R. David · H. Alla

Discrete, Continuous, and Hybrid Petri Nets

René David · Hassane Alla

Discrete, Continuous, and Hybrid Petri Nets

With 402 figures

Springer

Dr. René David
rene.david@lag.ensieg.inpg.fr
Pr. Hassane Alla
hassane.alla@inpg.fr
INP Grenoble
Laboratoire d'Automatique de Grenoble
ENSIEG
B.P. 46
38402 Saint Martin d'Heres Cedex
France

ISBN 3-540-22480-7 Springer Berlin Heidelberg New York

Library of Congress Control Number: 2004110950

Springer. Part of Springer Science+Business Media
springeronline.com

Typesetting: data delivered by authors
Cover design: design&production, Heidelberg
Printed on acid free paper 62/3020/M - 5 4 3 2 1 0

Foreword

Petri nets do not designate a single modeling formalism. In fact, newcomers to the field confess sometimes to be a little puzzled by the diversity of formalisms that are recognized under this "umbrella". Disregarding some extensions to the theoretical modeling capabilities, and looking at the *level of abstraction* of the formalisms, Condition/Event, Elementary, Place/Transition, Predicate/Transition, Colored, Object Oriented... net systems are frequently encountered in the literature. On the other side, provided with appropriate *interpretative extensions,* Controled Net Systems, Marking Diagrams (the Petri net generalization of State Diagrams), or the many-many variants in which time can be explicitly incorporated –Time(d), Deterministic, (Generalized) Stochastic, Fuzzy...– are defined. This represents another way to define practical formalisms that can be obtained by the "cross-product" of the two mentioned dimensions. Thus Petri nets constitute a modeling paradigm, understandable in a broad sense as "the total pattern of perceiving, conceptualising, acting, validating and valuing associated with a particular image of reality that prevails in a science or a branch of science" (Thomas S. Kuhn).

As a modeling paradigm, Petri nets define a conceptual framework in which particular formalisms can be obtained from a few principles and basic concepts (reduced syntactical/semantics "apparatus"). This means that there exist an underlying structure or set of relations among the different formalisms, what easies (if possible) the displacement among them. As a clear consequence, both at the conceptual and operational level, a substantial economy, coherence and synergy among models for different phases of the life cycle has been gained. In this context continuous and hybrid Petri nets are introduced in this useful book, expanding the existing perspective.

Let us point out that Petri nets are sound mathematical models that can be approached in several ways, showing centrality in the basic concepts of communication and locality in which are based. In particular Petri nets can be "found" in an Axiomatic way (as Carl Adam Petri does himself), extending Automata Theory in some sense, looking for Vector Addition Systems, considering a fragment of Linear Logic or using Theory of Regions in graphs, for example. The important question from a practical point of view is that Petri nets allows the modeling of intricate interleaving of cooperation and competition relationships between concurrent subsystems, what transpire through conflicts and synchronizations between local actions of the different implied subsystems, as can be seen in some of the proposed modeling exercises of this text.

Like Discrete Event Dynamic Systems, Petri nets have been developed at the intersection of other fields, although all of them are also very young from an historical perspective: Systems Theory and Automatic Control, Computer Science/Engineering and Operational Research, essentially. Introduction of continuous Petri nets goes back less than two decades ago, during the *8th European Workshop on Applications and Theory of Petri Nets* (Zaragoza, June, 1987), now *International Conference on Applications and Theory of Petri Nets.* The source of inspiration for the introduction of such formalism was the fluidification of Queuing Networks in Operations Research. In fact, most Queuing Networks can be seen as timed Petri nets provided with particular interpretative extensions. It is just a simple coincidence, that at the same meeting, with José Manuel Colom, we proposed the systematic use of Linear Programming for the structural analysis of (discrete) Petri nets, working with the fundamental or state equation of the net system. In fact our proposal can be rephrased as just relaxing Integer Programming Problems into Linear Programming Problems (or sets of them) in order to obtain necessary or sufficient conditions for the verification of some qualitative properties, or bounds (even exact values in some cases) for some quantitative measures. The main difference is that David and Alla fluidify at the net level, while we do that at some equation level. In perspective, the advantage of the approach of our colleagues from Grenoble was the possibility of studying the transient behavior. Anyhow, as the Monsieur Jourdain from the *Bourgeois Gentilhomme* of Molière, instead of prose "we were speaking in *continuous* Petri nets", without knowing it. In fact, I was fully aware of it only some years ago, when preparing a talk I was honored to give at the scientific meeting organised to celebrate the retirement of René David.

Baltasar Gracián says that "good things, when short, are twice as good". Even if I cannot guarantee quality for this foreword, I will try to be short. Anyhow I should explain why I feel so happy when I was invited to write this foreword. On one hand, René David was my Ph.D. advisor, in a work that introduced Petri nets and Programmable Logic Controlers in the Laboratoire d´Automatique de Grenoble some twenty five years ago. On the other hand, Hassane Alla enjoyed some stages in our Department in Zaragoza when preparing his Ph.D. Thus, in some sense, I am a "lost link" in this chain.

Last but more importantly, let me point out that this book provides a comprehensive view of developments in the field in the last decades. In an integrated way you can found the basics of discrete, continuous and hybrid variants of net formalisms. It is not simply a practically oriented book, because a significant number of theoretical concepts and results are presented, many of them are recent, some just introduced in this work. In the same sense, this is not simply a research book, given that much material related to applications is provided, not only many small academic exercises, but even several case studies. In other words, the authors show that this is a deep, practical and alive field. We are grateful to them for their time and efforts to provide such an integrated perspective.

Manuel Silva
University of Zaragoza
Royal Academy of Engineering of Spain

Preface

Carl Adam Petri is a contemporary German mathematician. In the early sixties, he defined a general purpose mathematical model for describing relations existing between conditions and events [Pe 62]. Then, he continued working on the net theory [Pe 76 & 79]. Many other researchers have provided a large number of theoretical results concerning these nets.

Engineers have appropriated this model, usually called Petri net, for modeling the discrete event dynamic systems they have to handle, particularly for computer science and automatic control applications. Various extensions have been proposed. The European Workshop on the theory and applications of Petri nets, rechristened the International Conference in 1989, which has been held every year since 1980, is certainly the international conference where the largest number of results have been presented on the subject. However, there are many conferences in which one or several sessions are devoted to the use of Petri nets for the domain treated in the conference.

In the early eighties, one of the authors was involved in a study of performance evaluation of production lines. Following [Zi 56], [GeSc 80], and [DuFo 82], the production of a machine was modeled by a continuous flow in [TeDa 87] and further works (although the number of parts processed by a machine was discrete). The idea of continuous Petri nets came from this study: since a continuous model was a good approximation for these discrete systems (formally shown later in [DaXiDa 90][MaCh 91]), why not for Petri nets? The other author, involved in the study of colored Petri nets [Si *et al.* 85], joined the first one to prepare the first paper on the subject [DaAl 87]. Without betraying the confidentiality of the program committee discussions, the authors were told that a kind of "battle" occured in Zaragoza: some members thought that the models called continuous Petri nets were not Petri nets because they were not "discrete" and hence not relevant to the conference; others thought that it was a really innovative approach for Petri nets. Finally, the paper was accepted and, in the next paper, the authors have proved that their model was a limit case of discrete Petri nets [DaAl 90].

Further studies have been conducted by the team of the authors, particularly the definition of hybrid Petri nets including discrete components and continuous components in the same model [LeAlDa 91], and in other teams. In 2001, a special issue of the *Journal of Discrete Event Dynamic Systems* was devoted to hybrid Petri nets [*SI* 01]. This means that these models have been accepted without problem by the community.

When writing this book, the authors were guided by two main goals.

The first one is to start from the basics of Petri nets and to reach an accurate understanding of continuous and hybrid Petri nets, while preserving the consistency of basic concepts throughout the book. For example, the notions of multiple firing and enabling degree are "similar" for discrete and continuous (hence, hybrid) Petri nets, and, when time is involved, the concepts of immediate transition and simultaneous firing are "similar" in all the models (discrete, continuous and hybrid). This homogeneity of concepts provides a *unified framework for all the models presented.*

The second goal of the authors is to present a *didactic tutorial* which is easy to understand due to many simple examples and detailed figures.

Informally, two parts may be considered: Chapters 1 to 3 are devoted to regular *discrete Petri nets*, while *continuous and hybrid Petri nets* are studied in Chapters 4 to 7.

Chapters 1 and 2 present the *basic concepts and properties of Petri nets*. The content of this part was used by both authors for introductory courses on Petri nets at post-graduate level. These two chapters contain what should be known by a student or an engineer who wants to use Petri nets. Except for an accurate presentation of conflicts (more accurate than usual), these chapters contain the classical basics of autonomous Petri nets, i.e. in which time is not involved.

Chapter 3 presents *non-autonomous discrete Petri nets*. The concept of synchronized Petri net, with infinite server semantics, is the basic one. Then, interpreted Petri nets, a model for specification of logic controlers, theoretical basis of Grafcet model, as well as timed and stochastic Petri nets appear to be particular cases of synchronized Petri nets.

Chapter 4 presents *autonomous continuous and hybrid Petri nets*, i.e. models in which time is not involved. In this chapter, the behavior of these nets is studied from a qualitative point of view (as in Chapters 1 and 2 for regular discrete Petri nets).

Timed continuous and hybrid Petri nets are defined and studied in Chapters 5 to 7. A detailed explanation of the behavior of these models, taking into account various kind of conflicts, is proposed.

In addition to the seven chapters, fifteen *appendices* provide details on specific points. These points are not dealt with in the chapter for the following reason: although they may be useful for an accurate understanding of some points, they are not absolutely necessary to understand the contents of chapters.

There is *continuity* between most of the successive chapters. Except Chapter 4 which is quite independent from Chapter 3, to read a chapter you need some knowledge of the previous ones. However, a reader who has some basic knowledge of the domain could read a chapter without reading the previous ones: he (or she) is helped by many *cross references* to sections, definitions, properties, figures, etc., where some useful notions are defined, discussed, or illustrated.

A large part is devoted to exercises with solutions. The proposed exercises are classed into eight parts. The first seven parts correspond to exercises relative to Chapters 1 to 7, respectively. Then, the eighth part contains three case studies: three examples of hybrid systems found in the literature are presented (the first of them was explicitely presented as a benchmark).

For simplified reading, only a few bibliographical references are given in the body of the chapters. Each chapter ends with "Notes and References".

All the main *notations* are given at the beginning of the book. At the end, a complete *index* allows a reader to easily locate where a particular concept is defined or discussed.

Acknowledgements

The authors wish to express their gratitude to the Master and Ph.D. students who have contributed by their scientific work and helped in many, many ways. We cannot mention all of them, but we would like to name the main contributors to the topic: Manuel Silva, Hugues Deneux, Nouredine Zerhouni, Jean Le Bail, Eric Dubois, François Charbonnier, Mohamed Allam, and Calin Munteanu who programmed the algorithms which will be available at the internet address `http://sirphyco.lag.ensieg.inpg.fr/`.

The book has benefited from collaboration with researchers of other teams with whom we had the opportunity to work and write papers in relation with this book: Mohamed Moalla, Jean-Baptiste Cavaillé, Gérard Bel, and Simona Caramihai. It has also benefited from a pertinent problem asked by Paul Caspi. Researchers from various other teams have made very useful contributions to the development of the topic. They are cited in the body of the book. Let us mention here Angela Di Febbraro, Alessandro Giua, and Guiseppe Menga, Guest Editors of the Special Issue already cited.

The presentation of the three application examples in part 8 of Exercises was permitted and helped by: Hermès Science, publisher, for Ronan Champagnat and co-authors (gas storage); Nathalie Audry, Isabel Demongodin, and François Prunet (bottling chain); Andrea Balluchi and co-authors (four-stroke engine).

The authors would like to thank the colleagues who helped them to improve the manuscript by their invaluable comments and suggestions: Isabel Demongodin, Alessandro Giua, Zdenek Hanzalek, Jean Le Bail, Mohamed Moalla, Stéphane Mocanu, Manuel Silva, and Janan Zaytoon.

We owe a special acknowledgement to Jean Le Bail who provided very creative ideas when he was working in our team, then when reading the manuscript.

We would like to emphasize the enthusiastic support of Manuel Silva who has shown his *continuous* interest in our work, provided excellent new results with his team, suggested to complete the book with some "substantial" application examples (already evoked), and who wrote the foreword for this book.

Last but not least, we would like to thank the members of *Laboratoire d'Automatique de Grenoble* for their assistance in very many areas, as well as Julia Summerton who corrected a lot of mistakes and clumsy turns of phrase in our initial draft.

René David
Emeritus director of research at the Centre National de la Recherche Scientifique
Hassane Alla
Professor at the University Joseph-Fourier

Contents

Notation

General

$\mathbf{A}$	Matrix.
$\mathbf{a}$	Vector.
$a(i)$ or a_i	ith component of vector $\mathbf{a}$.
$\mathbf{A}^{T}$; $\mathbf{a}^{T}$	Transposed matrix of $\mathbf{A}$; transposed vector of $\mathbf{a}$.
$(a, b, \ldots, v)$	Vector $[a \;\; b \;\; \ldots \;\; v]^{T}$.
$\mathbf{0}$	Vector (0, 0, …, 0).
$\mathbf{1}$	Vector (1, 1, …, 1).
$\mathbf{a} \geq \mathbf{b}$	For every i: $a(i) \geq b(i)$.
$\mathbf{a} \gneq \mathbf{b}$	$\mathbf{a} \geq \mathbf{b}$ and at least one i such that $a(i) > b(i)$.
$\mathbf{a} > \mathbf{b}$	For every i: $a(i) > b(i)$.
D^*	Monoid built over the alphabet D.
ε	Word of length zero.
D^+	D^* except ε.
$\overline{\mathcal{L}}$	Set of prefixes of language $\mathcal{L}$.
$a \cdot b$	Product (algebraic, Boolean, of events…). Not used for concatenation.
$\mathcal{N}$; $\mathcal{N}^*$	Natural numbers; except zero.
$\mathcal{Q}$; $\mathcal{Q}_+$; $\mathcal{Q}_+^*$	Rational numbers; positive or zero; positive.
$\mathcal{R}$; $\mathcal{R}_+$; $\mathcal{R}_+^*$	Real numbers; positive or zero; positive.
${}^*\mathcal{R}_+$	Hyperreal positive or zero numbers (non-standard).
${}^*x$	Variable x when it is measured by a number in ${}^*\mathcal{R}_+$.
$[a, b]$; $]a, b[$;	Interval from a to b including a and b; excluding a and b;
$[a, b[$; $]a, b]$	including a (if $a \neq b$) but not b; including b (if $a \neq b$) but not a.
$\lvert a, b \rvert$	Interval which is either $[a, b]$, or $]a, b[$, or $[a, b[$, or $]a, b]$.
$A \setminus B$	Set of components in set A but not in set B.

Petri nets

PN	Petri Net.
P_i; P	Place number i of a Petri Net; set of places.
T_j; T	Transition number j of a Petri Net; set of transitions.
$m(P_i)$ or m_i	Marking of place P_i.
$\mathbf{m}$ or $\mathbf{m}_k$	Marking of a Petri Net.

$\mathbf{m}_0$	Initial marking of a Petri Net.
$\mathbf{e}$ or $\mathbf{e}_k$	Enabling vector of a Petri Net.
Q	Unmarked Petri net.
R	Autonomous marked Petri net: $R = \langle Q, \mathbf{m}_0 \rangle$.
$\mathbf{W}$	Incidence matrix.
$\mathbf{W}^-$; $\mathbf{W}^+$	Input incidence matrix; output incidence matrix.
$\mathrm{Pre}(P_i, T_j)$	Component w_{ij}^- of $\mathbf{W}^-$.
$\mathrm{Post}(P_i, T_j)$	Component w_{ij}^+ of $\mathbf{W}^+$.
S or S_k	Firing sequence.
$\mathbf{s}$	Characteristic vector of firing sequence S.
$\mathbf{m}_1 \xrightarrow{S}$	Sequence S is a possible firing sequence from $\mathbf{m}_1$.
$\mathbf{m}_1 \xrightarrow{S} \mathbf{m}_2$	Firing sequence S leads from $\mathbf{m}_1$ to $\mathbf{m}_2$.
ω	Unbounded marking of a place.
$\mathcal{M}(\mathbf{m})$	Set of reachable markings from marking $\mathbf{m}$.
$\mathcal{L}(\mathbf{m})$	Language generated from marking $\mathbf{m}$.
$°T_j$; $T_j°$	Set of the input places of T_j; of the output places of T_j.
$°P_i$; $P_i°$	Set of the input transitions of P_i; of the output transitions of P_i.
T_1T_2	Successive firings of T_1 then T_2.
$[T_1T_2]$	Simultaneous firings of T_1 and T_2.
$\{T_1T_2\}$	Concurrent firings of T_1 and T_2.
$q(T_j, \mathbf{m})$	Enabling degree of T_j for marking $\mathbf{m}$.
$T_1 < T_2$	Transition T_1 takes priority over T_2.
$r(T_j, \mathbf{m})$	Allowing degree of T_j (in a priority PN) for marking $\mathbf{m}$.
$\langle P_1, \{T_1, T_2, ...\}\rangle$	Structural conflict.
$\langle P_1, \{T_1, T_2, ...\}, \mathbf{m}\rangle$	Effective or generalized conflict for marking $\mathbf{m}$

Conditions, Events, Actions, and Timings

a, a'	Boolean variable and its complement.
$\uparrow a$, $\downarrow a$	Rising edge and falling edge of variable a.
E^k; E	External event; set of external events.
e	Always occurring event.
E_j	Event associated with transition T_j (= E^k or e).
C_j	Condition associated with transition T_j.
R_j	Receptivity associated with transition T_j (i.e. $R_j = E_j \cdot C_j$).
R_S	Synchronized PN based on automomous PN R (in Section 3.2).
X or X_h	Set of compatible events (Section 3.2).
$\mathcal{L}_{\max}$; $\mathcal{L}_{\mathrm{ind}}$	Maximal set of sequences generated by the environment if all events in E are compatible; if they are independent from each other.
$T(X, \mathbf{m})$	Set of transitions receptive to events in the set X for marking $\mathbf{m}$.
EFS	Elementary firing sequence
$\mathbf{m} \xrightarrow{S/E^k} \mathbf{m}'$	Sequence S is fired on occurrence of E^k (or E^k replaced by a time).
'Event'	Description of an event.

$[Y]$	Predicate: $[Y] = 1$ if Y is true.
$[t/P_i/50\text{ s}]$	50 s elapsed since last marking of P_i ($\equiv$ 50 s/Xi in Grafcet).
O^*	Impulse action/operation (asterisk optional if no ambiguity).
$x := a$	Allocation of value a to variable x (or $x = a$ if no ambiguity).
T-timed PN	Transition-timed Petri Net.
P-timed PN	Place-timed Petri Net.
d_k	Timing associated with T_k (T-timed PN) or with P_k (P-timed PN).
P&T-timed PN	Petri Net with timings associated with both places and transitions.
T-time PN	Time Petri Net: firing time intervals associated with transitions.
P-time PN	Time Petri Net: sojourn time intervals associated with places.
$[a_k, b_k]$	Time interval associated with T_k (T-time) or with P_k (P-time).
$\mathbf{m}^a$; $\mathbf{m}^u$	Available marking of a P-timed PN; unavailable marking.
$\mathbf{m}_{0,\min}$, $\mathbf{m}_{0,\max}$	Initial markings for P-timed & T-timed PNs (Remarks 3.12, 3.13).

Specific to Continuous and Hybrid Petri nets

OG-firing	Simultaneous firing of one or several transitions at one go.
$[T_1]^\alpha$	Firing of T_1, quantity α at one go ($\alpha \in \mathcal{R}_+$).
$(T_1)^\alpha$	Firing of T_1, quantity α in one or more successive firings.
$[U_j]$	OG-firing of one or more transitions (discrete or/and continuous).
$\mathbf{m}_k^*$	Macro-marking defined by the set of marked C-places.
P^D; P^C	Set of D-places (discrete places); of C-places (continuous places).
T^D; T^C	Set of D-transitions; of C-transitions.
$\mathbf{W}^D$; $\mathbf{W}^C$; $\mathbf{W}^{CD}$	Part of $\mathbf{W}$ related to arcs among discrete nodes; among continuous nodes; among C-places and D-transitions.
$\mathcal{M}_L(\mathbf{m}_0)$	Linearized reachability space.
lim-$\mathcal{M}(\mathbf{m}_0)$	Set of limit reachable markings.
0^+	Ratio α / k where α is a finite positive number and $k \to \infty$.
$\tilde{m}_i$	Marking of P_i: may be 0 or 0^+ or a positive real number.
$\tilde{\mathbf{m}}$	Marking $(\tilde{m}_1, \tilde{m}_2, \ldots)$.
V_j; $\mathbf{V}$	Maximal firing speed of C-transition T_j; vector of values V_j.
v_j; $\mathbf{v}$	Instantaneous firing speed of C-transition T_j; vector of values v_j.
U_j; $\mathbf{U}$	Flow rate = maximal speed for a server for T_j; vector of values U_j.
I_i, O_i, B_i	Feeding speed, draining speed, balance of C-place P_i.
$D(T_j, \mathbf{m})$	D-enabling degree of C-transition T_j for marking $\mathbf{m}$.
$C(T_j, \mathbf{m})$	C-enabling degree of non-immediate C-transition T_j for marking $\mathbf{m}$.
$\mathbf{n}$	Vector of number of firings of D-transitions in a hybrid PN.
$\mathbf{m}^{C,\text{bef}}$	C-marking before firing of immediate C-transitions (unstable).
$\mathbf{m}^{C,\text{aft}}$	C-marking after firing of immediate C-transitions (stable).
R^C	Subjacent continuous PN for marked hybrid PN R.
$a[t_1, t_2[$	Enabling density a between minimal and maximal residual times.
ed	Enabling density vector.
$P(T_j)$; $\mathbf{P}$	Critical place of T_j; vector of critical places (for VHPN, AHPN).
m_i^*	Asymptotic marking of P_i (for VHPN, AHPN).

IB-state	Invariant behavior state: $\mathbf{v}$ is constant.
IB-phase	$\mathbf{v}(t)$ such that the functions $v_j(t) = f(t)$ do not change.
I-phase	Firings of immediate C-transitions leading to a stable marking.
Case 1 conflict	Between D-transitions.
Case 2 conflict	Between C-transitions for a common input C-place.
Case 3 conflict	Between a D-transition and a C-transition.
Case 4 conflict	Between C-transitions for a common input D-place.
$[\alpha_1 T_1, \alpha_2 T_2]$	Sharing aiming at $v_1 / \alpha_1 = v_2 / \alpha_2$.
C1-event	The marking of a marked C-place becomes zero.
C2-event	An unmarked place becomes marked.
C3-event	The vector $\mathbf{V}$ changes (if $\mathbf{V}(t)$ is piecewise constant).
C4-event	The marking of a C-place, origin of an inhibitor arc, reaches the weight of this arc.
C5-event	Change of critical place of a C-transition (for VHPN, AHPN).
C6-event	Critical place P_i reaches asymptotic value m_i^* (for AHPN).
CD-event	The continuous firing of a D-transition either begins or ends.
D1-event	Firing of a D-transition.
D2-event	Enabling degree of a D-transition change due to a C-place marking.
CHPN; CCPN	Constant speed hybrid PN; constant speed continuous PN.
VHPN; VCPN	Variable speed (C-marking dependent) hybrid PN; continuous PN.
AHPN; ACPN	Asymptotic hybrid PN; continuous PN.
DHPN	Differential hybrid PN.

For speed calculation (in Chapters 5 and 6)

LPP	Linear programming problem.
TOS	Time-ordered sequence.
C_1 to C_4	Sets of speed constraints, defined pp. 174, 175, and 180.
PL_k	Set of transitions in priority level k.
$T_{SF}(h)$	Set of surely firable transitions at passage h.
J_h	Sum of speeds to be maximized at passage h.
$p(T_j)$	Defined p. 173.
P_{ne}, T_{se}	Defined p. 175.
T_{pf}	Defined p. 181.
$T(J_h)$, T_{nc}, T_{ndc}, T_{sby}, T_{wait}	Defined p. 184.
$v_a^{(a)}$, $v_b^{(a)}$, $\mathbf{v}^{(a)}$, $B_c^{(a)}$	Defined p. 190.
$UP(K_F)$, $DOWN(K_F)$	Defined p. 191.
$TB_i^{(x)}$, $v_{a,\max}$	Defined p. 193.
G_a, MG_b, $\mathcal{G}_c$, $\mathcal{MG}_d$, $T(G_a)$	Defined pp. 199-200 and Appendix L.
$m_c^{(Ak)}$, V_j', $u(T_j)$	Defined p. 237.
$a(T_j)$	Defined p. 239.

Some other notations are not in this list because they are used *locally* (i.e. close to their definitions).

1

Bases of Petri Nets

The main users of Petri nets are computer and automatic control scientists. However, this tool is sufficiently general to model phenomena of extremely varying types.

Petri nets present two interesting characteristics. Firstly, they make it possible to model and visualize behaviors with parallelism, concurrency, synchronization and resource sharing. Secondly, the theoretical results concerning them are plentiful; the properties of these nets have been and still are extensively studied.

In this chapter, the bases of Petri nets, various classes of these nets and their modeling power are presented in a fairly intuitive manner.

1.1 BASIC CONCEPTS

The basic components of a Petri net, the concepts of marking and firing, the distinction between autonomous and non-autonomous Petri nets, and the essential characteristics of a Petri net are presented in this section.

1.1.1 Places, Transitions and Arcs

A Petri net (PN) has two types of nodes, namely places and transitions (Figure 1.1). A **place** is represented by a circle and a **transition** by a bar (certain authors represent a *transition* by a box). Places and transitions are connected by **arcs**. The number of places is *finite* and *not zero*. The number of transitions is also *finite* and *not zero* (in certain cases, we may have to consider *degenerate* PNs having no place or no transition). An arc is directed and connects either a place to a transition or a transition to a place.

In other words, a PN is a bipartite graph, i.e. places and transitions alternate on a path made up of consecutive arcs. It is compulsory for each arc to have a node at each of its ends. (From a node k, place or transition, to a node h, transition or place, there is at most one arc.)

Figure 1.1a represents a PN with 7 places, 6 transitions and 15 directed arcs. The set of places of a PN will be called P and its set of transitions will be called T. For the example in question, we thus have $P = \{P_1, P_2, P_3, P_4, P_5, P_6, P_7\}$ and $T = \{T_1, T_2, T_3, T_4, T_5, T_6\}$.

Place P_3 is said to be **upstream** or an **input** of transition T_3 because there is a directed arc from P_3 to T_3. Place P_5 is said to be **downstream** or an **output** of transition T_3 because there is a directed arc from T_3 to P_5. In a similar way, a transition is said to be an input (upstream) or an output (downstream) of a place. A transition without an input place is a **source transition.** A transition without an output place is a **sink transition.**

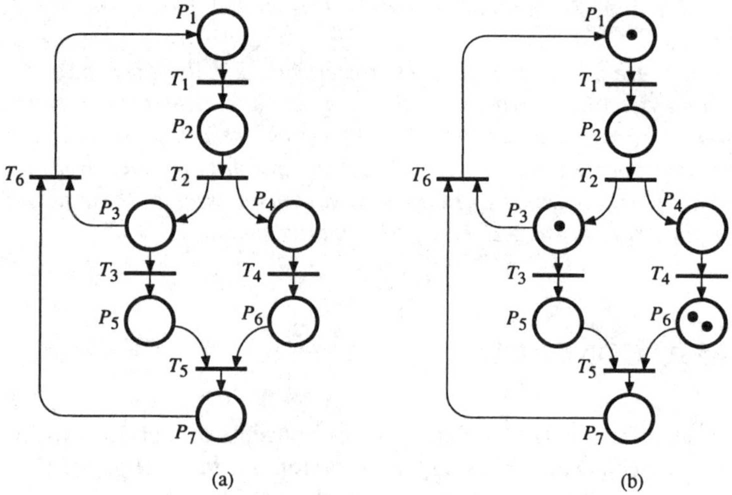

Figure 1.1 Petri Net. (a) Not marked. (b) Marked.

1.1.2 Marking

Figure 1.1b represents a marked Petri net. Each place contains an integer (positive or zero) number of **tokens** or **marks**. The number of tokens contained in a place P_i will be called either $m(P_i)$ or m_i. For the example in question, we have $m_1 = m_3 = 1$, $m_6 = 2$ and $m_2 = m_4 = m_5 = m_7 = 0$. The net marking, $\mathbf{m}$, is defined by the vector of these markings, i.e., $\mathbf{m} = (m_1, m_2, m_3, m_4, m_5, m_6, m_7)$. The marking of the PN in Figure 1.1b is thus $\mathbf{m} = (1, 0, 1, 0, 0, 2, 0)$. The marking defines the state of the PN, or more precisely the state of the system described by the PN. The evolution of the state thus corresponds to an evolution of the marking, an evolution which is caused by firing of transitions, as we shall see.

In this book, *marked* Petri nets are considered practically always. We shall call them quite simply *Petri nets*. On the other hand, **unmarked PNs** will be specified when necessary.

1.1.3 Firing of a Transition

A transition can only be fired if each of the input places of this transition contains at least one token[1]. The transition is then said to be **firable**, or **enabled** (*in this chapter,* these two words are synonymous). A source transition is thus always enabled. The transitions in Figures 1.2a, b and c (before firing) are enabled because in each case places P_1 and P_2 contain at least one token. This is not the case for the example of Figure 1.2d in which transition T_1 is not enabled, since P_1 does not contain any tokens.

Firing of a transition T_j consists of withdrawing a token from each of the input places of transition T_j and of adding a token to each of the output places of transition T_j. This is illustrated in Figure 1.2. In Figure b (when there is no ambiguity on the figure referred to, we shall write for example Figure b instead of Figure 1.2b, to make the text less cumbersome), we note that there are 2 tokens in place P_3 after firing because there was already one beforehand. In Figure c we observe that a token remains in place P_1 after firing.

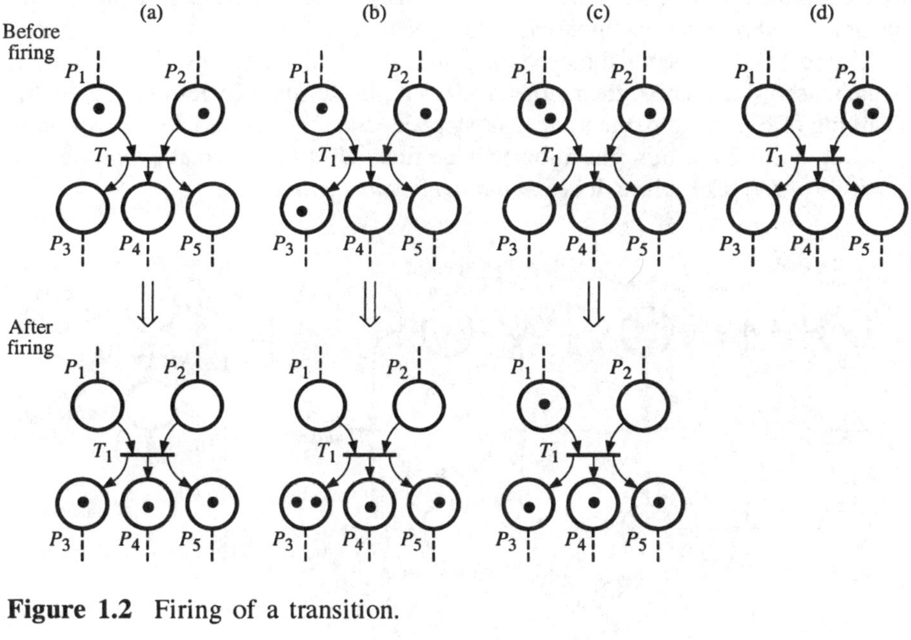

Figure 1.2 Firing of a transition.

---

[1] Note for the reader already having some knowledge of Petri nets: for the time being, we consider **ordinary** PNs, in which the arcs are not weighted (by default the weight of an arc is 1).

When a transition is enabled, this does not imply that it will be immediately fired. This only remains a possibility. In Figure 1.1b, two transitions are enabled, T_1 and T_3. Although we do not know when these transitions will be fired, we do know that the next evolution of the marking will correspond either to a firing of T_1 or to a firing of T_3 (thus that no other evolution is possible).

The firing of a transition is *indivisible*. Although there is no concept of duration in a PN (if it is neither timed nor synchronized), it is useful to consider that the firing of a transition has *a zero duration,* in order to facilitate understanding of the concept of indivisibility.

1.1.4 Autonomous and Non-autonomous Petri Nets

Figure 1.3a represents the seasonal cycle. A place is associated with each season, and the changeover from one season to the next with each transition. The marking of Figure 1.3a corresponds to spring. It can be seen on this PN that there is only one enabled transition T_1, and that this transition will therefore be the next one to be fired. The following marking will thus correspond to summer but we have no indication as to when the transition will take place. This PN describes the seasonal cycle in a *qualitative* manner. When a PN describes the functioning of a system evolving in an autonomous manner, i.e. whose firing instants are either unknown or not indicated, we may say that it is an *autonomous* Petri net. Although the word autonomous is not necessary, it allows this PN to be clearly distinguished from a non-autonomous Petri Net.

Figure 1.3b represents the cycle of states of a system, for example a motor, which stops, then starts, then stops again, etc. In the state corresponding to the marking of Figure 1.3b, the system is stopped, and the only enabled transition is T_1. But in this case, this transition will be fired *when* the external event, i.e. the *start-up order,* takes place. It is a *non-autonomous* Petri net.

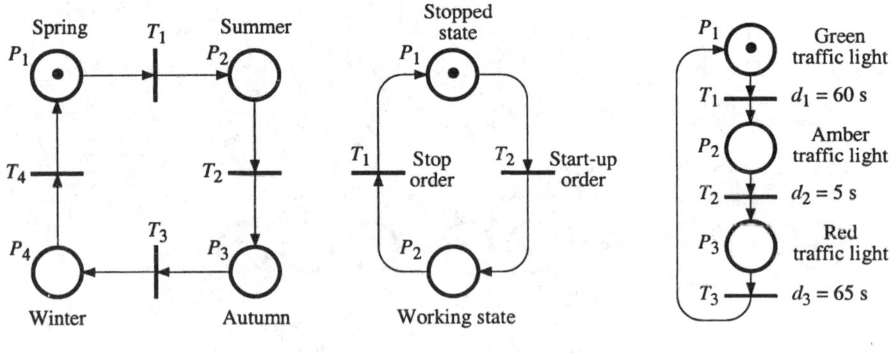

Figure 1.3 (a) Autonomous Petri net. (b) and (c) Non-autonomous Petri nets.

Figure 1.3b represents a system (a traffic light) whose behavior is *driven by time*. When there is a token in P_1, the traffic light is green and T_1 is enabled. This transition will be fired 60 s after the time when it became enabled. When T_1 is fired, the token in P_1 is taken out and a token is deposited in P_2: the traffic light becomes amber and T_2 becomes enabled . And so on. It is also a *non-autonomous* Petri net.

A **non-autonomous Petri net** describes the functioning of a system whose evolution is conditioned by external events and/or by time. A non-autonomous PN is *synchronized* and/or *timed.*

This chapter is dedicated to *autonomous PNs*. Non-autonomous PNs will be dealt with in Chapter 3 onwards.

1.1.5 The Essential Characteristics

A PN is used to describe and analyze a system. The properties will be given in detail in Section 1.3, but we can already define the essential characteristics which will be sought.

The dynamics of a system described by a PN is represented by the evolution of the markings. The concepts of *live* PNs (no transitions can become unfirable) and especially of *deadlock free* PNs, are important. Deadlock occurs if no transitions can be fired any longer. In almost every case, this corresponds to a system which is either badly designed or badly modeled.

A PN for which a return to the initial marking is always possible is *reversible*.

The concept of *conflict* frequently appears. A conflict structure exists when the same place is an input of *two* or *more* transitions. An example is given in Figure 1.1a, where place P_3 is an input of transitions T_3 and T_6. If both these transitions are enabled and there is only one token in P_3, only one of these transitions can be fired. There is thus a conflict between these transitions. This concept relates to a *decision* to be taken between two firings. It is found, for example, in the case of *sharing of a common resource*.

The number of tokens in a Petri net may increase and/or decrease (down to 0). A PN should, generally speaking, have the property of being *bounded* (finite number of tokens). An important particular case is that of *safe* nets (in which there is at most one token in each place).

1.2 SPECIAL PETRI NETS

This section presents some particular structures of PNs, followed by various abbreviations and extensions to the basic model.

1.2.1 Particular Structures

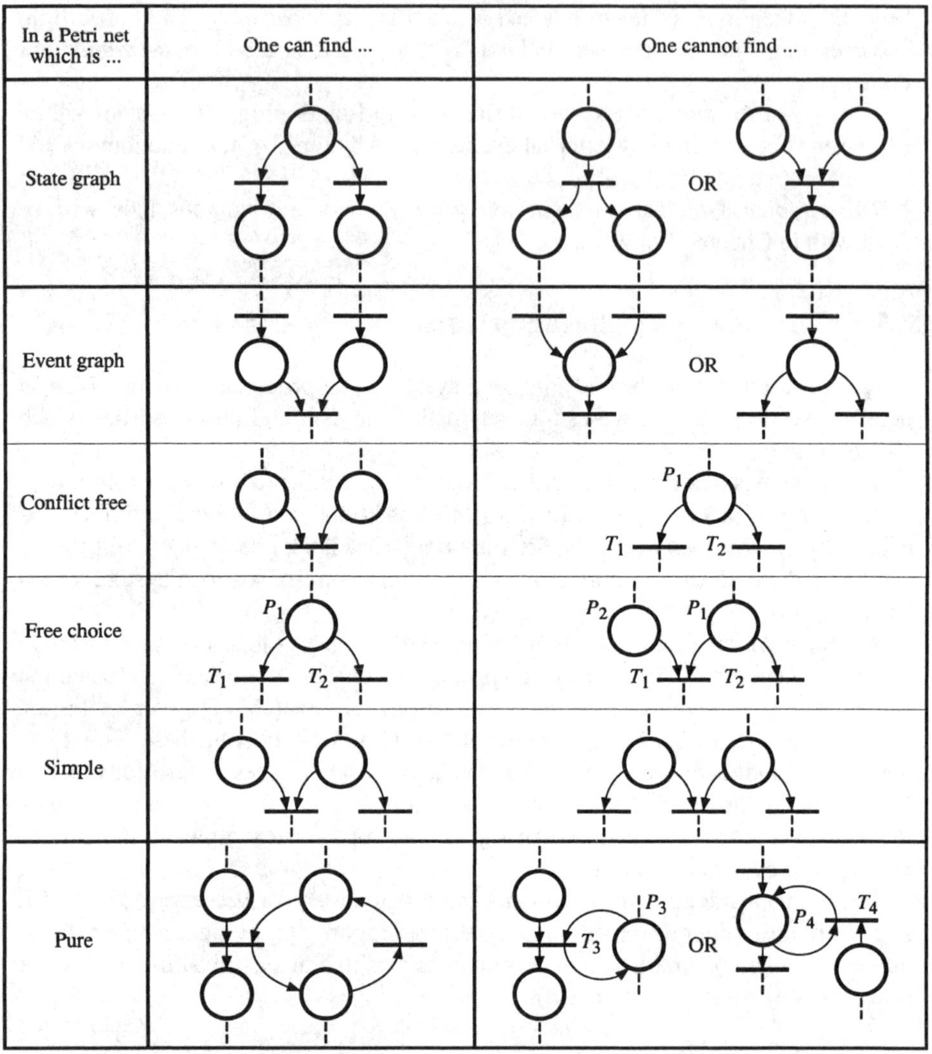

Figure 1.4 Illustration of particular structures.

Certain PNs have particular structures, i.e. they possess characteristics and properties not possessed by nets in the most general case. All the PNs presented in this section have *structural* characteristics, i.e. relating to *unmarked PNs*. They are illustrated in Figure 1.4.

1.2.1.1 State Graph

An unmarked PN is a *state graph* if and only if every transition has exactly one input and one output place.

Figure 1.5a represents an unmarked PN which is a state graph, in accordance with the above definition. Since each transition has only one input and one output place, the representation of a state graph in the classical sense (or state diagram, Figure 1.5b) contains the same information. In Figure 1.5b, a single directed arc (known as a transition) is made to correspond to the set of a transition with its input and output arc in Figure 1.5a.

It should be observed that if we *mark* a PN known as a *state graph,* its behavior will be equivalent to the state graph in the classical sense (representing an automaton which is in only one state at a time) *if and only if it contains exactly one token* (Figure 1.5c). The marking of Figure 1.5d, which may generate 6 different markings according to the firings of transitions, is clearly not equivalent to a 3-state automaton.

Generalized PNs, in which weights are associated with the arcs, will be introduced in Section 1.2.2. In a *state graph* (and also in an *event graph*, defined below), *the weight of all the arcs is* 1.

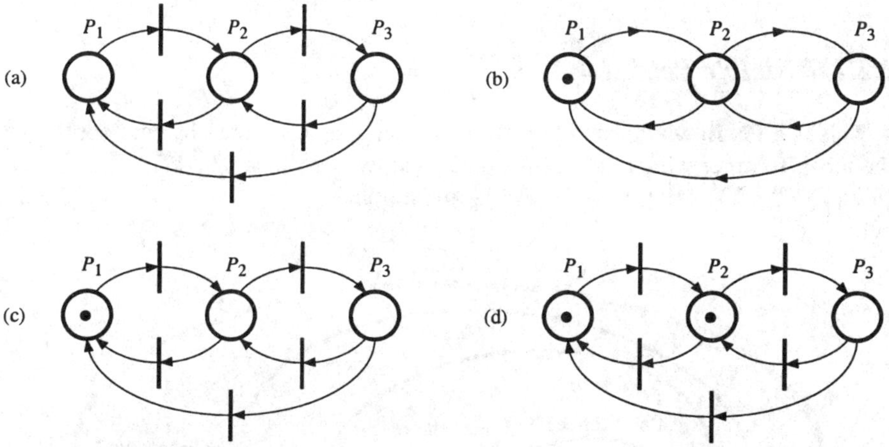

Figure 1.5 State graph. (a) Unmarked. (b) Representation of a state graph in the classical sense. (c) Marked with *one* token. (d) Marked with several tokens.

1.2.1.2 Event Graph

A PN is an *event graph* if and only if every place has exactly one input and one output transition. An *event graph* is also called *transition graph* or *marked graph* by some authors (the latter expression may be confusing with marked PN).

An event graph is thus the *dual* of a state graph.

1.2.1.3 Conflict Free Petri Net

This is a PN in which every place has at most one output transition.

A *conflict* (more precisely **structural conflict**) therefore corresponds to the existence of a place which has at least two output transitions. A structural conflict will be noted by the pair formed by one place and one set of output transitions of this place: $\langle P_1, \{T_1, T_2, ...\}\rangle$.

1.2.1.4 Free Choice Petri Net

Two definitions exist, which are distinguished by the appellations *free choice* and *extended free choice*.

A *free choice* PN is a PN in which for every conflict $\langle P_1, \{T_1, T_2, ...\}\rangle$ none of the transitions T_1, T_2, ... possesses an other input place than P_1.

An *extended free choice* PN is such that for every conflict $\langle P_1, \{T_1, T_2, ...\}\rangle$ all the transitions T_1, T_2, ... have the *same set of input places*. That is to say that if T_1 has P_1 and P_2 for input places, then T_2 has P_1 and P_2 for input places, etc.

In such a PN (free choice or extended free choice), if a transition involved in a conflict is enabled, all the transitions involved in the same conflict are also enabled.

1.2.1.5 Simple Petri Net

This is a PN in which each transition can only be concerned by one conflict at the most. In other words, if a transition T_1 and two conflicts $\langle P_1, \{T_1, T_2, ...\}\rangle$ and $\langle P_2, \{T_1, T_3, ...\}\rangle$ exist, then the PN is not simple.

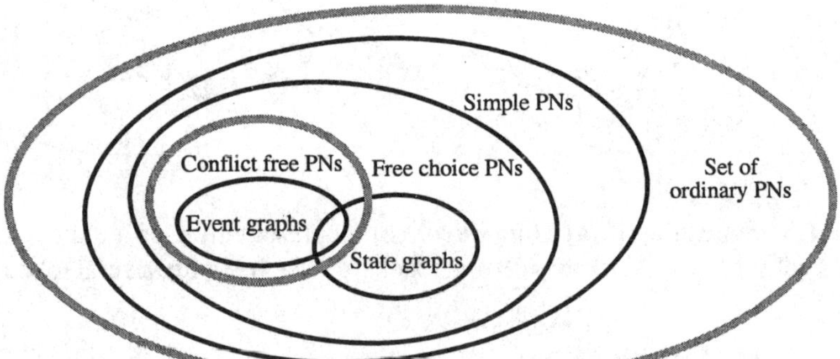

Figure 1.6 Relations among subsets of ordinary Petri nets.

Remark 1.1

a) It can be noted that the set of simple PNs includes the set of free choice PNs, which includes the set of conflict free PNs, which in turn includes the set of event graphs. The set of state graphs is included in the set of free choice PNs. These relations among subsets of **ordinary Petri nets** (i.e. *marked autonomous Petri nets* functioning according to the rules laid down in Section 1.1) are illustrated in Figure 1.6.

b) In a *state graph* there may be conflicts, but there is *no synchronisation* (i.e., a transition with at least two input places). In an *event graph* there may be synchronisations, but there is *no conflict*.

1.2.1.6 Pure Petri Net

A pair consisting of a place P_i and a transition T_j is called a *self-loop* if P_i is both an input and output place of T_j. A Petri net is *pure* if it has no self-loop.

Figure 1.4. shows two examples of impure nets. In one of them, place P_3 is both the input and the output place of T_3, and in the other P_4 is both the input and the output place of T_4. Transition T_3, in the first case, and transition T_4, in the second case, are said to be self-loop transitions or impure transitions. Places P_3 and P_4 are said to be self-loop places or impure places.

Property 1.1 All impure PNs can be converted into pure PNs. ❑

References where the conversion technique can be found are given in the Notes and References Section at the end of the chapter.

1.2.2 Abbreviations and Extensions

The types of Petri nets presented here do not correspond exactly to the functioning rules laid down above. The *abbreviations* correspond to simplified representations, useful for simplifying the graphical representation, but to which an *ordinary Petri net* can always be made to correspond. The *extensions* correspond to models to which functioning rules have been added, in order to enrich the initial model, thereby enabling a greater number of applications to be treated. It follows that all the *properties* of ordinary Petri nets *are maintained for the abbreviations* with a few adaptations, whereas these properties are *not all maintained for the extensions*.

1.2.2.1 Generalized Petri Nets

A generalized PN is a PN in which weights (strictly positive integers) are associated with the arcs. Figure 1.7a represents a generalized PN such that the arc $P_1 \rightarrow T_1$ has the weight 3, and arc $T_1 \rightarrow P_4$ has the weight 2. All the other arcs, whose weight is not explicitly specified, have a weight 1. When an arc $P_i \rightarrow T_j$

has a weight p, this means that transition T_j will only be enabled if place P_i contains at least p tokens. When this transition is fired, p tokens will be removed from place P_i. When an arc $T_j \rightarrow P_i$ has a weight p, this means that when T_j is fired, p tokens will be added to place P_i.

In Figure a, transition T_1 is enabled since P_1 contains 3 tokens and P_2 contains more than one token. Firing consists of removing three tokens from P_1 and one token from P_2, and of adding one token to P_3 and two tokens to P_4. The marking of Figure b is obtained.

Figure c is a simple application example of assembly. As soon as one car frame and four wheels are available, the four wheels may be mounted on the frame in order to obtain a car. In the figure, T_2 is not enabled because P_5 contains less than four tokens.

Property 1.2 All generalized Petri nets can be converted into ordinary PNs. ❑

References where the proof can be found are given in the Notes and References Section at the end of the chapter. From a theoretical point of view, it is interesting to know that this transformation is possible: the modeling power of the ordinary PNs is the same as the modeling power of the generalized PNs (i.e., the set of systems which can be modeled is the same for both models). However, the transformation rapidly becomes extremely complex. Fortunately, this transformation will not need to be carried out since the properties of ordinary PNs are easily adapted to the generalized PNs.

An *equal conflict* PN is a weighted generalization of the (extended) free choice PN. A generalized PN is *equal conflict* if, for any structural conflict $\langle P_i, \{T_j, T_k, ...\}\rangle$: 1) all the transitions T_j, T_k, ... have the same set of input places (as in an extended free choice PN, Section 1.2.1.4); 2) all the arcs $P_i \rightarrow T_j$, $P_i \rightarrow T_k$ ..., have *the same weight*. Hence, all the transitions involved in a conflict are enabled when one of them is enabled.

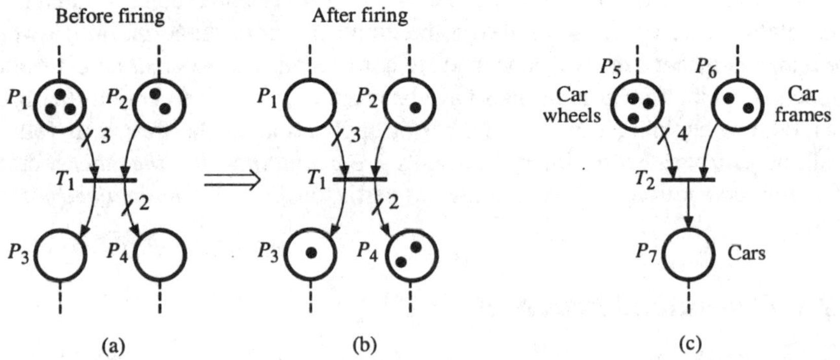

Figure 1.7 Parts of generalized Petri nets. (a) and (b) Firing. (c) Assembly example.

1.2.2.2 Finite Capacity Petri Nets

A finite capacity PN is a PN in which capacities (strictly positive integers) are associated with places. Firing of an input transition of a place P_i, whose capacity is Cap(P_i), is only possible if firing of this transition does not result in a number of tokens in P_i which exceeds this capacity (this definition works for any PN; however a somewhat different definition is used by some authors for impure PNs).

Figure 1.8 illustrates this notion. Figure a is a PN whose place P_2 has the capacity 2. Transition T_1 is enabled and its firing results in the marking of Figure b. In this figure, transitions T_1 and T_2 are enabled. Firing of transition T_1 results in Figure c. In this Figure c, transition T_1 can no longer be fired, although there is a token in P_1, because the marking of P_2 has reached its maximum capacity.

Property 1.3 All finite capacity PNs can be converted into ordinary PNs. ❑

The transformation is quite simple in the case of pure PNs, and is illustrated in Figures 1.8d to f. A *complementary place* is added to P_2, known as place P_2', whose marking is also complementary to the capacity of P_2. That is to say $m(P_2') = \text{Cap}(P_2) - m(P_2)$. Thus, when $m(P_2) = \text{Cap}(P_2)$, we have $m(P_2') = 0$ and transition T_1 is no longer enabled. In the general case, a place P_i' whose marking is complementary to this capacity is associated with a place P_i whose capacity Cap(P_i) is finite. All the input transitions of P_i are output transitions of P_i' and vice versa.

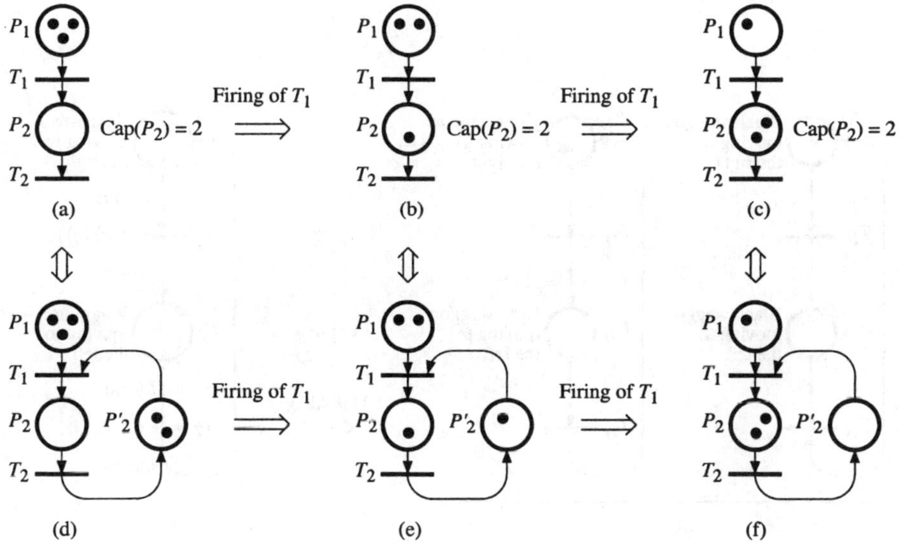

Figure 1.8 Illustration of a finite capacity PN.

1.2.2.3 Colored Petri Nets

Colored PNs have tokens to which *colors* are allocated. They form a category of nets whose intuitive perception is less clear than for the generalized and finite capacity PNs. Furthermore, they are of great value for the modeling of certain complex systems. There are several types of PNs which can be called colored, with variants in their definitions.

Property 1.4 All colored PNs with a finite number of colors can be converted into ordinary PNs. ❑

Here is an example illustrating Property 1.4. Figure 1.9 presents a system made of two waggons whose behavior is as follows. The black waggon moves continuously from left to right between spots L_b and R_b, then back to L_b, and so on. The grey waggon has a similar behavior between spots L_g and R_g. Figure b is a Petri net (not connected[2], made of two connected parts) modeling the behavior of the system.

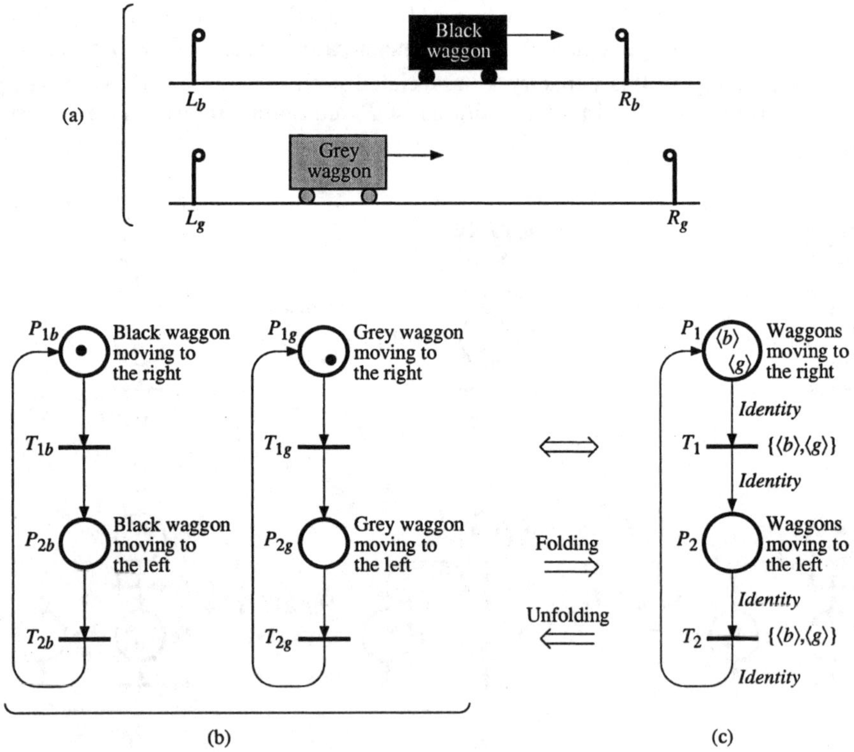

Figure 1.9 (a) System to be modeled. (b) Ordinary PN. (c) Colored PN.

[2] Readers not familar with the connected graph concept may consult Appendix C.

It can be observed that the two parts of Figure b have the same structure. Then, this structure can be represented only once by "superimposing" the two parts of the PN in Figure b. However, it is necessary to identify the tokens by labels called **colors**, in order to preserve the information concerning each waggon (for example, one must know which waggon is moving to the right and which wagon is moving to the left when they are not moving in the same direction). This is represented by the colored PN in Figure c. Colors $\langle b \rangle$ and $\langle g \rangle$ are associated with the black and grey waggons respectively. Place P_1, representing the set of places $\{P_{1b}, P_{1g}\}$, contains two tokens, one with the color $\langle b \rangle$ and the other with the color $\langle g \rangle$. Transition T_1, representing the set of transitions $\{T_{1b}, T_{1g}\}$, may be fired with respect to any color in the set $\{\langle b \rangle, \langle g \rangle\}$. The firing conditions are defined by functions associated with each arc. In our simple example, the *identity* function is associated with all the arcs. Let us consider the firing of transition T_1 with respect to color $\langle b \rangle$ (this transition could be fired with respect to colors $\langle b \rangle$ or $\langle g \rangle$). The *identity* functions associated with arcs $P_1 \rightarrow T_1$ and $T_1 \rightarrow P_2$ mean that, for firing color $\langle b \rangle$, the transition is enabled if there is at least one token $\langle b \rangle$ in P_1, and the firing consists of removing one token $\langle b \rangle$ from P_1 and adding one token of the same color to P_2 (this firing corresponds to a firing of T_{1b} in Figure b).

In the general case, a color may be a *n*-uple and the functions associated with the transitions may be complex (the tokens drawn and put down do not necessarily have the same color). A place (a transition) in a colored PN always corresponds to a set of places (of transitions) of an ordinary PN. The merging of a set of places and/or transitions is called **folding** and the reverse operation **unfolding**. The complete unfolding of a colored PN (Property 1.4) results in a unique ordinary PN.

1.2.2.4 Extended Petri Nets

An extended Petri net contains special arcs qualified as inhibitor. An *inhibitor arc* is a directed arc which leaves a place P_i to reach a transition T_j. Its end is marked by a small circle as shown in Figure 1.10. In this figure, the inhibitor arc between P_2 and T_1 means that transition T_1 is only enabled if place P_2 does not contain any tokens. Firing consists in removing a token from each input place of T_1, with the exception of P_2, and in adding a token to each output place of T_1.

In Figure 1.10a, transition T_1 is not enabled because P_1 does not contain any tokens. In Figure b, T_1 is not enabled because P_2 contains one token. In Figure c, the transition is enabled and the marking obtained after firing is shown in Figure d.

The following example is illustrated in Figure 1.11a. An administration lets customers in and then closes the entrance door before starting work. Once they have been served, the customers leave through another door. The entrance door will only be opened again when *all the customers* who came in have gone out. In

Figure a, the number of tokens in place P_2 represents the number of customers who have come in but not yet gone out and places P_1 and P_3 represent the state *entrance door open* and *entrance door closed,* respectively. We move from one state to the other by the firing of transitions T_3 and T_4. Firing of transition T_1 corresponds to the entrance of a customer. This firing adds a token to place P_2, while place P_1 does not change its marking. Since P_1 is the input and output place of transition T_1, the added token compensates the token removed. This operation is called "reading" of place P_1 (i.e. firing of T_1 is conditioned by the marking of P_1, but does not affect this marking). When the service begins, transition T_3 is fired, i.e. the door is closed. This door will only be opened again when *all the customers* have left, i.e. when place P_2 no longer contains any tokens. This constraint is taken into account by the inhibitor arc $P_2 \rightarrow T_4$ which only allows firing of T_4 if P_2 is empty. This firing cannot be represented by an ordinary PN, since the number of customers who may come in is not bounded.

Consider now a variant of the above example: after the entrance door has been closed, it may be open again only if there are no more than two customers waiting or being served. The corresponding system is shown in Figure 1.11b, where the weight of the inhibitor arc $P_2 \rightarrow T_4$ is 3. It means that firing of T_4 is inhibited if the number of tokens in P_2 is 3 or more (hence it is allowed if $m(P_2) = 0, 1,$ or 2).

In the general case, if the weight of an inhibitor arc $P_i \rightarrow T_j$ is s, then transition T_j is enabled only if $m(P_i) < s$; this is a **threshold test**. The case $s = 1$ is an important particular case called **zero test** since transition T_j is enabled only if $m(P_i) < 1$, i.e. if $m(P_i) = 0$.

Property 1.5 In the general case, extended PNs cannot be converted into ordinary PNs.

Property 1.6 If an extended PN is *bounded,* it can be converted into an ordinary PN.

❑

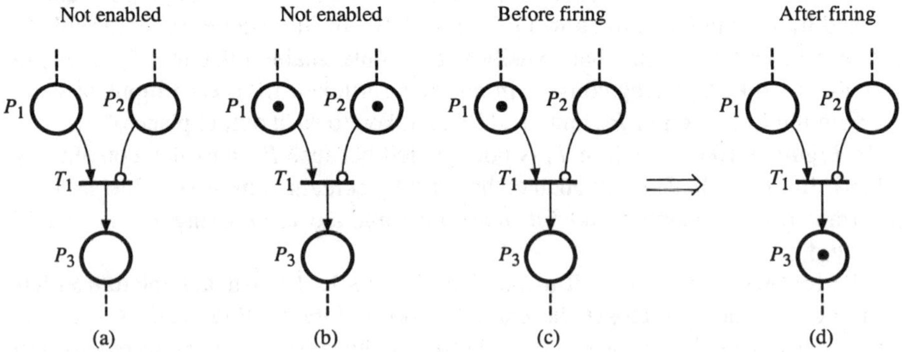

Figure 1.10 Transition with an inhibitor arc.

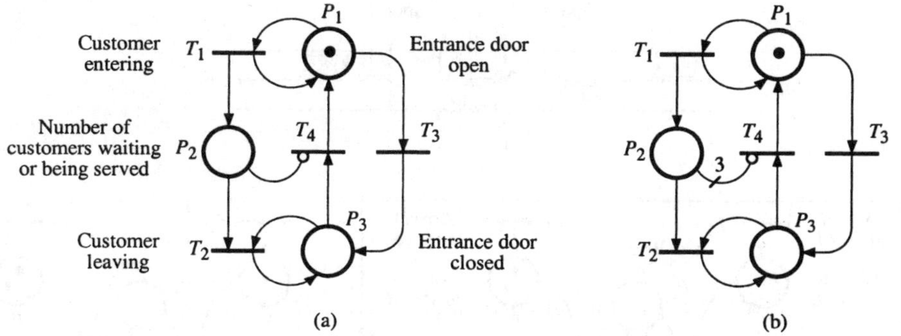

Figure 1.11 Examples of extended PNs. (a) Zero test. (b) General case.

This transformation takes place using complementary places. If there is a transition $P_i \rightarrow T_j$ with an inhibitor arc, an ordinary PN must be constructed containing a place P_i' which is marked when and only when place P_i is empty.

1.2.2.5 Priority Petri Nets

This type of net is used when we wish to choose between a number of enabled transitions. It is made up of 1) a Petri net and 2) a partial order relation on the net transitions.

Figure 1.12 presents a part of priority PN. There are two stuctural conflicts, namely $\langle P_1, \{T_1, T_2\}\rangle$ and $\langle P_2, \{T_2, T_3, T_4\}\rangle$. A strict order for the corresponding transitions is assigned for each conflict; i.e., $T_2 < T_1$ and $T_4 < T_2 < T_3$ for our example. The relation $T_j < T_k$ means that T_j has priority over T_k if both are enabled. For the marking $\mathbf{m}_1 = (1, 1, 0, ...)$ in Figure 1.12, there are two *effective conflicts* (this concept will be specified in Section 2.1.4), namely $\langle P_1, \{T_1, T_2\}, \mathbf{m}_1\rangle$ and $\langle P_2, \{T_2, T_3\}, \mathbf{m}_1\rangle$ because transitions T_1, T_2, and T_3, are enabled for this marking $\mathbf{m}_1$ (but not T_4). Since $T_2 < T_1$ and $T_2 < T_3$, transition T_2 will be fired. For the markings $\mathbf{m}_2 = (0, 1, 1, ...)$ and $\mathbf{m}_3 = (1, 1, 1, ...)$, transition T_4 will be fired. For $\mathbf{m}_4 = (0, 1, 0, ...)$, transition T_3 will be fired. A more general case of priority is presented in Appendix B (understanding of which requires concepts presented later in the book).

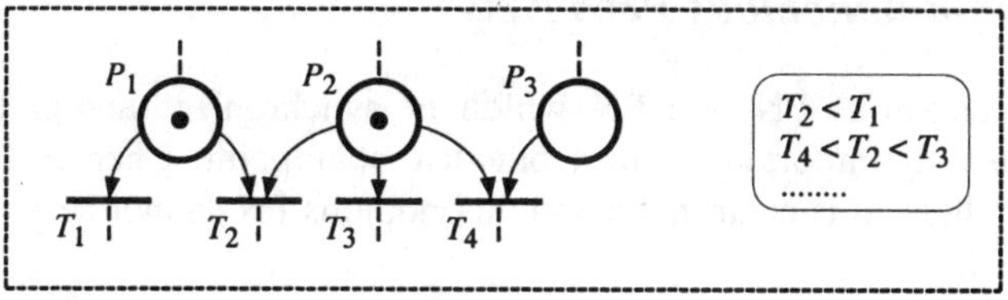

Figure 1.12 Part of priority Petri net.

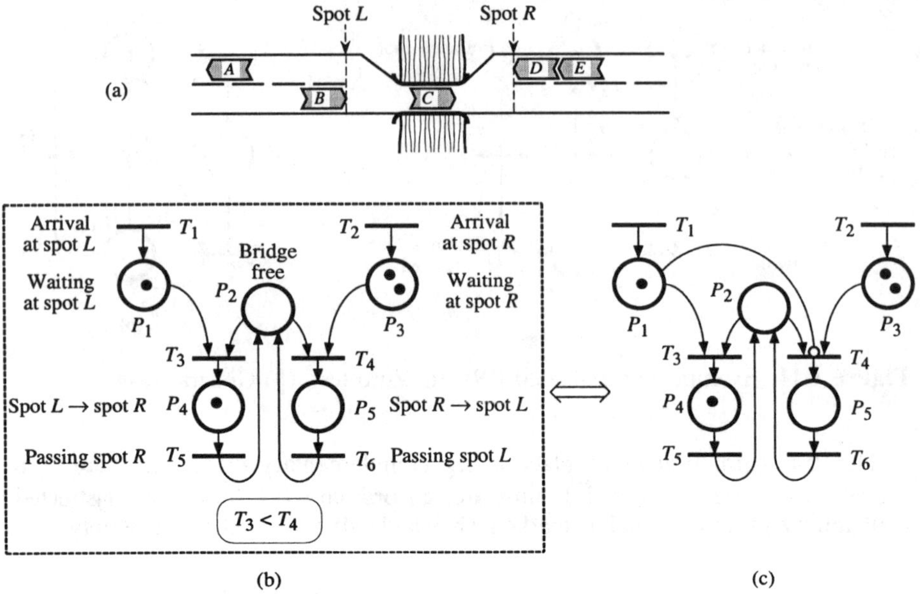

Figure 1.13 Example. (a) System to be modeled. (b) Priority PN. (c) Equivalent extended PN.

An application example is presented in Figure 1.13. The system to be modeled is shown in Figure a. Cars moving from left to right (like *B* and *C*) or in the opposite direction (like *A*, *D*, and *E*) must pass on a bridge. At most one car must be on the bridge (between spots *L* and *R*). If cars are waiting at both spots, priority is given to the cars waiting at spot *L*, i.e. cars moving from left to right. This system is modeled by the priority PN in Figure b, which consists of an ordinary PN and the order relation $T_3 < T_4$. An equivalent behavior is modeled by the extended PN in Figure c, since T_4 can be fired only if there is one token in P_2 and zero token in P_1.

Property 1.7 Priority PNs cannot be converted into ordinary PNs.

1.2.2.6 Non-autonomous Petri Nets

A non-autonomous PN is a PN which is synchronized and/or timed. This category of nets is important from a practical standpoint. Chapter 3 is entirely given over to them. It is clear that a non-autonomous PN cannot be converted into an ordinary PN.

1.2.2.7 Continuous and Hybrid Petri Nets

The distinguishing feature of continuous PNs is that the marking of a place is a real (positive) number and no longer an integer. Firing takes place like a continuous flow. These nets enable systems to be modeled which cannot be modeled by ordinary PNs, and a suitably close model to be obtained when the number of markings of an ordinary PN becomes too large. A hybrid PN is a combination of discrete (regular) PN and continuous PN. Chapters 4 to 6 are given over to continuous and hybrid PNs.

1.2.2.8 Conclusion

It appears that the *generalized*, *finite capacity* and *colored* PNs are all *abbreviations* of ordinary PNs (Properties 1.2 to 1.4). All the properties that we shall look at in the following sections can thus be adapted to these models.

The *extended, priority, non-autonomous* and *continuous* PNs are all *extensions* of the ordinary PNs. Some but not all the properties of the ordinary PNs are applicable to them.

The *extended* and *priority* PNs have the computational power of Turing machines. It follows that all priority PNs can be modeled by extended PNs, (an example is shown in Section 1.2.2.5).

1.3 MODELING OF SOME CONCEPTS

One of the key features of Petri nets is their capacity to graphically present certain relations and to *visualize* certain concepts. We shall give some examples of this below.

In Figure 1.14a, we clearly see that after transition T_1 has been fired and until transition T_2 is fired, there are concurrent evolutions, from place P_1 to place P_3 on the one hand and from place P_2 to place P_4 on the other. Each of these two evolutions can be carried out at its own pace.

In Figure 1.14b, place P_1 models the availability of a *resource* which can be used by the lefthand part as from the firing of T_1 and up to the firing of T_2 or by the righthand part in similar conditions, but not by both at the same time (*resource sharing*). In this figure, the lefthand part and the righthand part evolve concurrently except that there is a *mutual exclusion* between the activities from T_1 to T_2 on the one hand and from T_3 to T_4 on the other. For example, in computer science, the mutual exclusion of program parts which use the same resource, such as for example a common memory, can be modeled in this way.

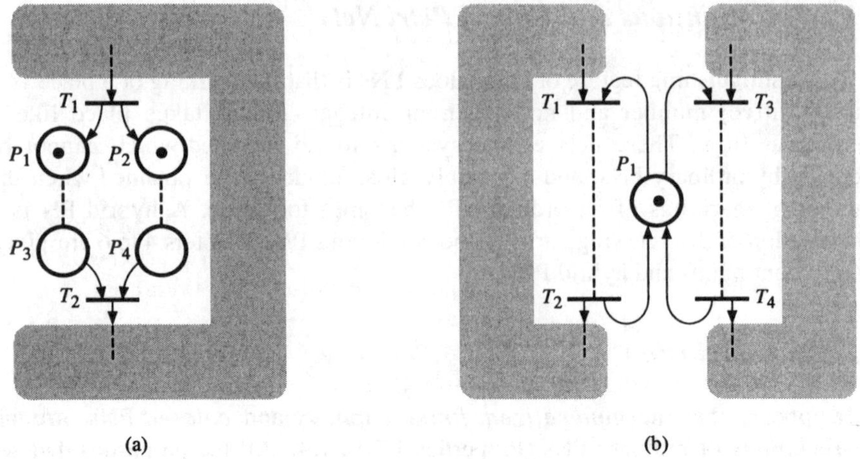

Figure 1.14 Concurrency. (a) Parallelism. (b) With resource sharing.

In a more general case, there may be several identical resources (several tokens in P_1). In this case, each part can use some resources simultaneously. For example, if there are three resources (initially three tokens in P_1), at some time the left hand part may use two resources while the right hand part uses one resource (or the three resources may be used by the same part).

Figure 1.15a represents the reciprocal synchronization of two evolutions. In the lefthand part, place P_1 is marked, but before proceeding with the evolution, we wait for place P_2 to also be marked. As soon as both places are marked, transition T_1 can be fired, thereby causing the marking of places P_3 and P_4 and the continuation of the evolution in each of the parts of the PN. Likewise, if place P_2 had been marked before place P_1, the righthand part would have waited in order to synchronize with the lefthand part (*rendez-vous* or *meeting* type).

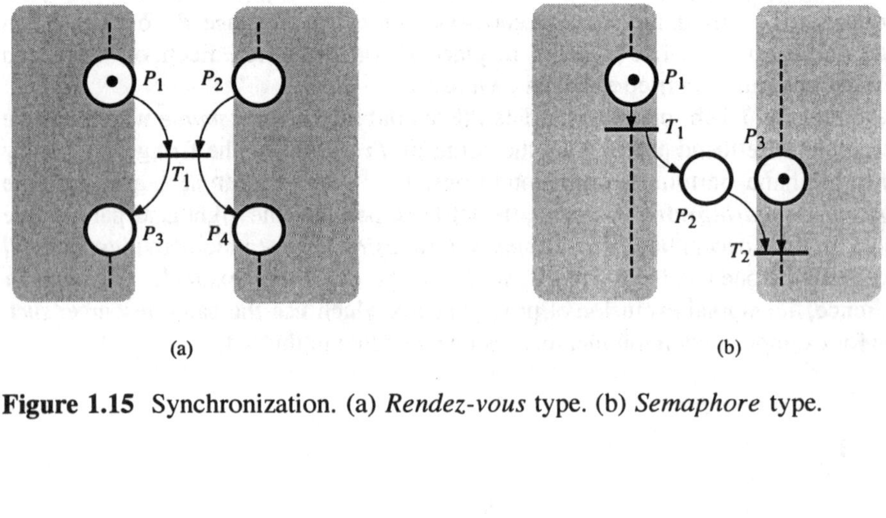

Figure 1.15 Synchronization. (a) *Rendez-vous* type. (b) *Semaphore* type.

Figure 1.15b illustrates a different case. The lefthand part evolves separately from the righthand part. On the other hand, transition T_2 must be fired after transition T_1. In the figure, place P_3 is marked before place P_2: as soon as transition T_1 is fired, transition T_2 can also be fired. If T_1 were fired when there is no token in P_3, place P_2 memorizes the authorization to fire T_2 (*semaphore* type).

Place P_1 of Figure 1.16a memorizes the fact that transition T_1 has been fired, thereby authorizing the future firing of T_2, provided that the other enabling conditions have been met (this case is fairly similar to that of Figure 1.15b).

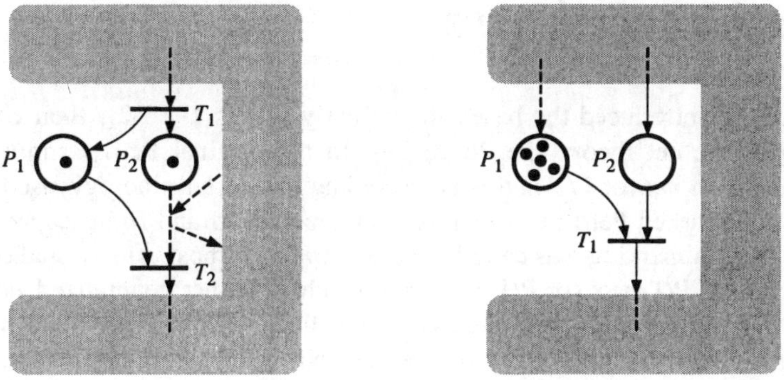

(a) (b)

Figure 1.16 Memorizing. (a) Of a firing. (b) Of a number.

A number can also be memorized (number of parts in a buffer for example) by the presence of several tokens in a place. There are 5 tokens in place P_1 of Figure 1.16b.

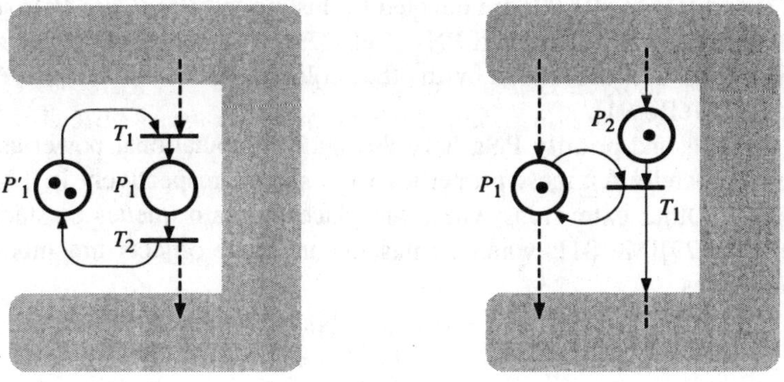

(a) (b)

Figure 1.17 (a) Limited capacity. (b) Reading.

Figure 1.17a represents a functioning such that firing of transition T_1 is only possible if place P_1 contains less than two tokens. We have already encountered similar cases of *limited capacity*. Each firing of T_1 decreases the number of tokens in P_1'.

In Figure 1.17b, the firing of T_1 is also conditioned by the presence of at least one token in P_1. However, this firing does not modify the marking of P_1. It corresponds to a *reading* of the marking of this place.

NOTES and REFERENCES

C. A. Petri introduced the bases in the early sixties [Pe 62], then continued working on the net theory [Pe 76 & 79]. In the original Petri terminology, a *place/transition net* (or *P/T-net*) was, according to the terminology used in this book, a non-marked Petri net which is both generalized and finite capacity. The same net with a marking was called a *P/T-system*. The most widely studied net is the subclass of P/T-nets (or P/T-systems) which is neither generalized nor finite capacity. In many papers this subclass (here called ordinary Petri nets) has been called place/transition net, or P/T-net, or just Petri nets by various authors.

Several methods can be used for converting a generalized PN into an ordinary PN [By 75][Ha 74][Si 79][KaMi 79]. Some of them are presented in the books [Br 83][Si 85][DaAl 89&92].

The concept of free choice was introduced in [Ha 72], while extended free choice and simple PNs appeared in [Co 72]. Equal conflict PNs were introduced in [TeSi 93] and widely studied in [TeSi 96].

A technique allowing the conversion of an impure PN into a pure PN can be found in [Br 83] or [DaAl 89&92] .

The model of colored PNs introduced by K. Jensen [Je 81] is widely used; parts of [Al 87] and [DaAl 89&92] are devoted to this model. *Predicate* Petri nets are also an interesting kind of colored PNs [GeLa 79]. The concepts of hierarchy and *high-level nets* are associated with the colored PNs [Si 85][HuJeSh 89] [MaMuSi 87][JeRo 91].

Extended PNs and priority PNs have the same computational power as Turing machines (Appendix A); these properties were shown respectively in [AgFl 73] and [Ha 75]. Other extensions where the places contain queues or stacks were proposed [Ro 79][Me 81]; when firings occur, these objects are modified by appropriate rules.

2

Properties of Petri Nets

The main properties possessed by certain PNs are presented in Section 2.1. These properties are linked to the concepts of boundedness, liveness and deadlocks, conservative components and repetitive components.

We must now look into how we can find out which PNs possess which properties. Seeking the properties is the goal of Section 2.2. The two basic methods are drawing up the graph of markings, on the one hand, and linear algebra on the other.

2.1 PRESENTATION OF THE MAIN PROPERTIES

This section deals mainly with ordinary Petri nets. When we wish to mention properties relating to a larger category, it will be specified.

After introduction of notations and definitions in Section 2.1.1, the concept of boundedness is presented in Section 2.1.2, liveness and deadlock in Section 2.1.3, conflicts in Section 2.1.4, and invariants in Section 2.1.5.

2.1.1 Notations and Definitions

Let us first introduce a few useful notations and concepts. Figure 2.1 represents a PN with its initial marking $\mathbf{m}_0$. The *marking* of a PN at a given moment is a *column vector* whose *i*th component is the marking of place P_i at this moment (see Figure 2.1). For simplicity's sake, we shall write the markings in the *transposed form* in the text. We shall use square brackets to represent a matrix and round brackets to represent the transposed form. For example:

$$\mathbf{m}_0 = (1, 0, 0, 0, 0) = \begin{bmatrix} 1 & 0 & 0 & 0 & 0 \end{bmatrix}^T = \begin{bmatrix} 1 \\ 0 \\ 0 \\ 0 \\ 0 \end{bmatrix}$$

For marking $\mathbf{m}_0$, there is an enabled transition which is T_1. The *firing* of T_1 *from* $\mathbf{m}_0$ *results in the marking* $\mathbf{m}_1 = (0, 1, 0, 1, 0)$. This will be given as:

$$\mathbf{m}_0 \xrightarrow{T_1} \mathbf{m}_1.$$

For the marking $\mathbf{m}_1$, there are two enabled transitions, T_2 and T_3. If the markings obtained by firing of these transitions are called $\mathbf{m}_2$ and $\mathbf{m}_3$ respectively (see Figure 2.2), we shall thus have:

$$\mathbf{m}_1 \xrightarrow{T_2} \mathbf{m}_2 = (0, 0, 1, 1, 0),$$

$$\mathbf{m}_1 \xrightarrow{T_3} \mathbf{m}_3 = (0, 1, 0, 0, 1).$$

For marking $\mathbf{m}_2$, only transition T_3 is enabled. Its firing results in $\mathbf{m}_4$:

$$\mathbf{m}_2 \xrightarrow{T_3} \mathbf{m}_4 = (0, 0, 1, 0, 1).$$

For marking $\mathbf{m}_3$, only transition T_2 is enabled. Its firing also results in $\mathbf{m}_4$. For marking $\mathbf{m}_4$, only transition T_4 is enabled. Its firing results in the initial marking $\mathbf{m}_0$. All possible evolutions starting from the initial marking have been seen.

When the PN under consideration is implicit, $\mathcal{M}(\mathbf{m}_0)$ denotes the **set of markings reachable** from marking $\mathbf{m}_0$ by a *finite* firing sequence. For our example, $\mathcal{M}(\mathbf{m}_0) = \{\mathbf{m}_0, \mathbf{m}_1, \mathbf{m}_2, \mathbf{m}_3, \mathbf{m}_4\}$. These reachable markings are illustrated in Figure 2.2.

Starting from marking $\mathbf{m}_0$, first T_1 and then T_2 can be fired. After these firings, the marking is $\mathbf{m}_2$. This can be written as:

$$\mathbf{m}_0 \xrightarrow{T_1T_2} \mathbf{m}_2.$$

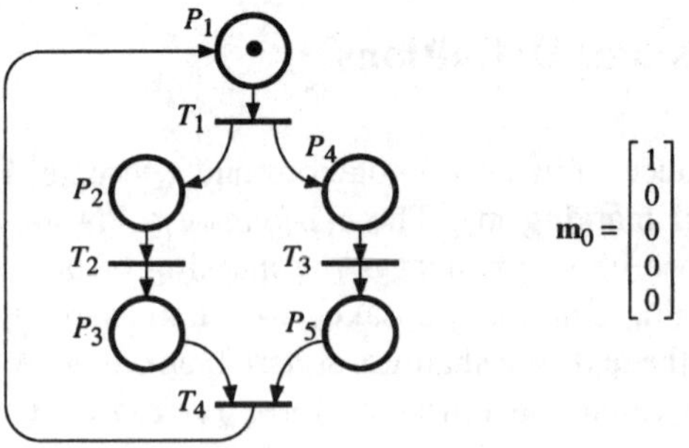

Figure 2.1 Petri net with its initial marking.

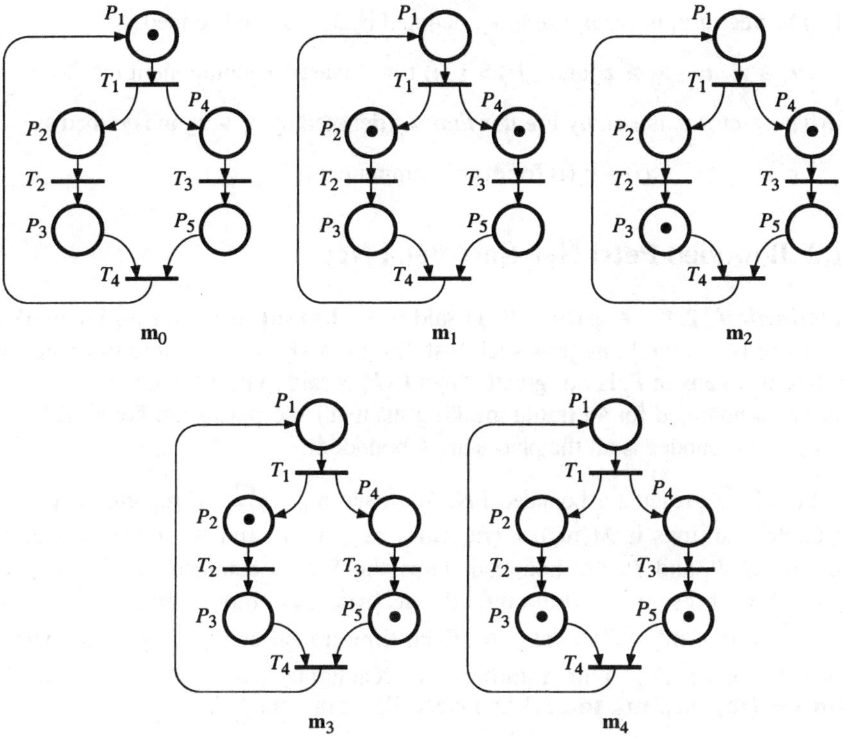

Figure 2.2 Reachable markings of a Petri net.

The string T_1T_2 is known as a **firing sequence.** The following can be written:

$$S = T_1T_2$$

and $\mathbf{m}_0 \xrightarrow{S} \mathbf{m}_2$.

A firing sequence is defined from a given marking. It is a series of transitions which are firable in turn (without other firing of transitions). For example, the sequence T_1T_4 is not a firing sequence from marking $\mathbf{m}_0$. Indeed, when T_1 has been fired, transition T_4 is not enabled and therefore cannot be the first transition fired after T_1.

Let us now introduce a notation which will be useful in the sequel.

Notation 2.1 Let $\mathbf{z}_1$ and $\mathbf{z}_2$ be two vectors of the same size (for example two markings of the same PN), and $z_j(i)$ denote the ith component of $\mathbf{z}_j$.

a) The vector $\mathbf{z}_1$ is *greater than or equal to* $\mathbf{z}_2$: denoted by $\mathbf{z}_1 \geq \mathbf{z}_2$ and defined as

$$\mathbf{z}_1 \geq \mathbf{z}_2 \Leftrightarrow z_1(i) \geq z_2(i) \text{ for every component } i.$$

b) The vector $\mathbf{z}_1$ is *greater than* $\mathbf{z}_2$: denoted by $\mathbf{z}_1 \gneq \mathbf{z}_2$ and defined as

$$\mathbf{z}_1 \gneq \mathbf{z}_2 \Leftrightarrow \mathbf{z}_1 \geq \mathbf{z}_2 \text{ and } z_1(i) > z_2(i) \text{ for at least one component } i.$$

c) The vector $\mathbf{z}_1$ is *strictly greater than* $\mathbf{z}_2$: denoted by $\mathbf{z}_1 > \mathbf{z}_2$ and defined as

$$\mathbf{z}_1 > \mathbf{z}_2 \Leftrightarrow z_1(i) > z_2(i) \text{ for every component } i.$$

2.1.2 Bounded Petri Net, Safe Petri Net

Definition 2.1 A **place** P_i is said to be **bounded** for an initial marking $\mathbf{m}_0$ if there is a natural integer k such that, for all markings reachable from $\mathbf{m}_0$, the number of tokens in P_i is not greater than k (P_i is said to be k-bounded).

A *PN is bounded* for an initial marking $\mathbf{m}_0$ if all the places are bounded for $\mathbf{m}_0$ (the PN is k-bounded if all the places are k-bounded).

❑

Figure 2.3a presents a bounded PN. We have $\mathbf{m}_0 \xrightarrow{T_1T_2} \mathbf{m}_0$ etc. The set of reachable markings is $\mathcal{M}(\mathbf{m}_0) = \{\mathbf{m}_0, \mathbf{m}_1, \mathbf{m}_2, \mathbf{m}_3\}$, and we observe that the markings of P_1 and P_2 are bounded. Thus the PN is bounded. It is 3-bounded. Figure 2.3b presents an unbounded net from $\mathbf{m}_0$. Indeed $\mathbf{m}_0 \xrightarrow{T_1} \mathbf{m}_1$, $\mathbf{m}_1 \xrightarrow{T_1} \mathbf{m}_2$, $\mathbf{m}_2 \xrightarrow{T_1} \mathbf{m}_3$ etc. Each time transition T_1 is fired, a token is added to place P_2. The number of reachable markings is unbounded: $\mathcal{M}(\mathbf{m}_0) = \{\mathbf{m}_0, \mathbf{m}_1, \mathbf{m}_2, \mathbf{m}_3, ...\}$ and place P_2 is not bounded.

Definition 2.2 A PN is said to be **safe** for an initial marking $\mathbf{m}_0$ if for all reachable markings, each place contains zero or one token.

❑

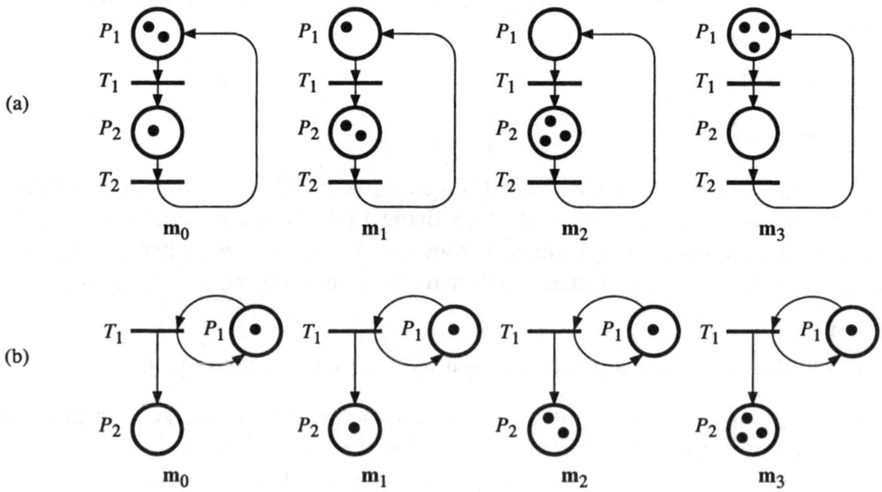

Figure 2.3 (a) Bounded PN. (b) Unbounded PN.

A safe PN is thus a particular case of bounded Petri net for which all the places are 1-bounded. The PN of Figure 2.3a is bounded but not safe. The PN of Figure 2.2 is safe. As a matter of fact we can see that for each of the reachable markings, there is always zero or one token in each place.

Remark 2.1 The properties of *safe* and *bounded* PNs *depend on the initial marking* $\mathbf{m}_0$. This is clear for the first of these properties. It is sufficient that $\mathbf{m}_0$ be such that one of the places contains 2 tokens for the PN not to be safe. As for the property of bounded PNs, it can be seen in Figure 2.3b that if the initial marking of P_1 is zero, transition T_1 is not and never will be enabled. The marking of P_2 will never change. If $\mathbf{m}_0 = \{0, N\}$ then $\mathcal{M}(\mathbf{m}_0) = \{\mathbf{m}_0\}$ whatever $N \geq 0$. This PN would thus be bounded with this initial marking. Figure 2.3a shows that the PN is bounded whatever $\mathbf{m}_0$ (irrespective of the evolution, the number of tokens remains constant). An *unmarked* PN is said to be **structurally bounded** if *for all initial finite marking,* the marked PN is bounded.

Property 2.1 If a PN is unbounded for $\mathbf{m}_0$, it is unbounded for $\mathbf{m}'_0 \geq \mathbf{m}_0$.

Remark 2.2. The concept of a *bounded* PN applies to all the *abbreviations* and *extensions*. The concept of a safe PN could apply to all the abbreviations and extensions (but with slight differences), with the exception of the continuous PNs since the place markings are not integers. A generalized PN which is safe would degenerate into an ordinary PN. This is illustrated in Figure 2.4. For the initial marking $\mathbf{m}_0$ of Figure a, the PN is safe, but transition T_3 will never be enabled since two tokens are needed in P_1, and functioning is equivalent to that of the ordinary PN in Figure b. For the initial marking $\mathbf{m}'_0$ of Figure c, the PN is not safe since firing of T_4 would add 2 tokens to P_2.

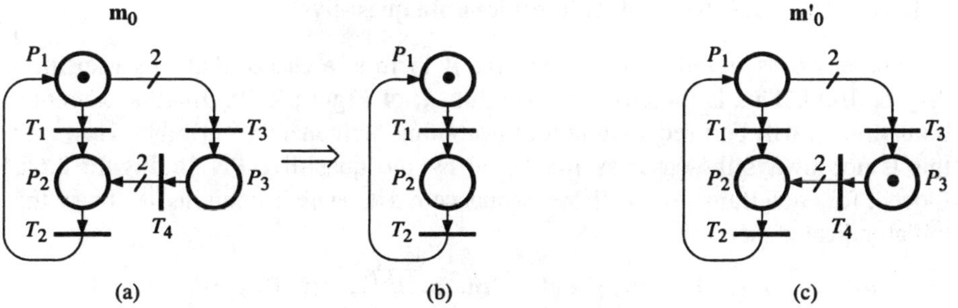

Figure 2.4 Degeneration of a generalized safe PN.

2.1.3 Liveness and Deadlock

The marking of a PN evolves by the firing of transitions. When certain transitions are no longer firable and when all or part of the PN no longer "functions", there is most probably a problem in the design of the system described. We shall now look at the concepts relating to this.

Definition 2.3 A **transition** T_j is **live** for an initial marking $\mathbf{m}_0$ if for every reachable marking $\mathbf{m}_i \in \mathcal{M}(\mathbf{m}_0)$ a firing sequence S from $\mathbf{m}_i$ exists, which contains transition T_j.

❑

In other words, whatever the evolution, a possibility always remains for firing T_j. This concept is illustrated in Figure 2.5. Figure a presents a PN which has 2 reachable markings. For each of these markings, a firing sequence exists which contains T_1. Indeed, $\mathbf{m}_0 \xrightarrow{T_1} \mathbf{m}_1$, and $\mathbf{m}_1 \xrightarrow{T_2T_1} \mathbf{m}_1$. Transition T_1 is thus live.

The PN of Figure b has 3 reachable markings from $\mathbf{m}_0$:

$$\mathcal{M}(\mathbf{m}_0) = \{\mathbf{m}_0, \mathbf{m}_1, \mathbf{m}_2\}.$$

We may observe that there is no firing sequence which contains T_1 from $\mathbf{m}_1$ (or from $\mathbf{m}_2$). That is to say that if marking $\mathbf{m}_1$ is reached, transition T_1 will never be enabled again. This transition is not live.

Definition 2.4 A *PN is live* for an initial marking $\mathbf{m}_0$ if all its transitions are live for $\mathbf{m}_0$.

❑

In other words, if a PN is live, this means that, regardless of the evolution, no transition will become unfirable on a permanent basis.

Let us now look at some weaker properties.

Definition 2.5

a) A **transition** T_j is **quasi-live**[1] for an initial marking $\mathbf{m}_0$, if there is a firing sequence from $\mathbf{m}_0$ which contains T_j.

b) A *PN is quasi-live* if all its transitions are quasi-live.

❑

In other words, a transition is quasi-live if there is a chance that this transition may be fired. This is the case of transition T_1 of Figure 2.5b. In this example, transition T_1 will be fired once before becoming permanently unfirable. However this is not always the case, as illustrated by the quasi-live PN in Figure 2.6a. Indeed, for each transition, a firing sequence exists which contains it, from the initial indicated marking:

$$\mathbf{m}_0 \xrightarrow{T_1} (0, 1, 0, 0), \quad \text{and} \quad \mathbf{m}_0 \xrightarrow{T_2T_3T_4} (0, 0, 1, 0).$$

However, there is a conflict $\langle P_1, \{T_1, T_2\}\rangle$. If the first transition fired is T_1, then the remaining transitions will never be fired, and if the first transition fired is T_2, then T_1 will never be fired.

---

[1] The term *non-dead* is also used. The authors' choice is commented in Notes & References at the end of this chapter.

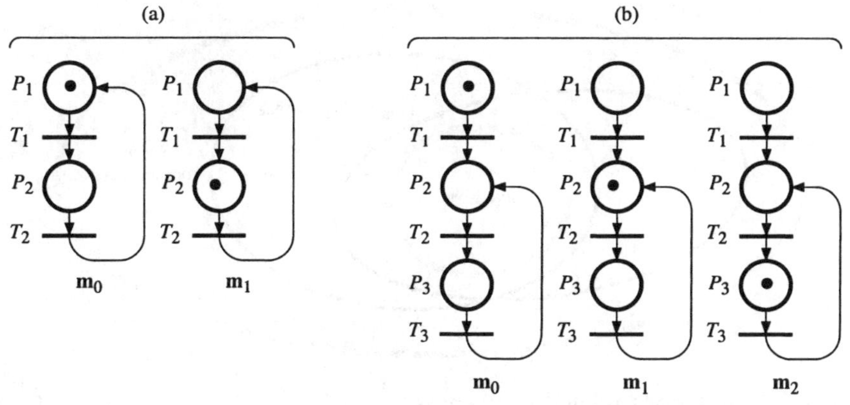

Figure 2.5 Illustration of liveness.

Definition 2.6 A **deadlock**[2] (or *sink state*) is a marking such that no transition is enabled.
❑

In Figure 2.6a, firing of transition T_1 results in the marking $\mathbf{m}_1 = (0, 1, 0, 0)$. This is a deadlock. Henceforward, the marking can no longer evolve.

Definition 2.7 A PN is said to be **deadlock free** for an initial marking $\mathbf{m}_0$ if no reachable marking $\mathbf{m}_i \in \mathcal{M}(\mathbf{m}_0)$ is a deadlock.
❑

Figure 2.6b is a *deadlock free* PN (certain authors call this *pseudo-live*). If T_1 is the first transition fired, the marking $\mathbf{m}_1 = (0, 1, 0, 0, 0)$ is obtained, and transitions T_3 and T_5 are live from this marking. If T_2 is the first transition fired, behavior is symmetrical. In other words, a deadlock free PN is such that there will always be a part "which functions".

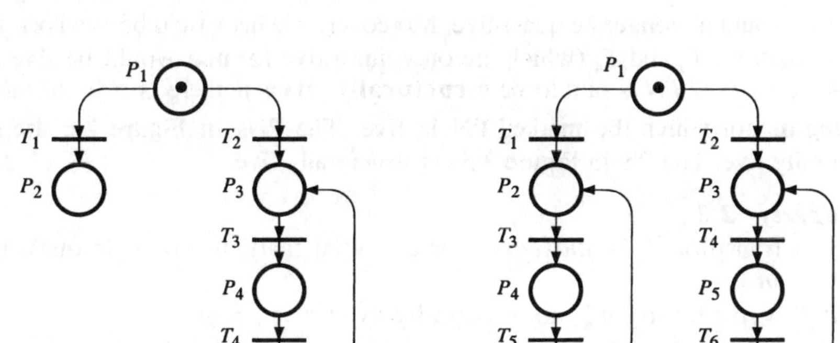

Figure 2.6 Quasi-live PNs. (a) With deadlock. (b) Deadlock free.

[2] Some authors use the word *deadlock* with another meaning (Definition 2.16 in Section 2.2.4).

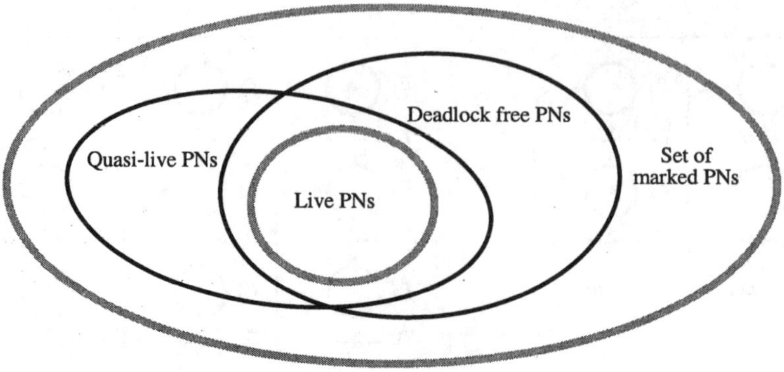

Figure 2.7 Relations related to liveness.

Remark 2.3 See Figure 2.7.

a) The properties of *quasi-liveness* and *deadlock* are independent. The two PNs of Figure 2.6 are quasi-live; the first has a deadlock and the second is deadlock free.

b) The properties of *liveness* and *deadlock* are not independent: a live PN has no deadlock, i.e., a PN with a deadlock cannot be live. As a matter of fact, a PN contains at least one transition (Definition 2.11 in Section 2.2.2.1), which cannot be fired if a deadlock is reached.

Remark 2.4 The properties of *liveness* and *deadlock* clearly *depend* on the *initial marking*. If the initial marking of the PN of Figure 2.5a is zero, i.e. $\mathbf{m}'_0 = (0, 0)$, this is a deadlock state and no transitions are enabled. Let us take a less trivial example. We have seen that the PN of Figure 2.6a was quasi-live and with a state of deadlock if the initial marking is $\mathbf{m}_0 = (1, 0, 0, 0)$. If the initial marking were $\mathbf{m}'_0 = (0, 0, 1, 0)$, the net would not be quasi-live since transitions T_1 and T_2 would no longer be quasi-live. Moreover, this net would be deadlock free since transitions T_3 and T_4 (which are only quasi-live for $\mathbf{m}_0$) would be live for $\mathbf{m}'_0$. An *unmarked PN* is said to be **structurally live** if there is a finite *initial marking* $\mathbf{m}_0$ for which the marked PN is live. The PNs in Figure 2.6 are not structurally live. The PN in Figure 2.5a is structurally live.

Property 2.2

a) If a transition T_j is *quasi-live* for an initial marking $\mathbf{m}_0$, it is quasi-live for $\mathbf{m}'_0 \geq \mathbf{m}_0$.

b) If T_j is *live* for $\mathbf{m}_0$, it is not necessarily live for $\mathbf{m}'_0 \geq \mathbf{m}_0$.

c) If a PN is *deadlock free* for $\mathbf{m}_0$, it is not necessarily so for $\mathbf{m}'_0 \gneqq \mathbf{m}_0$.

❑

Property 2.2a is quite clear since all firing sequences from $\mathbf{m}_0$ are also firing sequences from $\mathbf{m}'_0$. Figure 2.8 illustrates the fact that the properties of liveness and deadlock free are not maintained by increasing the marking. For marking

$\mathbf{m}_0 = (1, 0, 0)$, the reachable markings are (1, 0, 0) and (0, 1, 0); thus transitions T_1 and T_2 are live and T_3 is never enabled. For marking $\mathbf{m}'_0 = (2, 0, 0)$, we obtain $\mathbf{m}'_0 \xrightarrow{T_1T_3} (0, 0, 1)$ which is a deadlock state (thus T_1 and T_2 are not live). This example thus illustrates both Properties 2.2b and 2.2c. It also corresponds to a PN which is structurally not live. Indeed, irrespective of the initial marking (finite), firing of transition T_3 removes two tokens from the set of places $\{P_1, P_2\}$. At the end of a finite number of firings of T_3, only 0 or 1 token will remain in $\{P_1, P_2\}$, and transition T_3 will never be enabled.

Remark 2.5 The concepts of liveness, quasi-liveness and deadlock apply to all the *abbreviations* and may be generalized to all PN *extensions*. For continuous PNs, new concepts will be given in Section 4.3. ❑

The concepts relating to a marking which is always reachable and to the possibility to return to the initial marking, are defined below.

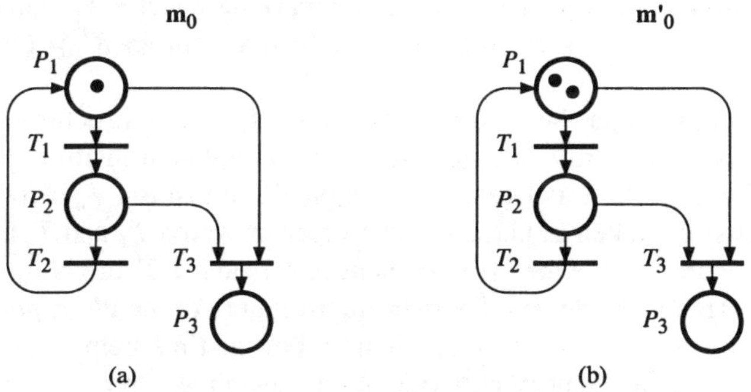

Figure 2.8 (a) Transitions T_1 and T_2 are live, and T_3 is not quasi-live. (b) Transitions T_1 and T_2 are not live, and T_3 is quasi-live.

Definition 2.8

a) A PN has a **home state** $\mathbf{m}_h$ for an initial marking $\mathbf{m}_0$ if for every reachable marking $\mathbf{m}_i \in \mathcal{M}(\mathbf{m}_0)$, a firing sequence S_i exists such that $\mathbf{m}_i \xrightarrow{S_i} \mathbf{m}_h$.

b) A PN is **reversible** for an initial marking $\mathbf{m}_0$ if $\mathbf{m}_0$ is a home state (the word *reinitializable* is also used). ❑

It is clear that the existence of a home state depends on the initial marking. For example, the PN of Figure 2.6a has no home state for $\mathbf{m}_0 = (1, 0, 0, 0)$. On the other hand, it has two home states for $\mathbf{m}'_0 = (0, 0, 1, 0)$: these two states are the markings $\mathbf{m}'_0$ and (0,0,0,1). The set of home states is the **home space**. This notion of home state may apply to all the PN abbreviations and extensions.

2.1.4 Conflicts

In this section, generalized PNs will be explicitly considered.

The concept of the structural conflict have already been presented in Section 1.2.1.3. We shall now specify the concepts of *effective conflict*, *general conflict*, *persistence*, and *concurrency*. Remember that a structural conflict corresponds to a set of at least 2 transitions T_1 and T_2 which have an input place in common. This is written as $K = \langle P_i, \{T_1, T_2, ...\}\rangle$.

In an *ordinary PN*, an *effective conflict* is the existence of a structural conflict K, and of a marking $\mathbf{m}$, such that the number of tokens in P_i is less than the number of output transitions of P_i which are enabled by $\mathbf{m}$.

An effective conflict is represented by a triple $K^E = \langle P_i, \{T_1, T_2, ...\}, \mathbf{m}\rangle$. This notion is illustrated in Figure 2.9. The structural conflict $K_1 = \langle P_2, \{T_1, T_2\}\rangle$ exists in Figures a to c. In Figure a, the current marking $\mathbf{m}_1$ only enables transition T_1 and there is thus no effective conflict. In Figure b, both conflict transitions are enabled but there are two tokens in P_2; thus there is no effective conflict since both transitions can be fired. In Figure c, the marking enables T_1 and T_2 but since there is only one token in P_2, there is an effective conflict $K^E = \langle P_2, \{T_1, T_2\}, \mathbf{m}_3\rangle$.

In Figure d, although there is an effective conflict, the structure is particular. Indeed, the marking only allows one of the enabled transitions to be fired. However, if one of these is fired, for example T_3, transition T_4 can then be fired since there is still a token in place P_5. The order in which T_3 and T_4 are fired may have no importance, since the firing sequences T_3T_4 and T_4T_3 are both possible and result in the same marking. For the marking in Figure d, the PN is *persistent*; this concept will be discussed later in this section. For the time being, let us define an effective conflict in the general case (i.e., for a generalized PN).

Definition 2.9 An **effective conflict**, denoted by $K^E = \langle P_i, \{T_1, T_2, ...\}, \mathbf{m}\rangle$ is the existence of a structural conflict $K = \langle P_i, \{T_1, T_2, ...\}\rangle$, and of a marking $\mathbf{m}$, such that the transitions in the set $\{T_1, T_2, ...\}$ are enabled by $\mathbf{m}$ and the number of tokens in P_i is less than the sum of the weights of the arcs $P_i \rightarrow T_1$, $P_i \rightarrow T_2$, ...

❑

For the marking $\mathbf{m}_2$ of Figure 2.9b, it is clear that the firings of T_1 and T_2 are independent since there are enough tokens to fire both. It follows that their concurrent firings can be contemplated. Let us denote $[T_1T_2]$ the **multiple firing** of T_1 and T_2. This means that transitions T_1 and T_2 are fired *in one step* (i.e. *simultaneously*). The double firing $[T_1T_2]$ leads from $\mathbf{m}_2$ to another marking, say $\mathbf{m}_4$, hence one can write: $\mathbf{m}_2 \xrightarrow{[T_1T_2]} \mathbf{m}_4$.

In this example, simultaneous firing of T_1 and T_2 is possible. However, it is not mandatory. Let us denote by $\{T_1T_2\}$ the **concurrent firing** of both transitions, i.e., they can be fired in any order, possibly simultaneously:

$$\{T_1T_2\} = T_1T_2 + T_2T_1 + [T_1T_2]. \qquad (2.1)$$

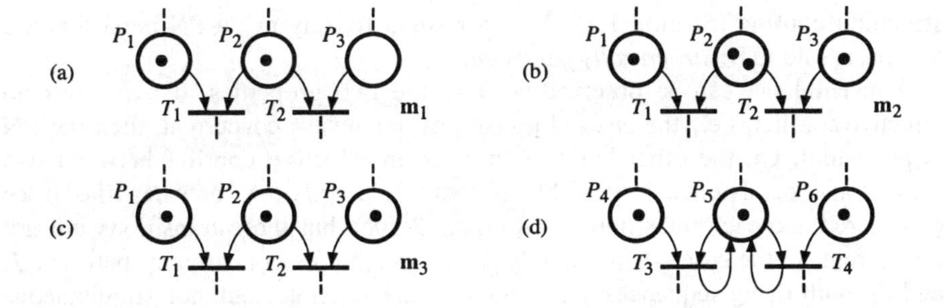

Figure 2.9 (a) and (b) No effective conflicts. (c) Effective conflict. (d) Effective conflict but persistence.

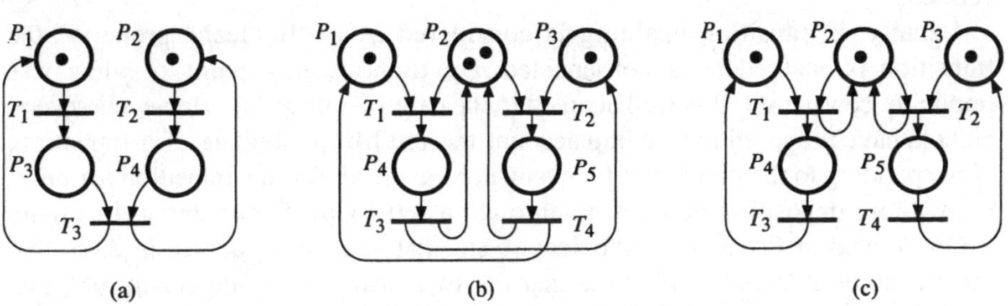

Figure 2.10 (a) and (b) Persistence and concurrency. (c) Only persistence.

It will be specified in Section 2.2.2.2 (Remark 2.11) that a firing sequence leads to a marking which depends on the transitions fired but not on their order in the sequence. Hence, the firing of T_1T_2 or T_2T_1 or $[T_1T_2]$ or $\{T_1T_2\}$ leads to the same marking $\mathbf{m}_4$; then $\mathbf{m}_2 \xrightarrow{\{T_1T_2\}} \mathbf{m}_4$. In a general case, the concurrent firing $\{T_aT_b\ldots T_u\}$ denotes the firing of all the transitions T_a, T_b, ..., and T_u, in any order, possibly including one or several simultaneous firings.

Two (or more) transitions are said to be **concurrent** if they are enabled and *causally independent*, i.e., a transition may fire before or after or simultaneously with the other. In most cases, these two transitions have no common input place; this is illustrated in Figure 2.10a. where T_1 and T_2 are concurrent. The transitions may have a common input place, but, in this case, there are enough tokens in this place to fire both transitions (i.e., there is no effective conflict), as in Figure 2.10b. If, for each reachable marking $\mathbf{m}_i \in \mathcal{M}(\mathbf{m}_0)$, there is no effective conflict, then at any time the enabled transitions are concurrent.

Let us now specify the concept of persistence. A PN is **persistent** if for every reachable marking $\mathbf{m}_i \in \mathcal{M}(\mathbf{m}_0)$, there is the following property: for every pair of transitions T_j and T_k enabled by marking $\mathbf{m}_i$, the sequence T_jT_k is a firing sequence from $\mathbf{m}_i$ (as well as T_kT_j by symmetry). In other words, this means that an enabled transition can only be disabled by its own firing. Note that a PN without structural conflict (Section 1.2.1.3) is persistent for any $\mathbf{m}_0$; a PN persistent for any $\mathbf{m}_0$ is said to be *structurally persistent.*

A resemblance can be observed between the two properties. If a PN has no effective conflict, i.e., the enabled transitions are always concurrent, then the PN is persistent. On the other hand, if there is an effective conflict between two transitions, persistence is possible if there is a self-loop between the place considered and each transition (as in Figure 2.10c), but these transitions are not concurrent. In the case where there is persistence but no concurrency between T_j and T_k, both firing sequences T_jT_k and T_kT_j are possible, but not simultaneous firing $[T_jT_k]$. The reader may verify that for all the reachable markings for the PNs in Figures 2.10a and b, the enabled transitions are always concurrent (hence, these PNs are persistent). The PN in Figure c is persistent, but concurrency is not verified.

Usually, a transition enabling is considered as a "Boolean" property: the transition is enabled or is not enabled. Up to now, this point of view was implicitly considered (it is well adapted to the case of safe PNs). However, several authors have proposed to take into account the **enabling degree** of a transition, corresponding to the number of firings which can occur for this transition (*at once*, i.e. without depositing new tokens through a self-loop). For example, in Figure 2.11a, transition T_1 can be fired twice; we say that its enabling degree is 2, or that this transition is 2-*enabled*[3]. Note that the two firings of T_1 are concurrent, i.e., the firing sequence may be $\{T_1T_1\} = T_1T_1 + [T_1T_1]$. According to Appendix A, T_1T_1 may be written as $(T_1)^2$, and, for example, $\{T_1T_1\} = \{(T_1)^2\}$.

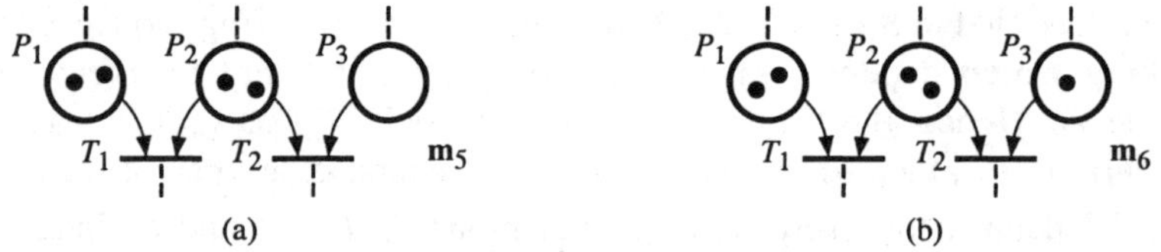

Figure 2.11 (a) Transition T_1 is 2-enabled. (b) General conflict.

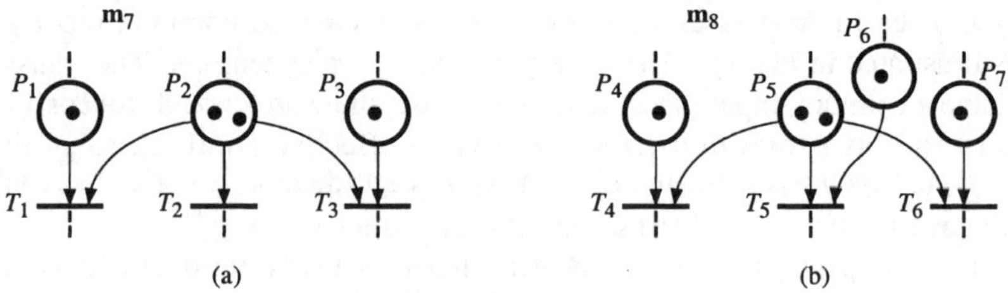

Figure 2.12 (a) No general conflict between T_1 and T_3. (b) General conflict without "subconflicts".

[3] These concepts will be formally defined in Section 3.1.

In Figure 2.11b, transitions T_1 and T_2 are enabled with enabling degrees 2 and 1 respectively. However, we cannot have two firings of T_1 *and* a firing of T_2, since there are only 2 tokens in P_2. This situation, which is not an effective conflict according to Definition 2.9, will be called a *general conflict*: the firing of T_2 decreases the enabling degree of T_1, and the double firing of T_1 disables T_2; the concurrent firings $\{T_1T_1\}$ and $\{T_1T_2\}$ are possible, but not $\{T_1T_1T_2\}$.

Definition 2.10 A **general conflict**, denoted by $K^G = \langle P_i, \{T_1, T_2, ...\}, \mathbf{m}\rangle$ is the existence of a structural conflict $K = \langle P_i, \{T_1, T_2, ...\}\rangle$, and of a marking $\mathbf{m}$, such that the number of tokens in P_i does not suffice to fire all the output transitions of P_i according to their *enabling degrees*[4].

Remark 2.6 According to the notation proposed, if $K_1 = \langle P_i, A\rangle$ is a structural conflict such that A is a set of at least 3 transitions, then, for any subset A' of A containing at least two transitions, $K_2 = \langle P_i, A'\rangle$ is a structural conflict. Let us now discuss general conflicts.

❑

Various authors define an effective conflict, or its generalization known here as a general conflict, between *two* transitions. We prefer that the definition *may consider a greater number of transitions*. Let us illustrate why from examples. According to Definition 2.10, in Figure 2.12a, $\langle P_2, \{T_1, T_3\}, \mathbf{m}_7\rangle$ is not a general conflict, although $\langle P_2, \{T_1, T_2, T_3\}, \mathbf{m}_7\rangle$, $\langle P_2, \{T_1, T_2\}, \mathbf{m}_7\rangle$, and $\langle P_2, \{T_2, T_3\}, \mathbf{m}_7\rangle$ are general conflicts. In Figure b, neither $\langle P_5, \{T_4, T_5,\}, \mathbf{m}_8\rangle$ nor $\langle P_5, \{T_4, T_6\}, \mathbf{m}_8\rangle$ nor $\langle P_5, \{T_5, T_6\}, \mathbf{m}_8\rangle$ is a general conflict, but $\langle P_5, \{T_4, T_5, T_6\}, \mathbf{m}_8\rangle$ is a general conflict. Hence, *a general conflict involving three transitions, or more, is not always induced from conflicts between two transitions* (conflicts between two transitions imply the others only for particular structures, such as Figure 2.12a or an equal conflict PN for example).

Remark 2.7 The concepts of effective conflict, general conflict, concurrency, and persistence, naturally apply to the *abbreviations* with a few adaptations in the definitions.

As it will be seen, these concepts also apply to the *extensions*, but the definitions must be complemented when non-autonomous PNs are considered. The concept of general conflict will be particularly useful for synchronized PNs and continuous PNs.

---

[4] I.e., $m(P_i) < \sum_j w_{\bar{ij}} \cdot q_j$, where $w_{\bar{ij}}$ is the weight of the arc $P_i \rightarrow T_j$ and q_j is the enabled degree of T_j, according to the notations which will be introduced in Sections 2.2.2.1 and 3.1. Note that, if an effective conflict exists, it is also a general conflict according to the definitions, but the reverse is not true. The name *effective conflict* may be used *synonymously* with *general conflict* when the enabling degrees of the transitions concerned are 1. Priority in case of general conflict is explained in Appendix B.

2.1.5 Invariants

Starting from an initial marking, the marking of a PN can evolve by the firing of transitions and, if there is no deadlock, the number of firings is unlimited. However, not just any marking can be reached, all the reachable markings have some properties in common; a property which does not vary when the transitions are fired is said to be *invariant*. Similarly, not just any transition sequence can be fired; some *invariant* properties are common to the possible firing sequences.

Hence, invariants enable certain properties of the reachable markings and firable transitions to be characterized, irrespective of the evolution.

2.1.5.1 Conservative Component

In Figure 1.3a, represented again in Figure 2.13a, we can see that, regardless of the evolution, there will always be one and only one token for all 4 places. Thus at all times, $m_1 + m_2 + m_3 + m_4 = 1$. This *invariant* property has an obvious meaning, namely that at all times we are at *one and only one season.*

The PN in Figure 1.8d is shown again in Figure 2.13b. We observe that $m_2 + m_2' = 2$ for all the reachable markings ($m_2' = m(P_2')$ represent the number of tokens in P_2'). This invariant is quite natural since place P_2' is *complementary* of P_2, in order to ensure that there will never be more than 2 tokens in P_2 whose capacity is limited. The number of tokens in P_2' corresponds to the number of "places available" in P_2.

Figure 2.2 presents a PN and all the reachable markings from an initial marking $\mathbf{m}_0$. The PN with the initial state $\mathbf{m}_0$ is shown again in Figure 2.13c. We can see that there is always one and just one token in the set of places $\{P_1, P_2, P_3\}$, i.e. $m_1 + m_2 + m_3 = 1$. Likewise, $m_1 + m_4 + m_5 = 1$. By calculating the sum of the two equations we have found, $2\,m_1 + m_2 + m_3 + m_4 + m_5 = 2$ is obtained. This is a new invariant which relates to all the places of the PN and which means that the *weighted sum* of tokens in the PN is constant (weight 2 for place P_1 and weight 1 for the rest).

Let R be a PN and P the set of its places. A **marking invariant** (also called *linear place invariant*, also simply called invariant) is obtained if there is a sub-set of places $P' = \{P_1, P_2, ..., P_r\}$ included in P and a weighting vector $(q_1, q_2, ..., q_r)$ for which all the weights q_i are positive integers such that:

$$q_1 \cdot m(P_1) + q_2 \cdot m(P_2) ... + q_r \cdot m(P_r) = \text{constant, for every } \mathbf{m} \in \mathcal{M}(\mathbf{m}_0). \quad (2.2)$$

The set of places P' is a **conservative component**. Net R is said to be **conservative** if P is a conservative component. The property of being a conservative component is independent from the initial marking (it is a structural property). On the other hand, the constant of the marking invariant depends on the initial marking. For example, in Figure 2.13b, $m(P_2) + m(P_2') = 2$. But if the initial marking were such that $m_0(P_2) + m_0(P_2') = N$, we would have $m(P_2) + m(P_2') = N$ for every reachable marking $\mathbf{m}$.

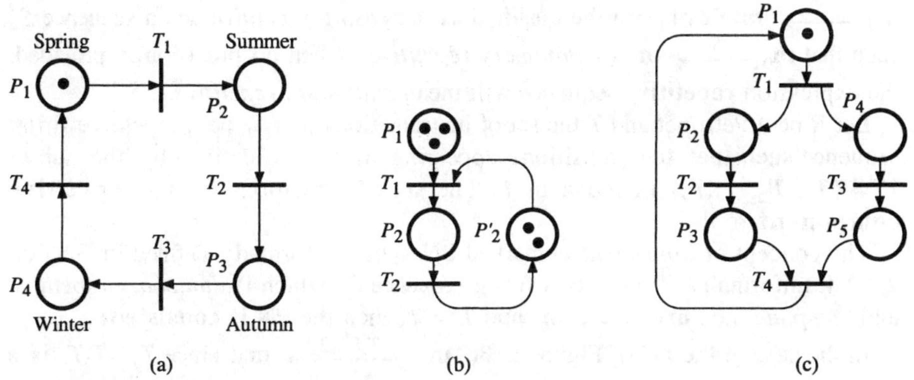

Figure 2.13 Illustration of conservative components.

Generally speaking, a conservative component has a physical meaning. For example, a system is in *one and only one state at a time* (Figure 2.13a), or a *number of entities* is maintained (Figure 2.13b). We shall look at further examples.

2.1.5.2 Repetitive Component

The firing sequences which are possible from the initial marking $\mathbf{m}_0$ of Figure 2.13a are as follows: T_1, T_1T_2, $T_1T_2T_3$, $T_1T_2T_3T_4$, $T_1T_2T_3T_4T_1$, etc. The firing sequence $T_1T_2T_3T_4$ is particular since $\mathbf{m}_0 \xrightarrow{T_1T_2T_3T_4} \mathbf{m}_0$. This sequence causes a return to the initial state. The sequence may thus be repeated. This is a **repetitive sequence.** A repetitive sequence which contains all the transitions (each at least once) is a **complete** repetitive sequence.

It is clear that if a transition sequence is a firing sequence from $\mathbf{m}_0$, then all its prefixes (see Appendix A) are also firing sequences from $\mathbf{m}_0$. For Figure 2.13a, $T_1T_2T_3T_4$ is the only *minimal* repetitive sequence, i.e., none of its proper prefixes are repetitive sequences (ε is not considered). In this example, the set of firing sequences is defined by the set of prefixes $\overline{\mathcal{L}}_1$ of $\mathcal{L}_1 = (T_1T_2T_3T_4)^*$, i.e., $\overline{\mathcal{L}}_1 = (T_1T_2T_3T_4)^* (\varepsilon + T_1 + T_1T_2 + T_1T_2T_3)$ where the iteration corresponds to the firing of the repetitive sequence any number of times, and the second part of the expression corresponds to all the proper prefixes of this sequence.

For the example of Figure 2.13c, there are two minimal repetitive sequences which are $S_1 = T_1T_2T_3T_4$ and $S_2 = T_1T_3T_2T_4$. The language generated is $\overline{\mathcal{L}}_2$ such that $\mathcal{L}_2 = (S_1 + S_2)^* = (T_1T_2T_3T_4 + T_1T_3T_2T_4)^*$.

For Figure 2.3b (Section 2.1.2), the case observed is slightly different from those we have just seen. The sequence $S_3 = T_1$ can be infinitely repeated, although it does not lead back to the initial marking $\mathbf{m}_0$, but to a marking $\mathbf{m}'_0 \gneq \mathbf{m}_0$. It is nevertheless a repetitive sequence. To be more precise, a sequence S_k such that

$\mathbf{m}_0 \xrightarrow{S_k} \mathbf{m}'_0 \gneq \mathbf{m}_0$ may be qualified as *increasing repetitive* and a sequence S_k such that $\mathbf{m}_0 \xrightarrow{S_k} \mathbf{m}_0$ as *stationary repetitive*. When no precision is provided, the expression **repetitive** sequence will mean *stationary repetitive*.

Let R be a Petri net and T the set of its transitions. Let S_k be a repetitive firing sequence such that the transitions appearing in S_k are defined by the subset $T' = \{T_1, T_2, ..., T_r\}$ included in T. The set of transitions T' is a **repetitive component**.

The concept of *consistent* unmarked PN will be formally defined in Section 2.2.2.4: unformally, if there is a firing sequence S_k which is *stationary repetitive* and *complete*, i.e., $\mathbf{m}_0 \xrightarrow{S_k} \mathbf{m}_0$ and $T' = T$, then the PN is **consistent**.

In the case of the PN of Figure 2.13c, $\mathbf{m}_0 \xrightarrow{T_1T_2T_3T_4} \mathbf{m}_0$; since $T_1T_2T_3T_4$ is a complete firing sequence, the PN is consistent. In the case of the PN of Figure 2.3b, $\{T_1\}$ is an increasing repetitive component, and the PN is increasing repetitive, i.e., it is not consistent.

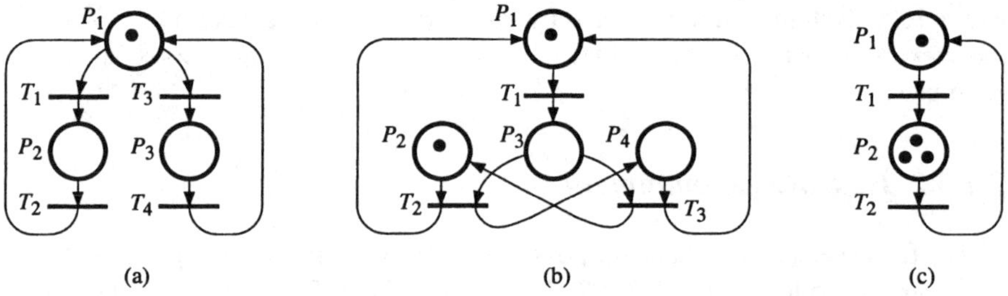

Figure 2.14 Illustration of repetitive sequences.

Let us now look at some slightly different examples.

In the case of the PN of Figure 2.14a, two repetitive sequences can immediately be seen, namely $S_1 = T_1T_2$ and $S_2 = T_3T_4$, since each of these sequences causes a return to the initial marking. There are two repetitive components which are $\{T_1, T_2\}$ and $\{T_3, T_4\}$. None of these components contains all the transitions of the PN. However, other repetitive components can be constructed using these components as a base. The set of all the repetitive sequences may be written as $\mathcal{L}_3 = (T_1T_2 + T_3T_4)^+$ (i.e., all the sequences in $(T_1T_2 + T_3T_4)^*$ except the sequence ε). The sequence $S_3 = T_1T_2T_3T_4$ is an element of $\mathcal{L}_3$. This is a *complete* repetitive sequence, hence the PN is consistent.

The example given in Figure 2.14b shows that a repetitive sequence can contain several firings of the same transition (while still being minimal). The initial marking $\mathbf{m}_0 = (1, 1, 0, 0)$ is such that $\mathbf{m}_0 \xrightarrow{T_1T_2T_1T_3} \mathbf{m}_0$. The sequence $S_4 = T_1T_2T_1T_3$ is thus repetitive. This sequence contains transition T_1 twice (and none of its proper prefixes is repetitive). Since S_4 is complete and stationary repetitive, the PN is consistent.

Property 2.3 Let $\mathcal{L}(\mathbf{m}_0)$ be the set of firing sequences from the initial marking $\mathbf{m}_0$. If $\mathbf{m}_0' \geq \mathbf{m}_0$ then $\mathcal{L}(\mathbf{m}_0') \supseteq \mathcal{L}(\mathbf{m}_0)$. In particular: if S is a repetitive sequence for $\mathbf{m}_0$, it is also a repetitive sequence for $\mathbf{m}_0'$.

❑

Here is an intuitive proof for Property 2.3. Let $\mathbf{m}_0' = \mathbf{m}_0 + \mathbf{m}_1$, where $\mathbf{m}_1$ is a vector $(m_1(P_1), \ldots, m_1(P_n))$ corresponding to the tokens added to the PN in order to obtain $\mathbf{m}_0'$ from $\mathbf{m}_0$. Assume a restricted behavior of the PN with the marking $\mathbf{m}_0'$ such that $m_1(P_i)$ tokens are "frozen" in each place P_i, i.e., these tokens do not move. This restricted behavior generates the language $\mathcal{L}(\mathbf{m}_0)$. Hence, if behavior is not restricted, i.e., if the tokens corresponding to $\mathbf{m}_1$ are not frozen, additional firing sequences may be obtained. Thus $\mathcal{L}(\mathbf{m}_0') \supseteq \mathcal{L}(\mathbf{m}_0)$.

The repetitive sequences give an idea of the "cyclical" behavior of a PN. However, additional information is given by the maximal *synchronic advance* of one transition on another. Consider for example the PN in Figure 2.14c. One can observe that T_1 can be fired at most once before a firing of T_2, since there is only one token in P_1; similarly, T_2 can be fired at most three times before a firing of T_2, since there are three tokens in P_2. More generally, let S_k be a firing sequence of this PN. If $N_k(T_j)$ is the number of firings of T_j in the sequence S_k, then

$$-3 \leq N_k(T_1) - N_k(T_2) \leq 1.$$

This means that, for any firing sequence of the PN in Figure 2.14c, the synchronic advance of T_1 on T_2 is at most 1 and the synchronic advance of T_2 on T_1 is at most 3.

2.2 SEEKING THE PROPERTIES OF PETRI NETS

Section 2.2.1 concentrates on drawing up the *graph of markings* or coverability root tree; this is the basic method. Section 2.2.2 presents the application of methods based on *linear algebra*; these results are powerful and elegant. An intuitive description of the *reduction* methods is briefly presented in Section 2.2.3. Finally, some other results are presented in Section 2.2.4.

2.2.1 Graph of Markings and Coverability Root Tree

As will be shown, the graph of markings can be used only if the PN is bounded and the coverability root tree may be used if the PN is not bounded (in fact for any PN). The reader can find the vocabulary related to graph theory in Appendix C.

2.2.1.1 Graph of Markings

The **graph of markings** (or **reachability graph**) is made up of vertices which correspond to reachable markings and of arcs corresponding to firing of transitions resulting in the passing from one marking to another.

Figure 2.15a presents a PN with its initial marking $\mathbf{m}_0 = (2, 0)$. Figure b is the graph of markings. As from $\mathbf{m}_0$, only transition T_1 can be fired. Marking $\mathbf{m}_1 = (1, 1)$ is then obtained. In the case of marking $\mathbf{m}_1$, both transitions are enabled. If T_1 is fired, $\mathbf{m}_2 = (0, 2)$ is reached, and if T_2 is fired, $\mathbf{m}_3 = (0, 1)$ is reached. Each of these two markings is a deadlock. Figure b thus presents all the reachable markings and all the single transition firings which exist between these markings. This graph of markings can be used to find the properties of the PN. We can see that it is bounded (no place has ever any more than 2 tokens), that it is not live and even that there are two deadlock states, that it is quasi-live (since from $\mathbf{m}_0$, T_1 can be fired at least once and T_2 likewise), that it has no home state and thus that it is not reversible.

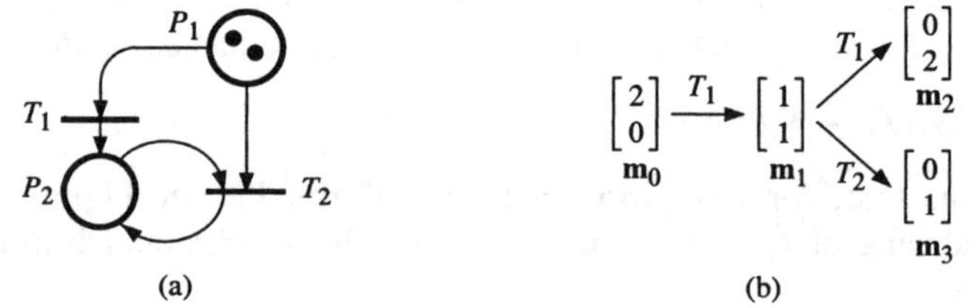

Figure 2.15 First example of a graph of markings.

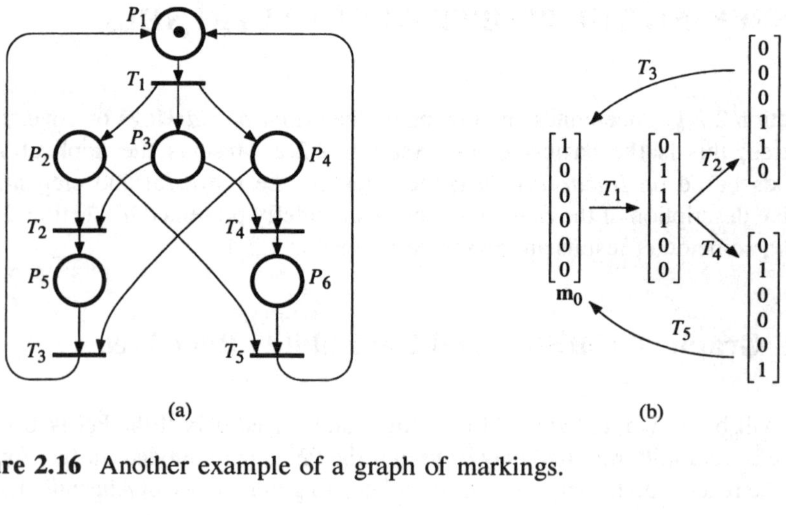

Figure 2.16 Another example of a graph of markings.

Remark 2.8 When drawing up a graph of markings (or coverability root tree), *it is usual not to represent the multiple firings*. For example, the double firing $[T_1T_1]$ is possible from $\mathbf{m}_0$ to $\mathbf{m}_2$ in Figure 2.15. However, as far as *autonomous PNs* are concerned (i.e. no time is involved, as in all this chapter), the set of *reachable markings* is the same with or without multiple firings.

❑

The graph of markings of the PN of Figure 2.16a is represented in Figure b. This graph can be used to find all the properties of this PN. We observe notably that it is safe, live, reversible and that $T_1T_2T_3$ and $T_1T_4T_5$ are repetitive sequences. The firing sequence $S = T_1T_2T_3T_1T_4T_5$ is complete and such that $\mathbf{m}_0 \xrightarrow{S} \mathbf{m}_0$; hence the PN is consistent.

2.2.1.2 Coverability Root Tree

In the case of the PN of Figure 2.17a, transition T_1 is enabled. If it is fired, there is one token in P_1. There are then two enabled transitions, T_1 and T_2. If T_1 is fired, there are two tokens in P1, and so on. Figure 2.17b represents the start of the construction of the graph of markings. However, this graph cannot be constructed since the PN is not bounded, i.e. the number of reachable markings is unbounded. We shall construct a root tree called a coverability root tree which possesses a finite number of vertices, by construction.

Let us construct the coverability root tree for the example in question (Figure 2.17c). Starting from the initial marking $\mathbf{m}_0 = (0)$ (the root), we have $\mathbf{m}_0 \xrightarrow{T_1} \mathbf{m}_1 = (1)$. Since $\mathbf{m}_1 \gneqq \mathbf{m}_0$, the firing of T_1 may be repeated as often as necessary. The symbolic marking ω is then associated with P_1, which means that the number of tokens in P_1 can reach an integer k as large as required. We shall call (ω) **a macro-marking** since it represents a *set of possible markings* (infinite set). Starting from macro-marking (ω), two transitions are enabled, T_1 and T_2. If there are k tokens in P_1, there will be k+1 once T_1 has been fired, or k-1 once T_2 has been fired. Since these two numbers k+1 and k-1 can be as large as required, the symbol ω is also associated with them. The coverability root tree shown in Figure 2. 17c is thus obtained. There is no point in continuing, since we have already indicated on this graph the possible firings from (ω).

Algorithm 2.1 **Construction of the coverability root tree.**
Every vertex corresponds to either a marking or a macro-marking. This is illustrated in Figures 2.18b for the PN in Figure a.

Step 1. From the initial marking $\mathbf{m}_0$, all the enabled transitions and the corresponding successive markings are indicated. If any of these markings is greater than $\mathbf{m}_0$, ω is placed for each of the components greater than the corresponding components of $\mathbf{m}_0$.

Step 2. For each new vertex $\mathbf{m}_i$ of the tree, either Step 2.1 or Step 2.2 is performed.

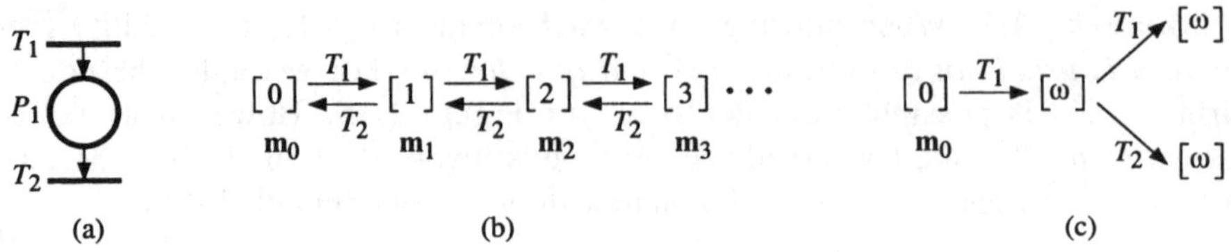

Figure 2.17 Example of a coverability root tree.

Step 2.1. If there is a vertex $\mathbf{m}_j = \mathbf{m}_i$ on the path from $\mathbf{m}_0$ to $\mathbf{m}_i$ (this latter not included), then $\mathbf{m}_i$ has no successor.

Step 2.2. If there is no vertex $\mathbf{m}_j = \mathbf{m}_i$ on the path from $\mathbf{m}_0$ to $\mathbf{m}_i$, the root tree is extended by adding all the successors of $\mathbf{m}_i$. For each successor $\mathbf{m}_k$ of $\mathbf{m}_i$: 1) A component ω of $\mathbf{m}_i$ remains a component ω for $\mathbf{m}_k$;

2) If there is a vertex $\mathbf{m}_j$ on the path from $\mathbf{m}_0$ to $\mathbf{m}_k$ such that $\mathbf{m}_k \gneqq \mathbf{m}_j$, then ω is placed for each of the components greater than the components of $\mathbf{m}_j$.

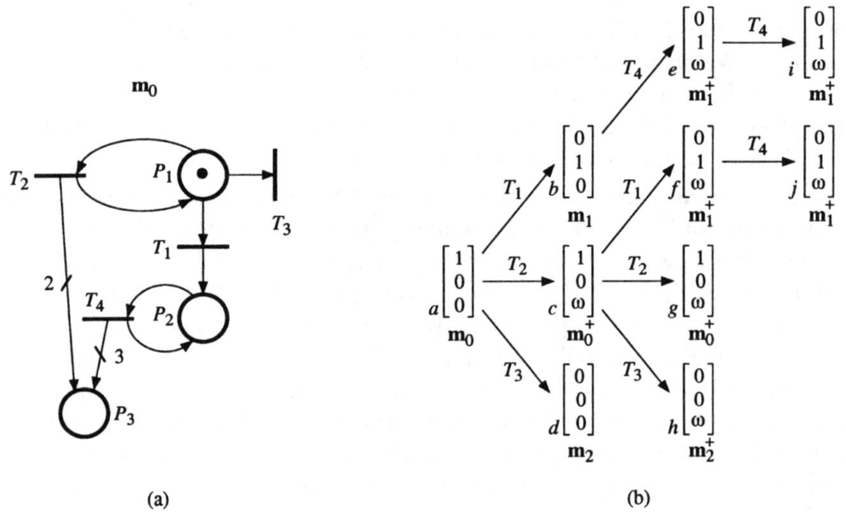

Figure 2.18 Another example of a coverability root tree.

Example (Figures 2.18a and b):

Step 1. $\mathbf{m}_0 \xrightarrow{T_1} \mathbf{m}_1$; $\mathbf{m}_0 \xrightarrow{T_2} (1, 0, 2)$, hence we write $\mathbf{m}_0 \xrightarrow{T_2} (1, 0, \omega) = \mathbf{m}_0^+$ since $(1, 0, 2) \gneqq (1, 0, 0) = \mathbf{m}_0$; $\mathbf{m}_0 \xrightarrow{T_3} \mathbf{m}_2$.

Step 2.

Step 2.2 for vertex *b*: $\mathbf{m}_1 \xrightarrow{T_4} (0, 1, 3)$, hence we write $\mathbf{m}_1 \xrightarrow{T_4} (0, 1, \omega) = \mathbf{m}_1^+$ since $(0, 1, 3) \gneqq (0, 1, 0) = \mathbf{m}_1$.

Step 2.2 for vertex *c*: $\mathbf{m}_0^+ \xrightarrow{T_1} \mathbf{m}_1^+$; $\mathbf{m}_0^+ \xrightarrow{T_2} \mathbf{m}_0^+$; $\mathbf{m}_0^+ \xrightarrow{T_3} \mathbf{m}_2^+$.

Step 2.2 for vertex *d*: no enabled transition, it is a deadlock.

Step 2.2 for vertex *e*: $\mathbf{m}_1^+ \xrightarrow{T_4} \mathbf{m}_1^+$.

Step 2.2 for vertex *f*: $\mathbf{m}_1^+ \xrightarrow{T_4} \mathbf{m}_1^+$.

Step 2.1 for vertex *g*: no successor since $(1, 0, \omega) = \mathbf{m}_0^+$ was already found on the path from vertex *a* to vertex *g* (vertex *c*).

Step 2.2 for vertex *h*: no enabled transition, it is a deadlock.

Step 2.1 for vertex *i*: no successor.

Step 2.1 for vertex *j*: no successor.

❑

The coverability root tree obtained by the above procedure always contains *a finite number of vertices.*

Using the coverability root tree of Figure 2.18b, a certain amount of information can be obtained: places P_1 and P_2 are bounded, but not P_3; there is an infinity of deadlocks which correspond to $\mathbf{m}_2$ and $\mathbf{m}_2^+$; the PN is quasi-live. However, for an unbounded PN, the *reachability problem* (i.e., given any marking, is it reachable?) and the *liveness problem* cannot be solved by the coverability root tree alone, because some information is lost by the use of symbol ω. The symbol ω represents a number which is *unbounded* but it *cannot be any number*: for example, for the node *e*, $m(P_3) = 3\,N$ where N is an integer, and for the node *f*, $m(P_3)$ is an even number.

For a bounded PN, the coverability root tree contains all possible reachable markings. It is called the reachability root tree. The reachability graph (i.e., graph of markings) is obtained by merging all the vertices of the reachability root tree corresponding to the same marking. It may be noticed that the coverability root tree can be very large, even for small PNs.

2.2.2 Linear Algebra

We shall introduce a certain formalism into the definition of Petri nets, along with certain of their properties. This section applies equally to *generalized Petri nets* and to ordinary PNs. The expression Petri net (without explicit precision) can thus apply to either type.

2.2.2.1 Notations and Definitions

Definition 2.11 An **unmarked ordinary PN** is a quadruple $Q = \langle P, T, \text{Pre}, \text{Post}\rangle$ such that:

$P = \{P_1, P_2, ..., P_n\}$ is a finite, not empty, set of places;

$T = \{T_1, T_2, ..., T_m\}$ is a finite, not empty, set of transitions;

$P \cap T = \emptyset$, i.e. the sets P and T are disjointed;

Pre: $P \times T \to \{0, 1\}$ is the input incidence application;

Post: $P \times T \to \{0, 1\}$ is the output incidence application.

□

$\text{Pre}(P_i, T_j)$ is the weight of the arc $P_i \to T_j$: weight 1 if the arc exists and 0 if not. For example, in Figure 2.19, $\text{Pre}(P_1, T_1) = 1$ and $\text{Pre}(P_1, T_4) = 0$.

$\text{Post}(P_i, T_j)$ is the weight of the arc $T_j \to P_i$. In Figure 2.19, $\text{Post}(P_4, T_1) = 1$ and $\text{Post}(P_4, T_2) = 0$. "Pre" and "Post" thus relate to transition T_j of the pair (P_i, T_j).

Definition 2.12 An **unmarked generalized PN** is defined as an unmarked ordinary PN, except that:

Pre: $P \times T \to \mathcal{N}$;

Post: $P \times T \to \mathcal{N}$.

□

For example, in Figure 1.7a (Section 1.2.2.1), $\text{Pre}(P_1, T_1) = 3$ and $\text{Post}(P_4, T_1) = 2$. These arcs have weights greater than 1.

The following notations will be used:

$°T_j = \{P_i \in P \mid \text{Pre}(P_i, T_j) > 0\}$ = set of input places of T_j;

$T_j° = \{P_i \in P \mid \text{Post}(P_i, T_j) > 0\}$ = set of output places of T_j;

$°P_i = \{T_j \in T \mid \text{Post}(P_i, T_j) > 0\}$ = set of input transitions of P_i;

$P_i° = \{T_j \in T \mid \text{Pre}(P_i, T_j) > 0\}$ = set of output transitions of P_i.

Using this notation, the *validation conditions* can now be expressed as in Definition 2.14. Before that, let us define a marked PN.

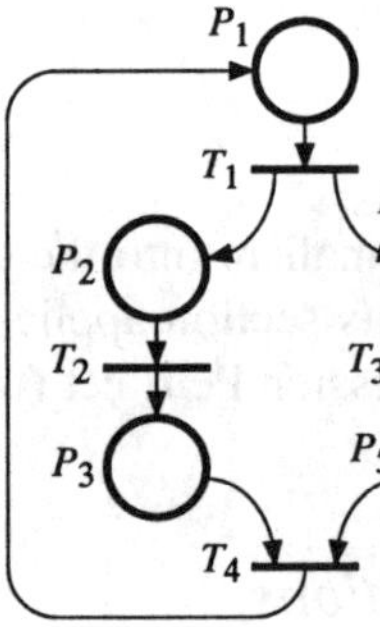

Figure 2.19 Example of a Petri net.

Definition 2.13 A **marked PN** is a pair $R = \langle Q, \mathbf{m}_0 \rangle$ in which Q is an unmarked PN and $\mathbf{m}_0$ an initial marking.

Definition 2.14 Transition T_j is enabled for a marking $\mathbf{m}_k$ if

$$m_k(P_i) \geq \text{Pre}(P_i, T_j) \text{ for every}^5\ P_i \in {}^\circ T_j. \quad (2.3)$$

❑

The following matrix is known as the **input incidence matrix**:

$$\mathbf{W}^- = [w_{ij}^-],\ \text{where } w_{ij}^- = \text{Pre}(P_i, T_j); \quad (2.4)$$

and the following matrix is known as the **output incidence matrix**:

$$\mathbf{W}^+ = [w_{ij}^+],\ \text{where } w_{ij}^+ = \text{Post}(P_i, T_j). \quad (2.5)$$

For example for the PN in Figure 2.19, we have:

$$\mathbf{W}^- = \begin{array}{c} \begin{matrix} T_1 & T_2 & T_3 & T_4 \end{matrix} \\ \begin{bmatrix} 1 & 0 & 0 & 0 \\ 0 & 1 & 0 & 0 \\ 0 & 0 & 0 & 1 \\ 0 & 0 & 1 & 0 \\ 0 & 0 & 0 & 1 \end{bmatrix} \end{array} \begin{matrix} P_1 \\ P_2 \\ P_3 \\ P_4 \\ P_5 \end{matrix} \quad \text{and} \quad \mathbf{W}^+ = \begin{array}{c} \begin{matrix} T_1 & T_2 & T_3 & T_4 \end{matrix} \\ \begin{bmatrix} 0 & 0 & 0 & 1 \\ 1 & 0 & 0 & 0 \\ 0 & 1 & 0 & 0 \\ 1 & 0 & 0 & 0 \\ 0 & 0 & 1 & 0 \end{bmatrix} \end{array} \begin{matrix} P_1 \\ P_2 \\ P_3 \\ P_4 \\ P_5 \end{matrix}.$$

The following matrix is known as the **incidence matrix** (or *flow matrix*):

$$\mathbf{W} = \mathbf{W}^+ - \mathbf{W}^- = [w_{ij}]. \quad (2.6)$$

For the PN of Figure 2.19, the incidence matrix is as follows:

$$\mathbf{W} = \begin{array}{c} \begin{matrix} T_1 & T_2 & T_3 & T_4 \end{matrix} \\ \begin{bmatrix} -1 & 0 & 0 & +1 \\ +1 & -1 & 0 & 0 \\ 0 & +1 & 0 & -1 \\ +1 & 0 & -1 & 0 \\ 0 & 0 & +1 & -1 \end{bmatrix} \end{array} \begin{matrix} P_1 \\ P_2 \\ P_3 \\ P_4 \\ P_5 \end{matrix}.$$

A column of this matrix corresponds to the marking modification made by the firing of the corresponding transition. For example, the first column of matrix **W** indicates that firing of transition T_1 consists of removing a token from place P_1 and adding a token to each of the places P_2 and P_4. This naturally provides no indication concerning the possibility of firing a transition since the *incidence matrix is independent from the marking*.

Remark 2.9 If a PN is *pure* (Section 1.2.1.6), its incidence matrix enables the unmarked PN to be constructed.

---

[5] If the inequality is true for every $P_i \in {}^\circ T_j$, it is true for every place in P since $\text{Pre}(P_i, T_j) = 0$ for $P_i \notin {}^\circ T_j$.

If a PN is *impure,* the unmarked PN cannot be reconstructed from its incidence matrix. Indeed, according to (2.4), (2.5), and (2.6), if the firing of a transition removes and adds tokens in the same place (self-loop), the incidence matrix only contains the difference (there is thus a loss of information on the PN). All the information on the unmarked PN is known if both the input and output incidence matrices are known.

2.2.2.2 Fundamental Equation

Let S be a firing sequence which can be performed from a marking $\mathbf{m}_i$, which can be written as $\mathbf{m}_i \xrightarrow{S}$. For example, for the marking $\mathbf{m}_1$ of Figure 2.20a, $\mathbf{m}_1 \xrightarrow{T_2}$ is obtained. The firing sequence $S_1 = T_2$ contains transition T_2 once and each of the other transitions zero times. The **characteristic vector** of sequence S, written as $\mathbf{s}$, is the m-component vector whose component number j corresponds to the number of firings of transition T_j in sequence S. For our example $\mathbf{s}_1 = (0, 1, 0, 0)$.

If the firing sequence S is such that $\mathbf{m}_i \xrightarrow{S} \mathbf{m}_k$, then a **fundamental equation** is obtained:

$$\mathbf{m}_k = \mathbf{m}_i + \mathbf{W} \cdot \mathbf{s}. \tag{2.7}$$

For our example:

$$\underset{\mathbf{m}_1}{\begin{bmatrix}0\\1\\0\\1\\0\end{bmatrix}} + \underset{\mathbf{W}}{\begin{bmatrix}-1 & 0 & 0 & +1\\+1 & -1 & 0 & 0\\0 & +1 & 0 & -1\\+1 & 0 & -1 & 0\\0 & 0 & +1 & -1\end{bmatrix}} \cdot \underset{\mathbf{s}_1}{\begin{bmatrix}0\\1\\0\\0\end{bmatrix}} = \begin{bmatrix}0\\1\\0\\1\\0\end{bmatrix} + \begin{bmatrix}0\\-1\\+1\\0\\0\end{bmatrix} = \underset{\mathbf{m}_2}{\begin{bmatrix}0\\0\\1\\1\\0\end{bmatrix}}.$$

The marking obtained from the marking $\mathbf{m}_1 = (0, 1, 0, 1, 0)$ by firing of T_2 is thus the marking $\mathbf{m}_2 = (0, 0, 1, 1, 0)$ which is represented in Figure 2.20b.

Another example: we have $\mathbf{m}_2 \xrightarrow{T_3T_4T_1T_3}$ and for $S_2 = T_3T_4T_1T_3$, we have $\mathbf{s}_2 = (1, 0, 2, 1)$. The carrying out of this firing sequence gives the marking $\mathbf{m}_3$ obtained by:

$$\underset{\mathbf{m}_2}{\begin{bmatrix}0\\0\\1\\1\\0\end{bmatrix}} + \underset{\mathbf{W}}{\begin{bmatrix}-1 & 0 & 0 & +1\\+1 & -1 & 0 & 0\\0 & +1 & 0 & -1\\+1 & 0 & -1 & 0\\0 & 0 & +1 & -1\end{bmatrix}} \cdot \underset{\mathbf{s}_2}{\begin{bmatrix}1\\0\\2\\1\end{bmatrix}} = \begin{bmatrix}0\\0\\1\\1\\0\end{bmatrix} + \begin{bmatrix}0\\+1\\-1\\-1\\+1\end{bmatrix} = \underset{\mathbf{m}_3}{\begin{bmatrix}0\\1\\0\\0\\1\end{bmatrix}}.$$

The marking corresponding to $\mathbf{m}_3$ is indicated in Figure 2.20c. If we now consider the carrying out of the sequence $S_1 \cdot S_2$ from the marking $\mathbf{m}_1$, we have $S_1 \cdot S_2 = T_2T_3T_4T_1T_3$ and $\mathbf{s}_1 + \mathbf{s}_2 = (1, 1, 2, 1)$, and the fundamental equation (2.7) gives $\mathbf{m}_1 + \mathbf{W} \cdot (\mathbf{s}_1 + \mathbf{s}_2) = \mathbf{m}_3$. This is illustrated in Figure 2.20.

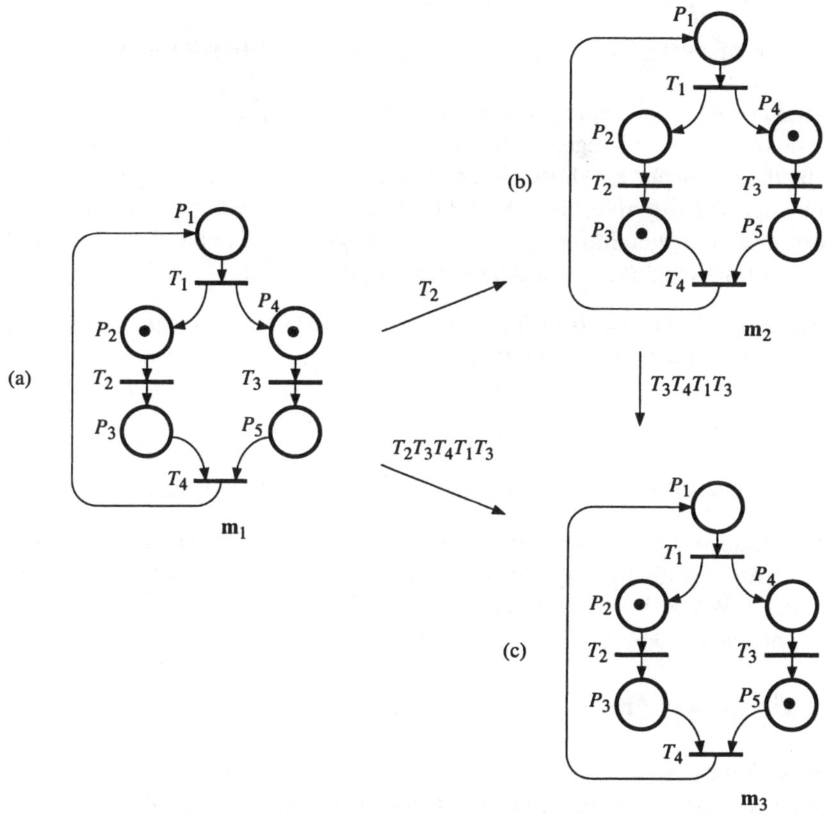

Figure 2.20 Illustration of firing sequences.

Remark 2.10 Vector **s** is said to be a "possible" characteristic vector if at least one firing sequence S corresponds to it, from a marking $\mathbf{m}_i$. Not all the **s** vectors whose components are positive or zero integers are possible. For example, $\mathbf{s} = (0, 1, 0, 1)$, is not possible from the marking $\mathbf{m}_1$ of Figure 2.20a. In fact, neither sequence T_2T_4 nor sequence T_4T_2 are possible from $\mathbf{m}_1$.

Remark 2.11 Several firing sequences may correspond to a possible characteristic vector **s**. For example, two possible S sequences correspond to $S = (0, 1, 1, 0)$ from the marking $\mathbf{m}_1$ of Figure 2.20a. These firing sequences are

T_2T_3 and T_3T_2. According to the fundamental equation (2.7), *two firing sequences with the same characteristic vector result in the same marking* (hence, the order of transitions in a firing sequence has no influence on the marking which will be reached). The reverse is not true, i.e. two sequences resulting in the same marking $\mathbf{m}_k$ from a marking $\mathbf{m}_j$ do not necessarily have the same characteristic vector.

2.2.2.3 Conservative Components & Marking Invariants

Let us consider a weighting vector for places $\mathbf{x} = (q_1, q_2, ..., q_n)$ such that each q_i is a positive or zero integer. Value q_i is the weight associated with place P_i. Let $P(\mathbf{x})$ be the set of places whose weight is not zero; $P(\mathbf{x})$ is thus a subset of P. Let us consider, for example, the PN of Figure 2.19. The vector $\mathbf{x} = (1, 1, 1, 0, 0)$ corresponds to a weighting of value 1 for places P_1, P_2 and P_3 and to a zero weighting for P_4 and P_5. We thus have $P(\mathbf{x}) = \{P_1, P_2, P_3\}$.

Property 2.4 The set B of places is a *conservative component* if and only if a weighting vector $\mathbf{x}$ exists such that

$$P(\mathbf{x}) = B \text{ and } \mathbf{x}^T \cdot \mathbf{W} = \mathbf{0}. \tag{2.8}$$

The vector $\mathbf{x}$ is a *P-invariant* (a non-negative P-invariant as explained below). ❑

Indeed, according to the fundamental equation (Equation (2.7) in Section 2.2.2.2), for any firing sequence S from $\mathbf{m}_0$, the marking reached is given by $\mathbf{m}_k = \mathbf{m}_0 + \mathbf{W} \cdot \mathbf{s}$. Left multiplication of both terms by $\mathbf{x}^T$ leads to: $\mathbf{x}^T \cdot \mathbf{m}_k = \mathbf{x}^T \cdot \mathbf{m}_0 + \mathbf{x}^T \cdot \mathbf{W} \cdot \mathbf{s}$. Thus, if $\mathbf{x}^T \cdot \mathbf{W} = \mathbf{0}$, we have

$$\mathbf{x}^T \cdot \mathbf{m}_k = \mathbf{x}^T \cdot \mathbf{m}_0 \tag{2.9}$$

whatever S may be, i.e., for every marking $\mathbf{m}_k \in \mathcal{M}(\mathbf{m}_0)$. Since $\mathbf{x}^T \cdot \mathbf{m}_0$ is a scalar quantity, $\mathbf{x}^T \cdot \mathbf{m}_k$ is a marking invariant. In other words, the number of tokens in the set of places $P(\mathbf{x})$ weighted by vector $\mathbf{x}$, is constant.

For the example considered, we have:

$$\mathbf{x}_1^T \cdot \mathbf{W} = \begin{bmatrix} 1 & 1 & 1 & 0 & 0 \end{bmatrix} \cdot \begin{bmatrix} -1 & 0 & 0 & +1 \\ +1 & -1 & 0 & 0 \\ 0 & +1 & 0 & -1 \\ +1 & 0 & -1 & 0 \\ 0 & 0 & +1 & -1 \end{bmatrix} = \begin{bmatrix} 0 & 0 & 0 & 0 \end{bmatrix}.$$

Therefore, the set of places $P(\mathbf{x}_1) = \{P_1, P_2, P_3\}$ is a conservative component. This is illustrated in Figure 2.21a. With the initial marking $\mathbf{m}_0 = (1, 0, 0, 0, 0)$, we obtain the *constant* of the marking invariant by:

$$\mathbf{x}_1^{\mathrm{T}} \cdot \mathbf{m}_0 = \begin{bmatrix} 1 & 1 & 1 & 0 & 0 \end{bmatrix} \cdot \begin{bmatrix} 1 \\ 0 \\ 0 \\ 0 \\ 0 \end{bmatrix} = 1.$$

Hence the marking invariant is $m_1 + m_2 + m_3 = 1$.

For the same PN, we also have $\mathbf{x}_2 = (1, 0, 0, 1, 1)$ such that $P(\mathbf{x}_2) = \{P_1, P_4, P_5\}$ is a conservative component. The marking invariant is $m_1 + m_4 + m_5 = 1$ (see Figure 2.21b).

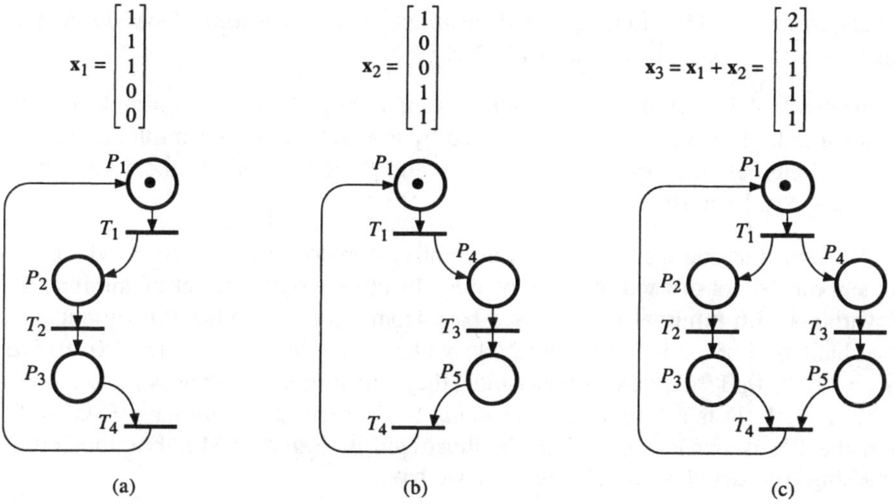

Figure 2.21 Conservative components.

For the time being, we are only concerned by P-invariants, but T-invariants have *similar properties* (Section 2.2.2.4).

A vector **x** which is a solution of $\mathbf{x}^{\mathrm{T}} \cdot \mathbf{W} = \mathbf{0}$, is known as a **P-invariant** (P standing for place). For our purpose, only *non-negative P-invariants*[6] are considered (i.e. $\mathbf{x} \geq \mathbf{0}$). Then, from now on, the positiveness of these invariants is *implicit.* The set of places $P(\mathbf{x})$ is known as the *support* of the P-invariant **x** (such a support is then a conservative component).

A P-invariant $\mathbf{x}_1$ is a **minimal support P-invariant** if there is no P-invariant $\mathbf{x}_2$ such that $P(\mathbf{x}_2) \subset P(\mathbf{x}_1)$. If two minimal support P-invariants have the same support, then they are linearly dependent. For example, if $\mathbf{x}_1$ and $\mathbf{x}_2$ are two minimal support P-invariants such that $P(\mathbf{x}_1) = P(\mathbf{x}_2)$, then $\alpha\,\mathbf{x}_1 = \beta\,\mathbf{x}_2$ where α

[6] Formally, a non-negative P-invariant may be called a P-semiflow.

and β are positive integers. A minimal support P-invariant whose non-negative components do not share a common factor except the value 1 is a *minimal P-invariant*. All minimal support P-invariants with the same support correspond to the same minimal P-invariant; for example, if $\mathbf{x}_1 = (0, 1, 2)$, $\mathbf{x}_2 = (0, 2, 4)$, and $\mathbf{x}_3 = (0, 3, 6)$, are three minimal support P-invariants, $\mathbf{x}_1$ will be called a minimal P-invariant ($\mathbf{x}_2 = 2\,\mathbf{x}_1$ and $\mathbf{x}_3 = 3\,\mathbf{x}_1$).

Property 2.5 Let $\mathbf{x}_1$ and $\mathbf{x}_2$ be two P-invariants.

a) If α and β are positive or zero integers, then $\alpha\,\mathbf{x}_1 + \beta\,\mathbf{x}_2$ is a P-invariant. If α and β are both positive, then $P(\alpha\,\mathbf{x}_1 + \beta\,\mathbf{x}_2) = P(\mathbf{x}_1) \cup P(\mathbf{x}_2)$.

b) If all the components of vector $\mathbf{x}_1 - \mathbf{x}_2$ are positive or zero integers, then $\mathbf{x}_1 - \mathbf{x}_2$ is a P-invariant.

Definition 2.15 Let $\mathbf{x}_1$ be a P-invariant. It is a **minimal P-invariant** if there is no P-invariant $\mathbf{x}_2$ such that $\mathbf{x}_1 \gneqq \mathbf{x}_2$.

Remark 2.12 A *minimal P-invariant* is not always a minimal support P-invariant. However, it can be obtained by a weighted sum of minimal support P-invariants, as in Property 2.5a where α and β are rational numbers instead of integers. See Exercise 2.11.

❑

We can thus see that minimal conservative components exist from which the others can be constructed by composition. In other words, the set of minimal P-invariants with minimal support is a base from which the other P-invariants can be obtained. For the PN of Figure 2.21, we have seen that $\mathbf{x}_1 = (1, 1, 1, 0, 0)$ and $\mathbf{x}_2 = (1, 0, 0, 1, 1)$ were P-invariants (they are minimal). Thus $\mathbf{x}_3 = \mathbf{x}_1 + \mathbf{x}_2 = (2, 1, 1, 1, 1)$ is a P-invariant according to Property 2.5a. We have $P(\mathbf{x}_3) = P$, i.e. the PN is conservative. This is illustrated in Figure 2.21c. For the initial marking considered, since $\mathbf{x}_3^{\mathrm{T}} \cdot \mathbf{m}_0 = 2$, we have:

$$2\,m_1 + m_2 + m_3 + m_4 + m_5 = 2.$$

Property 2.6 Let $P(\mathbf{x}) = \{P_1, P_2, ..., P_r\}$ be a conservative component and $(q_1, q_2, ..., q_r)$ be the corresponding weighting vector. All the places P_i of $P(\mathbf{x})$ are bounded and we have

$$m(P_i) \le \frac{\mathbf{x}^{\mathrm{T}} \cdot \mathbf{m}_0}{q_i}. \tag{2.10}$$

❑

The proof is straightforward since, from (2.2) in Section 2.1.5.1 and (2.9):

$$q_1 \cdot m(P_1) + ... + q_i \cdot m(P_i) + ... + q_r \cdot m(P_r) = \mathbf{x}^{\mathrm{T}} \cdot \mathbf{m}_0.$$

Remark 2.13 The conservative property is structural: an unmarked PN Q is **conservative** if and only if there is a P-invariant $\mathbf{x} > \mathbf{0}$ such that $\mathbf{x}^{\mathrm{T}} \cdot \mathbf{W} = \mathbf{0}$.

If there is $\mathbf{x} > \mathbf{0}$ such that $\mathbf{x}^{\mathrm{T}} \cdot \mathbf{W} \le \mathbf{0}$, then Q is *structurally bounded.*

2.2.2.4 Repetitive Components & Firing Invariants

The properties are similar to those examined in Section 2.2.2.3. A vector **y** which is a solution of $\mathbf{W} \cdot \mathbf{y} = \mathbf{0}$ is known as a **T-invariant**. The weighting vector which we shall now consider corresponds to a characteristic vector **s** associated with a firing sequence S. If $\mathbf{s} = \mathbf{y}$ is a T-invariant, the set of transitions appearing in S, i.e. corresponding to the non zero elements of **s**, is the *support* of the T-invariant. It is denoted by $T(\mathbf{s})$.

In the paragraph preceding Property 2.5, this property, Definition 2.15, and Remark 2.12, *the word P-invariant may be replaced by T-invariant.*

Property 2.7 Let D be a set of transitions. The set D is a *repetitive (stationary) component* if and only if a firing sequence S (whose characteristic vector is **s**) exists such that:

$$T(\mathbf{s}) = D \text{ and } \mathbf{W} \cdot \mathbf{s} = \mathbf{0}. \tag{2.11}$$

The vector **s** is a *T-invariant.*

If $\mathbf{W} \cdot \mathbf{s} > \mathbf{0}$, D is an increasing repetitive component.

❑

For the PN in Figure 2.14a (Section 2.1.5.2), $\mathbf{y}_1 = (1, 1, 0, 0)$ and $\mathbf{y}_2 = (0, 0, 1, 1)$ are minimal T-invariants corresponding to repetitive sequences $S_1 = T_1T_2$ and $S_2 = T_3T_4$ (i.e., $\mathbf{s}_1 = \mathbf{y}_1$ and $\mathbf{s}_2 = \mathbf{y}_2$). The sum of both minimal T-invariants is $\mathbf{y}_3 = \mathbf{y}_1 + \mathbf{y}_2 = (1, 1, 1, 1)$. There is a firing sequence $S_3 = T_1T_2T_3T_4$ such that $T(\mathbf{s}_3) = T(\mathbf{y}_3) = T$; the PN is consistent (Section 2.1.5.2).

For the PN in Figure 2.14b, $\mathbf{y}_4 = (2, 1, 1)$ is the only minimal T-invariant corresponding to the repetitive sequence $S_4 = T_1T_2T_1T_3$.

For the PN in Figure 2.14c, the only minimal T-invariant is $\mathbf{y} = (1, 1)$. This *does not mean that there is only one minimal repetitive sequence.* It means that, for any repetitive sequence, the number of firings of T_1 and T_2 is the same because they have the same weight in **y** (according to Property 2.5, if **y** is a T-invariant, then $\alpha\mathbf{y}$ is a T-invariant). For example, $S_5 = T_1T_2$, $S_6 = T_2T_2T_1T_1$ and $S_7 = T_2T_2T_1T_2T_1T_1$, are *minimal repetitive sequences* since none of their proper prefixes are repetitive. Their characteristic vectors are $\mathbf{s}_5 = (1, 1)$, $\mathbf{s}_6 = (2, 2)$, and $\mathbf{s}_7 = (3, 3)$.

Remark 2.14

a) If $\mathbf{W} \cdot \mathbf{s} = \mathbf{0}$, then **s** is a T-invariant. However, not all T-invariants necessarily correspond to a repetitive component, because at least one firing sequence must correspond to it. It is interesting to note that an algebraic calculation enables us to find the T-invariants (*independently of the marking*). According to Property 2.7, if there is a T-invariant **s** such that S is a firing sequence from marking **m**, then $\mathbf{m} \xrightarrow{S} \mathbf{m}$. In particular, the existence of such a T-invariant is a necessary condition for the PN to be reversible.

b) If S is a *repetitive sequence* from the marking $\mathbf{m}_1 \in \mathcal{M}(\mathbf{m}_0)$, and if S is a firing sequence from $\mathbf{m}_2 \in \mathcal{M}(\mathbf{m}_0)$, then S is also a repetitive sequence from the marking $\mathbf{m}_2$. This property is derived from the fundamental equation.

Remark 2.15 The consistency property is structural: an unmarked PN Q is defined as **consistent** if there is a T-invariant $\mathbf{y} > \mathbf{0}$ such that $\mathbf{W} \cdot \mathbf{y} = \mathbf{0}$.
If there is $\mathbf{y} > \mathbf{0}$ such that $\mathbf{W} \cdot \mathbf{y} \geq \mathbf{0}$, then Q is *structurally repetitive.*

2.2.2.5 Seeking P-invariants and T-invariants

An algorithm designed to find all the minimal P-invariants will first be presented, then we shall see how it can be adapted to look for the T-invariants. This Algorithm 2.2 is made up of several steps. After Step 1 and Step 2 have been performed, all the minimal support P-invariants[7] have been obtained, plus possibly some others which are not minimal. Step 3 consists of removing the non-minimal ones. The complexity of this algorithm is only polynomial.

Algorithm 2.2 **Search for minimal support P-invariants**

Step 1. Let $\mathbf{A}$ be the dimension n unit matrix (n = number of places) and $\mathbf{B} = \mathbf{W}$ (incidence matrix). Construct matrix $[\mathbf{A} \mid \mathbf{B}]$.

Step 2. For each index j of transition T_j.

Step 2.1. Add to matrix $[\mathbf{A} \mid \mathbf{B}]$ as many lines i as there are linear combinations of two lines, with positive integer coefficients, such that element (i, j) is zero.

Step 2.2. Eliminate from matrix $[\mathbf{A} \mid \mathbf{B}]$ all the lines k whose element (k, j) is not zero.

Step 3. Let $l_A \cdot l_B$ denote a line l of the matrix $[\mathbf{A} \mid \mathbf{B}]$ and $P(l_A)$ the support of l_A (i.e., the set of places for which the weight is not zero). If there are two lines p and q of the matrix such that $P(p_A) \supseteq P(q_A)$, then line p is removed.

Step 4. The minimal support P-invariants correspond to the non zero lines of $\mathbf{A}$.

Example (Figure 2.22)

Step 1. The matrix of 4 lines (P_1 to P_4) and 7 columns is obtained.

Step 2.

Step 2.1 for T_1: the line $P_1 + P_2$ is added which is the sum of the first two lines.

Step 2.2 for T_1: lines P_1 and P_2 are removed.

Step 2.1 for T_2: the line $P_3 + 2\,P_4$ is added (weight 1 for P_3 and weight 2 for P_4).

Step 2.2 for T_2: lines P_3 and P_4 are removed.

Step 2.1 for T_3: no lines are added.

Step 2.2 for T_3: line $P_1 + P_2$ is removed.

Step 3. No lines can be removed.

[7] More specifically, a P-invariant corresponding to every minimal support P-invariant.

Step 4. Only line $P_3 + 2\ P_4$ remains, which indicates that there is a conservative component $\{P_3, P_4\}$ with respective weights 1 and 2 for these two places, i.e., the only minimal P-invariant is $\mathbf{x} = (0, 0, 1, 2)$.

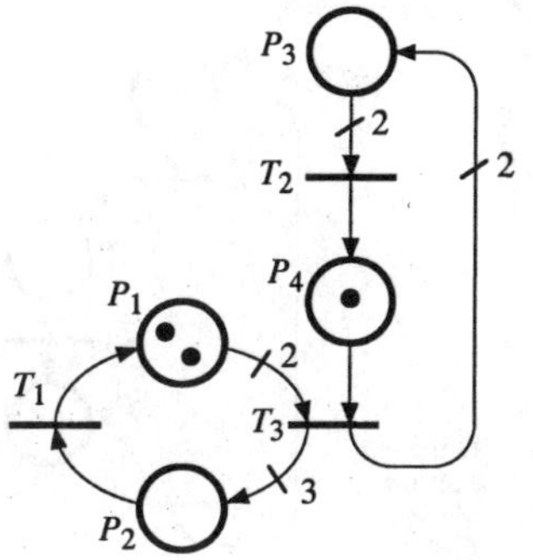

A				B			
P_1	P_2	P_3	P_4	T_1	T_2	T_3	
1	0	0	0	+1	0	-2	P_1
0	1	0	0	-1	0	+3	P_2
0	0	1	0	0	-2	+2	P_3
0	0	0	1	0	+1	-1	P_4
1	1	0	0	0	0	+1	$P_1 + P_2$
0	0	1	2	0	0	0	$P_3 + 2\,P_4$

Figure 2.22 Illustration of the algorithm giving the P-invariants.

Remark 2.16 The treatment in Step 3 may be added to Step 2 as a sub-step called Step 2.3. In this way, non-minimal or redundant P-invariants are eliminated after the treatment of each transition. In some cases this may avoid generation of many non-minimal P-invariants. While not useful in practice when Step 2.3 is added, Step 3 should theoretically be maintained to ensure that only minimal support P-invariants remain.

Adaptation of Algorithm 2.2 for searching for the T-invariants

Let us recall that a P-invariant $\mathbf{x}$ is defined by $\mathbf{x}^T \cdot \mathbf{W} = \mathbf{0}$ and a T-invariant $\mathbf{y}$ by $\mathbf{W} \cdot \mathbf{y} = \mathbf{0}$. Hence the T-invariants associated with $\mathbf{W}$ are found by looking for the P-invariants associated with the incidence matrix $\mathbf{W}^T$. In other words, if Q_1 is an unmarked PN, the T-invariants of Q_1 are the P-invariants of the *dual PN* Q_2 defined in this way: a transition in Q_2 is associated with each place of Q_1; a place in Q_2 is associated with each transition of Q_1; an arc directed in the other direction in Q_2 is associated with each arc of Q_1. See Exercise 2.10.

2.2.3 Reduction Methods Preserving Some Properties

Although building the graph of markings is certainly an efficient method for finding the properties of a PN of modest size, it may be very long and tedious for a PN with many reachable markings. The use of linear algebra also becomes difficult when the size of the studied PN increases. Reduction methods can be used to transform a PN into a simpler PN, while preserving some properties of the initial PN. But a word of warning, the simplified nets are *not* "equivalent" nets: do not look for any semantics in them.

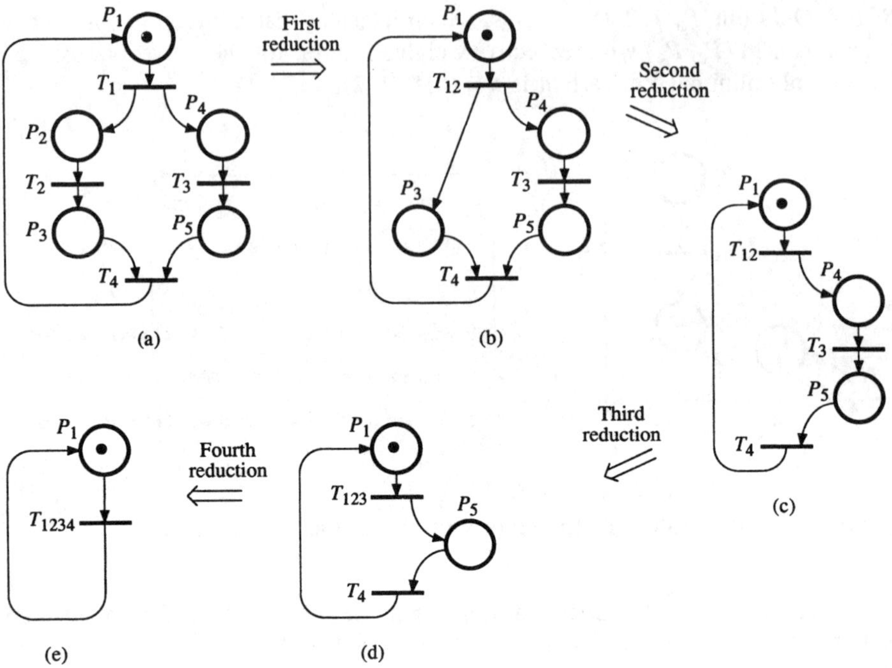

Figure 2.23 Illustration of the reduction process.

We will not explicitly describe these methods but simply illustrate their principle with an example. Two kinds of reduction are illustrated in Figure 2.23 (other kinds of reduction exist).

The reduction from Figure a to Figure b, called *substitution of place P_2*, is possible because some conditions are satisfied. The main condition is that the output transition of P_2 has no other input place: if a token is deposited in P_2, sooner or later firing of T_2 will occur and a token will be deposited in P_3.

The reduction from Figure b to Figure c consists of cancelling the *implicit place P_3*. The marking of P_3 never forms an obstacle to the firing of its output transition T_3 since $m_5 = 1$ implies $m_3 = 1$. Furthermore, the marking of P_3 can be deduced from the markings of the other places: $m_3 = m_4 + m_5$. Then, P_3 is "implicit" and may be cancelled.

Substitution of places P_4 then P_5 leads from Figure c to Figure e.

These reductions are made in such a way that some properties are preserved. In our example: the PN in Figure a is *live if and only if* the PN in Figure e is live; the PN in Figure a is *bounded if and only if* the PN in Figure e is bounded; the PN in Figure a *has a home state if and only if* the PN in Figure e has a home state. It is clear that the PN in Figure e is live, is bounded, and has a home state. Hence, one can deduce (thanks to the preservation of properties by the reductions) that the PN in Figure a is live, is bounded, and has a home state.

There is another kind of reduction method preserving the minimal support P-invariants (but not other properties like liveness and boundedness). It is a graphical adaptation of Algorithm 2.2.

2.2.4 Other Results

The results obtained concerning PN properties are very numerous. Let us state some of them which may possibly be the most useful. To begin with, we shall present some properties of *strongly connected event graphs*, then introduce the concepts of *siphon* and *trap*, and finally present some properties related to *liveness*.

2.2.4.1 Strongly Connected Event Graphs

An event graph was defined in Section 1.2.1.2 (subclass of ordinary PN); a strongly connected graph (and an elementary circuit) are defined in Appendix C. A **strongly connected event graph** has both properties.

Property 2.8 Let R be a *strongly connected event graph*, whose set of transitions is $T = \{T_1, T_2, ..., T_m\}$, and whose set of elementary circuits is

$$C = \{C_1, C_2, ... ,C_k, ...\}.$$

a) To each elementary circuit corresponds a minimal P-invariant such that, if $P(C_k) = \{P_1, P_2, ..., P_r\}$ is the set of places in the circuit C_k, then

$$m(P_1) + m(P_2) + ... + m(P_r) = m_0(P_1) + m_0(P_2) + ... + m_0(P_r) = m(C_k).$$

b) The only minimal T-invariant is the m-component vector $\mathbf{y} = (1, 1, ..., 1)$. It follows that if several minimal repetitive sequences exist, each of them contains each transition the same number of times.

c) The net R is live if and only if each elementary circuit contains at least one token, i.e. if $m(C_k) \geq 1$ for every C_k.

❑

Note that elementary circuits can be found in an event graph which is not strongly connected. Property 2.8a applies to these elementary circuits too.

Property 2.8 may be illustrated by the PN in Figure 2.21c (Section 2.2.2.3), which is a strongly connected event graph. There are two elementary circuits, namely $P_1T_1P_2T_2P_3T_4P_1$ and $P_1T_1P_4T_3P_5T_4P_1$, corresponding to invariants

$$m(P_1) + m(P_2) + m(P_3) = 1 \text{ and } m(P_1) + m(P_4) + m(P_5) = 1,$$

which are illustrated in Figures 2.21a and b (Property 2.8a). There are two minimal repetitive sequences, namely

$$S_1 = T_1T_2T_3T_4 \text{ and } S_2 = T_1T_3T_2T_4,$$

corresponding to the same minimal T-invariant $\mathbf{s}_1 = \mathbf{s}_2 = (1, 1, 1, 1)$ (Property 2.8b). Property 2.8c is easy to observe in this example. If there is no token in the circuit $P_1T_1P_2T_2P_3T_4P_1$, for example, transitions T_1, T_2 and T_4 will never be enabled. On the other hand, the PN is live if each circuit contains at least one token.

Strongly connected event graphs correspond to a class of PNs without conflicts, which is important from a practical point of view. Note that **strongly connected state graphs** have *dual properties*: there is a minimal T-invariant associated with each elementary circuit; only one minimal P-invariant for the whole net; the net is live if and only if there is at least one token in it (see Figure 1.5 in Section 1.2.1.1, for example).

2.2.4.2 Siphons and Traps

Let us now consider the concepts of siphon and trap, limited to the case of the ordinary PNs. Let $P' = \{P_1, P_2, ..., P_r\}$ be a set of places of a PN. The set of input transitions of the places of P' and the set of output transitions of the places of P', are denoted by $°P'$ and $P'°$, respectively. That is to say

$$°P' = °P_1 \cup °P_2 ... \cup °P_r \quad \text{and } P'° = P_1° \cup P_2° ... \cup P_r°.$$

Definition 2.16 A **siphon** (some authors use the words *deadlock*[8] or *lock*) is a set of places P' such that the set of input transitions of P' is included in the set of output transitions of P', i.e.,

$$°P' \subseteq P'°.$$

Definition 2.17 In an ordinary PN, a **trap** is a set of places P' such that the set of output transitions of P' is included in the set of input transitions of P':

$$°P' \supseteq P'°.$$

❑

According to these definitions, we see that the union of two siphons is a siphon, and that the union of two traps is a trap. Minimal siphons and minimal traps can thus be defined for a PN.

The set of places $P' = \{P_1, P_2\}$ of the Figures 2.24a is a siphon since $°P' = \{T_1, T_2\}$ and $P'° = \{T_1, T_2, T_3\}$. For the PN in this figure, transition T_2 is enabled and if it is fired, transition T_1 can then be fired and so on. There is a token circulating between P_1 and P_2 and when this token is in P_2, transition T_3 can be fired. If T_3 is fired, there are no tokens in P' and it follows that none of the output transitions of P' will ever be able to be fired. From this time on, the set P' is said to be *deficient* for the marking obtained, which means that the marking of P' is such that none of the transitions of $P'°$ will ever be enabled.

---

[8] But this word already has another meaning (Definition 2.6 in Section 2.1.3).

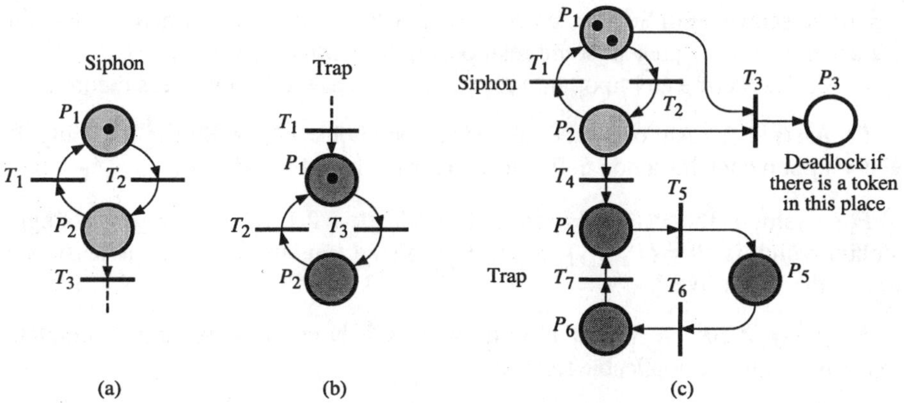

Figure 2.24 Siphons and traps.

In an informal manner, we can say that a siphon P' has the following property: if for a marking **m** none of the output transitions of P' are firable *because of the marking of the places of* P' (i.e. P' is deficient for marking **m**), then none of these transitions will ever be firable for any marking obtained from **m**.

The set of places $P' = \{P_1, P_2\}$ of Figure 2.24b is a trap since $°P' = \{T_1, T_2, T_3\}$ and $P'° = \{T_2, T_3\}$. A trap has a property which is dual that of a siphon. A trap can be said to have the following property: as soon as a transition of $°P'$ is fired, trap P' remains non deficient whatever the future evolution may be. The trap $P' = \{P_1, P_2\}$ of Figure b is non deficient since there is a token in P_1. If there were no token in P_1 and P_2, then P' would be deficient, but as soon as transition T_1 were fired, it would never be deficient again.

For the PN in Figure c, $\{P_1, P_2\}$ is a siphon, $\{P_4, P_5, P_6\}$ is a trap, and the marking **m** = (0, 0, 1, 0, 0, 0) is a deadlock. Note that the properties of siphon and trap are structural, while the deadlock depends on $\mathbf{m}_0$ (see Exercise 2.14).

Let us specify the notion of **deficient place**. For an ordinary PN, a place is non deficient if and only if it is marked. For a generalized PN, a place is non deficient if it is marked and/or: either 1) it has no output transitions, or 2) it contains enough tokens to enable at least one of its output transitions. **A set of places** P' **is non deficient** if at least one of its places is non deficient. For a *generalized PN, the definition of a siphon remains unchanged,* but *the definition of a trap is more complicated* in order to ensure the property of non deficiency.

Generally speaking, the search for siphons and traps is not a simple problem. As in the case of searching for invariants, it is possible to use linear algebra.

2.2.4.3 Liveness Related to Other Properties

After liveness (a basic property) was introduced in Section 2.1.3, various other concepts and properties were presented: siphons and traps in Section 2.2.4.2, repetitive sequence in Section 2.1.5.2. Equal conflict PN was defined in Section 1.2.2.1. Let us now present some relationship that might be useful.

Let us first define an important property, known as the Commoner's theorem.

Property 2.9 An ordinary extended free choice PN is live if and only if every siphon contains a non deficient trap for the initial marking $\mathbf{m}_0$.

❑

For example the PN of Figure 2.14c (Section 2.1.5.2) possesses a single siphon which is $P' = \{P_1, P_2\}$. Since P' is also a trap and contains at least one token, the PN is live.

Property 2.10 A PN is live if there is a home state $\mathbf{m}$ and a complete repetitive sequence applicable from $\mathbf{m}$.

Property 2.11 A persistent PN is live if and only if there is a reachable marking $\mathbf{m} \in \mathcal{M}(\mathbf{m}_0)$ and a complete repetitive sequence (whether increasing or not) applicable from $\mathbf{m}$.

Property 2.12 Let $R = \langle Q, \mathbf{m}_0 \rangle$ be a bounded *equal conflict PN* (Section 1.2.2.1). If $\langle Q, \mathbf{m}_0 \rangle$ is live, then $\langle Q, \mathbf{m}'_0 \rangle$ is live for any $\mathbf{m}'_0 \geq \mathbf{m}_0$.

❑

Since free choice and extended free choice PNs are particular cases of equal conflict PNs, Property 2.12 applies to them.

As stated above, liveness is a basic property, often required. Boundedness is necessary whenever the modeled system is to be implemented[9]. A marked PN $\langle Q, \mathbf{m}_0 \rangle$ is said to be **well-behaved** if it is both *live* and *bounded.* An unmarked PN, Q, is said to be **well-formed** if there exists an initial marking $\mathbf{m}_0$ such that $\langle Q, \mathbf{m}_0 \rangle$ is well-behaved.

2.2.5 Concluding Remarks

Modeling and analysis of real physical systems is a tricky problem, especially in the case of complex systems. In the latter case, the overall net may be broken down into a number of sub-nets. If the latter are interconnected in an anarchical manner, a large PN may be obtained whose size makes it unusable. Hence the need to structure when specifying a system. It is then possible to analyze during the various structuring stages, while still guaranteeing the validity of the results for the overall model. Some ideas are presented in Section 2.2.5.1.

A large number of PN analysis software programs have been developed; they are briefly described in Section 2.2.5.2.

[9] When modeling a production system, for example, a place may represent a counter of manufactured parts. Even if such a place is unbounded, its implementation will necessarily be bounded; however, the bound may be large enough for the purpose of the software, simulation related to a finite time for example.

2.2.5.1 Structuring

Two main approaches can be considered when specifying an application. The first of these is a descending approach, i.e. by successive refinings. When this approach is not possible, the system can be specified using partial descriptions and the overall net obtained by composition of the various sub-nets.

The approach by *successive refinings* first consists of roughly describing the behavior of the system by a relatively simple PN and of analyzing it. In this stage, the transitions represent complex operations. The description is then refined by replacing the transitions by PN parts known as "well-formed blocks" preserving boundedness, liveness, etc. A number of description levels can thus follow on from one another. The final model will exhibit the properties preserved by the well formed blocks. Figure 2.25 gives an example. Figure a is a simple PN which is live and safe. When the *well-formed blocks* of Figure b and c take the places of transitions T_1 and T_2 in Figure a, we obtain Figure d which is also live and safe.

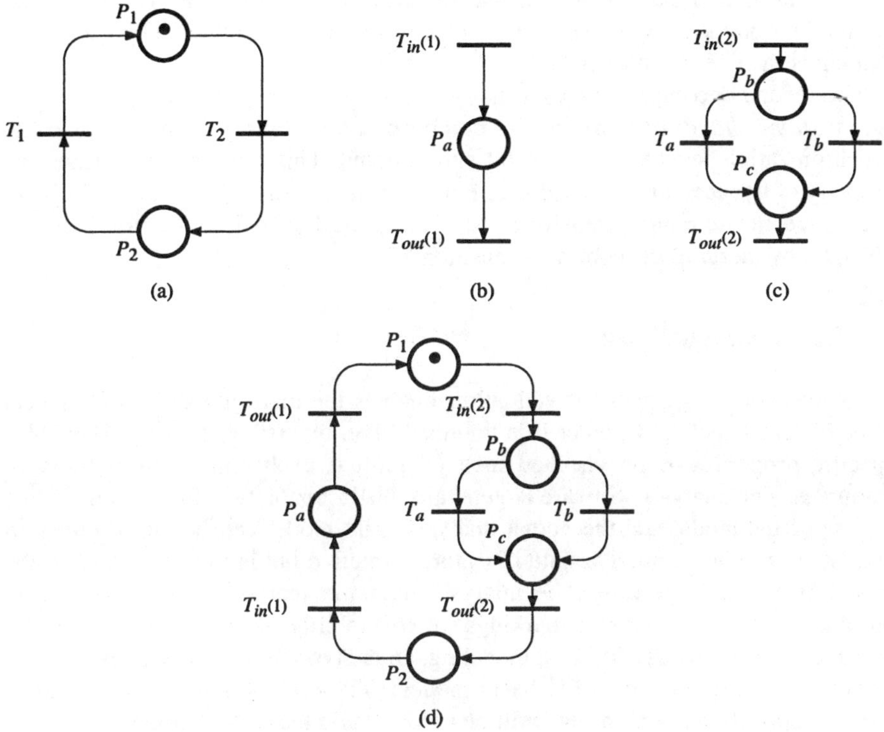

Figure 2.25 Successive refinings. (a) Basic PN. (b) and (c) Well-formed blocks "sequence" and "if then otherwise". (d) Resulting PN.

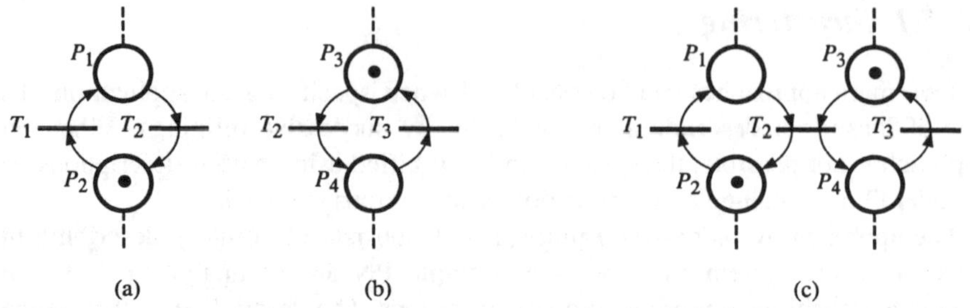

Figure 2.26 Composition by *rendez-vous.* (a) Sub-net 1. (b) Sub-net 2. (c) Overall net.

In the approach by *net composition*, partial specifications of the system studied are given. Overall synchronization is obtained by the composition of the various sub-nets describing these partial specifications.

An initial decomposition consists of constructing *sub-systems with shared places*. The shared place(s) will appear in each of the sub-systems. The overall model is obtained by these shared places merging together. This approach maintains the liveness property.

The second decomposition uses the *rendez-vous* mechanism. In this case, the *sub-systems have one or more common transitions* which achieve the synchronization between the different sub-systems. This approach maintains the property of liveness and boundedness. For example, the sub-nets a and b of Figure 2.26 have the common transition T_2. The overall model of Figure 2.26c is obtained by merging the common transition T_2.

2.2.5.2 Analysis Software

Analysis of a PN, known as validation, enables the main properties of a model to be checked such as whether it is bounded, live, reversible, etc. It also enables specific properties to be checked such as mutual exclusion or limitations of resources. The analysis software is generally made up of two parts, namely the editing of the model and the actual analysis. The model can be edited either in textual or graphic form. The latter is more attractive but is only possible if the model is not a large one. The analysis uses the methods which have been presented: graph of reachable markings (or coverability root tree), linear algebra, and reduction methods. To assist modeling, an analysis software may also have a library of already designed PN basic models. These models may form standard special application functions and will be used to build the overall model.

Considerable information on Petri net tools can be found at the following internet address: `www.daimi.au.dk/PetriNets/tools`.

NOTES and REFERENCES

Additional theoretical material can be found in some books [Pe 81][Br 83] [Si 85][Re 85][Re 89] or survey papers [Mu 89].

For the property defined in Definition 2.5a (Section 2.1.3) the term *quasi-live* is used by some authors, for example [Br 83] (in french: quasi-vivant). For the same meaning, the term *non-dead* is also used [Si 93]. The authors appreciate the term non-dead for a discrete transition. However, it is not so well adapted for a PN in which all the transitions have this property (Definition 2.5b). In order to have the same term for a transition and a PN, *quasi-live* is chosen. Furthermore, it will appear that this term (with the same definition) is pertinent for a continuous PN (Section 4.3).

Analysis of Petri nets by stepwise refinement was studied in [Va 79], then generalized in [SuMu 83].

The concept of *q-enabling* appears naturally in the contexts of stepwise refinement (as a limitation of the method [Va 79] [SuMu 83]), of timed PNs since a transition may be fired twice at the same time (various papers including [Zu 85] and [HoVe 87]), and in continuous PNs (where q corresponds to a maximum "*quantity of firing*" [DaAl 90]). For autonomous discrete PNs, a formal definition is given in [DaAl 92] and the name *enabling degree* is introduced in [CaSi 92]. The concept of conflict taking this enabling degree into account was explicitly introduced in [Ch 93b]: the *improved effective conflict* proposed is asymmetric (for example, in Figure 2.11b, G. Chiola considers that T_2 is in conflict with T_1 because the firing of T_2 reduces the enabling degree of T_1, although T_1 is not in conflict with T_2 because the firing of T_1 does not modify the enabling degree of T_2). With E. Teruel and M. Silva [TeSi 96], the authors prefer a symmetric relation; our Definition 2.10 differs from the definition in [TeSi 96] by the fact that it can treat conflict among more than two transitions (as explained in Remark 2.6). The concept expressed by "there is a conflict situation when the enabling vector is not an enabling step" [TeFrSi 98] is similar. This concept of general conflict was implicitely considered for timed PNs in [Ca *et al.* 89] (in which an "enabling bound" is considered), for synchronized and timed PNs in [DaAl 92].

The concept of synchronic advance is defined and studied in [KlLa 82]. It is generalized in [SiCo 89] where various synchronic concepts are presented.

Properties of the Petri nets developed from the linear algebra outlook were tackled in the seventies [LaSc 74][Li 76][Si 78][MeRo 80]. The fact that the P-invariants of a PN can be expressed as a linear combination of minimal support P-invariants with positive coefficients (similarly for T-invariants) was shown in [Si 78][MeRo 80]. The names P-semiflow and T-semiflow which may be used instead of P-invariant and T-invariant, were proposed in [Me 78][MeRo 80].

In certain cases, calculation complexity for seeking invariants can be exponential. Characterization by linear algebra has been enhanced by reducing the invariant calculations to a problem of linear programming. The complexity of the algorithms then becomes polynomial. Algorithm 2.2 was given in [MaSi 82] and, independently, in [To 82]. Improvements (for marking and synchronic invariants) were proposed in [CoSi 89][SiCo 89]. Structural analysis by mathematical programming techniques are developped in [SiTeCo 98].

The proof that the coverability root tree is finite was given in [KaMi 69].

The accessibility problem (given a marking, is it reachable?) was shown to be decidable [Ma 81][Ko 82][Re 89].

Proofs that some reduction methods preserve properties like liveness and boundedness, for example, were given by G. Berthelot [Be 83][Be 85]. Some reductions are also presented in [Da 76][Bo 78]. The reduction method preserving minimal support P-invariants is based on Algorithm 2.2. The most useful of these reduction methods are presented in [DaAl 89 &92].

The concepts of siphon and trap were introduced in [HoCo 70] and [Co 72]. They were generalized in [Ch 93b].

3

Non-Autonomous Petri Nets

Autonomous PNs which allow a qualitative approach were studied in Chapters 1 and 2. In this chapter, we shall present extensions of Petri nets which enable us to describe not only ***what*** *"happens" but also* ***when*** *"it happens". These Petri nets will enable systems to be modeled whose firings are synchronized on external events, and/or whose evolutions are time dependent.*

After an introduction in Section 3.1, Section 3.2 is devoted to synchronized PNs. Up to now, it was considered that, in a synchronized PN, a transition could be fired only once at a given time. In [DaAl 89, 2nd edition, & 92], a model called "extended synchronized PN" was briefly presented. In this chapter, this model is developed and rechristened "synchronized PN". This model, more general than the original synchronized PN, is based on the fact that a transition which is q-enabled may be fired q times at the same instant.

Then, interpreted PNs (for control *of discrete event systems) and timed PNs (for* performance evaluation*) are presented in Sections 3.3 and 3.4. These models appear to be special cases of synchronized PNs. As a consequence, all the properties shown for synchronized PNs are relevant to these models.*

3.1 INTRODUCTION

Assume transition T_1 in Figure 3.1 models an assembly operation: a component corresponding to a token in P_1 is assembled to a component corresponding to a token in P_2 and the resulting product corresponds to a token placed in P_3. Consider the model in Figure 3.1a; transition T_1 is 2-enabled (i.e., enabled twice) since there are two tokens in P_1 and more than two tokens in P_2. Is it possible to carry out two assemblies at the same time? If we assume that there is a single server (a single machine, or a single operator if the assembly is manual), then transition T_1 can be fired only once at a time. Two interpretations may be considered. For the **single-server** semantics (Figure a), it is always assumed that a transition cannot be fired more than once at a time, i.e., this *limitation* is *implicit*. For the **infinite-server** semantics (Figure b), q firings of a transition can be performed simultaneously if the transition is q-enabled; if there is a single server, the *limitation* to a single firing must be *explicit* as illustrated in Figure b. In this figure, transition T_1 is only 1-enabled due to the additional place P_4 marked with one token.

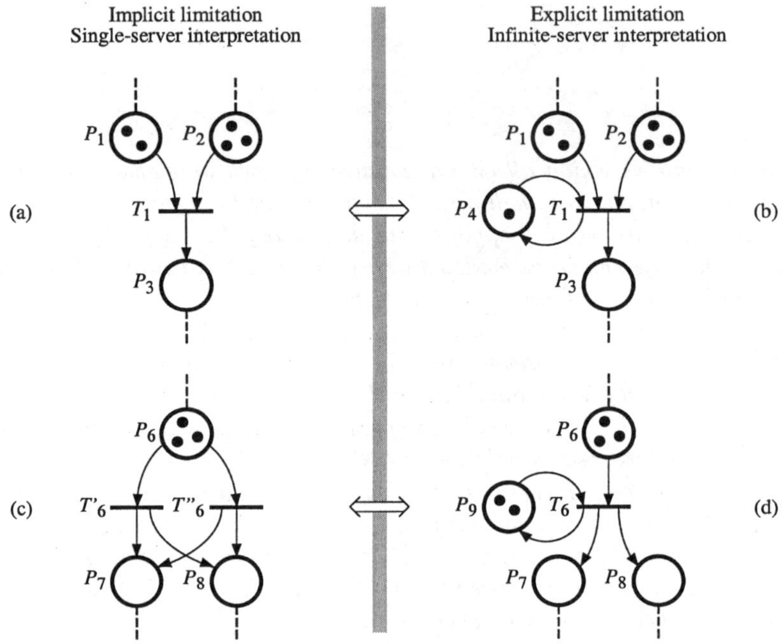

Figure 3.1 Single-server semantics versus infinite-server semantics.

Finite-server behavior can always be implemented by means of infinite-server semantics. Figure 3.1d shows a 2-server case (two tokens in P_9). As illustrated in Figure c, use of the single-server semantics requires a duplication of the transition.

Up to the early eighties, it was quite usual to adopt implicitely the single-server semantics, i.e., it was assumed that an enabled transition could be fired once only at one go. Below are several observations of this fact.

1) Enabling was considered to be a Boolean property. The enabling degree was not taken into account for possible conflicts. This point was widely discussed in Section 2.1.4.

2) The language generated by a marked PN is usually defined as a set of sequences of transitions. Neither the simultaneous firing of several transitions nor the multiple firing of a transition are modeled in this language. For the 2-transition PN in Figure A.1c (Appendix A), the language generated is built from

the alphabet $D_1 = \{T_1, T_2\}$. In this example, both T_1 and T_2 may be 1-enabled at the same time, and each one may be 2-enabled. According to the notation introduced in Section 2.1.4, the double firings $A_1 = [T_1T_1]$, $A_2 = [T_2T_2]$, and $A_3 = [T_1T_2]$ may be considered. A language taking into account these double firings should be built from the alphabet $D_2 = \{T_1, T_2, A_1, A_2, A_3\}$ and would be much more complicated.

3) In [FlNa 85], for example, the firing rates associated with the transitions of a stochastic PN are based on the single-server semantics.

4) The original definition of synchronized PNs is also based on the single-server semantics [MoPuSi 78].

Since the *infinite-server semantics is more general* than the single-server semantics (a finite-server behavior can always be implemented by means of infinite-server semantics as illustrated in Figure 3.1), in this chapter, *synchronized PNs will be defined on the basis of infinite-server semantics*. For this purpose, let us first formally define q-enabling.

Definition 3.1 The **enabling degree** of transition T_j for marking $\mathbf{m}$, denoted by q or $q(T_j, \mathbf{m})$ is the integer q such that

$$q \leq \min_{i:P_i \in {}^\circ T_j} \left(\frac{m(P_i)}{\mathrm{Pre}(P_i, T_j)} \right) < q + 1. \tag{3.1}$$

If $q > 0$, transition T_j is enabled; it is said to be ***q*-enabled**.

❑

Definition 3.1 applies to a generalized PN. For the particular case of an ordinary PN, in which the weight is 1 for all arcs, (3.1) may be simplified as:

$$\min_{i:P_i \in {}^\circ T_j} \left(m(P_i) \right) = q. \tag{3.2}$$

Synchronized PNs, explained in Section 3.2 form a basic model. Interpreted PNs, useful for modeling logic controlers and based on synchronized PNs, will then be discussed in Section 3.3. Various kinds of timed PNs, useful for performance evaluation, are presented in Section 3.4. In Appendix G, they are shown to be special cases of synchronized PNs. Appendix H presents the so-called time PNs.

3.2 SYNCHRONIZED PETRI NETS

In an autonomous PN, we know that a transition may be fired if it is enabled, but we do not know when it will be fired. In a synchronized Petri net, an event is associated with each transition, and the *firing* of this transition will occur:

if the transition is enabled,

when the associated event occurs.

The *external* events correspond to a change in state of the external world (including the time); by opposition, a change in internal state, a change in marking, could be called an *internal* event. *An event occurrence has no duration.*

Definition 3.2 A synchronized PN is a triple $\langle R, E, \text{Sync}\rangle$ such that:

R is a marked PN,

E is a set of **external events,**

Sync is a function from the set T of the transitions of R to $E \cup \{e\}$ in which e is the **always occurring event** (it is the neutral element of the monoid E^*, see Appendix A).

3.2.1 Principle

The set of external events is $E = \{E^1, E^2, ...\}$. The notation E^i (E super i) corresponds to the "name" of an external event. The notation E_j (E sub j) corresponds to the event associated with transition T_j.

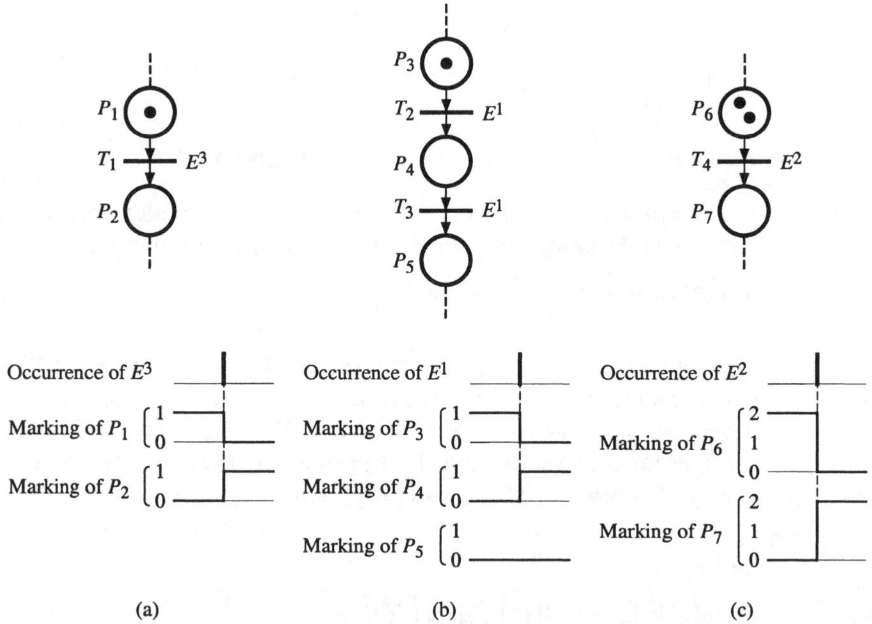

Figure 3.2 Firing principle of a synchronized transition.

The three examples given in Figure 3.2 illustrate the main concepts of a synchronized PN. The more complicated examples that we shall look at (for example, the case of a conflict) have an interpretation which is consistent with these concepts.

In Figure 3.2a, the external event E^3 is associated with transition T_1. In this figure, transition T_1 is said to be **receptive** to event E^3, because it is enabled. It will become **firable** when event E^3 occurs, and it will be fired immediately (see the corresponding timing diagram).

In Figure 3.2b, transition T_2 is *receptive* to event E^1, because it is enabled. It is fired when event E^1 occurs. On the other hand, transition T_3 is not fired, although it is synchronized on E^1 because it is not enabled when E^1 occurs (it is not receptive to this event).

In Figure 3.2c, transition T_4 is *receptive* to event E^2, because it is enabled. In fact, we can observe that transition T_4 is 2-enabled. It can be fired twice. Hence, when event E^2 occurs, it is fired twice at one go[1]. This is illustrated in the timing diagram.

Remark 3.1 In an autonomous PN, a transition could either be qualified indifferently by *enabled* or *firable*. This is no longer the case for a synchronized PN. A transition is *enabled* when each of its input places contains enough tokens (at least one token for an ordinary PN). If it is enabled, it is *firable on occurrence* of the event associated with it. It is then immediately fired (except possibly if there is a conflict not enabling all the firable transitions to be fired). ❑

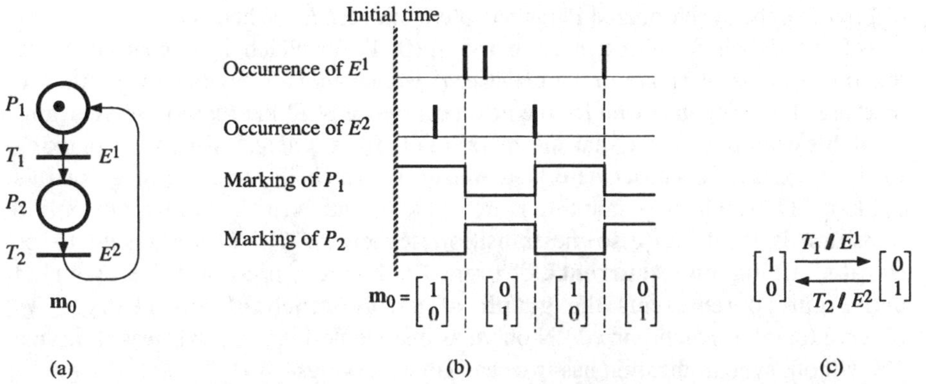

Figure 3.3 Example of behavior of a synchronized PN.

Figure 3.3a represents a synchronized PN with an initial marking $\mathbf{m}_0 = (1, 0)$. The *initial time* is represented by a *hachured line*. The set of external events is $E = \{E^1, E^2\}$. Figure 3.3b shows the evolution of the markings when the sequence of events is $Z = E^2E^1E^1E^2E^1$. For the initial marking, only transition T_1 is enabled. This transition is receptive to event E^1. The synchronized PN is thus only receptive to this event, which means that the occurrence of any other event

[1] In the case of single-server semantics, transition T_1 was fired only once [MoPuSi 78]. Both models have the same modeling power; this is shown in Appendix E which can be read after Section 3.2.2.

will not affect the marking. Thus the first event of sequence Z, which is E^2, does not alter anything. As soon as event E^1 occurs, T_1 is fired, which results in the marking (0, 1). The only transition then enabled is T_2 which is receptive to event E^2. When this event occurs (after the second occurrence of E^1 which had no effect), T_2 will be fired and so on. The set of possible evolutions is represented by the graph of markings in Figure 3.3c. For each firing of a transition, we have indicated after the slash / the event whose occurrence has caused the firing.

The *same event* may be associated with *several transitions* of a synchronized PN (Figure 3.2b already gave an example of this). Let us consider the synchronized PN of Figure 3.4. For the marking $\mathbf{m}_0 = (1, 0)$, transition T_1 is enabled and receptive to event E^3. When this event occurs, T_1 is fired which results in the marking (0, 1). It is then transition T_2 which is enabled and receptive to event E^3. Its firing on the 2nd occurrence of E^3 brings us back to the initial marking. This evolution is illustrated in Figure b, and the graph of markings is represented in Figure c.

We shall consider a new event which is not an *external* event. This is the *always occurring event,* which we shall write as *e*.

In Figure 3.5, event *e* is associated with transition T_2. This means that when transition T_2 is enabled, it will be receptive to this event and thus immediately firable since this event is "always occurring". For the initial marking $\mathbf{m}_0 = (1, 0)$ of Figure a, the synchronized PN is receptive to event E^3. When this event occurs, T_1 is fired, thereby resulting in the marking (0, 1) for which T_2 is enabled. Since this transition is receptive to the event *e*, it is *immediately fired* and we return to marking (1, 0). We then wait for the next occurrence of E^3 in order to evolve again.

In this example we see that the marking (1, 0) is *stable*, whereas the marking (0, 1) is *unstable* because there is a transition which is receptive to *e* for this marking. Thus, when the marking is $\mathbf{m}_0 = (1, 0)$ and event E^3 occurs, not only a transition is fired, but also the transition sequence T_1T_2. There is said to be **iterated firing on occurrence** of event E^3. Figure b illustrates this evolution and Figure c represents the **graph of stable reachable markings**. We observe that this synchronized PN only has one stable marking, whereas the same PN without synchronization has two reachable markings.

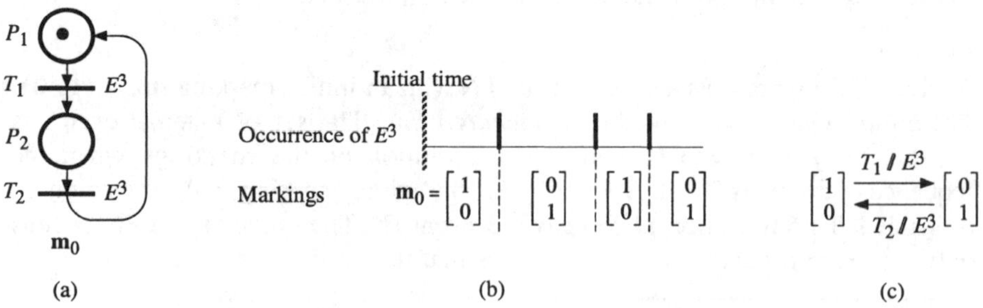

Figure 3.4 The same event associated with several transitions.

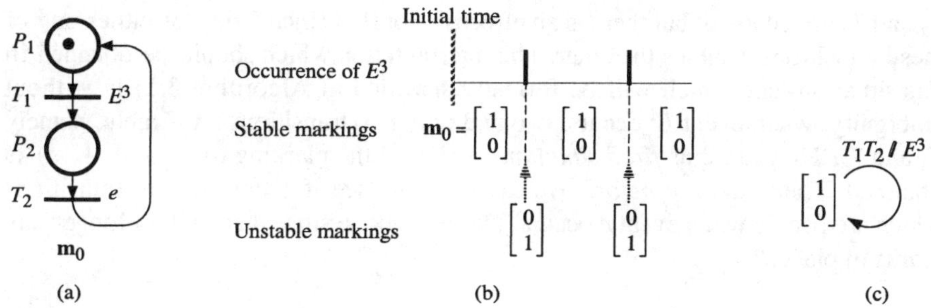

Figure 3.5 Always occurring event, *e*.

Let us now introduce the general concept of immediate transition.

Definition 3.3 In a non-autonomous[2] PN, an **immediate transition** is a transition which is *fired q times as soon as it is q-enabled,* for any $q > 0$ (or possibly less if there is an actual conflict among two or more immediate transitions).

❑

According to this definition, if T_j is an *immediate transition*, any marking $\mathbf{m}$ such that $q(T_j, \mathbf{m}) > 0$ is *unstable*. For a synchronized PN, it is equivalent to say that a *transition* is *immediate* or that it is *synchronized on event e*.

Remark 3.2 If there is a *conflict between an immediate transition T_a and a non-immediate transition T_b*, then *T_a takes priority over T_b*. This "hierarchy" is natural since firing of T_a is "immediate" whereas T_b must wait for an external event (synchronized PN) or the end of a timing (timed discrete PN).

❑

Let us now consider Figure 3.6a. There are two enabled transitions, T_2 and T_5, and both are receptive to event E^2. When this event occurs, both transitions are firable. Since there is no conflict between these transitions, both are fired simultaneously on occurrence of E^2. According to the notation introduced in Section 2.1.4, this double firing is denoted by $[T_2T_5]$ or $[T_5T_2]$ (the order of transitions in the square brackets is not significant since the firing is simultaneous). Let $\mathbf{m}_1 = (0, 1, 0, 0, 0, 1, 0)$ and $\mathbf{m}_2 = (0, 0, 1, 0, 0, 0, 1)$. We have $\mathbf{m}_1 \xrightarrow{[T_2T_5]} \mathbf{m}_2$.

The case shown in Figure 3.6b is more tricky (it is a part of a PN, the rest of the net is assumed to be without effect at the moment considered). The two enabled transitions T_1 and T_3 are receptive to the same event E^1, and the synchronized PN is persistent for the indicated marking $\mathbf{m}_1 = (1, 0, 1, 0, 0, ...)$. However, firing of T_1 (before T_3) would result in the marking $\mathbf{m}_2 = (0, 1, 1, 0, 0, ...)$ for which the net is not persistent. Indeed, for this marking, both transitions

---

[2] An *immediate transition* can be found in a synchronized PN (here), in a control interpreted PN (Section 3.3.1), in a timed discrete PN (Section 3.4.2.2), in a stochastic PN (Section 3.4.3.2), in a timed continuous PN (5.1.2.5), hence in an hybrid PN (Sections 6.1.2 and 6.1.5.2).

T_2 and T_3 are enabled, but there is an effective conflict since firing of either one of these transitions disables the other. The interpretation which should be obtained in this situation (and which will be formally specified in Algorithm 3.1) is without ambiguity: when event E^1 occurs, two and only two transitions are firable, namely T_1 and T_3. They are thus *fired simultaneously* and the marking (0, 1, 0, 0, 1, ..) is obtained. Then, and *not before,* we can look and see if there are any transitions which are firable when event *e* occurs. There are none since there are no longer any marks in place P_3.

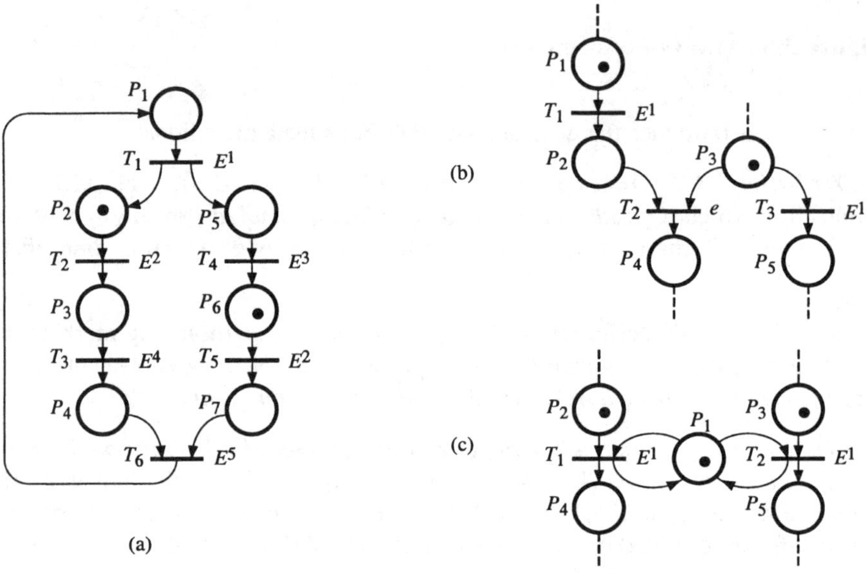

Figure 3.6 Two transitions firable at the same time.

In each of the two examples that we have seen above, there are two transitions which are receptive to the same event, but they are not in conflict. When two transitions are in *general conflict* (Definition 2.10 in Section 2.1.4) and, furthermore, they are *firable at the same moment,* there is said to be an **actual conflict.** This is the case for Figure 3.6c. The resolution of this actual conflict may be obtained by priority, or random drawing (see Appendix B).

Note that two transitions in general conflict which are synchronized on *independent* external events cannot be in actual conflict since these two events cannot occur simultaneously. In Appendix D, it is shown that two events which are independent cannot occur at the same time, while two events which are not independent can occur at the same time.

In the sequel, E^1 and E^2 denote two *independent events.*

In Figure 3.7a, there is a general conflict $K_1^G = \langle P_2, \{T_1, T_2\}, \mathbf{m}_6\rangle$ because there are not enough tokens in P_2 for a double firing of T_1 (which is 2-enabled) and a firing of T_2 (which is 1-enabled). However, *an actual conflict cannot occur* in this marking because E^1 and E^2 cannot occur simultaneously; if E^1 occurs first, there is a double firing of T_1, and if E^2 occurs first, there is a firing of T_2.

In Figure 3.7b, *an actual conflict occurs* when E^1 occurs because the two transitions in general conflict are firable at this time. The conflict resolution consists of the choice between the double firings $[T_1T_2]$ and $[T_1T_1]$.

In Figure 3.7c, there is a general conflict $K_2^G = \langle P_5, \{T_4, T_5, T_6\}, \mathbf{m}_7\rangle$. However, for this marking, *no actual conflict can occur*. If E^1 occurs first, the double firing $[T_4T_5]$ is performed (for $\mathbf{m}_7$, there is no general conflict between T_4 and T_5). If E^2 occurs first, T_6 is fired.

In Figure 3.7d, transition T_6 is not enabled but both T_4 and T_5 are 2-enabled. There is a general conflict $K_3^G = \langle P_5, \{T_4, T_5\}, \mathbf{m}_8\rangle$. *An actual conflict occurs* when E^1 occurs because there are not enough tokens in P_5 to fire T_4 twice and T_5 twice. There is a choice among $[T_4T_5]$, $[(T_4)^2]$, and $[(T_5)^2]$.

Based on the previous explanation, we will now specify when an actual conflict occurs. Let us first denote by X a **set of simultaneous events**. For example, if E^1, E^2, E^3, and E^4 are independent events (more specifically they are pairwise independent), the events[3] $E^5 = (E^1 + E^2)$ and $E^6 = (E^1 + E^3)$ are not independent. They are **compatible**, i.e., *they may occur simultaneously*; for the set of events $E = \{E^1, E^2, E^3, E^4, E^5, E^6\}$, the following sets of simultaneous events can be obtained: $X_1 = \{E^1, E^5, E^6\}$, $X_2 = \{E^2, E^5\}$, $X_3 = \{E^3, E^6\}$, $X_4 = \{E^4\}$ (see Notation 3.1). Other example: let E^7 denote an event occurring after each minute and E^8 occurring after each hour; they obviously are not independent, and for the set of events $E = \{E^7, E^8\}$, the sets $X_5 = \{E^7\}$ and $X_6 = \{E^7, E^8\}$ of simultaneous events can be obtained.

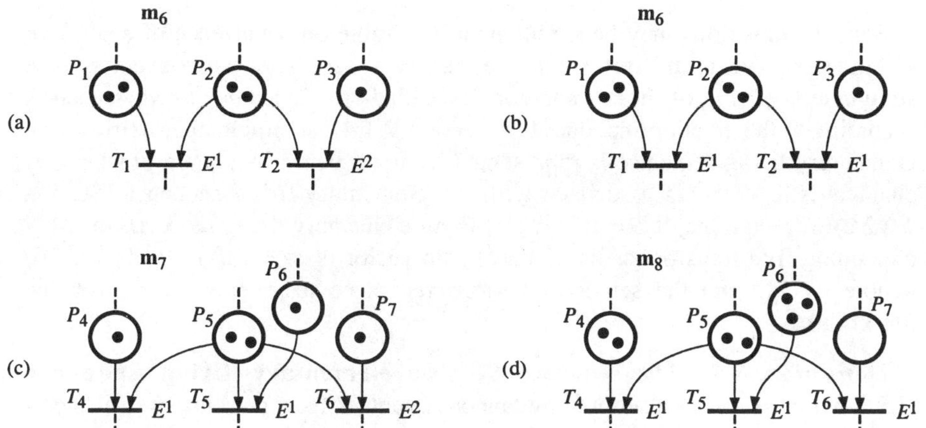

Figure 3.7 Actual conflict versus general conflict.

[3] According to Appendix D, the sum of two events is an event.

Notation 3.1 **Event and set of events**
Let $X_j = \{E^1, E^2\}$ be a set of two compatible events. According to Appendix D, stating that the product of two events is an event, it is synonymous to say "*the set X_j of events* occurs" or "*the event* $E^1 \cdot E^2$ occurs". Hence, when a set of events contains only one event, there is no ambiguity if we say "the event" instead of "the set of events", if we write $X = E^i$ instead of $X = \{E^i\}$, $X = e$ instead of $X = \{e\}$.

Definition 3.4 An **actual conflict occurs when** the set of simultaneous events X_h occurs if

1) the marking of the synchronized PN is $\mathbf{m}$,

and 2) there is a general conflict $K^G = \langle P_i, \{T_j\}, \mathbf{m}\rangle$ such that all the transitions in the set $\{T_j\}$ are synchronized on events in X_h.

❑

A synchronized PN is such that either an external event or the always occurring event (i.e., an element of $E \cup \{e\}$) is associated with each transition. A synchronized PN is said to be **totally synchronized** if none of its transitions are associated with the element e.

3.2.2 Iterated Firing On Occurrence of an External Event

Section 3.2.2.1 presents the concept of elementary firing sequence (EFS), corresponding to one or more simultaneous firings of transitions. Then, Section 3.2.2.2 explains how a string of EFS can be obtained by iteration.

3.2.2.1 Elementary Firing Sequence

Several transitions may be simultaneously firable on occurrence of a set X of simultaneous events (this set X may be either a subset of E or the event e). The simultaneous firing of these transitions is called an *elementary firing sequence*; according to the notation introduced in Section 2.1.4 for simultaneous firings, an elementary firing sequence is represented by a sequence in square brackets. A characteristic vector is associated with an elementary firing sequence (Section 2.2.2.2); for example, if $S = [(T_1)^2 T_2 T_4]$ is an elementary firing sequence in a PN containing five transitions, its characteristic vector is $\mathbf{s} = (2, 1, 0, 1, 0)$. We denote by $T(X,\mathbf{m})$ the set of transitions receptive to events in X, for the marking $\mathbf{m}$.

Definition 3.5 The sequence S_k is an **elementary firing sequence (EFS)** with respect to a set of simultaneous events X for a marking $\mathbf{m}$, if it meets the following three conditions[4].

1) All the transitions in S_k belong to $T(X, \mathbf{m})$.

---

[4] The notation $\mathbf{W}^-$ corresponds to the input incidence matrix (Section 2.2.2.1).

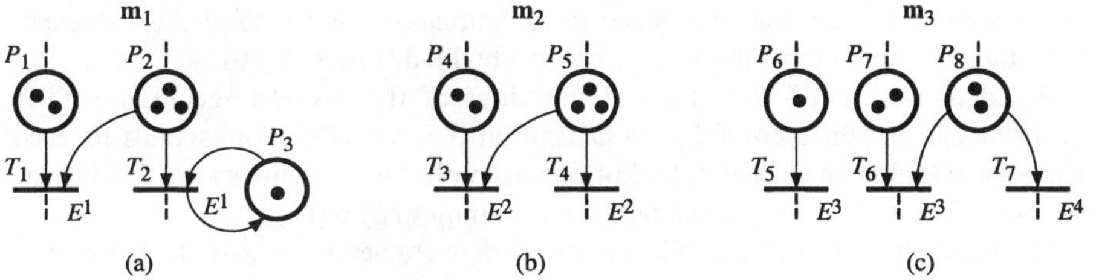

Figure 3.8 Illustration for elementary firing sequences.

2) $\mathbf{W}^- \cdot \mathbf{s}_k \le \mathbf{m}$. (3.3)

3) There is no sequence S_h meeting conditions 1 and 2 such that $\mathbf{s}_h \gneq \mathbf{s}_k$.

Remark 3.3 Condition 3 in Definition 3.5 is a generalization and a formalization of the principle intuitively presented in Figure 3.2c (Section 3.2.1). ❑

Figures 3.8a, b, and c, are parts of synchronized PNs in which it is assumed that the other transitions are not receptive to the considered events.

Figure 3.8a. For the marking $\mathbf{m}_1 = (2, 3, 1, ...)$ of the figure, T_1 is 2-enabled and T_2 is 1-enabled. On occurrence of the event E^1, both transitions are firable, hence $T(E^1, \mathbf{m}_1) = \{T_1, T_2\}$. Since there are enough tokens to fire these transitions according to their enabling degree, i.e., there is no actual conflict, the only EFS is $S_1 = [(T_1)^2T_2]$.

Figure 3.8b. On occurrence of the event E^2, the set of firable transitions is $T(E^2, \mathbf{m}_2) = \{T_3, T_4\}$. Transition T_3 is 2-enabled and T_4 is 3-enabled. Since there are not enough tokens in P_5 to fire these transitions according to their enabling degree, i.e., there is an actual conflict, there are several possible EFS, namely $S_2 = [(T_3)^2T_4]$, $S_3 = [T_3(T_4)^2]$, and $S_4 = [(T_4)^3]$.

Figure 3.8c. On occurrence of E^3, the set of firable transitions is $T(E^3, \mathbf{m}_3) = \{T_5, T_6\}$. There is no actual conflict: the only EFS is $S_5 = [T_5(T_6)^2]$. However, if events E^3 and E^4 occur simultaneously (which is possible if they are not independent), there is an actual conflict. For the set of simultaneous events $X_1 = \{E^3, E^4\}$, the set of firable transitions is $T(X_1, \mathbf{m}_3) = \{T_5, T_6, T_7\}$. The possible EFS are $S_6 = [T_5(T_6)^2]$, $S_7 = [T_5T_6T_7]$, and $S_8 = [T_5(T_7)^2]$; transition T_5 is in each EFS because it is not involved in the actual conflict between T_6 and T_7.

Let us now illustrate (Figure 3.9) an algorithm allowing all EFS to be found, given a marking $\mathbf{m}$ and a set of simultaneous events X. Let $\mathbf{m}_4$ be the marking of the synchronized PN in Figure 3.9a. What are the possible EFS, given the set of simultaneous events $X_2 = \{E^1, E^2\}$? Let R_1 denote the initial PN (Figure a).

According to the first condition in Definition 3.5, all the transitions in an EFS belong to $T(X_2, \mathbf{m}_4)$ which is a subset of the transitions synchronized by E^1 or E^2; then, all the transitions which are not synchronized by E^1 or E^2, are cancelled, as are also their input and output arcs. According to the second condition in Definition 3.5, the EFS depend only on the input incidence matrix $\mathbf{W}^-$; then all the arcs from a transition to a place (thus corresponding to $\mathbf{W}^+$) are cancelled. After these cancellations, the Petri net R_2 is obtained (Figure 3.9b).

We shall now build the graph of markings of the PN R_2: see Figure 3.9c. According to the third condition in Definition 3.5, an EFS corresponds to each deadlock marking in Figure 3.9c. For our example, two deadlocks are found: $\mathbf{m}_a$ corresponding to $S_a = [T_1T_2]$ and $\mathbf{m}_b$ corresponding to $S_b = [(T_2)^3]$.

The algorithm illustrated in Figure 3.9 always converges in a finite number of steps. As a matter of fact, by construction, the output incidence matrix of the PN R_2 is empty. Hence, each transition firing $\mathbf{m} \xrightarrow{T_j} \mathbf{m}'$ in R_2 is such that $\mathbf{m} \gneq \mathbf{m}'$; since the initial marking of R_2 is finite, its graph of markings is necessarily finite (and does not contain any circuit).

If the graph of markings of R_2 contains a *single* deadlock, the corresponding elementary firing sequence is said to be a **maximal EFS**.

Property 3.1

a) If $T_j \in T(X, \mathbf{m})$ is q_j-enabled, there is at least one EFS containing q_j times transition T_j.

b) An EFS S_k is a *maximal EFS* if and only if every transition $T_j \in T(X,\mathbf{m})$ appears q_j times in S_k (this EFS is unique, give or take a permutation).

Proof

a) Assume $T_j \in T(X, \mathbf{m})$ is q_j-enabled. If no other transition is fired, it is possible to perform q_j firings of T_j; in other words, in the marking graph of R_2, there is a path beginning with q_j firings of T_j. For example in Figure 3.9, T_1 is 1-enabled and T_2 is 3-enabled: in Figure c, from $\mathbf{m}_4$ there is a path beginning with T_1 and a path beginning with $T_2T_2T_2$.

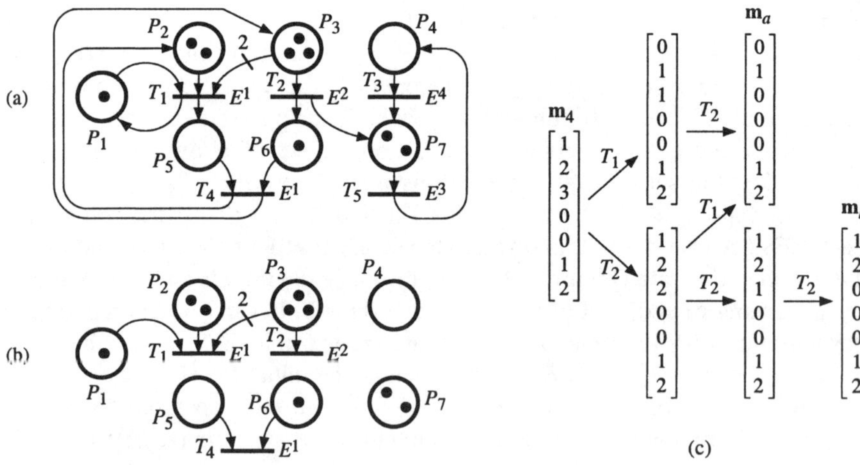

Figure 3.9 Research of elementary firing sequences. (a) PN R_1 with marking $\mathbf{m}_4$. (b) Modified PN R_2 given $X_2 = \{E^1, E^2\}$. (c) Graph of markings for R_2.

b) In the graph of markings of the net R_2, all the paths leading to some deadlock $\mathbf{m}_k$ correspond to the same characteristic vector $\mathbf{s}_k$, i.e., to the same EFS. Then, if there is a single EFS, according to Property a, every transition T_j is fired q_j times. If there are two or more EFS, there is an actual conflict: for any EFS there is at least one transition which is not fired up to its enabling degree.

Remark 3.4 If, given a set of simultaneous events X and a marking $\mathbf{m}$, there are two or more EFS, the resolution consists of choosing the EFS which is fired. This choice may be based, for example, on a priority order of the transitions involved in the actual conflict.

If a synchronized PN is such that no actual conflict can occur, it is said to be **deterministic**. The "deterministic or not" property depends on the initial marking, on the events associated with the transitions, and on the possible occurrence of events.

3.2.2.2 Iterated Firing

In the sequel, we shall consider the *iterated firing on occurrence of an external event* E^i. In a general case, instead of an event E^i, a set of simultaneous events X_j may be considered. However, in order to simplify reading, *we shall assume implicitly that all external events are independent*, hence that two of these events cannot occur simultaneously. We shall specify when we wish to mention the case of possibly simultaneous events.

An **iterated firing on occurrence of an external event** E^i consists of the firing of an EFS on occurrence of E^i, possibly followed by the firing of one of more EFS on occurrence of e. This is illustrated in Figure 3.10.

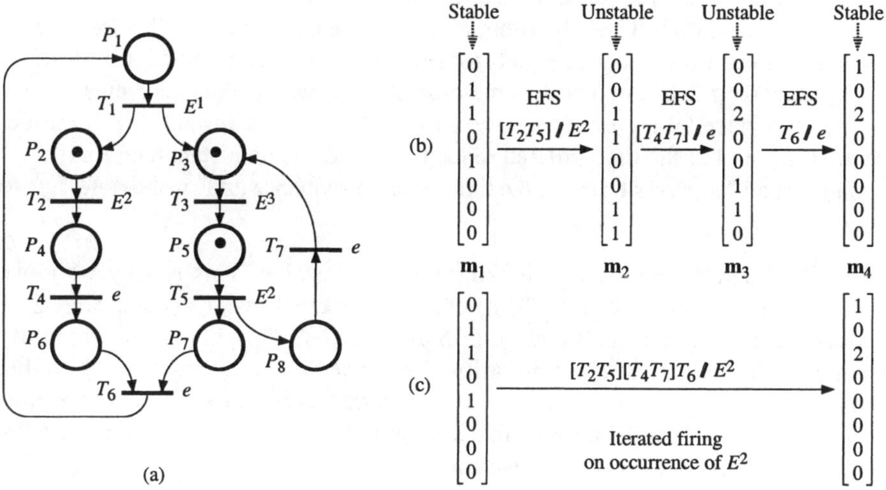

Figure 3.10 Iterated firing on occurrence of E^2.

The current marking is $\mathbf{m}_1 = (0, 1, 1, 0, 1, 0, 0, 0)$. The synchronized PN is receptive to the two external events E^2 (transitions T_2 and T_5 are receptive to it) and E^3 (transition T_3 is receptive to it). On occurrence of event E^2, the elementary firing sequence $[T_2T_5]$ is fired, thereby resulting in the marking $\mathbf{m}_2 = (0, 0, 1, 1, 0, 0, 1, 1)$. For this marking, the synchronized PN is receptive to event e (transitions T_4 and T_7 are receptive to it), which means that this marking is unstable. Once the EFS $[T_4T_7]$ has been fired, we obtain the marking $\mathbf{m}_3 = (0, 0, 2, 0, 0, 1, 1, 0)$ for which the synchronized PN is still receptive to event e (transition T_6). Once the EFS T_6 has been fired, we obtain the marking $\mathbf{m}_4 = (1, 0, 2, 0, 0, 0, 0, 0)$ for which the synchronized PN is not receptive to e. Thus this marking is stable. The synchronized PN is now receptive to the external events E^1 (transition T_1) and E^3 (transition T_3). No evolution of the marking will take place before one of these events occurs. The iterated firing on occurrence of E^2 has thus resulted in a changeover from marking $\mathbf{m}_1$ to marking $\mathbf{m}_4$. The firing sequence was $[T_2T_5][T_4T_7]T_6$, i.e., T_2 and T_5 simultaneously, then T_4 and T_7 simultaneously, then T_6. We may write $\mathbf{m}_1 \xrightarrow{[T_2T_5][T_4T_7]T_6/E^2} \mathbf{m}_4$ or, because there is only one possible string of EFSs: $\mathbf{m}_1 \xrightarrow{E^2} \mathbf{m}_4$.

Algorithm 3.1 **Interpretation of a synchronized PN**

Step 1. Initialization of the marking. Let $X = e$. Go to *Step 3*.

Step 2. Wait for the next external event. Let t be the time when this event occurs and $X(t)$ the set of simultaneous events occurring at this instant. Let $X = X(t)$,

Step 3. Determine the set of transitions firable on occurrence of X. If this set is empty, go to *Step 2*.

Step 4. Carry out an[5] EFS (elementary firing sequence). Let $X = e$. Go to *Step 3*.

Remark 3.5 **Variant of Algorithm 3.1**

Algorithm 3.1 explains how the real-time behavior of a synchronized PN should be understood. If the algorithm is to be used to simulate a behavior given a sequence of occurring external events (or set of events), it is modified as follows.

Step 1: add "Initialization of the time-ordered sequence of external events".

Step 2: replace "Wait for the next external event ... this instant" by "Consider the first instant t of the time-ordered sequence (if there is no instant then *End*)".

Step 3: add "suppress time t in the time-ordered sequence and" just before "go to *Step 2*".

❑

In Algorithm 3.1, the initial marking may be unstable. This is why we move to *Step 3* at the end of *Step 1*. Then, when an external event occurs, *Step 2* is performed, followed by an iteration of *Steps 3* and *4*. This algorithm implicitly assumes that this number of iterations is finite, i.e. that from every stable reachable marking, every occurrence of an external event (or set of simultaneous events) results in a stable marking in a *finite* number of EFS. A synchronized PN with this property is said to be *stable* (or *prompt*).

---

[5] If there are several possible EFS (i.e., there is an actual conflict), one of them is chosen.

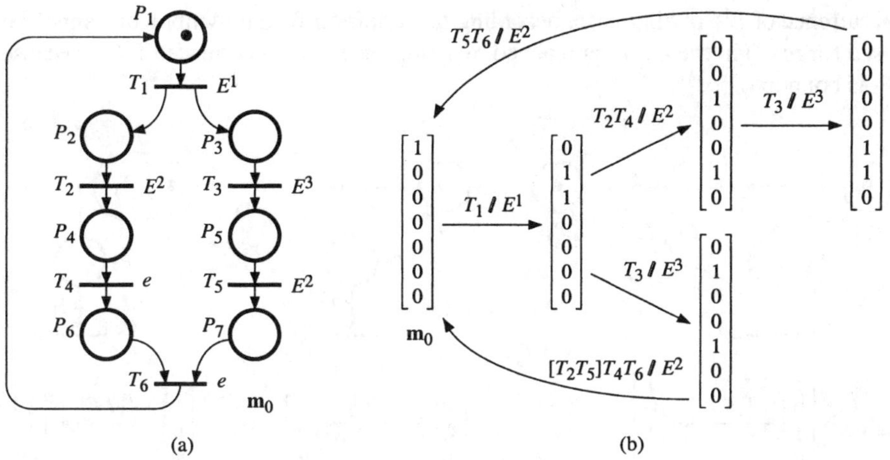

Figure 3.11 (a) Synchronized PN. (b) Graph of stable reachable markings.

Algorithm 3.1 enables the evolution of the markings of a synchronized PN to be deduced, given a sequence of external events. Its application to all the possible cases allows the **graph of stable reachable markings** to be obtained. This is illustrated in Figure 3.11. Figure a presents a synchronized PN with its initial marking $\mathbf{m}_0$. Figure b presents the graph of stable reachable markings. When the synchronized PN is not bounded, the graph of reachable markings cannot be constructed, since the number of these markings is unbounded. The coverability root tree of the stable reachable markings will then be constructed.

Algorithm 3.2 **Coverability root tree** *of the stable reachable markings*
This root tree is constructed according to Algorithm 2.1 in Section 2.2.1.2 with the two following differences.

1) Instead of the move to a graph node caused by the firing of a transition, it takes place by the iterated firing on occurrence of an external event.

2) When moving from $\mathbf{m}$ to $\mathbf{m}' \gneq \mathbf{m}$ on occurrence of E^a, a value ω is entered in the component $m'(i)$ if

α) $m'(i) > m(i)$,

and β) there is no positive integer k such that the marking $\mathbf{m} + k \cdot (\mathbf{m}' - \mathbf{m})$ enables a transition synchronized on E^b compatible with E^a.

❑

Let us illustrate the condition β in Algorithm 3.2 with the examples in Figure 3.12. In Figure a, transition T_2 is synchronized on E^1 like T_1; however, after any number of firings of T_1, T_2 remains not enabled since P_4 is empty (hence, the number of tokens in P_2 is not bounded: it is denoted by ω). In Figure b, T_2 is enabled after three firings of T_1; however, firing of T_2 cannot occur simultaneously with a firing of T_1 if E^2 and E^1 are not compatible (hence, P_2 is not bounded). In Figure c, after three firings of T_1, T_2 and T_1 are fired simultaneously on the fourth occurrence of E^1; in this case, according to condition β, the symbol ω cannot be used for $m'(2)$ in the marking reached by firing of T_1 on occurrence of E^1 (because P_2 is bounded).

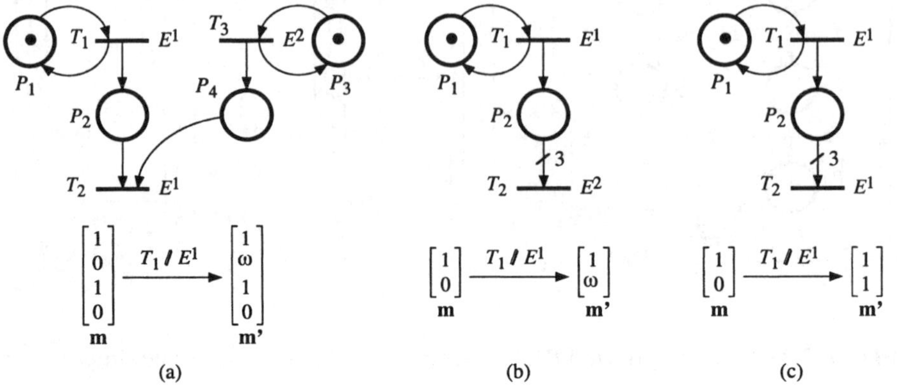

Figure 3.12 Illustration of condition β in Algorithm 3.2.

Examples of complete coverability root trees of stable markings are given in Figures 3.13a and 3.14b.

In this section, it is assumed that any sequence of events (or sets of compatible events) can occur. However, according to Section 3.2.3.3, some sequences of events might be impossible.

Notation 3.2 **Synchronization by the always occurring event** e is assumed for all the transitions without explicit event in a synchronized PN, i.e., if no explicit event is associated with T_j, then $E_j = e$ (i.e., T_j is an immediate transition). This simplified notation will be used in Figure 3.13 for example.

No confusion with a non-synchronized PN is possible since there is *at least one transition synchronized on an external event* in a synchronized PN.

3.2.3 Properties of the Synchronized PNs

3.2.3.1 Promptness or Stability

The first feature expected from a synchronized PN is that it be *prompt* (or *stable*), i.e., for every stable reachable marking, and for every external event E^i (or set of compatible events), the iterated firing on occurrence of E^i contains a finite number of EFS. If the number of EFS is always less than or equal to k, the synchronized PN is said to be k-**prompt** (or k-**stable**). The word *prompt* was introduced by the authors of the first paper on synchronized PNs [MoPuSi 78]. At the end of this section, the concept will be generalized to all the non-autonomous PNs. For some of them, the word *stable* is better adapted. Then, from now on, the word *stable* will be used instead of *prompt.*

Property 3.2 If a synchronized PN is such that for every elementary circuit $P_1T_1P_2T_2 \ldots P_qT_qP_1$, there is at least one transition out of T_1, T_2, ... T_q which is synchronized on an external event, then it is stable.

❑

This property is illustrated in Figure 3.13a. Each time E^1 occurs, a token is placed in P_2. Assume there are q tokens in P_2 at some instant: if E^3 occurs, place P_2 is emptied; but if E^2 occurs before E^3, the q tokens pass simultaneously in P_3 then return to P_2 (according to Notation 3.2 in Section 3.2.2.2, T_3 is synchronized by event $E_3 = e$). This synchronized PN is stable because it satisfies the condition in Property 3.2: for the elementary circuit $P_1T_1P_1$, transition T_1 is synchronized on E^1; for the elementary circuit $P_2T_2P_3T_3P_2$, transition T_2 is synchronized on E^2.

If T_2 were synchronized on e instead of E^2, after the first firing of T_1, T_2 and T_3 would be fired in turn *ad infinitum*: the firing sequence would be $S_1 = T_1T_2T_3T_2T_3\ldots$ If T_1 were synchronized on e, the firing sequence from the initial time would be $S_2 = T_1T_1T_1\ldots$

Note that a *source transition* must not exist in a synchronized PN. For example, P_1 cannot be removed in Figure 3.13a; otherwise, T_1 would be ∞-enabled and an infinite number of tokens would be placed in P_1 at the first occurrence of E^1. On the other hand, a *sink transition* may exist (e.g., T_4).

Condition of Property 3.2 ensures that there cannot be a circuit for which all the transitions are fired in turn on occurrence of e. However, this sufficient condition is not necessary, as shown in Figure 3.13b. The circuit $P_1T_1P_2T_2P_1$ is such that event e is associated with T_1 and T_2, i.e. with all the transitions of this circuit. Nevertheless, there is no instability, since after firing of T_1 on occurrence of e, transition T_2 is not enabled and vice versa. Furthermore, although circuits $P_3T_1P_3$ and $P_4T_2P_4$ are self-loops whose transitions are synchronized by event e, there is no instability.

Property 3.2 is interesting because it is easy to check. However, a stronger property can be highlighted. For a synchronized PN to be stable, it is *necessary and sufficient for every repetitive sequence* S_k, stationary or increasing, (from every reachable marking whatsoever) *to contain at least one transition synchronized on an external event* (will be shown and generalized in Property 3.4). For example, for the PN in Figure 3.13b, it is necessary and sufficient for one of the two transitions T_3 or T_4 to be synchronized on an external event (from $\mathbf{m}_0$, the repetitives sequences are T_3T_4 and $T_3T_2T_4T_1$). If we assume that T_3 is synchronized on E^1, while T_1, T_2 and T_4 are synchronized on e, the occurrence of E^1 from the marking $\mathbf{m}_0 = (0, 1, 1, 0)$ would result in the iterated firing $T_3T_2T_4T_1$ or T_3T_4 (after firing of T_3, there is an actual conflict between T_2 and T_4) bringing us back to the marking $\mathbf{m}_0$; the behavior is not deterministic but the synchronized PN is *stable.*

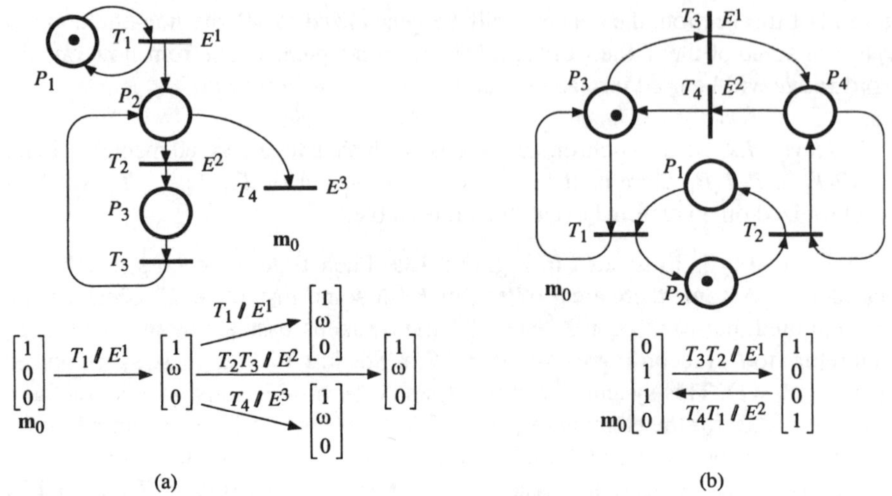

Figure 3.13. Stable PNs (using Notation 3.2). (a) Illustration of Property 3.2. (b) Stable PN although it has circuits without external event.

Property 3.3 A *totally synchronized* PN is 1-stable.

❑

As a matter of fact, a synchronous PN is totally synchronized if all the transitions are synchronized on external events (end of Section 3.2.1). Since there is no transition synchronized on e, every firing of an EFS on occurrence of an external event (or set of compatible events) results in a stable marking.

Remark 3.6 According to Definition 3.3 in Section 3.2.1, *immediate transitions* can be found in any kind of non-autonomous PNs, and a marking enabling such a transition is *unstable*. According to Appendix G, timed PNs (deterministic or stochastic) are special cases of synchronized PNs, and according to Section 5.1.1, a timed continuous PN is a limit case of discrete timed PN. It follows that *the behaviors related to immediate transitions in these various models can be deduced from the behaviors of immediate transitions* (i.e. synchronized on e) *in a synchronized PN.*

❑

Hence, a generalization to any kind of non-autonomous PN is presented in the sequel. It may be applied to *timed discrete* PNs (deterministic or stochastic timings), *continuous* PNs, and *hybrid* PNs. Although these models have not yet been defined in this book, it is sufficient to know that immediate transitions may exist in all these models.

Definition 3.6 A *non-autonomous PN* is **stable** if there is no reachable marking $\mathbf{m}$ such that $\mathbf{m} \xrightarrow{S}$ where S is an infinite sequence of *immediate transitions* (or sequence of EFS made up of immediate transitions).

Property 3.4 A non-autonomous PN is *stable* if and only if, for *any repetitive sequence* S_k of the corresponding autonomous PN, there is *at least one transition* in S_k *which is not immediate*.

Proof

Necessary condition. Let S_k denote a repetitive sequence, i.e. there is a reachable $\mathbf{m}$ such that $\mathbf{m} \xrightarrow{S_k} \mathbf{m}'$, $\mathbf{m}' \geq \mathbf{m}$. If S_k contains only immediate transitions, then $\mathbf{m} \xrightarrow{(S_k)^N}$, for any integer $N \geq 0$, hence the PN is not stable.

Sufficient condition. If, for any reachable marking $\mathbf{m}$ and any finite length sequence S_h such that $\mathbf{m} \xrightarrow{S_h}$, either S_h is not repetitive or it contains at least one non-immediate transition, then the PN is stable according to Definition 3.6.

Property 3.5 If every elementary circuit contains at least one non-immediate transition, then the non-autonomous PN is *stable.*

❑

Property 3.5 is a generalization of Property 3.2. Its truth implies that the condition in Property 3.4 is verified.

Let us emphasis that the stability property (for any non-autonomous PN) is very important. An unstable non-autonomous PN can easily be built, but it has no practical usefulness. From now on, it will be implicitely assumed that the models considered are stable.

3.2.3.2 Boundedness, Safeness, and Liveness

The definitions of a *bounded* PN, *safe* PN, *live* PN, are easily extended to stable synchronized PNs. A stable synchronized PN is bounded if, for every stable reachable marking, all the places are bounded (this implies that the synchronized PN is also bounded for all the transient markings). A safe PN is a particular case of bounded PN in which every place is 1-bounded. A transition T_j of a stable synchronized PN is live if, for every stable reachable marking, there is a sequence of external events enabling T_j to be fired in onc of the firings on occurrence of the events of this sequence. The concepts of a live synchronized PN, of quasi-live transition and deadlock, may be generalized in the same manner. The synchronized PNs of Figures 3.5, 3.11 and 3.13b are bounded and live. The synchronized PN in Figure 3.13a is not bounded.

We shall call *R an autonomous PN* and R_S *the same PN with a synchronization* (i.e. R_S is obtained by associating either an external event or an event e with each transition of R). This synchronization generates constraints on the evolution of R_S, which did not exist for net R. These constraints are due to the fact that, in a synchronized PN, each transition firing is not independent from the other firings: there are elementary firing sequences (an EFS is a set of transitions which *must be fired simultaneously*) and there are *iterated firings* (an iterated firing corresponds to a sequence of EFS which must be carried out before any other external event is taken into consideration). It follows that 1) the set of stable reachable markings of

R_S (and even the set of all the reachable markings) is included in the set of reachable markings of R, and that 2) the set of possible firing sequences in R_S is included in the set of possible firing sequences in net R (the language of R_S is included in that of R). The consequence is that, *as a rule, the properties of an autonomous PN are not preserved when this same net is synchronized.*

Property 3.6 The condition that an *autonomous PN R* be *bounded* for an initial marking $\mathbf{m}_0$ *is sufficient but is not necessary* for the *synchronized PN* R_S to be *bounded* for the same initial marking.

❑

It is obvious that this condition is sufficient since the set of possible evolutions of the synchronized PN is included in the set of possible evolutions of the autonomous PN.

Figure 3.14 shows that the condition is not necessary. Figure a represents an unbounded autonomous PN, and Figure b the same net with a synchronization, both with the corresponding coverability root tree. This synchronized PN is bounded because transitions T_1 and T_2 are synchronized on the same external event E^1. When there are two tokens in place P_1, the firings of T_1 and T_2, are simultaneous (in the same EFS).

Property 3.7 The condition that an *autonomous PN R* be *safe* for an initial marking $\mathbf{m}_0$ *is sufficient but is not necessary* for the *synchronized PN* R_S to be *safe* for the same initial marking.

❑

Sufficiency of the condition in Property 3.7 is obviously true for the same reason as for Property 3.6.

The synchronized PN of Figure 3.15a may represent a feedback shift register: a token in P_i corresponds to the Boolean value 1 in cell i; event E^1 corresponds to the rising edge of the clock. This synchronized PN is safe as shown by its graph of stable reachable markings (Figure 3.15b): from the marking $\mathbf{m}_0$, for example, the token already present in P_2 is taken out *exactly when* another token is placed in P_2. Without synchronization, the same net would not be safe, as shown in Figure 3.15c.

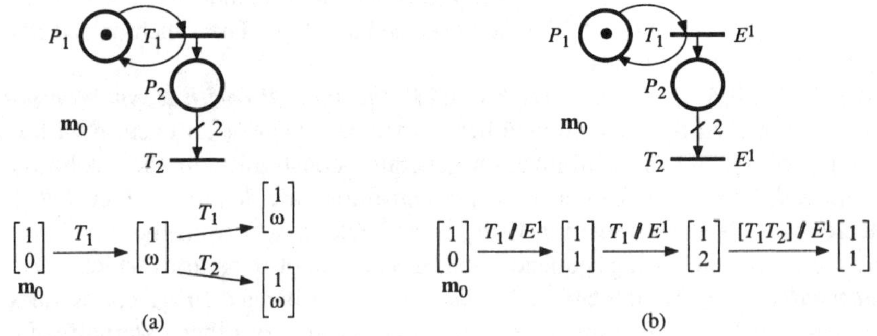

Figure 3.14 R is not bounded, and R_S is bounded.

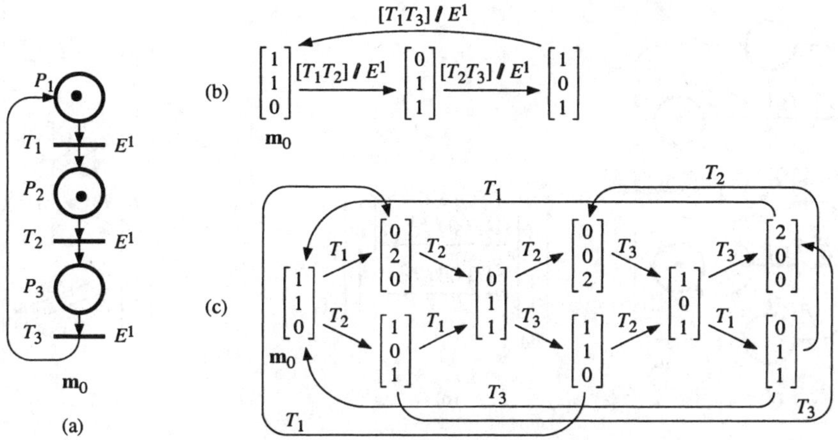

Figure 3.15 R is not safe and R_S is safe. (a) Synchronized PN R_S. (b) Graph of reachable markings of R_S. (c) Graph of reachable markings of R.

Property 3.8 The condition that an *autonomous PN R* be *live* for an initial marking $\mathbf{m}_0$ is *neither necessary nor sufficient* for the *synchronized PN* R_S to be *live* for the same initial marking. ❑

The synchronized PN of Figure 3.16a is live as shown by its graph of stable reachable markings (Figure 3.16b). Without synchronization, the same net would not be live, as shown in Figure 3.16c. Indeed, the firing sequence T_1T_1 from $\mathbf{m}_0$ results in a deadlock. The same is true for sequence T_2T_2. These sequences are not possible for R_S due to the synchronization.

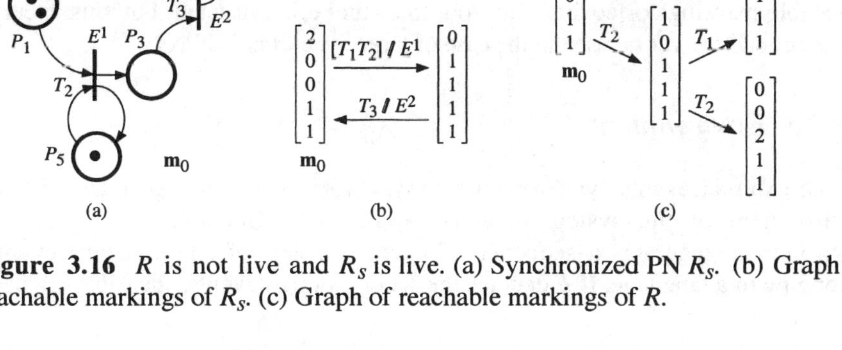

Figure 3.16 R is not live and R_S is live. (a) Synchronized PN R_S. (b) Graph of reachable markings of R_S. (c) Graph of reachable markings of R.

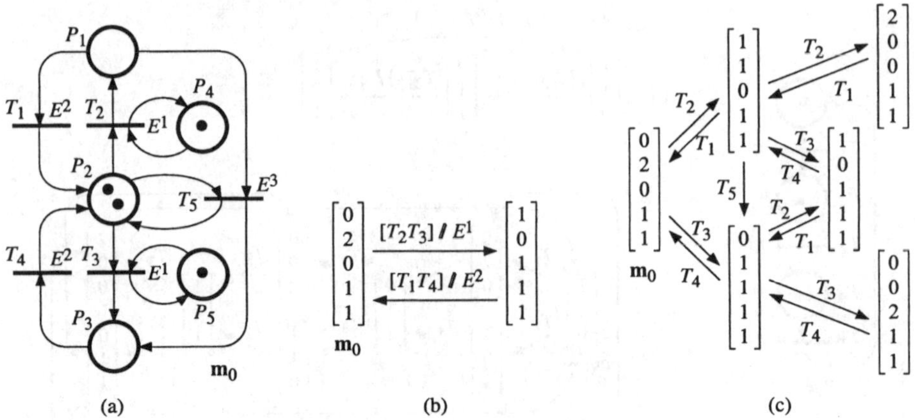

Figure 3.17 R is live and R_S is not live. (a) Synchronized PN R_S. (b) Graph of reachable markings of R_S. (c) Graph of reachable markings of R.

The synchronized PN of Figure 3.17a is not live. Indeed, transition T_5 is never fired as is shown in Figure 3.17b, because the marking (1, 1, 0) which enables this transition is never reached. For the same PN without synchronization, this marking is obtained by firing of T_2 from $\mathbf{m}_0$ (Figure 3.17c). For the synchronized PN, T_2 and T_3 are in the same EFS because E^1 is associated with both these transitions, and T_1 and T_4 are in the same EFS.

We note that the PN of Figure 3.17a is not *simple* (Section 1.2.1.5) because transition T_5 is concerned by two structural conflicts, $\langle P_1, \{T_1, T_5\}\rangle$ and $\langle P_2, \{T_2, T_3, T_5\}\rangle$. The simple PNs possess an interesting property described below.

Property 3.9 If an ordinary *simple* PN is live for an initial marking $\mathbf{m}_0$, then every *totally synchronized* PN constructed from R is live for $\mathbf{m}_0$.

Remark 3.7 We can see that *the properties of an autonomous PN are not necessarily preserved* when the same PN has a synchronization. However, the majority of methods which enable the properties of a Petri net to be discovered apply to autonomous PNs (linear algebra, reductions). We thus do not have many means at our disposal to find the properties of a synchronized PN. The graph of reachable markings or coverability root tree can be constructed, but this method can practically never be used in the case of large synchronized PNs.

3.2.3.3 Environment

The external events synchronizing a synchronized PN are generated by the environment of the system modeled by the PN. Generally speaking, this environment generates a sequence of events, or sets of simultaneous events, belonging to a language. If E denotes the set of external events, then the language generated by the environment is a subset of $(\mathcal{P}(E) - \varnothing)^*$, where $\mathcal{P}(E)$ is the set of subsets of E and $\varnothing$ is the empty set.

Consider for example the set $E = \{E^1, E^2, E^3\}$. Let us denote by $X_{ij...} = E^i \cdot E^j \cdot ...$ the simultaneous occurrence of the events in the subset $\{E^i, E^j, ...\}$ of E. Then, all the event sequences generated by the environment are in

$$\mathcal{L}_{\text{max}} = (E^1 + E^2 + E^3 + X_{12} + X_{13} + X_{23} + X_{123})^*. \tag{3.4}$$

If all the events in E are independent from each other, all the event sequences generated are in

$$\mathcal{L}_{\text{ind}} = (E^1 + E^2 + E^3)^*. \tag{3.5}$$

The language will always be a subset of $\mathcal{L}_{\text{max}}$ and sometimes a subset of $\mathcal{L}_{\text{ind}}$ (adapted according to the number of events in E). We give some examples below:

First example. Let $E = \{E^1, E^2, E^3, E^4\}$ be the set of events generated by the system in Figure D.1 in Appendix D, where $E^1 = \uparrow x_1$ (event *'the level pass from low to middle'*), $E^2 = \downarrow x_1$ (event *'the level pass from middle to low'*), $E^3 = \uparrow x_2$ (event *'the level pass from middle to high'*), and $E^4 = \downarrow x_2$ (event *'the level pass from high to middle'*). It is clear that these events cannot occur simultaneously. Assume the initial level is low. Obviously the sequence of events $Z = E^1E^4$ cannot occur. As a matter of fact, after the event E^1, the level is middle, hence the event E^4 = *'the level pass from high to middle'* cannot occur. The language corresponding to the possible sequences of events is the set $\overline{\mathcal{L}_1}$ of prefixes of $\mathcal{L}_1 = (E^1(E^3E^4)^*E^2)^*$.

Second example. Let $E = \{E^1, E^2\}$ such that E^1 denotes an event occurring after each minute and E^2 occurring after each hour. Event E^2 cannot occur by itself; the corresponding language is a subset of $(E^1 + X_{12})^*$, where X_{12} corresponds to the simultaneous occurrence of E^1 and E^2, i.e., $X_{12} = E^1 \cdot E^2$. More precisely, the language is the set $\overline{\mathcal{L}_2}$ of prefixes of $\mathcal{L}_2 = ((E^1)^{59}X_{12})^*$.

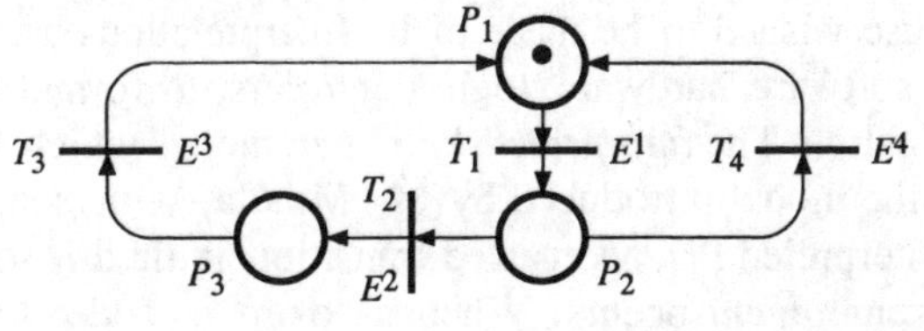

Figure 3.18 Totally synchronized PN, with a single transition per event.

Let us assume that a synchronized PN is *totally synchronized* and that *each external event* is *associated with a single transition*. This is for example the case of the PN in Figure 3.18: an external event E^i is associated with each transition T_i. If the event sequences correspond to the language $\mathcal{L}_{\text{ind}} = E^*$, then this synchronized PN has all the properties (bounded, live, etc.) of the corresponding autonomous PN.

On the other hand, if there is *a correlation among certain events,* i.e. the language containing all the possible event sequences is a proper subset of E^*, then the properties may not be preserved. Let us assume for example that every occurrence of event E^1 is followed by at least one occurrence of event E^2 before occurrence of E^4. Then transition T_4 would not be live because every firing of T_1 would be followed by a firing of T_2. More generally the following properties are obtained.

Property 3.10 Consider a PN R_S synchronized on events in $E = \{E^1, E^2, \ldots\}$. Let $\mathcal{L}_{\text{ind}} = E^*$ and $\mathcal{L}_1$ be a proper subset of E^*.

a) If R_S is bounded for $\mathcal{L}_{\text{ind}}$, it is bounded for $\mathcal{L}_1$.
b) The PN R_S may be unbounded for $\mathcal{L}_{\text{ind}}$ but bounded for $\mathcal{L}_1$.
c) If R_S is safe for $\mathcal{L}_{\text{ind}}$, it is safe for $\mathcal{L}_1$.
d) The PN R_S may not be safe for $\mathcal{L}_{\text{ind}}$ but safe for $\mathcal{L}_1$.
e) The PN R_S may be live for $\mathcal{L}_{\text{ind}}$ but not live for $\mathcal{L}_1$.
f) The PN R_S may not be live for $\mathcal{L}_{\text{ind}}$ but live for $\mathcal{L}_1$.

❑

When moving from an autonomous to a synchronized PN, only the affirmative boundedness and safeness properties were preserved (Properties 3.6, 3.7, and 3.8 in Section 3.2.3.2). Property 3.10 is true for the same reason: the synchronization generates constraints restricting the possible behaviors; similarly, the language $\mathcal{L}_1 \subset E^*$ generates more constraints than E^*. Property 3.10 reinforces the comment in Remark 3.7.

3.3 INTERPRETED PETRI NETS

The expression "interpreted Petri nets" can be applied to various interpretations according to the use wished to be made of it. Interpretations are found adapted to the description of software, hardware, logic controlers, to formal languages and to performance evaluation. The *interpreted Petri net* model which we shall present here is based on the model introduced by M. Moalla, with some modifications. Basically, in an interpreted PN, an enabled transition is fired *if* some *condition* is satisfied, *when* some *event* occurs. When a token is added to a place, some operation is performed and the token remains unavailable for some time (for enabling of the output transitions).

The model we shall define, called "control interpreted PN", presents both extensions and restrictions with respect to the Moalla's model. Let us explain why we have made these choices.

The model defined by M. Moalla was based on synchronized PNs with a single-server semantics. However, we have chosen, in this book, the *infinite-server semantics*; if q tokens are simultaneously added to a place (an input transition of which was q-enabled), shall we perform q times the operation associated with this place? This question does not arise if we restrict our attention to *safe* PNs. We also limit our interest to PNs whose behavior is *deterministic*. These choices are justified because the modeling power of safe PNs is sufficient (theoretically and practically) for logic controlers (the main target) and, furthermore, for standard applications this control must be deterministic.

In Moalla's model, delays were associated with places. In fact, timed models are more useful for performance evaluation (Section 3.4) than for control. Hence, the model we shall present is not timed. If some timing is necessary, it can be modeled using an operation *'launching of timing'* and a condition [*time elapsed*].

After these restrictions, we consider extensions: Boolean conditions may stem from the environment; operations may act punctually on the environment and Boolean variables may be "permanent" actions on the environment. With these additional possibilities, the inputs and outputs of the control interpreted PN are *homogeneous with those of the Grafcet model.* See Appendix E.

3.3.1 Definition of a Control Interpreted Petri Net

Before giving a definition, let us introduce progressively the main concepts. Consider Figure 3.19a. The control interpreted Petri net is a model of logic controler based on a synchronized PN.

The control interpreted PN receives information *from the* **environment** (made up of the controled system and, where applicable, human operators and other control interpreted PNs). This information consists in Boolean variables (C_j^e in the figure) and events (E_j in the figure). In fact, according to Appendix D, the events may be deduced from Boolean variables: for example, if x is a Boolean variable, the knowledge of $x(t)$ for $t \geq 0$ implies the knowledge of the occurrence times of events $\uparrow x$ and $\downarrow x$.

The control interpreted PN sends outputs *to the environment*: Boolean outputs depending on the marking (A_i in the figure), impulse outputs, i.e., events depending on a change of marking (B_i^* in the figure), and variables resulting from a calculation (V_k in the figure, numerical or Boolean).

Inside the control interpreted PN, the control part sends operation orders (O_i^* in the figure, event type) and receives Boolean information from the data processing part (C_j^o in the figure). The model assumes that a calculation has no duration: the result is *immediately available*.

Figure 3.19b presents a part of control interpreted PN. Roughly speaking, the *inputs are associated with the transitions* (changes of states, i.e., transition firings dependent on them) and *outputs are associated with places*. We can see that event E_j and condition C_j are associated with transition T_j. Condition C_j is a Boolean function depending on both the data processing part and the environment. Event E_j is either an external event derived from the environment or the always occurring event e (as explained in Section 3.2).

Transition T_j *will be fired:*

if transition T_j is enabled

and *if* condition C_j is true,

when event E_j occurs.

The product $R_j = E_j \cdot C_j$ is called the **receptivity** of transition T_j. If T_j is enabled, it is *receptive* to R_j. If transition T_j is enabled and if condition C_j is true, transition T_j is *receptive* to event E_j. Let us note that T_j is an *immediate transition* (Definition 3.3, Section 3.2.1) if and only if $C_j = 1$ and $E_j = e$.

As we can see in Figure 3.19b, actions denoted O_i^*, B_i^*, and A_i are associated with place P_i. When a token is deposited in place P_i at instant t, the operation O_i^* is carried out and the impulse action B_i^* is sent to the environment. The Boolean output A_i has the Boolean value 1 as long as there is a token in P_i. We have marked with an asterisk the event-type outputs (O_i^* and B_i^*) because, in this case, they are abstract. However, when there is no ambiguity the asterisk may not be necessary; examples will be given in the sequel.

***Notation 3.3* Simplified notation.** The various values will be as follows by default:

a) If E_j is not specified, then $E_j = e$ (as in Notation 3.2, Section 3.2.2.2).

b) If C_j is not specified, then $C_j = 1$, i.e., the logical condition C_j is true.

c) If O_i^* is not specified, then O_i^* is the *identity operator,* i.e. there is no modification in the state of the variables of the data processing part.

d) If B_i^* is not specified, there is no impulse action.

e) If A_i is not specified, all the Boolean outputs have the Boolean value 0 (provided that there is no other marked place with Boolean outputs associated).

❑

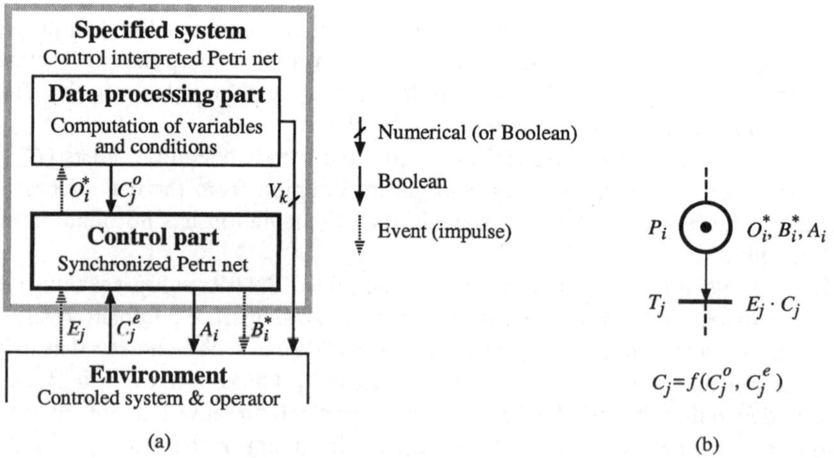

Figure 3.19 Control interpreted Petri net.

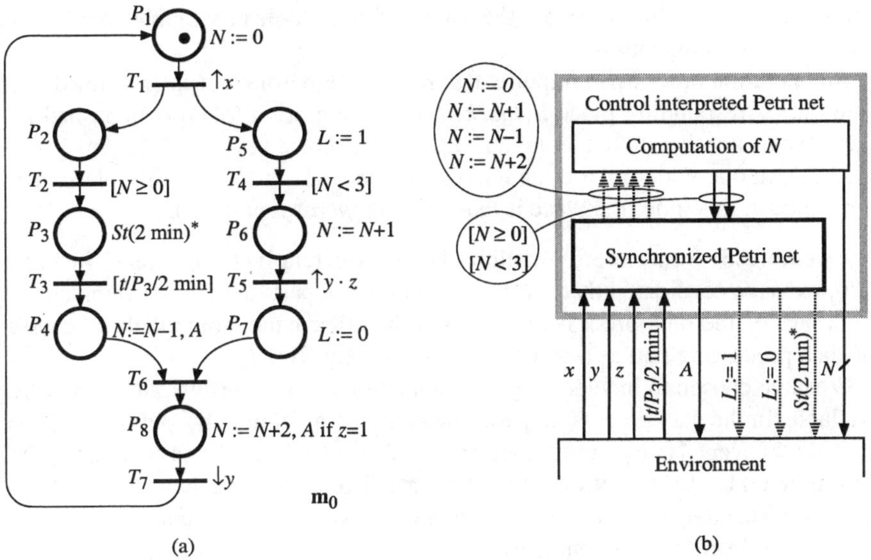

Figure 3.20 (a) Example of control interpreted PN. (b) Inputs and outputs.

An example of control interpreted PN is given in Figure 3.20a and the corresponding inputs and outputs are illustrated in Figure b. The behavior of the control interpreted PN is determined by the three Boolean variables, x, y, and z, derived from the environment.

According to Figure a, the external events synchronizing the PN are $\uparrow x$, $\uparrow y$, and $\downarrow y$: $E_1 = \uparrow x$ (i.e., T_1 is synchronized on $\uparrow x$); $E_5 = \uparrow y$; $E_7 = \downarrow y$; $E_2 = E_3 = E_4 = E_6 = e$.

The conditions associated with the transitions are: $C_2 = [N \geq 0]$ (which means that, when there is a token in P_2, T_2 is fired if and only if the predicate $[N \geq 0]$ is true (i.e., if $N \geq 0$); $C_4 = [N < 3]$; $C_5 = z$; $C_3 = [t/P_3/2 \text{ min}]$, where $[t/P_3/2 \text{ min}]$ is a Boolean variable whose value is 1 if [*a time at least equal to 2 min has elapsed since the last time that* P_3 *was marked*]; $C_1 = C_6 = C_7 = 1$.

Consider now the operation orders. When P_1 is marked, the value N is reset (operation $N := 0$). Other operations on the value N are associated with places P_4, P_6, and P_8.

Consider finally the *actions on the environment*. There is a level action, A, and three impulse actions, $L := 1$, $L := 0$, and $St(2 \text{ min})^*$. The Boolean value of A is 1 when there is a token in P_4, and when there is a token in P_8 if the Boolean value of z is 1 (a level action may be *conditional*: this is the case for action A when is is associated with P_8). The impulse actions may activate some device whose state is Boolean. Assume for example that L is the state of a light: $L := 1$ switches on the light (the Boolean variable L) and $L := 0$ switches off the light. A special action is associated with place P_3 which can be expressed as *'start a timing for 2 minutes'* written $St(2\text{ min})^*$ in Figure a; the value of the Boolean variable $[t/P_3/2\text{ min}]$ results from this operation[6].

An operation order or an impulse action on the environment is performed every time the corresponding place *is marked.* In some cases a token is deposited in a place when another token is taken out of this place (Figure 3.15 in Section 3.2.3.2). An operation order or an impulse action associated with this place is not performed at this time (the place is *marked without discontinuity*).

Remark 3.8 Some representation choices are relatively arbitrary. In Figure 3.20, we have considered that the 2-min timer was a part of the environment; why not a part of the data processing part? On the other hand, some authors consider the data processing part as a part of the environment [Br 83].

We have chosen to include only the calculations, whose results are immediately available, in the data processing part. The effect of a timing, by definition, is not immediate; event $\uparrow[t/P_3/2\text{ min}]$ does not result immediately from a calculation: it should be added to the set of external events. The choice we have made implies that *no event stems from the data processing part* but only conditions (all the events stem from the environment).

❑

The behavior of the model in Figure 3.20 will be specified in Section 3.3.2. For the time being, and prior to defining a control interpreted PN, we shall see under what conditions a safe interpreted PN can be deterministic.

An interpreted PN describes a controler. This controler is deterministic if, for every input sequence, its output sequence is deterministic (abuse of language since the word *controler* is used incorrectly: we should really say *sequential machine* if we do not only refer to the input/output behavior). The state of an interpreted PN is defined by two components: the marking, on the one hand, and the state of the data processing part, on the other. Thus, for an interpreted PN to be deterministic, both these components must also be so.

For the **marking** to be **deterministic**, it is necessary and sufficient that there is no actual conflict (Definition 3.4 in Section 3.2.1), i.e., at each firing of an elementary firing sequence, this EFS is unique (and consequently maximal).

An operation O_j^* which would allocate a random value to a variable V_k, for example, would not be deterministic. We assume that all operations O_j^* are deterministic. Given this assumption, in order for the state of the **data processing part** to be **deterministic**, it is necessary and sufficient that for every pair of operations O_i^* and O_j^* which could be carried out at the same time, these two operations are *compatible*. The operations O_i^* and O_j^* are *strongly compatible* if the set of variables transformed by O_i^* is disconnected from the set of variables transformed by O_j^*, for example $O_i^*: X := X + 1$, and $O_j^*: Y := 0$. Operations O_i^* and O_j^* are *weakly compatible* if they transform the same variable but their execution order does not affect the final result, for example

[6] Note that there is some redundancy in this specification. The existence of the Boolean variable $[t/P_3/2\text{ min}]$ *implies* the timing starting $St(2\text{ min})^*$ associated with place P_3.

$O_i^*: X := X + 1$, and $O_j^*: X := X + 2$; the final result is $X := X + 3$. In the other cases, these operations are said to be *incompatible*. For example $O_i^*: X := X + 1$, and $O_j^*: X := X^2$ are incompatible. Likewise $O_i^*: X := 0$ and $O_j^*: X := 1$ are incompatible.

Property 3.11 **A safe control interpreted PN** is **deterministic** if and only if it satisfies the following two conditions.

1) There is no actual conflict.

2) For every pair of places P_i and P_j such that O_i^* and O_j^* are incompatible, we have $m_a(P_i) + m_a(P_j) \leq 1$ for every reachable marking $\mathbf{m}_a$ (stable or unstable). ❑

Note that the existence of a marking *invariant* $m(P_i) + m(P_j) + ... = 1$ is a *sufficient condition* to satisfy Condition 2 of Property 3.11. This condition ensures that if operation O_i^* is carried out, O_j^* is not carried out, and vice versa. Note that it is in our interest to ensure that the operations simultaneously carried out are *strongly compatible,* in order to minimize the risk of error...

Let us now define the control interpreted PNs.

Definition 3.7 A **control interpreted Petri net** exhibits the following five characteristics (1 to 3, necessarily, 4 and 5, possibly). See Figure 3.19.

1) It is *synchronized* on external events and *stable* (Section 3.2.3.1).

2) It is *safe*.

3) It is *deterministic*.

4) It has *a data processing part* whose state is defined by a set of variables $V = \{V_1, V_2, ...\}$. This state is modified by operations O_i^* which are associated with the places. It determines the value of the predicates C_j^o.

5) It receives Boolean information C_j^e *from the environment*. It sends level actions A_i (Boolean) and impulse actions B_i^* (event type), associated with the places, *to the environment*.

3.3.2 Interpretation Algorithm of a Control Interpreted PN

The presentation of the interpretation algorithm of a control interpreted PN will have recourse to notions which have been fully studied for synchronized PNs: the elementary firing sequence (EFS) and the iterated firing (Section 3.2).

Algorithm 3.3 **Interpretation of a control interpreted PN**

Step 1. Initialization of the marking; initialization of all the level actions at value 0; execution of the operations and impluse actions associated with the marked places. Go to *Step 5*.

Step 2. *Wait for the next external event*. When a new external event occurs (or set of compatible events), determine the set $\mathcal{T}$ of the transitions receptive to this event (or set of events). If $\mathcal{T}$ is empty, go to[7] *Step 6*.

Step 3. Carry out the EFS (made up of all transitions in $\mathcal{T}$).

[7] The value of a conditional output may change even if the state does not change.

Step 4. Carry out all the operations and impulse actions associated with the places which just became marked in *Step 3*.
Step 5. Determine the set $\mathcal{T}$ of the transitions receptive to event e (always occurring). If $\mathcal{T}$ is not empty, go to *Step 3*.
Step 6. The marking is stable.
Step 6.1. Determine the set $\mathcal{A}_0$ of the level actions which must be inactivated (actions associated with the places which were marked at *Step 2* and which are not marked now, and conditional actions associated with places still marked for which the conditions are no longer verified).
Step 6.2. Determine the set $\mathcal{A}_1$ of the level actions which must be activated (actions associated with the places which were not marked at *Step 2* and which are now marked, possibly under certain conditions, and conditional actions associated with the places still marked for which the conditions are verified whereas they were not at *Step 2*)[8].
Step 6.3. Set to 0 all the actions which belong to $\mathcal{A}_0$ and do not belong to $\mathcal{A}_1$. Set to 1 all the actions which belong to $\mathcal{A}_1$. Go to *Step 2*.

❑

Let us now illustrate this algorithm with an example. Consider the control interpreted PN in Figure 3.20a (the marking shown is $\mathbf{m}_0$) and the timing diagram of x, y, and z, presented in Figure 3.21. The projection of the considered input sequence of events in $(\uparrow x + \downarrow x + \uparrow y + \downarrow y + \uparrow z + \downarrow z)^*$ is $S = \uparrow x \downarrow y \uparrow y \uparrow z \downarrow y \uparrow y \downarrow z$ (in fact another event must be taken into account: $\uparrow[t/P_3/2 \text{ min}]$; but we do not yet know where it will take place in the preceding sequence).

From application of Algorithm 3.3 presented in Figure 3.21, the following observations can be made:

a) On occurrence of $\uparrow x$, the stable marking $\mathbf{m}_2 = (0, 0, 1, 0, 0, 1, 0, 0)$ is reached. The 2-min timing starts at this time since P_3 became marked. For the example in the figure, the event $\uparrow[t/P_3/2 \text{ min}]$ will occur two minutes later, between the first occurrence of $\uparrow y$ and the first occurrence of $\uparrow z$.

b) A level action with the same value (0 or 1) in two successive stable markings, keeps the value without discontinuity: this value can change only when the marking is stable (*Step 6*). Example: $A = 1$ for $\mathbf{m}_3$ and $\mathbf{m}_5$.

c) An operation or impulse action is performed when the corresponding place becomes marked, even if the marking is not stable (in *Step 4*). Example: $L := 1$ performed when the instable marking $\mathbf{m}_1$ is reached (between $\mathbf{m}_0$ and $\mathbf{m}_2$).

Remark 3.9

a) As illustrated in Figure 3.19b, the receptivity of a transition may always be modeled by the product of an event and a condition: $R_i = E_i \cdot C_i$. According to Notation 3.3, the "always true" condition is associated with receptivities which depend only on an external event; example: $R_1 = \uparrow x$ (examples are taken from Figure 3.20a), then $C_1 = 1$, i.e., $R_1 = \uparrow x \cdot 1$. Similarly, the "always occurring" event is associated with receptivities which depend only on a condition; example:

---

[8] Reference to *Step 2* is obviously not significant at initial time.

$R_3 = [t/P_3/2 \text{ min}]$, then $E_3 = e$, i.e., $R_3 = e \cdot [t/P_3/2 \text{ min}]$.

b) When T_3 becomes enabled, the condition $C_3 = [t/P_3/2 \text{ min}]$ is not true, by definition. The event $\uparrow[t/P_3/2 \text{ min}]$ occurs exactly when this condition becomes true. Then, in this case, the behavior would be the same if

$$E_3 \cdot C_3 = \uparrow[t/P_3/2 \text{ min}] \cdot 1 \quad \text{instead of} \quad E_3 \cdot C_3 = e \cdot [t/P_3/2 \text{ min}].$$

In other words, in some cases, a control interpreted PN behaves similarly if the receptivity of some transition is the event $\uparrow x$ or the condition x (if $x = 0$ when the transition becomes enabled).

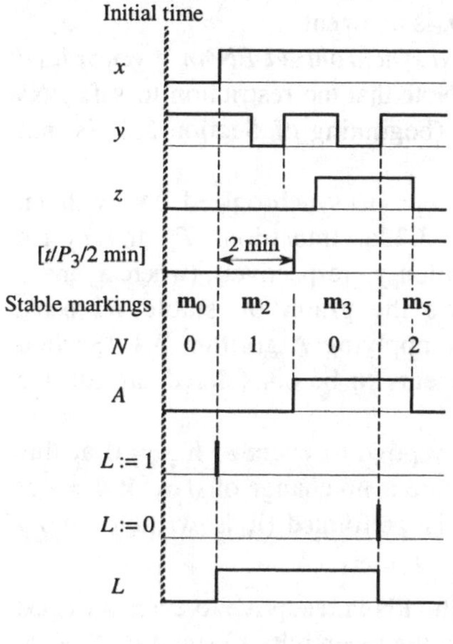

Step 1. $\mathbf{m}_0 = (1, 0, 0, 0, 0, 0, 0, 0)$; operation $N := 0$.
Step 5. $\mathcal{T} = \varnothing$.
Step 6. $\mathcal{A}_0 = \varnothing$; $\mathcal{A}_1 = \varnothing$; no change.
Step 2. Event $\uparrow x$; $\mathcal{T} = \{T_1\}$.
Step 3. Firing of T_1; $\mathbf{m}_1 = (0, 1, 0, 0, 1, 0, 0, 0)$.
Step 4. Impulse action $L := 1$.
Step 5. Event e; $\mathcal{T} = \{T_2, T_4\}$.
Step 3. Firing of $[T_2T_4]$; $\mathbf{m}_2 = (0, 0, 1, 0, 0, 1, 0, 0)$.
Step 4. $St(2 \text{ min})^*$; $N := N+1$.
Step 5. Event e; $\mathcal{T} = \varnothing$.
Step 6. $\mathcal{A}_0 = \varnothing$; $\mathcal{A}_1 = \varnothing$; no change.
Step 2. Event $\downarrow y$; $\mathcal{T} = \varnothing$.
Step 6. $\mathcal{A}_0 = \varnothing$; $\mathcal{A}_1 = \varnothing$; no change.
Step 2. Event $\uparrow y$; $\mathcal{T} = \varnothing$ (because z=0).
Step 6. $\mathcal{A}_0 = \varnothing$; $\mathcal{A}_1 = \varnothing$; no change.
Step 2. Event $\uparrow[t/P_3/2 \text{ min}]$; $\mathcal{T} = \{T_3\}$.
Step 3. Firing of T_3; $\mathbf{m}_3 = (0, 0, 0, 1, 0, 1, 0, 0)$.
Step 4. $N := N - 1$.
Step 5. Event e; $\mathcal{T} = \varnothing$.
Step 6. $\mathcal{A}_0 = \varnothing$; $\mathcal{A}_1 = \{A\}$; A takes the value 1.
Step 2. Event $\uparrow z$; $\mathcal{T} = \varnothing$.
Step 6. $\mathcal{A}_0 = \varnothing$; $\mathcal{A}_1 = \varnothing$; no change.
Step 2. Event $\downarrow y$; $\mathcal{T} = \varnothing$.
Step 6. $\mathcal{A}_0 = \varnothing$; $\mathcal{A}_1 = \varnothing$; no change.
Step 2. Event $\uparrow y$; $\mathcal{T} = \{T_5\}$ (condition z=1 is now true).
Step 3. Firing of T_5; $\mathbf{m}_4 = (0, 0, 0, 1, 0, 0, 1, 0)$.
Step 4. Impulse action $L := 0$.
Step 5. Event e; $\mathcal{T} = \{T_6\}$.
Step 3. Firing of $[T_6]$; $\mathbf{m}_5 = (0, 0, 0, 0, 0, 0, 0, 1)$.
Step 4. $N := N+2$.
Step 5. Event e; $\mathcal{T} = \varnothing$.
Step 6. $\mathcal{A}_0 = \{A\}$; $\mathcal{A}_1 = \{A\}$; no change.
Step 2. Event $\downarrow z$; $\mathcal{T} = \varnothing$.
Step 6. $\mathcal{A}_0 = \{A\}$; $\mathcal{A}_1 = \varnothing$; A returns to value 0.

Figure 3.21 Illustration of Algorithm 3.3 (related to Figure 3.20a).

Remark 3.10 *Grafcet* is a tool inspired from Petri nets, whose purpose is the specification of logic controlers. It is the basis of the Sequential Function Chart (SFC), an International Standard in 1987. References are given in Notes & References at the end of the chapter, and Appendix E is devoted to Grafcet.

3.3.3 Interpreted PN Without Outputs: Generalization of the Concept of Synchronized PN

According to Appendix D, the product of a condition and an event is an event. Then, from Section 3.3.1 (Figure 3.19b) and Remark 3.9a, *a receptivity can always be considered as an event.* According to Notation 3.3a and Remark 3.9b, if a receptivity is represented by the condition C^a, the corresponding event may be developed as $e \cdot C^a$. In other words, $C^a = e \cdot C^a$ as far as receptivities are concerned, i.e., *a condition may be considered as an event* by extension: this *event is always occurring as long as* $C^a = 1$.

It follows that a receptivity which would be the sum of an event E^b and a condition C^a is the sum of both events E^b and $C^a = e \cdot C^a$, i.e., an event. More generally, *any function of events and conditions* using the *sum*, the *product*, and the *symbols* $\uparrow$ and $\downarrow$, presented in Appendix D, is an event.

Hence, *an interpreted PN without outputs is a synchronized PN in a generalized meaning* (since every receptivity is an event). Note that the restriction to safe PNs which was used for control interpreted PNs (beginning of Section 3.3) is not useful here since there are no outputs.

Let us now illustrate this generalized concept of synchronized PN with an example (Figure 3.22). As shown in Figure 3.22a, transitions T_1 and T_2 are synchronized by the event $\uparrow x$ and by the condition y, respectively (where x and y are Boolean variables). Figure 3.22b shows the graph of stable reachable *significant markings*[9]. It may be obtained by applying Algorithm 3.1 (Section 3.2.2.2) to all possible cases; the external events to be considered are all the changes of the Boolean variables x and y.

At initial time, for $\mathbf{m}_0 = (1, 1)$, only T_2 is receptive to event e. If $y = 0$ at this time, the firing condition is not satisfied and there is no change of state. If $y = 1$ at this time, the firing $(1, 1) \xrightarrow{T_2 / e \cdot y} (2, 0)$ is performed (it is written simply $(1, 1) \xrightarrow{T_2 / y} (2, 0)$).

If the marking is $\mathbf{m}_0$, *after the initial time*, the PN is receptive to events $\uparrow x$ and $\uparrow y$ (according to Remark 3.9b in Section 3.3.2, the receptivity to y and to $\uparrow y$ are equivalent if $y = 0$; furthermore $y = 0$ if $\mathbf{m}_0$ is stable). If $\uparrow x$ occurs first, firing $(1, 1) \xrightarrow{T_1 / \uparrow x} (0, 2)$ is performed. If $\uparrow y$ occurs first, $(1, 1) \xrightarrow{T_2 / y} (2, 0)$ occurs.

If the marking is $(2, 0)$, the PN is only receptive to event $\uparrow x$. If $y = 0$ when $\uparrow x$ occurs, there is a single firing of T_1 denoted by $(2, 0) \xrightarrow{T_1 / \uparrow x \cdot y'} (1, 1)$ (since T_1 is 1-enabled because of the single token in P_3). If $y = 1$ when $\uparrow x$ occurs, the firing of T_1 is immediately followed by a firing of T_2 (iterated firing): $(2, 0) \xrightarrow{T_1 T_2 / \uparrow x y} (2, 0)$.

---

[9] According to Appendix E, the **significant marking** is the marking restricted to the places which *do not have the same marking in all the stable states.*

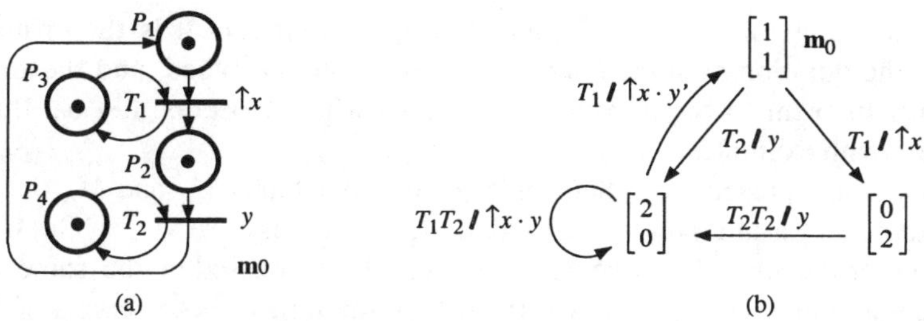

Figure 3.22 (a) Petri net synchronized by an event and a condition. (b) Graph of reachable stable markings (restricted to the significant marking).

If the marking is (0, 2), $y = 0$ (otherwise the marking would not be stable) and the PN is only receptive to the condition y (becoming true when event $\uparrow y$ occurs). When this event occurs, there is an iterated firing: first a firing of T_2 due to $\uparrow y$, then a new firing of T_2, on occurrence of e because the condition y is true: $(0, 2) \xrightarrow{T_2T_2/y} (2, 0)$. Hence, *we have pointed out a new concept: synchronization by a condition* (instead of an event).

3.4 TIMED PETRI NETS

Timed Petri nets are useful for *performance evaluation* (computer systems, manufacturing systems, etc.). A timing may be associated with the duration of an operation or with an expected time before some event occurrence such as a failure for example. *Constant* timings, on one hand, and *stochastic* timings with exponential distribution, on the other hand, are commonly used models since they allow performance evaluation thanks to *analytical methods*.

After some general information in Section 3.4.1, constant timings and stochastic timings are presented in Sections 3.4.2 and 3.4.3. Appendix G explains that the timed PNs correspond to special cases of synchronized PN (and vice-versa). This presentation does not simplify the use of timed PNs but it allows these PNs to inherit all the properties of synchronized PNs presented in Section 3.2.

Models in which a timing has a minimal and a maximal value, called time PNs, are briefly presented in Appendix H.

3.4.1 General Information

Basically, two models of timed PNs can be used; time is associated with the places or with the transitions. Transfers are possible from one model to another (their modeling powers are analyzed in Appendix O). In a PN, it is natural to associate with a place a state which has some duration and to associate with a transition a change of state, this change having no duration. It is then natural to associate the duration of some operation or state with a place, and the time of waiting for an event to the transition to be fired when it occurs. Let us illustrate these ideas with examples.

Figure 3.23a represents a system made of two machines M_A and M_B on which four customers pass alternatively. Machine M_A has a single server, while M_B is a double-server machine (i.e., two customers can be processed at the same time). The processing time of M_A is d_A, and the processing time of both servers of M_B is d_B. In Fig. 3.23a, the state of the system is as follows: the tokens in P_1 represent customers waiting for an operation on the machine M_A; a customer is processed by the machine M_A (token in P_2), hence the machine is not available (no token in P_2') and the transition T_1 cannot be fired. A customer is processed by M_B (token in P_4); a server of this machine remains idle (token in P_4') since there is no token in P_3. Operation times d_A and d_B are naturally associated with the places P_2 and P_4; this means that a token arriving in P_2 must remain during d_A before allowing firing of transition T_2 (during this time, the token is said to be *unavailable*). Similarly, a token placed in P_4 remains unavailable for d_B. It may be *convenient not to specify the delay associated with a place if this delay is zero*; in our example, any token placed in P_1, P_2', P_3, or P_4', is immediately available. In Fig. 3.23a, the times d_A and d_B may be either deterministic (P-timed PN) or stochastic (exponential distribution or any other distribution).

Remark 3.11 A transition is enabled if there are available tokens in all its input places. It can then be fired. *In a general case, the transition could be fired later*: firing of T_2 in Figure 3.23a corresponds to removing the part from the machine, *possibly after* the end of processing. If *transitions are fired at any time* (not specified) after their enabling, this non-deterministic behavior will be called at **free speed**. *Free speed behavior* may be used if some freedom is useful for improving some performance (for scheduling problems, for example).

However, since performance evaluation of completely specified systems is mainly targeted when timed PNs are used, we are mainly concerned with functioning at **maximal speed**. I.e., unless otherwise specified, *transitions are fired as soon as possible.*

❑

Figure 3.23b represents the state of a machine M_C which can fail and be repaired. The token in P_5 means that the machine is operational: transition T_5 can be fired and d_F represents the time when the failure will occur. Similarly, d_R represents the repair time. In this example, the times d_F and d_R are stochastic, with any distribution; if they are exponentially distributed, the rates (to failure and to repair) may be represented instead of the times (usual in stochastic PNs).

In Fig. 3.23a, the durations are naturally associated with the places modeling the corresponding operations. In Fig. 3.23b, the durations are naturally associated with the transitions fired when the corresponding events occur. Now, if we want to associate all the delays with the transitions, how can we modify Fig. 3.23a to meet this requirement? Two solutions are presented in the sequel.

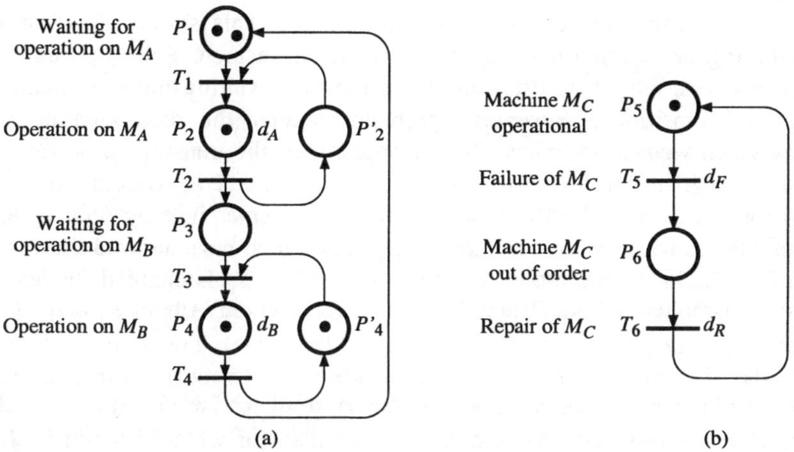

Figure 3.23 Natural modeling. (a) Timed operations. (b) Waited events.

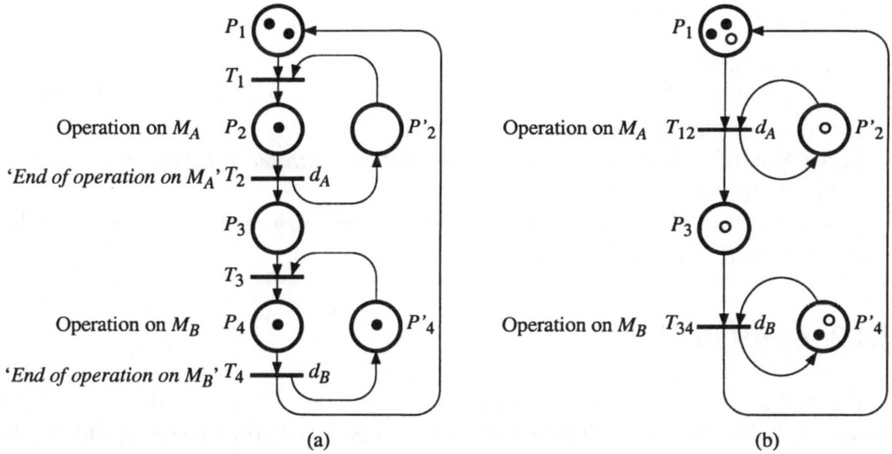

Figure 3.24 Time associated with transitions. (a) Time to events. (b) Duration.

The first solution consists of preserving the structure[10] of the PN and of associating the delays with the transitions corresponding to the events *'End of operation on M_A'* and *'End of operation on M_B'*. This solution is illustrated in Fig. 3.24a (*all the transitions without explicit delay are immediate*, i.e., fired as soon as they are enabled).

The second solution consists of associating an operation with a transition. For our example, the sub-PN made of T_1, P_2, T_2, and the corresponding arcs, in Fig. 3.23a, are replaced by the single transition T_{12} in Fig. 3.24b. Similarly, transition

[10] In some cases additional places and transitions are necessary, as explained in Section 3.2.4.1.

T_{34} in Fig. 3.24b represents the operation on M_B. This solution consisting of representing an operation by a transition in a discrete PN, although it is intuitively less natural, is often used because the resulting model contains less places and transitions. However, a problem remains: the representation of the tokens which were in the places that disappeared in the transformation from Fig. 3.23a to Fig. 3.24b. The token in P_2 on Fig. 3.23a should be "in the transition T_{12}" in Fig. 3.24b. In this figure, this is represented by *reserved* tokens (represented as white tokens) in the input places which have allowed the firing of T_1 in Fig. 3.23a. In the model in Fig. 3.24b, when T_{12} is enabled (at least one non-reserved token in both P_1 and P_2'), a token is reserved in both P_1 and P_2' (this reservation corresponds to the firing of T_1 in Fig. 3.23a). The tokens are reserved for d_A, then T_{12} is fired (this firing corresponds to the firing of T_2 in Fig. 3.23a): the reserved tokens are taken out and non-reserved tokens are placed in P_3 and P_2'. Informally, the reserved tokens correspond to a token which "should be in the transition", hence they are not available for re-enabling a transition.

In the rest of the book, *the time associated with a transition corresponds to a time to event* (as in Figure 3.24a), unless otherwise specified. In other words, there are *no reserved tokens* as in Figure 3.24b.

Notation 3.4

a) A PN with timings associated with *places* will be called **P-timed** (P standing for place, see Section 3.4.2.1).

b) A PN with timings associated with *transitions, without reserved tokens,* will be called **T-timed** (see Section 3.4.2.2).

c) A PN with timings associated with *transitions, with reserved tokens,* will be called **T-timed with reserved tokens**.

3.4.2 Constant Timing

P-timed and T-timed PNs are presented in Sections 3.4.2.1 and 3.4.2.2. The stationary behavior will be presented for the T-timed case (the most frequent in the literature on Petri nets) in Section 3.4.2.3. Adaptation to a P-timed PN is easy. In addition, it is always possible to change-over from a P-timed to a T-timed PN (Figure 3.25 and Appendix O).

3.4.2.1 P-Timed Petri Nets

A timing d_i, of zero value if not specified, is associated with each place P_i.

Definition 3.8 A **P-timed PN** is a pair $\langle R, \text{Tempo}\rangle$ such that:

R is a marked PN;

Tempo is a function from the set P of places to the set of positive or zero rational numbers. Tempo(P_i) = d_i = timing associated with place P_i.

❑

The P-timed PNs have been defined with timing values which are rational numbers, in order to have periodical functionings. However it is possible to define timed PNs with real timings (comment also relevant to T-timed PNs).

When a token is deposited in place P_i, this token must remain in this place at least for a time d_i. This token is said to be **unavailable** for this time. When the time d_i has elapsed, the token then becomes **available**.

Operating of a P-timed PN. At any time t, the present marking $\mathbf{m}$ is the sum of two markings $\mathbf{m}^a$ and $\mathbf{m}^u$, such that $\mathbf{m}^a$ is the marking made up of the available tokens and $\mathbf{m}^u$ is the marking made up of the unavailable tokens. A transition is enabled for the marking $\mathbf{m} = \mathbf{m}^a + \mathbf{m}^u$ if it is enabled for the marking $\mathbf{m}^a$. Firing of a transition is carried out as for an untimed PN, by only removing available tokens from the input places. This firing has a zero duration. If a token is deposited in a place P_i during a firing carried out at instant t_1, then this token is unavailable in the interval $[t_1, t_1 + d_i[$.

Definition 3.9 If a token is deposited in P_i at time t_1, for t_2 such that $t_1 \leq t_2 < t_1 + d_i$, the duration $(t_1 + d_i) - t_2$ is called the **residual time to availability**.

Remark 3.12 *Usually, the initial marking is made up of available tokens.* This initial marking will be referred as $\mathbf{m}_{0,\min}$ since all the *residual times to availability* are zero, hence minimal.

If all the residual times to availability are maximal at initial time (i.e. d_i for a token in P_i), the initial marking is denoted by $\mathbf{m}_{0,\max}$.

If another initialization is used, all the residual times must be specified.

❑

Figures 3.23a and 3.24a illustrate a way of changing over from a P-timed to a T-timed PN. In this example, every output transition of a timed place (i.e., whose timing is not zero) has a single input place. Figure 3.25 illustrates the transformation when a timed place has an output transition common to two input places. A timed place is replaced by two places separated by a timed transition which is fired when the time is elapsed. Consider the token in P_2: as long as it is unavailable in Figure a, there is a token in P_2 in Figure b; when the token becomes available in Figure a, T_2' is fired in Figure b. When there are a token in P_1' and in P_2', transition T_2 can be immediately fired.

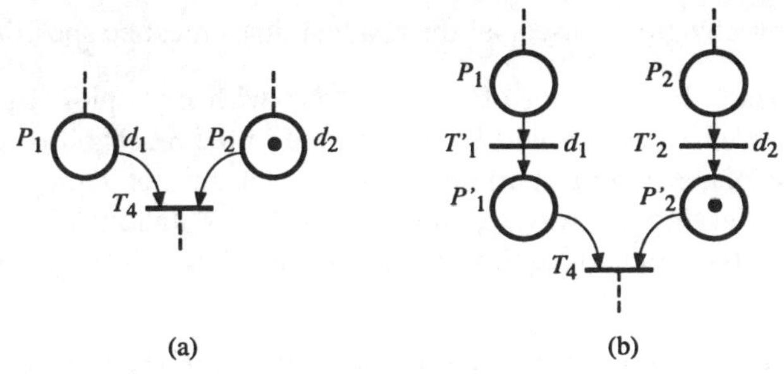

Figure 3.25 Change-over from a P-timed to a T-timed PN.

3.4.2.2 T-Timed Petri Nets

A timing d_j, of zero value if not specified, is associated with transition T_j.

Definition 3.10 A **T-timed PN** is a pair $\langle R, \text{Tempo}\rangle$ such that:

R is a marked PN;

Tempo is a function from the set T of transitions to the set of positive or zero rational numbers. Tempo$(T_j) = d_j$ = timing associated with T_j.

❑

When a transition T_j becomes enabled, it is not fired immediately. It will be fired when the time d_j has elapsed after enabling[11], except in case of conflict.

Operating of a T-timed PN. 1) Assume first that an enabled transition will not be disabled, except by its own firing. In this case: if a transition T_j is enabled at time t_1, it will be fired at $t_1 + d_j$; if T_j is enabled at t_1, then becomes 2-enabled at t_2, $t_1 \leq t_2 \leq t_1 + d_j$, it will be fired at $t_1 + d_j$ then at $t_2 + d_j$ (may be generalized to q-enabling). Note that T_j is an *immediate transition* (Definition 3.3, Section 3.2.1) if and only if $d_j = 0$.

2) Now, in case of a *general conflict* (Definition 2.10, Section 2.1.4), the first transition, involved in the conflict, which becomes firable is fired: there is *no actual conflict*[12] except if two transitions in conflict become firable at the same time. Hence, in case of a general conflict a transition may be disabled before becoming firable.

Definition 3.11 An enabling of T_j at time t_1 should lead to a firing at $t_1 + d_j$. For t_3 such that $t_1 \leq t_3 < t_1 + d_j$, if this enabling was not canceled by another firing, the duration $(t_1 + d_j) - t_3$ is called the **residual time to firing**.

Hence, for a transition whose enabling degree is q, the *vector of residual times to firings* contains q components.

Remark 3.13 *Usually, the initial marking is such that the residual times to firings are maximal* (i.e. d_j for every enabling of T_j). It will be referred as $\mathbf{m}_{0,\max}$ (this initialization is assumed if it is not specified).

If all the residual times to firing are zero at initial time, the initial marking is denoted by $\mathbf{m}_{0,\min}$.

If another initialization is used, all the residual times must be specified.

❑

Let us illustrate the behavior of T-timed PNs with examples. In the sequel, only the *significant markings* will be represented in the reachability graph. This means that the places whose markings are constant are not represented (such a place P_i whose marking is always m_i models a m_i-server station; its only effect is the limitation of the enabling degree of the corresponding transition to value m_i).

---

[11] According to note 10 in Section 3.4.1, behavior at *maximal speed* is considered.

[12] This concept was introduced in the context of synchronized PNs (Definition 3.4, Section 3.2.1). According to Appendix G, it is relevant to timed PNs.

This loss of information is compensated by the knowledge of the vector of enabling degrees.

As will be shown in the examples, a **node** N_k *of the reachability graph* contains the following information: 1) the *marking* $\mathbf{m}_k$, 2) the *enabling vector* $\mathbf{e}_k$, and 3) a list of q *residual times to firings* (at the moment when this node is reached) for a q-enabled transition.

(a)

(b)

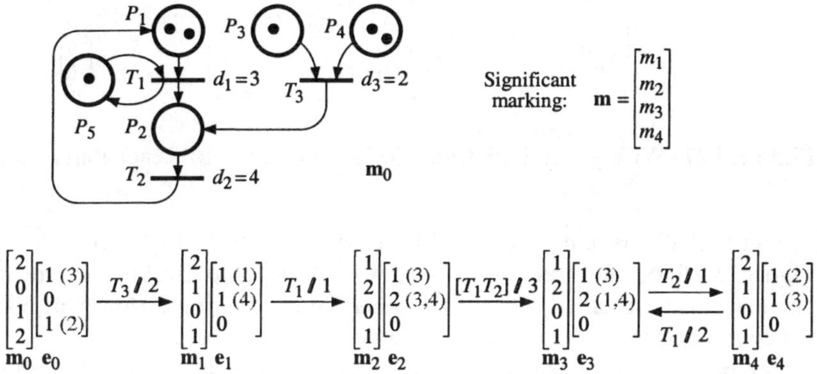

Figure 3.26 Without general conflict. (a) T-timed PN. (b) Reachabilty graph.

Figure 3.26a is a T-timed PN without general conflict. Its reachability graph is shown in Figure b. Let $\mathbf{m}_0$ denote the initial significant marking, i.e. $\mathbf{m}_0 = (m_1(0), m_2(0), m_3(0), m_4(0)) = (2, 0, 1, 2)$. For this marking, the *enabling vector*, is $\mathbf{e}_0 = (q(T_1, \mathbf{m}_0), q(T_2, \mathbf{m}_0), q(T_3, \mathbf{m}_0)) = (1, 0, 1)$. The effect of the constant marking $m_5(t) = 1$ for any t, is that $q(T_1, \mathbf{m}) \leq 1$ for any $\mathbf{m}$. This information, not represented in $\mathbf{m}_0$, is taken into account in $\mathbf{e}_0$ (i.e., $q(T_1, \mathbf{m}_0) = 1$ although $m_1(0) = 2$).

According to $\mathbf{e}_0$, at initial time T_1 and T_3 are 1-enabled. According to Remark 3.13, $\mathbf{m}_{0,\max}$ is assumed, hence the *residual times* to firings are 3 for T_1 and 2 for T_3 (corresponding to $d_1 = 3$ and $d_3 = 2$, respectively). These values are written in grey characters in Figure 3.26b. These residual times to firings decrease when the time increases. At $t = 2$, the residual time to firing of T_3 reaches the value 0, then T_3 is fired, resulting in $\mathbf{m}_1$. When this marking is reached, T_1 is still 1-enabled but its residual time to firing is reduced to 1 (since two time units have been spent since enabling); T_2 becomes 1-enabled and its residual time to firing is $d_2 = 4$. Hence the next firing will be the firing of T_1 which will occur one time unit after $\mathbf{m}_1$ has been reached (value shown after the /) and result in $\mathbf{m}_2$. For this marking, T_2 is 2-enabled and the residual times to firings are 3 for the first enabling and 4 for the second one. Three time units after $\mathbf{m}_2$ was reached, both T_1 and T_2 are fired. The double firing $[T_1T_2]$ leads to $\mathbf{m}_3$, and so on.

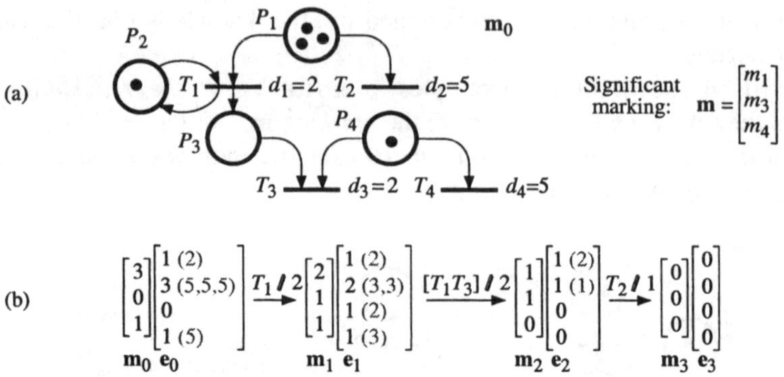

Figure 3.27 With general conflicts. (a) T-timed PN. (b) Reachabilty graph.

Figure 3.27 presents a case with *general conflicts*. In Figure 3.27 b, as in Figure 3.26 b, the transitions which will become firable (if not disabled beforehand) are specified with their residual times to firings. However, because of the conflicts, some firings will not occur.

For $\mathbf{m}_0$, T_1 is 1-enabled and T_2 is 3-enabled; there is a general conflict since $m_1(0)) = 3 < q(T_1, \mathbf{m}_0) + q(T_2, \mathbf{m}_0) = 4$. The firing of T_1 occurs at $t = 2$, resulting in $\mathbf{m}_1$. For this marking, T_1 is again 1-enabled, whereas the enabling degree of T_2 is reduced to 2 (and the corresponding residual times are reduced to 3 units). A token was deposited in P_3 at firing of T_1, resulting to another general conflict between T_3 and T_4. And so on. Finally, the conflict between T_1 and T_2 results in two firings of T_1 (at $t = 2$ and $t = 4$) and one firing of T_2 (at $t = 5$); the conflict between T_3 and T_4 results in a firing of T_3 (although T_4 was enabled before T_3, its was disabled at $t = 4$).

In the example in Figure 3.27, there is no actual conflict. However, if the timing associated with T_2 were $d_2 = 6$ instead of $d_2 = 5$, T_1 and T_2 would become firable at the same time $t = 6$, while $m_1 = 1$: this corresponds to an *actual conflict* since only one of them could be fired.

Remark 3.14 According to the end of Section 3.4.1, in this book, tokens are never reserved. It follows that the behavior of a T-timed PN is different from the behavior of a T-timed PN with reserved tokens presented in [DaAl 89 &92]. According to Section 3.4.1 (Figures 3.23a and 3.24b), a T-timed PN with reserved tokens is nothing but an *abreviation* of a P-timed PN. Then, it is clear that both have the same modeling power [DaAl 89 &92]. On the other hand, the modeling power of T-timed PNs is at least equal to the modeling power of P-timed PNs: see Appendix O in which models both P-timed and T-timed (called P&T-timed PNs) are presented.

Remark 3.15 In all the Section 3.4.2 (except this remark), it is assumed that the timings associated with places or transitions are constant. However, *timings functions of time*, $d_j(t)$ for T_j, or functions of other *time-dependent* variables, may be used. Examples, useful for modeling, appear in Appendix G (Section G.3) and in Solutions to Exercises 6.8, 6.9, 7.9, and 8.2

3.4.2.3 Stationary Behavior

We note that the functioning represented in Figure 3.26b reaches a periodical functioning after a certain time. This is a general property.

Property 3.12 Let R_T be a T-timed PN (or P-timed) whose timings are rational numbers. If there are no actual conflicts, functioning at maximal speed results in a *stationary functioning*[13], at the end of a finite time, for every initial marking such that R_T is bounded.

❑

In fact, a *node* in the reachability graph is characterized by the *number of tokens* in every place and the *residual time to firing* of every transition enabling, at the moment when this node is reached (Figure 3.26b or 3.27b). Then, the number of nodes is finite because the PN is bounded and the set of timings and residual times (obtained by differences between residual times and timings) is finite (hence, they allow a lowest common multiple). In order to study stationary behavior, we shall look at the firing frequency of transitions (basic performance).

Definition 3.12 The **firing frequency** F_j, of a transition T_j is the mean number of firings of T_j per time unit, when *stationary behavior* is established.

❑

Let us consider the circuit $P_1T_1P_2T_2P_3T_3P_1$ in the T-timed PN of Figure 3.28. At every firing of T_1, there was at least one token in P_1 for a duration $d_1 = 3$ (by definition of functioning). Hence, the average number of tokens in P_1, in stationary behavior, is at least $d_1 \cdot F_1$, i.e., $m_{av}(P_1) \geq d_1 \cdot F_1$. Similarly, $m_{av}(P_2) \geq d_2 \cdot F_2$ and $m_{av}(P_3) \geq d_3 \cdot F_3$.

Since $\mathbf{x}_1 = (1, 1, 1, 0)$ is a P-invariant, the number of tokens in the set of places $\{P_1, P_2, P_3\}$ is constant. Then $m_{av}(P_1) + m_{av}(P_2) + m_{av}(P_3) = m_0(P_1) + m_0(P_2) + m_0(P_3) = 2$, and we can write the inequality:

$$d_1 \cdot F_1 + d_2 \cdot F_2 + d_3 \cdot F_3 \leq m_0(P_1) + m_0(P_2) + m_0(P_3). \tag{3.6}$$

Similarly, for the P-invariant $\mathbf{x}_2 = (0, 1, 0, 1)$:

$$d_1 \cdot F_1 + d_2 \cdot F_2 \leq m_0(P_4) + m_0(P_2). \tag{3.7}$$

Now, from the T-invariant $\mathbf{y} = (1, 1, 1)$, it can be determined that all the transitions have the same frequency when stationary behavior is reached, i.e.,

$$F_1 = F_2 = F_3. \tag{3.8}$$

---

[13] A *stationary functioning* (also obtained for a *deterministic resolution* of actuals conflicts) is either *periodical* (Figure 3.26) or degenerated in a single state, i.e., a *deadlock* (Figure 3.27).

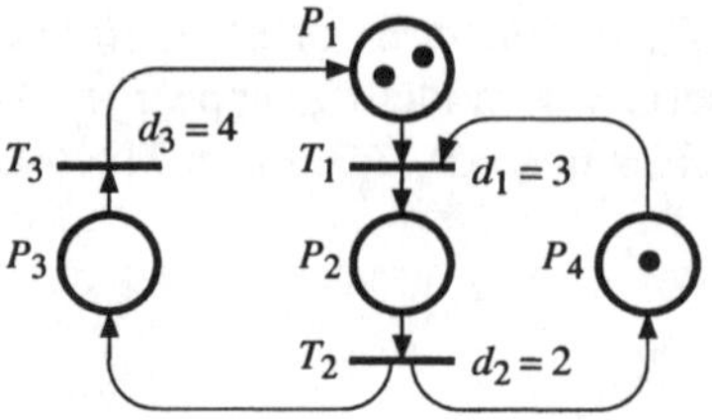

Figure 3.28 Firing frequencies.

Given $d_1 = 3$, $d_2 = 2$, $d_3 = 4$ and the intial marking in Figure 3.28, from (3.6) to (3.8) one obtains $9\,F_1 \leq 2$ and $5\,F_1 \leq 1$. Hence $F_1 = \min\left(\frac{2}{9}, \frac{1}{5}\right) = \frac{1}{5} = 0.2$.

In the circuit $P_4T_1P_2T_2P_4$, a token deposited in a place is immediately used for enabling a transition; this behavior is a **proper speed functioning**.

Generally speaking, the following relations are obtained.

1) A *relation* connecting timings, frequencies and the initial marking which is associated with every *P-invariant*:

$$\mathbf{x}^{\mathrm{T}} \cdot \mathbf{W}^{-} \cdot \mathbf{D} \cdot \mathbf{f} \leq \mathbf{x}^{\mathrm{T}} \cdot \mathbf{m}_0, \tag{3.9}$$

in which $\mathbf{x}$ is a P-invariant, $\mathbf{D}$ a diagonal matrix such that $D_{jj} = d_j$ is the timing associated with transition T_j, $\mathbf{W}^{-}$ is the input incidence matrix, $\mathbf{f}$ is the firing frequency vector and $\mathbf{m}_0$ the initial marking[14].

2) A *relation* between the firing frequencies. Let $\{\mathbf{y}_1, \ldots, \mathbf{y}_k, \ldots\}$ be the set of T-invariants and F_k the firing frequency associated with $\mathbf{y}_k$ (existence of several T-invariants is due to existence of conflicts). The firing frequency vector is such that

$$\mathbf{f} = \sum_k F_k \cdot \mathbf{y}_k\,. \tag{3.10}$$

3) From all these relations, the firing frequencies are determined corresponding to functioning at maximal speed (when the problem can be solved).

❑

In performance evaluation, systems which are modeled by *event graphs* are often encountered. It is then interesting to point out the particular properties of timed strongly connected event graphs (see Section 2.2.4.1).

Property 3.13 Let R_T be a n-place, m-transition *T-Timed* (or P-timed) *strongly connected event graph*, whose set of elementary circuits is $C = \{C_1, C_2, \ldots, C_{n\,m\,1}\}$ (see Appendix C). Let $C_k = P_1T_1P_2T_2 \ldots P_rT_rP_1$ be an elementary circuit, $d(C_k) = d_1 + d_2 + \ldots + d_r$, and $m(C_k) = m_0(P_1) + m_0(P_2) + \ldots + m_0(P_r)$.

a) The firing frequency corresponding to the *proper speed* functioning of the elementary circuit C_k is

---

[14] For a P-timed PN, Inequality (3.9) would be replaced by $\mathbf{x}^{\mathrm{T}} \cdot \mathbf{D} \cdot \mathbf{W}^{+} \cdot \mathbf{f} \leq \mathbf{x}^{\mathrm{T}} \cdot \mathbf{m}_0$, in which $\mathbf{D}$ is a diagonal matrix such that $D_i = d_i$ is the timing associated with place P_i.

$$F(C_k) = \frac{m(C_k)}{d(C_k)} \tag{3.11}$$

(also true for an elementary circuit in an event graph not strongly connected).

b) All the transitions have the same firing frequency, i.e.,

$$F_1 = F_2 = ... = F_m = F(R_T). \tag{3.12}$$

c) This *firing frequency* is

$$F(R_T) = \min_{k=1}^{n-m+1}\big(F(C_k)\big). \tag{3.13}$$

❑

Properties 3.13a and b are direct consequences of Properties 2.8a and b in Section 2.2.4.1. Part c of Property 3.13 follows from parts a and b. Property 3.13 may be applied to Figure 3.28, which is a strongly connected event graph.

3.4.3 Stochastic Petri Nets

The basic stochastic PN model and the so-called generalized stochastic PN are presented in Sections 3.4.3.1 and 3.4.3.2, then analyzed in Section 3.4.3.3.

3.4.3.1 Basic Model

Basically, a stochastic PN may be considered as a timed PN in which the timings have stochastic values.

Figure 3.29a illustrates a transition in such a PN: firing of transition T_1 will occur when a time d_1 has elapsed after its enabling and this time is a random value. In this basic model, usually called stochastic PN, the random variable d_1 follows an *exponential law of rate* μ. This means that

$$\Pr[d_1 \le t + \mathrm{d}t \mid d_1 > t] = \mu \cdot \mathrm{d}t. \tag{3.14}$$

The probability density and the distribution function of this law are, respectively,

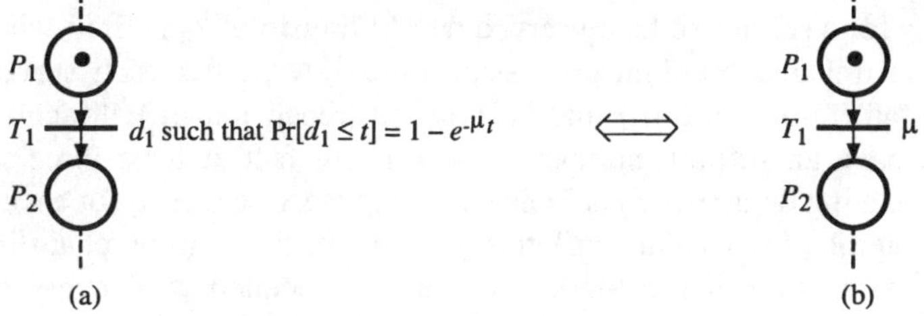

Figure 3.29 Notation for a stochastic PN. (a) Specified. (b) Usual.

$$h(t) = \mu \cdot e^{-\mu t}, \text{ and} \tag{3.15}$$

$$H(t) = \Pr[d_1 \le t] = 1 - e^{-\mu t}. \tag{3.16}$$

The average value of this law is $1 / \mu$ and its variance $1 / \mu^2$. It is clear, from the previous equations, that this law is completely defined by the parameter μ. Hence, it is usually denoted by this parameter as illustrated in Figure 3.29b.

Definition 3.13 A **stochastic PN** is a pair $\langle R, \text{Rate}\rangle$ such that:

R is a marked PN;

Rate is a function from the set T of transitions to the set of finite positive real numbers. Rate(T_j) = μ_j = firing rate associated with T_j.

❑

A fundamental feature of an exponential law is the *memoryless property*, i.e.:

$$\Pr[d_1 \le t_0 + t \mid d_1 > t_0] = \Pr[d_1 \le t]. \tag{3.17}$$

This property may be intrepreted in the following way: let d_j be a random variable exponentially distributed, representing for example the service time of a customer. The service of this customer begins at time $t = 0$. If at time t_0 the service is not yet completed, the distribution law of the residual service time is exponential with the same rate as the distribution law of d_j.

This property is important since it implies the following one.

Property 3.14 If transition T_j, whose firing rate is μ_j, is q-enabled at time t (Definition 3.1 in Section 3.1), then

$$\Pr[T_j \text{ will be fired between } t \text{ and } t + \mathrm{d}t] = q \cdot \mu_j \cdot \mathrm{d}t, \tag{3.18}$$

independently of the times when the q enabling occurred (simultaneously or not).

The product $q \cdot \mu_j = \mu_j(\mathbf{m})$ is the *firing rate associated with T_j for marking* $\mathbf{m}$.

❑

It results from Property 3.14 that the marking $\mathbf{m}(t)$ of the stochastic PN is an homogeneous Markovian process, and thus an homogeneous Markov chain can be associated with every stochastic PN.

Let us present the example in Figure 3.30. Figure a represents a production system made up of two machines whose processings are modeled by T_1 and T_2. Machine M_1 is a two-server machine (two tokens in P_3); the tokens in P_1 model parts waiting for a service or being served by M_1; transition T_1 is fired when a part processing is finished (random processing time, rate μ_1 for each server). Both servers can fail (failure rate μ_3) and be repaired (repair rate μ_4). Machine M_2 is assumed to have an infinite number of servers (in fact at least three since the maximum enabling degree of T_2 is 3) never failing (service rate μ_2 for each).

From the graph of reachable markings in Figure b, the Markov chain in Figure c is obtained. A state of the Markov chain is associated with every reachable marking and the transition rates of the Markov chain are obtained from Property 3.14. For example, for $\mathbf{m}_0$, both transitions T_1 and T_3 are enabled: T_1 is 2-enabled, then there is a transition to $\mathbf{m}_1$ whose rate is $2 \cdot \mu_1$ (because two parts are processed); T_3 is 2-enabled, then there is a transition to $\mathbf{m}_4$ whose rate is $2 \cdot \mu_3$ (because there are two servers which can fail). Transition T_3 will be fired ($\mathbf{m}_4$) if a processor fails before its end of service; otherwise, T_1 will be fired ($\mathbf{m}_1$).

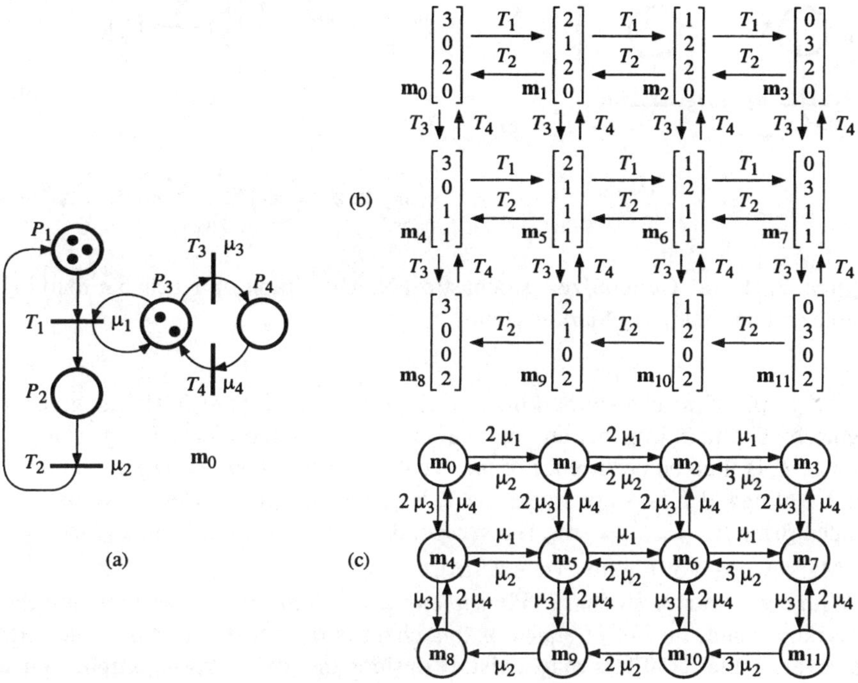

Figure 3.30 (a) Stochastic PN. (b) Graph of markings. (c) Markov chain.

Note that there is *no actual conflict in a stochastic PN*. For example, the probability that a failure of M_1 (transition T_3) occurs simultaneously with the end of service (transition T_1) is zero since continuous time is considered.

3.4.3.2 Generalized Stochastic Petri Net

In a generalized stochastic Petri net (GSPN), in addition to exponentially timed transitions, there are *immediate transitions*. According to Definition 3.3 in Section 3.2.1, an immediate transition is fired as soon as it is enabled (equivalent to an *infinite* firing rate). *Our notation*: if a transition[15] has no associated rate, it is immediate (T_1 in Figure 3.31a); this is consistent with Notation 3.2 (Section 3.2.2.2) and Appendix G.

[15] Some authors represent an immediate transition by a rectangular box.

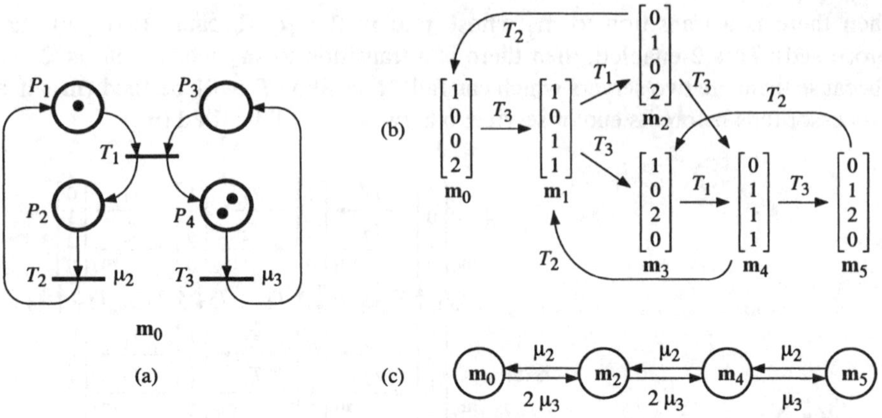

Figure 3.31 (a) Generalized stochastic PN. (b) Graph of reachable markings (stable and unstable). (c) Markov chain.

The graph of reachable markings of the GSPN in Figure 3.31a is given in Figure b. The markings $\mathbf{m}_1$ and $\mathbf{m}_3$ are unstable. Assume the marking is $\mathbf{m}_0$: as soon as T_3 is fired on occurrence of the event *'end of random timing for one of the two enablings of* T_3*'* occurs, transition T_1 is immediately fired. Hence, we have the iterated firing $\mathbf{m}_0 \xrightarrow{T_3T_1} \mathbf{m}_2$; it corresponds to the transition from $\mathbf{m}_0$ to $\mathbf{m}_2$ in the Markov chain of Figure c (rate $2 \cdot \mu_3$).

With a generalized stochastic PN *without actual conflict*, we can then associate a Markov chain; the set of states of this chain is the subset of stable markings. However, actual conflicts may exist. Consider the following example: a part (represented by a token) may be processed by M_1 or by M_2. The choice of one of them may be modeled by two immediate transitions in actual conflict. If branching probabilities are associated with these logical actions, independently of the timing specifications, a Markov model may be obtained.

3.4.3.3 Analysis and Simulation of Stochastic Petri Nets

Two complementary approaches may be used to analyze a stochastic PN. The first approach concerns the conservation properties in a PN (deduced from the P-invariants and T-invariants). The second approach consists of analyzing a continuous time, discrete state space Markov process (*bounded* PN).

Let $T(\mathbf{m})$ denote the set of transitions enabled by $\mathbf{m}$. If $T_k \in T(\mathbf{m})$, the *conditional firing probability* of T_k from $\mathbf{m}$ is:

$$\Pr[T_k \text{ will be fired} \mid \mathbf{m}] = \frac{\mu_k(\mathbf{m})}{\sum\limits_{j:T_j \in T(\mathbf{m})} \mu_j(\mathbf{m})}; \tag{3.19}$$

the *dwelling time* $\mu(\mathbf{m})$ follows an exponential law, and the mean dwelling time in marking $\mathbf{m}$ is $1 / \mu(\mathbf{m})$ (in Figure 3.31, $\mu(\mathbf{m}_1) = \infty$ and $\mathbf{m}_1$ is unstable):

$$\mu(\mathbf{m}) = \sum_{j:T_j \in \mathcal{T}(\mathbf{m})} \mu_j(\mathbf{m}). \tag{3.20}$$

Let $\mathbf{n}(t)$ denote the *random firing vector* (i.e., the jth component $n_j(t)$ is the random number of firings of T_j in the interval $[0, t[$). The fundamental Equation (2.7) can be rewritten as $\mathbf{m}(t) = \mathbf{m}(0) + \mathbf{W} \cdot \mathbf{n}(t)$. If we consider the expected values of $\mathbf{m}(t)$ and $\mathbf{n}(t)$, $\mathrm{E}[\mathbf{m}(t)] = \mathbf{m}(0) + \mathbf{W} \cdot \mathrm{E}[\mathbf{n}(t)]$ is obtained.

Similarly, if $\mathbf{x}$ is a P-invariant, $\mathbf{x}^{\mathrm{T}} \cdot \mathrm{E}[\mathbf{m}(t)] = \mathbf{x}^{\mathrm{T}} \cdot \mathbf{m}(0)$ is obtained from (2.9).

Let us assume that the marking process is ergodic and stationary, i.e., it converges towards the same limit $\mathbf{m}^*$ in temporal and in probabilistic mean value, and the firing frequency vector converges towards $\mathbf{f}^*$.

The *mean marking vector in stationary behavior*, verifies

$$\mathbf{x}^{\mathrm{T}} \cdot \mathbf{m}^* = \mathbf{x}^{\mathrm{T}} \cdot \mathbf{m}(0), \tag{3.21}$$

and the *firing frequency vector in stationary behavior*, verifies

$$\mathbf{W} \cdot \mathbf{f}^* = \mathbf{0}. \tag{3.22}$$

Vector $\mathbf{f}^*$ is a T-invariant. Given $\mathbf{W} = \mathbf{W}^+ - \mathbf{W}^-$ (Equation (2.6)),

$$\mathbf{W}^+ \cdot \mathbf{f}^* = \mathbf{W}^- \cdot \mathbf{f}^* \tag{3.23}$$

is obtained from (3.22): the flow of tokens entering a place equals the flow leaving this place.

The mean dwelling time of a token in place P_i is given by Little's formula:

$$D^*(P_i) = \frac{m^*(P_i)}{\mathbf{W}_i^+ \cdot \mathbf{f}^*}, \text{ where } \mathbf{W}_i^+ \text{ is the } i\text{th line of } \mathbf{W}^+.$$

The mean marking vector, the firing frequency vector, and the mean dwelling times of tokens in places can be obtained from the probabilities of the states of the Markov chain (presented in the previous sections), resulting from the standard analysis of Markov chains. An example is proposed in Exercise 3.15.

Simulation of a stochastic PN (not GSPN). Let us present two methods.

Resampling. At initial time and after each firing: 1) draw a random time to firing for each enabled transition, according to its firing rate (Property 3.14); 2) Fire (once) the transition with the shorter time (since the drawn numbers are practically rational although they should be real, one could obtain the same shorter time for two transitions: one of them may be randomly chosen for firing).

Enabling memory. A set of q enablings is associated with a q-enabled transition. After a firing, a time to firing is drawn for each new enabling while the others do not change. This algorithm may be used for any distribution of d_j.

Simulation of a GSPN. Similar, except that a transition firing may be replaced by an *iterated firing*, and that a random drawing must be performed in case of actual conflict between two immediate transitions.

NOTES and REFERENCES

For the concepts presented in Section 3.1, various authors have used the terms single-server and infinite-server semantics, for example [ReSi 00], or implicit and explicit limitation, for example [DaAl 89 &92].

The synchronized PNs were studied by M. Moalla and co-authors [MoPuSi 78][Mo 81][Mo 85]. The model they have defined assumes the single-server semantics. The concept of EFS presented here (based on the infinite-server semantics) was introduced by J. LeBail [Le 92], who proposes an algebraic method for finding the various EFS if the PN is not generalized. In [DaAl 89, 2nd Edition, & 92], the model called here "synchronized PN" was called "extended synchronized PN". In the last book, EFS stood for "extended firing sequence"; in this book, the same entity is still called EFS but stands for "elementary firing sequence". Synchronized PNs with single-server and infinite-server semantics have the same modeling power; this is shown in Appendix F.

The concept of immediate transition is usual in generalized stochastic PNs [AjBaCo 84]. It is explicitely used in hybrid stochastic PNs proposed in [BaGiMe 98]. Although the name "immediate transition" was not usual, the concept was available in timed PNs [Ra 73][Si 78] and in synchronized and interpreted PNs [MoPuSi 78][Mo 81]. The general definition of an immediate transition and the subsequent concept of stability, relevant to discrete, continuous, and hybrid non-autonomous PNs, are introduced in this book (Definitions 3.3 and 3.6).

Properties 3.3, 3.6, 3.8, and 3.9 are shown in [MoPuSi 78] in the context of single-server semantics; the proofs may be adapted to infinite-server semantics. Property 3.1 is given in [Le 92]. Property 3.2 is adapted from a property given in [DaAl 89 & 92] in the context of single-server semantics. Properties 3.4, 3.5, 3.7, and 3.10, are introduced in this book. The generalized concept of synchronized PN presented in Section 3.3.3 was implicitly contained in interpreted PNs; it is explained explicitely in this book.

Interpreted PNs were studied by M. Moalla [Mo 81][Mo 85]. Grafcet is a tool for modeling logic controlers [AF 77][Da 95]. It was the basis of an International Standard. The International Electrotechnic Commission (IEC) Standard on Programmable Controlers [IE 92] includes only some elements of the Sequential Function Chart (SFC) and refers to [IE 88], then the second edition [IE 02] which is the basic reference. The first edition [IE 88] was presented in a way emphasizing the practical side. Many improvements about formal behavioral aspects and structured and hierarchical descriptions are brought in the second edition [IE 02] (let us note that the word Grafcet appears in the title of this second edition).

Formal presentations can be found in [Mo 85][DaAl 89 & 92] [Da 95]; comparisons between interpreted PNs and grafcets can be found in these references and in [DaDe 83a]. In the publications prior to this book, the conditions for a safe interpreted PN to be deterministic were somewhat less restrictive than in this book, because the single-server semantics was assumed. Nevertheless, the control interpreted PN presented in Section 3.3, which is "similar" to a grafcet (see Appendix E), is independent from the semantics because it is safe and without actual conflict; it corresponds to the modeling constraints recommended for specification of a logic controler. The stability of a grafcet (or of a control interpreted PN) was analyzed in [DaDe 83b].

Other modelings for controled discrete event systems were proposed [HoKrGi 97].

The concepts of strongly compatible and weakly compatible operations considered in Section 3.3.1 were introduced in [DaAl 89 & 92]; in [Br 83], only operations qualified here as *strongly compatible* were considered to be compatible.

Interpreted PNs and grafcets are adapted to representation of real-time systems. However, for large-size systems, more structured programming languages exist, such as *Signal, Esterel* or *Lustre* (synchronous languages) [Le *et al.* 86][BeCoGo 87][Ca *et al.* 87][BeBe 91].

Timings associated with transitions and places were introduced respectively by C. Ramchandani [Ra 73] and J. Sifakis [Si 77]. In this book, it is considered that the firing of a timed transition has no duration (timing is associated with waiting for an event). Various authors have contemplated alternative approaches in which the firing process is split into two phases: tokens are removed from input places at start of firing and tokens are deposited in output places at end of firing after some time has elapsed [Zu 80][RaPh 84][HoVe 87]; the model considered in [DaAl 89 &92] corresponds to this category (tokens are "reserved" between start and end of firing: Figure 3.24b in Section 3.4.1).

Performance evaluation of PNs with constant timings was studied in [Ra 73] [Si 77][RaHo 80][ChCa 83]. The polynomial time computation based on these methods, through a linear programming problem, was introduced in [Ca *et al.* 89].

Stochastic PNs were introduced by S. Natkin and G. Florin [Na 80][FlNa 85]. Their analysis is studied in [Mo 82][Du *et al.* 84][Aj *et al.* 89]. Generalized stochastic PNs were proposed in [AjBaCo 84]. Analysis of deterministic and stochastic Petri nets (DSPN) working in some particular cases is proposed in [AjCh 84]. Phase-type distributions [Ne 81][MoCo 99] allow modeling or approximation of any probability distribution by a combination of exponential laws. They can be used for constructing Markov chains for various timing problems (however, except in some particular cases, their use may be tedious). Various behaviors of stochastic PNs, depending on how past history is taken into account, are studied in [Aj *et al.* 89]: resampling, enabling memory, as well as a preemptive-resume algorithm.

Analysis of Markov chains can be found in [KeSn 76] for example.

Some uncertainty about a system state may be modeled by fuzzy PNs [CaVaDu 96].

4

Autonomous Continuous and Hybrid Petri Nets

The marking of a place in a PN may correspond to the state of a device, e.g. a machine is or is not available. This marking can be compared to a Boolean variable. A marking can also be associated with an integer, e.g. the number of parts in the input buffer of a machine. In this second case, the number of tokens may be a large number. This may result in such a large number of reachable markings that a limit is formed for use of PNs. A number of authors studying production systems have modeled a number of parts by a real number, an approximation which generally proves very satisfactory. Why not then in a PN?

The continuous Petri net is a model in which the number of marks in the places are real numbers instead of integers. This model is presented in Section 4.1. Then, hybrid PNs containing a "discrete part" and a "continuous part" are defined in Section 4.2. Properties of continuous and hybrid PNs are presented in Section 4.3. Finally, Section 4.4 is devoted to a model called extended *Hybrid PN.*

All the models in this chapter are autonomous, i.e., not dependent on time.

4.1 AUTONOMOUS CONTINUOUS PETRI NETS

In Section 4.1.1, continuous PNs are shown to be a limit case of discrete (regular) PNs.

After a formal definition of continuous PNs in Section 4.1.2, reachability and conflicts are considered in Section 4.1.3.

4.1.1 From Discrete Petri Net To Continuous Petri Net

Let us consider a PN R (i.e., autonomous, discrete, ordinary or generalized) defined by its graph Q (places, transitions, arcs) and its marking $\mathbf{m}$, and let us apply a transformation which consists of dividing each mark into k equal parts (without any other modification of the PN). This new discrete PN and its marking will be denoted by $R'_{(k)}$ and $\mathbf{m}'_{(k)}$ (or more simply R' and $\mathbf{m}'$ when there is no ambiguity).

Normally, *token* and *mark* are synonymous. We shall use the word *mark* for the marking of the initial PN. Each **mark** is divided into k and the new unit which is one kth of mark is called a **token**. This designation is consistent with the usual meanings, since there is equivalence for $k = 1$.

See Figure 4.1 for example. The considered transformation applied to the PN of Figure b gives the PN of Figure c in which the markings are expressed in tokens, i.e., $\mathbf{m}'_{(k)} = (2k, 0)$. The new PN (represented for $k = 4$ in Figure c) possesses all the characteristics of a regular discrete PN.

For the PN R of Figure b, the firing of T_1 consists of *removing a mark* from place P_1 and *adding a mark* to place P_2.

For the PN $R'_{(k)}$ of Figure c, the firing of T_1 consists of *removing a token* from place P_1 and *adding a token* in place P_2. The marking of a place can thus be expressed in tokens (integer) or in marks (rational number if k is finite). Let m'_i be the *marking* of the place P_i *expressed in tokens* of the PN $R'_{(k)}$. For the same PN, we shall write

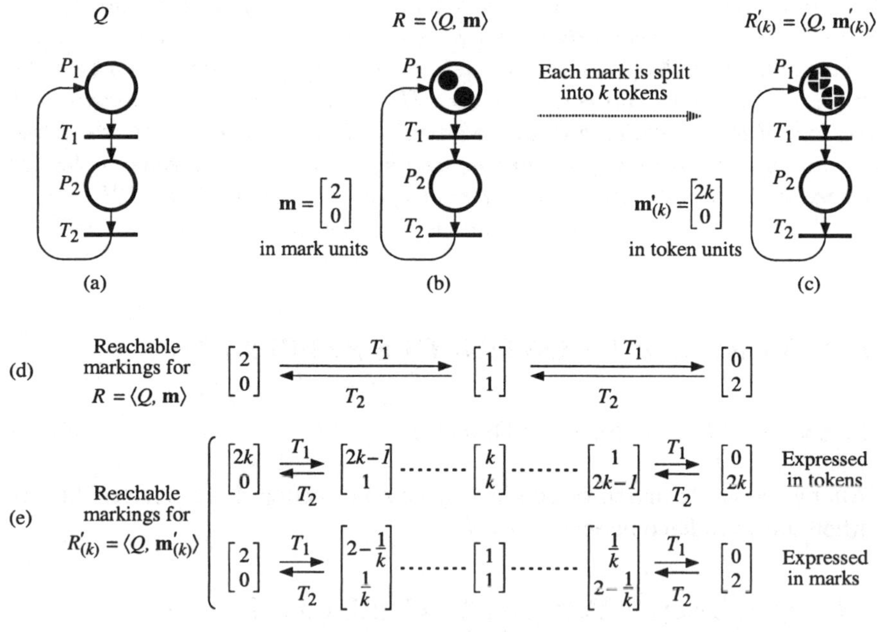

Figure 4.1 Transformation of a PN. (a) Unmarked PN Q. (b) Marked PN $R = \langle Q, \mathbf{m} \rangle$. (c) Transformed PN $R'_{(k)} = \langle Q, \mathbf{m}'_{(k)} \rangle$, represented for $k = 4$. (d) Graph of markings for R. (e) Graph of markings for $R'_{(k)}$.

$$m_i = \frac{m'_i}{k}$$

the *marking* of place P_i *expressed in marks.*

If we compare the markings expressed in unit *marks* for the PNs R and $R'_{(k)}$ (Figure 4.1d and e), we see that the reachable markings of the PN of Figure 4.1b are included in those of Figure 4.1c.

We shall now introduce a new notation. The notation $[T_1T_2]$ corresponding to the simultaneous firing of both T_1 and T_2 (once each) was presented in Section 2.1.4. Since $(T_1)^2 = T_1T_1$, a double firing of T_1 can be denoted by $[(T_1)^2] = [T_1T_1]$. We now propose a simpler notation, which will be useful in the sequel.

Notation 4.1 Let $[T_j]^\alpha = [(T_j)^\alpha]$, where α is a non-negative number, denote the firing of T_j, α times simultaneously (i.e., *at one go*).

❑

In other words, $[T_j]^\alpha$ represents α firings of T_j *at one go*, whereas $(T_j)^\alpha$ represents α *successive* firings of T_j. For continuous PNs, non-integer values of α will be considered.

Figure 4.2a shows the set of possible markings and the corresponding transition firings for the PN R in Figure 4.1b, in the plane defined by m_1 and m_2. In addition to the single transition firings $(2, 0) \xrightarrow{T_1} (1, 1)$, $(1, 1) \xrightarrow{T_2} (2, 0)$, $(1, 1) \xrightarrow{T_1} (0, 2)$, and $(0, 2) \xrightarrow{T_2} (1, 1)$, we have represented all the multiple transition firings, namely $(2, 0) \xrightarrow{[T_1]^2} (0, 2)$, $(1, 1) \xrightarrow{[T_1T_2]} (1, 1)$, and $(0, 2) \xrightarrow{[T_2]^2} (2, 0)$.

The possible markings of $R'_{(k)}$ are shown in Figure 4.2b for $k = 4$. There are very many possible multiple transition firings. Only two of them are illustrated:

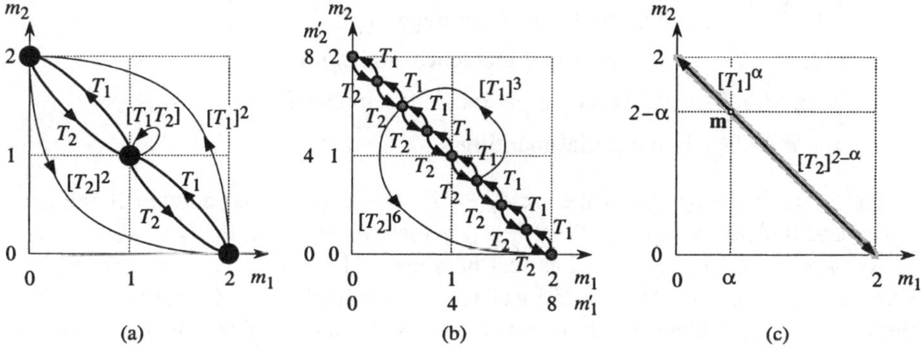

Figure 4.2 From discrete to continuous Petri net. (a) Graph of reachable markings for PN R in Figure 4.1b. (b) Graph of reachable markings of $R'_{(k)} = \langle Q, \mathbf{m}'_{(k)} \rangle$ for $k = 4$. (c) Reachable markings of $R'_{(\infty)}$.

$$(5,3) \xrightarrow{[T_1]^3} (2,6) \text{ and } (2,6) \xrightarrow{[T_2]^6} (8,0), \text{ in } \textit{tokens}, \text{ i.e.,}$$

$$(1.25, 0.75) \xrightarrow{[T_1]^{0.75}} (0.5, 1.5) \text{ and } (0.5, 1.5) \xrightarrow{[T_2]^{1.5}} (2,0), \text{ in } \textit{marks}.$$

When k tends to infinity, the set of reachable markings becomes infinite. It can be represented by a segment of line between (2, 0) and (0, 2) as illustrated in Figure 4.2c. The marking can no longer be expressed in tokens (since m_i' may become infinite). We use the marking expressed in marks. For marking $\mathbf{m} = (\alpha, 2-\alpha)$, where α is any real number in the range [0, 2], enabling degrees of T_1 and T_2 are, respectively, α and $2-\alpha$. Figure c illustrates the possible firings of these transitions *according to their enabling degrees*. In fact, from $\mathbf{m}$, T_1 can be fired β times at one go (β is called the **firing quantity**), such that $0 \le \beta \le \alpha$:

$$(\alpha, 2-\alpha) \xrightarrow{[T_1]^\beta} (\alpha-\beta, 2-\alpha+\beta).$$

Similarly, the firing of $[T_2]^\gamma$, $0 \le \gamma \le (2-\alpha)$ is possible. Finally, the multiple firing of $[(T_1)^\beta (T_2)^\gamma]$ is possible at one go from $\mathbf{m}$.

4.1.2 Definition

In the sequel, some formalism is introduced. After definitions of a continuous PN and of enabling in such a PN, events changing the set of enabled transitions are pointed out.

Definition 4.1 A marked **autonomous continuous PN** is a 5-uple $R = \langle P, T, \text{Pre}, \text{Post}, \mathbf{m}_0 \rangle$ such that:

$P = \{P_1, P_2, ..., P_n\}$ is a finite, not empty, set of places;

$T = \{T_1, T_2, ..., T_m\}$ is a finite, not empty, set of transitions;

$P \cap T = \emptyset$, i.e. the sets P and T are disjointed;

Pre: $P \times T \rightarrow \mathcal{Q}_+$ is the input incidence application[1];

Post: $P \times T \rightarrow \mathcal{Q}_+$ is the output incidence application;

$\mathbf{m}_0$: $P \rightarrow \mathcal{R}_+$ is the initial marking.

❑

Pre(P_i, T_j) is the weight of the arc $P_i \rightarrow T_j$: positive rational number if the arc exists and 0 if not. Similarly, Post(P_i, T_j) is the weight of the arc $T_j \rightarrow P_i$.

Arc weights could be defined as real numbers[2]. However, since a weight has a fixed value, its definition as a rational value, practically, is not a restriction. On the other hand, a place marking must be a real number since it may change continuously.

As for a discrete PN, $R = \langle Q, \mathbf{m}_0 \rangle$ where $Q = \langle P, T, \text{Pre}, \text{Post} \rangle$ represents the

[1] Notation $\mathcal{Q}_+$ and $\mathcal{R}_+$ correspond respectively to the sets of non-negative rational numbers and non-negative real numbers.

[2] In previous presentations, including [DaAl 89 & 92], arc weights were defined as real numbers.

unmarked PN (Definitions 2.11 to 2.13 in Section 2.2.2.1). In a continuous PN, places and transitions are represented by a double line (Figure 4.3a for example).

Definition 4.2 In a continuous PN, the **enabling degree** of transition T_j for marking $\mathbf{m}$, denoted by q or $q(T_j, \mathbf{m})$ is the real number q such that

$$q = \min_{i:P_i \in {}^\circ T_j} \left(\frac{m(P_i)}{\mathrm{Pre}(P_i, T_j)} \right). \tag{4.1}$$

If $q > 0$, transition T_j is enabled; it is said to be **q-enabled**.

❑

Definition 4.2 applies to a generalized PN. For the particular case of a PN in which the weight is 1 for all arcs, (4.1) may be simplified as:

$$q = \min_{i:P_i \in {}^\circ T_j} \left(m(P_i) \right). \tag{4.2}$$

We shall now introduce concepts which will be useful for studying reachability. Since the number of markings in a continuous PN may be infinite, we define *macro-markings* whose number is finite.

Definition 4.3 Let $\mathbf{m}_k$ be a marking. The set P of places may be divided into two subsets: $P^+(\mathbf{m}_k)$ the set of places P_i such that $m_k(P_i) > 0$, and $P^0(\mathbf{m})$ the set of places P_i such that $m_k(P_i) = 0$.

A **macro-marking** is the union of all markings $\mathbf{m}_k$ with the same set $P^+(\mathbf{m}_k)$ of marked places. A *macro-marking* will be denoted by $\mathbf{m}_j^*$ (or possibly $\mathbf{m}_j$ if it contains a single marking). It may be specified by its set of marked places $P^+(\mathbf{m}_j^*)$.

Property 4.1 The number of reachable macro-markings of a n-place continuous PN is less than or equal to 2^n.

❑

This property is a direct consequence of Definition 4.3, since each macro-marking is based on the Boolean state of every place: marked or not marked. The number of macro-markings is necessarily finite (even if the continuous PN is unbounded) because the number of places is finite.

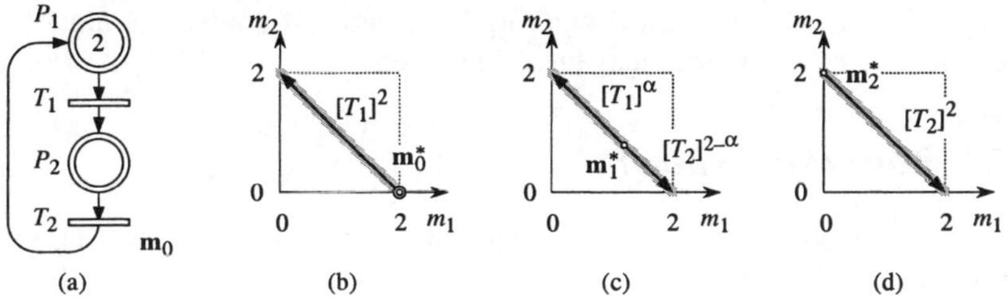

Figure 4.3 (a) Continuous Petri net. (b) to (d) Illustration of its macro-markings.

The continuous PN in Figure 4.3a has three macro-markings (illustrated in Figures b to d), namely $\mathbf{m}_0^*$, $\mathbf{m}_1^*$, and $\mathbf{m}_2^*$, such that $P^+(\mathbf{m}_0^*) = \{P_1\}$, $P^+(\mathbf{m}_1^*) = \{P_1, P_2\}$, and $P^+(\mathbf{m}_2^*) = \{P_2\}$. The fourth macro-marking, which would be possible according to Property 4.1, is $\mathbf{m}_3^*$, such that $P^+(\mathbf{m}_3^*) = \varnothing$.

In this example, $\mathbf{m}_0^*$ corresponds to the single marking $\mathbf{m}_0 = (2, 0)$; $\mathbf{m}_1^*$ corresponds to an infinite number of markings $\mathbf{m}_\alpha = (\alpha, 2 - \alpha)$, $0 < \alpha < 2$; $\mathbf{m}_2^* = (0, 2)$. The macro-marking $\mathbf{m}_3^* = (0, 0)$ is not reachable.

Given a marking, the set of enabled transitions is known; this is true for any PN, discrete or continuous. An interesting feature of continuous PNs is that *knowledge of the set of marked places* (i.e. knowledge of the macro-marking) is sufficient to know the set of enabled transitions[3]. This is a direct consequence of Definition 4.2. Hence, a reachability graph whose vertices are the macro-markings, can be built according to the following property.

Property 4.2 In a continuous PN, a change of macro-marking, hence a change of set of enabled transitions, can occur only if an **event** belonging to one of the followings types occurs (C in their names stands for continuous) .

C1-event: the *marking* of a marked place *becomes zero.*

C2-event: an unmarked place *becomes marked.*

4.1.3 Reachability and Conflicts

When a continuous PN is obtained from a discrete PN, such as the PN in Figure 4.3a which is obtained from the PN in Figure 4.1b (same Pre and Post functions and same initial marking), the initial PN is called the **discrete counterpart** of the continuous PN and vice-versa. For example, the PN in Figure 4.1b is the *discrete counterpart* of the PN in Figure 4.3a, and the PN in Figure 4.3a is the *continuous counterpart* of the PN in Figure 4.1b.

When a continuous PN is built from scratch, it may be converted into a continuous PN with a discrete counterpart if its initial marking is a vector of rational numbers. As a matter of fact, arc weights (rational according to Definition 4.1) and initial place markings can be multiplied by the least common multiple of their denominators. Note that, even if the marking was defined as a vector of real numbers (Definition 4.1), *initial* marking with rational numbers is not a big restriction since next markings may be real numbers.

4.1.3.1 Reachability Graph

Let us analyze the reachability of the continuous PN in Figure 4.4a. The set of reachable markings will be shown to correspond to all the markings in the grey

---

[3] This property is *true for an ordinary* discrete PN but *not for a generalized* discrete PN. Assume $°T_1 = \{P_1\}$ and $\mathrm{Pre}(P_1, T_1) = 2$: if $m_1 = 1$, T_1 is not enabled; if $m_1 = 2$, T_1 is enabled.

triangle in Figure 4.4b. For the discrete counterpart of the considered PN, the reader may verify that the reachable markings correspond to the four black points in this figure and that the language generated is $\overline{\mathcal{L}}$ such that $\mathcal{L} = T_1(T_2T_3)^*T_1$. One can observe that *the set of markings of the discrete counterpart is included in the set of markings of the continuous PN*; this is a general property.

According to Property 4.1, the number of reachable macro-markings of the 3-place continuous PN in Figure 4.4a cannot be more than $2^3 = 8$. In fact, there are only 7 macro-markings because the marking (0, 0, 0) is not reachable. These macro-markings, illustrated in Figure 4.4c, correspond to: three vertices of the triangle ($\mathbf{m}_0^*$, $\mathbf{m}_5^*$, and $\mathbf{m}_6^*$), each one corresponding to one marked place; three sides of the triangle, except the adjacent vertices, ($\mathbf{m}_1^*$, $\mathbf{m}_3^*$, and $\mathbf{m}_4^*$), each one corresponding to two marked places; all the area of the triangle, except the sides, $\mathbf{m}_2^*$ for which all the places are marked. Figure 4.4c shows the transitions enabled for every macro-marking.

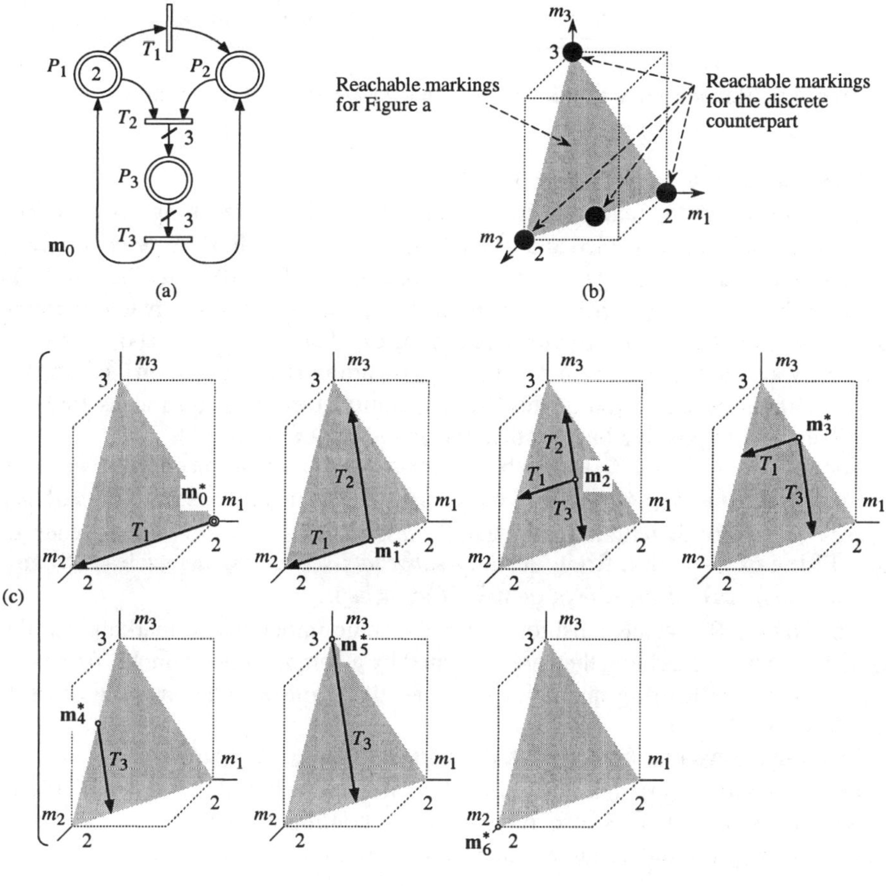

Figure 4.4 (a) Continuous Petri net. (b) Reachable markings for Figure a and for its discrete counterpart. (c) Illustration of the macro-markings.

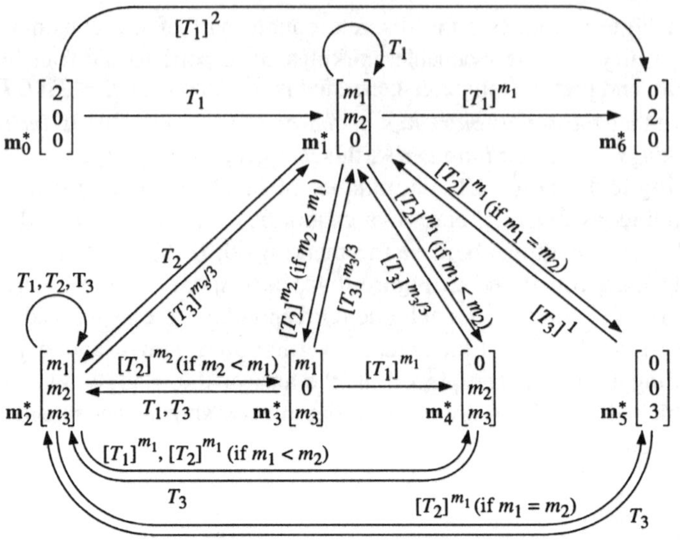

Figure 4.5 Reachability graph for the continuous PN in Figure 4.4a.

The reachability graph is shown in Figure 4.5.

The initial macro-marking corresponds to a single marking, $\mathbf{m}_0^* = \mathbf{m}_0 = (2, 0, 0)$; only T_1 is enabled and its enabling degree is 2. If T_1 is *fired according to its enabling degree*, i.e. $[T_1]^2$, macro-marking $\mathbf{m}_6^* = (0, 2, 0)$ is reached. If T_1 is fired with a *firing quantity less than its enabling degree*, macro-marking $\mathbf{m}_1^* = (m_1, m_2, 0)$ is reached (this notation means that, in $\mathbf{m}_1^*$, m_1 and m_2 have a positive value and that P_3 is not marked). The arrow from $\mathbf{m}_0^*$ to $\mathbf{m}_1^*$ is labelled by T_1 without specification of the firing quantity; this notation means that the firing quantity is positive but less than the enabling degree.

For $\mathbf{m}_1^*$, both T_1 and T_2 are enabled. Firing of T_1 leads to $\mathbf{m}_6^*$ (firing quantity *equal to enabling degree*, i.e. m_1) or to $\mathbf{m}_1^*$ (firing quantity *less than enabling degree*). Firing of T_2 with a firing quantity *less than its enabling degree* leads to $\mathbf{m}_2^*$. Firing of T_2 with a firing quantity *equal to its enabling degree* leads to $\mathbf{m}_3^*$ (if $m_2 < m_1$), $\mathbf{m}_4^*$ (if $m_1 < m_2$), or $\mathbf{m}_5^*$ (if $m_1 = m_2$).

And so non. If several transitions have the same initial macro-marking and the same final macro-marking, they are separated by a comma. For example, firings of T_1, T_2, or T_3 (with firing quantities less than the corresponding enabling degrees) lead from $\mathbf{m}_2^*$ to itself.

Macro-marking $\mathbf{m}_6^*$ is a deadlock (as in the discrete counterpart).

For the PN in Figure 4.4a, there is a P-invariant (Section 2.2.2.3; since this is a structural property it is also relevant to a continuous PN): $\mathbf{x} = (3, 3, 2)$. Accordingly, given the initial marking $\mathbf{m}_0 = (2, 0, 0)$, the marking invariant

$$3\, m_1 + 3\, m_2 + 2\, m_3 = 6 \tag{4.3}$$

is obtained. The set of reachable markings (represented by the grey triangle in Figure 4.4b) corresponds to all markings satisfying Equation (4.3) and such that $m_1 \geq 0$, $m_2 \geq 0$, and $m_3 \geq 0$. Figure 4.5 shows that all the macro-markings are reachable. In the next section, it will be shown that all markings satisfying (4.3) can effectively be reached.

4.1.3.2 Firing Sequence and Reachability Space

Consider the marking $\mathbf{m}_a = (\alpha, \beta, \gamma)$, such that $\gamma = (6 - 3\alpha - 3\beta) / 2$, according to Equation (4.3). This marking can be reached from $\mathbf{m}_0$ by the firing sequence $S_a = [T_1]^x[T_2]^y$ such that $x = (2 - \alpha + \beta) / 2$ and $y = (2 - \alpha - \beta) / 2$. Hence, the reachability space, $\mathcal{M}(\mathbf{m}_0)$ according to the notation introduced in Section 2.1.1, corresponds to all the grey triangle in Figure 4.4b. Both firing quantities x and y are positive, except for $\mathbf{m}_a = \mathbf{m}_0$ $(x = y = 0)$ and $\mathbf{m}_a = (0, 2, 0)$ $(x = 2, y = 0)$. For example, if $\mathbf{m}_a = (0.3, 0.7, 1.5)$, the firing quantities are $x = 1.2$ and $y = 0.5$. As illustrated in Figure 4.6a,

$$\mathbf{m}_0 \xrightarrow{[T_1]^{1.2}[T_2]^{0.5}} \mathbf{m}_a. \tag{4.4}$$

Marking $\mathbf{m}_a$ cannot be reached from $\mathbf{m}_0$ at one go (only T_1 is enabled for $\mathbf{m}_0$). We obtain $\mathbf{m}_0 \xrightarrow{[T_1]^{1.2}} \mathbf{m}_b$ and $\mathbf{m}_b \xrightarrow{[T_2]^{0.5}} \mathbf{m}_a$. Obviously, there are many other firing sequences than $S_a = [T_1]^{1.2}[T_2]^{0.5}$ leading from $\mathbf{m}_0$ to $\mathbf{m}_a$ (their number is infinite), for example $S_2 = [T_1]^{0.6}[T_2]^{0.25}[T_1]^{0.6}[T_2]^{0.25}$, illustrated in Figure 4.6a.

As for a discrete PN, if the firing sequence S is such that $\mathbf{m}_i \xrightarrow{S} \mathbf{m}_k$, then the *fundamental equation* (Equation (2.7) in Section 2.2.2.2)

$$\mathbf{m}_k = \mathbf{m}_i + \mathbf{W} \cdot \mathbf{s} \tag{4.5}$$

is true for a continuous PN; the only difference is that $\mathbf{s}$ is a vector of real numbers. For our example, $\mathbf{s}_a = \mathbf{s}_2 = (1.2, 0.5, 0)$ and Equation (4.5) applied to (4.4) gives:

$$\begin{bmatrix} 0.3 \\ 0.7 \\ 1.5 \end{bmatrix} = \begin{bmatrix} 2 \\ 0 \\ 0 \end{bmatrix} + \begin{bmatrix} -1 & -1 & +1 \\ +1 & -1 & +1 \\ 0 & +3 & -3 \end{bmatrix} \cdot \begin{bmatrix} 1.2 \\ 0.5 \\ 0 \end{bmatrix}.$$

Let us now introduce some notations.

Notation 4.2

a) Let $[U_j]$ or $[U_j]^\alpha$ denote an **at one go firing (OG-firing)**, i.e., the simultaneous firing of one or several transitions. This is a generalization of Notation 4.1. For example $[U_1] = [T_1]$, $[U_1]^\alpha = [T_1]^\alpha$, or $[U_2] = [(T_1)^{2.6}(T_2)^{0.2}]$.

The same OG-firing may have several expressions, for example:

$$[(T_1)^x(T_2)^y]^\alpha = [(T_1)^{x\alpha}(T_2)^{y\alpha}]. \tag{4.6}$$

Then, from (4.6), any OG-firing may be written as $[U_k]$: $[U_j]^\alpha = [(U_j)^\alpha] = [U_k]$. For example $[T_1]^\alpha = [(T_1)^\alpha] = [U_3]$.
Then, without loss of generality, a **firing sequence** may be written:

$$S = [U_1][U_2]\ldots\ [U_k]. \tag{4.7}$$

The number k of OG-firings is the *length* of the firing sequence.

b) $\alpha S = [U_1]^\alpha[U_2]^\alpha\ldots\ [U_k]^\alpha,\ \alpha \in \mathcal{R}_+.$ (4.8)

c) $S^\alpha = ([U_1][U_2]\ldots\ [U_k])^\alpha,\ \alpha \in \mathcal{N}$ (4.9)

(or $\alpha \in \mathcal{R}_+$ if the length of S is less than or equal to 1, see Remark 4.1b).

Remark 4.1
a) Given a continuous PN, the set $\mathcal{L}(\mathbf{m}_0)$ of finite firing sequences from $\mathbf{m}_0$ is an infinite set (except if $\mathbf{m}_0$ is a deadlock) which is a subset of $(\sum U_j)^*$. Note that the sequence ε of length 0 has the property $\varepsilon^\alpha = \varepsilon$ for any $\alpha \geq 0$.
b) From Notation 4.2 and Remark 4.1a, $\alpha S = S^\alpha$ if the length of S is 0 or 1:

$$\alpha\varepsilon = \varepsilon^\alpha = \varepsilon \text{ (length 0) and } \alpha[U_1] = [U_1]^\alpha \text{ (length 1).}$$

Remark 4.2 While not usual, the concept of *OG-firing could be used for discrete PNs*. For example, assume a marking $\mathbf{m}$ (in a discrete PN) such that T_2 is 1-enabled and T_3 is 2-enabled, and there is no conflict between T_2 and T_3. The possible OG-firings from $\mathbf{m}$ are T_2, T_3, $[T_3]^2$, $[T_2T_3]$, and $[T_2(T_3)^2]$. An elementary firing sequence (EFS, Definition 3.5 in Section 3.2.2.1), is nothing but an OG-firing provoked by some event. In some sense, the concept of OG-firing is more general than the concept of simultaneous (multiple) firing, since in $[T_1]^\alpha$, for example, α may be any positive value, possibly *infinitely small.*

Property 4.3 Let $S = [U_1][U_2]\ldots\ [U_k]$ be such that $\mathbf{m}_0 \xrightarrow{S} \mathbf{m}_k$.
a) For any $\mathbf{m}'_0 \geq \mathbf{m}_0$, S is firable from $\mathbf{m}'_0$.
b) For any $\alpha \geq 0$, αS is firable from $\mathbf{m}'_0 = \alpha\mathbf{m}_0$ and $\alpha\mathbf{m}_0 \xrightarrow{\alpha S} \alpha\mathbf{m}_k$.
c) If $0 \leq \alpha \leq 1$, then αS is firable from $\mathbf{m}_0$.

Proof Let $\mathbf{m}_1$, $\mathbf{m}_2$, ... denote the markings reached from $\mathbf{m}_0$ after firings of $[U_1]$, $[U_2]$, ... , and let $\mathbf{u}_j$ denote the characteristic vector associated with $[U_j]$.
a) Similar to the discrete case (Property 2.3 in Section 2.1.5.2).
b) If $[U_1]$ is firable from $\mathbf{m}_0$, $[U_1]^\alpha$ is firable from $\alpha\mathbf{m}_0$, since, from Definition 4.2 in Section 4.1.2, $q(T_i, \alpha\mathbf{m}_0) = \alpha \cdot q(T_i, \mathbf{m}_0)$ for every transition T_i. After firing of $[U_1]^\alpha$ from $\mathbf{m}'_0 = \alpha\mathbf{m}_0$, $\mathbf{m}'_1$ is reached:

$$\mathbf{m}'_1 = \alpha\mathbf{m}_0 + \mathbf{W} \cdot \alpha\mathbf{u}_1. \tag{4.10}$$

Since $\mathbf{m}_1 = \mathbf{m}_0 + \mathbf{W} \cdot \mathbf{u}_1$, $\mathbf{m}'_1 = \alpha\mathbf{m}_1$ and the property is shown by iteration.
c) Direct consequence of Properties a and b.

Property 4.4 *The reachability space* of a marked continuous PN *is a convex set.*

Proof Let S_1 and S_2 such that $\mathbf{m}_0 \xrightarrow{S_1} \mathbf{m}_1$ and $\mathbf{m}_0 \xrightarrow{S_2} \mathbf{m}_2$, and let $\alpha \in [0, 1]$. We have to show that any marking on the segment of line between $\mathbf{m}_1$ and $\mathbf{m}_2$, i.e., any marking $\mathbf{m} = \alpha\mathbf{m}_1 + (1-\alpha)\mathbf{m}_2$ is reachable from $\mathbf{m}_0$.

This property is a direct consequence of Property I.1b in Appendix I, since $\beta=(1-\alpha)$ satisfies the condition $0 \le \alpha + \beta \le 1$.

For example, the concatenation of αS_1 and $(1-\alpha)S_2$ leads the continuous PN to the targeted marking $\mathbf{m}$:

$$\mathbf{m}_0 \xrightarrow{(\alpha S_1)((1-\alpha)S_2)} \alpha\mathbf{m}_1 + (1-\alpha)\mathbf{m}_2. \quad (4.11)$$

4.1.3.3 Conflicts

Consider the marking $\mathbf{m} = (m_1, m_2, m_3)$ in Figures 4.6b to d. According to Figure 4.4a, the enabling degrees of the three transitions are $q(T_1, \mathbf{m}) = m_1$, $q(T_2, \mathbf{m}) = \min(m_1, m_2)$, and $q(T_3, \mathbf{m}) = m_3 / 3$.

Since there is no conflict between T_1 and T_3, they can be fired concurrently. All the OG-firings $[(T_1)^x(T_3)^y]$ such that $x \le m_1$ and $y \le m_3 / 3$ are possible: the set of markings reachable at one go from $\mathbf{m}$ is represented by the dark parrallelogram in Figure 4.6b.

Consider now simultaneous firings of T_1 and T_2. Their *firing quantities are not independent* since there is a *general conflict* $\langle P_1, \{T_1, T_2\}, \mathbf{m}\rangle$. All the OG-firings $[(T_1)^x(T_2)^y]$ such that $x \le m_1$, $y \le \min(m_1, m_2)$, *and* $x + y \le m_1$ are possible: the set of markings reachable at one go from $\mathbf{m}$ is represented by the dark trapezoid in Figure c. It is not a parallelogram because of the conflict (i.e., $x + y \le m_1$).

All the possible OG-firings $[(T_1)^x(T_2)^y(T_3)^z]$ from $\mathbf{m}$ correspond to the union of the previous parallelogram and trapezoid, as shown in Figure d. If $x = 0$, i.e., for an OG-firing $[(T_2)^y(T_3)^z]$, the reachable markings correspond to the segment of line at the right side of the dark pentagon in Figure d.

Note that $\mathbf{y} = (0, 1, 1)$ is a T-invariant. It follows that $\mathbf{m} \xrightarrow{[(T_2)^x(T_3)^x]} \mathbf{m}$ and that $\mathbf{m} \xrightarrow{[(T_1)^x(T_2)^{y+\alpha}(T_3)^\alpha]} \mathbf{m}_1$ if $\mathbf{m} \xrightarrow{[(T_1)^x(T_2)^y]} \mathbf{m}_1$, for example (if the corresponding OG-firings are possible).

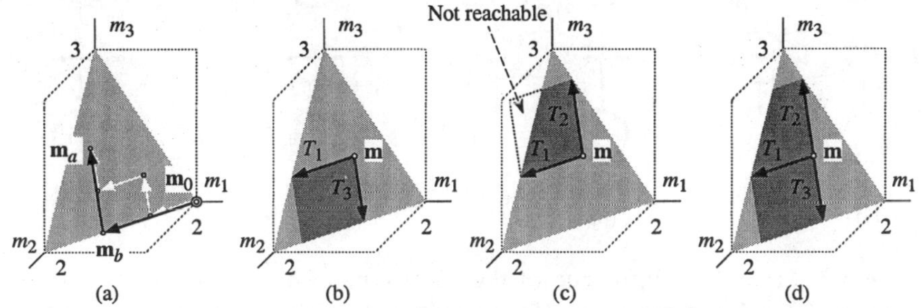

Figure 4.6 (a) Reachability of $\mathbf{m}_a$. (b) OG-firings of T_1 and T_3. (c) Conflict between T_1 and T_2. (d) OG-firings from $\mathbf{m}$.

4.2 AUTONOMOUS HYBRID PETRI NETS

Continuous PNs are particularly suitable for modeling flows: liquid flow or continuous production of a machine. However, a flow may be suddenly interrupted: closing a valve or machine breakdown for example. This is equivalent to suddenly having another continuous PN. This situation can be modeled by a **hybrid PN** containing *continuous places and transitions* (**C-places** and **C-transitions**) and *discrete places and transitions* (**D-places** and **D-transitions**). In addition, in a hybrid PN, a discrete marking may be converted into a continuous marking and vice-versa.

The marking of a C-place is represented by a real number, whose unit is called a *mark*, and the marking of a D-place is represented by dots, called *tokens* (or marks when a common word is useful).

4.2.1 Intuitive presentation

The basic concepts which can be modeled thanks to a hybrid PN are illustrated in Figure 4.7.

In Figure a, the *behavior of the continuous part is influenced by the discrete part.* Transition T_1 is enabled only if there is at least one token in P_1. However, the marking of P_1 is not modified by the firing of T_1. Assume that T_1 is fired with the firing quantity α ($\alpha \le m_1 = 1$): it corresponds to removing the quantity α from P_1 (input place) and adding the quantity α to P_1 (output place), hence the marking of P_1 does not change (similar to *reading*, Figure 1.17b in Section1.3).

In Figure b, *the behavior of the discrete part is influenced by the continuous part.* Transition T_4 is enabled only if there are at least 1.2 marks in P_6. However, the marking of P_6 is not modified by the firing of T_4 since arcs $P_6 \rightarrow T_4$ and $T_4 \rightarrow P_6$ have the same weight (*reading* a C-place marking).

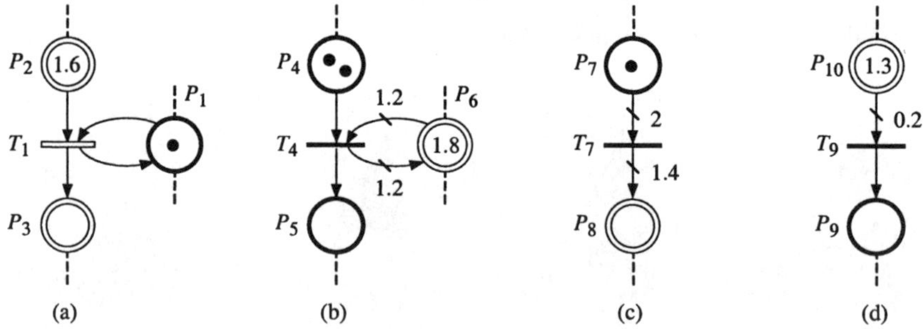

Figure 4.7 (a) and (b) Influence of the discrete part on the continuous part and vice-versa. (c) and (d) Conversion of a discrete marking into a continuous marking and vice-versa.

In Figure c, *a discrete marking is converted into a continuous marking*. If $m(P_7) \geq 2$, a firing of T_7 can be performed: two tokens are removed from P_9 and 1.4 marks are deposited in P_8.

In Figure d, *a continuous marking is converted into a discrete marking*. If $m(P_{10}) \geq 0.2$, a firing of T_9 can be performed: 0.2 marks are removed from P_{10} and one token is deposited in P_9. Note that conversions (from continuous to discrete marking or vice-versa) are made through D-transitions (T_7 and T_9).

All the concepts presented in Figure 4.7 are illustrated with concrete examples in Figure 4.8 (Figures 4.7a to c in Figure 4.8a, and Figure 4.7d in Figure 4.8b).

Figure 4.8a represents a production system in which the parts, arriving by batches, are assumed to be processed continuously. When a batch arrives, T_1 is fired and a token is deposited in P_1. Each batch contains 60 parts which are poured into the input buffer (firing of T_2, converting an integer number of batches into a real number of parts). Transition T_2 can be fired only if there is enough space available in the input buffer, i.e., $m(P_5) \geq 60$. Part processing, i.e., continuous firing of T_5 can be performed only if the machine is working (token in P_2). When the output buffer contains 500 parts, the machine may be stopped (firing of T_3). The self-loops $P_2 \rightarrow T_5 \rightarrow P_2$ and $P_6 \rightarrow T_3 \rightarrow P_6$ illustrate respectively the influence of a D-place on a C-transition and the influence of a C-place on a D-transition, without modification of their markings.

Figure 4.8b illustrates bottling. A barrel containing 57 liters of wine[4] is placed in bottles containing 0.75 liter each. This is modeled by transition T_2: this transition is enabled if there are at least 0.75 marks in P_4 and a token in P_1. Place P_3 shows that several bottles cannot be filled simultaneously.

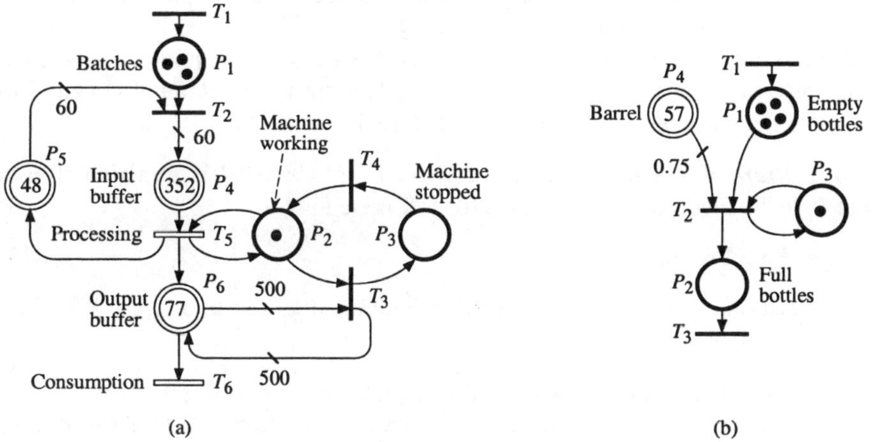

Figure 4.8 (a) Production system. (b) Bottling.

[4] In France, the usual capacity of a large barrel of wine is 228 liters. A smaller barrel corresponding to a quarter of large barrel contains 57 liters.

4.2.2 Definition

After definitions of a hybrid PN and of enabling in such a PN, events changing the set of enabled transitions are pointed out.

Definition 4.4 A marked **autonomous hybrid Petri net** is a sextuple $R = \langle P, T, \text{Pre}, \text{Post}, \mathbf{m}_0, h\rangle$ fulfilling the following conditions:

$P = \{P_1, P_2, ..., P_n\}$ is a finite, not empty, set of places;

$T = \{T_1, T_2, ..., T_m\}$ is a finite, not empty, set of transitions;

$P \cap T = \emptyset$, i.e. the sets P and T are disjointed;

h: $P \cup T \rightarrow \{D, C\}$, called "hybrid function", indicates for every node whether it is a discrete node (sets P^D and T^D) or a continuous one (sets P^C and T^C);

Pre: $P \times T \rightarrow \mathcal{Q}_+$ or $\mathcal{N}$ is the input incidence application;

Post: $P \times T \rightarrow \mathcal{Q}_+$ or $\mathcal{N}$ is the output incidence application;

$\mathbf{m}_0$: $P \rightarrow \mathcal{R}_+$ or $\mathcal{N}$ is the initial marking.

In the definitions of Pre, Post, and $\mathbf{m}_0$, $\mathcal{N}$ corresponds to the case where $P_i \in P^D$, and $\mathcal{Q}_+$ or $\mathcal{R}_+$ corresponds to the case where $P_i \in P^C$.

Pre and Post functions must meet the following criterion: if P_i and T_j are such that $P_i \in P^D$ and $T_j \in T^C$, then $\text{Pre}(P_i,T_j) = \text{Post}(P_i,T_j)$ must be verified.

❑

As for a discrete or a continuous PN, the structure (implying the incidence matrix) is defined by the quadruple $\langle P, T, \text{Pre}, \text{Post}\rangle$. An *unmarked hybrid PN* is a 5-uple $Q = \langle P, T, \text{Pre}, \text{Post}, h\rangle$, i.e., in addition to the structure, the nature of each node (discrete or continuous) is specified in Q.

The last condition of Definition 4.4 states that an arc must join a C-transition to a D-place as soon as a reciprocal arc exists. This ensures marking of D-places to be an integer whatever evolution occurs (Figure 4.7a and 4.8a).

Definition 4.5 A **discrete transition** in a hybrid PN is **enabled** if each place P_i in $°T_j$ meets the condition (as for a discrete PN):

$$m(P_i) \geq \text{Pre}(P_i,T_j). \tag{4.12}$$

The *enabling degree* of a D-transition is defined as for a discrete PN, i.e., Definition 3.1 in Section 3.1.

❑

Note that this definition does not distinguish the case where P_i is a D-place from the case where P_i is a C-place. Since marking evolution of a C-place may be continuous, an arc joining a C-place to a D-transition can be interpreted as being an enabling condition relating to a threshold that must overtake C-place marking. See P_5 and P_6 in Figure 4.8a. See also Figures 4.9a to d.

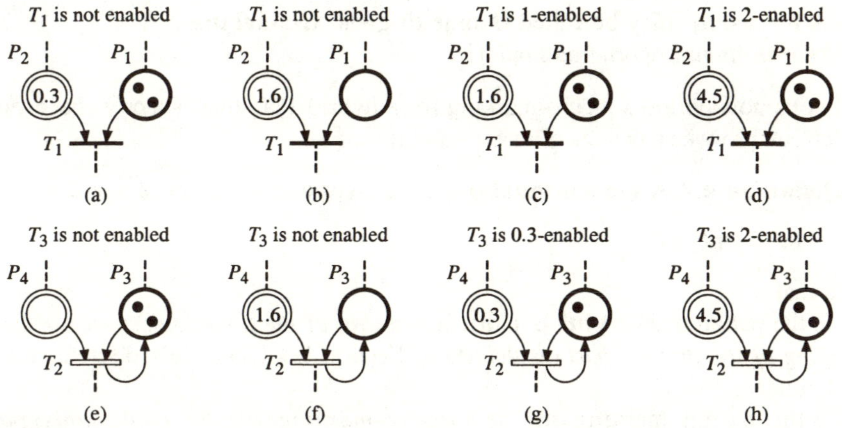

Figure 4.9 Enabling (a) to (d) D-transition. (e) to (h) C-transition.

Definition 4.6 A **continuous transition** in a hybrid PN is **enabled** if each place P_i in $°T_j$ meets the condition:

$$m(P_i) \geq \mathrm{Pre}(P_i,T_j), \text{ if } P_i \text{ is a D-place,} \tag{4.13}$$

or

$$m(P_i) > 0, \text{ if } P_i \text{ is a C-place.} \tag{4.14}$$

The *enabling degree* of a C-transition is defined as for a continuous PN, i.e., Definition 4.2 in Section 4.1.2.

❑

For a C-transition, the kind of place preceding the transition must be specified because the enabling conditions are different according to whether it is a D-place or a C-place. Note that the enabling condition of a C-transition with regard to a C-place is similar to a continuous Petri net. See Figures 4.9e to h. In Figure 4.9g, for example, T_2 is enabled because $m(P_4) > 0$ (even if $m(P_4) < \mathrm{Pre}(P_4, T_2)$).

Remark 4.3 In this chapter, only autonomous PNs are concerned. Readers already familar with some authors' work may recall that, *when time is involved*, i.e., when transition firings correspond to firing speeds, an "empty" place may be fed and emptied at the same speed. Chapter 5 shows that the marking of this place can be modeled by an infinitely small value, hence formally satisfying Equation (4.14). This remark is relevant to Definition 4.2 too.

❑

Let us now give some helpful notations.

Notation 4.3 Let $\mathbf{m}^D$ and $\mathbf{m}^C$ denote the markings of the discrete part (places in P^D) and of the continuous part (places in P^C), respectively. The *places* in the hybrid PN *may be ordered* in such a way that the index i of any D-place P_i is always lower than the index k of any C-place P_k. In that way,

$$\mathbf{m} = (\mathbf{m}^D, \mathbf{m}^C), \tag{4.15}$$

where $\mathbf{m}^D$ and $\mathbf{m}^C$ may be called **D-marking** and **C-marking**.

The *transitions are ordered similarly.*

❑

Let us now define a macro-marking for a hybrid PN, then specify the events which can provoke a change of macro-marking.

Definition 4.7 A **macro-marking** for a hybrid PN is a set of markings

$$\mathbf{m}^* = (\mathbf{m}^D, \mathbf{m}^{C*}),$$

such that:

1) the partial marking $\mathbf{m}^D$ is either a *marking of the discrete part*, or a *macro-marking of the discrete part* (as defined in Section 2.2.1.2) if this discrete part is not bounded;

2) the partial marking $\mathbf{m}^{C*}$ is a *macro-marking of the continuous part* (Definition 4.3 in Section 4.1.2).

Property 4.5 In a hybrid PN, a change of macro-marking, hence a change of set of enabled transitions, or of enabling degrees of D-transitions, can occur only if an **event** belonging to one of the following types occurs.

C1-event: the marking of a marked C-place becomes zero.

C2-event: an unmarked C-place becomes marked.

D1-event: *Firing* of a D-transition.

D2-event: enabling degree of a D-transition changes because of the marking of a C-place.

❑

The C1-events and C2-events are related to the continuous part (similar to events in Property 4.2, Section 4.1.2). A D1-event is clearly an event similar to the events considered in a discrete PN. A D2-event is specific to hybrid PNs (its name contains a D because this event modifies the enabling degree of a D-transition).

Remark 4.4 If *all the nodes* of the Petri net are *discrete* nodes, the hybrid PN is degenerated into a classical discrete PN. If *all the nodes* of the PN are *continuous* nodes, the hybrid PN is degenerated into a continuous PN.

4.2.3 Reachability and conflicts

One could say that the reachability for a hybrid PN can be studied thanks to tools used for discrete PNs on the one hand, and for continuous PNs on the other. It will appear that obtaining a reachability graph is sometimes tedious, although there is no new theoretical difficulty.

New kinds of conflict occur, but they are easily understood when the concept of conflict is well understood for a discrete PN and for a continuous PN.

4.2.3.1 Reachability Graph

We shall use several examples in turn, in order to present progressively the difficulties in drawing the reachability graph.

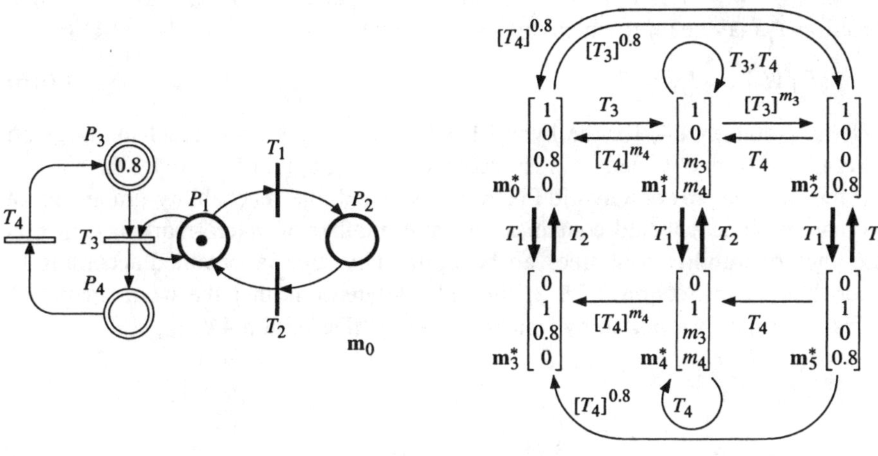

(a) (b)

Figure 4.10 (a) A hybrid PN. (b) Its reachability graph.

The marking of the hybrid PN in Figure 4.10a is a macro-marking, according to Definition 4.7: $\mathbf{m}_0^* = \mathbf{m}_0 = (\mathbf{m}^D, \mathbf{m}^C) = (1, 0, 0.8, 0)$, where $\mathbf{m}^D = (1, 0)$ and $\mathbf{m}^C = (0.8, 0)$. Consider first that $\mathbf{m}^D$ does not change. Then, the partial graph made up of $\mathbf{m}_0^*$, $\mathbf{m}_1^*$, and $\mathbf{m}_2^*$, and the related transitions is obtained (Figure 4.10b); it is exactly the reachability graph of the continuous PN corresponding to the continuous part of Figure 4.10a (except that $\mathbf{m}^D$ is added into the marking). If the discrete firings of T_1 and T_2 (bold arrows) are added, the reachability graph in Figure 4.10b is obtained: T_3 cannot be fired when $m_1 = 0$. This example illustrates C1 and C2-events (among $\mathbf{m}_0^*$, $\mathbf{m}_1^*$, and $\mathbf{m}_2^*$, for example) and D1-events (between $\{\mathbf{m}_0^*, \mathbf{m}_1^*, \mathbf{m}_2^*\}$ and $\{\mathbf{m}_3^*, \mathbf{m}_4^*, \mathbf{m}_5^*\}$)

Assume now that $\mathbf{m}^C = (2, 0)$ instead of $(0.8, 0)$ in Figure 4.10a: Figure 4.11a is obtained. For the PN in this figure, T_3 is 1-enabled, hence at one go, $[T_3]^x$, $x \leq 1$, can be performed. Thus, place P_3 cannot be emptied by a single OG-firing: as illustrated in Figure 4.11b, $\mathbf{m}_2^*$ cannot be reached from $\mathbf{m}_0^*$ by a single OG-firing. Furthermore, a transition from $\mathbf{m}_1^*$ to $\mathbf{m}_2^*$, for example, can be performed only if $m_3 \leq 1$. The reachability graph in Figure 4.11b is admissible. However, it is not isomorphic to the graph in Figure 4.10b. In order to obtain a graph isomorphic to Figure 4.10b, let us introduce a new notation.

Consider again the initial marking in Figure 4.11a. After the OG-firing $[T_3]^{0.7}$, for example, $[T_3]^x$, $x \leq 1$, can still be performed since $m_1 = 1$ and $m_3 \geq 1$. It appears that from $\mathbf{m}_0^* = (1, 0, 2, 0)$, $\mathbf{m}_2^* = (1, 0, 0, 2)$ can be reached by a sequence of two or more OG-firings; for example, $[T_3]^1[T_3]^1$, or $[T_3]^{0.7}[T_3]^1[T_3]^{0.3}$, or $[T_3]^{0.7}[T_3]^{0.5}[T_3]^{0.8}$. We propose to denote all these sequences of OG-firings by $(T_3)^2$, according to the following notation.

Notation 4.4 Let T_j be a C-transition, a sequence of OG-firings of a single transition T_j may be represented by a single notation

$$[T_j]^x[T_j]^y\ldots[T_j]^z = (T_j)^{x+y+\ldots+z}\,. \tag{4.16}$$

❑

Using Notation 4.4, for the hybrid PN in Figure 4.11a, the reachability graph in Figure 4.11c is obtained; it is isomorphic to the graph in Figure 4.10b.

If the *discrete part* of a hybrid PN is not bounded, the reachability graph cannot be obtained since it would contain an infinite number of macro-markings. In this case, the coverability root tree can be built. This tree is obtained according to Algorithm 2.1 in Section 2.2.1.2; the only difference is that the word "marking" in this algorithm is replaced by "macro-marking" (Definition 4.7).

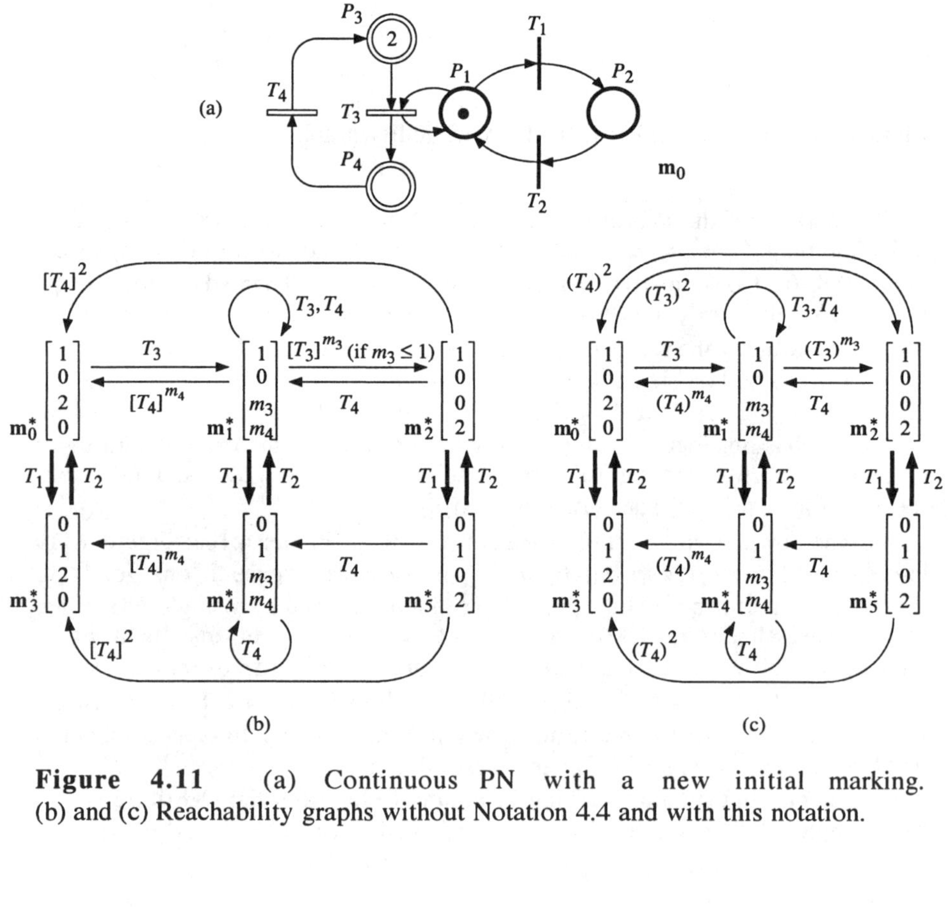

Figure 4.11 (a) Continuous PN with a new initial marking. (b) and (c) Reachability graphs without Notation 4.4 and with this notation.

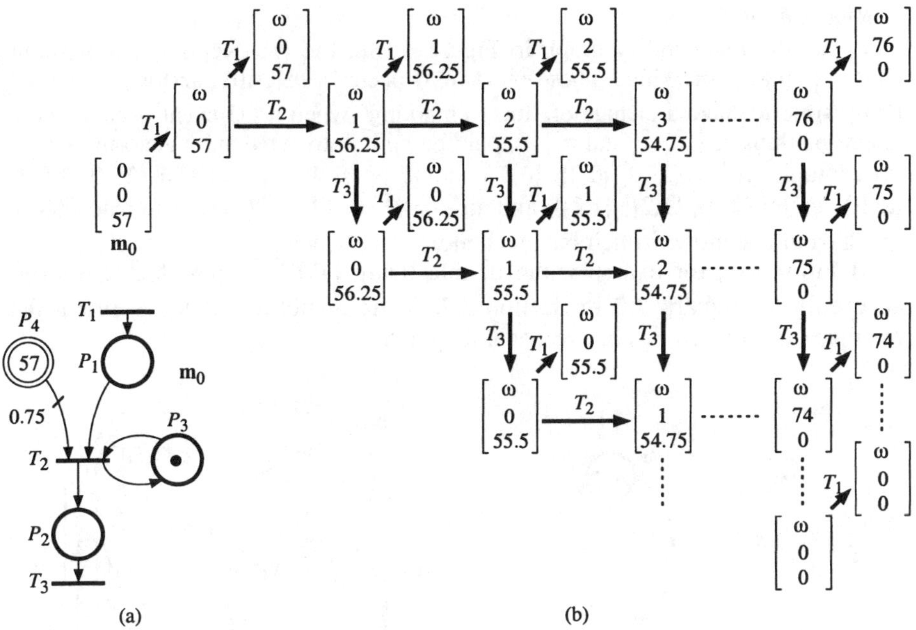

Figure 4.12 Coverability root tree (significant marking (m_1, m_2, m_4)).

This is illustrated in Figure 4.12. Figure a presents the initial marking of the bottling example in Figure 4.8b. Figure b shows the coverability root tree, restricted to the significant marking (m_1, m_2, m_4) because $m_3 = 1$ at any time. From $\mathbf{m}_0 = (0, 0, 57)$, only T_1 can be fired. The marking (1, 0, 57) is reached. Since $(1, 0, 57) \gneqq (0, 0, 57)$, $(\omega, 0, 57)$ is written instead of (1, 0, 57). From the new macro-marking $(\omega, 0, 57)$, T_1 and T_2 can be fired: $(\omega, 0, 57) \xrightarrow{T_1} (\omega, 0, 57)$ and $(\omega, 0, 57) \xrightarrow{T_2} (\omega, 1, 56.25)$. After the new firing of T_1 the process stops since the same macro-marking is reached. After firing of T_2 the process continues on since $(\omega, 1, 56.25)$ is not greater than or equal to $(\omega, 0, 57)$: m_2 has increased but m_4 has decreased. Then, the process continues up to the final macro-marking $(\omega, 0, 0)$ which is a deadlock.

In the hybrid PNs of Figures 4.10 to 4.12, only C1-events, C2-events and D1-events can occur. For the hybrid PN in Figure 4.13a, D2-events can also occur. The reachability graph is presented in Figure b. The transitions among $\mathbf{m}_0^*$, $\mathbf{m}_1^*$, and $\mathbf{m}_2^*$ correspond to the evolution of the continuous part when the marking of the discrete part remains unchanged (Notation 4.4 is used). Assume that, from the initial marking, only T_4 is fired. Then m_3 increases (while m_4 decreases); when $m_3 \geq 1$, T_1 becomes 1-enabled, and when $m_3 \geq 2$, T_1 becomes 2-enabled. These changes of the enabling degree of T_1 correspond to D2-events, according to Definition 4.5 in Section 4.2.2.

Remark 4.5

a) For the reachability graph in Figure 4.13b, the D2-events occur without change of macro-markings. These events may occur in $\mathbf{m}_1^*$, $\mathbf{m}_4^*$, and $\mathbf{m}_7^*$. It would be possible to have a change of "macro-marking" when a D2-event occurs: every macro-marking $\mathbf{m}_1^*$, $\mathbf{m}_4^*$, and $\mathbf{m}_7^*$, should be split into three sub-macro-markings, for example, $\mathbf{m}_1^* = (2, 0,]0,3[,]0,3[)$ should be split into $\mathbf{m}_{11}^* = (2, 0,]0,1[,]2,3[)$, $\mathbf{m}_{12}^* = (2, 0, [1,2[,]1,2])$, and $\mathbf{m}_{13}^* = (2, 0, [2,3[,]0,1])$. The reachability graph would be more explicit but much more complicated.

b) From $\mathbf{m}_7^*$, for example, the double firing $[T_2T_2]$ is possible. However, according to Remark 2.8 in Section 2.2.1.1, it is not usual to represent the multiple (discrete) firings in a reachability graph.

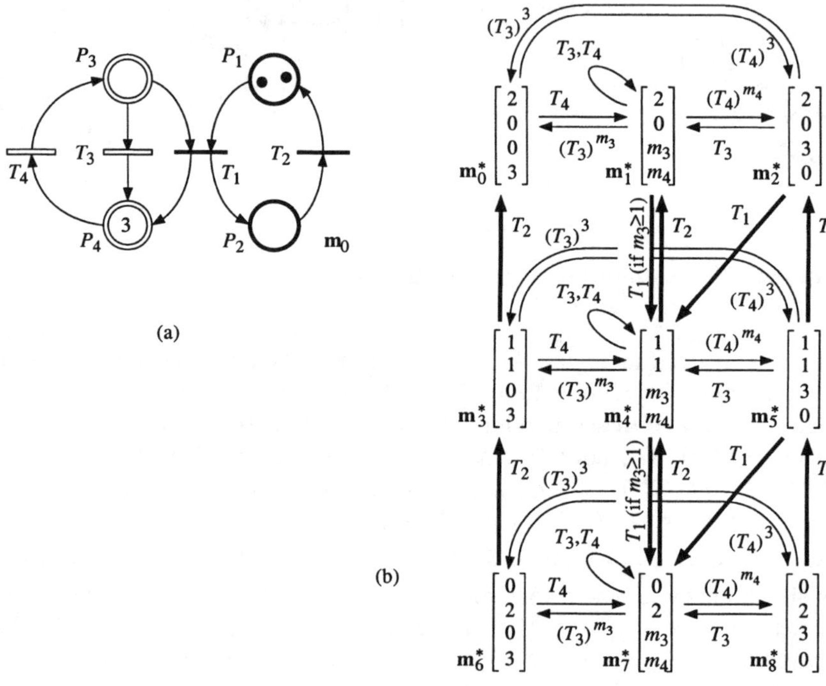

Figure 4.13 Hybrid PN where a D2-event may occur.

4.2.3.2 Firing Sequence and Reachability Space

A *firing sequence* is a string of OG-firings, each OG-firing corresponding to firings of C-transitions, or to firings of D-transitions, or both. For example, according to Notation 4.2 in Section 4.1.3.2, from the marking $\mathbf{m}_0 = (2, 0, 0, 3)$ in Figure 4.13a, the firing sequence

$$S_1 = [U_1][U_2][U_3][U_4],$$

where $[U_1] = [T_4]^{0.5}$, $[U_2] = [(T_3)^{0.3}(T_4)^{2.1}]$, $[U_3] = T_1$, and $[U_4] = [T_1T_2(T_3)^{0.1}]$, i.e.,

$$S_1 = [T_4]^{0.5}[(T_3)^{0.3}(T_4)^{2.1}]T_1[T_1T_2(T_3)^{0.1}], \tag{4.17}$$

can be performed. This firing sequence is made up of four OG-firings. The first one corresponds to a firing of the C-transition T_4 (the only transition enabled for $\mathbf{m}_0$). The second one is a simultaneous firing of T_3 and T_4 (firing quantities less than their enabling degrees which are 0.5 and 2.5, respectively). The third one is a single firing of the D-transition T_1. The fourth OG-firing corresponds to a simultaneous firing of the D-transitions T_1 and T_2 (once each) and of the C-transition T_3 (quantity 0.1).

As for a discrete PN and a continuous PN, if the firing sequence S is such that $\mathbf{m}_i \xrightarrow{S} \mathbf{m}_k$, then the *fundamental equation* (Equation (2.7) in Section 2.2.2.2)

$$\mathbf{m}_k = \mathbf{m}_i + \mathbf{W} \cdot \mathbf{s} \tag{4.18}$$

is true for a hybrid PN; the only difference is that $\mathbf{s}$ is a vector of $|T^D|$ integers and $|T^C|$ real numbers corresponding to the D-transitions and the C-transitions, respectively. For our example, $\mathbf{s}_1 = (2, 1, 0.4, 2.6)$ and Equation (4.18) applied to (4.17) gives:

$$\begin{bmatrix} 1 \\ 1 \\ 0.2 \\ 2.8 \end{bmatrix} = \begin{bmatrix} 2 \\ 0 \\ 0 \\ 3 \end{bmatrix} + \begin{bmatrix} -1 & 1 & 0 & 0 \\ 1 & -1 & 0 & 0 \\ -1 & 0 & -1 & 1 \\ 1 & 0 & 1 & -1 \end{bmatrix} \cdot \begin{bmatrix} 2 \\ 1 \\ 0.4 \\ 2.6 \end{bmatrix}. \tag{4.19}$$

In other words, $(2, 0, 0, 3) \xrightarrow{[T_4]^{0.5}[(T_3)^{0.3}(T_4)^{2.1}]T_1[T_1T_2(T_3)^{0.1}]} (1, 1, 0.2, 2.8)$.

Property 4.6 *The reachability space* of a marked hybrid PN is a union of disjoint *convex sets.*

❑

Property 4.6 follows from the previous results. Take a discrete marking $\mathbf{m}_1^D$. If it does not change, the continuous marking $\mathbf{m}^C$ may reach any value in a convex set $A(\mathbf{m}_1^D)$, according to Property 4.4. Now, if the discrete marking changes to reach the value $\mathbf{m}_2^D$, the new marking cannot be in $A(\mathbf{m}_1^D)$. And so on.

Note that a *necessary and sufficient condition* to have *at least one convex set* which is not degenerated into a single marking, is that there are at least one C-place and one C-transition linked by an arc, and that this C-transition is *quasi-live* (Definition 2.5 in Section 2.1.3). For example, for the hybrid PN in Figure 4.12, every convex set corresponds to a single marking (this hybrid PN does not contain any C-transition).

4.2.3.3 Conflicts

For a hybrid PN as for a discrete or a continuous PN, the 3-uple $\langle P_i, \{T_j, T_k, \ldots\}, \mathbf{m}\rangle$ is a *general conflict* if the number of marks in P_i does not suffice to fire all the output transitions of P_i according to their enabling degrees, i.e., $m(P_i) < q(T_j, \mathbf{m}) + q(T_k, \mathbf{m}) + \ldots$ (Definition 2.10 in Section 2.1.4). All the cases corresponding to two transitions in conflict are illustrated in Figure 4.14 (markings of the other places are assumed to be sufficient not to add other constraints on the concerned firings). Figure c and d are similar to conflicts in a continuous PN and in a discrete PN, respectively. The four other cases can be found only in a hybrid PN.

The behavior of Figure a is similar to the behavior of Figure d since the two transitions are discrete: the two possible OG-firings are T_1 and T_2 is Figure a, and T_7 and T_8 in Figure d. More generally, the OG-firings are similar in both figures if $m(P_4) \leq m(P_1) < m(P_4) + 1$.

In Figure b, the firing quantity for T_4 may be less than the weight of $P_2 \rightarrow T_4$ because T_4 is a C-transition. The possible OG-firings are: $[T_3(T_4)^\alpha]$, $0 \leq \alpha \leq 0.3$ and $[T_4]^\beta$, $0 < \beta \leq 1.3$.

In Figure e, OG-firings T_9 and $[T_{10}]^\alpha$, $0 < \alpha \leq 1$, are possible. However, after $[T_{10}]^\alpha$, firing of T_9 is still possible, while after firing of T_9, T_{10} is no longer enabled.

For Figure f, the possible OG-firings are: $[T_{11}]^\alpha$, $0 < \alpha \leq 1.25$ (because $m(P_6) / 0.8 = 1.25$); $[T_{12}]^\beta$, $0 < \beta \leq 1$; and $[(T_{11})^\alpha (T_{12})^\beta]$ such that $0 < \alpha < 1.25$, $0 < \beta < 1$, and $0.8\alpha + \beta \leq 1$.

The *conflict resolution* is possible thanks to priorities or sharing (Appendix B). This subject will be specified for timed hybrid PNs, in Sections 6.1.3 and 6.2.1.

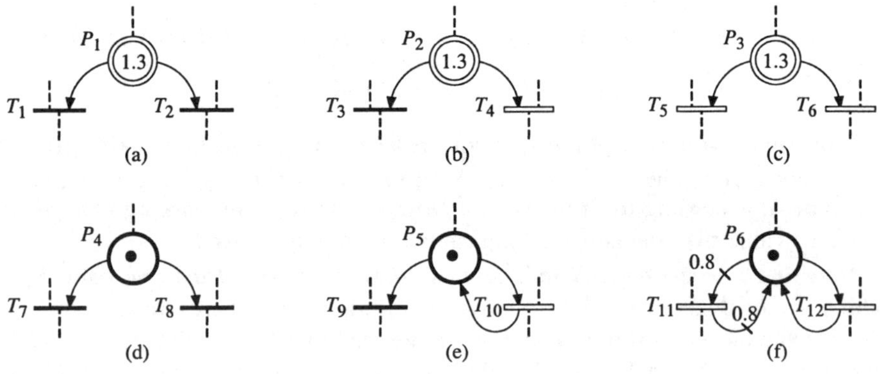

Figure 4.14 Conflicts in a hybrid PN.

4.3 PROPERTIES OF AUTONOMOUS CONTINUOUS AND HYBRID PETRI NETS

Some definitions and properties of autonomous PN are true for both discrete and continuous PNs; they are given in Section 4.3.1. New concepts about reachability and liveness appear for continuous PNs; they are presented in Sections 4.3.2 to 4.3.4. Properties of hybrid PNs, extensively based on the previous ones, are then given in Section 4.3.5.

4.3.1 Definitions and Properties Similar for Discrete and Continuous Petri Nets

Any *definition or property related to the structure* is true both for a discrete PN and a continuous PN. This is obviously true since the structure does not involve the marking. Now, some features related to the markings may also be similar for a discrete PN and a continuous PN.

4.3.1.1 Definitions

All the names given to the particular structures presented in Section 1.2.1, i.e., *state graph*, *event graph*, *conflict free PN*, *free choice* or *extended free choice PN*, *simple PN*, and *pure PN*, can also be given to continuous PNs. Let us recall that all the arc weights are 1 in a state graph or an event graph. This condition is not required for the other structures; the weighted generalization of a (extended) free choice PN is called an *equal conflict PN* (Section 1.2.2.1).

A continuous PN may also be *colored*, since this operation is similar to a folding (Section 1.2.2.3).

The following definitions, initially given in the context of discrete PNs, can also be used in the context of continuous PNs: Definitions 2.1 (*bounded* place and *bounded* PN), 2.3 (*live* transition), 2.4 (*live* PN), 2.5 (*quasi-live* transition and *quasi-live* PN), 2.6 (*deadlock*), 2.7 (*deadlock free* PN), 2.8 (*home state* and *reversible* PN), 2.10 (*general conflict*), 2.15 (*siphon*), 2.16 in which "ordinary" is understood as "the weight of all arcs is 1" (*trap*).

It will appear that additional definitions concerning liveness and deadlocks may be useful. Nevertheless, the basic definitions listed above remain meaningful.

Let us recall and emphasize some concepts, relevant to continuous as well as discrete PNs, which will be useful in the following sections. They are illustrated by the continuous PN in Figure 4.15. Although this section is entitled "Definitions", the sequel is a mix of definitions and properties.

An unmarked PN Q is *conservative* if there is a P-invariant $\mathbf{x} > \mathbf{0}$ such that $\mathbf{x}^T \cdot \mathbf{W} = \mathbf{0}$; if there is $\mathbf{x} > \mathbf{0}$ such that $\mathbf{x}^T \cdot \mathbf{W} \leq \mathbf{0}$, then Q is *structurally bounded* (Sections 2.1.5.1 and 2.2.2.3, Remark 2.12). From $\mathbf{W}_1$ in Figure b, the P-invariants $\mathbf{x}_1 = (1, 5, 0)$ and $\mathbf{x}_2 = (0, 5, 2)$ can be found. Since $\mathbf{x}_3 = \mathbf{x}_2 + \mathbf{x}_1$

= (1, 10, 2) is such that $\mathbf{x}_3 > \mathbf{0}$ (i.e., all its components are positive), the PN in Figure a is conservative.

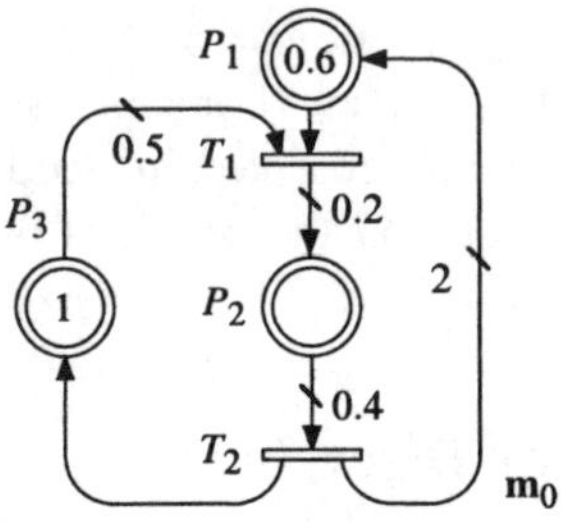

$$\mathbf{W}_1 = \begin{bmatrix} -1 & 2 \\ 0.2 & -0.4 \\ -0.5 & 1 \end{bmatrix} \begin{matrix} P_1 \\ P_2 \\ P_3 \end{matrix} \quad (\text{columns } T_1, T_2) \qquad \mathbf{x}_3 = \begin{bmatrix} 1 \\ 10 \\ 2 \end{bmatrix} \qquad \mathbf{y}_1 = \begin{bmatrix} 2 \\ 1 \end{bmatrix}$$

(b)

Figure 4.15 (a) Continuous PN. (b) Incidence matrix and invariants.

An unmarked PN Q is *consistent* if there is a T-invariant $\mathbf{y} > \mathbf{0}$ such that $\mathbf{W} \cdot \mathbf{y} = \mathbf{0}$; if there is $\mathbf{y} > \mathbf{0}$ such that $\mathbf{W} \cdot \mathbf{y} \geq \mathbf{0}$, then Q is *structurally repetitive* (Sections 2.1.5.2 and 2.2.2.4, Remark 2.14). From $\mathbf{W}_1$, the T-invariants $\mathbf{y}_1 = (2, 1)$ can be found. Since $\mathbf{y}_1 > \mathbf{0}$, the PN in Figure a is consistent.

Given that the (unmarked) PN in Figure a is consistent, a marking $\mathbf{m}$ can be found such that a *complete repetitive* firing sequence can be performed from $\mathbf{m}$. For example, from the marking $\mathbf{m}_0$ in Figure a, $S_1 = [T_1]^{0.6}[T_2]^{0.3}$ can be fired. This sequence is complete since every transition is fired with some positive quantity. It is repetitive since its characteristic vector is $\mathbf{s}_1 = 0.3 \cdot \mathbf{y}_1$.

4.3.1.2 Properties

The following Properties, initially shown in the context of discrete PNs, are also true in the context of continuous PNs: Properties 2.1 (unbounded for $\mathbf{m}_0$ implies *unbounded* for $\mathbf{m}'_0 \geq \mathbf{m}_0$), 2.2a (quasi-live for $\mathbf{m}_0$ implies *quasi-live* for $\mathbf{m}'_0 \geq \mathbf{m}_0$), 2.3 ($\mathcal{L}(\mathbf{m}'_0) \supseteq \mathcal{L}(\mathbf{m}_0)$ if $\mathbf{m}'_0 \geq \mathbf{m}_0$), 2.4 (*P-invariant* and *conservative component*), 2.5 (*composition* of P-invariants), 2.6 (*bounds* of places in a conservative component), 2.7 (*T-invariant* and *repetitive component*), 2.8 (*strongly connected event graph*), 2.10 (*live* if there is a *home state* ...), 2.11 (bounded *equal conflict, live* for $\mathbf{m}'_0 \geq \mathbf{m}_0$ if live for $\mathbf{m}_0$).

Search for minimal P-invariants is possible using Algorithm 2.2 in Section 2.2.2.5. For a continuous PN, the same algorithm can be used, as well as the adaptated algorithm for searching for the T-invariants (in the same section).

According to Algorithm 2.2, the P-invariants and T-invariant obtained are vectors of integer numbers. This is always possible since the arc weights are rational numbers (Definition 4.1 in Section 4.1.2): there is a least common multiple of their denominators, and arc weights and initial place markings can be multiplied by this value. Since the initial marking may be a vector of real numbers, a marking invariant could need a real number. For example, if the initial marking of the PN in Figure 4.15a were ($\sqrt{2}$, 0, 1), $m_1 + 5\,m_2 = \sqrt{2}$ would be a marking invariant.

4.3.2 Reachability and Limit Reachability for a Continuous Petri Net

According to (4.7) in Section 4.1.3.2, a firing sequence is a string of OG-firings $S = [U_1][U_2]\ldots[U_k]$, where each U_i involves *one or more* transitions.

Before giving some useful properties, let us present some new concepts.

Consider the continuous PN in Figure 4.16a. Each time T_1 then T_2 are fired according to their enabling degrees, the marking of P_1 is cut by half. The sum $m_1 + m_2$ decreases exponentially but never reaches zero[5] (remember the Zenon's paradox). The sequence

$$S_2 = [T_1]^{\frac{1}{2}}[T_2]^{\frac{1}{2}}[T_1]^{\frac{1}{4}}[T_2]^{\frac{1}{4}}[T_1]^{\frac{1}{8}}[T_2]^{\frac{1}{8}}[T_1]^{\frac{1}{16}}[T_2]^{\frac{1}{16}}\ldots \tag{4.20}$$

is firable from $\mathbf{m}_0$. The reachability graph is shown in Figure 4.16b. Sequence S_2 in (4.20) corresponds to transitions between $\mathbf{m}_0^*$ and $\mathbf{m}_2^*$ in this graph. Other firing sequences passing by $\mathbf{m}_1^*$ are possible. Figure c illustrates the set of reachable markings. All the markings in the grey triangle are reachable by a finite firing sequence, except the marking (0, 0). A marking is *reachable* if there is a *finite* sequence leading to it (Section 2.1.1). Let us now present the concepts of *limit reachability* and *linear reachability*.

Definition 4.8 Let $R = \langle Q, \mathbf{m}_0\rangle$ be an n-place continuous PN. The marking $\mathbf{m} \in (\mathcal{R}_+)^n$ is **limit reachable** if there is a sequence of reachable markings $\{\mathbf{m}_i\}$, $i \geq 1$, such that

$$\mathbf{m}_0 \xrightarrow{S_1} \mathbf{m}_1 \xrightarrow{S_2} \mathbf{m}_2 \ldots \mathbf{m}_{i-1} \xrightarrow{S_i} \mathbf{m}_i \ldots \tag{4.21}$$

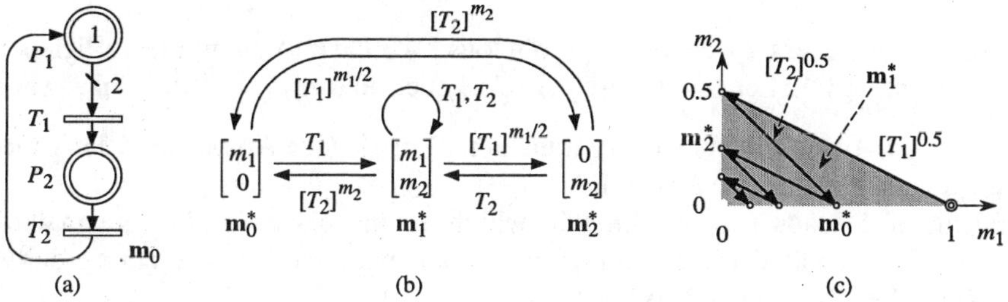

Figure 4.16 (a) Continuous PN. (b) Reachability graph. (c) Set of markings.

[5] Similar behavior is obtained with an *impure* PN containing a single place P_1 and a single transition T_1: $\text{Pre}(P_1, T_1) = 2$ and $\text{Post}(P_1, T_1) = 1$.

and $\lim_{i \to \infty} \mathbf{m}_i = \mathbf{m}$. Since one may have $S_i = \varepsilon$ for i greater than some finite value, a reachable marking is a particular case of limit reachable marking.

Let lim-$\mathcal{M}(\mathbf{m}_0)$, if Q is implicit, denote the set of limit reachable markings.

❑

For the example in Figure 4.16, $\mathbf{m} = (0, 0)$ is limit reachable: the set lim-$\mathcal{M}(\mathbf{m}_0)$ is the grey triangle including $\mathbf{m} = (0, 0)$. Note that this concept is not meaningful for a discrete PN.

Definition 4.9 Let $R = \langle Q, \mathbf{m}_0 \rangle$ be an n-place m-transition continuous PN. The set of all markings $\mathbf{m} \in (\mathcal{R}_+)^n$ such that there is a vector $\mathbf{s} \in (\mathcal{R}_+)^m$ fulfilling the fundamental equation $\mathbf{m} = \mathbf{m}_0 + \mathbf{W} \cdot \mathbf{s}$ is called the **linearized reachability space** and denoted by $\mathcal{M}_L(\mathbf{m}_0)$.

❑

All the reachable markings fulfill the fundamental equation: $\mathcal{M}(\mathbf{m}_0) \subseteq \mathcal{M}_L(\mathbf{m}_0)$. On the other hand, a marking in $\mathcal{M}_L(\mathbf{m}_0)$ may be not reachable (for a discrete as well as for a continuous PN).

According to Definition 4.3 in Section 4.1.2, a macro-marking for a continuous PN is specified by the set of places whose marking is not zero. Let the **middle macro-marking** (or **M-macro-marking**) be the one in which *all the places have a positive marking*. For example, $\mathbf{m}_1^*$ in Figure 4.16. The following algorithm determines whether the M-macro-marking is reachable.

Algorithm 4.1 **Reachability of the M-macro-marking.**

Step 1. Initialization: $P^{(0)} = \emptyset$; $P^{(1)} = \{P_k \mid m_0(P_k) > 0\}$; $i := 1$.

Step 2. **While** $P^{(i)} \neq P^{(i-1)}$ AND $P^{(i)} \neq P$ **do**

$T^{(i)} = \{T_j \mid {}^\circ T_j \subseteq P^{(i)}\}$

$P^{(i+1)} = P^{(i)} \cup T^{(i)\circ}$

$i := i + 1$

end_while

Step 3. $P^{(i)}$ is the set of places which can be marked (from $\mathbf{m}_0$). If $P^{(i)} = P$, the M-macro-marking is reachable (see Property 4.7).

Property 4.7 The *M-macro-marking is reachable* if and only if $P^{(i)}$ obtained in Algorithm 4.1 is equal to the set P of places.

Proof The necessary condition is obvious. We have to show the sufficiency. Let $N = \max_{k: P_k \in P} |P_k{}^\circ|$. Let $S = [U_1][U_2]\ldots[U_{i-1}]$ such that $\mathbf{m}_{l-1} \xrightarrow{[U_l]} \mathbf{m}_l$, where $[U_l]$ is an OG-firing in which every transition T_j in $T^{(l)}$ (see Algorithm 4.1) is fired by the quantity $q(T_j, \mathbf{m}_{l-1}) / (N + 1)$.

Firing of S leads to a marking in which all the places in $P^{(i)}$ are marked: because of the small firing quantities, no marked place can become empty during the firing of S. Property 4.7 follows.

❑

Let us illustrate Algorithm 4.1 and Property 4.7 with the example in Figure 4.17. Algorithm 4.1 applied to the marked continuous PN in Figure a is shown in Figure b. The set $P^{(3)} = P$ is obtained, hence the M-macro-marking is reachable.

Firing sequence S_3 in Figure c (defined in the proof of Property 4.7) leads to $\mathbf{m}_2$, a component of this M-macro-marking. Note that $N = |P_1^\circ| = 3$.

Property 4.8 Let Q be an unmarked continuous PN in which *each place is linked to at least one transition.* The PN $R = \langle Q, \mathbf{m}_0\rangle$ is *quasi-live* (Definition 2.5 in Section 2.1.3) *if and only if the M-macro-marking is reachable* from $\mathbf{m}_0$.

Proof *Sufficient condition.* For any marking $\mathbf{m}$ in the M-macro-marking, all the transitions are enabled (according to Equation (4.1) in Section 4.1.2).

Necessary condition (illustrated in Figure 4.18 which is a quasi-live PN). Note that it is always possible to fire new transitions without emptying a marked place (as in the proof of Property 4.7). If the PN is quasi-live: 1) each place *following* a transition may become marked since the transition may be fired (P_1 following T_1 and P_5 following T_4); 2) each place *preceding* a transition must be (P_2 preceding T_2) or must become (P_5 preceding T_5) marked since the transition must be or must become firable. The only place which may be never marked in the figure is P_3 which has no input transition and no output transition. If such a place does not exist in a quasi-live continuous PN, Algorithm 4.1 leads to $P^{(i)} = P$. Hence, the M-macro-marking is reachable, according to Property 4.7.

Property 4.9 Let $R = \langle Q, \mathbf{m}_0\rangle$ be a *quasi-live* continuous PN. If R is *bounded* for the initial marking $\mathbf{m}_0$, then Q is *bounded for any initial marking* (i.e., it is structurally bounded: there is $\mathbf{x} > \mathbf{0}$ such that $\mathbf{x}^{\mathrm{T}} \cdot \mathbf{W} \leq \mathbf{0}$).

❑

Proof Let $\mathbf{m}_M$ denote a marking in the M-macro-marking, reachable from $\mathbf{m}_0$. For any reachable marking $\mathbf{m}_1$, α exists such that $\alpha\mathbf{m}_1 \leq \mathbf{m}_M$. If there were $\mathbf{m}_1$ such that $\langle Q, \mathbf{m}_1\rangle$ is not bounded, then $\langle Q, \alpha\mathbf{m}_1\rangle$ would not be bounded (Property 4.3b in Section 4.1.3.2). Since $\mathbf{m}_M \geq \alpha\mathbf{m}_1$ is reachable from $\mathbf{m}_0$, $\langle Q, \mathbf{m}_0\rangle$ would not be bounded .

❑

Example: $\langle Q, \mathbf{m}_0\rangle$ in Figure 4.17a is quasi-live and bounded, hence Q is structurally bounded.

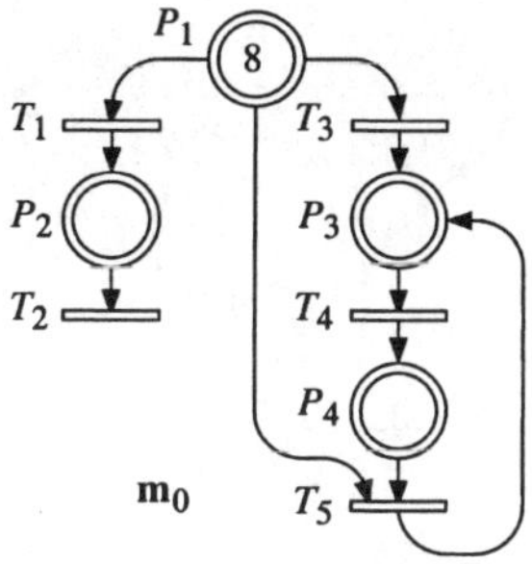

(a)

Step 1. $P^{(0)} = \varnothing$; $P^{(1)} = \{P_1\}$.
Step 2. $T^{(1)} = \{T_1, T_3\}$.
$P^{(2)} = \{P_1, P_2, P_3\}$.
Step 2. $T^{(2)} = \{T_1, T_2, T_3, T_4\}$.
$P^{(3)} = \{P_1, P_2, P_3, P_4\}$.
Step 3. $P^{(3)} = P$.

(b)

$N = 3 \Rightarrow N + 1 = 4$.
$S_3 = [U_1][U_2]$.

$\mathbf{m}_0 = (8, 0, 0, 0)$.
$[U_1] = [(T_1)^2(T_3)^2]$
$\mathbf{m}_1 = (4, 2, 2, 0)$.
$[U_2] = [(T_1)^1(T_2)^{0.5}(T_3)^1(T_4)^{0.5}]$.
$\mathbf{m}_2 = (2, 2.5, 2.5, 0.5)$.

(c)

Figure 4.17 (a) Continuous PN. (b) Algorithm 4.1. (c) Firing sequence.

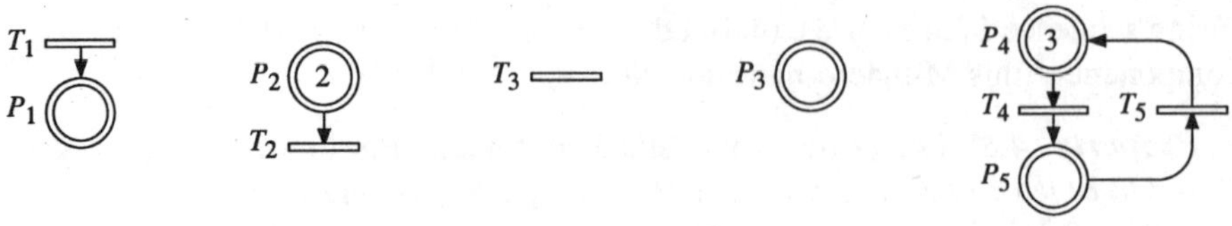

Figure 4.18 Illustration for the necessary condition of Property 4.8.

Property 4.10 Let $R = \langle Q, \mathbf{m}_0 \rangle$ be a *quasi-live* and *consistent* continuous PN. Then, lim-$\mathcal{M}(\mathbf{m}_0) = \mathcal{M}_L(\mathbf{m}_0)$, i.e., the set of *limit reachable markings* is the *linearized reachability space*.

❑

The proof of this property may be found in [ReTeSi 99]. Figure 4.15a is an example.

4.3.3 ε-Liveness for a Continuous Petri Net

According to Definition 2.3, in Section 2.1.3, accepted for a continuous PN too, transition T_1 in Figure 4.16a is live. However, the enabling degree may become very small, close to zero. In this section we present a new concept: ε-liveness.

Definition 4.10 A **transition** T_j is **ε-live** for an initial marking $\mathbf{m}_0$ if $\varepsilon > 0$ exists such that: for every reachable marking $\mathbf{m}_i \in \mathcal{M}(\mathbf{m}_0)$, there is $\mathbf{m}_k \in \mathcal{M}(\mathbf{m}_i)$ such that the enabling degree of T_j is at least ε, i.e., $q(T_j, \mathbf{m}_k) \geq \varepsilon$.

A continuous PN is *ε-live* if all its transitions are ε-live.

❑

It is clear that T_1 in Figure 4.16a is not ε-live (although it is live): for any value $\varepsilon > 0$ a marking $\mathbf{m}_i$ such that $m_i(P_1) + m_i(P_2) < \varepsilon$ can be reached; then no marking $\mathbf{m}_k$ such that $q(T_j, \mathbf{m}_k) \geq \varepsilon$ can be reached from $\mathbf{m}_i$.

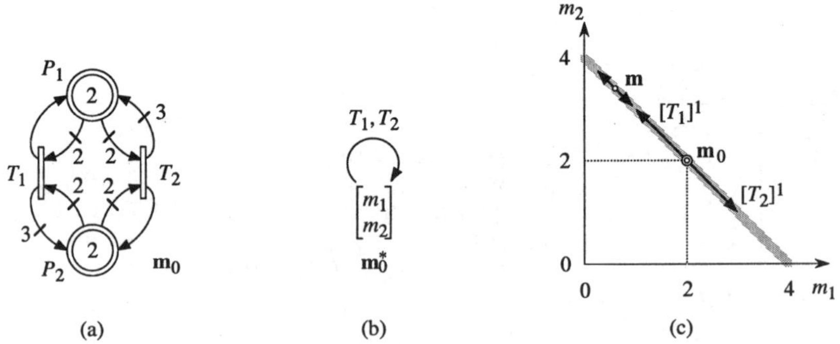

Figure 4.19 Example of ε-live continuous PN.

Consider now the continuous PN in Figure 4.19a. From the initial marking $\mathbf{m}_0 = (2, 2)$, the sequence $S_4 = [T_1]^1[T_1]^{0.5}[T_1]^{0.25}[T_1]^{0.125}\ldots$ can be fired. Hence, $m(P_1)$ may decrease exponentially towards zero. However, $m(P_1) = 0$ will never be reached. If $m(P_1) = \alpha$ is a very small value, $[T_2]^{\alpha/2}$ may be fired, thus increasing the marking in P_1: $(\alpha, 4 - \alpha) \xrightarrow{[T_2]^{\alpha/2}} (1.5\alpha, 4 - 1.5\alpha)$. The process may continue, the sequence

$$S_5 = [T_2]^{\alpha/2}[T_2]^{1.5\alpha/2}[T_2]^{(1.5)^2\alpha/2}[T_2]^{(1.5)^3\alpha/2}\ldots$$

can be fired. Hence, the marking P_1 may increase. Finally, from any reachable marking, the initial marking, for which $q(T_1, \mathbf{m}_0) = q(T_2, \mathbf{m}_0) = 1$, may be reached. The continuous PN is *reversible* (Definition 2.8, Section 2.1.3) and *ε-live*. Note that $m_i(P_1) + m_i(P_2)$ is constant in Figure 4.19a, whereas $m_i(P_1) + m_i(P_2)$ decreases to 0 in Figure 4.16a.

One can say that a continuous PN **ε-dealocks** if no transition is ε-live. Formally, if for any $\varepsilon > 0$ a marking $\mathbf{m}_i$ exists such that $q(T_j, \mathbf{m}_k) < \varepsilon$ for every $\mathbf{m}_k \in \mathcal{M}(\mathbf{m}_i)$ and every transition T_j, then the PN ε-deadlocks.

To conclude, roughly speaking, a transition is ε-live if it can "*come to live again*" *after any marking for which it was* "*almost dead*". Note that ε-liveness implies liveness, but not conversely. The PN in Figure 4.16a is not ε-live while the PN in Figure 4.19a is ε-live. Both are live (according to the definition) but the first one may become "almost dead" for ever.

4.3.4 Lim-Liveness for a Continuous Petri Net

Lim-liveneness defined below is stronger than ε-liveness defined above.

Definition 4.11 A **transition** T_j is **lim-live** for an initial marking $\mathbf{m}_0$ if for any marking in $\mathbf{m}_i \in$ lim-$\mathcal{M}(\mathbf{m}_0)$, there is $\mathbf{m}_k \in \mathcal{M}(\mathbf{m}_i)$ such that $q(T_j, \mathbf{m}_k) > 0$.

A continuous PN is *lim-live* if all its transitions are lim-live.

❑

In other words, a transition is not lim-live if there is a string of successively reachable markings converging to a marking such that none of its successors enables the transition.

For example, the marking of Figure 4.16a (Section 4.3.2) converges to $\mathbf{m}_1 = (0, 0)$ if the sequence S_2 (4.20) is fired. From $\mathbf{m}_1$, T_1 and T_2 can no longer be fired; hence they are not lim-live. Other example: if the sequence $S_4 = [T_1]^1[T_1]^{0.5}[T_1]^{0.25}[T_1]^{0.125}\ldots$ is fired for the PN in Figure 4.19a, the marking converges to $\mathbf{m}_2 = (0, 0)$. Since, from this marking, T_1 and T_2 can no longer be fired, they are not lim-live.

Figure 4.20 shows two examples of lim-live continuous PNs, built from the previous examples.

The part made up of P_1, P_2, T_1, and T_2 in Figure 4.20a is similar to Figure 4.16a. As in this PN, the marking converges to $\mathbf{m}_3 = (0, 0, 1)$ if the sequence S_2

(4.20) is fired. However, from $\mathbf{m}_3$, T_1 and T_2 can be fired again after firing of T_3 by any positive quantity.

The part made up of P_1, P_2, T_1, and T_2 in Figure 4.20b is similar to Figure 4.19a. As in this PN, the marking converges to $\mathbf{m}_4 = (0, 4, 1, 0)$ if the sequence $S_4 = [T_1]^1[T_1]^{0.5}[T_1]^{0.25}[T_1]^{0.125}$... is fired. However, from $\mathbf{m}_4$, T_1 and T_2 can be fired again after firing of T_5 by any positive quantity. In addition, the continuous PN in Figure 4.20b is relevant to the following property.

Property 4.11 If a continuous PN $R = \langle Q, \mathbf{m}_0\rangle$ is *consistent*, *bounded*, and *lim-live* for initial marking $\mathbf{m}_0$, then:

a) The set of *limit reachable markings* is the *linearized reachability space*, i.e., $\text{lim-}\mathcal{M}(\mathbf{m}_0) = \mathcal{M}_L(\mathbf{m}_0)$.

b) The unmarked PN Q is structurally bounded.

c) $R = \langle Q, \mathbf{m}_0\rangle$ is reversible with respect to the set of limit reachable markings $\text{lim-}\mathcal{M}(\mathbf{m}_0)$.

Proof

a) Since the PN is lim-live, a firing sequence containing all the transitions exists. Thus the property is a consequence of Property 4.10.

b) Obvious from Property 4.9.

c) Let $\mathbf{m} \in \text{lim-}\mathcal{M}(\mathbf{m}_0)$. Since the PN is consistent, $\mathbf{m}_0 \in \mathcal{M}_L(\mathbf{m})$. Then applying Property a to $\langle Q, \mathbf{m}\rangle$, $\mathbf{m}_0 \in \text{lim-}\mathcal{M}(\mathbf{m})$ is obtained.

❑

One can say that a continuous PN **lim-dealocks** if there is a marking $\mathbf{m} \in \text{lim-}\mathcal{M}(\mathbf{m}_0)$ such that $q(T_j, \mathbf{m}) = 0$ for every T_j.

Note that lim-liveness implies ε-liveness, but not conversely; let us recall that ε-liveness implies liveness (which is a weak property for a continuous PN). These properties are summarized in Figure 4.21.

Property 4.12 Consider the continuous PN $R = \langle Q, \mathbf{m}_0\rangle$ and any positive value α. If $\langle Q, \mathbf{m}_0\rangle$ is live, then $\langle Q, \alpha\mathbf{m}_0\rangle$ is live; if $\langle Q, \mathbf{m}_0\rangle$ is ε-live, then $\langle Q, \alpha\mathbf{m}_0\rangle$ is ε-live; if $\langle Q, \mathbf{m}_0\rangle$ is lim-live, then $\langle Q, \alpha\mathbf{m}_0\rangle$ is lim-live.

❑

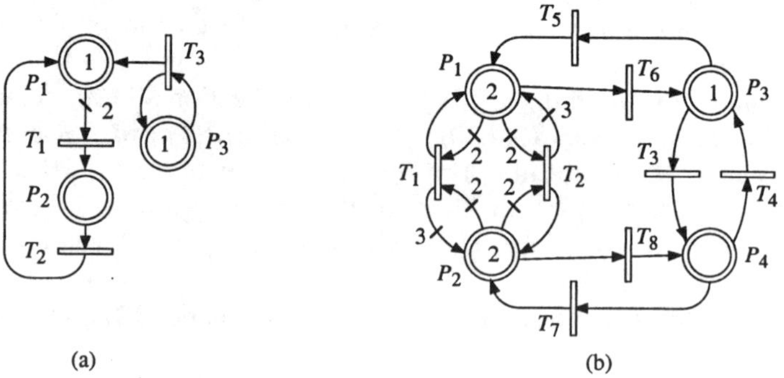

Figure 4.20 Examples of lim-live continuous PNs.

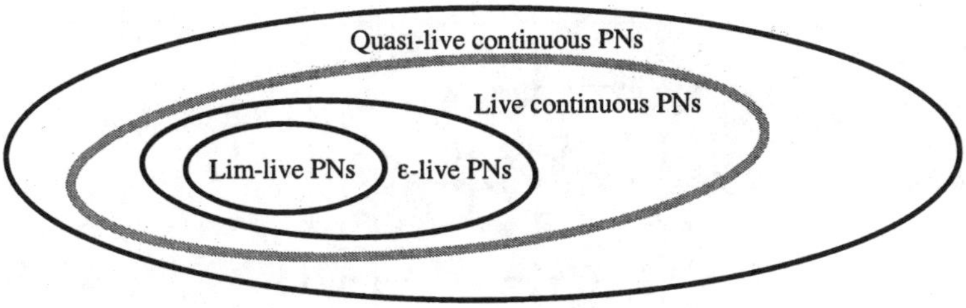

Figure 4.21 Relations among subsets of quasi-live continuous PNs.

Intuitively, this *monotonicity* property of continuous PNs is true because the *unit measuring a continuous value is arbitrary.* For liveness, the property is a direct consequence of Property 4.3b (Section 4.1.3.2). For the proof related to ε-liveness and lim-liveness, see [ReTeSi 99].

Finally, note that the discrete counterparts of Figures 4.16a, 4.19a, and 4.20b may deadlock, while the continuous PNs are respectively, live, ε-live, and lim-live. Conversely, a continuous PN may deadlock while its discrete counterpart does not: Exercise 4.7.

4.3.5 Properties for a Hybrid Petri Net

Basically, hybrid PNs inherit the properties of both discrete PNs and continuous PNs.

4.3.5.1 Similar Definitions and Properties

Any definition or property related to the structure is true for a discrete PN, a continuous PN and hence for a hybrid PN. All the definitions, related to the structure or to other features, presented in Section 4.3.1.1 as true for both a discrete PN and a continuous PN, are also true for a hybrid PN. Similarly, the properties presented in Section 4.3.1.2 as true for both a discrete PN and a continuous PN, are also true for a hybrid PN.

4.3.5.2 Reachability and Liveness

According to Sections 4.3.2 to 4.3.4, new concepts of reachability and liveness are necessary for continuous PNs (while they are not significant for a discrete PN). Let us comment on these new concepts for the hybrid PN in Figure 4.22. The marking is $\mathbf{m} = (\mathbf{m}^D, \mathbf{m}^C)$ with $\mathbf{m}^D = (m_1, m_2, m_3, m_4)$ and $\mathbf{m}^C = (m_5, m_6)$.

From the initial marking $\mathbf{m}_0$ in the figure, if the firing sequence $S_6 = T_1[T_6]^1[T_7]^1[T_6]^{0.5}[T_7]^{0.5}[T_6]^{0.25}[T_7]^{0.25}\ldots$ is performed, $\mathbf{m}_1 = (0, 1, 0, 0, 0, 0)$ is *limit reachable* since $(m_5 + m_6)$ decreases to the limit zero for an infinite length.

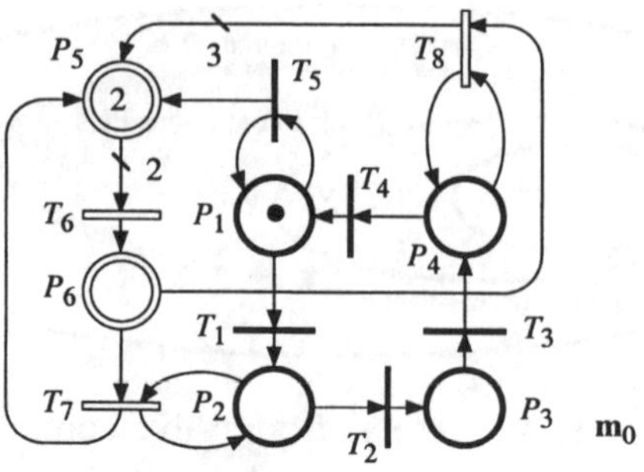

Figure 4.22 Example of hybrid PN.

The five D-transitions, T_1, T_2, T_3, T_4, and T_5, are *live*.

The three C-transitions, T_6, T_7, and T_8, are *lim-live*. As a matter of fact (see Definition 4.11), from the limit reachable continuous marking $\mathbf{m}_1^C = (0, 0)$, $\mathbf{m}_2^C = (1, 0)$ is reached after the first firing of T_5. From $\mathbf{m}_2^C$, firing of $[T_6]^{0.2}$, for example, leads to $\mathbf{m}_3^C = (0.6, 0.2)$. From $\mathbf{m}_3^C$, T_7 and T_8 will be enabled after the first firing of T_1 and the first firing of T_3, respectively.

4.3.5.3 Incidence Matrix

According to Notation 4.3 in Section 4.2.2, the incidence matrix of a hybrid PN may be written

$$\mathbf{W} = \begin{bmatrix} \mathbf{W}^D & \mathbf{0} \\ \mathbf{W}^{CD} & \mathbf{W}^C \end{bmatrix}, \tag{4.22}$$

where $\mathbf{W}^D$ corresponds to arcs among discrete nodes, $\mathbf{W}^C$ to arcs among continuous nodes, and $\mathbf{W}^{CD}$ to arcs among C-places and D-transitions. Arcs among D-places and C-transitions correspond to the submatrix $\mathbf{0}$, according to the restriction about these arcs in Definition 4.4 (Section 4.2.2). An example of incidence matrix is given in (4.19), Section 4.2.3.2.

When $\mathbf{W}^{CD} = \mathbf{0}$, the net is called an **elementary hybrid PN**. Figure 4.10a in Section 4.2.3.1 presents an elementary hybrid PN with an incidence matrix:

$$\mathbf{W} = \begin{bmatrix} -1 & 1 & 0 & 0 \\ 1 & -1 & 0 & 0 \\ 0 & 0 & -1 & 1 \\ 0 & 0 & 1 & -1 \end{bmatrix}. \tag{4.23}$$

In such a net, there is a decoupling between the discrete and the continuous parts (one part may influence the behavior of the other, but there is no conversion of discrete marking into continuous marking or vice-versa). A *minimal P-invariant* or

T-invariant of an elementary hybrid PN is related either to the discrete part or to the continuous part. The P-invariants (T-invariants) of an elementary hybrid PN are the linear combinations of these minimal invariants.

4.4 EXTENDED HYBRID PETRI NETS

According to Section 1.2.2.4, an extended (discrete) PN contains inhibitor arcs, allowing *threshold tests* of place markings and a *zero test* as a particular case. This section shows how these concepts can be modeled in a hybrid PN. In addition, a special marking is proposed.

4.4.1 Threshold Test

A threshold test in a hybrid PN is illustrated in Figure 4.23. The place whose marking is tested may be discrete or continuous, and the transition whose enabling depends on this marking may also be discrete or continuous; hence four cases must be considered.

For every example in Figure 4.23, there are a D-place and a C-place linked to the transition by "normal" arcs (P_1 and P_3 in Figure a). The markings of these places are sufficient for enabling the corresponding transitions. Hence, in every case, the enabling depends only on the place linked by an inhibitor arc. Let P_i, T_j, and s, denote the place, the transition, and the inhibitor arc weight. Transition T_j is enabled only if $m(P_i) < s$ (as in Section 1.2.2.4).

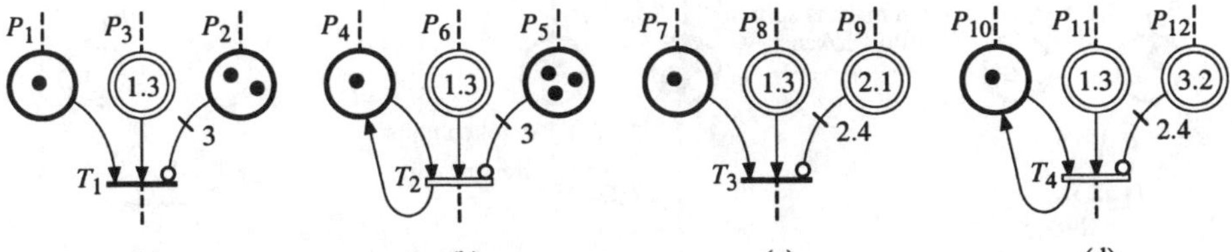

Figure 4.23 Threshold test in a hybrid PN. (a) and (b) D-place. (c) and (d) C-place.

In Figures a and b, the inhibitor arc starts from a D-place. The arc weight s is then an *integer*. For the marking in Figure a, T_1 is enabled since $m(P_2) = 2 < 3 = s$. In Figure b, T_2 is not enabled since $m(P_5) = 3 \geq 3 = s$.

In Figures c and d, the inhibitor arc starts from a C-place. The arc weight s is then a *rational number*. In Figure c, T_3 is enabled since $m(P_9) = 2.1 < 2.4 = s$. In Figure d, T_4 is not enabled since $m(P_{12}) = 3.2 \geq 2.4 = s$.

4.4.2 Zero Test and Arc Weight 0^+

The zero test of a D-place is similar to the zero test in a discrete PN, i.e., the arc weight is $s = 1$: since the arc weight is an *integer*, $m(P_i) < 1$ is verified only if $m(P_i) = 0$. The transition whose enabling is concerned may be discrete or continuous: see Figure 4.24.

The zero test of a C-place is different. For any rational number s, a marking of P_i such that $0 < m(P_i) < s$ may exist. Even if s is a very small value, this situation does not correspond to a zero test.

In order to model the zero test, the new symbol 0^+ is introduced.

Notation 4.5 The ratio

$$\frac{\alpha}{k}, \text{ where } \alpha \text{ is a finite positive number and } k \to \infty, \text{ is denoted by } 0^+.$$

❑

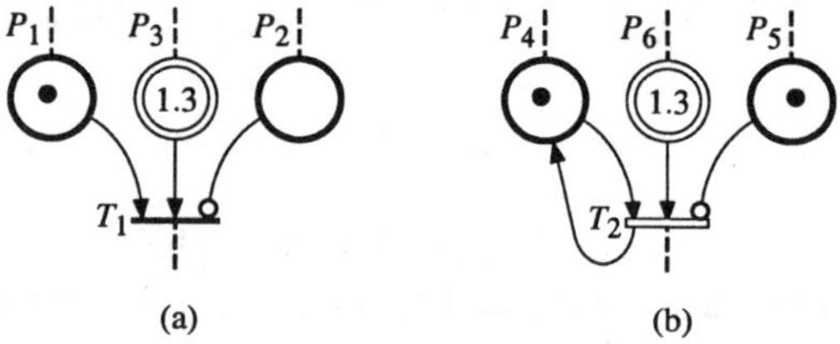

Figure 4.24 Zero test of a D-place. (a) T_1 is enabled. (b) T_2 is not enabled.

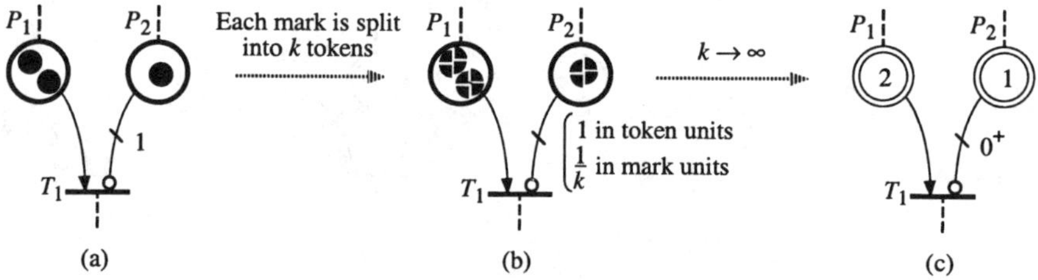

Figure 4.25 Zero test: from discrete to continuous PN.

From this notation, we may write that $0 < 0^+$ and there is no positive real number x such that $x < 0^+$. This feature is supported by the non-standard analysis, as explained in Appendix J where some algebra involving the symbol 0^+ is presented.

Figure 4.25 illustrates the conversion from a discrete to a continuous PN when a zero test is involved. Changing from Figure a to Figure b is similar to the conversion presented in Figure 4.1 (Section 4.1.1): each mark is split into k tokens (Figure b is represented for $k = 4$). In Figure b, the weight of the inhibitor arc is 1 in token units or $1 / k$ in mark units. According to Notation 4.5, this weight may be written 0^+ when $k \to \infty$; this is shown in Figure c.

The zero test of a C-place is illustrated in Figure 4.26. The enabling condition is similar for both a D-transition and a C-transition. Transition T_3 in Figure a is enabled because $m(P_9) = 0 < 0^+$ (and, obviously, the other markings are sufficient). Transition T_4 in Figure b is not enabled because $m(P_{12}) = 0.1 \geq 0^+$.

Various kinds of tests of a C-place marking are illustrated in Figure 4.27. In Figure a, when the production of a machine (corresponding to firing of T_2) is stopped (token in P_1), production will be allowed (token in P_2) when the level in the output buffer is less than 12; this model corresponds to a threshold test of P_3. In Figure b, production will be allowed when the output buffer is empty; it corresponds to a zero test of P_3. A special case of reading a C-place marking (Figure 4.7b, Section 4.2.1) is shown in Figure c: the firing of T_4 is possible only if P_5 is not empty, i.e., if $m(P_5) \geq 0^+$.

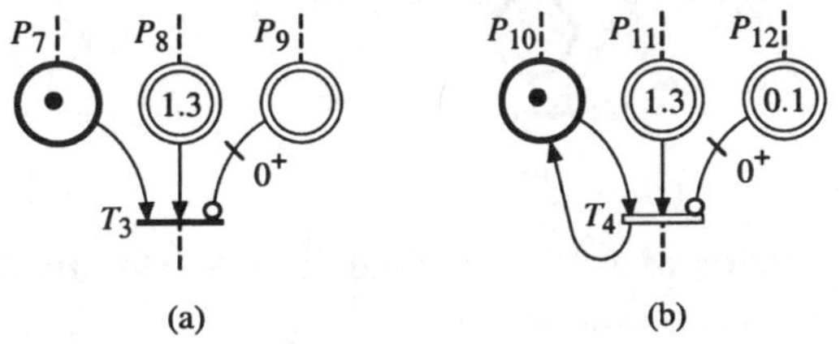

Figure 4.26 Zero test of a C-place. (a) T_3 is enabled. (b) T_4 is not enabled.

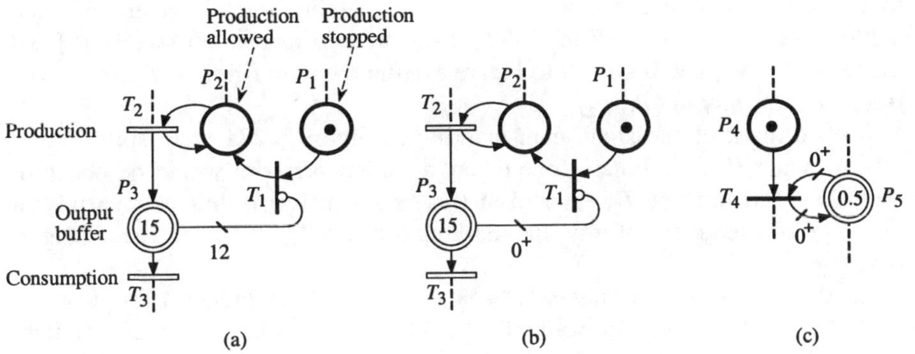

Figure 4.27 Test of a C-place marking. (a) Threshold test. (b) Zero test. (c) T_4 enabled only if P_5 is not empty.

4.4.3 Marking 0^+

Consider the discrete PN in Figure 4.28a. Tokens turn in the circuit $P_1T_1P_2T_2P_1$ with, from time to time, a firing of T_3 decreasing $m_1 + m_2$. This is a priority PN: T_2 has priority over T_3. The role of P_3 is only to limit the effect of the priority: without this place, the *allowing degree* of T_2 (Definition B.1 in Appendix B) would be $m(P_2)$, hence no firing of T_3 could be performed. Thanks to P_3, the allowing degree of T_2 is limited to 1: $r(T_2, \mathbf{m}) = \min(m(P_2), 1)$, and $r(T_3, \mathbf{m}) = m(P_2) - r(T_2, \mathbf{m}) = m(P_2) - \min(m(P_2), 1)$. In other words, T_3 can be fired only if $m(P_2) \geq 2$. The reachability graph, limited to the significant marking (m_1, m_2) is shown in Figure 4.29a.

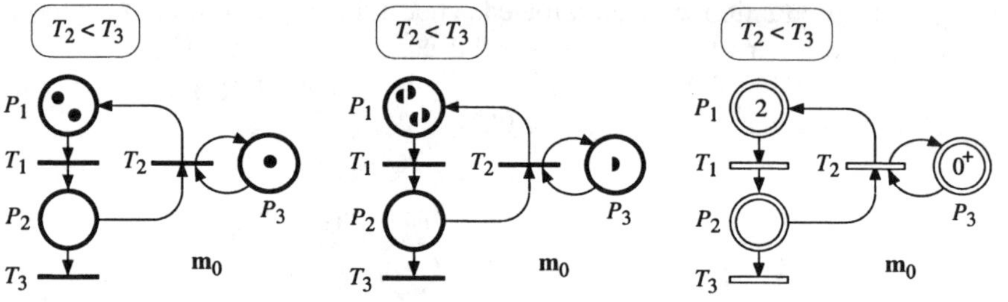

(a) (b) (c)

Figure 4.28 Conversion of a PN. (a) Initial discrete. (b) Significant marks split into 2 tokens. (c) Limit continuous PN.

Figure 4.28b is a conversion of the PN in Figure a: each mark of the significant marking is split into two tokens, while P_3 limits the priority of T_2 to one firing. In Figure b as in Figure a, T_3 may be fired only if the number of tokens in P_2 is at least 2. The corresponding reachability graph is given in Figure 4.29b. Note that, after two firings of T_3, there are two tokens left in $\{P_1, P_2\}$ and the reachability graph is similar to Figure a. After the third firing of T_3, one token is left permanently in $\{P_1, P_2\}$.

If each mark of the significant marking in Figure 4.28a were split into k tokens, while P_3 still contains one token, a similar behavior would be obtained: after $2k - 1$ firings of T_3, one token (i.e., $1 / k$ mark) is left permanently in $\{P_1, P_2\}$. If k tends to infinity, the marking left in $\{P_1, P_2\}$ is 0^+, according to Notation 4.5.

In other words, the marking $m(P_1) + m(P_2)$ in the PN of Figure 4.28c decreases to 0^+, but never reaches the value 0. The difference between 0^+ (value infinitely small but not nil) and 0, may be useful for enabling a transition (Figure 4.27c for example). Various applications will be presented for *timed* continuous and hybrid PNs.

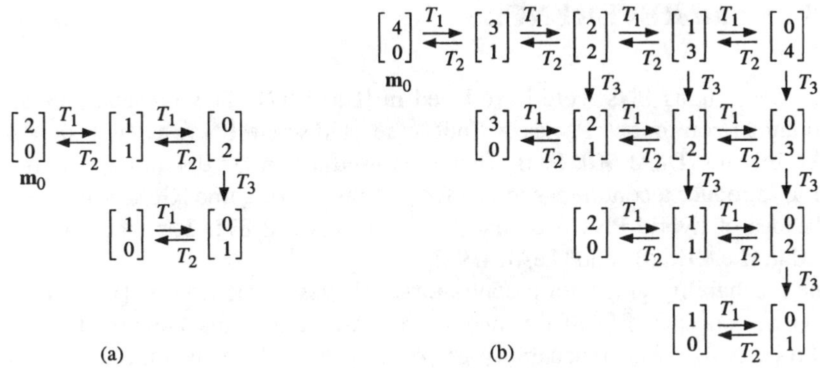

Figure 4.29 Reachability graphs for PNs in Figure 4.28a and b (significant markings).

Remark 4.6 Note that decreasing of $m(P_1) + m(P_2)$ from 2 to 0^+, in the PN of Figure 4.28, is linear in the quantity of firing of T_3: if each significant mark is split into k tokens, $m(P_1) + m(P_2) = 1 / k$ after $2k - 1$ firings of T_3. For the continuous PN in Figure 4.28c, the significant marking $(0, 0^+)$ may be reached by a sequence of two OG-firings: $S = [T_1]^2[(T_2)^{0^+}(T_3)^{2^-}]$, where 2^- means 2 minus 0^+.

The exponential decreasing of $m(P_1) + m(P_2)$ in Figure 4.16 (Section 4.3.2) is not of the same kind.

4.4.4 Definition

On the basis of the explanations in Sections 4.2.1 to 4.2.3, let us now define an extended hybrid PN.

Definition 4.12 The definition of an *autonomous* **extended hybrid PN** is similar to the definition of an *autonomous hybrid PN* (Definition 4.4, Section 4.2.2), except that:

1) one can have, in addition, *inhibitor arcs* (from D-places or C-places, to D-transitions or C-transitions);

2) the *weight of an arc* (inhibitor or ordinary) whose origin is a C-place, takes its value in $\mathcal{Q}_+ \cup \{0^+\}$ instead of $\mathcal{Q}_+$;

3) the *marking of a C-place* takes its value in[6] $\mathcal{R}_+ \cup \{0^+\}$ instead of $\mathcal{R}_+$.

❑

The concepts added in extended hybrid PNs will appear to be useful in the following chapters where time is introduced for continuous and hybrid PNs.

---

[6] Formally, one could say that the marking of a C-place takes its value in the set of non-negative hyperreal numbers $^*\mathcal{R}_+$. Similarly, the weight of an arc is in the subset of values x in $^*\mathcal{R}_+$ such that $st(x) \in \mathcal{Q}_+$. See Appendix J.

NOTES and REFERENCES

The continuous PNs were introduced in [DaAl 87]. The presentation of an autonomous continuous PN as a limit case of discrete PN was first shown in [DaAl 89] and [DaAl 90]. S. B. Gershwin used a similar division (of parts) in order to introduce a continuous model for a manufacturing line [GeSc 80].

The idea of hybrid PNs was first presented in [DaAl 87]. The model was then studied in [LeAlDa 91] and [LeAlDa 92].

The reachability graph for a continuous PN was introduced in [DaAl 03], as well as the concepts of OG-firing and macro-markings for this kinds of Petri nets, and Property 4.1. The reachability graph for a hybrid PN is introduced in this book, as well as the concepts of OG-firing and macro-markings for this kinds of PNs, and Properties 4.6 and 4.8. The conflicts for continuous or hybrid PNs are specified in greater detail than in the previous publications.

The concepts of linearized reachability space, ε-liveness, and lim-liveness were introduced in [ReTeSi 99], from which examples in Figures 4.16a and 4.19a are drawn. Properties 4.4 and 4.9 to 4.12 are shown in this reference, in which various other results concerning a continuous PN and its discrete counterpart can be found.

Properties 4.2, 4.3, 4.5, and 4.7 were partially presented in previous papers.

Extended hybrid PNs were introduced in [DaCa 00] and [DaAl 01]. The second reference, while published later, was written before the other. In these papers, extended hybrid PNs were introduced mainly to increase modeling power of hybrid PNs; various applications to modeling will be presented in the following chapters (including modeling of pure delays which is developed in [DaCa 00]). In this book, we intend to show that the arc weight and marking conventionally denoted by 0^+ (Notation 4.5 and Appendix J) correspond to concepts occurring naturally in hybrid PNs.

An overview of integrality relaxations in PNs is presented in [SiRe 02]. It is focused on the relationship between the properties of (discrete) PNs and the corresponding properties of their continuous approximations.

5

Timed Continuous Petri Nets

The basic model is an autonomous continuous PN, as defined in Chapter 4, plus a maximal speed associated with every transition[1]*. Firing of a transition is continuous, like a flow limited by the maximal speed. For the models considered in this chapter, the maximal speeds do not depend on the marking.*

For the basic model, the maximal speeds are constant. *In Section 5.1, the behavior of such a PN is presented as a limit case of timed discrete PN behavior. Various basic behaviors are then analyzed, in order to understand clearly the semantics of the model.*

After a presentation of the conflicts in a timed continuous PN and their resolutions in Section 5.2, an algorithm used to calculate the behavior for such a PN is proposed in Section 5.3. Section 5.4 presents some applications and properties.

The basic model is then generalized in Section 5.5. The maximal speeds may depend on time but remain independent from the marking (models in which the speeds may depend on the marking are presented in Chapter 7).

In this chapter, the expression "continuous PN" means "timed continuous PN" (with constant maximal speeds up to the end of Section 5.4). Whenever an autonomous PN is concerned, this is specified.

5.1 DEFINITION OF THE MODEL

In Section 5.1.1, the timed continuous PN model is presented as a limit case of timed discrete PN. On the basis of this presentation, elementary behaviors are analyzed in Section 5.1.2.

[1] This is the basic model proposed by the authors. However, from the autonomous continuous PN in Chapter 4, other kinds of timings may be considered. Some of them are presented in Notes & References at the end of the chapter. See also Chapter 7, Appendix O and Postface.

5.1.1 Limit Case of a Discrete Timed Petri Net

Consider the timed discrete PN in Figure 5.1a. Marks[2] flow from P_1 to P_2, and P_3 specifies the single-server behavior: every time d_1 is spent, a mark is removed from P_1 and deposited in P_2, as long as a mark remains in P_1. In other words, the flow rate from P_1 to P_2 is $1/d_1$, as long as P_1 is not empty.

Figure b is obtained from Figure a in the following way: each "moving" mark, i.e. involved in the *significant marking* (Appendix F), is split into two tokens, and the timing associated with the transition is divided by two, i.e., $d_1/2$; the single-server behavior is preserved. Now, the token flow rate is $2/d_1$. Since the number of tokens is twice the number of marks in Figure a, the time for emptying place P_1 is the same. This is illustrated in Figure 5.1d.

If each mark in P_1 is split into k tokens, and the timing associated with T_1 is d_1/k, the flow rate becomes k/d_1, and the time for emptying place P_1 is still the same. When k tends to infinity, the *mark unit* may be used instead of the *token unit*: the marking of P_1 is expressed in marks, and the *token flow rate* k/d_1 is replaced by the *mark flow rate* $1/d_1$, denoted by V_1. We obtain the timed continuous PN in Figure 5.1c. The single-server hypothesis is no longer explicit: it is included in the definition of the **maximal firing speed** V_1, as will be specified in the next example. Speed V_1 is qualified as *maximal*, because it cannot be obtained when P_1 is empty. Let v_1 denote the **instantaneous firing speed** (*maximal firing speed* is noted with a capital V and *instantaneous firing speed* with a small v). As long as $m_1 > 0$, $v_1 = V_1$, and when $m_1 = 0$, $v_1 = 0$. The word "firing" will often be omitted: *maximal speed* and *instantaneous speed.*

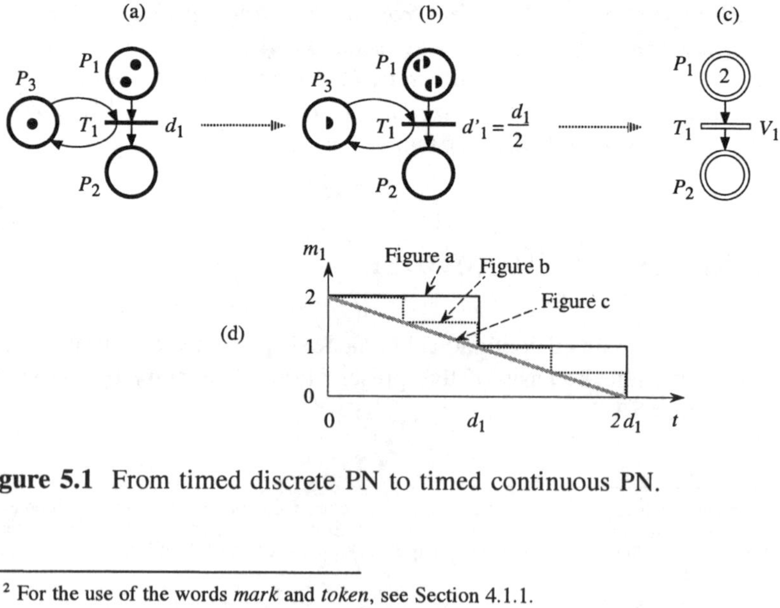

Figure 5.1 From timed discrete PN to timed continuous PN.

[2] For the use of the words *mark* and *token*, see Section 4.1.1.

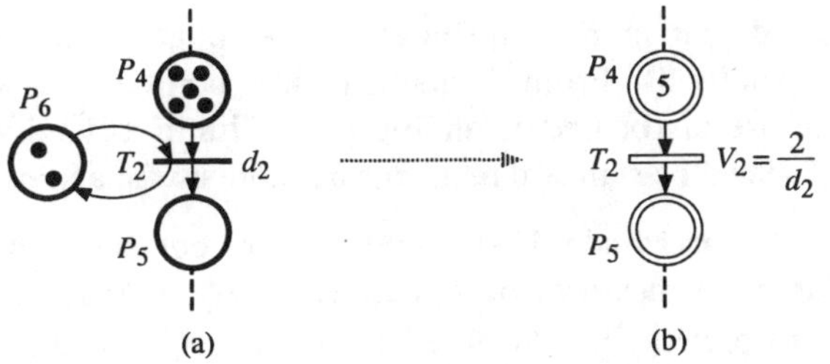

Figure 5.2 (a) Two-server machine. (b) Corresponding continuous PN.

Transition T_2 in Figure 5.2a illustrates a 2-server machine (two tokens in P_6). As long as there are at least two tokens in P_4, transition T_2, which is 2-enabled, is fired twice every d_2. For example, from the initial time, T_2 may be fired twice at time $t = d_2$, then twice at $t = 2\,d_2$, ... If there are *always* tokens in P_4, the average time between two firings is $d_2 / 2$ and the flow rate is $2 / d_2$. A transformation as in Figure 5.1 leads from Figure 5.2a to the timed continuous PN in Figure 5.2b, where the maximal firing speed V_2 is

$$V_2 = \frac{2}{d_2}: \tag{5.1}$$

the maximal firing speed V_2 is the product of the *flow rate* $1 / d_2$ *associated with a server* by the number of servers (i.e. number of tokens in P_6 in Figure 5.2a).

5.1.2 Analysis of Some Basic Behaviors

In this section, the behaviors of some basic timed continuous PNs are analyzed, by reference to the behaviors of "similar" timed discrete PNs. As in Section 4.1.1, a mark may be split into k tokens; a place marking is denoted by m_i *in mark units* (or m_i' in token units, rarely used). This analysis ensures an accurate understanding of the model. The algorithm calculating the evolution of a continuous PN (presented in Section 5.3) will be based on the behaviors explained in the sequel.

Remark 5.1 According to Definition 4.1 in Section 4.1.2, the place markings are (non-negative) real numbers. However, *during the analysis* of a behavior, the non-standard marking 0^+ will be used (see Appendix J). As a matter of fact, according to Definition 4.2 (Section 4.1.2), a non-immediate[3] transition is enabled if all its input places have a positive marking, and 0^+ is a positive marking. In some cases, other non-standard numbers may be used for an accurate analysis.

Nevertheless, *after* a detailed analysis, *all the results* (markings, time intervals, instantaneous speeds) *are expressed with real numbers*. This is possible because

[3] An immediate transition will appear to be a special case since a marking enabling it is unstable.

there is a standard real number infinitely close to any non-standard number [Ro 66] (see Appendix J). From a mathematical point of view, non-standard numbers may be useful for understanding some "limit" behaviors, but from a practical point of view, the standard numbers are more useful and pertinent.

Notation 5.1 A non-standard number is in $^*\mathcal{R}$ but not in $\mathcal{R}$ (see Appendix J). When a variable x is measured by a number in $^*\mathcal{R}$, it may be denoted by $^*x$, while the standard part of $^*x$ is denoted by x, i.e. $st(^*x) = x$. Notation $^*x$ is redundant when a non-standard number is denoted by β^+ or β^-. It will be used only if both $^*x$ and x are used; otherwise $x = 0^+$, for example, clearly means that x is measured by a number in $^*\mathcal{R}$.

❑

In order to simplify the presentation, the following notation will be used:

Notation 5.2 The interval $|a, b|$ denotes either $[a, b]$, or $[a, b[$, or $]a, b]$, or $]a, b[$. The values a and b are, respectively, the *lower limit* and the *upper limit*. This notation will be used in order to specify when some properties related to the markings and the speeds are true. If property A_1 is true in $|a, b|$, this means that A_1 is certainly true in $]a, b[$ but does not specify whether or not it is true for the limits. If A_1 and A_2 are true in $|a, b|$, A_1 may be true for the upper limit b whereas A_2 is not true for b, for example.

5.1.2.1 Sequences of Transitions, Same Maximal Speeds

Let us consider the timed continuous PN in Figure 5.3a; the initial marking is (2, 0) and the maximal speed associated with T_1 is $V_1 = 2$. According to Figure 5.1 presenting a timed continuous PN as a limit case of timed discrete PN, the discrete counterpart of the PN in Figure 5.3a is the PN in Figure 5.3b, and the behavior of the continuous PN corresponds to the behavior of the discrete PN in Figure 5.3c when k tends to infinity.

The behavior of the PN in Figure c is illustrated in Figures d and e. At $t = 0$, $m_1 = 2$ and $m_2 = 0$. The first firing of T_1 occurs at $t = d_1 = 1/2k$. At this time, the marking $(2 - 1/k,\ 1/k)$ is reached. And so on. The dot at the beginning of a line segment illustrating a constant marking, means that the corresponding value is true for the lower limit of the interval, i.e., $m_1 = 2$ for $t \in [0,\ 1/2k[$, $m_1 = (2 - 1/k)$ for $t \in [1/2k,\ 2/2k[$, etc.

Figure 5.3d shows that m_1 decreases by steps from 2 to 0 between $t = 0$ and $t = 1$. There are $2k$ successive firings of T_1. Each firing $[T_1]$ in token units corresponds to an OG-firing $[T_1]^{1/k}$ in mark units. These OG-firings occur at $t = 1/2k$, $t = 2/2k$, ..., $t = 1$.

When k tends to infinity, m_1 decreases linearly from 2 to 0 according to the grey line segment. Hence, for the continuous PN in Figure a, the following markings are obtained:

$$m_1(t) = 2 - 2t \text{ for } t \in [0, 1] \tag{5.2}$$

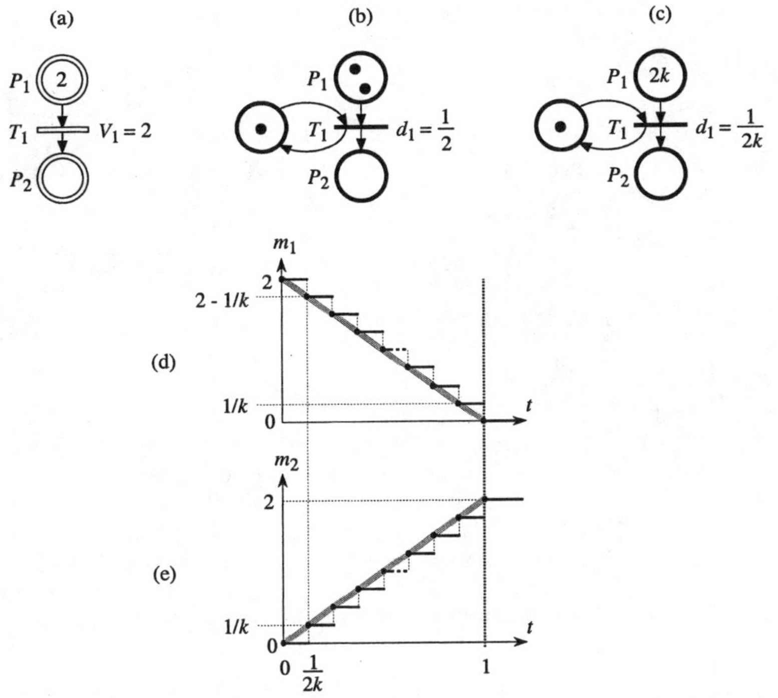

Figure 5.3 Continuous flow from P_1 to P_2.

and $\quad m_1(t) = 0$ for $t \in [1, \infty]$. (5.3)

As long as $m_1 > 0$, i.e. for $t \in [0, 1[$ according to (5.2), the instantaneous speed v_1 is the maximal speed V_1 (Section 5.1.1), then:

$$v_1(t) = V_1 = 2 \text{ for } t \in [0, 1[\tag{5.4}$$

and $\quad v_1(t) = 0$ for $t \in [1, \infty]$. (5.5)

Marking m_2 can be deduced from (5.2) and (5.3) since there is a marking invariant: $m_1 + m_2 = 2$. Hence,

$$m_2(t) = 2t \text{ for } t \in [0, 1] \tag{5.6}$$

and $\quad m_2(t) = 2$ for $t \in [1, \infty]$. (5.7)

According to Notation 5.2, the detailed results in (5.2) to (5.7) may be summarized as follows:

$$\text{For } t \in |\, 0, 1\, | : v_1(t) = 2,\ m_1(t) = 2 - 2t,\ m_2(t) = 2t. \tag{5.8}$$

$$\text{For } t \in |\, 1, \infty\, | : v_1(t) = 0,\ m_1(t) = 0,\ m_2(t) = 2. \tag{5.9}$$

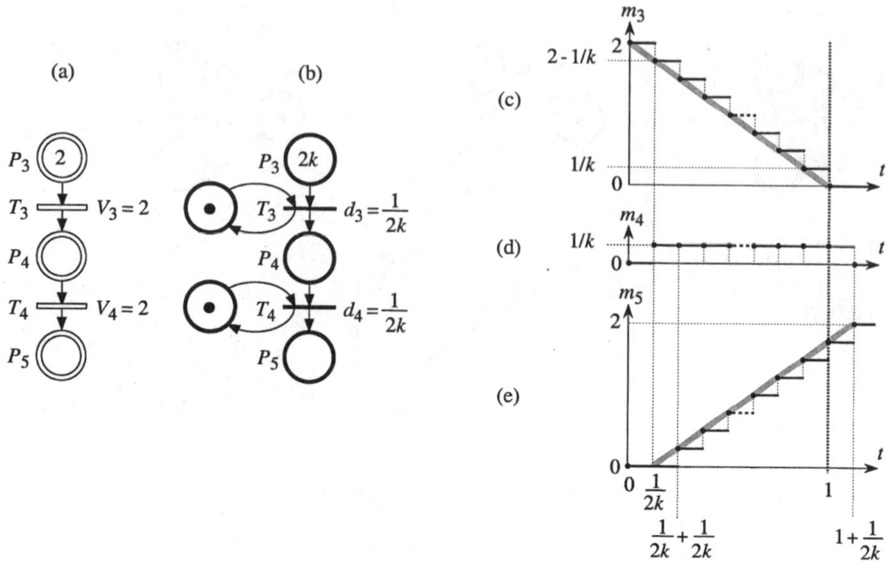

Figure 5.4 Flow through two transitions with the same maximal speed.

Let us now consider Figure 5.4. As before, the behavior of the PN in Figure a corresponds to the behavior of Figure b when k tends to infinity. The marking m_3 evolves similarly to m_1 in Figure 5.3.

Let us analyze the marking of P_4 in Figure b (see also Figure d). At time $t = 1/2k$, a token is deposited in P_4. This token enables T_4 which will be fired at $t = 2/2k$. At the same time $t = 2/2k$, the second firing of T_3 occurs. Hence, simultaneously, a token is deposited in and a token is removed from P_4. The marking remains unchanged: $m_4 = 1/k$ (corresponding to $m_4' = 1$ in token units). As shown in Figure d, $m_4 = 1/k$ from $t = 1/2k$ to $t = 1 + (1/2k)$. When k tends to infinity, ${}^*m_4 = 0^+$ (Notation 5.1, and Appendix J) for $t \in |0, 1|$. Let us now specify the values of m_4 for the limit values $t = 0$ and $t = 1$. According to Figure d, ${}^*m_4(0) = 0$ and ${}^*m_4(1) = 0^+$. Hence:

$$ {}^*m_4(t) = 0^+ \text{ for } t \in]0, 1]. \tag{5.10}$$

In this example, the marking ${}^*m_4 = 0^+$ lasts for some time interval (i.e. not for an infinitely short time). During this time interval, the marking of P_4 is positive. Transition T_4 is fired at the maximal speed $v_4(t) = V_4 = 2$. This is clearly observed in Figure 5.4e showing that m_5 increases at this speed.

A formal specification of enabling for continuous timed PNs will be given later (Section 5.1.3.2). For the time being, the following remark is sufficient.

Remark 5.2 For the examples in this section, all the maximal speeds are the same and each transition T_j has a single input place P_j. Then, according to the previous comments and Section 5.1.1, $v_j(t) = V_j$ if and only if ${}^*m_j(t) > 0$ (i.e., ${}^*m_j(t) \geq 0^+$) and $v_j(t) = 0$ if ${}^*m_j(t) = 0$.

❑

From (5.10) and Remark 5.2, the following results are obtained (for the notation t^-, see (J.7) in Appendix J):

$$v_4(t) = V_4 = 2 \text{ for } t \in]0, 1], \text{ and } v_4(t) = 0 \text{ for other values of } t; \tag{5.11}$$

$$^*m_5(t) = 2t^- \text{ for } t \in]0, 1], {}^*m_5(t) = 0 \text{ before and 2 after.} \tag{5.12}$$

Thus, from this analysis, the variables related to the continuous timed PN in Figure 5.4a, are expressed as follows *in standard numbers*:

$$\text{For } t \in |0, 1|: v_3(t) = v_4(t) = 2,\ m_3(t) = 2 - 2t,\ m_4(t) = 0,\ m_5(t) = 2t. \tag{5.13}$$

$$\text{For } t \in |1, \infty|: v_3(t) = v_4(t) = 0,\ m_3(t) = 0,\ m_4(t) = 0,\ m_5(t) = 2. \tag{5.14}$$

Finally, let us consider a last example in which all the maximal speeds are the same: the PN in Figure 5.5a. The markings of P_6 and P_7 in Figure 5.5 are similar to the markings of P_3 and P_4 in Figure 5.4. The instantaneous speeds of T_6 and T_7 in Figure 5.5 are similar to the instantaneous speeds of T_3 and T_4 in Figure 5.4.

As illustrated in Figures 5.5e and f, the following are obtained:

$$^*m_8(t) = 0^+ \text{ for } t \in]0, 1]; \tag{5.15}$$

$$v_8(t) = V_8 = 2 \text{ for } t \in]0, 1], \text{ and } v_8(t) = 0 \text{ for other values of } t; \tag{5.16}$$

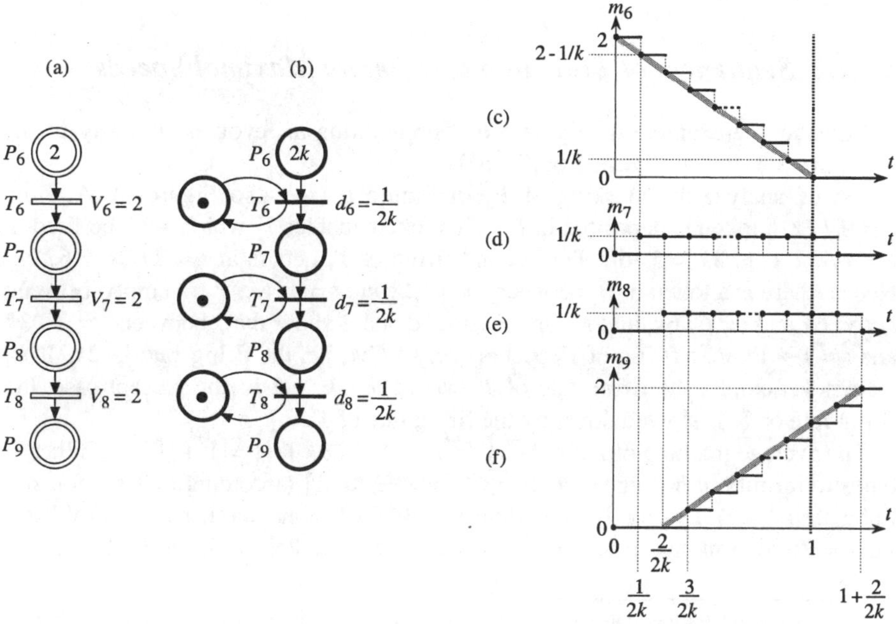

Figure 5.5 Flow through three transitions with the same maximal speed.

$$^{*}m_9(t) = 2t^- \text{ for } t \in]0, 1], \; ^{*}m_9(t) = 0 \text{ before and 2 after.} \tag{5.17}$$

Thus, from this analysis, the variables related to the continuous timed PN in Figure 5.5a, are expressed as follows *in standard numbers*:

$$\text{For } t \in |0, 1| : v_6(t) = v_7(t) = v_8(t) = 2,$$
$$m_6(t) = 2 - 2t, \; m_7(t) = m_8(t) = 0, \; m_9(t) = 2t. \tag{5.18}$$

$$\text{For } t \in |1, \infty| : v_6(t) = v_7(t) = v_8(t) = 0,$$
$$m_6(t) = m_7(t) = m_8(t) = 0, \; m_9(t) = 2. \tag{5.19}$$

Remark 5.3 For every example in this section, there is a place whose marking decreases linearly while the marking of another place increases linearly. As long as the *speed vector remains constant*, the PN is said to be in the same **IB-state** (standing for *invariant behavior state*). An IB-state lasts for a finite or infinite time interval. By abuse of language, this time interval is sometimes called[4] IB-state (instead of "time interval related to an IB-state").

During the IB-state, there is a continuous flow of marks at the maximal speeds. For example, in Figure 5.5, there is a *continuous flow* from P_6 to P_9 while P_7 and P_8 remain "almost" empty (markings 0^+). Throughout the IB-state, transitions T_6, T_7, and T_8, are fired at the same speed $v_j = 2$, except that $v_6(0) = 2$ and $v_6(1) = 0$ whereas $v_7(0) = v_8(0) = 0$ and $v_7(1) = v_8(1) = 2$.

The differences related to the lower and upper limit of the time interval have no practical interest for most of the timed continuous PNs. They have no *quantitative* effect, but their *qualitative* effect will be shown for some examples.

5.1.2.2 Sequences of Transitions, Different Maximal Speeds

Consider the example in Figure 5.6. The marking m_1 evolves similarly to m_1 in Figure 5.3, but in Figure 5.6a, $V_2 > V_1$.

Let us analyze the marking of P_2 in Figure b (see also Figure d). At time $t = 1/2k$, a token is deposited in P_2. This token enables T_2 which will be fired at $t = 1/2k + 1/3k = 5/6k$. The second firing of T_1 occurs at $t = 2/2k = 6/6k$. Hence, there is a token in P_2 between $t = 1/2k$ and $t = 5/6k$, P_2 is empty between $t = 5/6k$ and $t = 6/6k$, and so on. Figures d and e show that, between $t = 1/3k$ and $t = 1 + 1/3k$, a firing of T_2 occurs every $1/2k$, i.e. the firing rate is $2k$. If P_2 were never empty, the firing rate of T_2 would be $3k$, but in the present case, the firing rate of T_2 is slowed down by the firing rate of T_1.

The average marking of P_2 is $2/3k$ from $t = 1/3k$ to $t = 1 + 1/3k$. When k tends to infinity, it becomes $^{*}m_2(t) = 0^+$ for $t \in]0, 1]$ (according to Notation 4.5 in Section 4.4.2). During this time interval, the instantaneous firing speed of T_2 is $v_2(t) = 2$ (corresponding to the firing rate $2k$ for the PN in Figure b. The firing

[4] In previous publications, this time interval was called a *phase* by the authors. The name IB-state is now chosen for homogeneity with hybrid PNs: an IB-state for a continuous PN is a particular case of IB-state for a hybrid PN (Definition 6.5 in Section 6.1.5.3).

speed of T_2 is slowed down by the firing speed of T_1: $v_1(t) = V_1 = 2$ and $v_2(t) = v_1(t) = 2$.

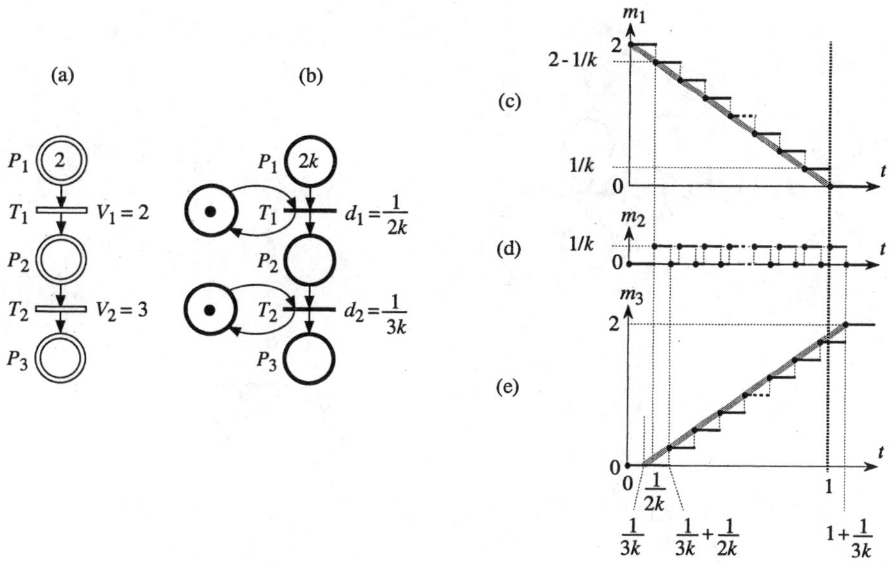

Figure 5.6 Flow with $V_2 > V_1$.

Thus, the variables related to the continuous timed PN in Figure 5.6a, are expressed as follows *in standard numbers*:

$$\text{For } t \in \left]0, 1\right] : v_1(t) = v_2(t) = 2,\ m_1(t) = 2 - 2t,\ m_2(t) = 0,\ m_3(t) = 2t. \quad (5.20)$$

$$\text{For } t \in \left]1, \infty\right[: v_1(t) = v_2(t) = 0,\ m_1(t) = 0,\ m_2(t) = 0,\ m_3(t) = 2. \quad (5.21)$$

Remark 5.4 In a *timed* continuous PN, the marking 0^+ is associated with a place which is (practically) empty and which is *both fed and emptied at the same speed.* This is true for P_4 in Figure 5.4, P_7 and P_8 in Figure 5.5, and P_2 in Figure 5.6.

The corresponding output transitions T_4 in Figure 5.4, T_7 and T_8 in Figure 5.5, have a maximal speed equal to the feeding speeds of the places concerned. It follows that these transitions are fired at their maximal speeds. On the other hand, in Figure 5.6, the feeding speed of P_2, $v_1(t) = 2$, is lower than the maximal speed $V_2 = 3$ of the output transition T_2. It follows that the speed of T_2 is slowed down to satisfy the equality $v_2(t) = v_1(t)$, i.e. $v_2(t) < V_2$. As a matter of fact, since P_2 is (practically) empty, marks cannot be removed from this place faster than they are deposited in it.

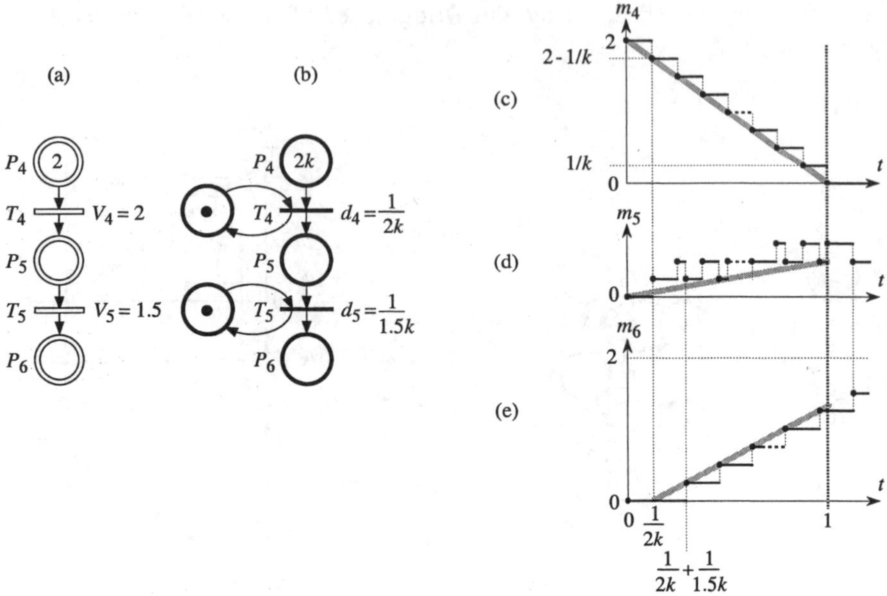

Figure 5.7 Flow with $V_5 < V_4$.

Two kinds of enablings may be considered. During the time interval $\lvert 0, 1 \rvert$, a transition like T_1 in Figure 5.6a is **strongly enabled** ($m_1(t) > 0$, measured by a standard number), and a transition like T_7 in Figure 5.5a or T_2 in Figure 5.6a is **weakly enabled** (${}^*m_2(t) = 0^+$). A formal definition will be given in Section 5.1.3.2 (Definitions 5.3 and 5.3').

❑

Let us now consider the example in Figure 5.7 in which $V_5 < V_4$.

Let us analyze the behavior of the discrete timed PN in Figure b (see also Figure d). At time $t = 1/2k$, a token is deposited in P_5. This token enables T_5 which will be fired at $t = 1/2k + 1/1.5k \approx 1.17/k$. The second firing of T_4 occurs at $t = 2/2k = 1/k$. Hence, the second firing of T_4 occurs before the first firing of T_5: there are two tokens in P_5 between $t = 1/k$ and $t = 1.17/k$. Since $d_5 > d_4$, the number of tokens in P_5 tends to increase (Figure d).

When k tends to infinity, $m_5(t)$ increases linearly from $t = 0$. Between $t = 0$ and $t = 1$, both T_4 and T_5 can be fired at their maximal speeds since $m_4 > 0$ and $m_5 > 0$, i.e. $v_4(t) = V_4 = 2$ and $v_5(t) = V_5 = 1.5$. It follows that, for this time interval:

$$m_5(t) = m_5(0) + 2t - 1.5t = 0.5t. \tag{5.22}$$

Thus, for the *first IB-state*:

$$\text{For } t \in \lvert 0, 1 \rvert : v_4(t) = 2,\ v_5(t) = 1.5, \quad m_4(t) = 2 - 2t,\ m_5(t) = 0.5t,\ m_6(t) = 1.5t. \tag{5.23}$$

At time $t = 1$, the *second IB-state* begins from the marking $(m_4, m_5, m_6) = (0, 0.5, 1.5)$. It is clear that P_4 remains empty and that $v_4(t) = 0$ for $t > 1$. From $t = 1$, the behavior of the subnet made up of P_5, T_5, and P_6, is qualitatively similar to the behavior of the PN in Figure 5.3a from $t = 0$. The second and third IB-states are as follows:

$$\text{For } t \in |1, 4/3|: v_4(t) = 0,\ v_5(t) = 1.5, \quad m_4(t) = 0,\ m_5(t) = 0.5 - 1.5(t-1),\ m_6(t) = 1.5 + 1.5(t-1). \tag{5.24}$$

$$\text{For } t \in |4/3, \infty|: v_4(t) = 0,\ v_5(t) = 0, \quad m_4(t) = 0,\ m_5(t) = 0,\ m_6(t) = 2. \tag{5.25}$$

5.1.2.3 Synchronization

Figure 5.8 illustrates synchronization. Transition T_1 in Figure b is fired every $d_1 = 1/4k$ as long as there are tokens in both P_1 and P_2. When T_1 has been fired $3k$ times, k tokens remain in P_1 but P_2 is empty. Then, T_1 is no longer enabled. The behavior of the PN in Figure a corresponds to k tending to infinity, i.e. T_1 is fired at the maximal speed $v_1(t) = V_1 = 4$ up to $t = 0.75$. At this time $m_2 = 0$ and the transition is no longer enabled. Hence, the behavior is made up of two IB-states:

$$\text{For } t \in |0, 0.75|: v_1(t) = 4,\ m_1(t) = 4 - 4t,\ m_2(t) = 3 - 4t,\ m_3(t) = 4t. \tag{5.26}$$

$$\text{For } t \in |0.75, \infty|: v_1(t) = 0,\ m_1(t) = 1,\ m_2(t) = 0,\ m_3(t) = 3. \tag{5.27}$$

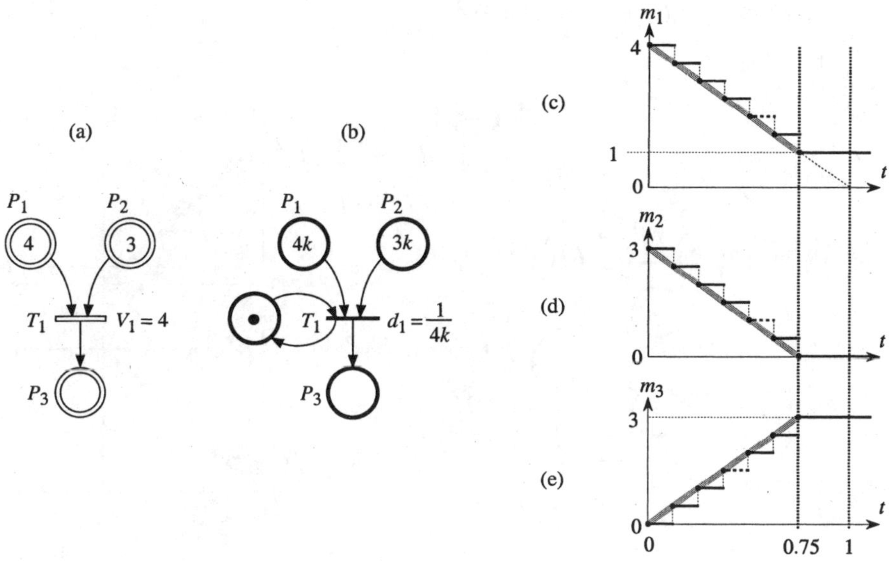

Figure 5.8 Synchronization.

5.1.2.4 Timed Continuous Petri Net With a Circuit

The example in Figure 5.9 contains a circuit.

Let us analyze the behavior of the discrete timed PN in Figure b. The first firing of T_1 occurs at time $t = 1/4k = 0.25/k$: a token is removed from P_1 while a token is deposited in P_2. The token in P_2 enables T_2 which will be fired at $t = 1/4k + 1/1.5k \approx 0.917/k$. The second and third firings of T_1 occur at $t = 0.5/k$ and $0.75/k$. At the first firing of T_2 (between the third and the fourth firing of T_1) a token is removed from P_2 while a token is deposited in P_1. And so on. This is illustrated in Figures c and d. Since the firing rate of T_1 is greater than the firing rate of T_2 (i.e., $d_1 < d_2$), the marking of P_1 tends to decrease while the marking of P_2 tends to increase. Hence, after some time, P_1 will become empty whereas there are $2k$ tokens in P_2. From this time, a token will be deposited in P_1 every $1/1.5k \approx 0.667/k$ time units (firing of T_2); this token remains $0.25/k$ time units in P_1 (up to firing of T_1). And so on (Figures c and d).

When k tends to infinity, i.e. for the behavior of the continuous PN in Figure a, two IB-states have to be considered: during the first IB-state, $m_1 > 0$, and during the second IB-state $m_1 = 0$.

First IB-state: $m_1(t) > 0$ for $t \geq 0$, and $m_2(t) > 0$ for $t > 0$ since $V_2 < V_1$. Then:

$$v_1(t) = V_1 = 4 \text{ and } v_2(t) = V_2 = 1.5. \tag{5.28}$$

Since, P_1 is simultaneously emptied by the continuous firing of T_1 and fed by the continuous firing of T_2:

$$m_1(t + \mathrm{d}t) = m_1(t) + (v_2(t) - v_1(t))\,\mathrm{d}t. \tag{5.29}$$

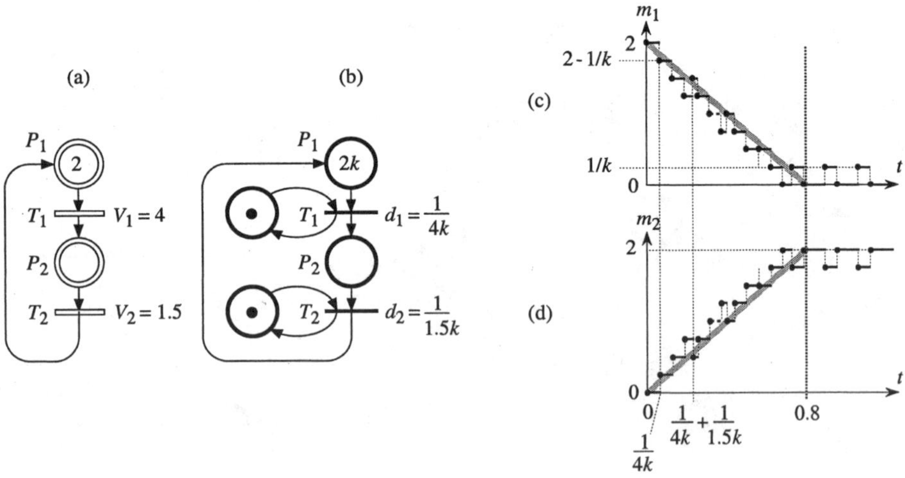

Figure 5.9 Timed continuous PN with a circuit.

From (5.28) and (5.29), and given $m_1(0) = 2$,

$$m_1(t) = 2 - 2.5t \tag{5.30}$$

is obtained. Similarly, $m_2(t) = 2.5t$ is obtained ($m_1 + m_2 = 2$). The first IB-state lasts as long as $m_1 > 0$, i.e. up to $t = 0.8$ according to (5.30).

Second IB-state: at $t = 0.8$, the marking of the PN in Figure a is (0, 2). Since $m_2(t) > 0$, transition T_2 is fired at its maximal speed: $v_2(t) = V_2 = 1.5$. Place P_1 is fed by the firing of T_2. Since the maximal firing speed of the output transition T_1 is greater than the instantaneous speed of the input transition T_2 (i.e. $V_1 > v_2(t)$), P_1 remains (almost) empty. In this IB-state, the behavior related to P_1 is similar to the behavior of place P_2 in Figure 5.6: ${}^*m_1(t) = 0^+$. This is shown in Figure 5.9c, where the average marking of P_1 in Figure b is $d_1 / d_2 = 0.375$ in token units, i.e. $d_1 / d_2 k = 0.375 / k$ in mark units.

Thus, the variables related to the continuous timed PN in Figure 5.9a, are expressed as follows *in standard numbers*:

$$\text{For } t \in \,]0, 0.8]: v_1(t) = 4, v_2(t) = 1.5, m_1(t) = 2 - 2.5t, m_2(t) = 2.5t. \tag{5.31}$$

$$\text{For } t \in \,]0.8, \infty[: v_1(t) = v_2(t) = 1.5, m_1(t) = 0, m_2(t) = 2. \tag{5.32}$$

5.1.2.5 Infinite Maximal Speed

The behavior of the continuous PN in Figure 5.10a corresponds to the behavior of the discrete PN in Figure 5.10b when k tends to infinity. In both cases, T_2 is an *immediate transition* (Definition 3.3 in Section 3.2.1), modeled by $d_2 = 0$ in the discrete case and $V_2 = \infty$ in the continuous case. A marking enabling an immediate transition is unstable. This means that at most one of the places P_1 or P_2 can remain marked; this is illustrated in Figures c and d. For the discrete model, as long as there are tokens in P_1, as soon as a token is deposited in P_2, T_2 is immediately fired; hence P_2 remains empty. When P_1 is empty, then m_2 can increase. The behavior of the continuous model is "similar".

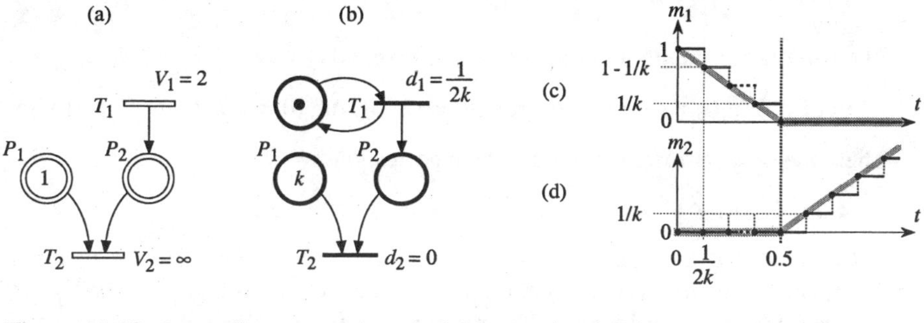

Figure 5.10 Immediate transition: infinite maximal firing speed.

The *two IB-states* are as follows:

$$\text{For } t \in \,]0, 0.5]: v_1(t) = 2, v_2(t) = 2, m_1(t) = 1 - 2t, m_2(t) = 0. \tag{5.33}$$

$$\text{For } t \in \,]0.5, \infty]: v_1(t) = 2, v_2(t) = 0, m_1(t) = 0, m_2(t) = 2(t - 0.5). \tag{5.34}$$

According to the behavior explained above, the instantaneous firing speed of an immediate transition depends on the other speeds. For the example in Figure 5.10a, $v_2(t) = V_1(t) = 2$ as long as $m_1(t) > 0$, then $v_2(t) = 0$ when $m_1(t) = 0$. If the PN is *stable* (Definition 3.6 in Section 3.2.3.1), for any T_j such that $V_j = \infty$, $v_j(t) < \infty$ for any $t > 0$. However, at $t = 0$, in some cases $v_j(0) = \infty$.

Roughly speaking, for both discrete and continuous PNs, a marking is stable if it does not change, immediately, in a *discrete* way. Then, let us now specify the concept of *stable marking* for a continuous PN.

Definition 5.1 In a *timed continuous PN*, a marking $\mathbf{m}(t)$ is **stable** if $\mathbf{m}(t + \mathrm{d}t) = \mathbf{m}(t) + \mathbf{v} \cdot \mathrm{d}t$, where $\mathbf{v}$ is a vector of *finite* or zero numbers.

❑

For the marking $\mathbf{m}_0 = (2, 0)$ in Figure 5.11a, T_1 is 2-enabled at $t = 0$, i.e., $q(T_1, \mathbf{m}_0) = 2$. According to the definition of an immediate transition, the quantity of firing of transition T_1 is 2 at $t = 0$. Hence, at $t = 0$, the marking changes immediately from (2, 0), *unstable*, to (0, 2) which is *stable*. This is illustrated in Figures c and d. Afterwards, T_1 behaves similarly to any other transition; for our example, $v_1(t) = v_2(t) = 2$ for $t > 0$.

General case: if a timed continuous PN contains one or more immediate transitions, $\mathbf{m}_0$ may be *unstable*. If $\mathbf{m}_0$ is unstable, let us denote it by $\mathbf{m}^{bef}(0)$; a marking $\mathbf{m}^{aft}(0)$ (*bef* and *aft* standing for *before* and *after*) is instantaneously reached: there is a discontinuity between $\mathbf{m}^{bef}(0)$ and $\mathbf{m}^{aft}(0)$: the speed vector of marking variation corresponds to a vector of Dirac functions. Then, from the *stable* marking $\mathbf{m}^{aft}(0)$, the marking evolves without discontinuity.

The *discrete firing of the continuous transition* T_1 in Figure 5.11a, at time 0, does not correspond to an IB-state (because the evolution is not continuous). Let us call this evolution an **I-phase** (I standing for initialization). The duration on an I-phase is zero and the firing speed of every transition is either ∞ (strongly enabled immediate transition) or 0 (other transitions). The firing quantity corresponding to an infinite speed during a nul time is obtained from the initial marking.

The *I-phase* for the PN in Figure 5.11a corresponds to:

$$\text{At } t = 0: v_1(t) = \infty, v_2(t) = 0, \mathbf{m}^{bef}(0) = (2, 0), \mathbf{m}^{aft}(0) = (0, 2). \tag{5.35}$$

Then, the only *IB-state* for this example corresponds to:

$$\text{For } t \in \,]0, \infty]: v_1(t) = 2, v_2(t) = 2, m_1(t) = 0, m_2(t) = 2. \tag{5.36}$$

Let us note that the marking of a place continuously emptied by the firing of an immediate transition is exactly 0 but not 0^+. For the PN in Figure 5.10a, ${}^*m_2(t) = 0$ for $t \in \,]0, 0.5]$, and for the PN in Figure 5.11a, ${}^*m_1(t) = 0$ for $t \in \,]0, \infty]$.

This is consistent with the meaning of 0^+ (Notation 4.5 in Section 4.4.2, and Appendix J). It follows that, *for some features* (particularly weak enabling), *an immediate transition cannot be considered as an ordinary* (non-immediate) *transition.*

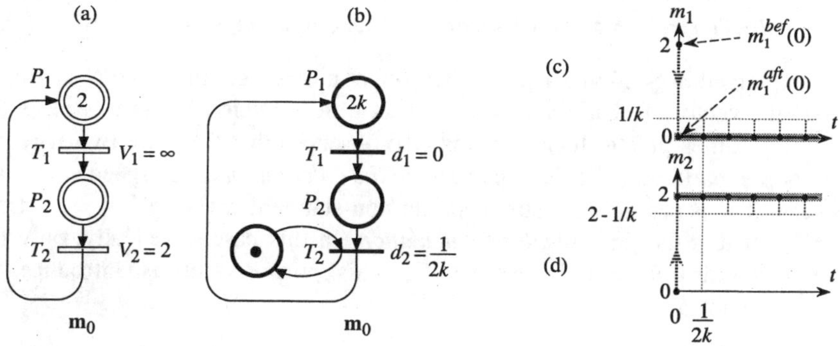

Figure 5.11 Infinite speed at t = 0.

Remark 5.5 In a general case, moving from $\mathbf{m}^{bef}(0)$ to $\mathbf{m}^{aft}(0)$ may be obtained by an *iterated firing sequence* (as in Section 3.2.2.2).

For example, consider a loop such as in Figure 5.11a with additional place P_3 and transition T_3, such that $V_1 = V_2 = \infty$ and $V_3 = 2$. If $\mathbf{m}_0 = \mathbf{m}^{bef}(0) = (2, 0, 0)$, immediate firing of T_1 leads to the marking (0, 2, 0) which is unstable, then the immediate firing of T_2 leads to the marking $\mathbf{m}^{aft}(0) = (0, 0, 2)$ which is stable. For this I-phase, $v_1 = \infty$, $v_2 = \infty$, and $v_3 = 0$.

Remark 5.6 As explained in Section G.3 of Appendix G, it is possible to define the behavior of a transition T_j whose speed is infinite when some Boolean condition C_j is true, or whose speed is infinite when some event E_j occurs.

In both cases, T_j is a *synchronized continuous transition* (see Section 3.3.3 for the synchronized discrete transition). Examples of *synchronized continuous PN* (all transitions synchronized) and continuous PNs with synchronized transitions and timed transitions are considered in Exercise 5.4.

5.1.3 Definitions

In this section, the concepts introduced in Section 5.1.2 are formalized. Conflicts, whose analysis and resolution are based on the content of this section, will be studied in Section 5.2.

5.1.3.1 Definition and Notation

Definition 5.2 A **timed continuous PN** is a pair $\langle R, \text{Spe}\rangle$ such that:

R is a marked autonomous continuous PN (Definition 4.1, Section 4.1.2);

Spe is a function from the set T of transitions to $Q_+ \cup \{\infty\}$. For T_j, $\text{Spe}(T_j) = V_j$ = maximal speed associated with transition T_j.

❑

As explained in Section 5.1.2, the *marking* of a timed continuous PN at time t *implies* the vector of *instantaneous speeds* at time t, which in turn *implies* the *marking* at times greater than t. It was also observed that there are two ways to consider the marking: either it is the *basis of the calculation* of the speed vector (in this case, it is necessary to distinguish the non-standard marking ${}^*m_i = 0^+$ from ${}^*m_i = 0$); or it is the *final result of calculation* (in this case, practically, only the standard marking $m_i = 0$ is useful). For this purpose, let us introduce the following notation.

Notation 5.3 The *marking used for calculation* of the speed vector is denoted by $\tilde{\mathbf{m}} = (\tilde{m}_1, \tilde{m}_2, \ldots)$. The marking $\tilde{m}_i$ may be 0, or 0^+, or a positive real number. The *marking finally useful* for an engineer is denoted by $\mathbf{m} = (m_1, m_2, \ldots)$. The marking m_i is a standard positive or nul real number.

❑

For example, during the second IB-state for the PN in Figure 5.9a (Section 5.1.2.4), we have $\tilde{\mathbf{m}} = (0^+, 2)$ and $\mathbf{m} = (0, 2)$, according to (5.32).

Remark 5.7 In $\tilde{\mathbf{m}}$, the only non-standard number used is 0^+. For the previous example, a non-standard marking would be ${}^*\mathbf{m} = (0^+, 2^-)$ (according to (J.7), Appendix J), since there is a marking invariant $m_1 + m_2 = 2$, but it is not useful for our purpose.

5.1.3.2 Enabling

According to Section 5.1.2.5, enabling of an immediate transition will be defined separately (Definition 5.3').

According to Definition 4.2 (Section 4.1.2) and Notation 5.3, enabling of a non-immediate transition in a timed continuous PN may be defined as follows.

Definition 5.3 Non-immediate transition T_j in a timed continuous PN is **enabled** at t if

$$\tilde{m}_i(t) > 0 \text{ for every } P_i \in {}^\circ T_j. \tag{5.37}$$

It is **strongly enabled** if

$$m_i(t) > 0 \text{ for every } P_i \in {}^\circ T_j, \tag{5.38}$$

and **weakly enabled** otherwise (i.e., if (5.37) is satisfied but not (5.38).

❑

All the concepts defined below are significant *after the stable marking* $\mathbf{m}^{aft}(0)$ has been reached, if an I-phase exists: Definitions 5.3', 5.4, and 5.5, as well as Properties 5.1 to 5.4.

Definition 5.4

a) The **feeding speed** of a place P_i in a timed continuous PN is:

$$I_i = \sum_{T_j \in {}^\circ P_i} \text{Post}(P_i, T_j) \cdot v_j . \tag{5.39}$$

If $I_i > 0$, place P_i is said to be **fed**.

b) The **draining speed** of a place P_i in a timed continuous PN is:

$$O_i = \sum_{T_k \in P_i^\circ} \text{Pre}(P_i, T_k) \cdot v_k . \tag{5.40}$$

❑

For the PN in Figure 5.6a (Section 5.1.2.2), for example, at $t = 0.5$, T_1 is strongly enabled because $m_1(0.5) > 0$, and T_2 is weakly enabled because $\tilde{m}_2(0.5) = 0^+$. At the same time, both P_2 and P_3 are fed: P_2 is fed because T_1 is fired at $v_1 = V_1 = 2$; P_3 is fed because T_2 is fired at $v_2 = 2$. It is clear in this example that T_1, strongly enabled, is fired at its maximal speed, and that T_2, weakly enabled, is fired at the feeding speed of P_2. Before giving a property specifying the firing speed of a weakly enabled transition, let us analyze various behaviors, then define weak enabling for an immediate transition.

Transitions T_3 in Figures 5.12a, b, c, and d, are weakly enabled at $t > 0$ (not at $t = 0$ but just after) for a duration which is greater than 1 in every case. Hence we look at the instantaneous speeds at time $t = 1$, for example.

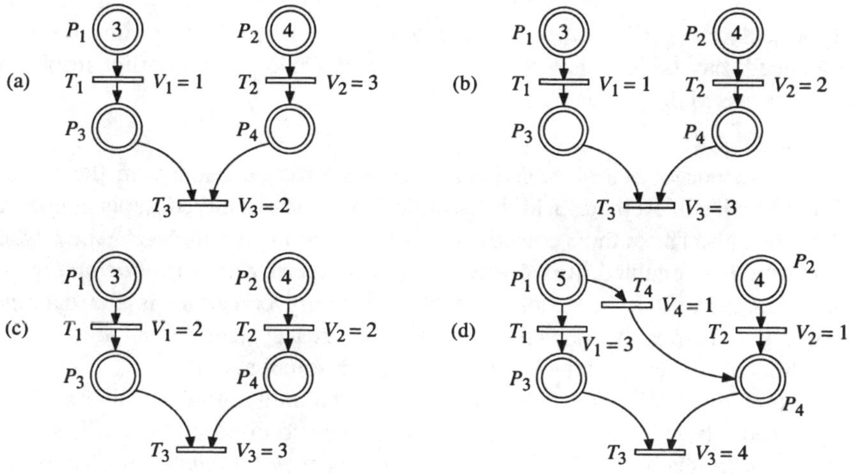

Figure 5.12 Weak enabling and synchronization.

In Figure a, T_1 and T_2 are strongly enabled, hence $v_1 = V_1 = 1$ and $v_2 = V_2 = 3$. Since $V_3 < v_2$, P_4 cannot remain empty: m_4 increases. On the other hand, $V_3 > v_1$, hence, v_3 is limited by the feeding speed of P_3: $v_3 = v_1 = 1$. The draining speed of P_3 ($O_3 = v_3$) is equal to its feeding speed ($I_3 = v_1$), and during this IB-state $\tilde{m}_3 = 0^+$.

In Figure b, both $v_1 = V_1 = 1$ and $v_2 = V_2 = 2$ are smaller than $V_3 = 3$. Then, due to synchronization, v_3 takes the smaller of these two speeds: $v_3 = \min(v_1, v_2) = v_1 = 1$. Since $v_2 > v_1$, m_4 increases and $\tilde{m}_3 = 0^+$.

In Figure c, both $v_1 = V_1 = 2$ and $v_2 = V_2 = 2$ are smaller than $V_3 = 3$. Then, $v_3 = \min(v_1, v_2) = v_1 = v_2 = 2$. Since $v_1 = v_2 = v_3$, $\tilde{m}_3 = \tilde{m}_4 = 0^+$.

In Figure d, P_4 is fed by two transitions: T_2 and T_4. Its feeding speed is $I_4 = v_2 + v_4 = 2$. Since $V_3 > v_1$ and $V_3 > v_2 + v_4$, $v_3 = \min(v_1, v_2 + v_4) = v_2 + v_4 = 2$. Then m_3 increases and $\tilde{m}_4 = 0^+$.

Definition 5.3' An *immediate transition* T_j in a timed continuous PN is **weakly enabled** at t if all the places P_i in $°T_j$ such that $m_i(t) = 0$ (at least one exists since an immediate transition cannot be strongly enabled after initial time) are such that $I_i(t) > 0$.

Property 5.1

a) If transition T_j is *strongly enabled* it is fired at its *maximal speed*: $v_j(t) = V_j$.

b) Let T_j be a *weakly enabled* transition and $Q_j(t)$ denote the subset of places P_i in $°T_j$ such that[5] $m_j(t) = 0$. If T_j is not involved in an actual conflict related to a place in $Q_j(t)$ (or if T_j takes priority over the other transitions involved in the conflict), the instantaneous firing speed of T_j is:

$$v_j(t) = \min_{P_i \in Q_j(t)} \left(\frac{1}{\text{Pre}(P_i, T_j)} \sum_{T_k \in °P_i} \text{Post}(P_i, T_k) \cdot v_k(t),\ V_j \right). \tag{5.41}$$

❑

If a weakly enabled transition is involved in an actual conflict, its instantaneous firing speed may be less than the value given by (5.41). The conflict resolution will be discussed in Section 5.2.

Remark 5.8

a) The concepts of *free speed* and *maximal speed* were explained, in the context of discrete PNs, in Remark 3.11 in Section 3.4.1. The same concepts applies to continuous PNs. The authors consider the *behavior at maximal speed* [6], i.e., when a transition T_j is enabled, it is always fired at a speed which is *maximal given some constraints*: if it is strongly enabled, the only constraint is the maximal speed V_j; if it is weakly enabled, $v_j(t)$ cannot be greater than the value in (5.41), and additional constraints may be due to conflict resolution.

b) In previous publications, the authors proposed an iterative calculation of the firing speeds, based on (5.41) and priorities; the proposed algorithm has some weaknesses. In this book, algorithms based on *linear programming problems* will be given (*maximization of some criteria, given some constraints*).

---

[5] The marking $\tilde{m}_i(t) = 0$ if T_j is *immediate* and $\tilde{m}_i(t) = 0^+$ otherwise, but $m_i(t) = 0$ for both cases.

[6] Behaviors at *free speed* are considered in [Ha 03] (where the word is introduced).

5.1.3.3 Balance

Definition 5.5 The **balance** of the marking of a place P_i (balance of P_i for short) in a timed continuous PN is:

$$B_i = I_i - O_i,$$

i.e. the *difference* between the *feeding* and the *draining* speeds (Definition 5.4).

❑

The balance of P_i corresponds to the time derivative of its marking, i.e.,

$$\frac{\mathrm{d}m_i(t)}{\mathrm{d}t} = B_i(t), \tag{5.42}$$

and $m_i(t + \mathrm{d}t) = m_i(t) + B_i(t) \cdot \mathrm{d}t.$ (5.43)

Property 5.2 Let $|\, t_1, t_2 \,|$ be a time interval corresponding to an IB-state (this interval is finite or infinitely long but not infinitely short).

a) $B_i(t) > 0$ for $t \in |\, t_1, t_2 \,| \Rightarrow m_i(t) > 0$ for $t \in |\, t_1, t_2 \,|$. (5.44)

b) $B_i(t) < 0$ for $t \in |\, t_1, t_2 \,| \Rightarrow m_i(t) > 0$ for $t \in |\, t_1, t_2 \,|$. (5.45)

c) $B_i(t) = 0$ for $t \in |\, t_1, t_2 \,| \Leftrightarrow m_i(t) =$ constant for $t \in |\, t_1, t_2 \,|$. (5.46)

Proof Let us first observe that Properties 5.2a, b, and c, are not specified for the lower and upper limits of the interval.

a) If the *balance is positive*, the *marking increases* according to (5.43). Since $m_i(t_1) \geq 0$, the property is shown. Examples: $m_2(t)$ in Figure 5.3, $m_5(t)$ in Figure 5.4, $m_9(t)$ in Figure 5.5, $m_5(t)$ and $m_6(t)$ in Figure 5.7, all these examples for $t \in |\, 0, 1 \,|$; $m_2(t)$ for $t \in |\, 0, 0.8 \,|$ in Figure 5.9.

b) If the *balance is negative*, the *marking decreases.* Since the marking cannot be negative, the negative balance is only possible if the marking is positive. Examples: $m_1(t)$ for $t \in |\, 0, 1 \,|$ in Figure 5.3; $m_1(t)$ and $m_2(t)$ for $t \in |\, 0, 0.75 \,|$ in Figure 5.8; $m_1(t)$ for $t \in |\, 0, 0.8 \,|$ in Figure 5.9.

c) This property, expressed in standard numbers (Notation 5.3), is a direct consequence of Equation (5.43). Examples in Figure 5.5: $m_7(t)$ and $m_8(t)$ for $t \in |\, 0, 1 \,|$; $m_6(t)$, $m_7(t)$, $m_8(t)$, and $m_9(t)$, for $t \in |\, 1, \infty \,|$. Examples in Figure 5.8: $m_1(t)$, $m_2(t)$, and $m_3(t)$, for $t \in |\, 0.75, \infty \,|$.

Property 5.3 Let $|\, t_1, t_2 \,|$ be a time interval corresponding to an IB-state. The marking of place P_i at the upper limit t_2 is $\tilde{m}_i(t_2) = 0^+$ if and only if:

1) $\tilde{m}_i(t) = 0^+$ for $t \in |\, t_1, t_2 \,|$, (5.47)

or 2) $m_i(t)$ decreases to 0 AND $I_i(t) > 0$ for $t \in |\, t_1, t_2 \,|$. (5.48)

Proof Let us first recall that, in general (according to Notation 5.2 in Section 5.1.2), if some property is known to be true for $t \in |\, t_1, t_2 \,|$, it may be true or false at t_2.

Sufficient condition

1) If $\tilde{m}_i(t) = 0^+$ for $t \in |t_1, t_2|$, then $B_i(t) = 0$ for $t \in |t_1, t_2|$, according to Property 5.2. After some time t_a, $t_1 > t_a > t_2$, $B_i(t)$ cannot change before $t = t_2$ (upper limit of the interval). Since $\tilde{m}_i(t)$ is a consequence of $B_i(t)$, it can change only *after* $t = t_2$. Hence $\tilde{m}_i(t_2) = 0^+$. Examples in Figure 5.5: $\tilde{m}_7(t) = 0^+$ and $\tilde{m}_8(t) = 0^+$ for $t \in |0, 1|$, hence $\tilde{m}_7(1) = \tilde{m}_8(1) = 0^+$ (whereas $\tilde{m}_6(1) = 0$).

2) Similarly to case 1, $I_i(t)$ cannot change before $t = t_2$, hence $\tilde{m}_i(t_2) > 0$. Since $m_i(t)$ decreases to 0 between t_1 and t_2, the only possible value at t_2 is $\tilde{m}_i(t_2) = 0^+$. Example in Figure 5.9: in $t \in |0, 0.8|$, $\tilde{m}_1(t)$ decreases to 0 AND $I_1(t) = 1.5$, hence $\tilde{m}_1(0.8) = 0^+$.

Necessary condition

If $\tilde{m}_i(t) = \alpha$ for $t \in |t_1, t_2|$, where α is a constant value different from 0^+, then $\tilde{m}_i(t_2) = \alpha$ (similar to case 1 of the sufficient condition). If $m_i(t)$ increases or decreases to a value β different from 0, the result is obvious since $\tilde{m}_i(t_2) = \beta \neq 0^+$.

Property 5.4 Let $|t_{s-1}, t_s|$ be a time interval corresponding to an IB-state. The marking of place P_i is $\tilde{m}_i(t) = 0^+$ for $t \in |t_{s-1}, t_s|$ if and only if:

1) $m_i(t) = 0$, $I_i(t) > 0$, and P_i is not emptied by an immediate transition (5.49)

or 2) $\tilde{m}_i(t_{s-1}) = 0^+$ and no transition in P_i° is enabled at time t_{s-1}. (5.50)

❑

Case 1 in Property 5.4 was already explained in Remark 5.4 (Section 5.1.2.2): there is a "dynamic" production of the marking 0^+ ($\tilde{m}_1$ in Figure 5.9a in Section 5.1.2.4, $t \in |0.8, \infty|$). In case 2, the value 0^+ is "static": this value is maintained during the IB-state, except if *an enabled transition in P_i° empties the place completely*. For the example in Figure 5.6 (Section 5.1.2.2), $\tilde{m}_2(1) = 0^+$ and $\tilde{m}_2(t) = 0$ for $t > 1$ because T_2 is enabled at time $t = 1$. An example of "static" 0^+ will be given in Figure 6.17 (Section 6.2.4). These comments lead to the following definition.

Definition 5.6 A marking 0^+ during a time interval $|t_1, t_2|$ is:

a) a **dynamic 0^+** if its corresponds to case 1 of Property 5.4;

b) a **static 0^+** if its corresponds to case 2 of Property 5.4.

❑

From previous results, one can conclude that *if a place* (without immediate output transition) *is fed during a time interval*, it cannot be "completely " empty at the upper limit of this time interval, i.e.:

$$I_i(t) > 0 \text{ for } t \in |t_1, t_2| \Rightarrow \tilde{m}_i(t_2) \geq 0^+. \tag{5.51}$$

As a matter of fact: 1) if $B_i(t) > 0$, the result is obvious; 2) if $B_i(t) < 0$, then $m_i(t_2) = \alpha \geq 0$, and property (5.51) is true even if $\alpha = 0$ according to (5.48); 3) if $B_i(t) = 0$, the result is a consequence of Properties 5.3 (case 1) and 5.4 (case 1).

5.1.3.4 Evolution Graph

The behavior of a timed continuous PN can be represented by an **evolution graph** in time, given in Figure 5.13c. This graph is represented in the form of a Petri net. Each place corresponds to an *IB-state*, i.e. a *constant instantaneous speed state* $\mathbf{v} = (v_1, v_2)$, and with each transition the event (change of marking) is associated, the occurrence of which produces a change from one speed state to another. If there is an I-phase, the markings *before* and *after* the I-phase are specified (see Appendix M and Exercise 5.3).

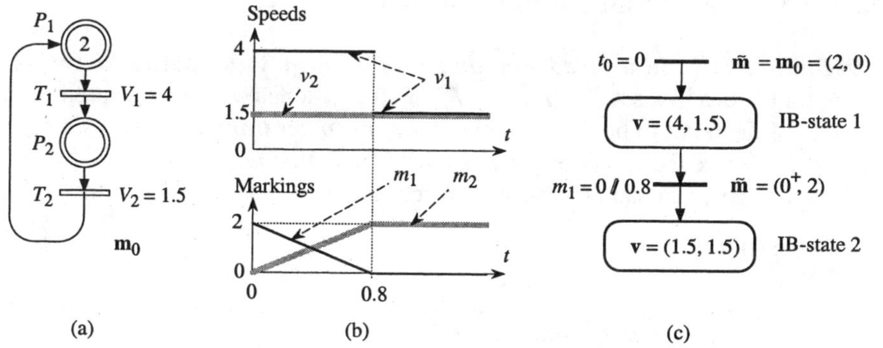

Figure 5.13 (a) Timed continuous PN. (b) Evolution of markings and speeds. (c) Evolution graph.

According to the detailed study of the behavior of the PN in Figure 5.13a (Section 5.1.2.4: Figure 5.9, Equations (5.31) and (5.32)), the evolution contains two IB-states which are presented in Figure 5.13c. At $t = 0$, the marking is $\mathbf{m}_0 = (2, 0)$ which implies $v_1 = 4$ and $v_2 = 1.5$, i.e. $\mathbf{v} = (4, 1.5)$. Hence, during IB-state 1, the marking evolves linearly: $m_1(t) = 2 - 2.5t$ and $m_2(t) = 2.5t$.

The change of IB-state is provoked by the event '*m_1 reaches the value* 0'. This is denoted $m_1 = 0$ near the corresponding transition; this transition will occur 0.8 time units after IB-state 1 has begun (value shown after the /). At $t = 0.8$, according to Notation 5.3 and Property 5.3b, the marking is $\tilde{\mathbf{m}} = (0^+, 2)$, and the vector speed $\mathbf{v} = (1.5, 1.5)$ of IB-state 2 is calculated from this marking (in this example, the calculation from $\mathbf{m} = (0, 2)$ gives the same result, but we will see examples where the results would be different).

Remark 5.9 The evolution graph can be built only if the *behavior is deterministic*. If there is no conflict, the behavior is deterministic. If there are actual conflicts, the resolution of these conflicts is necessary for the behavior to be deterministic. Conflicts are studied in Section 5.2 and algorithms leading to the building of the evolution graph are given in Section 5.3.

5.2 CONFLICTS

If an actual conflict exists, the instantaneous speed calculation requires the conflict resolution. Section 5.2.1 explains when an actual conflict exists, and Section 5.2.2 presents various methods for conflict resolution. Algorithms for this resolution will be presented in Section 5.3; it is assumed that *a resolution rule is proposed for each structural conflict*, since it is not known *a priori* if an actual conflict may exist or not, given that a structural conflict exists.

5.2.1 Existence of an Actual Conflict

Definition 5.7 In a *timed continuous PN*, there is an **actual conflict** among transitions in a set $\{T_1, T_2, \ldots, T_s\}$, if the instantaneous firing speed of at least one of them has to be slowed down *due to the other transitions in this set.*

Let $Q_j(t)$ denote the subset of places P_h in $°T_j$ such that $m_h(t) = 0$.

Formally, given **m** and $I_i(t)$, there is an actual conflict related to P_i at t if:

1) $m_i(t) = 0$,

and

$$2)\ I_i(t) < \sum_{T_j \in P_i^{\circ}} \min_{P_h \in Q_j(t)} \left(\frac{1}{\mathrm{Pre}(P_h, T_j)} \sum_{T_k \in {}^{\circ}P_h} \mathrm{Post}(P_h, T_k) \cdot v_k(t),\ V_j \right). \qquad (5.52)$$

In other words, the feeding speed of place P_i is not sufficient for firing all its output transitions at the firing speed in (5.41) (Property 5.1b, Section 5.1.3.2). ❑

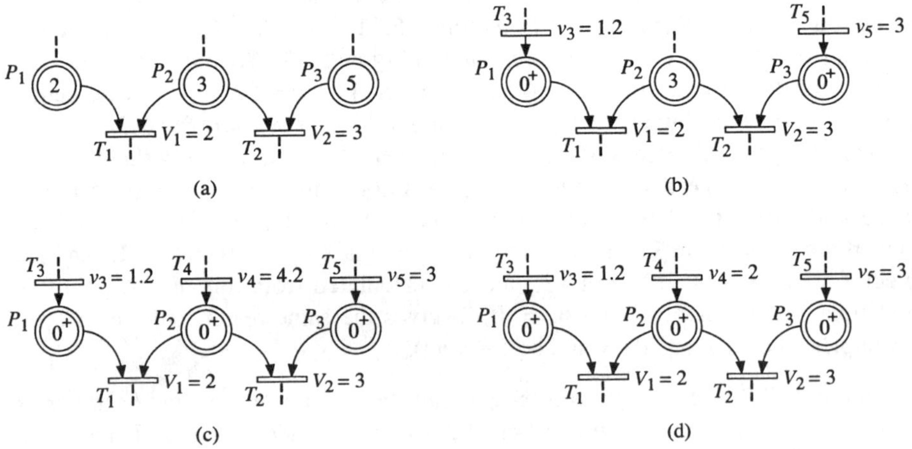

Figure 5.14 (a), (b) and (c) No actual conflict. (d) Actual conflict.

Figure 5.14 presents four cases corresponding to the same *structural conflict* $K_1 = \langle P_2, \{T_1, T_2\}\rangle$. Let us analyze which cases correspond to an actual conflict. According to (5.49) in Property 5.4 (Section 5.1.3.3), for a place P_i which is not emptied by an immediate transition, $\tilde{m}_i(t) = 0^+$ if $m_i(t) = 0$ and $I_i(t) > 0$: this marking 0^+ is explicitly written in Figure 5.14.

Figure a: no actual conflict. Both T_1 and T_2 are strongly enabled: T_1 is strongly enabled because $m_1 > 0$ and $m_2 > 0$; T_2 is strongly enabled because $m_2 > 0$ and $m_3 > 0$. Accordingly, both transitions can be fired at their maximal speeds, i.e. $v_1 = V_1 = 2$, and $v_2 = V_2 = 3$ (Property 5.1a).

If P_2 is not fed, i.e. if $I_2 = 0$, then $B_2 = -O_2 = -(v_1 + v_2) = -5$. According to Equation (5.43) in Section 5.1.3.3:

$$m_2(t + \mathrm{d}t) = m_2(t) - 5\ \mathrm{d}t. \tag{5.53}$$

As long as $m_2 > 0$, both transitions can be fired at their maximal speeds. Simply, this situation does not last very long (3 / 5 = 0.6 time unit) because P_2 is emptied by both T_1 and T_2.

Figure b: no actual conflict. Both T_1 and T_2 are weakly enabled but $m_2 > 0$ does not limit their instantaneous firing speeds. The instantaneous firing speed of T_1 is limited by the feeding speed of P_1, i.e. $v_1 = \min(V_1, v_3) = v_3 = 1.2$. Similarly, $v_2 = \min(V_2, v_5) = 3$.

Figure c. Both T_1 and T_2 are weakly enabled and $m_2 = 0$. The instantaneous firing speed of T_1 is limited by the feeding speeds of both P_1 and P_2, i.e. $v_1 \leq \min(V_1, v_3, v_4) = v_3 = 1.2$. Similarly, $v_2 \leq \min(V_2, v_4, v_5) = v_5 = 3$. Hence, $v_1 + v_2 \leq 1.2 + 3 = 4.2$. Since the feeding speed of P_2, i.e. $I_2 = v_4 = 4.2$, is sufficient to allow both $v_1 = 1.2$ and $v_2 = 3$, there is *no actual conflict.*

Then, the balances $B_1 = v_3 - v_1 = 1.2 - 1.2 = 0$, $B_2 = v_4 - (v_1 + v_2) = 4.2 - 4.2 = 0$, and $B_3 = v_5 - v_2 = 3 - 3 = 0$, are obtained.

Figure d. As in Figure c, the firing speeds of T_1 and T_2 are limited by the feeding speeds of P_1, P_2 and P_3: $v_1 \leq \min(V_1, v_3, v_4) = v_3 = 1.2$ and $v_2 \leq \min(V_2, v_4, v_5) = v_4 = 2$. Hence, $v_1 + v_2 \leq 1.2 + 2 = 3.2$. Since the feeding speed of P_2, i.e. $I_2 = v_4 = 2$, is not sufficient to allow both $v_1 = 1.2$ and $v_2 = 2$, *there is an actual conflict.* Resolution of this conflict is studied in the next section.

5.2.2 Conflict Resolution

When there is an actual conflict between two or several transitions, there are two *basic rules*[7] for resolution: the *priority* and the flow *sharing*[8].

[7] These rules express an *objective which is not necessarily reached.* For example, even if $T_a < T_b$, in some cases $v_a = 0$ and $v_b > 0$ (because T_a is not enabled).

[8] The flow sharing is possible because the flow is continuous. For a discrete PN, a "sharing" is obtained by probabilities assigned to every transition or by alternate firings (Appendix B).

Let us consider the example in Figure 5.14d. According to the analysis in Section 5.2.1, the conflict resolution consists of finding instantaneous speeds v_1 and v_2 such that

$$v_1 \leq 1.2, \tag{5.54}$$

$$v_2 \leq 2, \tag{5.55}$$

and $$v_1 + v_2 = 2. \tag{5.56}$$

Remark 5.10 Let us note that Equation (5.56) is $v_1 + v_2 = 2$ but not $v_1 + v_2 \leq 2$. This is consistent with Remark 5.8a: a transition which is enabled is fired at a speed which is *maximal given some constraints*. The conflict resolution will add new constraints in order to obtain a solution. For any resolution, Equation (5.56) must be satisfied. A solution satisfying (5.54) to (5.56) (for our example), is said to be **acceptable**. If $v_1 + v_2 \neq 2$, the solution is *not acceptable*; this idea will be illustrated in the sequel.

Another way to tackle conflict resolution is proposed in [Ha 03]. See Notes and References at the end of this chapter.

❑

In the sequel, four solutions are presented. Each solution is the consequence of a resolution rule (which may be chosen by the designer).

First resolution rule: priority of T_1 *over* T_2 (i.e., $T_1 < T_2$). According to (5.54), $v_1 = 1.2$ is chosen. Then, from (5.56), $v_2 = 2 - v_1 = 2 - 1.2 = 0.8$ is obtained[9].

Second resolution rule: priority of T_2 *over* T_1 (i.e., $T_2 < T_1$). According to (5.55), $v_2 = 2$ is chosen. Then, from (5.56), $v_1 = 2 - v_2 = 2 - 2 = 0$ is obtained.

Third resolution rule: sharing between T_1 *and* T_2 *aiming at*

$$v_1 = v_2. \tag{5.57}$$

Various kinds of sharing could be imagined. **Aiming at** (5.57) means that there is an equal flow through each transition, *if this solution is acceptable*. From (5.54) to (5.57), $v_1 = v_2 = 1$ is obtained.

Fourth resolution rule: sharing between T_1 *and* T_2 *aiming at*

$$v_1 = 4\,v_2. \tag{5.58}$$

There is no solution satisfying (5.54), (5.55), (5.56), and (5.58).

From (5.54), (5.55), and (5.58), the values $v_1 = 1.2$ and $v_2 = 0.3$ are obtained. *This solution is not acceptable* because $v_1 + v_2 = 1.5$ does not satisfy (5.56).

From (5.56) and (5.58), the values $v_1 = 1.6$ and $v_2 = 0.4$ are obtained. *This solution is not acceptable* because (5.54) is not satisfied. However, it will be used

[9] In our example, v_4 does not depend on v_1 and v_2. Examples of resolutions by priorities such that there is a feedback loop from the transitions to the place involved in the conflict, appear in Sections 5.3.2.1, 5.3.3.1 and Appendix K.

for finding the solution, as explained in the sequel. The *total flow* through T_1 and T_2 is $v_1 + v_2 = 2$ according to (5.56). If this total flow (corresponding to v_4) could be shared according to (5.58), the **parts** assigned to T_1 and T_2 (denoted by $p(T_1)$ and $p(T_2)$) would be

$$p(T_1) = 1.6 \text{ and } p(T_2) = 0.4. \tag{5.59}$$

According to (5.54) and (5.59), $v_1 < p(T_1)$, then the value of v_1 is 1.2 (maximal value given (5.54)). Since T_1 "*cannot use its part completely*", the difference may be "*added to the part*" of T_2. The solution is then:

$$v_1 = \min(p(T_1), 1.2) = 1.2; \tag{5.60}$$

$$v_2 = 2 - 1.2 = 0.8. \tag{5.61}$$

The result in (5.60) is obtained from (5.54) and (5.59), and (5.61) is obtained from (5.56) and (5.60).

Notation 5.4 The priority between two transitions T_a and T_b is denoted by $T_a < T_b$. The *sharing between these transitions aiming at* $v_a / \alpha_a = v_b / \alpha_b$ *will be denoted by* $[\alpha_a T_a, \alpha_b T_b]$. This sharing implies, obviously, that T_a and T_b have the same level of priority.

For example, assume a structural conflict $K_2 = \langle P_1, \{T_1, T_2, T_3, T_4\}\rangle$. In the sequel are some possible resolution rules, denoted by LR_i (LR stands for **local resolution**). Generally speaking, in a regular discrete PN, a transition firing is always the consequence of *local information* related to the transition and its input places and related conflicts (expressed by the "*concept of locality*" in the Foreword by M. Silva). For continuous PNs, the authors want to preserve the same "philosophy", i.e. a local conflict resolution):

LR_1: $T_1 < T_2 < T_3 < T_4$: four priority levels (complete priority ordering).

LR_2: $T_1 < [2T_3, 3T_4] < T_2$: three priority levels, with a sharing between T_3 and T_4 at the second level.

LR_3: $[2T_2, 3T_3] < [T_1, T_4]$: two priority levels, with a sharing between T_2 and T_3, then, at a second priority level a sharing between T_1 and T_4.

5.3 SPEED CALCULATION ALGORITHMS

First, algorithms of speed calculation *for one IB-state* are presented in Sections 5.3.1 to 5.3.3. Then, a complete algorithm, for all the IB-states, will be given in Section 5.3.4. The difficulty consists of obtaining an algorithm for one IB-state (the complete algorithm is simply an iteration of the first one).

An algorithm for a PN *without structural conflict* is first presented in Section 5.3.1 (Algorithm 5.1).

Then, algorithmic resolution of conflicts is tackled. The *basic algorithm is only based on priorities*, i.e., if N transitions are involved in the same conflict, there are N levels of priority among them (see LR_1 in Notation 5.4). This case, analyzed in Section 5.3.2, leads to Algorithm 5.4. The possibility of *sharings among transitions* will be studied in Section 5.3.3: Algorithm 5.7 allows mixing of priorities and sharings.

5.3.1 There is No Structural Conflict

If there is no structural conflict, the speed vector $\mathbf{v}_s$ related to IB-state s is obtained from the marking $\mathbf{m}_{s-1}$ at the beginning of this IB-state (whereas it is obtained from $\tilde{\mathbf{m}}_{s-1}$ in the general case). The reason for which the use of $\tilde{\mathbf{m}}_{s-1}$ is not necessary (nevertheless possible) will be explained in Remark 5.14 in Section 5.3.4.

The explanations and algorithm are illustrated with the example in Figure 5.15. Let us present the various data used to automatically calculate the result for the PN in Figure 5.15a. Three kinds of data are presented. They correspond to sets of constraints denoted by C_1, C_2, and C_3.

1) *Speed limits*. According to the definition of the model, an instantaneous speed must be positive and not greater than its specified maximal value:

$$C_1 = \{0 \le v_1 \le 3,\ 0 \le v_2 \le 3,\ 0 \le v_3 \le 1,\ 0 \le v_4 \le 2,\ 0 \le v_5 \le 2,\ 0 \le v_6 \le 1\}. \tag{5.62}$$

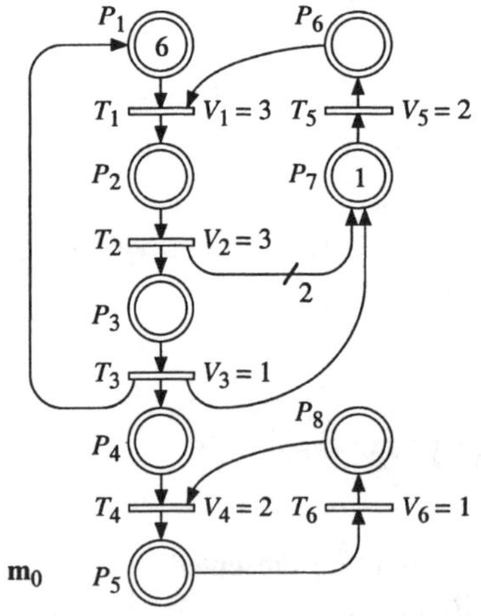

(a)

Step 0. $\mathbf{m} = \mathbf{m}_0 = (6, 0, 0, 0, 0, 0, 1, 0)$.
$C_1 = \{0 \le v_1 \le 3, 0 \le v_2 \le 3, ..., 0 \le v_6 \le 1\}$.

Step 1. $C_2 = \{v_1 - v_2 \ge 0,\ v_2 - v_3 \ge 0,\ v_3 - v_4 \ge 0,$
$v_4 - v_6 \ge 0,\ v_5 - v_1 \ge 0,\ v_6 - v_4 \ge 0\}$.

Step 2. $T_{SF} = \{T_1, T_2, T_3, T_5\}$.

Step 3. $C_3 = \{v_4 = 0, v_6 = 0\}$.

Step 4. $\mathbf{v} = (2, 2, 1, 0, 2, 0)$.

Step 5.1. $m_1 = 6 - t,\ m_2 = m_5 = m_6 = m_8 = 0,$
$m_3 = t,\ m_4 = t,\ m_7 = 1 + 3\,t$.

Step 5.2. $\Delta t = 6$ (m_1 becomes 0); $t_1 = 6$.

Step 5.3. $\mathbf{m}_1 = \mathbf{m}(6) = (0, 0, 6, 6, 0, 0, 19, 0)$.

(b)

Figure 5.15 Illustration of Algorithm 5.1 (no structural conflict).

2) *Non-negative balances.* Since a marking cannot be negative, the balance of a place P_i whose marking is $m_i = 0$ at the beginning of an IB-state, must be non-negative. For our example, B_2, B_3, B_4, B_5, B_6, and $B_8 \geq 0$, i.e.:

$$C_2 = \{v_1 - v_2 \geq 0, v_2 - v_3 \geq 0, v_3 - v_4 \geq 0, v_4 - v_6 \geq 0, v_5 - v_3 \geq 0, v_6 - v_4 \geq 0\}. \quad (5.63)$$

3) *Transitions whose instantaneous speed is zero and surely firable transitions.* From the initial marking in Figure 5.15a, it is known that, during the first IB-state, the instantaneous speed of T_5 is certainly positive (T_5 is said to be **surely firable**) because this transition is strongly enabled. Since T_5 is fired, place P_6 is fed, hence T_1 is surely firable since $m_1 \geq 0$. Since T_1 is fired, place P_2 is fed, hence T_2 is surely firable. Since T_2 is fired, place P_3 is fed, hence T_3 is surely firable. The **set of surely firable transitions**, denoted by T_{SF} is then:

$$T_{SF} = \{T_1, T_2, T_3, T_5\}. \quad (5.64)$$

By construction of T_{SF}, all the transitions in this set have a positive instantaneous speed, whereas all the transition out of this set, i.e. the transitions in $T \setminus T_{SF} = \{T_4, T_6\}$ have a zero instantaneous speed. As a matter of fact, T_4 could be fired only if P_8 were fed, i.e. if T_6 were fired, and T_6 could be fired only if P_5 were fed, i.e. if T_4 were fired. But these firings cannot occur since all the places in the circuit $P_8T_4P_5T_6$ have a zero marking. This information provides the third set of constraints, denoted by C_3:

$$C_3 = \{v_4 = 0, v_6 = 0\}. \quad (5.65)$$

❑

Then, calculation of the instantaneous speeds v_1 to v_6 consists of solving the following *linear programming problem* (LPP): maximize the criterion[10]

$$J_1 = v_1 + v_2 + v_3 + v_4 + v_5 + v_6, \quad (5.66)$$

given the set of constaints C_1, C_2, and C_3.

Let us note that, given the set C_3 of constraints (i.e., $v_j = 0$ for every $T_j \notin T_{SF}$), the criteria $\sum_{T_j \in T_{SF}} v_j$ and $\sum_{T_j \in T} v_j$ are equivalent. The set C_1 of constraints is independent from the marking, whereas C_2, T_{SF}, and C_3 depend on the marking.

From the above explanations, Algorithm 5.1 is obtained. Notations $P_{ne}^{(k)}$, $T_{SF}^{(k)}$, and T_{se} are used: $P_{ne}^{(k)}$ is the set of *non-empty* places (i.e. marked or fed) and $T_{SF}^{(k)}$ is the value of T_{SF}, at the kth iteration; T_{se} is the set of *strongly enabled* transitions.

Algorithm 5.1 **Speed vector if there is no conflict**
Step 0. Initialization: $\mathbf{m}$, and set C_1 of constraints corresponding to $0 \leq v_j \leq V_j$.
Step 1. Set up the constraints C_2 based on the balances ($B_i \geq 0$ if $m_i = 0$).

[10] If J_1 is maximized, every v_j ($j = 1, \ldots, 6$, in our example) is maximized. Since there is no conflict among the transitions in T, a criterion $J_1' = \sum_{T_j \in T} \alpha_j v_j$, $\alpha_j > 0$, would lead to the same vector $\mathbf{v}$.

Step 2. Let $T_{SF}^{(0)} = \varnothing$, $P_{ne}^{(1)} = \{P_i \mid m_i > 0\}$, $T_{SF}^{(1)} = T_{se} = \{T_j \mid {}^\circ T_j \in P_{ne}^{(1)}\}$, and $k = 1$.

While $T_{SF}^{(k)} \neq T_{SF}^{(k-1)}$ and $T_{SF}^{(k)} \neq T$ **do**

$P_{ne}^{(k+1)} = P_{ne}^{(k)} \cup (T_{SF}^{(k)})^\circ$

$T_{SF}^{(k+1)} = \{T_j \mid {}^\circ T_j \in P_{ne}^{(k+1)}\}$

$k = k + 1$

end_while

$T_{SF} = T_{SF}^{(k)}$

Step 3. Set up the constraints C_3: $v_j = 0$ for $T_j \notin T_{SF}$.

Step 4. Solve the LPP: maximize $J_1 = \sum_{T_j \in T} v_j$ given C_1, C_2, and C_3 (**v** is obtained).

Step 5.

Step 5.1. Calculate $\mathbf{m}(t)$ for the IB-state.

Step 5.2. Calculate the duration Δt of the IB-state (shorter time such that a component of $\mathbf{m}(t)$ becomes 0 as will be specified in Property 5.5), then time t_s corresponding to the end of this IB-state s.

Step 5.3. Calculate $\mathbf{m}(t_s)$. END.

❑

This algorithm is illustrated in Figure 5.15b. In *Step 2*, $T_{SF} = \{T_1, T_2, T_3, T_5\}$ is obtained. In this step, the last value of $P_{ne}^{(k)}$, $P_{ne}^{(5)} = \{P_1, P_2, P_3, P_4, P_6, P_7\}$, shows that P_5 and P_8 remain empty and that T_4 and T_6 are never fired.

5.3.2 Resolution By Priorities

Algorithmic resolution of conflicts is now tackled. In this section, the following hypothesis is made.

Hypothesis 5.1 For any pair of transitions T_a and T_b involved in the same structural conflict, a priority rule $T_a < T_b$ or $T_b < T_a$ is specified.

❑

The problems to be solved are illustrated in Section 5.3.2.1. Solutions are progressively presented in Sections 5.3.2.2, and 5.3.2.3.

5.3.2.1 Expected Results And Problems To Be Solved

The goal of Section 5.3.2.3 will be to calculate the vector **v** of instantaneous speeds during an IB-state, given 1) the PN under consideration, 2) the local conflict resolutions chosen for every structural conflict, and 3) marking $\tilde{\mathbf{m}}$ (Notation 5.3) at the beginning of the IB-state. Before giving an algorithm, let us emphasize some problems to be solved, through the analysis of two examples.

According to Section 5.1.1, a timed continuous PN is a limit case of timed discrete PN. Then, expected behaviors of timed continuous PNs will be explained from timed discrete PNs.

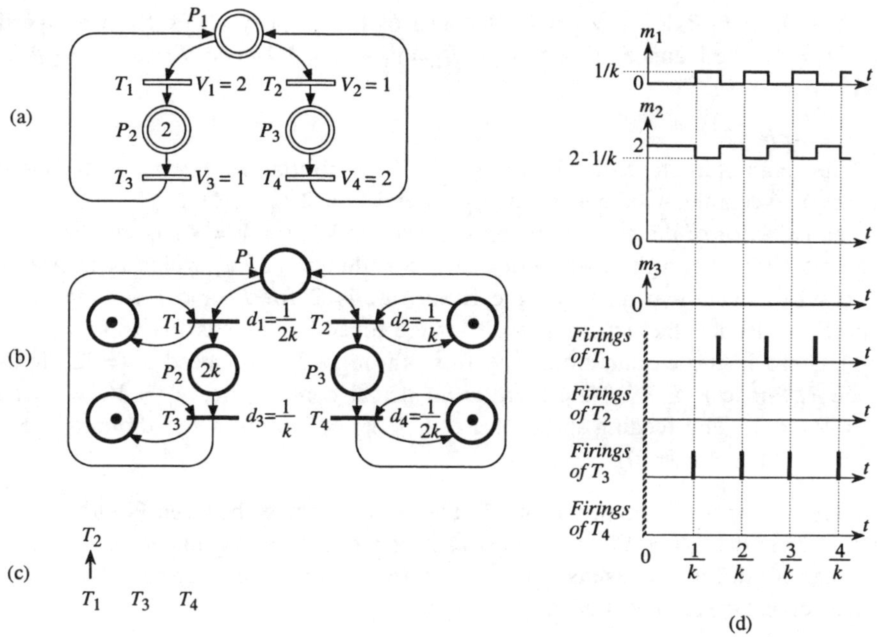

Figure 5.16 Example 1. (a) Timed continuous PN. (b) Discrete model. (c) Priority graph for both. (d) Behavior of the discrete model for $k = 4$.

Example 1

Consider the timed continuous PN in Figure 5.16a. According to Section 5.1, its behavior corresponds to the behavior of the discrete PN in Figure 5.16b when k tends to infinity. Let us assume (for both), the priority $T_1 < T_2$: the priority graph (see Appendix B) is given in Figure 5.16c.

The behavior of the discrete PN in Figure b is illustrated in Figure d for $k = 4$. Transition T_3 is 1-enabled at time $t = 0$; it will be fired at $t = 1 / k$. When T_3 is fired, a token is removed from P_3 and a token is deposited in P_1. Hence T_1 and T_2 become enabled. Because of the priority, T_1 will be fired and this firing will occur at $t = 1.5 / k$.

Since there are $2k$ tokens in P_2, T_3 will be fired again at $t = 2 / k$, $t = 3 / k$, etc., and T_1 will be fired again at $t = 2.5 / k$, $t = 3.5 / k$, etc. Since $d_1 < d_3$, the number of tokens in P_1 is at most 1, then T_2 will never be fired. Hence, the flow rate for T_1 and T_3 is k in token units, i.e. 1 in mark units, and the flow rate is 0 for T_2 and T_4.

Since a flow rate for the PN in Figure 5.16b corresponds to an instantaneous firing speed for the PN in Figure 5.16a, the expected result for this PN is $v_1 = v_3 = 1$ and $v_2 = v_4 = 0$. Intuitively, this is easy to understand. Since T_3 is strongly enabled, it is fired at its maximal speed $v_3 = V_3 = 1$. The feeding speed of place P_1 (Definition 5.4 in Section 5.1.3.2) is $I_1 = v_3 = 1$, then there is an actual conflict since $I_1 < V_1 + V_2 = 4$. Because of the priority, T_1 is fired at speed $v_1 = I_1 = 1$. The balance of place P_1 is $B_1 = v_3 - v_1 = 0$. Hence, T_2 *cannot be fired* and $v_2 = v_4 = 0$.

Example 2

This example, presented in Figure 5.17, is similar to Figure 5.16 (same priority), except the maximal speeds $V_1 = 1$ and $V_3 = 2$.

The behavior of the corresponding discrete model (for $k = 4$) is presented in Figure 5.17d. After a short transitory behavior (duration $2 / k$, which is infinitely short when k tends to infinity), the firing rate is 2 for T_3 and 1 for the other transitions. Hence, the expected result for the timed continuous PN is $\mathbf{v} = (1, 1, 2, 1)$. Here is the explanation. The flow through T_3, at speed $v_3 = 2$, feeds place P_1. Since $I_1 \geq V_1 + V_2$, there is no actual conflict, i.e. $v_1 = V_1 = 1$ and $v_2 = V_2 = 1$. The feeding speed of P_3, $I_3 = v_2 = 1$, allows the draining speed $O_3 = v_4 = 1$ (i.e. $B_3 = 0$).

❑

Let us compare Examples 1 and 2. The only difference between Figures 5.16a and 5.17a is that $V_1 > V_3$ in the first case, whereas $V_1 < V_3$ in the second case, and that this difference results in $v_2 = v_4 = 0$ in the first case whereas $v_2 = v_4 = 1$ in the second case. This is due to the priority $T_1 < T_2$.

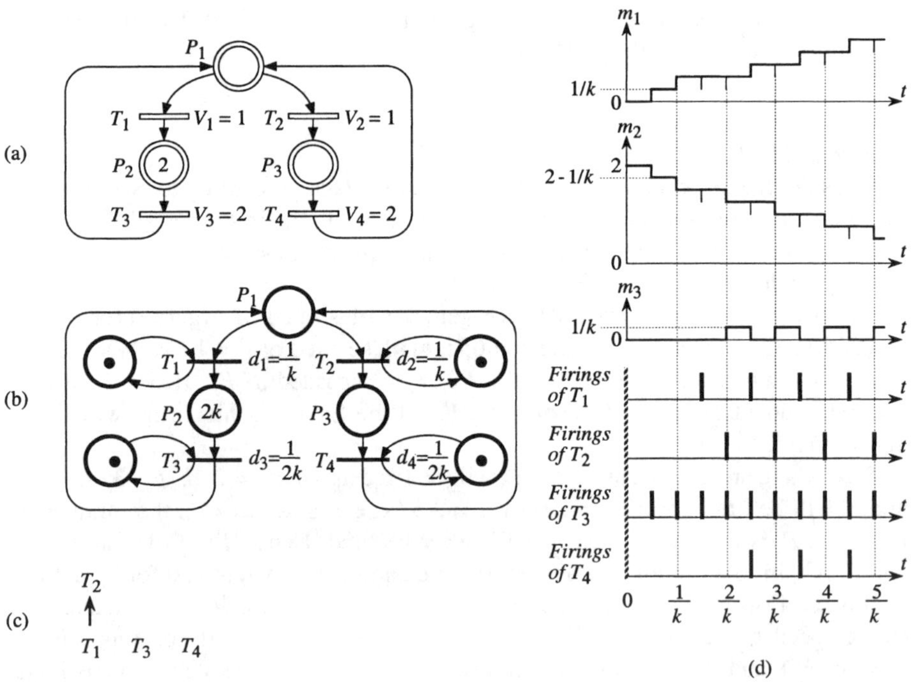

Figure 5.17 Example 2.

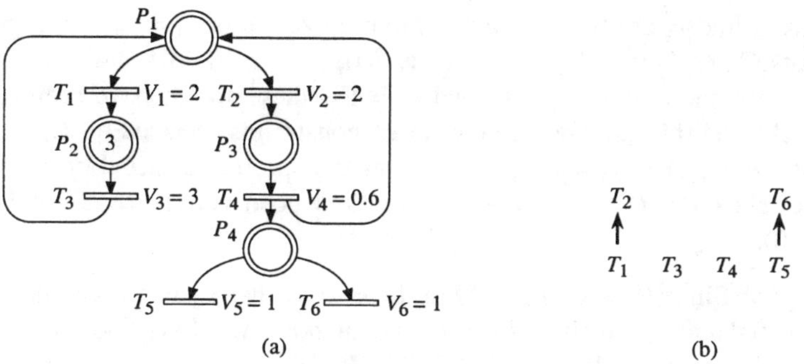

Figure 5.18 Illustration for successive passages for speed calculations.

If all the input flow I_1 to P_1 flows through T_1, then v_2 remains at value 0. This is the case for Example 1: $I_1 = v_3 = V_3 = 1$; then, since $V_1 > v_3$, $v_1 = \min(v_3, V_1) = v_3 = 1$, and $B_1 = v_3 - v_1 = 0$.

If all the input flow I_1 to P_1 does not flow through T_1, then v_2 has a positive value. This is the case[11] for Example 2: $I_1 > v_3 = V_3 = 2$; then, since $V_1 < v_3$, $v_1 = \min(v_3, V_1) = V_1 = 1$.

At the very beginning of the calculation process (in both Examples 1 and 2), it is known that $v_3 > 0$ since T_3 is strongly enabled. It follows that $I_1 > 0$ and $v_1 > 0$ since $T_1 < T_2$. In other words, it is known that T_3 and T_1 are in the set T_{SF} of *surely firable transitions* (see Section 5.3.1). From this *qualitative* analysis, one cannot know whether or not transition T_2 will be fired. This information will be available only *after calculation* of v_1 and v_3. Hence, if there are conflicts in the PN, the firing speeds will be calculated progressively and not all at once as in Algorithm 5.1 (Section 5.3.1) but in *several passages.* In the sequel, this process will be intuitively explained with an example containing two structural conflicts (one for which both transitions have a positive firing speed and one for which only the transition with the highest priority is fired).

Let us consider the example in Figure 5.18. The PN and the priority graph (corresponding to $T_1 < T_2$ and $T_5 < T_6$) are given in Figures a and b, respectively. The sets of constraints C_1 (*speed limits*) and C_2 (*non-negative balances* for empty places) are defined as in Section 5.3.1 (i.e. these constraints do not depend on the conflicts). On the other hand, the set T_{SF} of *surely firable transitions* and the set C_3 of constraints (transitions whose *speeds must remain zero*) change progressively, at each *passage*. The concept of **passage** is illustrated in the sequel; let $T_{SF}(h)$ and $C_3(h)$ denote the sets T_{SF} and C_3 at passage h.

Passage 1. Transition T_3 is strongly enabled, then $v_3 > 0$. It follows that P_1 is fed and that $v_1 > 0$ since $T_1 < T_2$. Right now, we don't know if T_2 will be fired or not, or T_4, T_5, and T_6. Hence, $T_{SF}(1) = \{T_1, T_3\}$.

[11] In the next equation, $I_1 > v_3$ will be written instead of $I_1 = v_3$ because we know that $v_2 > 0$.

Speeds v_1 and v_3 are then calculated: maximization of $J_1 = v_1 + v_3$, given the constraints C_1, C_2, and $C_3(1) = \{v_2 = 0,\ v_4 = 0,\ v_5 = 0,\ v_6 = 0\}$. Values $v_1 = 2$ and $v_3 = 3$ are obtained. Since $v_1 = V_1$ and $v_3 = V_3$, these speeds have been calculated once for all, and this provides a new set of constraints, denoted by $C_4(1)$, for the next passage: $C_4(1) = \{v_1 = 2,\ v_3 = 3\}$. From the speeds calculated up to now, the balance of place P_1, $B_1 = v_3 + v_4 - v_1 - v_2 = 1$ is positive. Hence, T_2 will be fired (i.e. $v_2 > 0$).

Passage 2. Since $B_1 > 0$, T_2 is fired. It follows that P_3 is fed and that $v_4 > 0$. Then P_4 is fed and $v_5 > 0$ since $T_5 < T_6$. Right now, we don't know if T_6 will be fired or not. Hence, $T_{SF}(2) = \{T_1, T_2, T_3, T_4, T_5\}$.

Speeds v_2, v_4, and v_5 are then calculated: maximization of $J_2 = v_2 + v_4 + v_5$, given the constraints C_1, C_2, $C_3(2) = \{v_6 = 0\}$, and $C_4(1) = \{v_1 = 2,\ v_3 = 3\}$. Values $v_2 = 1.6$ $v_4 = 0.6$, and $v_5 = 0.6$ are obtained. From the speeds calculated up to now, the balance of place P_4, $B_4 = v_4 - v_5 - v_6 = 0$. Hence, T_6 will not be fired (i.e. $v_6 = 0$).

The speed calculation is finished: $\mathbf{v} = (2, 1.6, 3, 0.6, 0.6, 0)$ at the end of passage 2. If B_4 were greater than 0 at the end of passage 2, a third passage would be necessary for calculation of v_6.

5.3.2.2 Setting Up the Set of Surely Firable Transitions

As explained in the previous section, calculation of the speed vector may require several *passages*. These successive passages correspond to increasing sets of surely firable transitions. This section shows how these sets are calculated. In the sequel, up to and including Algorithm 5.2, the first passage is mainly concerned.

(a)

$T_1 < T_2 < T_3$

$T_2 < T_4 < T_5$

$T_6 < T_7$

$T_8 < T_7$

T_9

T_{10}

(b)

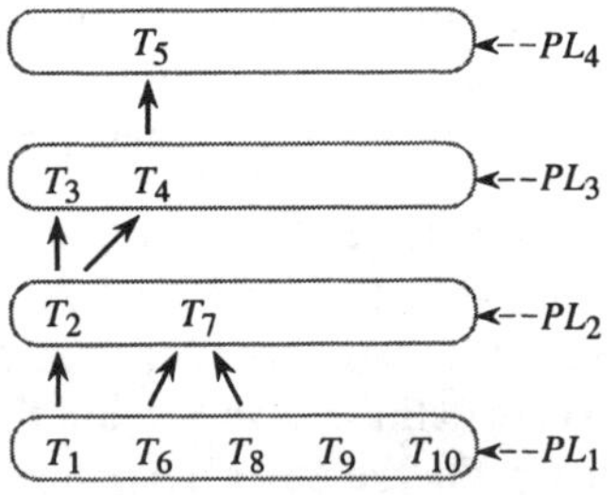

Figure 5.19 (a) Set of local resolutions. (b) Priority graph and levels.

Let us first define the **priority levels** and illustrate this concept with an example. A set of local resolutions is given in Figure 5.19a. It corresponds to a PN with the following structural conflicts: $K_1 = \langle P_1, \{T_1, T_2, T_3\}\rangle$, $K_2 = \langle P_2, \{T_2, T_4, T_5\}\rangle$, $K_3 = \langle P_3, \{T_6, T_7\}\rangle$, $K_4 = \langle P_4, \{T_7, T_8\}\rangle$; in addition, transitions T_9 and T_{10} are not involved in conflicts. The corresponding priority graph is given in Figure 5.19b. From this graph, the *sets of priority levels* $PL_1 = \{T_1, T_6, T_8, T_9, T_{10}\}$, $PL_2 = \{T_2, T_7\}$, $PL_3 = \{T_3, T_4\}$, and $PL_4 = \{T_5\}$, are obtained as follows: a transition T_j is in PL_1 if there is no transition T_k such that $T_k < T_j$; after deleting all the transitions in PL_1, PL_2 is defined similarly on the remaining set of transitions, and so on.

Consider now the example in Figure 5.20. The priority levels and several sets of transitions or places are illustrated with this example. The sets T_{se} of strongly enabled transitions, T_{SF} of surely firable transitions, and $P_{ne}^{(k)}$ of places non-empty (or fed) at *k*th step, were already used in Algorithm 5.1 (Section 5.3.1). The new **set T_{pf} of possibly firable transitions** is introduced now. Initially, it corresponds to the transitions in the first priority level which are not strongly enabled: $T_{pf} = PL_1 \setminus T_{se}$.

Algorithm 5.2, given below, leads to the set T_{SF} for the first passage[12]. It works like *Step 2* in Algorithm 5.1 (Section 5.3.1) with two differences: 1) all the transitions added to T_{SF}, after the initial value $T_{SF} = T_{se}$, belong to the first priority level, i.e. PL_1; 2) places with a 0^+ marking at the beginning of IB-state are in $P_{ne}^{(1)}$ (according to Remark 5.4 in Section 5.1.2.2, and Definitions 5.3 and 5.4 in Section 5.1.3.2, the transitions all of whose input places are in $P_{ne}^{(k)}$ are enabled).

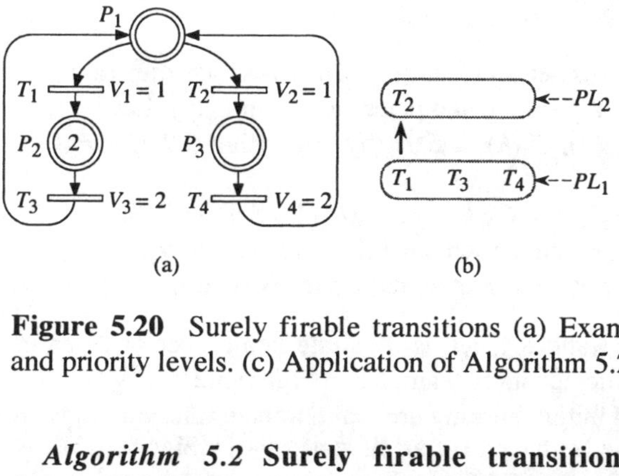

(a) (b)

Step 0. $T_1 < T_2; \tilde{\mathbf{m}} = (0, 2, 0)$.
Step 1. $T_{SF} = \{T_3\}$; $T_{pf} = \{T_1, T_4\}$. $P_{ne}^{(1)} = \{P_2\}$; $k = 1$.
Step 2. $P_{(1)} = \{P_1\}$; $P_{(4)} = \{P_3\}$.
Step 3. $k = 2$; $P_{ne}^{(2)} = \{P_1, P_2\}$.
Step 4.1. $P_{(1)} = \varnothing$; $P_{(4)} = \{P_3\}$.
Step 4.2. $T_{SF} = \{T_1, T_3\}$; $T_{pf} = \{T_4\}$.
Step 3. $k = 3$; $P_{ne}^{(3)} = \{P_1, P_2\}$.
Step 5. $J_1 = v_1 + v_3$.

(c)

Figure 5.20 Surely firable transitions (a) Example of PN. (b) Priority graph and priority levels. (c) Application of Algorithm 5.2 (leading to criterion J_1).

***Algorithm 5.2* Surely firable transitions: first criterion**

Step 0. Initialization: local conflict resolutions and marking $\tilde{\mathbf{m}}$.
Step 1. $T_{SF} = T_{se}$, $T_{pf} = PL_1 \setminus T_{se}$, $P_{ne}^{(1)} = \{P_i \mid \tilde{m}_i > 0\}$, and let $k = 1$.
Step 2. **If** $T_{pf} = \varnothing$ **then** go to *Step 5* **else** build $P_{(j)} = {}^\circ T_j$ for every T_j in T_{pf}.
Step 3. Let $k = k + 1$ and $P_{ne}^{(k)} = P_{ne}^{(k-1)} \bigcup_{T_j \in T_{SF}} T_j^\circ$.

If $P_{ne}^{(k)} = P_{ne}^{(k-1)}$ AND $k > 2$: go to *Step 5*.

Step 4.

Step 4.1. Delete all the places in $P_{ne}^{(k)}$ from the sets $P_{(j)}$.

[12] Value T_{SF} at the end of Algorithm 5.2 corresponds to $T_{SF}(1)$ according to the notation used in the previous section. But notation T_{SF} is sufficient here.

Step 4.2. For every T_j such that $P_{(j)} = \varnothing$: add T_j to T_{SF} and delete T_j from T_{pf}.
If T_{SF} has just been modified AND $T_{pf} \neq \varnothing$: go to *Step 3*.

Step 5.

$$J_1 = \sum_{T_j \in T_{SF}} v_j . \tag{5.67}$$

END.

❑

Figure 5.20c is an illustration of Algorithm 5.2 for the example in Figure 5.20a and b. Place P_1, fed by the firing of the strongly enabled transition T_3, is added to $P_{ne}^{(1)}$ to give $P_{ne}^{(2)}$ (in *Step 3*). In *Step 4*, T_1 is added to T_{SF} and deleted from T_{pf}. Since $P_{ne}^{(3)} = P_{ne}^{(2)}$ (*Step 3* again), the algorithm ends: $T_{SF} = \{T_1, T_3\}$ and $J_1 = v_1 + v_3$.

From maximization of J_1, the following result is obtained: $v_1 = V_1 = 1$ and $v_3 = V_3 = 2$. After this first passage, the second passage will conclude that T_2 and T_4 may be added to T_{SF}. This modification of T_{SF} will be obtained by Algorithm 5.3 (presented later) which gives $J_2 = v_2 + v_4$ (leading to $v_2 = v_4 = 1$). Let us now specify the *structure* of Algorithm 5.4 (which will be detailed in Section 5.3.2.3).

Let us recall that there are four sets of constraints (see the previous section): C_1 (*speed limits*) and C_2 (necessary *positive balances*) do not change during the speed calculation process; C_3 (*zero speed* for some transitions) and C_4 (*known speed values*) change at each passage.

Structure of Algorithm 5.4.

Step A. Initialization for IB-state s. Set up C_1 and C_2. Then find the transitions T_j such that $v_j = 0$ is implied by C_1 and C_2, and place '$v_j = 0$' in $C_4(0)$. Let $h = 1$.

Step B. Passage h. Calculate $T_{SF}(h)$, $C_3(h)$, and J_h (by Algorithm 5.2 if $h = 1$ or Algorithm 5.3 if $h > 1$).
Maximize J_h, given C_1, C_2, $C_3(h)$, and $C_4(h-1)$. Set up $C_4(h)$.
If another passage is required **then** $h = h + 1$ and iterate *Step B*.

Step C. Ending. Calculate the various parameters related to IB-state s.

❑

Before giving Algorithms 5.3 and 5.4, let us illustrate some special cases of behavior which must be taken into account: examples in Figure 5.21.

Even if the components of an initial marking are real numbers, the markings in Figure 5.21a and d can be obtained at the end of an IB-state: $\tilde{m}_2$ in Figure 5.21a is similar to $\tilde{m}_4$ in Figure 5.4a (Section 5.1.2.1); the way to reach the marking in Figure 5.21d will be explained in Section 5.4.1.2. Let us analyze first the example in Figure d, then the others. Algorithm 5.3 has not yet been given; however, the results expected from this algorithm (increase of T_{SF}) are presented in the sequel. For all examples in Figure 5.21 except Figure a, there is a structural conflict $K_c = \langle P_c, \{T_a, T_b\}\rangle$ and a priority $T_a < T_b$.

Figure d. Algorithm 5.2 gives $T_{SF} = \{T_6, T_a\}$ and $J_1 = v_6 + v_a$. Given the constraints denoted by C_1, C_2, and $C_3 = \{v_5 = 0, v_b = 0\}$, the values $v_6 = 1$ and $v_a = 1$ are obtained. All the balances are zero. This is the final result.

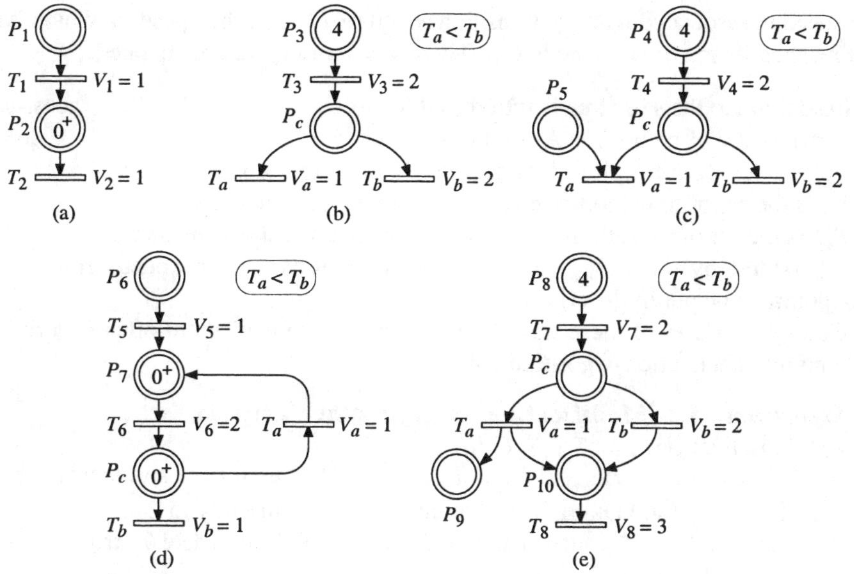

Figure 5.21 Illustrative examples.

Figure a. Given the constraints $v_1 \geq 0$ in C_1, and $B_1 = - v_1 \geq 0$ and $B_2 = v_1 - v_2 \geq 0$ in C_2, the values $v_1 = v_2 = 0$ are obtained in *Step A* presented in the structure of Algorithm 5.4. All the balances are zero and this is the final result. However, Algorithm 5.2 gives $T_{SF} = \{T_2\}$ and $J_1 = v_2$. This is a special case where T_2 is enabled and belongs to T_{SF}, whereas $v_2 = 0$! At the very beginning of the IB-state, T_2 is enabled. But P_2 *is immediately emptied by this transition*: if t denotes the time when the system passes from the previous IB-state to the present IB-state, the marking changes from $\tilde{m}_2(t) = 0^+$ to $\tilde{m}_2(t^+) = 0$. This is consistent with Property 5.4 (Section 5.1.3.3) and the comments which follow.

Figure b. From Algorithm 5.2, $T_{SF} = \{T_3, T_a\}$ and $J_1 = v_3 + v_a$. The values $v_3 = 2$ and $v_a = 1$ are obtained in passage 1. In Algorithm 5.3, T_{SF} is extended to $T_{SF} = \{T_3, T_a, T_b\}$ and $J_2 = v_b$ is the second criterion to be maximized. Given the constraints, particularly $C_4(1) = \{v_3 = 2, v_a = 1\}$, $v_b = 1$ is obtained in passage 2.

Figure c. Algorithm 5.2: $T_{SF} = \{T_4\}$ and $J_1 = v_4$ lead to $v_4 = 2$ (T_a is not enabled since $\tilde{m}_5 = 0$, and T_b is not yet in T_{SF} since $T_a < T_b$). In passage 2, $T_{SF} = \{T_4, T_b\}$ and $J_2 = v_b$, then $v_b = 2$, are obtained (although $T_a < T_b$).

Figure e. Algorithm 5.2: $T_{SF} = \{T_7, T_a, T_8\}$ and $J_1 = v_7 + v_a + v_8$ lead to $v_7 = 2$, $v_a = 1$, $v_8 = 1$. Then $B_c > 0$, Algorithm 5.3 will be used. Note that $v_7 = 2 = V_7$, $v_a = 1 = V_a$, but $v_8 = 1 < V_8$. Then, $C_4(1) = \{v_7 = 2, v_a = 1, v_8 \geq 1\}$: at the first passage, the minimum value $v_8 = 1$ was obtained but this value could be increased. Algorithm 5.3 will give $T_{SF} = \{T_7, T_a, T_8, T_b\}$ and $J_2 = v_b + v_8$ (sum of the speed v_b associated with the transition added to T_{SF} and the speed v_8 which has not reached its maximal value in C_4). Then, $v_b = 1$ and $v_8 = 2$ are obtained.

❑

In addition to the sets already introduced (C_1 to C_4, T_{SF}, T_{pf}, P_{ne}, PL_h), new sets of transitions will be used in Algorithms 5.3 and 5.4:

$T(J_h)$ is the set of transitions whose speeds are in criterion J_h;

T_{nc} is the set of transitions whose speeds are *not yet calculated*;

T_{ndc} is the set of transitions whose speeds are *not definitely calculated*;

T_{sby} (stand by) and T_{wait} are sets of transitions whose speeds must not (temporarily) be put in the criterion J_h.

Examples of uses of these sets will be presented, some of them after Algorithm 5.3 and the others after Algorithm 5.4.

***Algorithm 5.3* Modification of T_{SF}: other criteria**

Step 0. Initialization: T_{SF}, T_{pf}, h, $C_4(h)$.

Step 1. Let $T_{pf} = \{T_{pf} \cup PL_{h+1} \cup T_{sby}\} \setminus T_{SF}$, $P_{ne}^{(1)} = P_{ne}^{(k)}$, then $k = 1$ and empty T_{sby}.

Step 2. **If** $T_j \in T_{pf}$ (and taking into account the speeds currently in $C_4(h)$):

1) **If** there is $P_i \in {}^{\circ}T_j$ such that $I_i > 0$ and $B_i = 0$ **then** delete T_j from T_{pf} and from T_{nc} and add it to T_{sby}.

2) **If** there is $P_i \in {}^{\circ}T_j$ such that $\tilde{m}_i = 0$ and $B_i = 0$ **then** delete T_j from T_{nc}.

Steps 3, 4, and *5.* Similar to *Steps 2, 3,* and *4* in Algorithm 5.2.

Step 6. Let $T(J_{h+1}) = \{T_{SF} \cap T_{ndc}\} \setminus T_{wait}$. Then, **if** several transitions in $T(J_{h+1})$ are involved in the same structural conflict, **then** delete all these transitions from $T(J_{h+1})$ except the one corresponding to the lower index c of PL_c. **If** T_j has just been deleted from $T(J_{h+1})$, **then** delete from $T(J_{h+1})$ each transition whose firing is conditioned by firing of T_j, and iterate.

Step 7.

$$J_{h+1} = \sum_{T_j \in T(J_{h+1})} v_j . \tag{5.68}$$

END.

❑

Let us comment on Algorithm 5.3.

In *Step 1*, the transitions corresponding to the next priority level are added.

Step 6 defines the set of transitions which must be taken into account in the next criterion. For the example in Figure 5.21e, after maximization of $J_1 = v_7 + v_a + v_8$, T_8 is in $T(J_2)$ because its value may still increase (it is out of T_{nc} but in T_{ndc} since $v_8 \geq 1$ in $C_4(1)$). In some cases, several transitions in conflict may appear in $T(J_{h+1})$: only one is kept. For example, if Figure 5.21b were such that $V_3 < V_a$, $T(J_2) = \{T_a, T_b\}$ would be found at the beginning of *Step 6* in Algorithm 5.3; then T_b would be deleted. See Exercise 5.5.

Remark 5.11 All the transitions in the set T_{SF} obtained by Algorithm 5.2 are qualified "surely firable": their speeds have a positive value during the IB-state, or they are at least firable during an infinitely short time (like T_2 in Figure 5.21a). However, for the transitions in T_{SF} extended by Algorithm 5.3, the expression "surely firable" is an abuse of language: in some cases, the calculated firing speed may be zero because of priorities.

5.3.2.3 Algorithm And Application

Algorithm 5.4, corresponding to one IB-state, will be presented in the sequel. An application of this algorithm will be given further on. The heart of the algorithm is the resolution of a *linear programming problem* (*Step 4*) which is the maximization of a criterion J_h, given the four sets of constraints denoted by C_1 (*speed limits* related to the model), C_2 (*necessary positive balances*), C_3 (*nul speed for a transition, as long as it is not surely firable*), and C_4 (*speeds whose values have already been calculated*).

The *balances are important* for two purposes.

First purpose. If $m_i = 0$ (i.e., $\tilde{m}_i = 0$ or $\tilde{m}_i = 0^+$), it is necessary that $B_i \geq 0$ because the marking of place P_i cannot become negative (constraints C_2).

Second purpose. If $B_i < 0$, then m_i decreases. If the IB-state lasts long enough, place P_i will become empty and a new IB-state will start. Hence, the change of IB-state occurs when the first place whose balance is negative becomes empty. This is specified in the following property.

Property 5.5 In a timed continuous PN, a *change of IB-state* can occur only when an **event** of the following type occurs: the *marking* of a marked place P_i *becomes zero*, i.e. this *C1-event* occurs at t_1 if $m_i(t_1) = 0$ while $m_i(t_1 - dt) > 0$.
❑

A change of IB-state in a *timed continuous PN* corresponds to a change of macromarking in the corresponding *autonomous continuous PN*. According to Property 4.2 (Section 4.1.2), a macromarking can change if a C1-event or a C2-event occurs. A C1-event occurs if the *marking* of a marked place *becomes zero*. It corresponds to an event defined in Property 5.5. A C2-event occurs when an unmarked place *becomes marked*. In a *timed* continuous PN this event cannot occur alone: it may occur at t_1^+ if another marking became zero at t_1 or if $t_1 = 0$ is the initial time, hence the C2-event is not the cause of the beginning of the IB-state (illustratory examples are given for timed hybrid PNs in which a similar property is observed: Section 6.1.2). Let us now specify Algorithm 5.4.

Algorithm 5.4 **Resolution by priorities**

Step 0. Initialization: $\tilde{\mathbf{m}}$, local conflict resolutions, H = number of priority levels, $h = 1$; set of constraints C_1 corresponding to $0 \leq v_j \leq V_j$.

Step 1.

Step 1.1. Set up the constraints C_2 based on the balances ($B_i \geq 0$ if $\tilde{m}_i \leq 0^+$), $C_3 = \{v_j = 0 \mid \text{for every } T_j \in T\}$, $C_4(0) = \varnothing$, $T_{nc} = T_{ndc} = T$, $T_{sby} = T_{wait} = \varnothing$.

Step 1.2. Find the transitions T_j such that $v_j = 0$ from C_1 and C_2. Add '$v_j = 0$' to $C_4(0)$ and delete T_j from T_{nc} and from T_{ndc}.

Step 2. Obtaining T_{SF} and J_1 by Algorithm 5.2 (without its *Step 0*).

Step 3. If '$v_j = 0$' is in C_3, delete it if $T_j \in T_{SF}$.

Step 4.

Step 4.1. Solve the LPP: maximize J_h given C_1, C_2, C_3, and $C_4(h-1)$.

Step 4.2. For every T_j in $T(J_h)$, delete T_j from T_{nc}.

Step 5.

Step 5.1. For every T_j in T_{SF} such that $v_j = 0$:

1) for every $P_i = {}^\circ T_j$ such that $\tilde{m}_i = 0^+$ AND $V_j > 0$ AND T_j was not deleted in *Step 5* of Algorithm 5.3, replace $\tilde{m}_i = 0^+$ by $\tilde{m}_i = 0$ and delete P_i from P_{ne};
2) delete T_j from T_{SF} and add T_j to T_{pf}.

Step 5.2. Set up the set $C_4(h)$:

1) every component of $C_4(h-1)$ corresponding to a transition in $T(C_4(h-1)) \setminus T(J_h)$ is kept in $C_4(h)$;
2) for each $T_j \in T(J_h)$, **if** the value $v_j = x_j$ calculated in *Step 4.1* is such that $x_j = V_j$ **then** '$v_j = x_j$' is placed in $C_4(h)$ **else** '$v_j \geq x_j$' is placed in $C_4(h)$;
3) for each $T_j \notin T(J_h)$, **if** $v_j = x_j > 0$ is obtained **then** '$v_j \geq x_j$' or '$v_j = x_j$' (if $v_j = V_j$) is placed in $C_4(h)$.

Step 6.

Step 6.1. Delete from T_{ndc} transitions T_j such that '$v_j = x_j$' in $C_4(h)$.

Step 6.2. **If** v_j is in the criterion for the second time AND $C_4(h) = C_4(h-1)$ **then** place T_j in T_{wait}.

Step 6.3. **if** $C_4(h) \neq C_4(h-1)$ **then** $T_{wait} = \varnothing$.

Step 6.4. **if** (T_{pf} AND T_{ndc} are the same as during the last passage in this step AND $h \geq H$ AND T_{nc} is empty) **then** go to *Step 8*..

Step 7.

Step 7.1. Calculation of T_{SF} and J_{h+1} by Algorithm 5.3 (without its *Step 0*).

Step 7.2. Let $h = h + 1$.

Step 7.3. **if** ($T(J_h) \neq T(J_{h-1})$ AND $T(J_h) \neq \varnothing$) OR $h - 1 < H$ **then** go to *Step 3* **else** **if** $T_{ndc} = \varnothing$ **then** go to *Step 8*.

Step 7.4. Let $C_4(h) = C_4(h-1)$ and go to *Step 6*.

Step 8. Set the instantaneous speed vector $\mathbf{v}$ for the IB-state: **if** $T_j \in T(C_4(h))$ **then** $v_j = x_j$ **else** $v_j = 0$.

Step 9.

Step 9.1. Calculate $\tilde{\mathbf{m}}(t)$ and $\mathbf{m}(t)$ for the IB-state.

Step 9.2. Calculate the duration Δt of the IB-state (shorter time such that a component of $\mathbf{m}(t)$ becomes 0), then time t_s of the end of this IB-state s.

Step 9.3. Calculate $\tilde{\mathbf{m}}(t_s)$. END.

❑

Steps A to *C* of the *structure* presented in Section 5.3.2.2 correspond to Algorithm 5.4 in the following way. *Step A* corresponds to *Steps 0* and *1*. *Step B* for $h = 1$ begins with *Step 2* and ends with *Step 6*. *Step B* for $h > 1$ begins with *Step 7*, then *Steps 3* to *6*. *Step C* is performed by *Steps 8* and *9*.

For the example in Figure 5.21a, speeds $v_1 = v_2 = 0$ are obtained in *Step 1.2*. Other consequences of such a case are drawn in *Step 5.1*, particularly emptying P_2: $\tilde{m}_2$ becomes 0. For the examples in Figures 5.21b to e, the reader may verify that Algorithm 5.4 satisfies the explanations (given in Section 5.3.2.2) related to these examples.

Algorithm 5.4 ends in *Steps 8* and *9* if either nothing changes and T_{nc} is empty (test in *Step 6.4*) OR all the speeds have been definitely calculated, i.e. T_{ndc} empty (test in *Step 7.3*). The values $\tilde{m}_i(t_s) = 0^+$, for some places P_i, at the end of the IB-state, are established in *Step 9.1*. The way to assign these values 0^+ is specified in Property 5.4 (Section 5.1.3.3).

Example. This example is presented in Figure 5.22: IB-state 1, from $\mathbf{m}_0$, given the local resolutions $T_4 < T_3 < T_2$ for the structural conflict $\langle P_2, \{T_2, T_3, T_4\}\rangle$ and $T_4 < T_3$ for $\langle P_3, \{T_3, T_4\}\rangle$ (priority graph in Figure b). The results obtained at each step of Algorithm 5.4 are given in Figure c (steps where nothing is done are omitted).

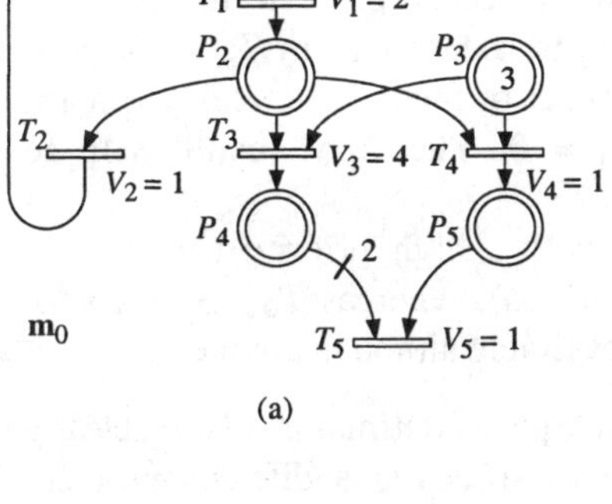

(a)

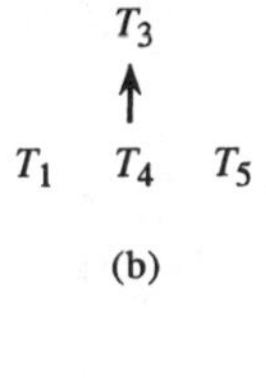

(b)

Step 0. $\tilde{\mathbf{m}} = \mathbf{m}_0 = (6, 0, 3, 0, 0)$; $T_4 < T_3 < T_2$; $h = 1$.
$C_1 = \{0 \le v_1 \le 2, 0 \le v_2 \le 1, ..., 0 \le v_5 \le 1\}$.
Step 1. $C_2 = \{v_1 - v_2 - v_3 - v_4 \ge 0, v_3 - 2v_5 \ge 0, v_4 - v_5 \ge 0\}$;
$C_3 = \{v_1 = 0, v_2 = 0, v_3 = 0, v_4 = 0, v_5 = 0\}$; $C_4(0) = \varnothing$.
Step 2. $T_{SF} = \{T_1, T_4\}$; $T_{pf} = \{T_5\}$; $J_1 = v_1 + v_4$.
Step 3. $C_3 = \{v_2 = 0, v_3 = 0, v_5 = 0\}$;
Step 4. $v_1 = 2, v_4 = 1$; $T_{nc} = \{T_2, T_3, T_5\}$.
Step 5.2. $C_4(1) = \{v_1 = 2, v_4 = 1\}$.
Step 6.1. $T_{ndc} = \{T_2, T_3, T_5\}$.
Step 7.1. $T_{SF} = \{T_1, T_3, T_4, T_5\}$; $T_{pf} = \varnothing$; $J_2 = v_3 + v_5$.
Step 7.2. $h = 2$.
Step 7.3. Go to *Step 3*.
Step 3. $C_3 = \{v_2 = 0\}$.
Step 4. $v_3 = 1, v_5 = 0.5$; $T_{nc} = \{T_2\}$.
Step 5.2. $C_4(2) = \{v_1 = 2, v_3 \ge 1, v_4 = 1, v_5 \ge 0.5\}$.
Step 6.1. $T_{ndc} = \{T_2, T_3, T_5\}$.
Step 7.1. $T_{SF} = \{T_1, T_3, T_4, T_5\}$; $T_{pf} = \varnothing$; $T_{sby} = \{T_2\}$; $J_3 = v_3 + v_5$.
Step 7.2. $h = 3$.
Step 7.3. Go to *Step 3*.
Step 3. $C_3 = \varnothing$.
Step 4. $v_3 = 1, v_5 = 0.5$; $T_{nc} = \{T_2\}$.
Step 5.2. $C_4(3) = \{v_1 = 2, v_3 \ge 1, v_4 = 1, v_5 \ge 0.5\}$.
Step 6.1. $T_{ndc} = \{T_2, T_3, T_5\}$.
Step 6.2. $T_{wait} = \{T_3, T_5\}$.
Step 7.1. $T_{SF} = T$; $T_{pf} = \varnothing$; $J_4 = v_2$.
Step 7.2. $h = 4$.
Step 7.3. Go to *Step 3* (then *Step 4*).
Step 4. $v_2 = 0$; $T_{nc} = \varnothing$.
Step 5.2. $C_4(4) = \{v_1 = 2, v_2 \ge 0, v_3 \ge 1, v_4 = 1, v_5 \ge 0.5\}$.
Step 6.1. $T_{ndc} = \{T_2, T_3, T_5\}$.
Step 6.3. $T_{wait} = \varnothing$.
Step 6.4. Go to *Step 8*.
Step 8. $\mathbf{v} = (2, 0, 1, 1, 0.5)$.
Step 9.1. $\tilde{m}_1 = m_1 = 6 - 2\,t$.
$\tilde{m}_2 = 0^+, m_2 = 0$.
$\tilde{m}_3 = m_3 = 3 - 2\,t$.
$\tilde{m}_4 = 0^+, m_4 = 0$.
$\tilde{m}_5 = m_5 = 0.5\,t$.
Step 9.2. $\Delta t = 1.5$ (m_3 becomes 0); $t_1 = 1.5$.
Step 9.3. $\tilde{\mathbf{m}}_1 = \tilde{\mathbf{m}}(1.5) = (3, 0^+, 0, 0^+, 0.75)$.

(c)

Figure 5.22 Illustration of Algorithm 5.4: resolution by priorities.

The calculation requires four passages.

The *first passage*, from *Step 2* to *Step 6.1* (first time), is performed with the set of *surely firable transitions* $T_{SF} = \{T_1, T_4\}$ and $J_1 = v_1 + v_4$.

The *second passage*, from *Step 7.1* (first time) to *Step 6.1* (second time), is performed with $T_{SF} = \{T_1, T_3, T_4, T_5\}$ and $J_2 = v_3 + v_5$.

The *third passage*, from *Step 7.1* (second time) to *Step 6.1* (third time), is performed with $T_{SF} = \{T_1, T_3, T_4, T_5\}$ and $J_3 = v_3 + v_5$.

The *fourth passage*, from *Step 7.1* (third time) to *Step 6.1* (fourth time), is performed with $T_{SF} = T$ and $J_4 = v_2$.

Let us note that $J_3 = J_2$ and that the same results ($v_3 = 1$ and $v_5 = 0.5$) are obtained at the second and third passages. The usefulness of T_{wait} appears in this case: T_3 and T_5 are put into T_{wait} and the speed of T_2 (with a lower priority than T_3) can then be calculated at the next passage. In *Step 6.3* of the fourth passage, T_{wait} is emptied; in some cases (but not in this example) a greater speed can be obtained later for v_3 or v_5. In the present example, the conditions in *Step 6.4* are satisfied at the *fourth passage*, particularly $T_{nc} = \varnothing$ (all speeds have been calculated), hence the speed calculation is over.

For the example in Figure 5.23, at the first passage, $T_{SF} = \{T_1, T_3\}$ gives $J_1 = v_1 + v_3$ which leads to $v_1 = 1$ and $v_3 = 1$. At the second passage, in *Step 7.1*, Algorithm 5.3 is performed. In this algorithm, T_2 is added to T_4 in T_{pf} in *Step 1*, then transfered from T_{pf} to T_{sby} in *Step 2*. As a matter of fact, T_2 is weakly enabled but cannot be fired because $T_1 < T_2$ and $B_1 = 0$. The final result will be $\mathbf{v} = (1, 0, 1, 0)$.

Let us note that T_{wait} is used for transitions with a high priority (in order to calculate speeds of transitions with lower priorities), whereas T_{sby} is used for transitions with a low priority (which may be not firable although enabled).

Remark 5.12 Although the behavior of a timed continuous PN is clearly defined, automatic calculation of the instantaneous speeds is a difficult problem when there are actual conflicts. Thanks to the various sets of places and transitions used in Algorithm 5.4, this algorithm is able to calculate the speeds in usual cases. Examples in which the priorities are *not exactly* satisfied by this algorithm can be built. However, the result obtained is always consistent with the definition of a timed continuous PN. This is illustrated in Exercise 5.10.

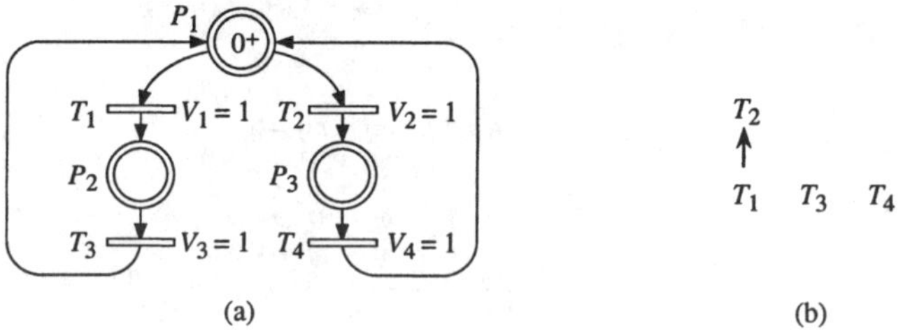

Figure 5.23 Example using T_{sby}.

5.3.3 Resolution By Sharings And Priorities

In this section, resolutions mixing sharings and priorities will be presented. A resolution by sharing is easy to understand (Section 5.2.2). However, it is more difficult to calculate automatically than the resolution by priorities.

Sharing between two transitions, anywhere in the PN, is first presented in Section 5.3.3.1. Single or multiple sharing among three or more transitions, if some restrictive conditions are satisfied, is then proposed in Section 5.3.3.2. Finally, an algorithm for a more general case, using several times the cases studied in Sections 5.3.3.1 and 5.3.3.2, is given in Section 5.3.3.3.

5.3.3.1 Single Sharing Between Two Transitions

In this section, it is assumed that there is a single sharing between two transitions, anywhere in the PN, whereas all the other conflicts are solved by priorities. A case of expected behavior is illustrated by Example 3.

Example 3

This example is presented in Figure 5.24. There is a sharing between T_1 and T_2 as shown by the priority graph in Figure c. The behavior of the discrete PN in Figure b is given in Figure d (for $k = 4$), assuming that the sharing in the discrete model is performed thanks to an alternate firing of T_1 and T_2.

In Figure d, one can observe that, after a short transitory behavior, the firing rate is 1 for every transition. Hence, the expected result for the timed continuous PN is $v_1 = v_2 = v_3 = v_4 = 1$. Here is the explanation. The flow through T_3, at speed $v_3 = 1$, feeds place P_1 and is shared between T_1 and T_2. Since P_3 is fed at speed v_2, T_4 can be fired at speed $v_4 = \max(v_2, V_4)$. Now, the feeding speed of P_1 is $I_1 = v_3 + v_4$, which is shared such that $v_1 + v_2 = I_1 = v_3 + v_4$. Since $v_3 = V_3 = 1$ and $v_4 \leq V_4 = 1$, the solution satisfying all the constraints is $v_1 = v_2 = v_3 = v_4 = 1$. (Note that $I_1 = 2$ in this case, whereas $I_1 = 1$ if $T_1 < T_2$, i.e. I_1 depends on the resolution because there are loops from P_1 to itself.)

❑

The hypothesis in this section is that there is a *single* sharing between *two transitions*. Then, if the concerned structural conflict implies N transitions ($N \geq 2$), $N - 1$ priority levels are necessary for this structural conflict. For example, for $K_1 = \langle P_1, \{T_1, T_2, T_3\}\rangle$ ($N = 3$), $[T_1, T_2] < T_3$ and $T_3 < [2T_1, T_2]$ are two possible 2-priority level local resolution rules.

The speed vector will be calculated by Algorithm 5.5 given below. The basic idea is as follows. If there is sharing between T_a and T_b (with the common place P_c), speeds vectors are first calculated, successively, assuming that $T_a < T_b$ then assuming that $T_b < T_a$. Afterwards, from the obtained results, the speed vector $\mathbf{v}$ is calculated. Let us now present notations used in Algorithm 5.5.

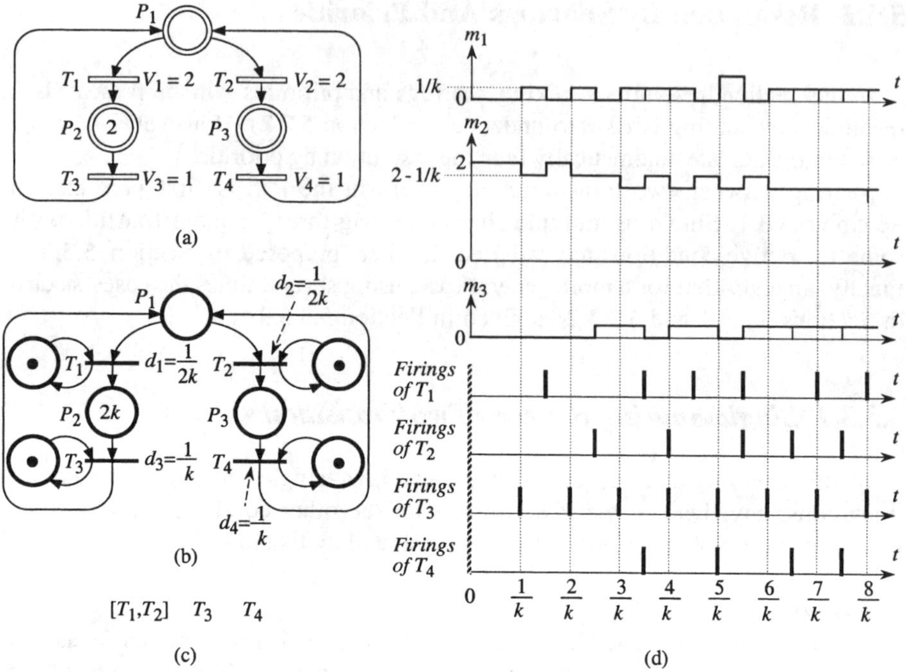

Figure 5.24 Example 3: sharing $[T_1, T_2]$.

Notation 5.5 Let $[\alpha_a T_a, \alpha_b T_b]$ be a sharing corresponding to the structural conflict $K = \langle P_c, \{T_a, T_b\}\rangle$.

1) For the priority $T_a < T_b$, the instantaneous speeds of T_a and T_b, the speed vector, and the balance obtained are denoted by $v_a^{(a)}$, $v_b^{(a)}$, $\mathbf{v}^{(a)}$, and $B_c^{(a)}$.

2) For the priority $T_b < T_a$, the instantaneous speeds of T_a and T_b, the speed vector, and the balance obtained are denoted by $v_a^{(b)}$, $v_b^{(b)}$, $\mathbf{v}^{(b)}$, and $B_c^{(b)}$.

❑

Once the values $v_a^{(a)}$, $v_b^{(a)}$, $B_c^{(a)}$, $v_a^{(b)}$, $v_b^{(b)}$, and $B_c^{(b)}$ are known, v_a and v_b are obtained from Algorithm 5.5. It seems reasonable that the instantaneous speed associated with a transition is maximal when this transition takes priority over the other: $v_a^{(a)} \geq v_a^{(b)}$, for example. However, as explained in Appendix K, this is not always true. The various cases considered in Algorithm 5.5 are explained in this appendix.

***Algorithm 5.5* One sharing between T_a and T_b**

Step 0. Input data: the timed continuous PN, $\tilde{\mathbf{m}}$, C-transitions T_a and T_b and common input C-place P_c, coefficients α_a and α_b. Priorities between other transitions in conflict.

Step 1.

Step 1.1. Modify the resolution rule: $T_a < T_b$.

Step 1.2. Algorithm 5.4 up to *Step 8*: calculation of $v_a^{(a)}$, $v_b^{(a)}$, $\mathbf{v}^{(a)}$, $B_c^{(a)}$.
If $m_c > 0$ **then** $\mathbf{v} = \mathbf{v}^{(a)}$ and go to *Step 7.*

Step 2.
Step 2.1. Modify the resolution rule: $T_b < T_a$.
Step 2.2. Algorithm 5.4 up to *Step 8*: calculation of $v_a^{(b)}$, $v_b^{(b)}$, $\mathbf{v}^{(b)}$, $B_c^{(b)}$.

Step 3.
If $v_b^{(a)} > v_b^{(b)}$ **then** $\mathbf{v} = \mathbf{v}^{(a)}$ and go to *Step 7*
else if $v_a^{(b)} > v_a^{(a)}$ **then** $\mathbf{v} = \mathbf{v}^{(b)}$ and go to *Step 7.*

Step 4.
If $v_a^{(a)} / \alpha_a \leq v_b^{(a)} / \alpha_b$ OR ($v_b^{(b)} = v_b^{(a)}$ AND ($v_a^{(b)} = v_a^{(a)}$ OR $B_c^{(a)} > 0$))
then $\mathbf{v} = \mathbf{v}^{(a)}$ and go to *Step 7*
else if $v_b^{(b)} / \alpha_b \leq v_a^{(b)} / \alpha_a$ OR ($v_a^{(a)} = v_a^{(b)}$ AND $B_c^{(b)} > 0$)
then $\mathbf{v} = \mathbf{v}^{(b)}$ and go to *Step 7*
else if $v_a^{(a)} = v_a^{(b)}$ **then** $v_a = v_a^{(a)}$ and $v_b = \alpha_b v_a / \alpha_a$, and go to *Step 6*
else if $v_b^{(b)} = v_b^{(a)}$ **then** $v_b = v_b^{(b)}$ and $v_a = \alpha_a v_b / \alpha_b$, go to *Step 6.*

Step 5. Place T_a and T_b in the same priority level. Then, Algorithm 5.4 up to *Step 8* with the additional constraint $v_a / \alpha_a = v_b / \alpha_b$. Go to *Step 7.*

Step 6. Algorithm 5.4 up to *Step 8* with the additional constraint: values v_a and v_b are fixed and put into $C_4^{(0)}$; $v_a = 0$ and $v_b = 0$ are deleted from C_3; T_a and T_b are deleted from T_{ndc} and T_{nc}.

Step 7. Perform *Step 9* of Algorithm 5.4. END.

5.3.3.2 One or Several Sharings Among Transitions

In this section, an algorithm will be proposed to solve a conflict among several transitions, for *any resolution rule*, if some hypotheses are verified. Informally, the main hypothesis is that the *data for the resolution do not depend on the result of the resolution.* In other words, there is no feedback from the transitions involved in the conflict to themselves: it is then possible to calculate the speeds upstream from these transitions, then the speeds of these transitions, and finally the speeds downstream from these transitions. This is formalized in Hypothesis 5.2a illustrated in Figure 5.25. In addition, according to Hypothesis 5.2b, no transition is involved in two sharings.

Hypothesis 5.2
a) The whole PN can be divided into three subnets $UP(K_F)$, K_F, and $DOWN(K_F)$, such that (illustration in Figure 5.25):

1) $K_F = \langle P_i, T(K_F)\rangle$ denotes a structural conflict, where $T(K_F) = \{T_1, T_2, \ldots, T_s\}$ contains at least two C-transitions.
2) There is no arc from a component of K_F to a component of $UP(K_F)$. There is no arc from a transition in $T(K_F)$ to P_i, and no arc from P_i to $DOWN(K_F)$.

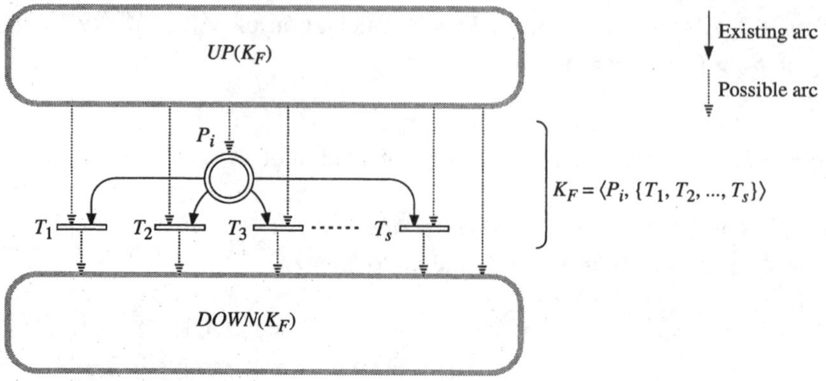

Figure 5.25 Illustration of Hypothesis 5.2a.

3) There is no arc from a component of $DOWN(K_F)$ either to a component of K_F or to a component of $UP(K_F)$.

b) Let $P(K_F) = {}^\circ T(K_F)$ denote the set of input places of the transitions in $T(K_F)$. The only place in $P(K_F)$ whose feeding speed is split by sharing[1] is P_i.

❑

Let us now introduce the concept of **temporary balance** of a place P_i, denoted by TB_i, and present intuitively the calculation process. A temporary balance has no meaning during an IB-state, it is used *during the calculation* of the instantaneous speeds of transitions involved in a conflict.

According to Notation 5.4 in Section 5.2.2, a sharing among the transitions in $T(K) = \{T_1, T_2, ..., T_s\}$ is denoted by $[\alpha_1 T_1, \alpha_2 T_2, ..., \alpha_s T_s]$; this sharing *aims at* $v_1 / \alpha_1 = v_2 / \alpha_2 = ... = v_s / \alpha_s$. Let us consider the example in Figure 5.26. There is an actual conflict since $V_1 + V_2 + V_3 + V_4 < I_5 = 3$. Several possible resolution rules will be successively considered, namely, LR_1: $T_1 < [T_2, T_3, T_4]$; LR_2: $[T_1, T_2] < [3T_3, T_4]$; LR_3: $[T_1, T_2, T_3] < T_4$; LR_4: $T_1 < T_4 < [T_2, T_3]$.

The *calculation process* for obtaining the instantaneous firing speeds is as follows. Initially $v_1 = v_2 = v_3 = v_4 = 0$. Since P_5 is fed, its *temporary balance* (as long as the value 0 is assigned to the speeds v_1 to v_4) is $TB_5^{(0)} = I_5 = 3$.

Resolution rule LR_1: $T_1 < [T_2, T_3, T_4]$. Since T_1 takes priority over the other transitions and $V_1 < TB_5^{(0)}$, $v_1 = V_1 = 1$ is assigned to this transition. Now, given $v_1 = 1$, the temporary balance is calculated again: $TB_5^{(1)} = I_5 - v_1 = 2$. Transitions T_2, T_3, and T_4 belong to the second priority level and a solution aiming at $v_2 = v_3 = v_4$ is sought. It follows that $v_2 = v_3 = v_4 = TB_5^{(1)} / 3 = 0.67$ is obtained.

Resolution rule LR_2: $[T_1, T_2] < [3T_3, T_4]$. Since T_1 and T_2 take priority over the other transitions and $V_1 + V_2 < TB_5^{(0)}$, $v_1 = V_1 = 1$ and $v_2 = V_2 = 1$ are assigned to these transitions. Now, given $v_1 = v_2 = 1$, the temporary balance is calculated again:

[1] Other places in $P(K)$ may have feeding speeds split by priorities.

$TB_5^{(1)} = I_5 - v_1 - v_2 = 1$. Transitions T_3 and T_4 belong to the second priority level and a solution aiming at $v_3 / 3 = v_4$ is sought. It follows that $v_3 = 0.75\ TB_5^{(1)} = 0.75$ and $v_4 = 0.25\ TB_5^{(1)} = 0.25$ are obtained.

Resolution rule LR_3: $[T_1, T_2, T_3] < T_4$. Since T_1, T_2, and T_3 belong to the first priority level and $V_1 + V_2 + V_3 \leq TB_5^{(0)}$, $v_1 = V_1 = 1$, $v_2 = V_2 = 1$, and $v_3 = V_3 = 1$, are assigned to these transitions. Now, given $v_1 = v_2 = v_3 = 1$, the temporary balance is calculated again: $TB_5^{(1)} = I_5 - v_1 - v_2 - v_3 = 0$. It follows that T_4 belonging to the second priority level cannot be fired, i.e. $v_4 = 0$ is obtained.

Resolution rule LR_4: $T_1 < T_4 < [T_2, T_3]$. The value $v_1 = V_1 = 1$ is assigned to transition T_1. Now, given $v_1 = 1$, the temporary balance is calculated again: $TB_5^{(1)} = I_5 - v_1 = 2$. The value $v_4 = 1$ can then be assigned to T_4. Given $v_1 = v_4 = 1$, $TB_5^{(2)} = I_5 - v_1 - v_4 = 1$ is obtained. Now, $TB_5^{(2)}$ is not sufficient to fire both T_2 and T_3 at their maximal speeds. There is a sharing: $v_2 = v_3 = 0.5$. Finally, $TB_5^{(3)} = I_5 - v_1 - v_2 - v_3 - v_4 = 0$ is the balance B_5 for the IB-state.

❑

In the last example (LR_4), when the calculation process for the sharing $[T_2, T_3]$ begins, the "available" temporary balance $TB_5^{(2)}$ is less than the initial one $TB_5^{(0)} = I_5$, because transitions with a higher priority have been "served" before. Let us introduce some notations, definitions and hypotheses necessary for sharing Algorithm 5.6. In the sequel, $TB_k^{(B)}$ denotes the value TB_k at the beginning of a sharing calculation (corresponding to $TB_5^{(2)}$ in the example above).

Notation 5.6

a) Let $P^+(K_F) \subseteq P(K_F)$ denote the subset of places P_k in $P(K_F)$ such that $m_k > 0$. Given Hypothesis 5.2 is verified,

$$D(K_F) = \langle P^+(K_F), \{TB_k^{(B)} \mid P_k \in P(K_F) \setminus P^+(K_F)\}\rangle \tag{5.69}$$

is the set of data necessary for the resolution of a possible conflict (by priority or sharing): set of places in $P(K_F)$ whose markings are positive, and temporary balances for the other places in $P(K_F)$.

b) Let $v_{a,\max}$ denote the maximal value of v_a, given $D(K_F)$.

c) Let $v_a^{(a)}$ denote the value of v_a, given $D(K_F)$ and T_a takes priority over all the other transitions in $T(K_F)$.

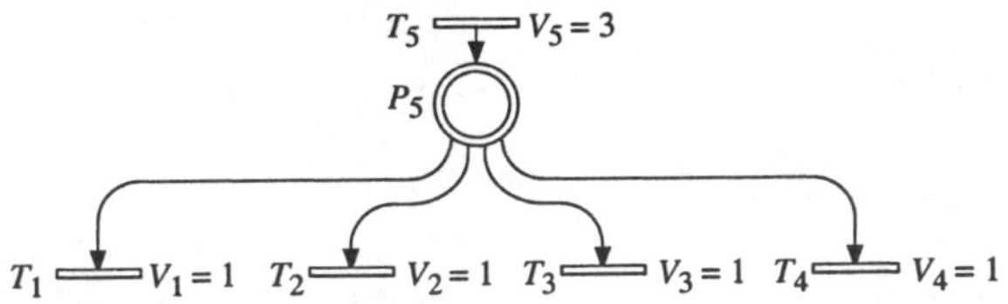

Figure 5.26 Illustration for temporary balance.

Property 5.6 If Hypothesis 5.2 is verified, then $v_{a,\max} = v_a^{(a)}$.

Proof Obvious, since neither $TB_i^{(B)}$ nor $TB_k^{(B)}$ for $P_k \in P(K_F)$, depend on the conflict resolution according to[14] Hypothesis 5.2b.

Property 5.7 There is an actual conflict among the transitions in $T(K_F)$ if and only if

$$m_i = 0 \quad \text{AND} \quad TB_i^{(B)} < \sum_{T_a \in T(K_F)} v_{a,\max}\,. \tag{5.70}$$

Proof
Sufficient condition: if $m_i = 0$, the transitions in $T(K_F)$ must share $TB_i^{(B)}$, which is less than the sum of the maximal values $v_{a,\max}$.
Necessary condition: if $m_i > 0$, there is obviously no conflict (examples in Figures 5.14a and b in Section 5.2.1); if $TB_i^{(B)} \geq \sum_{T_a \in T(K_F)} v_{a,\max}$, every transition T_a may be fired at $v_{a,\max}$.

❑

Before specifying Algorithm 5.6, let us consider the example, easy to understand, in Figure 5.27a. There is a structural conflict $K_4 = \langle P_4, \{T_1, T_2, T_3\}\rangle$, satisfying Hypothesis 5.2, and for the marking $\mathbf{m}_0$ there is an actual conflict involving the three transitions T_1 to T_3. Imagine that these transitions T_j correspond to machines having to process the contents of the respective places P_j, and that they need the common resource modeled by P_4. One possibility would be to share the resource among the three machines: the resource does not allow all the machines to work at their maximal speeds since $v_4 = V_4 < V_1 + V_2 + V_3$.

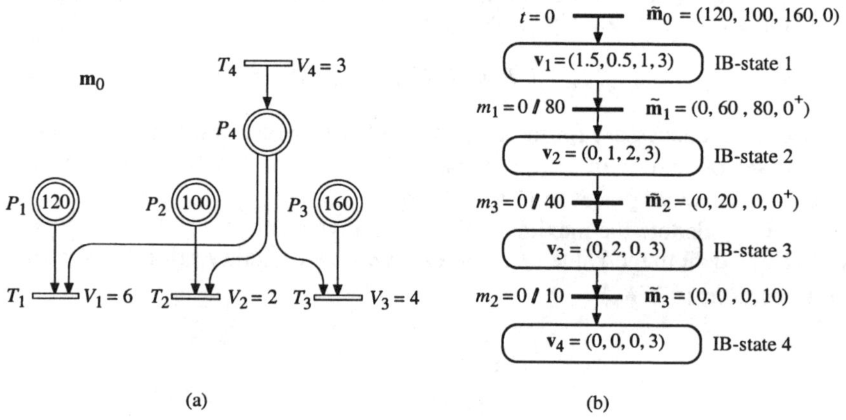

Figure 5.27 Sharing aiming at $\frac{v_1}{6} = \frac{v_2}{2} = \frac{v_3}{4}$.

[14] Property 5.6 is not always true if Hypothesis 5.2a is not satisfied. Examples are given in Appendix K.

For example, a sharing *aiming at*

$$\frac{v_1}{V_1} = \frac{v_2}{V_2} = \frac{v_3}{V_3}, \text{ i.e. } \frac{v_1}{6} = \frac{v_2}{2} = \frac{v_3}{4}, \tag{5.71}$$

may be sought, i.e. the common resource is shared proportionally to the maximal speeds of the transitions in this case, corresponding to the resolution rule $[6\,T_1,\ 2\,T_2,\ 4\,T_3]$. For the sharing rule in (5.71) and the marking in Figure 5.27a, the *parts* assigned to T_1, T_2, and T_3 are given in (5.72):

$$p(T_1) = I_4 \cdot \frac{V_1}{V_1 + V_2 + V_3} = 1.5\text{; similarly } p(T_2) = 0.5\text{, and } p(T_3) = 1. \tag{5.72}$$

The maximal possible speeds are $v_{1,\max} = \min(V_1, I_4) = 3$, $v_{2,\max} = \min(V_2, I_4) = 2$, and $v_{3,\max} = \min(V_3, I_4) = 3$. For the three transitions concerned, the parts assigned are smaller than the maximal possible speeds. It follows that the instantaneous firing speeds are the parts assigned, i.e., $v_1 = 1.5$, $v_2 = 0.5$, $v_3 = 1$. Hence, during IB-state 1, the markings are $m_1 = 120 - 1.5\,t$, $m_2 = 100 - 0.5\,t$, $m_3 = 160 - t$, and $m_4 = (3 - 1.5 - 0.5 - 1)\,t = 0$. This IB-state lasts up to $t = 80$, when P_1 becomes empty. At $t = 80$, $\mathbf{m}_1 = (0, 60, 80, 0)$; in fact, $\tilde{m}_4(80) = 0^+$, as specified in the evolution graph in Figure 5.27b (Properties 5.3a and 5.4a in Section 5.1.3.3).

At $t = 80$ a new sharing is calculated for the second IB-state. The parts (5.72) assigned to the three transitions, as well as $v_{2,\max}$ and $v_{3,\max}$, are the same as in IB-state 1. But $v_{1,\max} = 0$ because T_1 is no longer enabled. Hence, $v_1 = 0$ and the flow I_4 is shared between T_2 and T_3 during IB-state 2. The parts assigned are then:

$$p(T_2) = I_4 \cdot \frac{V_2}{V_2 + V_3} = 1 \text{ and } p(T_3) = I_4 \cdot \frac{V_3}{V_2 + V_3} = 2. \tag{5.73}$$

It follows that $v_2(t) = 1$ and $v_3(t) = 2$ are found for $t \in \left|80, 120\right|$. And so on.

The complete evolution is given in Figure 5.27b. During IB-state 3, there is no actual conflict since T_1 and T_3 are no longer enabled. During IB-state 4, only T_4 is enabled.

Algorithm 5.6 will be given below. It corresponds to the calculation of the speed vector $\mathbf{v}$ for a continuous timed PN illustrated in Figure 5.25 (let us recall that this algorithm should be used *only if Hypothesis 5.2 is satisfied*). Any local resolution rule $LR(K_F)$ may be used. Each priority level in $LR(K_F)$ is processed from *Step 3* to *Step 7*, *after* calculation of the speeds in $UP(K_F)$ in *Step 1* (which are data for *Steps 3* to *7*), and *before* calculation of speeds in $DOWN(K_F)$ performed in *Step 8*. Notation $T(K_f)$ corresponds to a set of transitions among which a sharing is performed (i.e. $T(K_f)$ is a subset of $T(K_F)$). When the processing related to a priority level begins, the temporary balance is $TB_i^{(B)}$, which becomes $TB_i^{(E)}$ during this processing.

***Algorithm 5.6* One or several sharings among transitions**

Step 0. Input data: the timed continuous PN, $\tilde{\mathbf{m}}$, $UP(K_F)$, K_F, $DOWN(K_F)$, $LR(K_F)$ (includes α_a for every T_a in each $T(K_f)$), priorities among other transitions in conflict.

Step 1. Calculate the speeds v_j of transitions T_j in $UP(K_F)$: Algorithm 5.4 up to *Step 8* applied to this subnet.

Step 2. Let $v_j = 0$ for all T_j in $T(K_F)$. Calculate $TB_k^{(B)}$ for every P_k in $P(K_F)$ (includes $TB_i^{(B)}$), from the speeds already known.

Step 3. **If** the first priority level in $LR(K_F)$ contains a single transition **then** let T_l denote this transition **else** go to *Step 5.*

Step 4. Calculate $v_l = v_{l,\max}$, then calculate the new values $TB_k^{(B)}$ for every P_k in $P(K_F)$ and go to *Step 7.*

Step 5.

Step 5.1. **For** every T_a in $T(K_f)$, calculate $v_{a,\max}$ and delete T_a from $T(K_f)$ if $v_{a,\max} = 0$ **end**

Step 5.2. **For** every T_a in $T(K_f)$, calculate $p(T_a)$ = function of $\mathbf{m}$ and the values α_j and $TB_k^{(B)}$ **end**

Step 6. Let $TB_i^{(E)} = TB_i^{(B)}$.

For every T_a in $T(K_f)$ such that $v_{a,\max} \leq p(T_a)$

$v_a = v_{a,\max}$

$TB_i^{(E)} = TB_i^{(E)} - \mathrm{Pre}(P_i, T_a) \cdot v_a$

delete T_a from $T(K_f)$

end

If $TB_i^{(E)} = TB_i^{(B)}$ **then** **for** every T_a in $T(K_f)$

$v_a = p(T_a)$ and delete T_a from $T(K_f)$

end

$TB_i^{(B)} = TB_i^{(E)}$.

If $T(K_f) \neq \varnothing$ AND $TB_i^{(E)} > 0$, go to *Step 5.*

Step 7. **If** $TB_i^{(B)} > 0$ OR $m_i > 0$, delete the first priority level in $LR(K_F)$. Then, if $LR(K_F)$ is not empty go to *Step3.*

Step 8. Calculate the speeds v_j of transitions T_j in $DOWN(K_F)$: Algorithm 5.4 up to *Step 8* applied to this subnet.

Step 9. Perform *Step 9* of Algorithm 5.4 for the whole net.

❑

The values $v_{a,\max}$ are obtained thanks to Property 5.6, and the general expression of $p(T_a)$ is

$$p(T_a) = TB_i^{(B)} \cdot \frac{\mathrm{Pre}(P_a, T_j) \cdot \alpha_a}{\sum_{T_j \in T(K_f)} \mathrm{Pre}(P_i, T_j) \cdot \alpha_j}. \tag{5.74}$$

For all the sharings in the example in Figure 5.27, the value v_a is always zero or $p(T_a)$. An example such that $0 < v_a < p(T_a)$ is given in Figure 5.28.

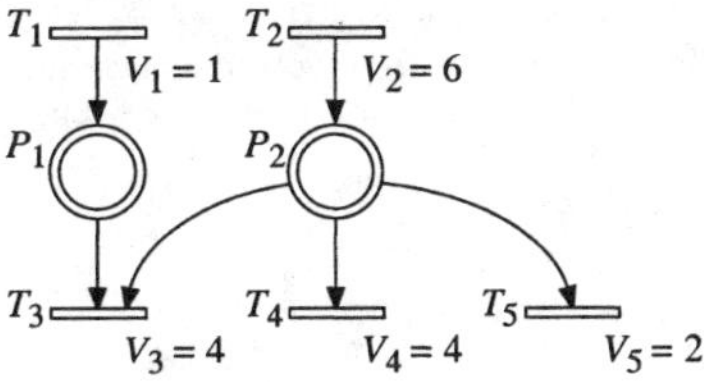

(a)

Step 0. $\mathbf{m} = (0,0)$; $UP(K_F)=\{T_1, T_2\}$; $DOWN(K_F)=\varnothing$; $R(K_F) = [T_3, T_4, T_5]$.

Step 1. $v_1 = 1, v_2 = 6$.

Step 2. $v_3 = v_4 = v_5 = 0$; $TB_1^{(B)} = 1$, $TB_2^{(B)} = 6$.

Step 3. $T(K_f) = \{T_3, T_4, T_5\}$; go to *Step 5.*

Step 5.1. $v_{3,\max} = 1$, $v_{4,\max} = 4$, $v_{5,\max} = 2$.

Step 5.2. $p(T_3) = p(T_4) = p(T_5) = 2$.

Step 6. $TB_2^{(E)} = 6$; $v_3 = 1$, $v_5 = 2$; $TB_2^{(E)} = 3$; $T(K_f) = \{T_4\}$; $TB_2^{(B)} = 3$.

Step 5.1. $v_{4,\max} = 3$.

Step 5.2. $p(T_4) = 3$.

Step 6. $TB_2^{(E)} = 3$; $v_4 = 3$; $T(K_f) = \varnothing$; $TB_2^{(B)} = 0$.

Step 9. $\mathbf{v} = (1, 6, 1, 3, 2)$.
$m_1 = m_2 = 0$; $\tilde{m}_1 = \tilde{m}_2 = 0^+$.
$\Delta t = \infty$. END

(b)

Figure 5.28 Illustration of Algorithm 5.6.

Figure 5.28b is an illustration of Algorithm 5.6. In *Step 5* (first time), $v_{3,\max} = 1$ and $p(T_3) = 2$ are found, then $v_3 = v_{3,\max} = 1$ (*Step 6*, first time). Similarly, since $v_{5,\max} = p(T_5) = 2$, $v_5 = v_{5,\max} = 2$. Since $v_{4,\max} > p(T_4)$ (*Step 5*, first time), the values $v_{4,\max}$ and $p(T_4)$ will be calculated again in a second iteration (*Steps 5* and *6*, second time).

5.3.3.3 Algorithm

Even if there is no restriction in the definition of the model, our algorithm of automatic calculation may require Hypothesis 5.2 to be verified. In this complex case, involving at least three transitions in conflict, all the speeds upstream from the conflict are calculated beforehand and all the speeds downstream from the conflict are calculated afterwards.

A way to verify whether or not Hypothesis 5.2a is satisfied is presented in Appendix L. The process is illustrated by the example in Figure L.1 which is represented again in Figure 5.29. Figure 5.29a shows a continuous PN (in Figure L.1a, only the structure of the PN was presented since neither the marking nor the maximal speed vector are necessary for obtaining Figure b), and Figure b is the corresponding *graph of relations among conflicts*. This graph is built as follows: it contains two kinds of nodes, illustrated in Figure b: the first kind is associated with every transition not involved in a conflict and the second kind is associated with every set (maximal) of transitions involved in a structural conflict. There is an arc from node n_1 to node n_2 in Figure b if there is at least one path of length 2 in Figure a from a transition in n_1 to a transition in n_2. For example, the path $T_1 \rightarrow P_1 \rightarrow T_2$ in Figure a implies the arc $T_1 \rightarrow \{T_2, T_3\}$ in Figure b. According to Remark L.1, arcs are added if the PN is not simple.

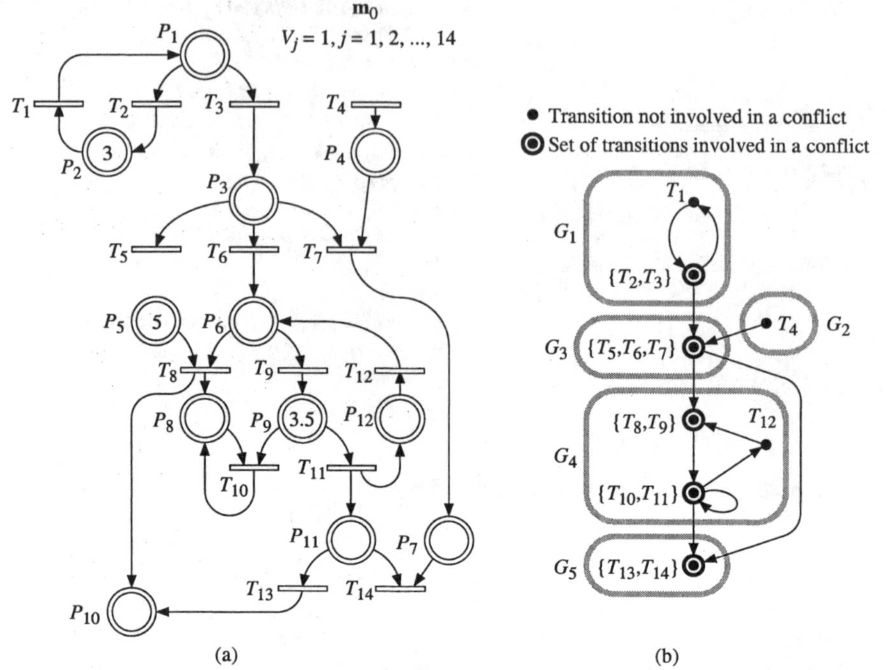

Figure 5.29 A timed continuous PN and its graph of relations among conflicts.

A *group* G_h of nodes in the *graph of relations among conflicts* is defined as follows: the subgraph made of the nodes in G_h and the arcs among them is a maximal strongly connected component. In Figure b, the groups G_1 to G_5 are obtained.

According to Property L.1: 1) a transition of the PN is *in exactly one group*; 2) the structural conflict $K = \langle P_i, T(K)\rangle$ satisfies *Hypothesis 5.2a if and only if there is a group* G_h such that $T(G_h) = T(K)$ and there is no self-loop from the node associated with $T(K)$ to itself. For the example in Figure 5.29b, only the conflicts $K_3 = \langle P_3, \{T_5, T_6, T_7\}\rangle$ and $K_{11} = \langle P_{11}, \{T_{13}, T_{14}\}\rangle$ satisfy this hypothesis.

The groups in a graph of relations among conflicts can always be ordered in a way such that the speeds of $T(G_h)$ do not depend on the speeds of $T(G_k)$ if G_k is after G_h. For our example,

$$\mathcal{G}_1 = (G_1, G_2, G_3, G_4, G_5) \tag{5.75}$$

and $$\mathcal{G}_2 = (G_2, G_1, G_3, G_4, G_5) \tag{5.76}$$

can be obtained. It follows from $\mathcal{G}_1$ that $\mathbf{v}(G_1)$, the speed vector related to the set of transitions $T(G_1)$, can be calculated first (independently from the speeds of the other transitions in the PN). Then, given $\mathbf{v}(G_1)$, $\mathbf{v}(G_2)$ can be calculated. Then, given $\mathbf{v}(G_1)$ and $\mathbf{v}(G_2)$, $\mathbf{v}(G_3)$ can be calculated, and so on.

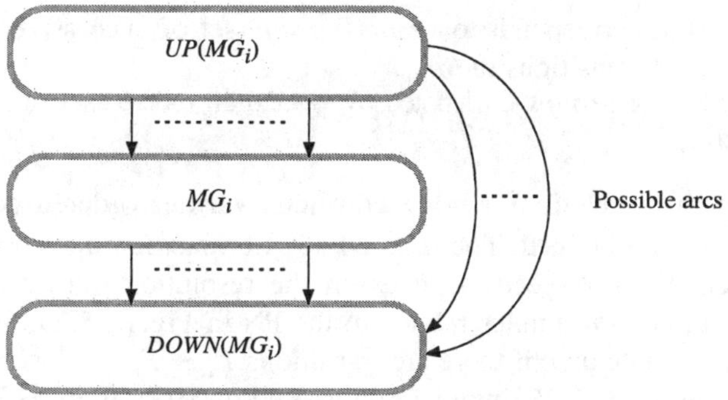

Figure 5.30 Macrogroup MG_i.

It follows that, for each successive group G_h, the speed vector $\mathbf{v}(G_h)$ can be calculated, practically, as for a complete timed continuous PN, with the algorithms already presented.

$\mathbf{v}(G_1)$: can be calculated by Algorithm[3] 5.4 (Section 5.3.2) if there is a priority between T_2 and T_3 ($T_2 < T_3$ or $T_3 < T_2$); can be calculated by Algorithm 5.5 (Section 5.3.3.1) if there is a sharing $[\alpha_2 T_2, \alpha_3 T_3]$.

$\mathbf{v}(G_2)$: obvious.

$\mathbf{v}(G_3)$: any conflict resolutions may be used since K_3 satisfies Hypothesis 5.2; can be calculated by Algorithm 5.6 (Section 5.3.3.2) in the general case.

$\mathbf{v}(G_4)$: can be calculated by Algorithm 5.5 if there is at most one sharing between two transitions ($[\alpha_8 T_8, \alpha_9 T_9]$ and priority between T_{10} and T_{11}, or priority between T_8 and T_9 and $[\alpha_{10} T_{10}, \alpha_{11} T_{11}]$).

$\mathbf{v}(G_5)$: can be calculated by Algorithm 5.5 if there is sharing between both transitions.

According to Appendix L, two or more *successive groups* in $\mathcal{G}$ can be gathered into a *macrogroup*. The concept of macrogroup is illustrated in Figure 5.30 in which only the arcs represented are possible, though not compulsory. Macrogroup MG_i corresponds to a set of successive groups in $\mathcal{G}$, $UP(MG_i)$ to a set of groups whose speeds must be calculated *before* the speeds of MG_i, and $DOWN(MG_i)$ to a set of groups whose speeds must be calculated *after* the speeds of MG_i. Examples: from (5.75), $MG_1 = \{G_3, G_4, G_5\}$ is such that $UP(MG_1) = \{G_1, G_2\}$ and $DOWN(MG_1) = \varnothing$; from (5.76), $MG_2 = \{G_1, G_3\}$ is such that $UP(MG_2) = \{G_2\}$ and $DOWN(MG_2) = \{G_4, G_5\}$.

Let us now specify some notations.

Notation 5.7

a) Notation G_h corresponds to a *group* in the graph of relations among conflicts and $T(G_h)$ is the set of transitions in G_h.

---

[3] This is true for each group if all its conflicts are solved by priorities.

b) Notation MG_p corresponds to a *macrogroup* (set of successive groups) and $T(MG_p)$ is the set of transitions in MG_p.

c) An ordered set of *groups* is denoted by $\mathcal{G}_r$ and an ordered set of *macrogroups* is denoted by $\mathcal{MG}_r$.

❑

Given a graph of relations among conflicts, various ordered sets $\mathcal{MG}_r$ of macrogroups can be obtained. The ordered set of macrogroups chosen for the successive calculations of speeds depends on the resolution rules chosen by the designer. Here are two examples related to the PN in Figure 5.29, illustrated in Figure 5.31 (let us note that, if there are transitions $T_a \rightarrow P_i \rightarrow T_b$ such that T_a is in MG_p and T_b is in MG_q, P_i may be represented in MG_p or in MG_q: only the transitions are significant).

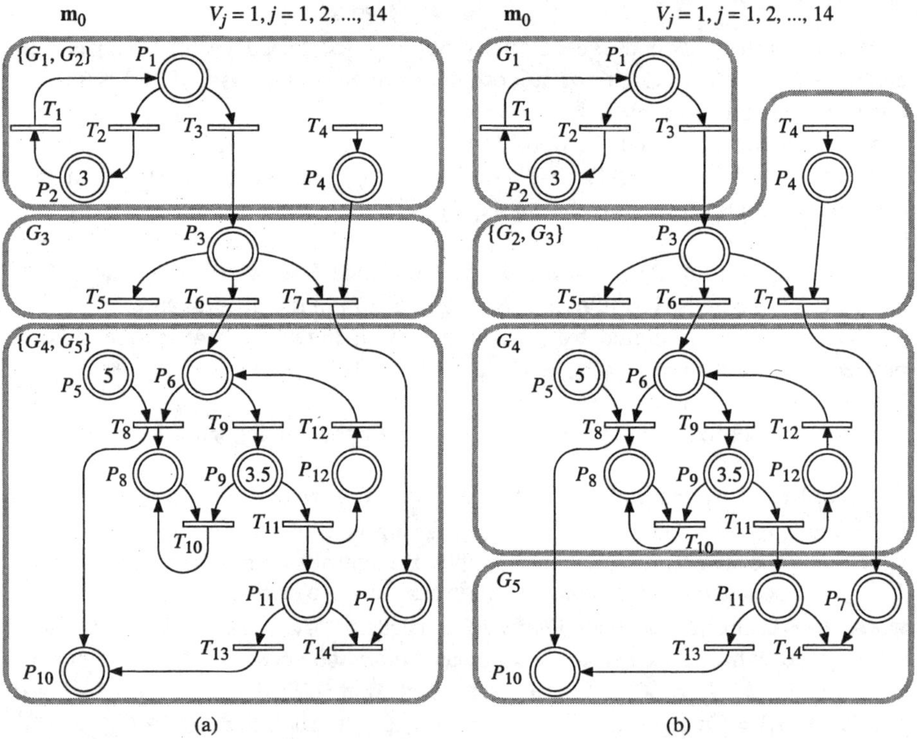

Figure 5.31 Two examples of ordered sets of macrogroups. (a) $\mathcal{MG}_1 = (\{G_1, G_2\}, \{G_3\}, \{G_4, G_5\})$. (b) $\mathcal{MG}_2 = (\{G_1\}, \{G_2, G_3\}, \{G_4\}, \{G_5\})$.

First example of local conflict resolutions: $T_3 < T_2$, $[T_5, T_6, 2T_7]$, $T_8 < T_9$, $T_{10} < T_{11}$, $[T_{13}, T_{14}]$. The speed calculations can be performed according to the ordered set shown in Figure 5.31a: $\mathcal{MG}_1 = (MG_1, MG_2, MG_3)$, in which

$MG_1 = \{G_1, G_2\}$, $MG_2 = \{G_3\}$, and $MG_3 = \{G_4, G_5\}$. After calculation of speeds of $T(MG_1)$ by Algorithm 5.4 (no sharing), speeds of $T(G_3)$ can be calculated by Algorithm 5.6 (Hypothesis 5.2 verified, the sharing among three transitions can be obtained); then, speeds of $T(MG_3)$ can be calculated by Algorithm 5.5 (only *one* sharing between *two* transitions among the three structural conflicts).

Second example of local conflict resolutions: $[T_2, T_3]$, $T_7 < [T_5, T_6]$, $T_9 < T_8$, $[T_{10}, T_{11}]$, $[3T_{13}, T_{14}]$. The speed calculations can be performed according to the ordered set $\mathcal{MG}_2$ shown in Figure 5.31b. The differences with $\mathcal{MG}_1$ are: 1) G_2 is gathered with G_3 (it could also be gathered with G_1), and this is possible because Algorithm 5.5 can be used (only one sharing between *two* transitions in $\{T_5, T_6, T_7\}$; 2) groups G_4 and G_5 cannot be gathered (one sharing in G_4 and one sharing in G_5).

Note that the limit case where there is a single macrogroup, $\mathcal{MG}_3 = (\{G_1, G_2, G_3, G_4, G_5\})$, could be used if all the conflicts are solved by priorities (Algorithm 5.4) or if there is at most one sharing between two transitions (Algorithm 5.5).

From the explanations above, Algorithm 5.7 allowing mixing of priorities and sharings can be obtained. This algorithm may be used *after the designer has chosen the local resolution rules*. Let us recall, in Remark 5.13, the constraints for these choices.

Remark 5.13 Except if an *immediate transition* T_a and a *non-immediate transition* T_b are involved in the same conflict (T_a must have a higher priority level than T_b according to Remark 3.2 in Section 3.2.1), *no restriction is due to the model*, i.e. any resolution rule could be defined in any case. However, for *automatic calculation of the speeds by Algorithm 5.7*, the following rules are applied for every macrogroup.

1) If a *conflict among transitions satisfies Hypothesis 5.2* (Section 5.3.3.2), there is no restriction related to the local resolution rule (except if both immediate and non-immediate transitions are involved in this conflict: any resolution among the immediate transitions, then any resolution among the other transitions).

2) If a *conflict among transitions does not satisfy Hypothesis 5.2*: at most one sharing between two transitions in the corresponding macrogroup (other resolutions by priorities).

Algorithm 5.7 **Resolution by priority and sharing**

Step 0. Input data: the timed continuous PN, $\tilde{\mathbf{m}}$, local conflict resolutions, $\mathcal{MG} = (MG_1, \ldots, MG_r)$.

Step 1. Let MG_g be the first term in $\mathcal{MG}$ and delete it from $\mathcal{MG}$.

Step 2. **If** there is no sharing among transitions in MG_g **then** go to *Step 3*
  else if there is sharing between two transitions in MG_g **then** go to *Step 4*
    else go to *Step 5*.

Step 3. Perform Algorithm 5.4 up to *Step 8* for subnet MG_g. Go to *Step 6*.

Step 4. Perform *Steps 1* to *6* of Algorithm 5.5 for subnet MG_g. Go to *Step 6*.

Step 5. Perform *Steps 2* to *7* of Algorithm 5.6 for subnet MG_g. Go to *Step 6*.

Step 6. If $\mathcal{MG}$ is not empty go to *Step 1*.
Step 7. Perform *Step 9* of Algorithm 5.4 for the whole net. END.

❑

Application of Algorithm 5.7 to the PN in Figure 5.29 for the first example of local conflict resolutions.

Step 0. $\tilde{\mathbf{m}}(0) = (0, 3, 0, 0, 5, 0, 0, 0, 3.5, 0, 0, 0)$; $T_3 < T_2$, $[T_5, T_6, 0.5\,T_7]$, $T_8 < T_9$, $T_{10} < T_{11}$, $[T_{13}, T_{14}]$; $\mathcal{MG} = (\{G_1, G_2\}, \{G_3\}, \{G_4, G_5\})$.
Step 1. $T(\{G_1, G_2\}) = \{T_1, T_2, T_3, T_4\}$; $\mathcal{MG} = (\{G_3\}, \{G_4, G_5\})$.
Step 2. Go to *Step 3*.
Step 3. Algorithm 5.4: $v_1 = 1$, $v_2 = 0$, $v_3 = 1$, $v_4 = 1$. Go to *Step 6*.
Step 6. Go to *Step 1*.
Step 1. $T(\{G_3\}) = \{T_5, T_6, T_7\}$; $\mathcal{MG} = (\{G_4, G_5\})$.
Step 2. Go to *Step 5*.
Step 5. Algorithm 5.6: $v_5 = 0.25$, $v_6 = 0.25$, $v_7 = 0.5$. Go to *Step 6*.
Step 6. Go to *Step 1*.
Step 1. $T(\{G_4, G_5\}) = \{T_8, T_9, T_{10}, T_{11}, T_{12}, T_{13}, T_{14}\}$; $\mathcal{MG} = \varnothing$.
Step 2. Go to *Step 4*.
Step 5. Algorithm 5.5: $v_8 = 1$, $v_9 = 0.25$, $v_{10} = 1$, $v_{11} = 1$, $v_{12} = 1$, $v_{13} = 0.5$, $v_{14} = 0.5$. Go to *Step 6*.
Step 6. $\mathcal{MG} = \varnothing$, then go to *Step 7*.
Step 7. $\mathbf{v} = (1, 0, 1, 1, 0.25, 0.25, 0.5, 1, 0.25, 1, 1, 1, 0.5, 0.5)$.
$m_i = 0$ and $\tilde{m}_i = 0^+$ for $i = 1, 3, 6, 7, 11, 12$; $m_2 = 3 - t$, $m_4 = 0.5\,t$, $m_5 = 5 - t$, $m_8 = t$, $m_9 = 3.5 - 1.75\,t$, $m_{10} = 1.5\,t$. Hence $\Delta t = 2$ $(m_9 = 0)$.
$\tilde{\mathbf{m}}(2) = (0^+, 1, 0^+, 1, 3, 0^+, 0^+, 2, 0^+, 3, 0^+, 0^+)$.

5.3.4 Complete Algorithm For All IB-states

Given an initial marking, the evolution of a timed continuous PN contains one or more successive IB-states, represented in an evolution graph. The analysis giving this evolution graph is specified in Algorithm 5.8.

Step 3.2 corresponds to the I-phase if it exists: the vector $\mathbf{m}_0$ is iteratively modified: the first one is $\mathbf{m}^{bef}(0)$ and the last one is $\mathbf{m}^{aft}(0)$. In this algorithm, the possibility of conflicts among immediate transitions involved in the I-phase has not been taken into account. The reason is that the authors do not want to make this algorithm cumbersome. However, if necessary, *Step 3.2* can be modified to take this possibility into account (according to Remark 5.14 below).

***Algorithm 5.8* For all IB-states**

Step 1. Initialization: timed continuous PN, initial marking $\mathbf{m}_0$, local conflict resolutions. Ordered set $\mathcal{MG}$ of macrogroups. Let $s = 1$.
Step 2. Setting the balances (literal expressions).
Step 3.
Step 3.1. If there is no T_j such that $V_j = \infty$ AND $q(T_j, \mathbf{m}_0) > 0$, let $\tilde{\mathbf{m}} = \mathbf{m}_0$ and go to *Step 4*.

Step 3.2. Let $T^{(\infty)}$ denote the set of transitions T_j such that $V_j = \infty$ AND $q(T_j, \mathbf{m}_0) > 0$. Let $\mathbf{m}_0 = \mathbf{m}_0 + \sum_{T_j \in T^{(\infty)}} q(T_j, \mathbf{m}_0) \cdot \mathbf{W}_{-,j}$, where $\mathbf{W}_{-,j}$ is the jth column of $\mathbf{W}$. Go to *Step 3.1*.

Step 4. Calculations for IB-state s: Algorithm[4] 5.7 without *Step 0*.

Step 5. **If** $t_s < \infty$ **then** let $\tilde{\mathbf{m}} = \tilde{\mathbf{m}}(t_s)$ and $s = s + 1$, and go to *Step 4*.
else END.

❑

Consider the example in Figure 5.32. The graph of relations among conflicts of the PN in Figure a is given in Figure b. It appears that the structural conflict $K_2 = \langle P_2, \{T_2, T_3, T_4\} \rangle$ does not satisfy Hypothesis 5.2 (Property L.1b in Appendix L). Hence, at most one sharing between two transitions can be automatically treated and the whole PN can be considered as a single macrogroup. Let us assume that the resolution rule $T_4 < [T_2, T_3]$ is chosen.

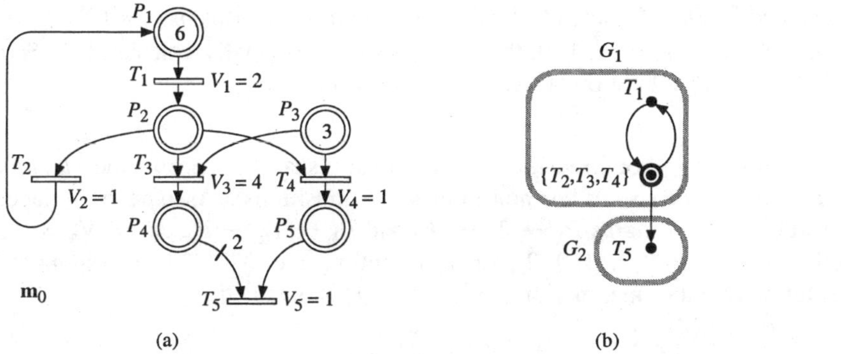

Step 1. PN in Figure 5.26a; $C_1 = \{0 \le v_1 \le 2, ...\}$; $s = 1$;
$\mathcal{MG} = (\{G_1, G_2\})$; $\mathbf{m}_0 = (6, 0, 3, 0, 0)$; $T_4 < [T_2, T_3]$.
Step 2. $B_1 = v_2 - v_1$, $B_2 = v_1 - v_2 - v_3 - v_4$,
$B_3 = -v_3 - v_4$, $B_4 = v_3 - 2v_5$, $B_5 = v_4 - v_5$.
Step 3.1. $\tilde{\mathbf{m}} = (6, 0, 3, 0, 0)$.
Step 4. $\mathbf{v} = (2, 0.5, 0.5, 1, 0.25)$; $m_1 = 6 - 1.5\,t$; $m_2 = 0$;
$m_3 = 3 - 1.5\,t$; $m_4 = 0$; $m_5 = 0.75\,t$; $\Delta t = 2$; $t_1 = 2$.
Step 5. $\tilde{\mathbf{m}} = (3, 0^+, 0, 0^+, 1.5)$; $s = 2$.
Step 4. $\mathbf{v} = (2, 1, 0, 0, 0)$; $m_1 = 3 - t$; $m_2 = t$; $m_3 = 0$;
$m_4 = 0$; $m_5 = 1.5$; $\Delta t = 3$; $t_2 = 5$.
Step 5. $\tilde{\mathbf{m}} = (0^+, 3, 0, 0, 1.5)$; $s = 3$.
Step 4. $\mathbf{v} = (1, 1, 0, 0, 0)$; $m_1 = 0$; $m_2 = 3$; $m_3 = 0$;
$m_4 = 0$; $m_5 = 1.5$; $\Delta t = \infty$; $t_3 = \infty$.

(c)

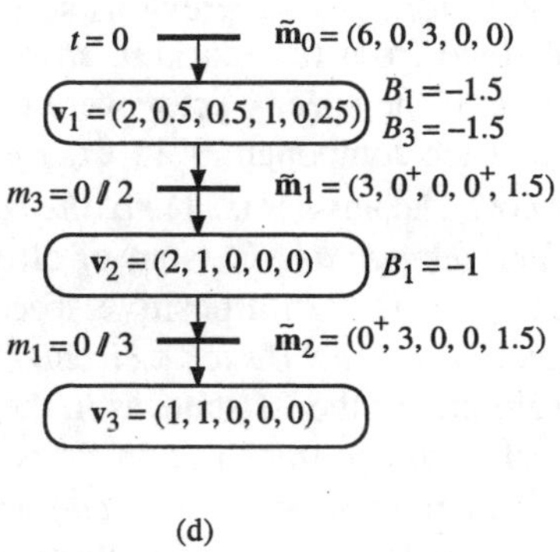

Figure 5.32 (a) Timed continuous PN. (b) Graph of relations among conflicts. (c) Illustration of Algorithm 5.8 to the PN in Figure a for the resolution rule $T_4 < [T_2, T_3]$. (d) Corresponding evolution graph.

---

[4] Algorithm 5.7 may possibly be replaced by Algorithm 5.4 if all the conflicts are solved by priorities, or Algorithm 5.1 if there is no conflict.

In Figure 5.32c, Algorithm 5.8 is applied to the timed continuous PN and its initial marking in Figure 5.32a. The corresponding evolution graph is given in Figure 5.32d. Although this information is redundant, the negative balances are shown for each IB-state. Let us comment on Figure 5.32c. The initial data, including $\mathcal{MG} = (\{G_1, G_2\})$, are given in *Step 1*.

In *Step 2*, balance equations $B_1 = v_2 - v_1$, $B_2 = v_1 - v_2 - v_3 - v_4$, ..., $B_5 = v_4 - v_5$ are set up.

Step 3.1: $\tilde{\mathbf{m}} = \mathbf{m}_0$ is the initial marking from which the calculations related to the first IB-state are performed (since there is no I-phase, *Step 3.2* is not performed).

Step 4 (first time) corresponds to the calculation related to IB-state 1. Since IB-state 1 ends at $t = 2 < \infty$, the new marking $\tilde{\mathbf{m}}_1 = (3, 0^+, 0, 0^+, 1.5)$ is set up in *Step 5* before returning to *Step 4*.

The vector of instantaneous speeds $\mathbf{v}_2 = (2, 1, 0, 0, 0)$ for IB-state 2 is then calculated in *Step 4* (second time). From the new marking $\tilde{\mathbf{m}}_2 = (0^+, 3, 0, 0, 1.5)$, the vector $\mathbf{v}_3 = (1, 1, 0, 0, 0)$ for IB-state 3 is finally calculated in *Step 4* (third time). The duration of this IB-state is infinite.

Remark 5.14 For *immediate transitions*, a sharing concerns the *quantity of firing* (instead of speeds for non-immediate transitions). Assume a 3-place, 2-transition PN, with arcs $P_1 \rightarrow T_1 \rightarrow P_2$ and $P_1 \rightarrow T_2 \rightarrow P_3$, $V_1 = V_2 = \infty$, an initial marking $\mathbf{m}^{\text{bef}} = (4, 0, 0)$, and a resolution rule $[3T_1, T_2]$. According to the resolution rule the final marking $\mathbf{m}^{\text{aft}} = (0, 3, 1)$ is obtained.

Remark 5.15

According to the previous sections, the speed vector $\mathbf{v}_s$ related to IB-state s is obtained from the marking $\tilde{\mathbf{m}}_{s-1}$ at the beginning of this IB-state. A question arises: if the speed vector were calculated from the vector $\mathbf{m}_{s-1}$ instead of $\tilde{\mathbf{m}}_{s-1}$ (i.e., each component 0^+ in $\tilde{\mathbf{m}}_{s-1}$ would be 0 in $\mathbf{m}_{s-1}$), would the result be correct or not? The answer is: 1) *no, the result would not be correct if there is a circuit in which the marking* 0^+ (sum of all the markings of the places in the circuit) turns in the circuit with a positive speed (Figure 5.21d for example); 2) *yes, the result would be correct if such a circuit does not exist* (a marking 0^+ may disappear at the beginning of the IB-state, as in the example in Figure 5.21a, or remain in a place P_i such that no transition in P_i° is enabled).

If there is *no structural conflict* in the PN, and given $\mathbf{m}_0$ is a vector of real numbers (Definition 4.1 in Section 4.1.2), a circuit in which the marking 0^+ turns with a positive speed cannot exist. This property will be explained in Section 5.4.2. It follows that, *if there is no structural conflict, the speed vector* $\mathbf{v}_s$ *may be calculated from* $\mathbf{m}_{s-1}$ (Algorithm 5.1 in Section 5.3.1).

5.4 PROPERTIES

The model studied from the beginning of this chapter and up to the end of Section 5.4, has the following feature: it is a timed continuous PN such that *every maximal firing speed is constant.* Let us call it **CCPN** (standing for Constant speed Continuous PN). Various properties related to the behavior of this model were given in the previous sections. Let us emphasize Property 5.1 (Section 5.1.3.1), defining the *maximal instantaneous speed associated with a transition as a function of enabling*, and Properties 5.3 and 5.4 (Section 5.1.3.3), defining *the conditions such that a marking* $\tilde{m}_i(t) = 0^+$ during an IB-state and at the end of an IB-state. Let us recall that the marking 0^+ does not appear in the definition of the basic model (Definition 5.2 in Section 5.1.3.1); it is a mathematical means for an accurate understanding of enabling conditions (Definition 5.3).

Some additional explanations will be given through illustrative examples in Section 5.4.1, then more general properties will be presented in Section 5.4.2. The modeling power of CCPN is emphasized in Section 5.4.3.

5.4.1 Illustratory Examples

5.4.1.1 A Simple Production System

The timed discrete PN in Figure 5.33a models the behavior of a production system made up of three single-server stations (a station may correspond to unloading/loading of parts on pallets and two stations to machining). Place P_i corresponds to the pallets waiting or being served in station M_i and a firing of T_i models an end of service in M_i. The service times are constant and given in the figure. The markings obtained from the initial marking $\mathbf{m}_0 = (12, 0, 0)$ are illustrated in Figure c.

The behavior of the CCPN in Figure b provides an approximation of the behavior of the discrete PN in Figure a. The processing in a station is represented as a continuous flow. According to the previous sections, we obtain:

IB-state 1. At $t = 0$, $\mathbf{m}_0 = (12, 0, 0)$. For $t \in |0, 4| : \mathbf{v}_1 = (4, 2, 1)$, (5.77)
$m_1 = 12 - 3t$, $m_2 = 2t$, $m_3 = t$.

IB-state 2. At $t = 4$, $\mathbf{m}_1 = (0, 8, 4)$. For $t \in |4, 12| : \mathbf{v}_2 = (1, 2, 1)$, (5.78)
$m_1 = 0$, $m_2 = 8 - (t - 4)$, $m_3 = 4 + (t - 4)$.

IB-state 3. At $t = 12$, $\mathbf{m}_2 = (0, 0, 12)$. For $t \in |12, \infty| : \mathbf{v}_3 = (1, 1, 1)$, (5.79)
$m_1 = 0$, $m_2 = 0$, $m_3 = 12$.

One can observe, in Figure 5.33c, that there is a transitory behavior up to $t = 12$, then a stationary behavior for $t \geq 12$.

Let us discuss the information which can be obtained from the discrete model and from the continuous model in both cases.

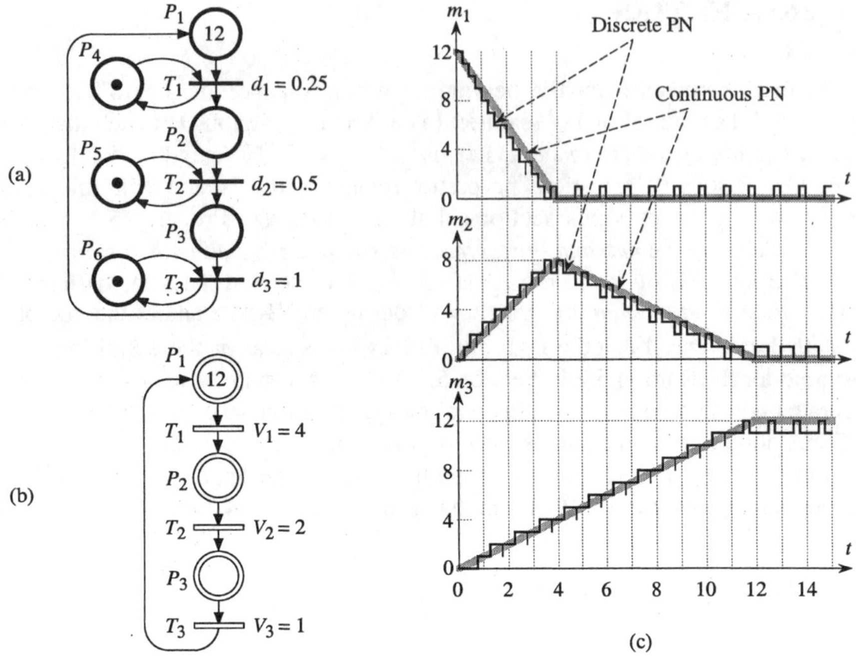

Figure 5.33 (a) Timed discrete PN modeling a production system. (b) Corresponding timed continuous PN (approximation of the behavior). (c) Evolution of the markings in both cases.

Stationary behavior. Since the PN in Figure a is a strongly connected event graph, the firing frequencies of the *discrete PN* can be calculated using Property 3.13 in Section 3.4.2.3. There are four circuits, namely $C_1 = P_1T_1P_2T_2P_3T_3P_1$, $C_2 = P_4T_1P_4$, $C_3 = P_5T_2P_5$, and $C_4 = P_6T_3P_6$. According to (3.11), $F(C_1) = 12/1.75 = 6.86$, $F(C_2) = 1/0.25 = 4$, $F(C_3) = 1/0.5 = 2$, $F(C_4) = 1/1 = 1$, and according to (3.12), $F_1 = F_2 = F_3 = \min(6.86, 4, 2, 1) = 1$. In other words, *the firing frequency is 1 for every transition* when the stationary behavior is reached (may be observed in Figure c: one token arrival and one token departure each time unit, for every place). For the *continuous PN* in Figure b, the stationary behavior is reached in IB-state 3; in this IB-state, according to (5.79), $\mathbf{v} = (1, 1, 1)$, i.e. the firing speed is 1 for every transition (corresponding to $v_j = F_j$ for each transition T_j).

Transitory behavior. For the *discrete model*, there is no analytical expression of the firing frequency or of the marking. A simulation can be performed in order to obtain the behavior appearing in Figure c; this simulation takes into account 55 events (i.e. 55 firings of transitions), and if the initial marking of P_1 were 10 000 instead of 12, the number of events during the transitory behavior would be about 50 000. For the *continuous model*, analytical expressions can be obtained for each

IB-state: (5.77) to (5.79). The speed vector and the marking equations are calculated three times (one at initial time and one at each event *'the marking of a place reaches the value 0'*). If the initial marking of P_1 were 10 000 instead of 12, the number of events to take into account would be the same (i.e. the number of IB-states would be the same: three).

There is no conflict in Figure 5.33b. Then, according to Section 5.3.1, speed may be calculated from markings $\mathbf{m}_i$ by Algorithm 5.1 (the markings $\tilde{\mathbf{m}}_i$ are not required). Many cases exist such that the marking 0^+ would not be necessary. Nevertheless, this marking must be used in some special cases; an example is presented in the next section.

5.4.1.2 About Marking 0^+

Figure 5.34a represents a timed continuous PN with a conflict whose resolution rule is the priority $T_4 < T_3$ (this example was proposed by Paul Caspi [Ca 87]). The evolution graph obtained from Algorithm 5.8 is in Figure b. Intuitively, the behavior can be explained as follows. At the very beginning, there is a flow (speed 1) through T_1 and T_2. Because of the priority, $v_4 = 1$ is obtained. Hence $I_2 = v_1 + v_4 = 2$, and $v_2 = 2$ is obtained; since $I_3 = 2 = V_3 + V_4$, there is no conflict, hence $v_3 = v_4 = 1$. In other words, during IB-state 1, the speed vector is $\mathbf{v} = (1, 2, 1, 1)$. During this IB-state, there are, roughly speaking, two juxtaposed flows: a flow at speed 1 in the path $P_1T_1P_2T_2P_3T_3$, plus a flow at speed 1 in the loop $P_2T_2P_3T_4P_2$. When P_1 becomes empty, at $t = 90$, the first flow stops whereas the flow in the loop continues (because of the priority). This behavior is effectively obtained by Algorithm 5.7 from $\tilde{\mathbf{m}}_1 = (0, 0^+, 0^+)$: as a matter of fact, this vector $\tilde{\mathbf{m}}_1$ is obtained since $\tilde{m}_2(90) = 0^+$ and $\tilde{m}_3(90) = 0^+$ according to Properties 5.4a and 5.3a (Section 5.1.3.3). It must be noted that, in this example, the correct behavior (which may be verified with a discrete PN) would not be obtained without the markings 0^+: it is clear that the speed vector $\mathbf{v} = (0, 0, 0, 0)$ would be obtained from $\mathbf{m}_1 = (0, 0, 0)$. Note that $\tilde{\mathbf{m}}_1$ contains some markings 0^+ whereas the initial marking $\tilde{\mathbf{m}}_0 = \mathbf{m}_0 = (90, 0, 0)$ contains only real numbers.

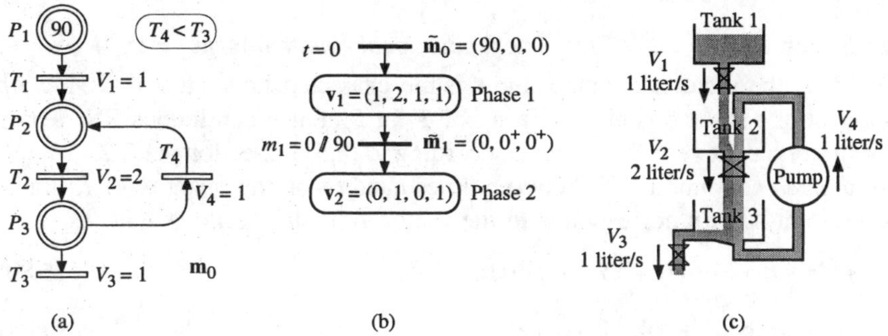

Figure 5.34 (a) Timed discrete PN. (b) Evolution graph. (c) Illustration.

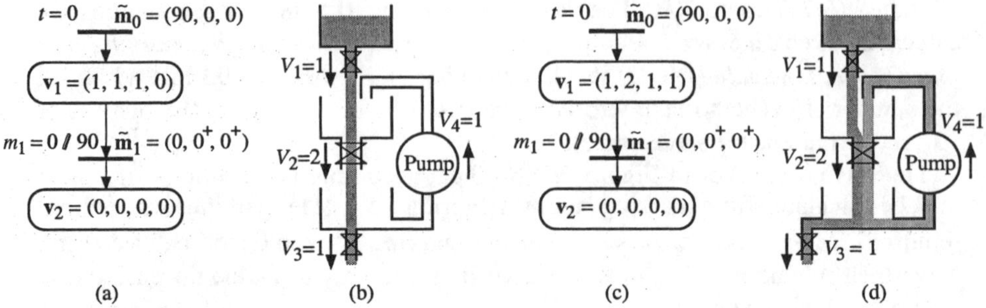

Figure 5.35 (a) and (b) Priority $T_3 < T_4$. (c) and (d) Sharing $[T_3, T_4]$.

Figure 5.34c is an illustration for the PN in Figure a. The contents of tanks i (i = 1, 2, 3) correspond to the markings of places P_i; the maximal speeds associated with the transitions follow. The figure shows the flows during IB-state 1. When tank 1 becomes empty, the flow through the loop continues; roughly speaking, the markings 0^+ correspond to the small quantity of liquid in the pipes.

Let us now assume other resolution rules for the PN in Figure 5.34a. They are illustrated in Figure 5.35.

1) The priority is $T_3 < T_4$ (instead of $T_4 < T_3$). The evolution graph is in Figure a. During the first IB-state, $\mathbf{v}_1$ = (1, 1, 1, 0): transition T_4 is not fired. When P_1 becomes empty, all the speeds become zero (and $\tilde{m}_2$ and $\tilde{m}_3$ become 0). Figure b provides an illustration of this behavior.

2) There is a sharing $[T_3, T_4]$. The evolution graph and an illustration are given in Figures c and d. During the first IB-state, the behavior is similar to the case $T_4 < T_3$. When P_1 becomes empty, all the speeds become zero. Because of the sharing $\tilde{m}_2 + \tilde{m}_3$ become 0 (the pipes become empty); this is consistent with (J.5) in Appendix J.

5.4.2 General Properties

Let S such that $\mathbf{m}_i \xrightarrow{S} \mathbf{m}_k$. The *fundamental equation* is $\mathbf{m}_k = \mathbf{m}_i + \mathbf{W} \cdot \mathbf{s}$, where $\mathbf{W}$ is the incidence matrix and $\mathbf{s}$ is the characteristic vector of S: the jth component of $\mathbf{s}$ is the quantity of firing of T_j in S. For a continuous PN, $\mathbf{s}$ is a vector of non-negative real numbers, according to (4.5) in Section 4.1.3.2.

For a *timed* continuous PN, between times t and t + dt, the quantity of firing of T_j is $v_j(t) \cdot \mathrm{d}t$, then $\mathbf{s}$ corresponds to the vector $\mathbf{v}(t) \cdot \mathrm{d}t$. It follows that:

$$\mathbf{m}(t + \mathrm{d}t) = \mathbf{m}(t) + \mathbf{W} \cdot \mathbf{v}(t) \cdot \mathrm{d}t, \qquad (5.80)$$

then $$\mathbf{m}(t_2) = \mathbf{m}(t_1) + \mathbf{W} \cdot \int_{t_1}^{t_2} \mathbf{v}(u) \cdot \mathrm{d}u \qquad (5.81)$$

is the **fundamental equation** for a timed continuous PN. It works for any $0 \le t1 \le t2$. Example for the PN in Figure 5.33b: according to (5.77) and (5.81),

$$\mathbf{m}(2.5) = \begin{bmatrix} 12 \\ 0 \\ 0 \end{bmatrix} + \begin{bmatrix} -1 & 0 & 1 \\ 1 & -1 & 0 \\ 0 & 1 & -1 \end{bmatrix} \cdot \int_0^{2.5} \begin{bmatrix} 4 \\ 2 \\ 1 \end{bmatrix} \cdot du = \begin{bmatrix} 4.5 \\ 5 \\ 2.5 \end{bmatrix}.$$

Let us now analyze the examples in Figure 5.36 (called Examples a, b, c, and d), before drawing more general properties. The incidence matrix are

$$\mathbf{W}_a = \begin{bmatrix} -1 & 1 \\ 1 & -1 \end{bmatrix}, \mathbf{W}_b = \begin{bmatrix} -1 & 1 \\ 2 & -1 \end{bmatrix}, \mathbf{W}_c = \begin{bmatrix} -2 & 1 \\ 1 & -1 \end{bmatrix}, \mathbf{W}_d = \begin{bmatrix} -1 & 1 \\ 1 & -1 \end{bmatrix}. \quad (5.82)$$

Example a. During the first IB-state, $B_1 = -1$. When m_1 becomes zero, the speed vector becomes definitively $\mathbf{v}_{2a} = (1, 1)$. Vector $\mathbf{x}_1 = (1, 1)$ is a P-invariant since $\mathbf{x}_1^T \cdot \mathbf{W}_a = \mathbf{0}$. It follows that $m_1 + m_2$ is constant (i.e., 3); m_1 decreases while m_2 increases during IB-state 1 (because $v_1 = V_1 > v_2 = V_2$), then $\mathbf{m} = (0, 3)$ during the second IB-state. During this *last IB-state*, $\mathbf{v}_{2a} = (1, 1)$ corresponds to the T-invariant $\mathbf{y}_1 = (1, 1)$: $\mathbf{W}_a \cdot \mathbf{y}_1 = \mathbf{0}$, i.e. the same quantity of firing of both transitions, $v_1 \cdot dt = v_2 \cdot dt$, does not change the marking.

Example b. There is neither P-invariant nor T-invariant. Since $\mathbf{W}_b \cdot \mathbf{v}_{1b} = (0, 1) > \mathbf{0}$, m_2 increases for the final speed vector $\mathbf{v}_{1b} = (1, 1)$. This PN is unbounded.

Example c. There is neither P-invariant nor T-invariant. Since $\mathbf{W}_c \cdot \mathbf{v}_{1c} = (-1, 0) < \mathbf{0}$, m_1 decreases for the speed vector $\mathbf{v}_{1c} = (1, 1)$. At $t = 3$, the marking is $\tilde{\mathbf{m}}(3) = \tilde{\mathbf{m}}_1 = (0^+, 0^+)$. Then, for $t > 3$, $\mathbf{v}_{2c} = (0, 0)$ and $\tilde{\mathbf{m}}(t) = (0, 0)$ (according to Property 5.4 and Algorithm 5.7).

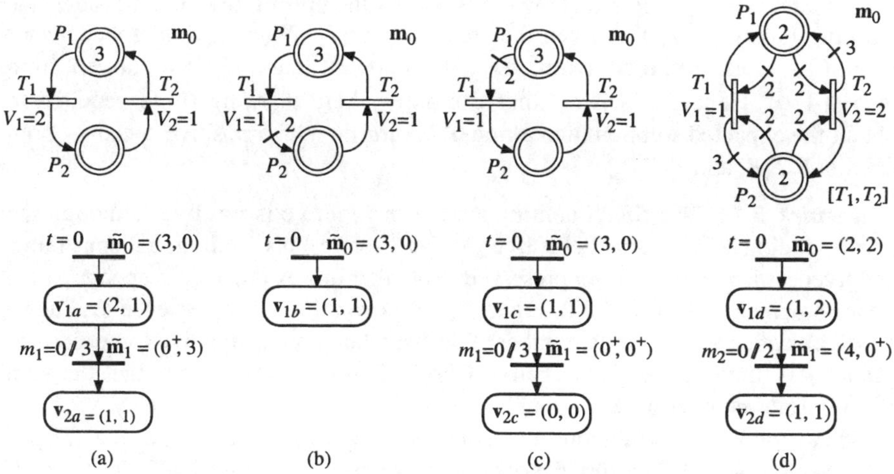

Figure 5.36 (a), (b), and (c) Various circuits. (d) Timed ε-live continuous PN.

For other values of maximal speeds ($V_1 = 2$ and $V_2 = 1$ for example) the evolution graphs of the PNs in Figures b and c could begin with different speed vectors. However, given the incidence matrix, the final speed vector (stationary behavior for examples a and c) would always be $\mathbf{v}_a = (v_\alpha, v_\alpha)$, $\mathbf{v}_b = (v_\beta, v_\gamma)$, and $\mathbf{v}_c = (0, 0)$, where v_α, v_β, $v_\gamma > 0$ depend on V_1 and V_2.

These observations may be generalized. Assume a PN consisting of a single circuit $P_1T_1P_2T_2 \ldots P_nT_nP_1$, and let

$$G = \prod_{P_i \in P} \frac{\mathrm{Post}(P_i, T_{i-1(\mathrm{mod}\, n)})}{\mathrm{Pre}(P_i, T_i)}. \tag{5.83}$$

If $G = 1$, the PN is conservative and there is a T-invariant $\mathbf{y} > \mathbf{0}$. The last IB-state is such that $\mathbf{m}$ is constant and the final speed vector is $\mathbf{v} = k \cdot \mathbf{y}$, where k is a positive constant (generalizes Example a). If $G > 1$, the PN is unbounded and the final speed vector is $\mathbf{v} > \mathbf{0}$ (generalizes Example b). If $G < 1$, the last IB-state is such that $\mathbf{m} = \mathbf{0}$ (as well as $\tilde{\mathbf{m}} = \mathbf{0}$) and $\mathbf{v} = \mathbf{0}$ (generalizes Example c).

Let us observe that, for Examples a, b, and c, the knowledge of $\tilde{\mathbf{m}}$ is not necessary for the calculation of speed vectors. The only case where $\tilde{\mathbf{m}} = (0^+, \ldots, 0^+)$ is reached is Example c (and its generalization) when the IB-state just before the last one ends: the speed vector $\mathbf{v} = \mathbf{0}$ of the last IB-state may be obtained from $\mathbf{m} = \mathbf{0}$ as well as from $\tilde{\mathbf{m}} = (0^+, \ldots, 0^+)$. In a PN without conflicts, circuits may have a behavior similar to Examples a, b, or c, plus, possibly, constraints due to synchronization and additional feedings due to input transitions to the places of the circuits. These modifications cannot lead to a circuit in which a marking 0^+ is turning with a positive speed: such a behavior can only be obtained if there is a conflict (as in Figure 5.34a). As a consequence, in a timed continuous PN without conflict, marking $\tilde{\mathbf{m}}$ is not required for calculation of the speed vector.

Consider now Example d (Figure 5.36d). At the end of the first IB-state, the marking is $\tilde{\mathbf{m}}_1 = (4, 0^+)$. Given the sharing rule $[T_1, T_2]$, the final speed vector $\mathbf{v}_{2d} = (1, 1)$ is obtained from $\tilde{\mathbf{m}}_1$. This result cannot be obtained from $\mathbf{m}_1 = (4, 0)$. Hence, this is another example where marking 0^+ is required to obtain the expected solution (a marking 0^+ is turning with a positive speed in both circuits $P_2T_1P_2$ and $P_2T_2P_2$).

Remark 5.16 The *timed* continuous PN in Figure c is not live, although the corresponding *autonomous* PN is live. As for discrete PNs, timing add constraints and liveness is not necessarily preserved. Note that this example corresponds to the Zenon's paradox (Section 4.3.2, Figure 4.16a); even if a firing sequence leading to deadlock contains an infinite number of terms, when time is involved, the time to deadlock is finite (five centuries before Christ, Zenon was not aware that the sum of an infinite number of terms could be finite!).

The *autonomous* continuous PN corresponding to Figure d is ε-live (Figure 4.19, Section 4.3.3). The *timed* model is live in case of sharing $[\alpha_1 T_1, \alpha_2 T_2]$ or if $V_1 \leq V_2$ and the priority rule is $T_1 < T_2$. It is not live only if $V_1 < V_2$ and the priority rule is $T_2 < T_1$ (or vice-versa), i.e., the priority given to the faster transition empties a place completely (for our example, $\tilde{m}_2$ becomes 0).

❑

Let us now give a property related to *speed decreasing*, then two properties and a conjecture concerning the *last IB-state*, i.e. the **final speed state**.

Property 5.8 If a timed continuous PN is *without conflict*, then:
a) If $v_j(t_1) = 0$ for some $t_1 > 0$, then $v_j(t) = 0$ for any $t \geq t_1$.
b) If $v_j(t_1) = \alpha$ for some $t_1 > 0$, then $v_j(t) \leq \alpha$ for any $t \geq t_1$.

Proof

a) If $v_j(t_1) = 0$ for some $t_1 > 0$, there is at least one path $P_1T_1P_2T_2 \ldots P_jT_j$ such that: 1) P_1 has no input transition or $P_1 = P_j$ (hence $T_1 = T_j$), and 2) $m_1(t_1) = m_2(t_1) = \ldots = m_j(t_1) = 0$ and no place in $\{P_1, P_2, \ldots, P_j\}$ is fed. It is clear that these places will never be fed since the PN is without conflict.

b) Speed $v_j(t)$ *can increase only* if there is a place P_j in $°T_j$ such that $m_j(t) = 0$ and there is T_k in $°P_j$ such that $v_k(t)$ increases. Similarly, speed $v_k(t)$ can increase only if there is a place P_k in $°T_k$ such that $m_k(t) = 0$ and there is T_h in $°P_k$ such that $v_h(t)$ increases. And so on. The process ends either with a place P_a which has no input transition or is P_j itself, or with a source transition T_a.

If the process ends with P_a, the instantaneous speed of every transition of the path from P_a to P_j is 0 and will remain 0 according to Property 5.8a. If the process ends with the source transition T_a, its speed cannot increase since $v_a(t) = V_a$ is constant. Thus, $v_j(t)$ cannot increase.

Property 5.9 A timed continuous PN *without conflict* converges to a final speed state in a finite time.

Proof Let $P^{(s)}$ denote the set of places P_i such that $m_i > 0$ and $B_i < 0$ during IB-state s. The duration of an IB-state, except the last one, is $\min_{P_i \in P^{(s)}} \frac{m_i}{B_i}$, i.e. finite. Given Property 5.8b, when the speed vector changes, it decreases: $\mathbf{v}_{s+1} < \mathbf{v}_s$. The number of speed states is necessarily finite, a change from IB-state s to IB-state $s+1$ occurring when a marking m_i becomes zero.

Conjecture 5.1 Property 5.9 is true also for a timed continuous PN *with conflicts, if the conflicts are solved in a deterministic way* (priority or sharing).

❑

The duration of an IB-state is finite for the same reason. Each component v_j of a speed vector $\mathbf{v}_s$ is the algebraic sum of maximal speeds V_k (or fractions of such a sum in case of sharing); since the values V_k are rational numbers, the number of possible vectors $\mathbf{v}_s$ is finite. For proving the conjecture, it remains to be shown that once IB-state s is completed, the same speed vector $\mathbf{v}_s$ cannot be obtained again.

Property 5.10

a) In the final IB-state, the *marking remains constant* if and only if the final firing speed is *either a T-invariant or the vector* **0**.

b) If the final firing speed is greater than **0** but is not a T-invariant, then the PN is unbounded.

Proof

a) According to Equation (5.80), the marking remains constant if and only if $\mathbf{W} \cdot \mathbf{v}(t) = \mathbf{0}$, i.e. if $\mathbf{v}(t)$ is a T-invariant (by definition) or if $\mathbf{v}(t) = \mathbf{0}$.

b) Assume that $\mathbf{v}(t)$ is not a T-invariant, i.e. $\mathbf{W} \cdot \mathbf{v}(t) \neq \mathbf{0}$. If a component of $\mathbf{W} \cdot \mathbf{v}(t)$ is less than 0, some place marking m_i is decreasing, then the current IB-state cannot be the last one. If $\mathbf{W} \cdot \mathbf{v}(t) \gneq \mathbf{0}$, the property is verified.

❑

Previous examples provide illustrations of the above properties. Property 5.9: Figures 5.33b, 5.36a, b, and c. Conjecture 5.1: Figures 5.32d, 5.34b, 5.35a and c, 5.36d. Property 5.10a: Figures 5.32d, 5.34b, and 5.36a for $\mathbf{v} \gneq \mathbf{0}$; Figures 5.35a and c, and 5.36c for $\mathbf{v} = \mathbf{0}$. Property 5.10b: Figure 5.36b.

5.4.3 Modeling Power

Naturally, a CCPN can model a *continuous system* in which the *speeds are constant*. It can also provide an approximation to the behavior of a *discrete event dynamic system*, if *large numbers* of entities (parts for example) are concerned. If the processing or waiting *times are constant*, the discrete system may be modeled by a T-timed PN (Section 3.4.2.2). A relatively good approximation of the behavior may be obtained from a timed continuous PN, as illustrated in Figure 5.33 (Section 5.4.1.1). If the processing or waiting *times are stochastic*, some behaviors may also be approximated, with a small error.

Consider the example in Figure 5.37. Figure a is a stochastic PN: T_1 models a single-server station whose firing rate is μ_1. At what time $t(N)$ will the marking $m_1 = N$ be reached? The first token will be deposited in P_1 at time τ_1: τ_1 is a random variable such that $\Pr[\tau_1 \leq t] = 1 - e^{-\mu_1 t}$ (Section 3.4.3). The second token will be deposited at $\tau_1 + \tau_2$ such that τ_2 has the same distribution function as τ_1, and so on. This process is modeled by the markov process in Figure b: $\mathbf{m}_i$ corresponds to the state such that $m_1 = i$, i.e. there are N tokens in P_1 when the state $\mathbf{m}_N$ is reached. As Figure b models an Erlang law, the mathematical expectation and the variance of the random variable $t(N)$ are:

$$E[t(N)] = \frac{N}{\mu_1} \quad \text{and} \quad \sigma^2[t(N)] = \frac{1}{N \cdot (\mu_1)^2}. \tag{5.84}$$

Consider now the continuous PN in Figure 5.37c. Since $m_1 = V_1 \cdot t$, $m_1 = N$ at the deterministic time

$$t(N) = \frac{N}{V_1}. \tag{5.85}$$

The comparison between (5.84) and (5.85) shows that the continuous model in Figure c, with $V_1 = \mu_1$, provides an approximation of the stochastic process in Figure a. The error is small if N is large (σ^2 decreases when N increases).

Let us now compare the behaviors of the stochastic PN in Figure 5.38a and the continuous PN in Figure b. The significant initial marking is the same for both and $V_j = \mu_i$ for $i = 1, 2$. The behaviors are illustrated in Figure c.

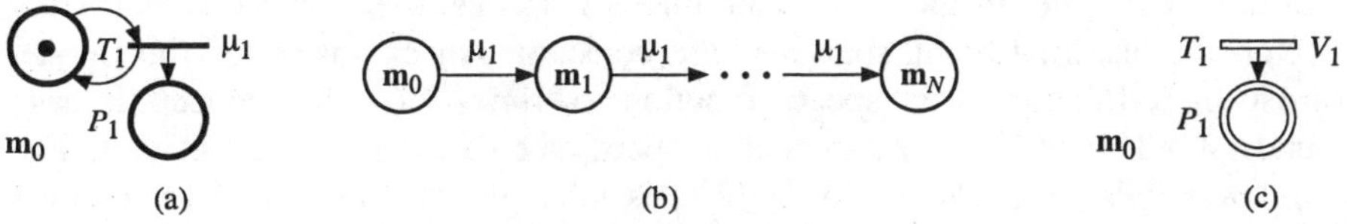

Figure 5.37 (a) Stochastic PN. (b) Markov process. (c) Continuous timed PN.

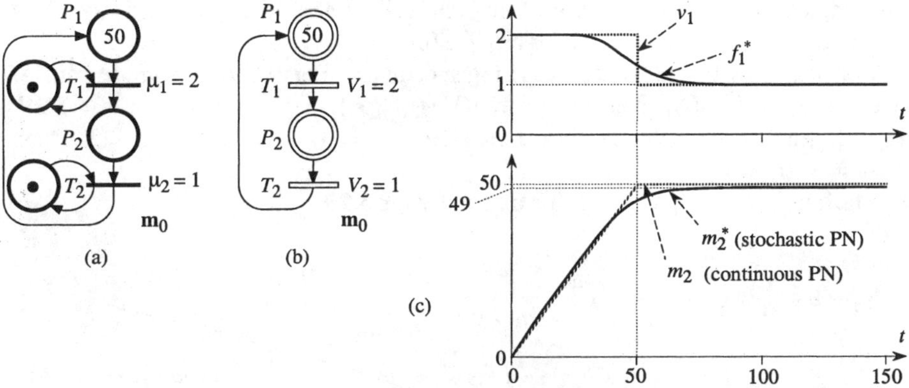

Figure 5.38 (a) Stochastic PN. (b) Continuous PN. (c) Both behaviors.

For the continuous PN: $v_1(t) = 2$ for $t \in \left|0, 50\right|$ and $v_1(t) = 1$ after, whereas $v_2(t) = 1$ for any $t \geq 0$; $m_1(t)$ decreases from 50 to 0 and $m_2(t)$ increases from 0 to 50 in the interval $t \in \left|0, 50\right|$ and $\mathbf{m}$ does not change afterwards.

For the stochastic PN, the *average firing frequencies* and the *average markings* may be calculated by standard Markov process analysis methods. For the stationary behavior (Section 3.4.3.3) $f_1^* = f_2^* \approx 1$ (error about $2^{-(N+1)} \approx 10^{-15}$), $m_1^* \approx 1$, and $m_2^* \approx 49$ are obtained. I.e., the stationary firing frequencies of the stochastic PN are practically equal to the stationary firing speeds of the continuous PN. Furthermore, the markings are close to each other if the initial marking is large enough. When the stationary behavior is reached, $m_2 = 50$ for the continuous PN whereas $m_2^* \approx 49$ for the stochastic PN (if $m_1(0) + m_2(0) = N$, then $\lim_{N\to\infty}(m_2^*/N)=1$). For large initial marking values, the approximation of the transitory behavior may also be acceptable (Figure 5.38c).

5.5 MAXIMAL SPEEDS FUNCTIONS OF TIME

In the CCPN, i.e. in the timed continuous PN considered up to now in this chapter, the maximal firing speeds V_j are constant. An extension of this model consists of defining maximal speeds functions of time, $V_j(t)$. The concept is easy to understand, even if the instantaneous speed calculation is more difficult. *The basic principles are the same*: 1) if T_j is strongly enabled at time t, then $v_j(t) = V_j(t)$; 2) if T_j is weakly enabled, then $v_j(t)$ may be less than $V_j(t)$; 3) the *fundamental equation* (5.81) in Section 5.3.3 is the same.

Let us consider the example in Figure 5.39. The maximal speeds are $V_1(t) = t$ for $0 \le t < 1.5$ and $V_1(t) = 1.5$ for $t \ge 1.5$, and $V_2(t) = t^{0.5}$ for $t \ge 0$. The initial marking is $m_1 = 0$. The behavior of this PN is illustrated in Figure b. Since T_1 is a source transition, $v_1(t) = V_1(t)$ for any $t \ge 0$.

Consider now the instantaneous firing speed $v_2(t)$. For $0 \le t \le 1$: m_1 remains 0 because $V_1(t) \le V_2(t)$, then $v_2(t) = \min(V_1(t), V_2(t)) = V_1(t)$. For $1 < t < 3.29$: $m_1 > 0$, then $v_2(t) = V_2(t)$. For $t \ge 3.29$: $m_1 = 0$ and $V_1(t) \le V_2(t)$, then $v_2(t) = V_1(t)$ again.

Marking m_1 is 0 up to $t = 1$. Then, for $1 \le t \le 3.29$,

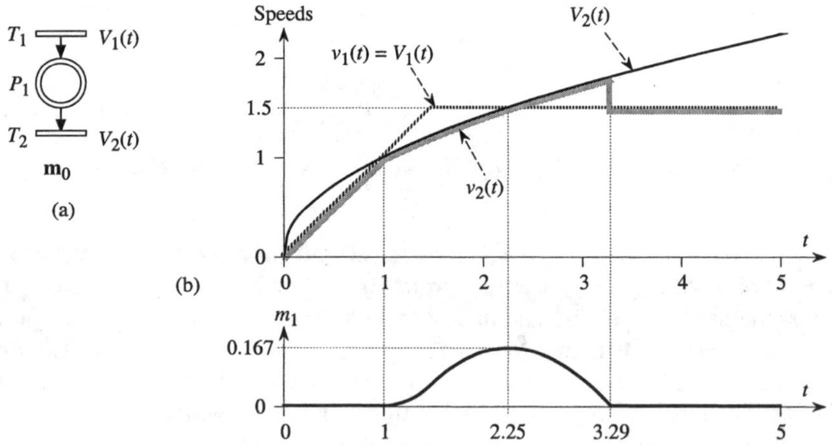

Figure 5.39 Continuous timed PN with maximal speeds functions of time.

$$m_1(t) = \int_1^t (v_1(u) - v_2(u)) \cdot du = \int_1^t (V_1(u) - V_2(u)) \cdot du.$$

For $1 < t < 2.25$, $B_1(t) = V_1(t) - V_2(t) > 0$, then $m_1(t)$ increases from 0 to 0.167. For $2.25 < t < 3.29$, T_2 remains strongly enabled, but $B_1(t) = V_1(t) - V_2(t) < 0$, then $m_1(t)$ decreases from 0.167 to 0. For $t \geq 3.29$, $m_1 = 0$.

Definition 5.8 The definition of a **continuous PN with maximal speeds functions of time** is similar to the definition of a *timed continuous PN* (Definition 5.2, Section 5.1.3.1), except that Spe is a function from the set T to the set of functions of time, defined for $t \in [0, \infty[$ and whose values are non-negative.

❑

The maximal speeds depending on time may be used for modeling the environment of a system modeled by a timed continuous PN. This is illustrated in Figure 5.40. Speed $V_1(t)$, associated with a *source transition* (Figure a) may model an *outside production.* Speed $V_2(t)$, associated with a *sink transition* (Figure b) may model a maximal *outside consumption.* For example, in Figure 5.39a, the marking $m_1(t)$ may model the available quantity of some product. The product is deposited in P_1 by firing of T_1. If some quantity of product is available at t, i.e. $m_1(t) > 0$, it is consumed at speed $v_2(t) = V_2(t)$. Otherwise, the consumption speed is limited by the production speed: $v_2(t) = v_1(t) = V_1(t)$. (In this model, unsatisfied demand is not memorized.)

Speed $V_3(t)$, associated with an "ordinary" transition (possibly, but not necessarily, a source or sink transition) may model an *outside control* (Figure c). An example is given in Figure d. Assume some product A can be transformed either into a product $A1$ or into a product $A2$. Assume that, at some time, it appears pertinent to reduce the production of product $A1$ (less demand or any other reason): the maximal speed $V_1(t)$ (in Figure d) may then be decreased.

Remark 5.17 Note that the reduction of $V_j(t)$ to 0 corresponds to a (temporary) modification of the structure (similar to cancelation of T_j).

❑

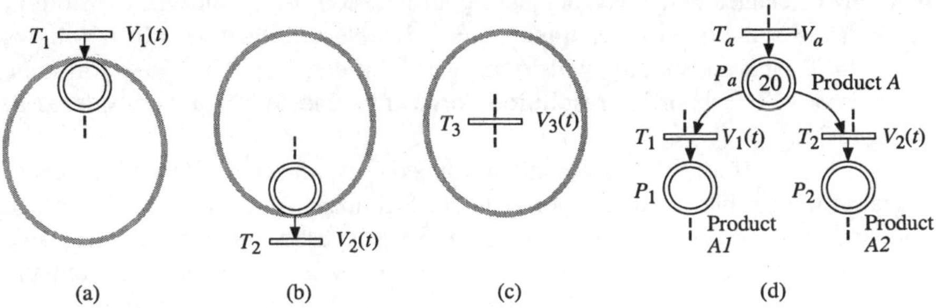

Figure 5.40 (a) Outside production. (b) Outside consumption. (c) Outside control. (d) Example of case c.

If the *maximal speeds depending on time are piecewise constant*, during a time interval where all the maximal speeds are constant, the PN corresponds to a CCPN (considered in this chapter except this section). A change of maximal speed vector then corresponds to a change of CCPN. An algorithm based on this simple idea is given in Appendix M.

NOTES and REFERENCES

Continuous PNs were introduced in [DaAl 87]. The model presented was the basic timed continuous PN, now called CCPN. The concepts of place fed, of weak and strong enabling, were the same as in this book. However, the algorithm seeking the enabled transitions was limited to the case where there is a "positive solution", i.e. the conflict resolutions are such that any enabled transition has a positive firing speed (for example, if all the conflicts are solved by sharing, a positive solution is obtained). In fact, all the transitions enabled if there is a positive solution are *potentially enabled* if the global conflict resolution (i.e., set of local conflict resolutions) is not specified. If T_j is potentially enabled, there is a global conflict resolution such that v_j is positive. On the other hand, v_j may be nil for another global conflict resolution. Assume, for example: 1) there is a path $P_1T_1P_2T_2$ and a conflict $\langle P_1, \{T_1, T_3\}\rangle$, and every transition T_1, T_2, and T_3, has a single input place; 2) $m_1 = m_2 = 0$ and $I_1 < V_3$. All the transitions concerned are potentially enabled: if there is a sharing $[\alpha_1 T_1, \alpha_3 T_3]$, v_1, v_2, and v_3 are positive. On the other hand if there is a priority $T_3 < T_1$, T_1 and T_3 are enabled, but T_1 is not fired (i.e., $v_1 = 0$) then T_2 is not enabled, according to Definition 5.3 in Section 5.1.3.2. In other words, enabling of some weakly enabled transitions depends on both the marking *and* the global conflict resolution; this is explicitly taken into account in the algorithms in Section 5.3.

In [DaAl 87] and [AlDa 98a], the speeds are calculated given the hypothesis that all the conflicts are solved by priorities (Hypothesis 5.1 in Section 5.3.2). The speeds are calculated iteratively, and during an iteration, v_a is calculated before v_b if $T_a < T_b$. For the previous example, $v_3 = I_1$ is obtained, then $v_1 = 0$; it follows that $v_2 = 0$. Sharing among transitions was considered in [Ch 93a] for some simple cases. The conflict resolution for a free-choice PN is considered in [BaGiMe 00].

The idea to find the instantaneous speeds as the solution of a linear programming problem was proposed by F. Balduzzi, A. Giua, and G. Menga [BaGiMe 98 & 00]. In this book, the same idea is used, but several successive LPPs are used in order to take into account the difficult problem of automatic conflict resolution (a relatively wide range of accepted solutions is proposed).

The algorithm proposed by the authors in 1987 was not able to obtain consistent continuity between IB-states for some particular cases of continuous

PNs. This problem was pointed out by P. Caspi [Ca 87]. Caspi's problem is presented in Figure 5.34a. It is now solved thanks to the marking 0^+ in $\tilde{\mathbf{m}}_1$ (the vector $\mathbf{v}_2$ in Figure 5.34b cannot be obtained from marking $\mathbf{m}_1 = (0, 0, 0)$).

Although the possibility of *infinite maximal speed* was explicitly presented in previous papers, its influence on behavior is studied for the first time in this book. A theoretical interest is the consistency of all the non-autonomous models in this book (immediate transitions exist in discrete, continuous, and hybrid PNs). A practical interest: even if the existence of an I-phase can always be avoided when a CCPN is built, such an I-phase can appear "naturally" in a PN whose maximal speeds are piecewise constant (Figure M.4 in Appendix M), or in a hybrid PN (see Algorithm 6.1 in Section 6.2.4 and Exercise 6.4).

In addition, the concept of *synchronized continuous transition* may be deduced from the infinite maximal speed notion (Remark 5.6, Section 5.1.2.5). This is explained in Section G.3 in Appendix G, and illustrated in Exercise 5.4.

Continuous PNs whose maximal speeds are piecewise constant have been studied by E. Dubois and co-authors [DuAlDa 94a & 94b][Du 95]. Approximation of the stochastic process in Figure 5.37 is drawn from [ReSi 01].

In [Ha 03], a model called free speed CCPN is studied. The model assumes that a transition T_j whose maximal speed is V_j may be fired at any speed v_j less than V_j (if T_j is strongly enabled; otherwise, the maximal possible speed may be reduced). Conceptually, this behavior is similar to the behavior of a timed discrete PN in which a transition may be fired at any time after it became firable (Remarks 3.11 in Section 3.4.1 and 5.8a in Section 5.1.3.2; see also Remark 5.10 in Section 5.2.2). The infinite set of possible behaviors is defined by a polytope. From this result, [Ha 03] proposes a conflict resolution in which the speed maximization takes priority over the priority between transitions. An almost similar approach, in the context of hybrid PNs, is proposed in [BaGiMe 00] (see Notes & References of Chapter 6).

The concepts of neutral, absorbing, and generating weighted marked graphs (i.e., PNs with a structure of event graph and weighted arcs) were known for discrete PNs [Te *et al.* 92][ChZhWa 93]. Properties of the corresponding continuous nets are studied in [MoDeSa 01]. Particularly, a method for directly calculating (i.e. without evolution graph) the final marking of a neutral net is proposed (a neutral net generalizes the example in Figure 5.36a, whereas the examples in Figures 5.36b and c are respectively generating and absorbing).

The *autonomous* continuous PNs presented in Chapter 4 is a basis from which *various kinds of timings* could be considered. The authors have chosen a maximal speed associated with every transition because this model is simple and powerful for modeling various systems. However, other timings have been proposed. In [BrBl 90], in addition to speeds associated with the transitions, constant timings are associated with the places (some marking deposited in P_i at t becomes available at $t + d_i$, for enabling the output transitions of P_i). In [Ol 93], a continuous PN in which constant timings are associated with the places and the transitions are immediate is proposed; various properties of event graphs are obtained with the help of max-plus algebra. The model in [CoGaQu 95 & 98] is similar while not limited to event graphs.

It was observed in Solution to Exercise 3.11 that a processing time and a transportation time are different in nature. For a continuous system, the same concepts may be illustrated with the following example. A liquid flows from tank 1 to tank 2 through a regulated valve and a long pipe. The flow through the valve can be modeled by a "speed" (3 liters/minute for example) and the time spent in the pipe can be modeled by a pure "delay" (5 minutes for example). This system is easily modeled with the model in [BrBl 90]. The model in [Ol 93] [CoGaQu 95 & 98] is well adapted for modeling the delay but not the speed. The model in [DaAl 87] (and in this chapter) is well adapted for modeling the speed but not the delay. This modeling will be possible with a timed extended hybrid PN (Section 6.4.3. See also Exercise 6.6).

6

Timed Hybrid Petri Nets

The basic model is an autonomous hybrid PN, as defined in Chapter 4, plus a timing associated with every discrete transition and a maximal speed associated with every continuous transition (or more formally a flow rate, as will be specified in Section 6.1.4).

For the basic model, the timings and the flow rates are constant. *The behavior and a formal definition of this model are presented in Section 6.1. An algorithm used to calculate the behavior for such a PN is proposed in Section 6.2.*

Other models, in which the maximal speeds or timings are not constant, are presented in Section 6.3. Finally, timed extended *hybrid Petri nets (based on the autonomous model introduced in Section 4.4) are presented with various application examples in Section 6.4.*

In this chapter, the expression "hybrid PN" means "timed hybrid PN" (with constant timings and maximal speeds up to the end of Section 6.2). Whenever an autonomous PN is concerned, this is specified.

6.1 DEFINITION OF THE MODEL

Generally speaking, the *discrete transitions* in a non-autonomous hybrid PN may be fired as the transitions in a discrete PN (i.e., they may be *synchronized*, or *timed* with *constant* or *stochastic* timings). Similarly, the *continuous transitions* in a non-autonomous hybrid PN may be fired as the transitions in a continuous PN (i.e., they can be *synchronized*, according to Appendix G, or maximal speeds may be *constant*, or *function of time*, or *function of the marking* as explained in the next chapter). However, the *basic model* is such that all the timings and maximal speeds are *constant*. This section is devoted to the presentation of this basic model, which will be called **CHPN**, C standing for constant and H for hybrid (when this "name" will be useful for comparison with other models).

6.1.1 Intuitive Presentation

Figure 6.1a presents a timed hybrid PN in which the set of places is $P = \{P_1, P_2, P_3, P_4\}$: the sets of discrete places and continuous places are $P^D = \{P_1, P_2\}$ and $P^C = \{P_3, P_4\}$, respectively. Similarly, the sets of transitions are $T = \{T_1, T_2, T_3, T_4\}$, $T^D = \{T_1, T_2\}$ and $T^C = \{T_3, T_4\}$. The initial marking is

$$\mathbf{m}_0 = (\mathbf{m}_0^D, \mathbf{m}_0^C) \text{ where } \mathbf{m}_0^D = (1, 0) \text{ and } \mathbf{m}_0^C = (180, 0).$$

The timings associated with the discrete transitions are $d_1 = 90$ and $d_2 = 60$ and the maximal speeds associated with the continuous transitions are $V_3 = 3$ and $V_4 = 2$ (another notation will be introduced in Section 6.1.4).

Note that there are two marking invariants: $m_1 + m_2 = 1$ and $m_3 + m_4 = 180$. The marking of the PN can thus be represented by (m_1, m_3) instead of (m_1, m_2, m_3, m_4) which is redundant. This enables us to represent the reachable space in the plane: see Figure 6.1b where A represents the initial marking. The evolution of the hybrid PN may be analyzed by the evolution graph in Figure 6.1c. This graph is made of IB-states (standing for *invariant behavior states*) and transitions among them. An IB-state is such that *the marking of the discrete part and the instantaneous speed vector of the continuous part remain constant* as long as the system is in the same IB-state. (More details are given in Section 6.1.5.3.)

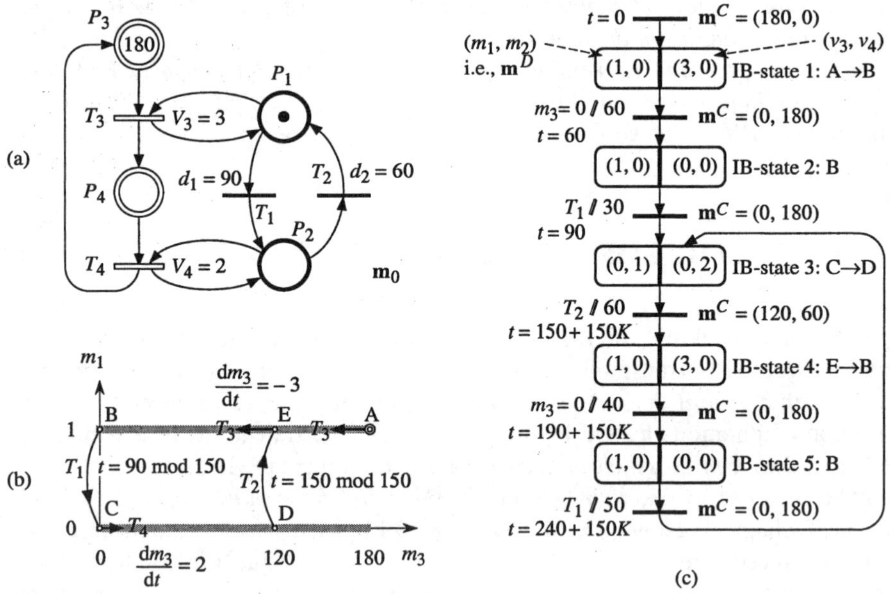

Figure 6.1 Example. (a) Timed hybrid Petri net. (b) Reachable space. (c) Evolution graph.

IB-state 1 in Figure c corresponds to the behavior from the initial state in Figure a: the marking of the discrete part is $\mathbf{m}^D = (m_1, m_2) = (1, 0)$ and the instantaneous speed vector is $(v_3, v_4) = (3, 0)$ (because T_3 is strongly enabled, whereas T_4 is not enabled since $m_2 = 0$). Continuous marking evolves continuously and linearly as long as the system is in this IB-state; at $t = 60$, m_3 becomes 0 and transition T_3 is no longer firable. During IB-state 1, $\mathbf{m}^C$ evolves continuously from (180, 0) to (0, 180) (from point A to point B in Figure b).

When m_3 reaches the value 0, at $t = 60$ (i.e. point B in Figure b), the marking keeps the same value since $v_3 = v_4 = 0$: IB-state 2.

The next event is the firing of T_1, occurring at $t = 90$. In Figure b, there is a jump from point B to point C. At $t = 90$, T_4 becomes (strongly) enabled and $v_4 = V_4 = 2$. This behavior corresponds to IB-state 3; passing from IB-state 2 to IB-state 3 is denoted by T_1 / 30 in Figure c, meaning that firing of T_1 occurs 30 time units after the time when IB-state 2 was reached. IB-state 3 corresponds to evolution from point C to point D in Figure b.

The next event is the firing of T_2, occurring at $t = 90 + 60 = 150$ (i.e. before emptying of m_4). In Figure b, there is a jump from point D to point E. At $t = 150$, T_4 is disabled while T_3 becomes enabled. This behavior corresponds to IB-state 4: evolution from point E to point B in Figure b. And so on.

The behavior is periodical, a period corresponding to IB-states 3, 4, and 5 in Figure c (K is the number of times this period has been performed). IB-state 5 is similar to IB-state 2, except it does not have the same duration (because the remaining time to firing of T_1 is not the same when both IB-states are reached).

For the *autonomous* hybrid PN corresponding to Figure 6.1a, i.e. without timings and firing speeds, the reachable space corresponds to the two grey segments of lines in Figure 6.1b. For the *timed* hybrid PN in Figure 6.1a, the reachable space is smaller: the segment of line on the right side of point D is never reached.

6.1.2 Events To Be Considered

According to Property 4.5 in Section 4.2.2, in a hybrid PN, a change of macro-marking, hence a change of the set of enabled transitions, or of enabling degrees of D-transitions, can occur only if an event of the following type occurs: C1-event, C2-event, D1-event, or D2-event. Let us illustrate these events in the case of *timed* hybrid PNs. For each part (a to f) of Figure 6.2, it is assumed that the (partial) marking corresponds to $\mathbf{m}_0$ and that the partial PN presented is not affected by another change.

Figure a. From $t = 0$, $v_1 = 1$. At $t = 3$, m_2 *becomes empty*: this is a C1-event.

Figure b. From $t = 0$, T_3 is enabled. This *D-transition will be fired* at $t = 2$: this is a D1-event.

Figure c. From $t = 0$, $v_6 = 0.1$ and T_5 is not enabled because $m_7 < \text{Pre}(P_7, T_5)$. At $t = 5$, m_7 *reaches the value* $1 = \text{Pre}(P_7, T_5)$ and T_5 *becomes enabled*: this is a

D2-event (in a more general case, a D2-event corresponds to a modification of the enabling degree of a D-transition, increasing or decreasing).

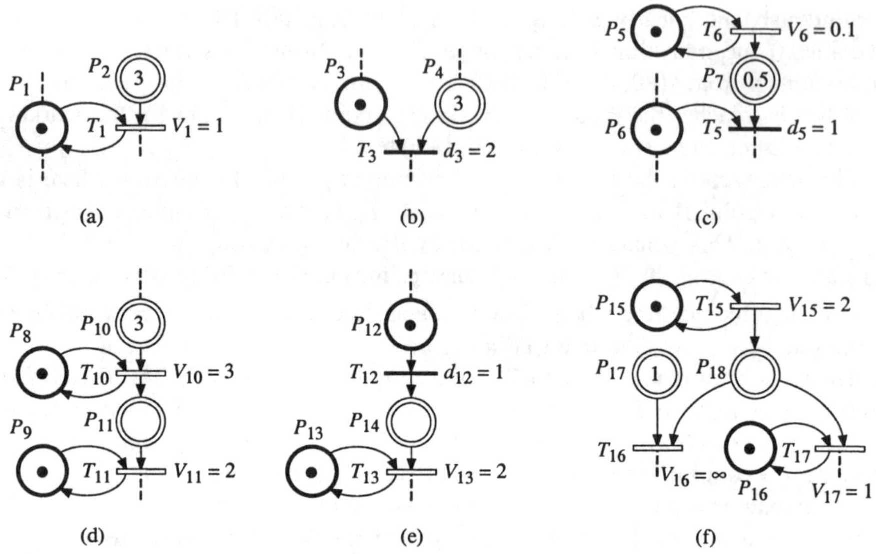

Figure 6.2 Illustrations. (a) C1-event. (b) D1-event. (c) D2-event. (d), (e), and (f) C2-events are immediate consequences of other events.

Figure d. From $t = 0$, $v_{10} = 3$, and at $t = 0$, $v_{11} = 0$ because $\tilde{m}_{11} = 0$. At $t = 0^+$, *the unmarked C-place* P_{11} *becomes marked* (as for P_4 in the continuous PN in Figure 5.4, Section 5.1.2.1): this is a C2-event, the immediate consequence of the initial time behavior.

Figure e. At $t = 0$, T_{12} is enabled. At $t = 1$, the D-transition T_{12} is fired and *the unmarked C-place* P_{14} *becomes marked*: this is a C2-event, the immediate consequence of a D1-event.

Figure f. From $t = 0$, $v_{15} = 2$, $v_{16} = 2$, and $v_{17} = 0$, because the immediate transition T_{16} takes priority over the non-immediate transition T_{17} (Remark 3.2 in Section 3.2.1). At $t = 0.5$, the C-place P_{17} becomes empty and T_{16} is no longer enabled; it follows that *the unmarked C-place* P_{18} *becomes marked*: this is a C2-event, the immediate consequence of a C1-event.

These examples illustrate that *changing from one IB-state to another can occur only* if a C1-event, a D1-event, or a D2-event occurs. Each time a C2-event occurs, it is an immediate consequence of either initialization (Figure 6.2d), or a C1-event (Figure f), or a D1-event (Figure e).

6.1.3 Conflict Resolutions

Conflicts in an *autonomous* hybrid PN were studied in Section 4.2.3.3. Let us now specify the behaviors when *time* is involved. Only *deterministic resolutions* are considered. In the sequel, examples corresponding to structural conflicts denoted by $K_c = \langle P_c, \{T_a, T_b\}\rangle$ will be presented. Four cases will be considered in turn (all the possibilities are considered in these four cases).

Case 1: $T_a, T_b \in T^D$.
Case 2: $T_a, T_b \in T^C$ and $P_c \in P^C$.
Case 3: $T_a \in T^D$ and $T_b \in T^C$, or vice-versa.
Case 4: $T_a, T_b \in T^C$ and $P_c \in P^D$.

Case 1

If the conflict is *between two D-transitions* T_a and T_b, the conflict resolution is *similar to the conflict resolution in a discrete timed PN* (P_c may be either a D-place or a C-place). The resolution may be a *priority* between T_a and T_b (or a deterministic alternate firing which can be explicitly modeled: Appendix B). However, since the first transition becoming firable is fired, an *actual conflict* can occur *only* if both transitions are firable exactly at the same time (Section 3.4.2.2). Hence, the priority is relevant only if this simultaneity occurs.

Case 2

If the conflict is *between two C-transitions* T_a and T_b, and P_c is a *C-place*, the conflict resolution is *similar to the conflict resolution in a continuous timed PN* (i.e., *priority* or *sharing* according to Sections 5.2 and 5.3).

Case 3

This conflict resolution is presented as a rule because it is a choice of the authors. Examples in Figure 6.3 illustrate the reasons of this choice.

Rule 6.1 If the conflict is *between a D-transition and a C-transition, the discrete transition takes priority over the continuous transition.*

❑

As before, it is assumed, without loss of generality, that the markings in the figures correspond to $\mathbf{m}_0$. Two cases have to be considered: either the place related to the conflict is a D-place, or it is a C-place.

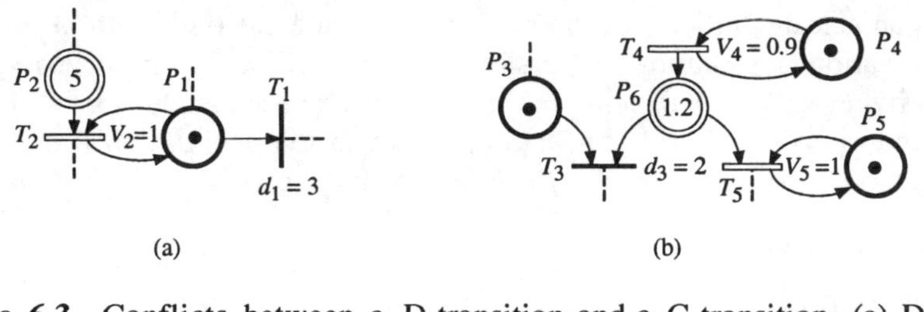

Figure 6.3 Conflicts between a D-transition and a C-transition. (a) D-place. (b) C-place.

In Figure 6.3a, the structural conflict is $K_1 = \langle P_1, \{T_1, T_2\}\rangle$, where P_1 is a *D-place*. The token in place P_1 may represent the availability of some resource for some production at speed $V_2 = 1$ represented by transition T_2 (continuous firing of T_2 may be the production of a machine whose availability is represented by a token in P_1, or the flow through a valve which is open when there is a token in P_1). At $t = 0$, both T_1 and T_2 are enabled. Continuous firing of T_2 begins immediately whereas T_1 will become firable at $t = d_1 = 3$. Hence, at $t = 3$ there is an actual conflict between T_1 and T_2. According to the meaning of the PN, *the priority of T_1 over T_2 is intuitively logical* (at $t = 3$, the machine becomes unavailable or the valve is closed). This is general: the firing of a C-transition corresponds to continuous working while the firing of a D-transition corresponds to a brutal change of state of the system.

In Figure 6.3b, the structural conflict is $K_2 = \langle P_6, \{T_3, T_5\}\rangle$, where P_6 is a *C-place*. At $t = 0$, both T_3 and T_5 are enabled. Continuous firing of T_5 begins immediately whereas T_3 will become firable at $t = d_3 = 2$. At $t = 2$ there is an actual conflict between T_3 and T_5 because both can be fired and the marking $m_6(2) = 1 = \text{Pre}(P_6, T_3)$, since $m_6(2) = 1.2 + 2\,(v_4 - v_5)$. This is a very special case because: if $m_6 > \text{Pre}(P_6, T_3)$ when T_3 can be fired, there is no conflict (T_3 is fired while T_5 continues to be fired); if $m_6 < \text{Pre}(P_6, T_3)$, there is no conflict because T_3 is no longer enabled. Hence, the priority of T_3 over T_5 specifies only the special case where $m_6 = \text{Pre}(P_6, T_3)$.

Remark 6.1 The authors have never encountered a practical example for which the priority of a D-transition over a C-transition would be a handicap for modeling. However, if it were necessary, the corresponding behavior could be modeled by a timed extended hybrid PN (Exercise 6.10).

Case 4

This case is illustrated in Figure 6.4. Typically, a machine (represented by the token in P_1), can be used to manufacture L type products (continuous firing of T_1) or R type products (continuous firing of T_2). If the machine is allocated to L type production, this production is performed at the maximal speed $v_1 = V_1 = 1$ (as long as $m_3 > 0$ or $I_3 \geq V_1$). Similarly, if the machine is allocated to R type production, this production is performed at the maximal speed $v_2 = V_2 = 2$ (as long as $m_4 > 0$ or $I_4 \geq V_2$).

Since the machine cannot be allocated *simultaneously to both* productions, there is an actual conflict. For the marking in Figure 6.4 (i.e., with $m_3 > 0$ and $m_4 > 0$), various resolutions are possible. a) Priority $T_1 < T_2$: $v_1 = 1$ and $v_2 = 0$. b) Priority $T_2 < T_1$: $v_1 = 0$ and $v_2 = 2$. c) Half time for each production: $v_1 = V_1 / 2 = 0.5$ and $v_2 = V_2 / 2 = 1$. d) Sharing such that the *proportion* β of time is allocated to L production and $(1 - \beta)$ of the time is allocated to R production:

$$v_1 = \beta\, V_1 \text{ and } v_2 = (1 - \beta)\, V_2, \tag{6.1}$$

hence, from (6.1):

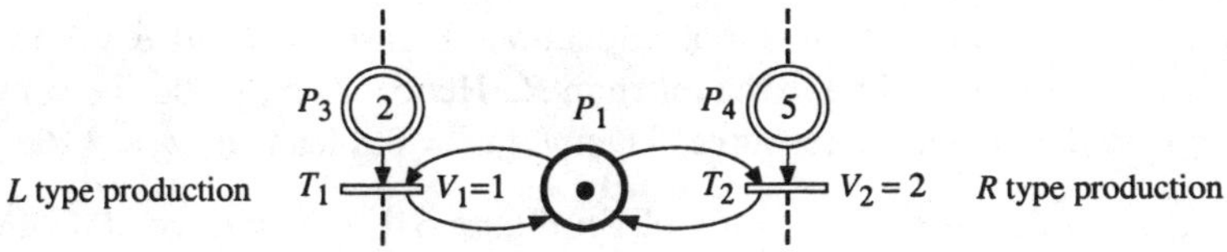

Figure 6.4 Two C-transitions share a common D-place.

$$\frac{v_1}{V_1} + \frac{v_2}{V_2} = \beta + (1 - \beta) = 1 = m_1. \tag{6.2}$$

In a general case, i.e. for any values of m_3, m_4, I_3, and I_4, the resolution must satisfy the following inequality: $\frac{v_1}{V_1} + \frac{v_2}{V_2} \leq 1 = m_1$.

As in a continuous PN, the resolution may be either a *priority*, $T_1 < T_2$ or conversely, or a *sharing*.

Remark 6.2 In a *case 2* conflict, similar to a conflict in a continuous PN, a sharing *shares a flow* (the feeding speed of the C-place involved in the conflict). In a *case 4* conflict, a sharing *shares available servers* (the number of tokens in the D-place involved in the conflict).

❑

From Remark 6.2, it seems pertinent that a sharing for a case 4 conflict be expressed in number of servers. For our example in Figure 6.4, a sharing denoted by $[\beta_1 T_1, \beta_2 T_2]$ aims at

$$\frac{1}{\beta_1} \cdot \frac{v_1}{V_1} = \frac{1}{\beta_2} \cdot \frac{v_2}{V_2}. \tag{6.3}$$

If $\beta_1 = \beta_2$, (6.3) corresponds to $v_1 / V_1 = v_2 / V_2$, i.e., the server is used half the time for *L* type production and half the time for *R* type production, since $v_1 / V_1 + v_2 / V_2 = 1$ (hence, $v_1 / V_1 = v_2 / V_2 = 0.5$).

From (6.2) and (6.3), the following resolution is obtained:

$$v_1 = \frac{\beta_1}{\beta_1 + \beta_2} \cdot V_1 \quad \text{and} \quad v_2 = \frac{\beta_2}{\beta_1 + \beta_2} \cdot V_2. \tag{6.4}$$

Given $V_1 = 1$ and $V_2 = 2$: for $\beta_1 = \beta_2$, $v_1 = V_1 / 2 = 0.5$ and $v_2 = V_2 / 2 = 1$ are obtained from (6.4); for $\beta_1 = 2$ and $\beta_2 = 1$, $v_1 = 2\,V_1 / 3 = 0.67$ and $v_2 = V_2 / 3 = 0.67$ are obtained.

Let us assume now that $m_3 = 0$, $I_3 = 0.2$, $m_4 = 5$, and that $\beta_1 = \beta_2$. Transition T_1 is weakly enabled, it can be fired at most at speed $v_1 = 0.2$. The coefficient of V_1 in (6.4) is no longer the *proportion of server* **used** by transition T_1, but the proportion of server initially **allocated** to T_1. For our example, the *allocations* to the transitions, denoted by $a(T_j)$, are $a(T_1) = a(T_2) = 0.5$. As $I_3 < a(T_1) \cdot V_1$, $v_1 = \min(a(T_1) \cdot V_1, I_3) = 0.2$. Since T_1 "*cannot use its allocation completely*",

T_2 may use more than its own initial allocation. Transition T_1 *uses* $v_1 / V_1 = 0.2$ of the resource corresponding to the token in P_1. Hence, T_2 may use the remaining part, i.e. $1 - 0.2 = 0.8$ of the resource. Then $v_2 / V_2 = 0.8$ leads to $v_2 = 1.6$.

From a generalization of this example in Figure 6.4, the concept of D-enabling degree will be introduced in Section 6.1.4. Then a *generalization of the conflict resolution by sharing* (if there are several tokens in P_1 or more than two transitions in $P_1°$), will be given in Section 6.2.1.

6.1.4 Flow Rate and Maximal Firing Speed

In this section, the notion of flow rate will be specified and the concept of D-enabling degree will be introduced.

Roughly speaking, a *flow rate* denoted by U_j, corresponds to the maximal speed provided by *a server* associated with transition T_j. In Figure 6.5a, the flow rate associated with T_1 is $U_1 = 2$; since the number of servers associated with T_1 is *always one* (one token in P_1), the maximal speed is $V_1 = U_1 \cdot m_1 = U_1 = 2$. Let us note that U_1 is homogeneous with the inverse of a time whereas V_1 is homogeneous with a marking divided by a time. This is consistent with the representation of a multi-server in a timed continuous PN, illustrated by Figure 5.2 in Section 5.1.1: V_2 is the product of the flow rate $1 / d_2$ by the number m_6 of servers.

In Figure 6.5b, the number of tokens in P_3, i.e. the number of servers corresponding to transition T_5, is not constant. The analysis of the subnet made up of P_3 to P_5, T_2 to T_4, and the related arcs, shows that P_3, presently empty, will contain sometimes one token and sometimes two tokens. Given $U_5 = 6$, $V_5 = 0$ when $m_3 = 0$, $V_5 = 6$ when $m_3 = 1$, and $V_5 = 12$ when $m_3 = 2$.

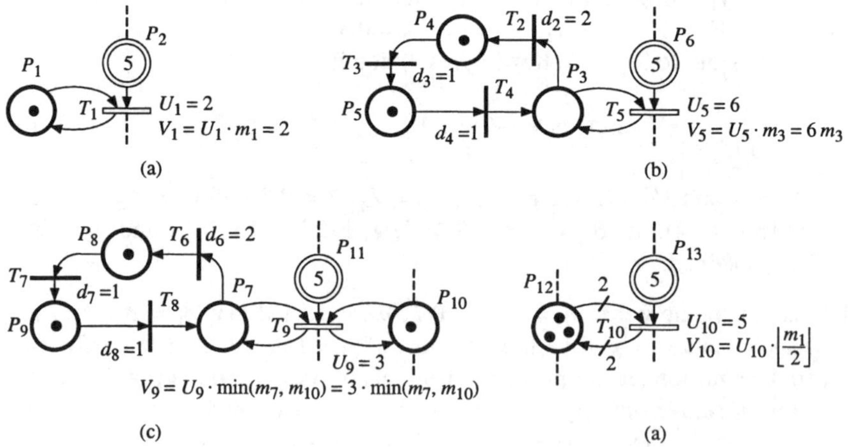

Figure 6.5 Maximal firing speed V_j versus flow rate U_j.

In Figure 6.5c, C-transition T_9 has two input D-places, namely P_7 and P_{10}. This means that firing of T_9 requires two kinds of "servers" represented respectively by the tokens in P_7 and in P_{10}. For example, a token in P_7 represents an available machine and a token in P_{10} represents an available workman able to use a machine. If either $m_7 = 0$ or $m_{10} = 0$, it means that either there is no machine available or no workman, hence T_9 cannot be fired. If $m_7 = 1$ and $m_{10} = 1$: there are a machine available and a workman, then T_9 can be fired at $V_9 = U_9 = 3$. If $m_7 = 2$ and $m_{10} = 1$: there are two machines available but only one workman which cannot use both machines; T_9 can be fired at $V_9 = U_9 = 3$. All these cases correspond to $V_9 = U_9 \cdot \min(m_7, m_{10})$. The term $\min(m_7, m_{10})$ will be called the *D-enabling degree* of T_9.

Before defining formally the D-enabling degree of a C-transition, let us present the following remark.

Remark 6.3 Consider a self-loop $P_i \rightarrow T_j \rightarrow P_i$ such that P_i is a D-place and T_j a C-transition.

1) *Case a.* If $\mathrm{Pre}(P_i, T_j) = \mathrm{Post}(P_i, T_j) = 1$, a token in P_i corresponds to a "server" as was already explained. If $m_i = 1$, there is one server, and if $m_i > 1$, there are several servers. If there is a one server, for example, it may be shared between two or more transitions because this "server sharing" models effectively a *time sharing*: the server is devoted to a transition for a *part of the time*.

2) *Case b.* If $\mathrm{Pre}(P_i, T_j) = \mathrm{Post}(P_i, T_j) = 2$, this means that *two tokens in P_i are required* for enabling T_j (Figure 6.5d). For example, if an operation can be performed by *two workmen together*, this operation cannot be performed if there is only one workman; unformally, a token in P_i is seen by T_j as a "partial server".

3) The D-enabling degree will be formally defined in the general case (Equation (6.5)). If the case b above does not exist, the definition is simplified (Equation (6.6)). Most of the results presented in the sequel are true in both cases, but the authors have never encountered a practical modeling problem in which case b (i.e. "partial servers") was required. In Section 6.2, presenting a calculation algorithm, only case a will be taken into account.

Definition 6.1 The **D-enabling degree** of a C-transition T_j for a marking $\mathbf{m}$, denoted by $D(T_j, \mathbf{m})$, is the enabling degree of T_j after *all the arcs from a C-place to a C-transition have been deleted.* Then[1],

$$D(T_j, \mathbf{m}) = \min_{P_i \in {}^{\circ}T_j \cap P^D} \left\lfloor \frac{m_i}{\mathrm{Pre}(P_i, T_j)} \right\rfloor. \tag{6.5}$$

❑

Equation (6.5) is illustrated in Figure 6.5d. For the very *important particular case* where all the arc weights linking a D-place to a C-transition are 1, (6.5) can be simplified as:

$$D(T_j, \mathbf{m}) = \min_{P_i \in {}^{\circ}T_j \cap P^D} m_i. \tag{6.6}$$

[1] The value $\lfloor x \rfloor$ is the greatest integer less than or equal to number x.

Definition 6.2 In a timed hybrid PN:

a) The **flow rate** U_j associated with transition T_j corresponds to its maximal speed if its D-enabling degree is 1.

b) The **maximal firing speed** of transition T_j is the *product* of its *flow rate* by its *D-enabling degree*:

$$V_j = U_j \cdot D(T_j, \mathbf{m}). \tag{6.7}$$

❑

Hence, given $0 \leq v_j \leq V_j$, it follows from (6.7) that $0 \leq v_j \leq U_j \cdot D(T_j, \mathbf{m})$.

Remark 6.4 For all the examples from the beginning of this chapter (except in this section), a value denoted by V_j was associated with a transition T_j. Formally, the values denoted by V_3 and V_4 in Figure 6.1 should have been denoted by U_3 and U_4 according to the notation introduced in this section. For all the other examples, $V_j = U_j$ because the D-enabling degree of every D-transition is one.

6.1.5 Formal Definitions

In this section, concepts introduced in Sections 6.1.1 to 6.1.4 are formalized and generalized.

6.1.5.1 Definition and Notations

Definition 6.3 A **timed hybrid PN** is a pair $\langle R, \text{tempo}\rangle$ such that:

R is a marked autonomous hybrid PN (Definition 4.4, Section 4.2.2);

Tempo is a function from the set T of transitions to the set of positive or zero rational numbers:

if $T_j \in T^D$, $d_j = \text{tempo}(T_j)$ = timing associated with T_j;

if $T_j \in T^C$, $U_j = \dfrac{1}{\text{tempo}(T_j)}$ = flow rate associated with T_j.

❑

From Definitions 6.2 and 6.3, for a C-transition T_j, U_j, then V_j, takes its value in $Q_+ \cup \{\infty\}$, just like the maximal speed of a transition in a timed continuous PN (Definition 5.2 in Section 5.1.3.1).

Let us recall that $\mathbf{m} = (\mathbf{m}^D, \mathbf{m}^C)$, where $\mathbf{m}^D$ and $\mathbf{m}^C$ are the markings related to places in P^D and P^C, respectively (Notation 4.3 in Section 4.2.2). Similarly to timed continuous PNs, the *marking used for calculation* of the speed vector is denoted by $\tilde{\mathbf{m}} = (\mathbf{m}^D, \tilde{\mathbf{m}}^C)$, in which the marking $\tilde{m}_i$ of a C-place may be 0, or 0^+, or a positive real number (Notation 5.3 in Section 5.1.3.1).

According to Section 4.2.2, the transitions in a hybrid PN, like the places, may be *ordered* in such a way that the index j of any D-transition T_j is always lower than the index h of any C-transition T_h. With this ordering, the following notations may be used.

Notation 6.1

a) The vector of instantaneous speeds may be denoted by $\mathbf{v} = (\mathbf{0}, \mathbf{v}^C)$. This vector whose dimension is $|T|$, contains $|T^D|$ 0's for the D-transitions; the dimension of $\mathbf{v}^C$ is $|T^C|$.

b) Let n_j denote the number of firings of D-transition T_j from the initial time. The vector of discrete firings may be denoted by $\mathbf{n} = (\mathbf{n}^D, \mathbf{0})$; this vector is made of the $|T^D|$-dimensional vector $\mathbf{n}^D$ corresponding to the numbers of firings of every D-transition, and $|T^C|$ zeros corresponding to the C-transitions.

c) By abuse of notation, *when there is no ambiguity*, $\mathbf{v}$ may be used instead of $\mathbf{v}^C$, and $\mathbf{n}$ may be used instead of $\mathbf{n}^D$. In other words, $\mathbf{v} = (\mathbf{0}, \mathbf{v}^C)$ or $\mathbf{v}^C$, and $\mathbf{n} = (\mathbf{n}^D, \mathbf{0})$ or $\mathbf{n}^D$, depending on the required dimension of the vector.

❑

Let S such that $\mathbf{m}_i \xrightarrow{S} \mathbf{m}_k$. The *fundamental equation* is $\mathbf{m}_k = \mathbf{m}_i + \mathbf{W} \cdot \mathbf{s}$, where $\mathbf{W}$ is the incidence matrix and $\mathbf{s}$ is the characteristic vector of S: the jth component of $\mathbf{s}$ is the quantity of firing of T_j in S. For a *hybrid* PN, $\mathbf{s}$ is a vector of $|T^D|$ *integers* and $|T^C|$ *real numbers*, according to (4.18) in Section 4.2.3.2.

We shall now present the *fundamental equation when time is involved.* In the sequel, time will be explicitly noted, for example $\mathbf{m}(t)$ denotes the marking $\mathbf{m}$ at time t. The characteristic vector $\mathbf{s}$ may be specified as:

$$\mathbf{s}(t) = \mathbf{n}(t) + \int_0^t \mathbf{v}(u) \cdot du\,, \tag{6.8}$$

where the first term of the sum corresponds to D-transitions and the second term to C-transitions. Hence,

$$\mathbf{m}(t) = \mathbf{m}(0) + \mathbf{W} \cdot \left(\mathbf{n}(t) + \int_0^t \mathbf{v}(u) \cdot du \right), \tag{6.9}$$

and
$$\mathbf{m}(t_2) = \mathbf{m}(t_1) + \mathbf{W} \cdot \left(\mathbf{n}(t_2) - \mathbf{n}(t_1) + \int_{t_1}^{t_2} \mathbf{v}(u) \cdot du \right) \tag{6.10}$$

is the **fundamental equation** for a *timed hybrid PN*. It works for any $0 \le t_1 \le t_2$. Equation (6.10) may be rewritten in a way pointing out the discrete part and the continuous part (using Notation 6.1):

$$\begin{bmatrix} \mathbf{m}^D(t_2) \\ \mathbf{m}^C(t_2) \end{bmatrix} = \begin{bmatrix} \mathbf{m}^D(t_1) \\ \mathbf{m}^C(t_1) \end{bmatrix} + \begin{bmatrix} \mathbf{W}^D & \mathbf{0} \\ \mathbf{W}^{CD} & \mathbf{W}^C \end{bmatrix} \cdot \left(\begin{bmatrix} \mathbf{n}^D(t_2) - \mathbf{n}^D(t_1) \\ \mathbf{0} \end{bmatrix} + \int_{t_1}^{t_2} \begin{bmatrix} \mathbf{0} \\ \mathbf{v}^C(u) \end{bmatrix} du \right). \tag{6.11}$$

Let us end this section with a property related to a periodical behavior.

Property 6.1 The existence of a T-invariant is a necessary condition for a periodical functioning of a timed hybrid PN.

Proof If a periodical behavior of period d has been reached, from (6.10) one obtains $\mathbf{m}(t + d) = \mathbf{m}(t) + \mathbf{W} \cdot \mathbf{p}(d) = \mathbf{m}(t)$, where $\mathbf{p}(d) = \mathbf{n}(t + d) - \mathbf{n}(t)$

$+ \int_t^{t+d} \mathbf{v}(u) \cdot du$ is the characteristic vector associated with the trajectory during a period. Since $\mathbf{W} \cdot \mathbf{p}(d) = \mathbf{0}$, $\mathbf{p}(d)$ is a T-invariant.

6.1.5.2 Enabling in Timed Hybrid Petri Nets

In Figure 6.6a, one machine (or server) is allocated to L type production and another is allocated to R type production. There is no conflict, both productions can be performed at their maximal speeds, i.e., $v_1 = V_1 = 1$ and $v_2 = V_2 = 2$, according to Definition 6.2b.

Assume now that *both servers are similar and gathered in a double server* as illustrated in Figure 6.6b. If one server is allocated to each production, the production speeds are as in Figure a, i.e. $v_1 = 1$ and $v_2 = 2$. However, in Figure 6.6b, both servers can be allocated to L type production, the production speed is then $v_1 = 2\,U_1 = 2$ while $v_2 = 0$ (similarly, if both servers are allocated to R type production, $v_2 = 2\,U_2 = 4$ while $v_1 = 0$).

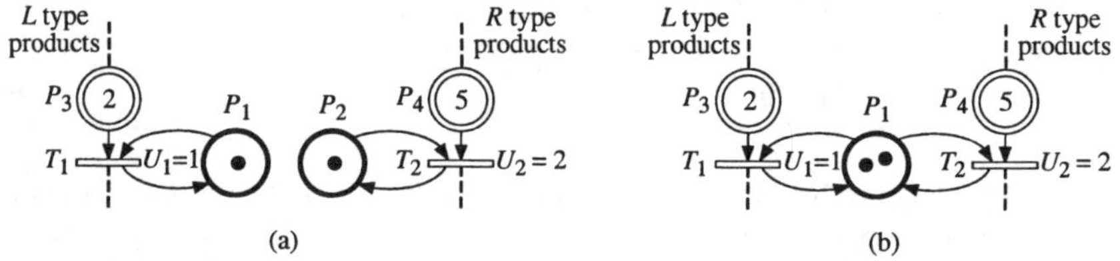

Figure 6.6 Two servers for two productions. (a) Independence. (b) Sharing.

According to Definitions 6.1 and 6.2, in Figure 6.6a, the D-enabling degrees $D(T_1, \mathbf{m}) = 1$ and $D(T_2, \mathbf{m}) = 1$, hence $V_1 = 1$ and $V_2 = 2$, are obtained. Similarly, in Figure 6.6b, $D(T_1, \mathbf{m}) = D(T_2, \mathbf{m}) = 2$, hence $V_1 = 2$ and $V_2 = 4$, are obtained.

Given Definitions 6.1 and 6.2, let us now compare the *semantics* of an *hybrid PN* and a *continuous PN*. This is illustrated in Figure 6.7.

Transition T_1 in Figure 6.7a, whose flow rate is $U_1 = 3$ and whose D-enabling degree is 2 may be fired at $v_1 = 2\,U_1 = 6$ (as long as $m_2 > 0$ or $I_2 \geq 6$; if $m_2 = 0$ and $I_2 < 6$, then $v_1 = I_2 < 6$). Transition T_1 in Figure b has the same behavior, since 1 server whose flow rate is $N \cdot U_j$ has the same maximal performance as N servers whose flow rates are U_j. In a timed *continuous* PN, each transition corresponds *implicitly* to a *single server*: see Figure c. This is consistent with modeling of multiple servers illustrated in Figure 5.2 in Section 5.1.1.

The case illustrated in Figures a, b, and c, can be generalized to N servers, for any finite positive integer N. The case of *infinite server* is illustrated in Figure d. For transition T_3 in Figure d, $D(T_3, \mathbf{m}) = \infty$. It follows that the behaviors of the infinite server in Figure d and the single server in Figure e are similar: T_3 is an *immediate transition* since $U_3 = \infty$, modeled as in Figure f in a timed continuous

PN. For any positive value of U_3 in Figure d, finite or infinite, the same Figures e and f are obtained. Hence, $U_3 = \infty$ can be written instead of $U_3 = 1$ (Figure g).

How can we distinguish the case in Figure c from the case in Figure d, i.e. the meaning of the number representing either a maximal speed *or* a flow rate (independently of the notation U_j or V_j)? If the PN contains *no* D-place *or* D-transition, it is a continuous PN (hence the timed *continuous* PN semantics is applied). If the PN contains *at least one* D-place *or one* D-transition, it is an hybrid PN (hence the timed *hybrid* PN semantics is applied). For an immediate C-transition in a hybrid PN, the model in Figure g should be prefered for avoiding errors).

Remark 6.5

a) The authors suggest to apply cautionly this semantics in the models in order to avoid semantic errors. However, if *every C-transition has a constant D-enabling degree* and there is *no case 4 conflict* (Section 6.1.3), an *abbreviation* associating maximal speeds V_j to every C-transition T_j (i.e. without D-place with a self loop) could be used. Be careful.

b) As in Chapter 3 (first time: Notation 3.2, Section 3.2.2.2), an *immediate transition* may be denoted without speed (continuous PN) or without flow rate (hybrid PN). Examples: $U_3 = \infty$ in Figure 6.7g or $V_3 = \infty$ in f could be not written. ❑

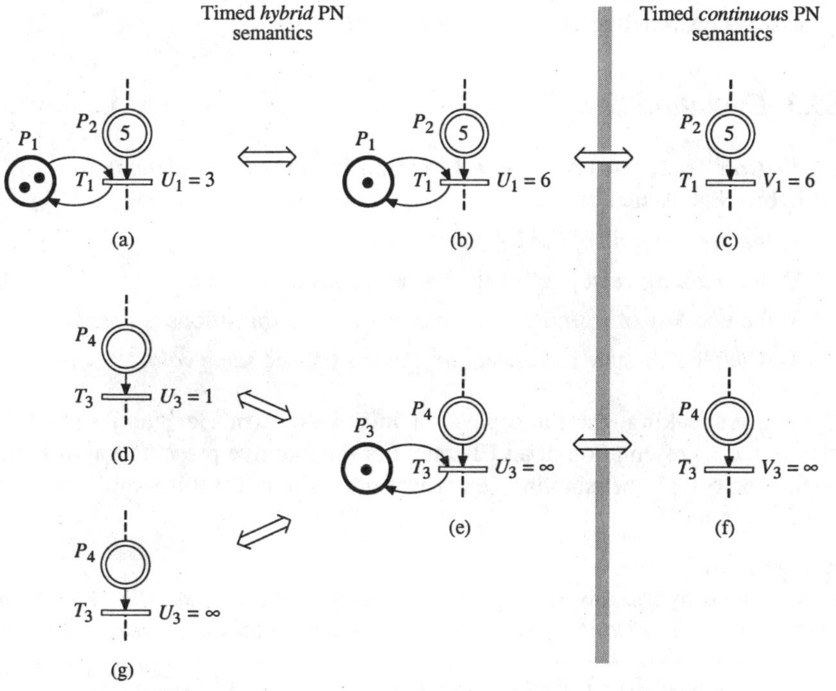

Figure 6.7 Timed hybrid PN semantics vs timed continuous PN semantics.

All the elements allowing specification of the *enabling concept in timed hybrid PNs* have now been presented.

Enabling of a *D-transition* in a timed hybrid PN is similar to enabling of a D-transition in an autonomous hybrid PN: Definition 4.5 in Section 4.2.2, and the enabling degree is defined in Definition 3.1 in Section 3.1.

Enabling of a *C-transition* in a timed hybrid PN is a *generalization* of enabling in a timed continuous PN (Definitions 5.3 and 5.3' in Section 5.1.3.2), taking into account the D-enabling degree related to D-places.

Definition 6.4

a) Non-immediate *C-transition* T_j in a timed hybrid PN is **enabled** at t if

$$D(T_j, \mathbf{m}(t)) > 0, \text{ and } \tilde{m}_i(t) > 0 \text{ for every } P_i \in {}^\circ T_j \cap P^C. \quad (6.12)$$

If T_j is enabled, it is **strongly enabled** if $m_i(t) > 0$ for every $P_i \in {}^\circ T_j \cap P^C$, and **weakly enabled** otherwise.

b) An *immediate C-transition* T_j in a timed hybrid PN is **weakly enabled** at t if $D(T_j, \mathbf{m}(t)) > 0$, and all the places P_i in ${}^\circ T_j \cap P^C$ such that $m_i(t) = 0$ are such that $I_i(t) > 0$.

❑

Remark 6.6 All the concepts defined for a *stable marking* in a timed *continuous* PN are relevant to a *C-place* P_i in a timed *hybrid* PN, in every IB-state (defined below): *feeding speed, draining speed, balance* (Definitions 5.4 and 5.5, with the sums restricted to ${}^\circ P_i \cap T^C$ instead of ${}^\circ P_i$).

6.1.5.3 Evolution Graph

Definition 6.5 An **IB-state** (standing for *invariant behavior state*) in a timed hybrid PN, is such that:

1) *the* marking $\mathbf{m}^D$ *of the D-places is* constant*;*

2) the *enabling vector*[2] $\mathbf{e}^D$ of the D-transitions is *constant*;

3) the vector $\mathbf{v}$ of *instantaneous speeds* of the C-transitions is *constant*.

4) when the IB-state is reached, $\mathbf{m}^C$ always has the *same value*[3].

❑

Generally speaking, the three piece of information ($\mathbf{m}^D$, $\mathbf{e}^D$, and $\mathbf{v}$) will be written in the representation of an IB-state. For the intuitive presentation in Figure 6.1c (Section 6.1.1), the enabling vector was not written: for this simple example, $\mathbf{e}^D = \mathbf{m}^D$ is found.

Remark 6.7

1) If a timed hybrid PN is degenerated into a timed *discrete* PN, an IB-state corresponds to a *constant marking* and a *constant enabling vector*. These two

---

[2] The enabling vector, for a T-timed discrete PN, is defined in Section 3.4.2.2.

[3] For example, in Figure 6.8b, IB-state 1 can be reached at initial time or at the end of IB-state 5. In both cases, $\mathbf{m}^C = (0, 0)$. This feature allows an IB-state to have always the same duration.

pieces of information were explicitly presented in the graphs corresponding to T-timed discrete PNs in Figures 3.26b and 3.27b (Section 3.4.2.2).

In a discrete PN, the *complete* marking implies the enabling vector; in Section 3.4.2.2, the enabling vector was necessary because only the *significant* marking was represented (i.e., not the complete one). In a *hybrid* PN, the *complete* $\mathbf{m}^D$ does not imply $\mathbf{e}^D$ because $\mathbf{e}^D$ depends also on $\mathbf{m}^C$.

In Figures 3.26b and 3.27b, the residual times to firings were specified. We shall not write this information in the evolution graphs for timed hybrid PNs. However, it can be deduced from some data (delays and priorities related to D-transitions) and, when the evolution graph is built, it can be taken into account in a time-ordered sequence of events.

2) If a timed hybrid PN is degenerated into a timed *continuous* PN, an IB-state is also an *IB-state* (Remark 5.3 in Section 5.1.2.1, and Section 5.1.3.4).

❑

According to the explanations in Section 6.1.2, Property 6.2 is obtained.

Property 6.2 In a timed hybrid PN, a *change of IB-state* can occur only if an *event* belonging to one of the following types occurs.

C1-event*: the marking of a marked C-place becomes zero.*

D1-event: *firing* of a D-transition.

D2-event: enabling degree of a D-transition changes because of the marking of a C-place.

❑

Let us now present two examples of evolution graphs.

Figure 6.8b presents the evolution graph corresponding to the timed hybrid PN in Figure 6.8a. The discrete marking is $\mathbf{m}^D = (m_1, m_2)$. The enabling vector for D-transitions is $\mathbf{e}^D = (q(T_1, \mathbf{m}), q(T_2, \mathbf{m}))$, and the instantaneous speed vector is $\mathbf{v} = (v_3)$.

At initial time, $\mathbf{m}^D = (2, 1)$, $\mathbf{e}^D = (1, 0)$, and $\mathbf{v} = (0)$ since $m_3 = 0$. This IB-state lasts up to firing of T_1 (D1-event), occurring at $t = 3$ and leading to IB-state 2. Firing of T_1 consists of removing two *tokens* from P_1 and placing 10 *marks* in P_3. Hence, there is a brutal (not continuous) change of the continuous marking $\mathbf{m}^C$, from $\mathbf{m}^{C,bef}(3) = (0, 0)$ to $\mathbf{m}^{C,aft}(3) = (10, 0)$ (notations similar to the case of infinite firing speed in a timed continuous PN, Section 5.1.2.5). Then $v_3 = 1$.

The next change, from IB-state 2 to IB-state 3, is due to the D2-event corresponding to m_4 reaching the value $5 = \mathrm{Pre}(P_4, T_2)$: T_2 becomes enabled. Transition T_2 will be fired two time units later (D1-event): changing from IB-state 3 to IB-state 4.

Three time units later, a C1-event (m_3 becomes 0) and a D2-event (m_4 reaches value 5) occur simultaneously. IB-state 5 is reached, then two time units later, T_2 is fired for the second time and IB-state 1 is reached again. The behavior is periodical.

Let us consider for example the evolution of the hybrid PN in Figure 6.8 from $t = 0$ to $t = 11.5$. During this time interval, T_1 is fired at $t = 3$, T_2 is fired at $t = 10$, T_3 is continuously fired at $v_3 = 1$ from $t = 3$ to $t = 11.5$. Hence,

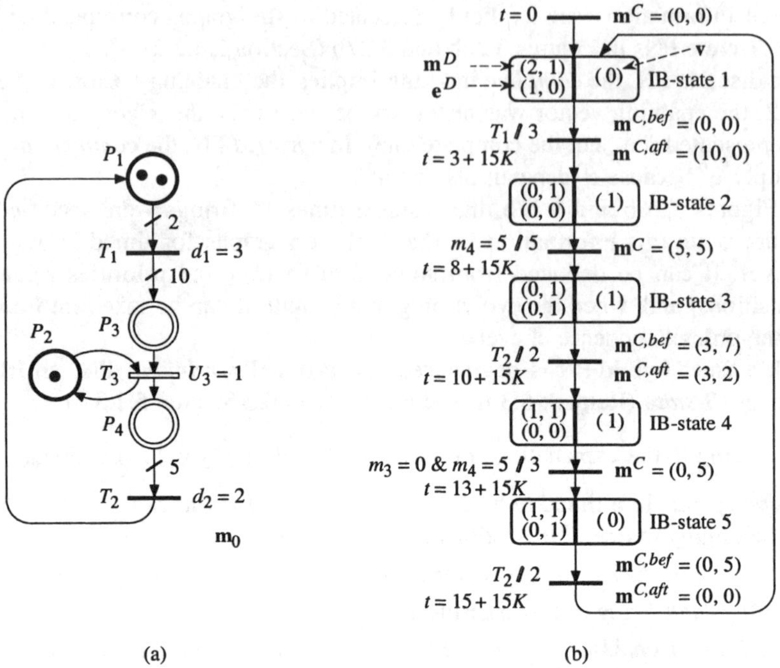

Figure 6.8 Example of timed hybrid PN and evolution graph.

$$\mathbf{m}(0)=\begin{bmatrix}2\\1\\0\\0\end{bmatrix};\quad \mathbf{W}=\begin{bmatrix}-2 & +1 & 0\\ 0 & 0 & 0\\ +10 & 0 & -1\\ 0 & -5 & +1\end{bmatrix};\quad \mathbf{n}(11.5)=\begin{bmatrix}1\\1\\0\end{bmatrix};\quad \int_0^{11.5}\mathbf{v}(t)\cdot dt=\begin{bmatrix}0\\0\\8.5\end{bmatrix}. \tag{6.13}$$

Then, $\mathbf{m}(11.5)$ is obtained from (6.9) and (6.13):

$$\mathbf{s}(11.5)=\mathbf{n}(11.5)+\int_0^{11.5}\mathbf{v}(t)\cdot dt=\begin{bmatrix}1\\1\\8.5\end{bmatrix},\text{ and }\mathbf{m}(11.5)=\begin{bmatrix}1\\1\\1.5\\3.5\end{bmatrix}. \tag{6.14}$$

For this example, let us note that: 1) according to the notation in Section 4.3.5.3, $\mathbf{W}^{CD}=\begin{bmatrix}+10 & 0\\ 0 & -5\end{bmatrix}\neq\mathbf{0}$, i.e. the hybrid PN is *not elementary*; 2) the periodic behavior appearing in Figure 6.8b, corresponds to the T-invariant $\mathbf{p}(15) = (1, 2, 10)$ (according to the notation used in the proof of Property 6.1 in Section 6.1.5.1).

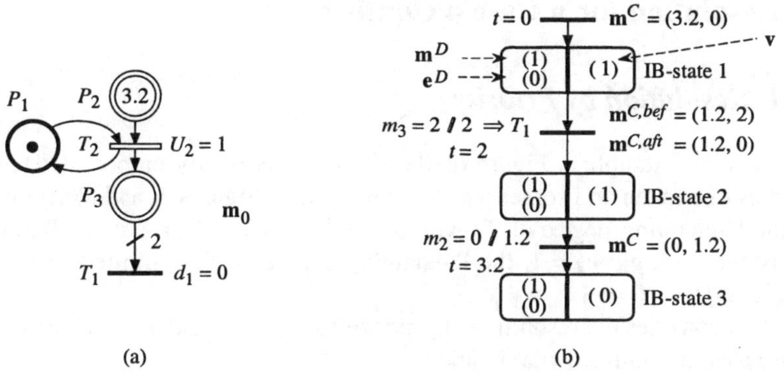

Figure 6.9 Another example of an evolution graph.

For the example in Figure 6.9, the discrete marking is $\mathbf{m}^D = (m_1)$, the enabling vector for the only D-transition is $\mathbf{e}^D = (q(T_1, \mathbf{m}))$, and the instantaneous speed vector is $\mathbf{v} = (v_2)$. In IB-state 1, $v_2 = 1$ since T_2 is strongly enabled. At $t = 2$, m_3 reaches value 2, and this D2-event results in an *immediate firing* of T_1 (which is an immediate transition). During IB-state 2, P_2 is emptied and IB-state 3 is a deadlock.

Remark 6.8 It could be possible to simplify the representation of an IB-state by representing only the significant discrete marking instead of the complete marking. For example, in Figure 6.8, the *significant* discrete marking (m_1) could replace the *complete* discrete marking (m_1, m_2) since m_2 is always 1.

6.2 ALGORITHM

In this section, it is assumed that the weights $\text{Pre}(P_i, T_j) = \text{Post}(P_i, T_j) = 1$ *for all the self-loops such that* $P_i \in P^D$ *and* $T_j \in T^C$. In other words, it is assumed that there is no "partial server". According to Remark 6.3 (Section 6.1.4), this restriction is not important from a practical point of view.

A generalization of the conflict resolution by sharing for case 4 (i.e. both transitions are continuous and the common place is discrete) is given in Section 6.2.1. The effects of C1-events, D1-events, and D2-events are analyzed in Section 6.2.2. An algorithm for building the evolution graph will then be presented in Section 6.2.3.

6.2.1 Resolution for a Case 4 Conflict

6.2.1.1 Resolution by Priority

Consider the example in Figure 6.10a. Place P_1, presently empty, will contain sometimes one token and sometimes two tokens (see Figures b and c). Hence, for $t > 0$, the D-enabling degree of T_3 will be $D(T_3, \mathbf{m}(t)) = 0$ or 1 or 2. Because of the constant marking $m_4 = 1$, the D-enabling degree of T_4 is limited to value 1, i.e., $D(T_4, \mathbf{m}(t)) = 0$ or 1.

We shall consider the resolutions by priorities $T_3 < T_4$ and $T_4 < T_3$ in turn for the examples in Figures 6.10a, b, and c.

Let us recall that, from (6.6) and (6.7) in Section 6.1.4:

$$V_j = U_j \cdot \min_{P_i \in {}^\circ T_j \cap P^D} m_i. \tag{6.15}$$

Priority $T_3 < T_4$ or $T_4 < T_3$ for Figure 6.10a. According to (6.15), $V_3 = V_4 = 0$, hence, $v_3 = v_4 = 0$.

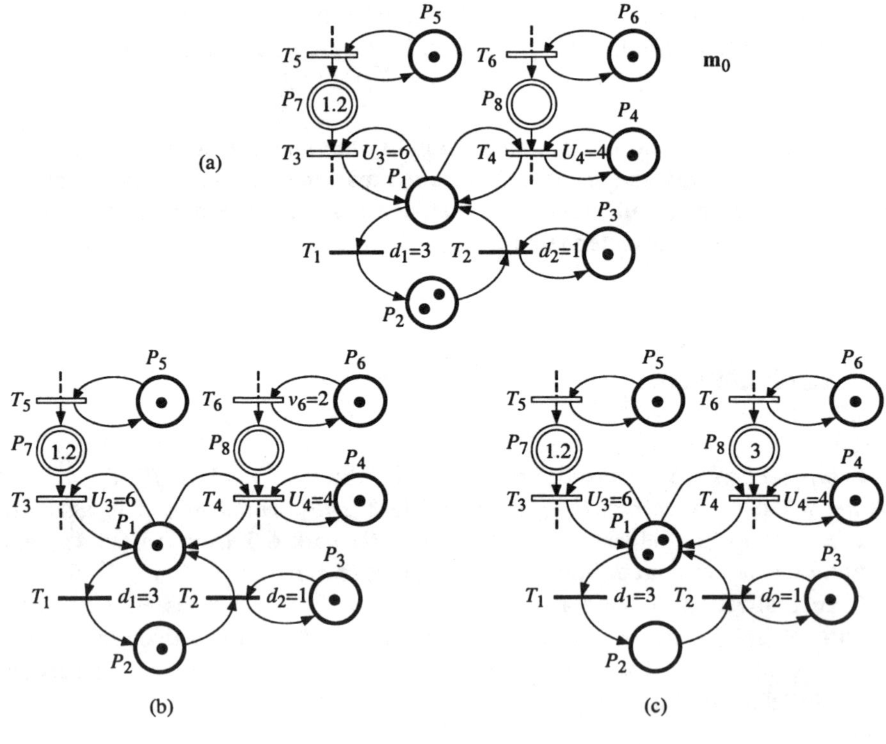

Figure 6.10 Illustration of sharing a common D-place (C-transitions).

Priority $T_3 < T_4$ for Figure 6.10b. According to (6.15), $V_3 = 6$. Hence, $v_3 = V_3 = 6$ since T_3 is strongly enabled. Transition T_3 uses $v_3 / U_3 = 1$ server.
In this case, there is no server available for T_4: $v_4 = 0$.

In the sequel, the concept of **available marking** will be illustrated. Roughly speaking, it is the number of servers represented by the marking in a D-place P_c involved in a case 4 conflict, which remain *available* for some transition T_j after all the servers *used by transitions with a higher priority* than T_j have been calculated.

Notation 6.2
a) *Available marking* in D-place P_c is denoted by $m_c^{(A)}$. It is a fictitious (not integer) marking only used during the calculation process. The value $m_c^{(Ak)}$ specifies the value of $m_c^{(A)}$ after the kth calculation of its value. By definition, $m_c^{(A0)} = m_c$. Marking $\mathbf{m}$ in which m_c is replaced by $m_c^{(A)}$ is denoted by $\mathbf{m}^{(A)}$.
b) Notation V_j' will be used to represent a *maximal firing speed when it is less than* V_j (due to additional constraints).
c) For a case 4 conflict $K = \langle P_c, T(K), \mathbf{m}\rangle$, where P_c is a D-place and $T(K)$ a set of C-transitions, $u(T_j)$ denotes the *quantity of servers used* by $T_j \in T(K)$. In other words, the conflict resolution consists of obtaining the values $u(T_j)$.
❑

Priority $T_4 < T_3$ for Figure 6.10b. According to (6.15), $V_4 = 4$. Since T_4 is weakly enabled, $v_4 = \min(V_4, v_6) = 2$. Hence T_4 uses $u(T_4) = v_4 / U_4 = 0.5$ server.
Since T_4 *does not use completely* the server (corresponding to the token in P_1), there is 0.5 *server available* for T_3. This available "quantity of servers" is the *available marking* of P_1: $m_1^{(A)} = 0.5$. The maximal firing speed of T_3 can be calculated by an expression similar to (6.15), in which m_1 is replaced by $m_1^{(A)}$. The general expression is (since, according to Notation 6.2a, m_i is a particular case of $m_i^{(A)}$):

$$V_j' = U_j \cdot \min_{P_i \in {}^\circ T_j \cap P^D} m_i^{(A)}, \tag{6.16}$$

and for our example: $V_3' = 0.5 \cdot U_3 = 3$. Since transition T_3 is strongly enabled, $v_3 = V_3' = 3$.

Priority $T_3 < T_4$ for Figure 6.10c. According to (6.15), $V_3 = 12$. Since T_3 is strongly enabled, $v_3 = V_3 = 12$. Transition T_3 uses $u(T_3) = v_3 / U_3 = 2$ servers.
In this case, there is no server available for T_4: $v_4 = 0$.

Priority $T_4 < T_3$ for Figure 6.10c. According to (6.15), $V_4 = 4$. Since T_4 is strongly enabled, $v_4 = V_4 = 4$. Transition T_4 uses $u(T_4) = v_4 / U_4 = 1$ server.
The *available marking* of P_1 is now $m_1^{(A)} = m_1 - u(T_4) = 1$. From (6.16), $V_3' = 6$ is obtained. Since T_3 is strongly enabled, $v_3 = V_3' = 6$.

Remark 6.9 In both Figures 6.10b and c, values $v_3 > 0$ are obtained even for the resolution by priority $T_4 < T_3$, i.e. when T_3 has the lowest priority. In the first case (Figure b), T_4 cannot use the server completely because it is *weakly enabled.*

In the second case (Figure c), T_4 *cannot use both servers completely* ($m_1 = 2$) *because its D-enabling degree cannot be greater than* $m_4 = 1$.

Similar phenomena will appear for resolution by sharing. When the servers modeled by the tokens in a D-place are shared between two or more C-transitions, it is possible that a C-transition cannot use completely the quantity of servers allocated to it.

6.2.1.2 Resolution by Sharing

Whereas sharing between two C-transitions in Section 5.3.2 was related to a common C-place, in this section, the common place is a D-place (generalization of the case 4 resolution presented in Section 6.1.3). In both cases, for a conflict $K = \langle P_c, T(K)\rangle$, it is assumed that *the only place in* $°T(K)$ *whose resource*[4] *is split by sharing is* P_c (Hypothesis 5.2b in Section 5.3.3.2).

Consider again the examples in Figure 6.10 and assume a sharing $[\beta_3 T_3, \beta_4 T_4]$ between T_3 and T_4. According to Section 6.1.3, this sharing aims at

$$\frac{1}{\beta_3}\cdot\frac{v_3}{U_3} = \frac{1}{\beta_4}\cdot\frac{v_4}{U_4}. \tag{6.17}$$

If both transitions were able to use their allocations completely, these allocations would be:

$$\frac{v_3}{U_3} = \frac{\beta_3}{\beta_3+\beta_4}\cdot m_1 \text{ and } \frac{v_4}{U_4} = \frac{\beta_4}{\beta_3+\beta_4}\cdot m_1. \tag{6.18}$$

Let us assume sharing $[T_3, 3\,T_4]$, i.e. $\beta_3 = 1$ and $\beta_4 = 3$. It *aims at* "the number of servers assigned to T_4 is three times the number of servers assigned to T_3". For these values, (6.17) and (6.18) become:

$$\frac{v_3}{U_3} = \frac{1}{3}\cdot\frac{v_4}{U_4}, \text{ and} \tag{6.19}$$

$$\frac{v_3}{U_3} = \frac{1}{4}\cdot m_1 \text{ and } \frac{v_4}{U_4} = \frac{3}{4}\cdot m_1. \tag{6.20}$$

Before the complete resolution of this example, let us present the general case.

Consider the *actual* conflit $K = \langle P_c, T(K), \mathbf{m}\rangle$, where P_c is a D-place and $T(K)$ a set of C-transitions. Let $T(K) = \{T_1, T_2, \ldots, T_s\}$, $s \geq 2$. Let us assume that this conflict is solved by sharing among the transitions denoted by $[\beta_1 T_1, \beta_2 T_2, \ldots, \beta_s T_s]$. In the general case, the number of servers used by T_j is $\mathrm{Pre}(P_c, T_j) \cdot v_j / U_j$. Hence, the sharing aims at:

$$\frac{\mathrm{Pre}(P_c,T_1)}{\beta_1}\cdot\frac{v_1}{U_1} = \frac{\mathrm{Pre}(P_c,T_2)}{\beta_2}\cdot\frac{v_2}{U_2} = \ldots = \frac{\mathrm{Pre}(P_c,T_s)}{\beta_s}\cdot\frac{v_s}{U_s}. \tag{6.21}$$

[4] The resource is a number of tokens in a D-place (as in the P_c considered) or a feeding speed for an empty C-place.

The quantity of servers *initially allocated* to T_j is

$$a(T_j) = \frac{\beta_j}{\sum\limits_{T_k \in T(K)} \beta_k} \cdot m_c. \tag{6.22}$$

The maximal firing speed in view of this allocation is then obtained by a modification of (6.16):

$$V_j' = U_j \cdot \min\left(\frac{\beta_j}{\sum\limits_{T_k \in T(K)} \beta_k} \cdot m_c^{(A)}, \min_{P_i \in \left(\left({}^{\circ}T_j \cap P^D\right) \setminus P_c\right)} m_i^{(A)} \right), \tag{6.23}$$

Let us now use this expression for sharing $[T_3, 3\,T_4]$ applied to Figure 6.10.

Sharing $[T_3, 3\,T_4]$ *for Figure 6.10b*. According to (6.23), $V_3' = 0.25 \times 6 = 1.5$ and $V_4' = 0.75 \times 4 = 3$. Since T_4 is weakly enabled, $v_4 = \min(V_4', v_6) = \min(3, 2) = 2$. Hence, $u(T_4) = v_4 / U_4 = 0.5$ server (T_4 cannot use its allocation completely). Then, it remains $m_1^{(A)} = 1 - u(T_4) = 0.5$ server for T_3, and $V_3' = 0.5 \times 6 = 3$ is obtained from (6.16). Finally, $v_3 = V_3' = 3$ is obtained.

Sharing $[T_3, 3\,T_4]$ *for Figure 6.10c*. According to (6.23), $V_3' = 0.5 \times 6 = 3$ and $V_4' = \min(1.5, 1) \times 4 = 4$. Let us note that, for T_4, the second term in the brackets of (6.23) (i.e., $m_4 = 1$) is *less than the first term* in the brackets (i.e., $3\,m_1 / 4 = 1.5$). This means that T_4 cannot use completely 1.5 servers which could be allocated to it, *even if it is strongly enabled.* Since T_4 is strongly enabled, $v_4 = V_4' = 4$ is obtained. Then, V_3' is recalculated and $v_3 = V_3' = 6$ is obtained.

Remark 6.10 Let us assume now that $m_8 > 0$ in Figure 6.10b. In this case, both T_3 and T_4 can use completely their *initial allocations*: $v_3 = 1.5$ and $v_4 = 3$.

According to Remark 6.9, there are *two reasons* for which a C-transition T_j cannot use completely its initial allocation (both can exist simultaneously for the same transition).

First reason: the *second term* in the brackets of (6.23) is *less than the first term* in the brackets (T_4 in Figure 6.10c). This does not depend on enabling the transition by C-places (strong, weak, or not enabled). Hence, new allocations can be **recalculated** immediately (i.e., *before* calculation of the instantaneous firing speeds) for the other transitions involved in the conflict.

Second reason: the transition is *weakly enabled* (T_4 in Figure 6.10b). In this case, the difference between the number of servers allocated and used is **transferred** to other transitions *after* the instantaneous firing speeds have been calculated. (We use two different words, *recalculation* and *transfer*, in order to specify the origin of the new calculation.)

The difference between the *recalculation* due to the first reason and the *transfer* due to the second reason is that the first one *depends only on the discrete marking* $\mathbf{m}^D$. This difference is not really important when a single conflict is concerned (example in Section 6.2.1.3), but is important when there are several conflicts among two or more subnets (Section 6.2.3).

6.2.1.3 Algorithmic Resolution

An algorithmic resolution of case 4 conflict is presented in the sequel. An example is given in Figure 6.11. The conflict *resolution rule* is specified by the three priority levels $T_1 < T_2 < [T_3, T_4]$ (i.e., with a sharing at the third level). It is assumed that the feeding speeds v_5, v_6, v_7, and v_8, *do not depend on the results* of the conflict resolution, i.e. v_1, v_2, v_3, and v_4.

The resolution consists of allocating some part of the three servers modeled by the tokens in P_1 to every transition involved in the conflict. The *calculation process* consists of allocating some "quantity of servers" to transition T_1 (higher priority level), then to T_2, and finally sharing the remaining quantity of servers between T_3 and T_4. The *available marking* $m_1^{(A,k)}$ of P_1 is calculated at each step. There are similarities between this kind of resolution and the resolution presented in Section 5.3.2.1 for timed continuous PNs. The *available marking* here and the *temporary balance* in Section 5.3.2.1 play similar roles.

At the beginning of the calculation process, the available marking is $m_1^{(A,0)} = m_1 = 3$. According to (6.15), $V_1 = 3$. Then, since T_1 is weakly enabled:

$$v_1 = \min(V_1, v_5) = 1.2. \tag{6.24}$$

The quantity of servers *used by* T_1 is:

$$u(T_1) = \frac{v_1}{U_1} = \frac{1.2}{3} = 0.4. \tag{6.25}$$

Hence, after this quantity has been assigned to T_1, the available marking of P_1 is $m_1^{(A,1)} = m_1^{(A,0)} - u(T_1) = 3 - 0.4 = 2.6$.

Among the remaining transitions, T_2 has the highest priority. From (6.16),

$$V_2' = U_2 \cdot m_1^{(A,1)} = 5.2 \tag{6.26}$$

is obtained. Since T_2 is weakly enabled:

$$v_2 = \min(V_2', v_6) = 0.4. \tag{6.27}$$

Then, $u(T_2) = v_2 / U_2 = 0.2$, and $m_1^{(A,2)} = 2.6 - 0.2 = 2.4$.

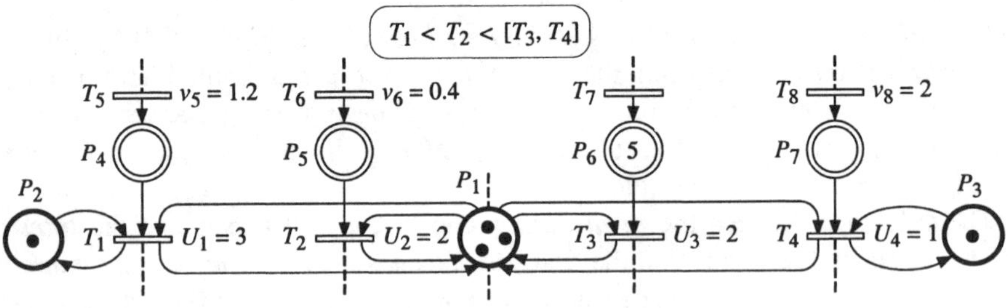

Figure 6.11 Case 4 conflict among four transitions.

Now, there is a sharing $[T_3, T_4]$ at the third priority level. In view of the available marking, $m_1^{(A2)} / 2 = 2.4 / 2 = 1.2$ servers are initially allocated to each of T_3 and T_4. Then,

$$V_3' = 2 \times 1.2 = 2.4 \quad \text{and} \quad V_4' = 1 \times \min(1.2, 1) = 1 \tag{6.28}$$

are obtained from (6.23). For V_4', the second term in the brackets is *less* than the first one: V_3' is *recalculated* as

$$V_3' = U_3 \times \left(m_1^{(A2)} - \frac{V_4'}{U_4}\right) = 2 \times \left(2.4 - \frac{1}{1}\right) = 2.8. \tag{6.29}$$

Since P_7 is empty,

$$v_4 = \min(V_4', v_8) = \min(1, 2) = 1. \tag{6.30}$$

Note that, although P_7 is empty, $v_4 = V_4'$ because $V_4' < v_8$. Hence, m_7 increases and T_4 becomes strongly enabled (if T_4 were weakly enabled, according to Remark 6.10 some quantity of server would be *transferred* from T_4 to T_3 after calculation of v_4).

Finally, since T_3 is strongly enabled:

$$v_3 = V_3' = 2.8. \tag{6.31}$$

As $u(T_3) = v_3 / U_3 = 1.4$ and $u(T_4) = v_4 / U_4 = 1$, we may verify that

$$u(T_1) + u(T_2) + u(T_3) + u(T_4) = 3 = m_1. \tag{6.32}$$

If $\sum_{T_j \in P_1^\circ} u(T_j) < m_1$, there were no conflict: the number of tokens in P_1 would be sufficient to fire all the transitions in P_1° at the maximal speeds in view of the other constraints.

6.2.2 Consequences of Various Events

Let t_s denote the time when IB-state s ends. In the algorithm for building the evolution graph, presented in the next section, the parameters related to IB-state s will be calculated. If these parameters are known, the time t_s is known. At time t_s, the parameters related to IB-state $s + 1$ must be calculated, either if t_s is the initial time t_0, or if $t_s < \infty$ and IB-state s is not equivalent to another IB-state k such that $k < s$ (periodical behavior).

During performance of the algorithm for building the evolution graph, all the events whose future occurrence is expected are listed in a *time-ordered sequence of events*. When the parameters related to IB-state s have been calculated, the time of the next events to be considered is t_s. One or more events (C1-events, D1-events, or D2-events, according to Property 6.2 is Section 6.1.5.3) are known to occur at t_s. This set of events is denoted by X_s. Let us analyze the possible effects of these events.

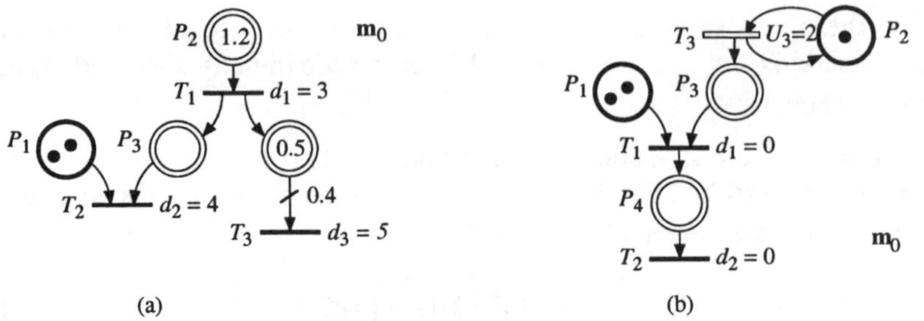

Figure 6.12 (a) A D1-event causes three D2-events. (b) A D2-event causes a string: D1-event then new D2-event then new D1-event.

Definition 6.6 Event E_1 **causes** event E_2 if E_2 is an *immediate consequence* of E_1 (i.e., E_2 occurs *after* E_1, but practically at the *same time*). ❑

C1-event (*the marking of a marked C-place becomes zero*)
A C1-event can cause a change of instantaneous speed vector (like in a timed continuous PN). Examples: $m_3 = 0$ at the end of IB-state 1, in Figure 6.1 (Section 6.3.1); $m_2 = 0$ at the end of IB-state 2, in Figure 6.9 (Section 6.1.5.3).

Obviously, a C1-event *cannot cause* a D1-event. A C1-event *cannot cause* a D2-event because there is no zero test in the basic timed hybrid PN (according to Section 4.4, a zero test can exist only in an *extended* hybrid PN). In Figure 6.8 (Section 6.1.5.3), $m_3 = 0$ and $m_4 = 5$ at the end of IB-state 4; the corresponding C1-event and D2-event occur *simultaneously*, but neither of them is *caused* by the other.

D1-event (*firing of a D-transition*)
It is clear that a D1-event can cause another D1-event, as in a timed discrete PN (because a D1-event modifies $\mathbf{m}^D$). For example, firing of T_1 can cause enabling of T_2 and firing of T_2 if it is an immediate transition.

A D1-event can also cause a C1-event (because a D1-event may modify $\mathbf{m}^C$). Examples: firing of T_2 at the end of IB-state 5, in Figure 6.8, empties P_4; firing of T_1, in Figure 6.9, empties P_3.

Since a D1-event may modify $\mathbf{m}^C$, it can also cause a D2-event. For example, each firing of T_2 in Figure 6.8 (ends of IB-states 3 and 5) decreases the enabling degree of this transition. In Figure 6.12a, the firing of T_1 at $t = 3$ decreases the enabling degree of T_1 (from 1 to 0) and increases the enabling degrees of T_2 (from 0 to 1) and T_3 (from 1 to 3); firing of T_3 at $t = 5$ decreases the enabling degree of T_3 from 3 to 2.

D2-event (*enabling degree of a D-transition changes because of the marking of a C-place*)
A D2-event cannot cause a C1-event, but a D2-event can cause a D1-event in case of immediate transition. Even if a D2-event cannot cause *directly* another D2-event, an iteration is possible as shown in the next example. In Figure 6.12b,

at $t = 0.5$, m_3 reaches the value 1; this D2-event causes the firing of T_1 (D1-event), which causes enabling of T_2 (D2-event), which causes in turn the firing of T_2 (D1-event).

The *conclusion* of this section is that, when the parameters of phase s are calculated at t_{s-1}, first a stable marking must be obtained, by iteration of D2-events and D1-events, then the vector of instantaneous speeds will be calculated.

6.2.3 Timed Hybrid PNs Automatically Treated in Algorithm 6.1

All the concepts defining how the behavior of a timed hybrid PN must be understood have now been presented. Algorithm 6.1 will present an automatic calculation of the successive IB-phases of such a hybrid PN, but only if some restrictive hypotheses are satisfied. These restrictions do not concern the model which has been defined but only the subclass which can be treated automatically by the proposed algorithm.

6.2.3.1 Hybrid PN Restricted to a Continuous PN

According to Section 6.2.2, when the parameters of an IB-state are calculated, the discrete marking $\mathbf{m}^D$ must be calculated before the speed vector $\mathbf{v}$. When $\mathbf{m}^D$ is known, the hybrid PN can be transformed into a continuous PN as explained in the sequel.

Definition 6.7 Given a timed hybrid PN $R(t_s)$, the **subjacent continuous PN**, denoted by $R^C(t_s)$, is obtained as follows.

1) *Replace* each flow rate U_j associated with T_j by the maximal firing speed V_j (Equation (6.7) or (6.15)).

2) *Add an equation* expressing the relation among instantaneous speeds v_j related to the same conflict, for every case 4 conflict.

3) *Delete* all the D-places and D-transitions and the adjacent arcs.

❑

Example: obtaining $R^C(t_s)$ in Figure 6.13b from $R(t_s)$ in Figure 6.13a.

1) Since $D(T_1, \mathbf{m}) = 1$ in Figure 6.13a, $U_1 = 4$ in this figure is replaced by $V_1 = U_1 \cdot D(T_1, \mathbf{m}) = 4$ in Figure 6.13b. Similarly, $U_2 = 1$ and $U_3 = 8$ in Figure a are replaced by $V_2 = 1$ and $V_3 = 8$ in Figure b. Naturally, $U_4 = \infty$ in Figure a is replaced by $V_4 = \infty$ in Figure b.

2) There is a *potential*[5] case 4 *actual* conflict $\langle P_1, \{T_1, T_3\}, \mathbf{m}\rangle$. The firing speeds v_1 and v_3 must be such that $v_1 / U_1 + v_3 / U_3 \leq m_1 = 1$, leading to:

$$2 v_1 + v_3 \leq 8. \tag{6.33}$$

[5] The conflict is *potential* as long as the speed calculation has not been performed: in some cases the marking may be sufficient to avoid an actual conflict.

3) D-places P_1 and P_2 are deleted as well as their adjacent arcs.

❑

A similar transformation is illustrated in Figures 6.13c and d. One can observe that the continuous PN in Figure b is made up of *two subnets* (A and B), whereas the continuous PN in Figure d is made up of a *single connected net.* The difference between these two cases is important for the conflict resolution. The first one may be calculated automatically as shown in the sequel. Let us note that the *subjacent continuous PN is independent from the resolution rule.* In the sequel, the instantaneous speeds of the continuous PN in Figure 6.13b will be calculated for three different *resolution rules*: $T_1 < T_3$, then $T_3 < T_1$, and finally $[T_1, T_3]$.

Priority $T_1 < T_3$.

Speeds v_1 and v_2 are calculated by Algorithm 5.7. Only the *connected subnet A* is concerned by this calculation. The values $v_1 = v_2 = 1$ are obtained.

Since $v_1 = 1$, $v_3 \leq 8 - 2\,v_1 = 6 = V_3'$ is obtained from (6.33). The firing speeds v_3 and v_4 are calculated by Algorithm 5.7 in which V_3' takes the place of V_3. Only the *connected subnet B* is concerned. Values $v_3 = v_4 = 6$ are obtained.

Remark 6.11 Another way to calculate V_3' is as follows. Since $v_1 = 1$, the quantity of server used by T_1 is $u(T_1) = v_1 / U_1 = 0.25$. Then T_3 can use the remaining part of server, i.e. $1 - 0.25 = 0.75$. The constraint $V_3 = 8$ is then replaced by $V_3' = U_3 \times 0.75 = 6$.

❑

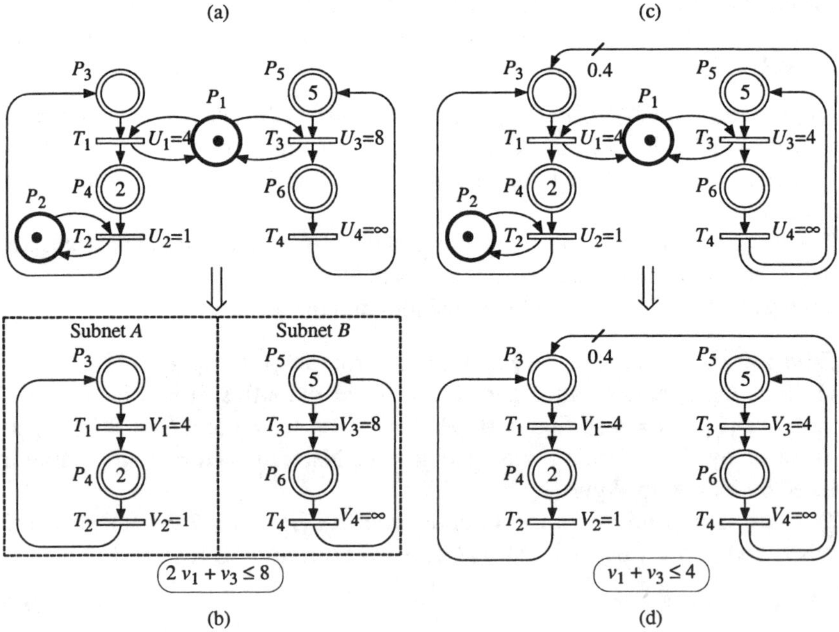

Figure 6.13 Transformation from the timed hybrid PN $R(t_s)$ to the subjacent continuous PN $R^C(t_s)$.

Priority $T_3 < T_1$.

Speeds v_3 and v_4 are calculated by Algorithm 5.7. Only the *subnet B* is concerned by this calculation. The values $v_3 = v_4 = 8$ are obtained.

According to (6.33), $v_1 = 0$ since $v_3 = 8$. With this constraint the calculation related to *subnet A* leads to $v_1 = 0$ and $v_2 = 1$.

Sharing $[T_1, T_3]$.

The constraints $V_1 = 4$ and $V_3 = 8$ are replaced by $V_1' = 2$ and $V_3' = 4$ ((6.23) in Section 6.2.1.2). The speed calculations (separately for subnets A and B) lead to: $v_1 = v_2 = 1$ and $v_3 = v_4 = 4$. Since T_1 does not use its allocation completely (i.e., $v_1 < V_1'$ because T_1 is not strongly enabled). A part of server is *transfered* to T_3. The new value of V_3' is 6. A new calculation for subnet B gives $v_3 = v_4 = 6$.

Let us now specify the first hypothesis which will be required for an automatic treatment by Algorithm 6.1. Hypothesis 6.1 is based on Property 6.3.

Property 6.3 The *structure* of the subjacent continuous PN *does not depend on the marking.*

Proof The structure of the subjacent continuous PN is obtained from step 3 in Definition 6.7, which is independent from the marking. (Only the values V_j and the relation among speeds v_j related to a conflict depend on $\mathbf{m}$.)

Hypothesis 6.1 Let T_a and T_b be any two C-transitions involved in a structural conflict $\langle P_i, \{T_a, T_b, \ldots\}\rangle$ such that P_i is a D-place. Transitions T_a and T_b *do not belong to the same connected subnet* of the subjacent continuous PN.

6.2.3.2 Consistency of Resolution Rules

Let us consider the timed hybrid PN in Figure 6.14. There are three structural case 4 conflicts, namely $K_1 = \langle P_4, \{T_4, T_8, T_{10}\}\rangle$, $K_2 = \langle P_5, \{T_5, T_9\}\rangle$, and $K_3 = \langle P_6, \{T_9, T_{12}\}\rangle$.

This hybrid PN satisfies Hypothesis 6.1, as shown in the sequel. After deleting D-places P_1 to P_6, D-transitions T_1 to T_3, and the adjacent arcs, three subnets remain: subnet A made up of P_7 to P_9, T_4 to T_6, and the arcs linking them; subnet B containing P_{10} to P_{12}, T_7 to T_9, and the arcs linking them; subnet C made up of P_{13} to P_{15}, T_{10} to T_{12}, and the arcs linking them. For any structural case 4 conflict K_1, K_2, or K_3, the transitions concerned belong to different subnets. For example, for $K_1 = \langle P_4, \{T_4, T_8, T_{10}\}\rangle$, T_4 is in subnet A, T_8 in subnet B, and T_{10} in subnet C.

One could imagine that each subnet corresponds to some continuous production. The three productions (subnets A, B, and C) use common resources modeled by the tokens in P_4 (one or two tokens, depending on time), P_5 (two tokens), and P_6 (one token). Then, it seems reasonable that there is some priorities *among these productions*, or resource sharing among some of them. For example, if subnet A takes priority over subnet B, then $T_4 < T_8$ and $T_5 < T_9$.

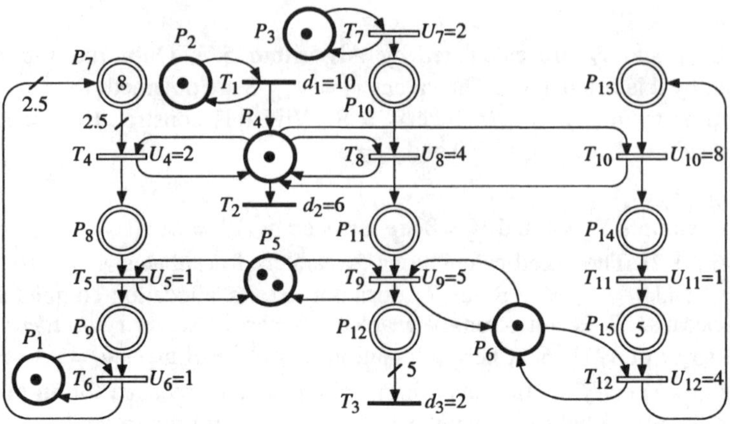

Figure 6.14 Illustration for consistency of resolution rules.

However, the resolution rules are local (i.e., specified for each conflict). A priority among the subnets is obtained if there is some *consistency* among the local resolution rules. For example, $T_4 < T_8$ is *not consistent* with $T_9 < T_5$.

Assume for example that the resolution rule for conflict K_1 is $T_4 < T_8 < T_{10}$. There is consistency if the resolution rules for K_2 and K_3 are $T_5 < T_9$ and $T_9 < T_{12}$.

For the resolution rule $T_4 < [2\, T_8, T_{10}]$, the consistency implies $T_5 < T_9$ and $[\beta_9 T_9, \beta_9 T_{12}]$: there is a sharing between T_8 and T_{10} (same priority level), then there is a sharing between T_9 and T_{12} (not necessarily with the same coefficients).

Note that the consistency with a sharing $[\beta_4 T_4, \beta_8 T_8, \beta_{10} T_{10}]$ would imply a sharing between T_5 and T_9, on the one hand, and a sharing between T_9 and T_{12}, on the other. This solution, corresponding to two sharings of places in $°T_9$ (P_5 and P_6) is not included in our hypotheses, according to the beginning of Section 6.2.1.2.

Let us now specify the hypothesis of consistency.

Definition 6.8 Given Hypothesis 6.1 is satisfied, there is **consistency** *among the local resolution rules* if: for any quadruple ($T_{a,X}$, $T_{b,X}$, $T_{a,Y}$, $T_{b,Y}$) such that transitions $T_{a,X}$ and $T_{b,X}$ belong to the same subnet X, transitions $T_{a,Y}$ and $T_{b,Y}$ belong to another subnet Y, $T_{a,X}$ and $T_{a,Y}$ are involved in structural case 4 conflict K_a, and $T_{b,X}$ and $T_{b,Y}$ are involved in structural case 4 conflict K_b, the local resolution rules are such that, either

1) $T_{a,X} < T_{a,Y}$ and $T_{b,X} < T_{b,Y}$, or $T_{a,Y} < T_{a,X}$ and $T_{b,Y} < T_{b,X}$,

or 2) $[\beta_1 T_{a,X}, \beta_2 T_{a,Y}]$ and $[\beta_3 T_{b,X}, \beta_4 T_{b,Y}]$.

Hypothesis 6.2 There is consistency among the local resolution rules. ❑

If the consistency is verified (Hypotheses 6.1 and 6.2 satisfied), *priority levels among the subnets* can be specified. For example, if the following local resolution rules for the hybrid PN in Figure 6.14 are $T_4 < [2\ T_8,\ T_{10}]$, $T_5 < T_9$, and $[T_9,\ 3\ T_{12}]$, the priority levels among the subnets can be denoted by

$$\text{Subnet } A < [\text{Subnet } B, \text{ Subnet } C]. \tag{6.34}$$

Subnet A takes priority over subnets B and C, and there is sharing among subnets B and C. Note that the sharing coefficients are not specified in (6.34). As a matter of fact, the coefficients may be different for each local resolution rule (for example, $[2\ T_8,\ T_{10}]$ for K_1 and $[T_9,\ 3\ T_{12}]$ for K_3.

Given the consistency is verified, the case 4 conflict resolution for the whole hybrid PN can be performed. This resolution is explained in the sequel, with the help of an example. Then, this algorithmic resolution will be included as a part of Algorithm 6.1.

Basically, the resolution follows the *same sequence of decisions* as in Section 6.2.1.3 (illustration in Figure 6.11). The difference is that the calculation for *a transition* in Section 6.2.1.3 is replaced by the calculation for *a subnet* in this case. First, the instantaneous speeds of transitions in the subnets with the higher priority level are calculated (one subnet or several in case of sharing among them), then the calculation for the second priority level is performed. And so on.

Consider for example the hybrid PN in Figure 6.15a. The subjacent continuous PN is shown in Figure 6.15b. For the conflict $K_1 = \langle P_1, \{T_1, T_3\}\rangle$, $v_1 / U_1 + v_3 / U_3 \leq m_1 = 3$, leads to $3\ v_1 + 2\ v_3 \leq 18$; for $K_2 = \langle P_2, \{T_2, T_4\}\rangle$, $v_2 / U_2 + v_4 / U_4 \leq m_2 = 1$, leads to $10\ v_2 + v_4 \leq 20$. In the sequel, resolutions in the cases Subnet $A <$ Subnet B, then [Subnet A, Subnet B], are presented.

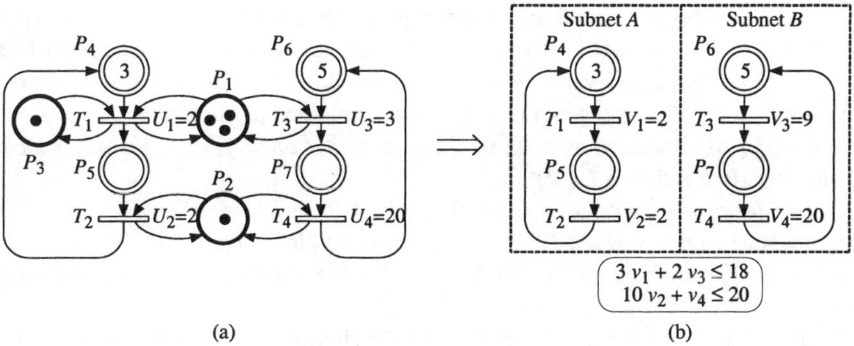

Figure 6.15 Case 4 conflict resolution between two subnets.

Example a: Subnet $A <$ Subnet B.

1) Speed calculation (Algorithm 5.7) for subnet A: values $v_1 = v_2 = 2$ are obtained.

2) Since T_1 uses $v_1 / U_1 = 1$ server, there remains $m_1 - 1 = 2$ servers for T_3; then $V_3' = 6$. Similarly, T_2 uses one server, then $V_4' = 0$. For subnet B, with $V_3' = 6$ and $V_4' = 0$ taking the places of V_3 and V_4 in Algorithm 5.7, values $v_3 = 6$ and $v_4 = 0$ are obtained.

Example b: [Subnet *A*, Subnet *B*] with $[T_1, T_3]$ and $[T_2, 3\,T_4]$.

1) From the coefficients assigned to T_1 to T_4, allocations $a(T_1) = 1.5$, $a(T_2) = 0.25$, $a(T_3) = 1.5$, and $a(T_4) = 0.75$ are obtained (from (6.22) in Section 6.2.1.2). Taking into account only the discrete marking, $V_1' = 2 \times \min(1.5, 1) = 2 = V_1$, $V_2' = 2 \times 0.25 = 0.5$, $V_3' = 3 \times 1.5 = 4.5$, and $V_4' = 20 \times 0.75 = 15$, are obtained from (6.23). For V_1', the second term in the brackets is less than the first one. Then, according to Remark 6.10 in Section 6.2.1.2, V_3' is recalculated: $V_3' = U_3 \times (3 - 1) = 6$ is obtained.

It follows that the speed calculation is performed with the maximal speed values $V_1 = 2$, $V_2' = 0.5$, $V_3' = 6$, and $V_4' = 15$. Instantaneous speeds $v_1 = 2$, $v_2 = 0.5$, $v_3 = 6$, and $v_4 = 6$, are obtained.

2) One can observe that T_4 does not use its allocation completely because it is weakly enabled: $u(T_4) = v_4 / U_4 = 0.3$. Hence, the part of allocation not used by T_4 may be transfered to T_2 (Remark 6.10). The new value $V_2' = U_2 \times (m_2 - u(T_4)) = 2 \times 0.7 = 1.4$ is obtained. Given $V_1 = 2$ and $V_2' = 1.4$, $v_1 = 2$ and $v_2 = 1.4$ are obtained (no change for subnet *B*).

❑

If every priority level contains exactly one subnet, there is no difficulty (Example a). Given $\mathbf{m}$, the speeds in the subnet with the *higher priority* are calculated. Then, given the available marking $\mathbf{m}^{(A,1)}$, the speeds for the subnet with the *second priority* level are calculated. And so on.

If there is sharing among two or more subnets in the same priority level, an iteration of the speed calculation may be required (Example b). After the first speed calculation ($v_1 = 2$, $v_2 = 0.5$, $v_3 = 6$, $v_4 = 6$), there is a **transfer** of allocation from T_4 to T_2. Note that T_4 is *weakly enabled* whereas T_2 is becoming *strongly enabled* (because $B_5 > 0$). This is a general property as explained in the sequel.

A *transfer* of allocation *from* T_a to T_b *is possible only if* T_a *is weakly enabled* (Remark 6.10, Section 6.2.1.2).

A *transfer* of allocation to T_b is *useful only* if T_b is *strongly enabled* or if it is *becoming strongly enabled* (like T_2 in our example). Otherwise, if T_b is and remains weakly enabled (or not enabled), a transfer to T_b could not increase its speed.

If more than two subnets are involved in the sharing, an allocation transfer may take place towards several transitions (according to the β_j coefficients). If, at some time during the calculation process, there are several possible transfers, the authors suggest that the first transfer chosen is made towards a transition whose input places are marked or fed from a strongly enabled transition. However, theoretically, several choices may exist.

Remark 6.12 If there is a complete ordering of the subnets, or each time there is an allocation transfer there is only one possible transfer (in case of sharing among subnets), the speed vector is unique. However, if a choice exists among possible transfers, the solution is not necessarily unique.

6.2.4 Algorithm for Building the Evolution Graph

All the concepts providing an algorithm[6] for building the evolution graph have now been presented. The known future events are in a time-ordered sequence, denoted by *TOS*. This time-ordered sequence is a finite string $(t_a, X_a)(t_b, X_b) \ldots (t_u, X_u)$, $t_a < t_b < \ldots < t_u$, in which X_s represents the events known (or expected) to occur at t_s. Each X_s, which may also be denoted by $X(t_s)$ is a triple

$$X_s = X(t_s) = \left(C1(t_s), D1(t_s), D2(t_s)\right), \tag{6.35}$$

in which $C1(t_s)$ is a set of C1-events, $D1(t_s)$ is a set of D1-events, and $D2(t_s)$ is a set of D2-events, all of them occurring at t_s.

Note that Algorithm 6.1 assumes that Hypotheses 6.1 and 6.2 are satisfied. After the presentation of this algorithm, an application example will be given.

Algorithm 6.1 **Timed hybrid Petri net**

Step 1. Initialization.

Step 1.1. Enter data related to the timed hybrid PN, initial marking, and local resolution rules for cases 1, case 2, and case 4 conflicts (Section 6.1.3). Set up the resolution rules among subnets

Step 1.2. Initialization of $\tilde{\mathbf{m}}^C$ (like *Step 3* in Algorithm 5.8). Initialization of the time-ordered sequence *TOS* (possibly D1-events). Let $t_s = t_0$.

Step 2. **If** $D2(t_s) \neq \varnothing$ **then** add the corresponding D1-events in *TOS*.

Step 3. **If** $D1(t_s) = \varnothing$ **then** go to *Step 4*

else fire the corresponding transitions (a subset of them if a case 1 conflict exists), then *update TOS* and go to *Step 2*.

Step 4.

Step 4.1. Build the subjacent continuous PN.

Step 4.2. Perform an I-phase if necessary (similar to *Step 3* in Algorithm 5.8).

Step 5. **If** at least one priority level has not been treated **then**

if the priority level contains a single subnet **then** calculate the values V_j' for its transitions, using (6.16) in Section 6.2.1.1.

else go to *Step 7*

else go to *Step 10*.

Step 6. Calculate the instantaneous firing speeds for the subnet (Algorithm 5.7 without its *Step 0*). Increase the priority level and go to *Step 5*.

---

[6] Let us recall that, according to the beginning of Section 6.2, it is assumed that there is no "partial server".

Step 7. Sharing: calculate values V_j' (using (6.23)) for the transitions in the subnets concerned; then recalculate some of them if necessary (Remark 6.10).

Step 8. Calculate the instantaneous firing speeds of the concerned subnets (Algorithm 5.7 without its *Step 0* applied to every subnet).

Step 9. **If** at least one weakly enabled transition does not use its allocation, which can be transfered to one or several strongly (or becoming strongly) enabled transitions, **then** perform this transfer (choice if several are possible) and go to *Step 8*

else increase the priority level and go to *Step 5*.

Step 10. Update TOS. For this purpose, calculate from $\mathbf{v}$ the times when a component of $\mathbf{m}(t)$ becomes 0 (C1-events) and the times of future D2-events.

Step 11. Delete the first time in *TOS*. Then, let t_s = first time in *TOS*.

If t_s does not exist (i.e., $t_s = \infty$) **then** END.

else Calculate $\tilde{\mathbf{m}}^C(t_s)$.

Step 12. **If** the parameters of IB-state s are similar to the parameters of IB-state k, $k < s$, **then** END (periodical behavior)

else go to *Step 2*.

❑

One can see that the *TOS* is updated in *Step 2*, i.e., after each *elementary firing sequence* of D-transitions (Definition 3.5 in Section 3.2.2.1), and in *Step 10*, i.e., after the speed vector of the C-transitions has been calculated.

Assume a C-transition T_j whose maximal speed is $V_j = \infty$, and such that all its input C-places are marked while it is not D-enabled (Definition 6.1, Section 6.1.4). It may become enabled when a D-transition is fired. Then, an I-phase must be performed in *Step 4.2* (Exercise 6.4).

Consider the timed hybrid PN in Figure 6.16a. All the cases of conflicts are present in this small example. For the case 1 conflict $\langle P_9, \{T_2, T_3\}\rangle$, the priority $T_2 < T_3$ is chosen (only if they become firable at the same time, Section 6.1.3). For the case 2 conflict $\langle P_6, \{T_5, T_6\}\rangle$, $T_5 < T_6$ is chosen. According to Section 6.1.3, the priority is given to the D-transition for the case 3 conflicts $\langle P_3, \{T_1, T_6\}\rangle$ and $\langle P_3, \{T_1, T_7\}\rangle$, i.e., T_1 is fired as soon as it is firable. The sharing $[T_6, T_7]$ is chosen for the case 4 conflict $\langle P_3, \{T_6, T_7\}\rangle$.

Let us now apply Algorithm 6.1 to this example. We may omit writing a step in which nothing is done. The sets $C1(t)$, $D1(t)$, and $D2(t)$, are written when they are modified.

Application of Algorithm 6.1 to the timed hybrid PN in Figure 6.16a.

Step 1.1. Unmarked PN in Figure 6.16a, $\mathbf{m}^D(0) = (1, 1, 1, 1)$, $\mathbf{m}^C(0) = (2, 0, 0, 15, 0)$, $T_2 < T_3$, $T_5 < T_6$, $[T_6, T_7]$.

Step 1.2. $\tilde{\mathbf{m}}^C = (2, 0, 0, 15, 0)$, $C1(0) = D1(0) = D2(0) = \varnothing$, and $D1(7) = \{T_1\}$.

Step 3. Go to *Step 4*.

Step 4.1. Subnet $A = \{P_5, P_6, P_7, T_4, T_5, T_6$, and arcs linking them$\}$, Subnet $B = \{P_8, P_9, T_7$, and arcs linking them$\}$, $V_4 = 2.5$, $V_5 = 2$, $V_6 = 2$, $V_7 = 2$. According to $[T_6, T_7]$, there is sharing [Subnet A, Subnet B].

Step 5. Go to *Step 7*.
Step 7. $V_6' = U_6 \times 0.5 = 1$ and $V_7' = U_7 \times 0.5 = 1$.
Step 8. Speeds: $v_4 = 2.5$, $v_5 = 2$, $v_6 = 0.5$, $v_7 = 1$.
Step 9. Transfer: $u(T_6) = 0.25 < 0.5 \Rightarrow V_7' = 2 \times (1 - 0.25) = 1.5$.
Step 8. New speed: $v_7 = 1.5$.
Step 10. Speed vector $\mathbf{v} = (2.5, 2, 0.5, 1.5)$.

1) Negative balances: $B_5 = -0.5$ and $B_8 = -1.5$. Hence m_5 and m_8 could become 0 at $t = 4$ and $t = 10$, respectively. Then, $C1(4) = \{m_1 = 0\}$ and $C1(10) = \{m_8 = 0\}$.

2) Positive balance $B_9 = +1.5$. Marking m_9 could reach values 6 (enabling of T_3) and 15 (enabling of T_2) at $t = 4$ and $t = 10$, respectively. Then, $D2(4) = \{(m_9 = 6: T_3)\}$ and $D2(10) = \{(m_9 = 15: T_2)\}$.

Step 11. Next time $t_s = 4$; $\tilde{\mathbf{m}}^C(4) = (0^+, 0^+, 2, 9, 6)$.
Step 12. Go to *Step 2*.
Step 2. D2-event $(m_9 = 6: T_3)$ occurring at $t_s = 4 \Rightarrow D1(7) = \{T_1, T_3\}$ since $d_3 = 3$. Then $D2(4)$ is deleted.
Step 3. Go to *Step 4* since $D1(4)$ is empty.
Step 4.1. The subjacent continuous PN is unchanged.
Step 5. Go to *Step 7*.
Step 7. $V_6' = U_6 \times 0.5 = 1$ and $V_7' = U_7 \times 0.5 = 1$.
Step 8. Speeds: $v_4 = 2$, $v_5 = 2$, $v_6 = 0$, $v_7 = 1$.
Step 9. Transfer: $u(T_6) = 0 < 0.5 \Rightarrow V_7' = 2 \times (1 - 0) = 2$.
Step 8. New speed: $v_7 = 2$.
Step 10. Speed vector $\mathbf{v} = (2, 2, 0, 2)$.

1) $B_8 = -2$, then m_8 could become 0 at $t = 4 + 9/2 = 8.5$ instead of $t = 10$. Then, $C1(8.5) = \{m_8 = 0\}$ and $C1(10) = \{\}$.

2) $B_9 = +2$. Marking m_9 could reach values 15 (enabling of T_2) at $t = 4 + (15 - 6)/2 = 8.5$ instead of $t = 10$. Then, $D2(8.5) = \{(m_9 = 15: T_2)\}$ and $D2(10) = \{\}$.

Step 11. Next time $t_s = 7$; $\tilde{\mathbf{m}}^C(7) = (0^+, 0^+, 2, 3, 12)$.
Step 12. Go to *Step 2*, then *Step 3*.
Step 3. $D1(7) = \{T_1, T_3\}$: fire T_1 and $T_3 \Rightarrow \tilde{\mathbf{m}}^D(7) = (1, 1, 0, 0)$ and $\tilde{\mathbf{m}}^C(7) = (0^+, 0^+, 2, 3, 6)$.
Step 4.1. Same structure of the subjacent continuous PN but now $V_6 = V_7 = 0$.
Step 5. Go to *Step 7*.
Step 7. $V_6' = 0$ and $V_7' = 0$.
Step 8. Speeds: $v_4 = 2$, $v_5 = 2$, $v_6 = 0$, $v_7 = 0$.
Step 10. Speed vector $\mathbf{v} = (2, 2, 0, 0)$.

All the $B_i = 0$, then m_8 and m_9 do not change: $C1(8.5) = \{\}$ and $C1(\infty) = \{m_8 = 0\}$, and $D2(8.5) = \{\}$ and $D2(\infty) = \{(m_9 = 15: T_2)\}$.

Step 11. Next time $t_s = \infty$: END.

❑

The evolution graph in Figure 6.16b is drawn from the results in this application of Algorithm 6.1.

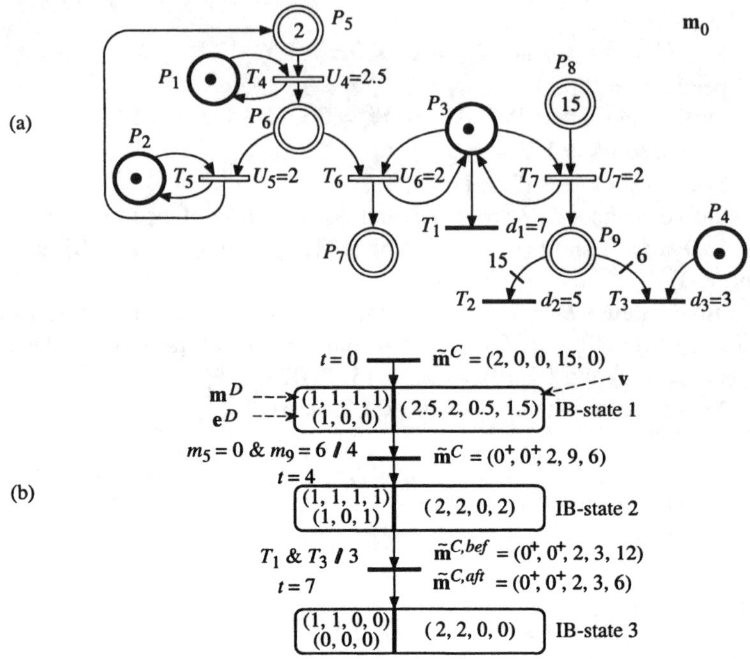

Figure 6.16 Illustration for Algorithm 6.1.

IB-state 1. Marking $\mathbf{m}^D = (1, 1, 1, 1)$ is given in *Step 1.1*. Enabling vector $\mathbf{e}^D = (1, 0, 0)$ because only T_1 is enabled ($D1(7) = \{T_1\}$ in *Step 1.2*). Speed vector $\mathbf{v} = (2.5, 2, 0.5, 1.5)$, according to *Step 10*, first time.

At $t = 4$ (end of IB-state 1), $C1(4) = \{m_1 = 0\}$ and $D2(4) = \{(m_9 = 6\text{: } T_3)\}$ (*Step 10*, first time), and $\tilde{\mathbf{m}}^C(4) = (0^+, 0^+, 2, 9, 6)$ (*Step 11*, first time).

IB-state 2. Transition T_3 became enabled: $D2(4) = \{(m_9 = 6\text{: } T_3)\}$. Speed vector $\mathbf{v} = (2, 2, 0, 2)$, according to *Step 10*, second time.

At $t = 7$ (end of IB-state 2), $\tilde{\mathbf{m}}^C(7) = (0^+, 0^+, 2, 3, 12)$ (*Step 11*, second time), then T_1 and T_3 are fired (*Step 3*, third time), resulting in $\mathbf{m}^D = (1, 1, 0, 0)$, $\mathbf{e}^D = (0, 0, 0)$, and $\tilde{\mathbf{m}}^C(7) = (0^+, 0^+, 2, 3, 6)$.

IB-state 3. Speed vector $\mathbf{v} = (2, 2, 0, 0)$, according to *Step 10*, third time. Infinite duration.

Remark 6.13

a) The time when an event is expected to occur may change. For example, in the previous example, the C1-event '$m_1 = 0$' is expected to occur at $t = 10$, as long as $v_7 = 1.5$; then it is expected to occur at $t = 8.5$ when v_7 becomes 2; finally it will never occur since v_7 becomes 0 before $t = 8.5$.

b) The conditions for an IB-state s to be equivalent to a previous IB-state k (*Step 12* in Algorithm 6.1) are given in Definition 6.5 (Section 6.1.5.3). For both IB-states, the *TOS* is the same except for a time shift $t_s - t_k$.

Remark 6.14 Figure 6.17 illustrates a case of *multiplication* of marking 0^+. Priority $T_4 < T_6$ is the resolution rule for the structural case 2 conflict $\langle P_6, \{T_4, T_6\}\rangle$. During IB-state 2, there is a continuous flow $v_3 = v_4 = v_5 = 1$ and $\tilde{m}_5 = \tilde{m}_6 = \tilde{m}_7 = 0^+$. Since these markings are generated by *flows through* "empty" places, they are *dynamic* 0^+ (Definition 5.6, Section 5.1.3.3). When T_1 is fired, at $t = 5$, V_3 hence v_3 becomes zero: P_6 and P_7 are no longer fed, then v_4 and v_5 become 0, and $\tilde{m}_6$ and $\tilde{m}_7$ become 0, whereas the infinitely small marking in the loop is gathered in P_5 (i.e., $\tilde{m}_5 = 0^+$) since T_3 is no longer enabled. As there is no flow through P_5, this marking is a *static* 0^+. When T_2 is fired, at $t = 8$, a reciprocal modification occurs. Since $m_5 + m_6 + m_7$ is constant when $v_6 = 0$, here is an illustration that $0^+ + 0^+ + 0^+ = 0^+$. This is consistent with (J.4) in Appendix J.

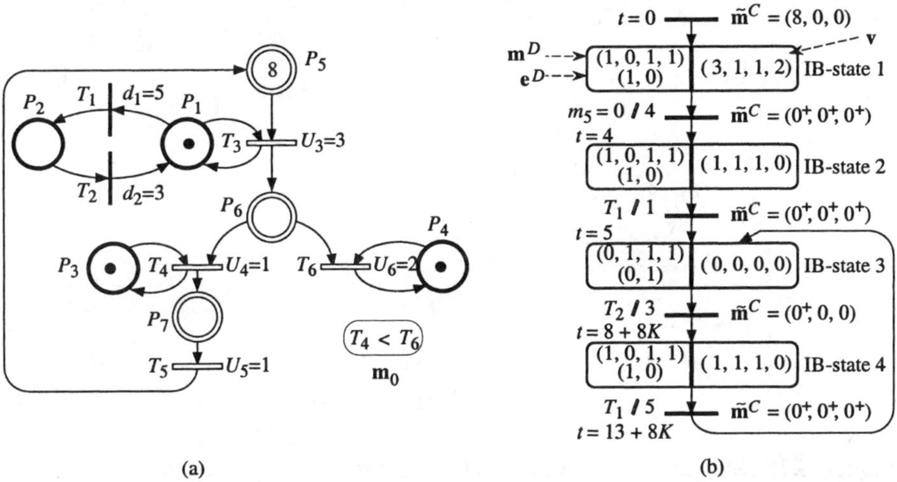

Figure 6.17 Illustration of the multiplication of 0^+.

Remark 6.15 Powerful analysis methods have been developed from a model called *hybrid automaton* (references related to hybrid automata are given in Notes & References at the end of this chapter). In Appendix N, an automatic transformation from a hybrid PN to a hybrid automaton is presented. This combines the *modeling power of hybrid PNs* with the *analysis power of hybrid automata.*

According to Definition 6.5 (Section 6.1.5.3), the number of IB-states may be unbounded if the hybrid PN is unbounded. However, thanks to the concept of *macro-IB-state*, an evolution graph can be built for *unbounded* hybrid PNs, and a hybrid automaton with a number of locations smaller than the number of macro-IB-states may be obtained. These results, as well as the concept of *quasi-periodicity*, are presented in Appendix N, Section N.2.2.

6.2.5 Resolution of a Case Not Treated by Algorithm 6.1

Hypothesis 6.1 is satisfied for the hybrid PN in Figure 6.13a, whereas it is not satisfied for the hybrid PN in Figure 6.13c (Section 6.2.3.1). However, the various instantaneous speeds for Figure 6.13d, represented again in Figure 6.18, can be calculated, as shown in this section.

Consider for example the *resolution rule* $T_1 < T_3$ for the continuous PN in Figure 6.18. It is clear that v_1 depends on v_2 and v_4, that v_4 depends on v_3 which depends in turns on v_1 .... A *general* algorithm taking into account these kinds of dependencies is difficult to find (could be a future research subject !). However, this *particular* problem is not difficult to solve.

Speed v_1 is limited by the bound $V_1 = 4$. Since $m_3 = 0$, transition T_1 is weakly enabled if $I_3 = v_2 + 0.4 \cdot v_4 < 4$. Then,

$$v_1 = \min(4, v_2 + 0.4 \cdot v_4). \tag{6.36}$$

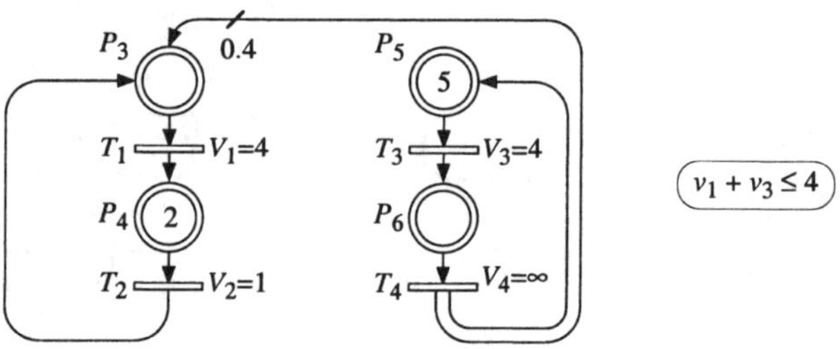

Figure 6.18 Subjacent continuous PN for hybrid PN in Figure 6.13c.

Since T_2 is strongly enabled,

$$v_2 = V_2 = 1. \tag{6.37}$$

According to equation $v_1 + v_3 \leq 4$ in Figure 6.18, and given T_3 is strongly enabled:

$$v_3 = 4 - v_1. \tag{6.38}$$

Since $V_4 = \infty$, P_6 remains empty, then

$$v_4 = v_3. \tag{6.39}$$

The set of four equations, (6.36) to (6.39), is a linear programming problem whose solution is $v_1 = 1.86$, $v_2 = 1$, $v_3 = 2.14$, and $v_4 = 2.14$.

6.3 VARIANTS OF THE MODEL

As explained at the beginning of Section 6.1, the *discrete transitions* in a non-autonomous hybrid PN may be fired as the transitions in a discrete PN (i.e., they may be *synchronized*, or *timed* with *constant* or *stochastic* timings). Similarly, the continuous transitions in a hybrid PN may be fired as the transitions in a continuous PN (i.e., they may be *synchronized*, according to Appendix G, or the flow rates may be *constant*, or *function of time*, or *function of the marking*).

The *basic model*, in which all the timings and flow rates are *constant*, was presented in Section 6.1. Variants of this model are presented in this section, with the help of examples: an example with *synchronized* D-transitions in Section 6.3.1, an example in which *stochastic* timings are associated with the D-transitions in Section 6.3.2, and an example in which the flow rates associated with the C-transitions are *function of time* (*piecewise constant*) in Section 6.3.3.

Examples of synchronized C-transitions are presented in Exercises 5.4 and 8.3 (Figures S 8.3**C**, **E**, and **F**).

6.3.1 Synchronized D-Transitions

The synchronized PNs were presented in Section 3.2.

The hybrid PN in Figure 6.19 contains D-transitions synchronized on external events (T_1, T_3, T_4, T_8, T_9, T_{11}, T_{12}, and T_{16}), D-transitions with constant timings (T_2, T_6, T_{10}, and T_{14}), and immediate D-transitions (T_5, T_7, T_{13}, and T_{15}) for which neither an event nor a delay is written in the figure. A constant flow rate is associated with every C-transition.

Let us now explain the behavior of the system represented in Figure 6.19.

Event E^1 is the deposition of a batch of 500 parts in buffer 1, represented by P_2. Six time units later, the 500 parts will be deposited in buffer 2 represented by P_{17} (if there is enough place, i.e. if $m_{18} \geq 500$). Then, the parts will be treated continuously (processings 1 and 2 represented by the continuous transitions T_{17} and T_{18}). Processing 1 is performed by two servers (two tokens in P_3). When event E^3 occurs, T_3 is fired and a server becomes unavailable (failure or maintenance). If P_4 is not empty, an unavailable server becomes available again when event E^4 occurs. Place P_7 contains a token representing a server which may be used either for processing 2 or for processing 4 (priority to processing 2, i.e. $T_5 < T_{13}$). As soon as 100 parts are in buffer 3 ($m_{19} \geq 100$), transition T_5 is fired if there is a token in P_7 (server not yet allocated). A token in P_8 means that the server is allocated to processing 2, but a set-up (5 time units corresponding to d_6) is necessary before processing a new batch. Processing 2 begins when there is a token in P_9. When the 500 parts in the batch have been completely processed, transition T_7 is immediately fired (because $m_{21} = 500$) and T_8 becomes enabled. When event E^2 occurs (corresponding to an external demand of a processed batch), transition T_8 will be fired.

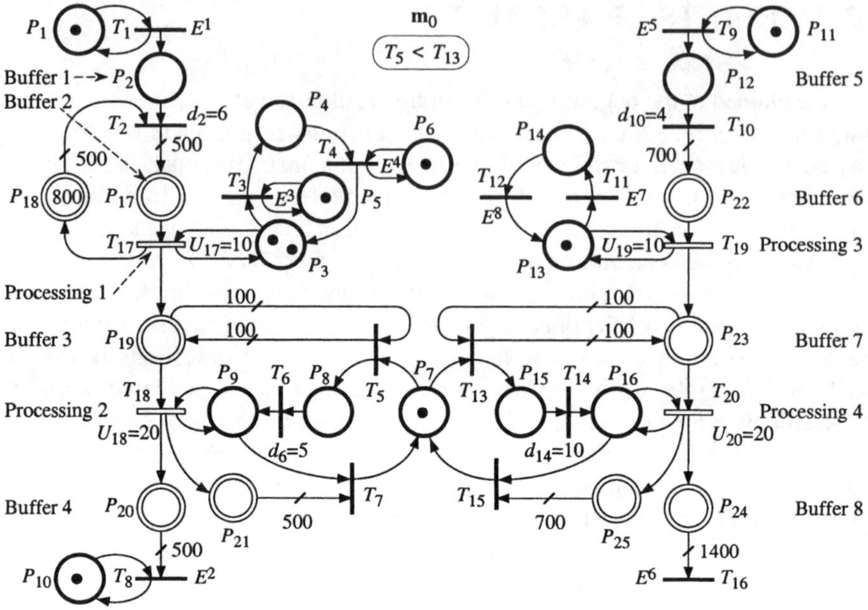

Figure 6.19 A production system with synchronized transitions.

The behavior corresponding to the right part of Figure 6.19 is almost similar. Here are the main differences: 1) whereas the capacity of buffer 2 is 800, buffer 6 is assumed to have an infinite capacity (as well as the other buffers in the model); 2) processing 3 is performed by a single server (i.e. v_{19} = 10 or 0 whereas v_{17} = 20 or 10 or 0); 3) an output batch contains 1400 parts, i.e. two input batches (700 parts each); 4) when event E^6 occurs, all the 1400-part batches which can be made are effectively drawn (for example, if m_{24} = 3000 when E^6 occurs, T_{16} is fired twice and m_{24} = 200 remains).

Given a time ordered sequence of external events, a single path simulation of the behavior can be performed. This behavior may be expressed as a string of IB-states. In addition to *C1-events*, *D1-events*, and *D2-events* (Property 6.2, Section 6.1.5.3), *external events* may cause a change of IB-state.

Let us illustrate the behavior of the hybrid PN in Figure 6.19 for the sequence of external events in Figure 6.20a (for t < 125): given the initial marking $\mathbf{m}_0$ in Figure 6.19 and the sequence of events in Figure 6.20a (i.e., E^1 occurs at t = 4 and t = 59, E^2 at t = 80, E^3 at t = 30, E^4 at t = 52.5, E^5 at t = 18.5, E^6 at t = 115, and E^7 and E^8 do not occur), the behavior illustrated in Figures 6.20 b, c, and d, is obtained.

At t = 0, $\mathbf{m}_0$ is stable, no C-transition is enabled. Nothing can change without an external event.

At t = 4, E^1 occurs and T_1 is fired. A token is deposited in P_2 and T_2 becomes enabled.

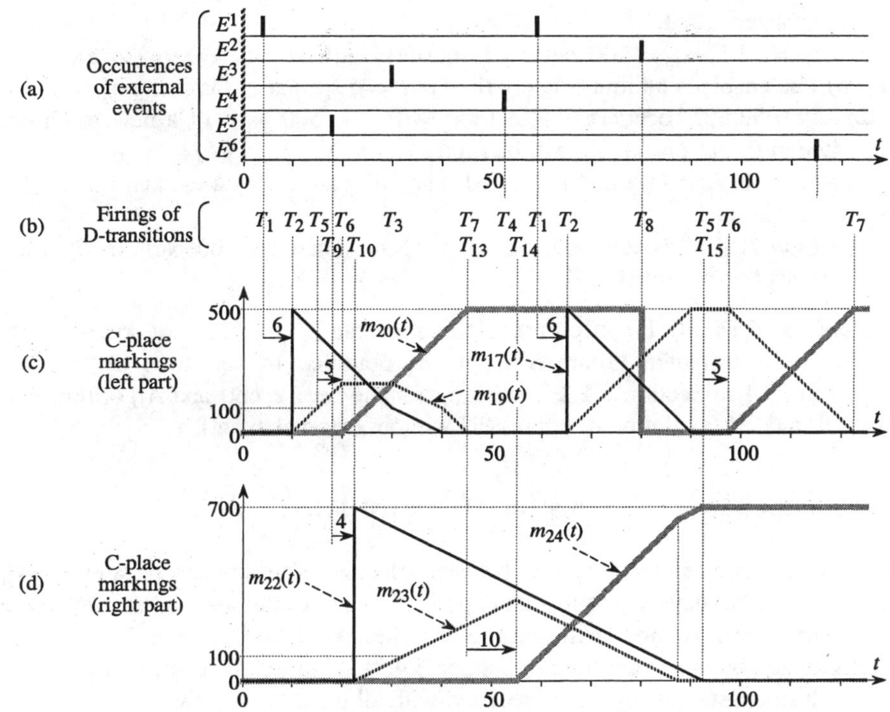

Figure 6.20 Example of behavior for the hybrid PN in Figure 6.19.

At t = 10, T_2 is fired (d_2 = 6 elapsed), resulting in m_2 = 0, m_{17} = 500, and m_{18} = 300 (in Figure 6.20c, m_{18} is not represented because m_{18} = 800 – m_{17}). From this time, T_{17} is continuously fired at speed v_{17} = 20, resulting in $B_{17} = -20$, $B_{18} = B_{19} = +20$.

At t = 15, m_{19} reaches value 100: T_5 becomes enabled and is immediately fired, then T_6 becomes enabled.

At t = 18.5, E^5 occurs and T_9 is fired. A token is deposited in P_{12} and T_{10} becomes enabled. This does not modify the behavior of the left part.

At t = 20, T_6 is fired (d_6 = 5 elapsed), resulting in m_9 = 1. From this time, T_{18} is continuously fired at speed v_{18} = 20, resulting in $B_{19} = 0$ and $B_{20} = B_{21} = +20$ (in Figure 6.20c, m_{21} is not represented: it increases as m_{20} and returns to 0 by firing of T_7 as soon as m_{21} = 500).

At t = 22.5, T_{10} is fired (d_{10} = 4 elapsed), resulting in m_{12} = 0 and m_{22} = 700. From this time, T_{19} is continuously fired at speed v_{19} = 10, resulting in $B_{22} = -10$, and $B_{23} = +10$.

At t = 30, E^3 occurs and T_3 is fired. Since only one token remains in P_3, the D-enabling of T_{17} becomes 1. From this time, T_{17} is fired at speed v_{17} = 10, resulting in B_{17} = -10 and B_{19} = -10.

And so on.

Let us observe that:

1) At $t = 32.5$, $m_{23} = 100$ but T_{13} is not enabled since there is no token in P_7 (T_{13} will be enabled and immediately fired at $t = 45$, just after the firing of T_7).

2) At $t = 52.5$, E^4 occurs and T_4 is fired. After the 500 parts of a new batch are deposited in P_{17}, at $t = 65$, T_{17} will be fired at $v_{17} = 20$.

3) At $t = 80$, E^2 occurs and T_8 is fired. The 500 parts in P_{20} are taken out of the system represented.

4) At $t = 115$, E^6 occurs but nothing happens. There are 700 parts in P_{24}, but 1400 are necessary for firing T_{16}.

Remark 6.16 If the occurrence times of the events are not known, the behavior illustrated in Figure 6.20 can be obtained in real-time by mixing Algorithm 3.1 in Section 3.2.2.2 (waiting for the next event) and Algorithm 6.1 in Section 6.2.4 (evolution of hybrid PN without external event).

6.3.2 Stochastic Timings for D-Transitions

Let us assume that the durations between successive occurrences of some event are unknown but that they follow an exponential law (or this law is an admissible approximation). A stochastic timing can be associated with a D-transition synchronized by this event (like in Section 3.4.3). Figure 6.21 shows an example in which stochastic timings are associated with all the D-transitions.

Consider a transfer line composed of three machines M_1, M_2 and M_3 in order, and two intermediate buffers B_1 and B_2 with respective finite capacities, C_1 and C_2 (Figure 6.21a). The parts move on the machines, and wait in the intermediate buffers if required. We assume that there are always unworked parts upstream M_1 and available space downstream M_3. Let us assume that the processing time on machine M_i is a constant S_i, that M_i can fail randomly with a *time dependent* failure rate λ_i and that the repair rate is μ_i. The failure rate is **time dependent** if the machine may fail at any time when it is operational (it may be working or blocked or starved). If a machine can fail only when the machine is working (hence it cannot fail if it is blocked or starved), the failure is **operation dependent**. The model of operation dependent failure is more pertinent for a manufacturing system. It will be considered in Section 7.3.2.

This system may be modeled by a discrete PN in which some transitions have a constant timing (processing time on a machine) and others have a stochastic timing (failures and repairs).

An approximation of its behavior may be obtained by a hybrid PN as shown in Figure 6.21b. The flow on a machine is represented by the continuous firing of a C-transition, for example T_7 corresponds to M_1 and $U_7 = 1 / S_1$. Transition T_7 is fired at its maximal speed $V_7 = U_7$ as long as M_1 is operational (a token in P_1) and B_1 is not full (i.e. $m_7 > 0$). If M_1 fails, T_1 is fired: the timing of this transition is stochastic, with firing rate λ_1. Similarly, the repair rate μ_1 is associated with T_4.

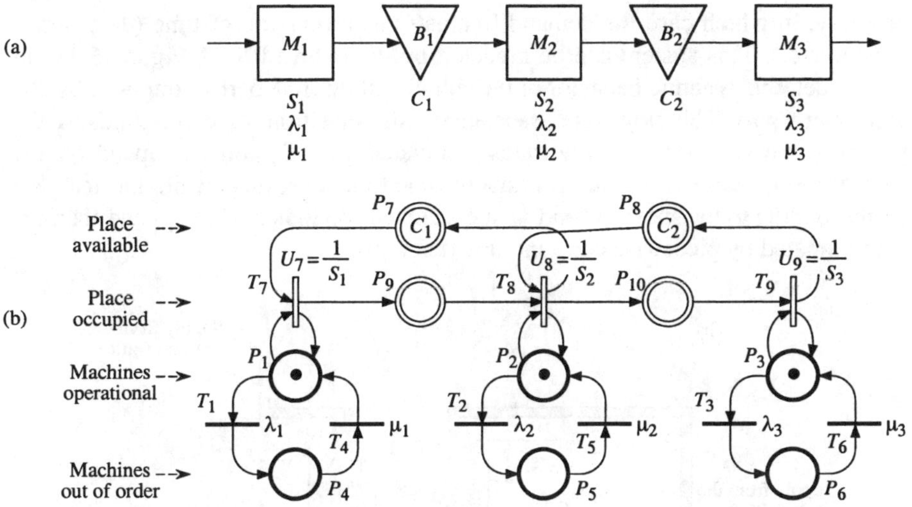

Figure 6.21 (a) Transfer line. (b) Hybrid PN with stochastic timings associated with D-transitions.

If M_1 and M_3 are operational while M_2 fails over a long period of time, M_1 may become blocked (B_1 full) and M_3 may become starved (B_2 empty). In Figure 6.21b, this situation corresponds to $m_7 = m_{10} = 0$: neither T_7 nor T_9 are enabled.

Remark 6.17 Naturally, a single path simulation of the behavior can be performed: the times to failures and to repairs are randomly drawn according to the corresponding distribution laws. However, the average performance of such a system can be obtained by analytical methods (obviously more complicated than in the cases where either all the timings are constant or all the timings are stochastic with exponential laws). References will be given in the Notes and References at the end of Chapter 7.

6.3.3 C-Transitions with Flow Rates Functions of Time

In the example presented in this section, some flow rates are functions of time.

Figure 6.22 presents part of a water supply system: more details about this system can be found in [Me 94]. This system consists of a tank which can be fed either from ground water or from a natural source. Water coming from ground water is pumped and water coming from the source arrives by gravity. When the volume of water corresponding to a level is less than L_{min}, then the pumps are started; they are stopped when this level reaches L_{max}. This requires electrical energy (energy is not required to supply the tank by the natural source). It can be noticed that the source supply depends on the rain and also on the season, while ground water is always available. The tank water can be used for the water supply of a city, in which case the demand fluctuates as a function of time (day, night, season, etc.). This system can be modeled by the hybrid PN of Figure 6.23. In this model, the dynamic behavior of the significant discrete part is imposed by the continuous part. The flow rate associated with transition T_5 corresponds to the pumping power while the flow rates associated with T_3 and T_4 depend on the environment. These speeds are continuous time functions, meanwhile the real data corresponding to the speeds v_3 and v_4 are sampled (periodical reading) and then are approximated by piecewise constant time functions.

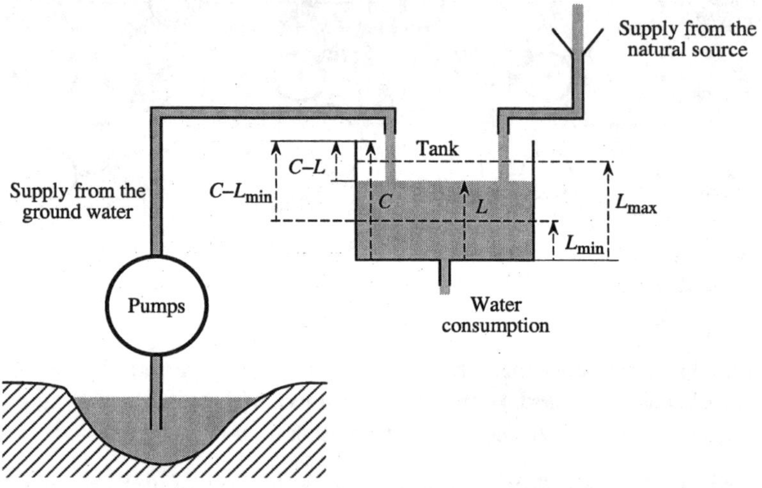

Figure 6.22 Water supply system.

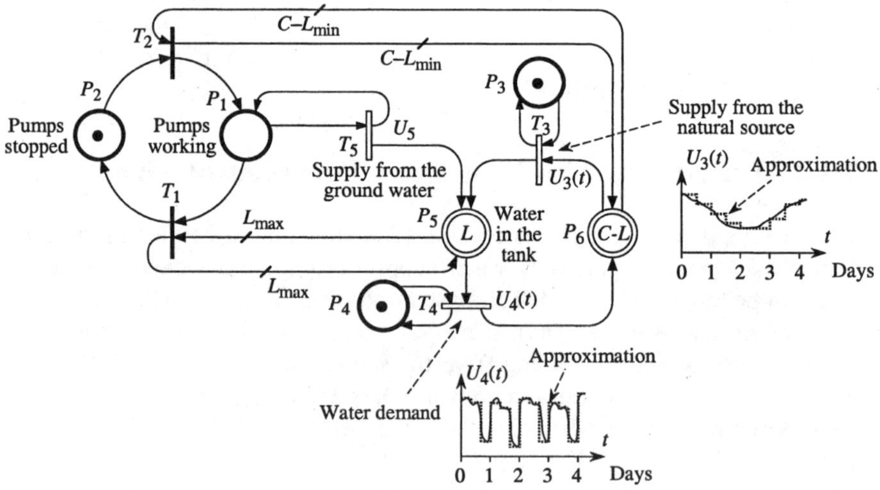

Figure 6.23 Hybrid PN with flow rates function of time.

The quantity of water in the tank is given by its level L (C-place P_5). Place P_6 corresponds to the empty part of the tank: if $m_6 \geq C - L_{\min}$, then $m_5 \leq L_{\min}$.

D-places P_3 and P_4 are necessary because of the hybrid PN semantics explained in Section 6.1.5.2. D-transitions T_1 and T_2 are immediate.

Given piecewise constant functions $U_3(t)$ and $U_4(t)$, a single path simulation of the behavior can be performed. This behavior may be expressed as a string of IB-states. In addition to *C1-events*, *D1-events*, and *D2-events*, *events* corresponding to changes of the vector $\mathbf{U}(t)$ (vector of flow rates associated with the C-transitions) may cause a change of IB-state (see Appendix M).

Another way to manage the volume in the tank is to ensure that the tank is full every morning at 6 AM (corresponding to another hybrid PN model).

Remark 6.18 In Appendix M, the behavior study of a timed continuous PN if $\mathbf{V}(t)$ is progressively known (Hypothesis M.1) or completely known (Hypothesis M.2) is proposed. The same kind of study can be performed for a timed hybrid PN if $\mathbf{U}(t)$ satisfies Hypothesis M.1 or M.2.

More generally, if the vector **tempo**(t) of the values tempo(T_j) (Definition 6.3 in Section 6.1.5.1) satisfies Hypothesis M.1 or M.2, a study as in Appendix M can be performed for a hybrid PN whose values tempo(T_j) are piecewise constant (i.e., the timing d_j associated with a D-transition and the flow rate associated with a C-transition are piecewise constant). Algorithms M.1 and M.2 can be adapted: **tempo**(t) takes the place of $\mathbf{V}(t)$ and Algorithm 6.1 takes the place of Algorithm 5.7.

Remark 6.19 All the variants may be mixed. For example, in Figure 6.19, T_1 could be a stochastic transition and U_{19} could be a function of time.

6.4 EXTENDED TIMED HYBRID PETRI NETS

In the context of autonomous PNs, extended hybrid PNs were presented in Section 4.4. In the present section, *time* is involved. According to Definition 4.12 in Section 4.4.4, an *inhibitor arc*, an *arc weight* 0^+, or an *initial marking* 0^+, can be found in an extended hybrid PN. These elements are proposed in order to extend the modeling power of timed hybrid PNs. However, these elements must be used carefully; *the authors do not claim* that there are semantics allowing the behavior of any hybrid PN to be understood in which arc weights and initial markings 0^+ are introduced randomly. The only claim is that some concepts, useful for modeling some physical systems, can be modeled with the help of these elements. Several examples will be presented and for each case the semantics will be specified.

Modeling of zero buffers by marking 0^+ will be explained in Section 6.4.1. In Section 6.4.2, arc weights 0^+ designed to test whether or not a C-place is empty are presented. Modeling the pure delay of a continuous flow by arc weights 0^+ is shown in Section 6.4.3.

Remark 6.20 A *hybrid* PN is degenerated into a *continuous* PN if there is neither a D-place nor a D-transition. Hence, some concepts and modelings presented for a hybrid PN can also be obtained in a continuous PN. It follows that *extended continuous PNs* are defined implicitly as a special case of extended hybrid PNs (Exercise 6.5).

6.4.1 Modeling of Zero Buffers

Although an initial marking 0^+ does not exist in a continuous or hybrid PN, timed or not (Definitions 4.4 and 6.3 in Sections 4.2.2 and 6.1.5.1), such a marking can appear "naturally" in a *timed* continuous or hybrid PN *after* initial time. This 0^+ marking may be transient (Figure 5.4, Section 5.1.2.1) or permanent (Figure 6.17, Section 6.2.4). A marking 0^+ can be permanent only if it is in a circuit. We shall propose introducing an *initial marking* 0^+ in a circuit, but first we shall analyze the example in Figure 6.24.

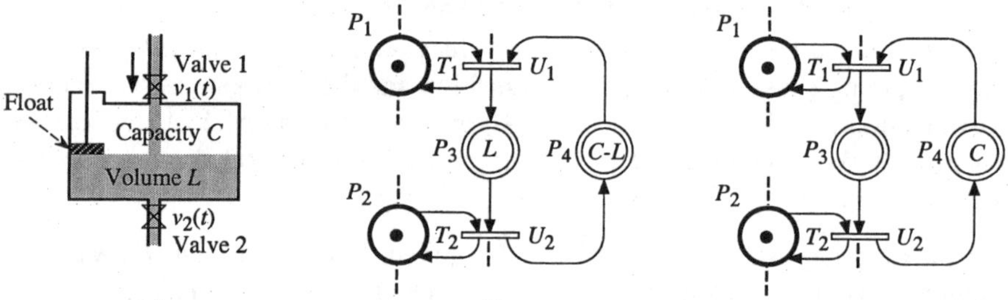

(a) (b) (c)

Figure 6.24 (a) Finite capacity buffer. (b) Corresponding hybrid PN. (c) Initial marking if the buffer is empty.

Figure 6.24a represents a tank which is used as a buffer (capacity C) between valves 1 and 2. When the quantity L of liquid in the buffer increases, the air in the tank may go out, but when $L = C$, the float shuts the tank. The liquid flows, $v_1(t)$ through valve 1 and $v_2(t)$ through valve 2, are relatively independent thanks to this buffer. However, if $L = 0$, $v_2(t) \leq v_1(t)$, and if $L = C$, $v_1(t) \leq v_2(t)$ (it is assumed that the delay between valve 1 and valve 2 is negligible).

The corresponding behavior may be modeled by the (partial) hybrid PN in Figure 6.24b. A token in P_1 means that valve 1 is open; in this case, $V_1 = U_1 \cdot m_1 = U_1$. It is possible that $v_1(t) < V_1$ if the input flow is less than V_1. Markings m_3 and m_4 correspond respectively to the quantity of liquid in the tank and to the empty part in this tank. The marking invariant $m_3 + m_4 = C$ models the *capacity* of the buffer. If the buffer is empty at initial time, the initial marking corresponds to Figure 6.24c. If the buffer were full at initial time, the initial marking would be $m_3 = C$ and $m_4 = 0$.

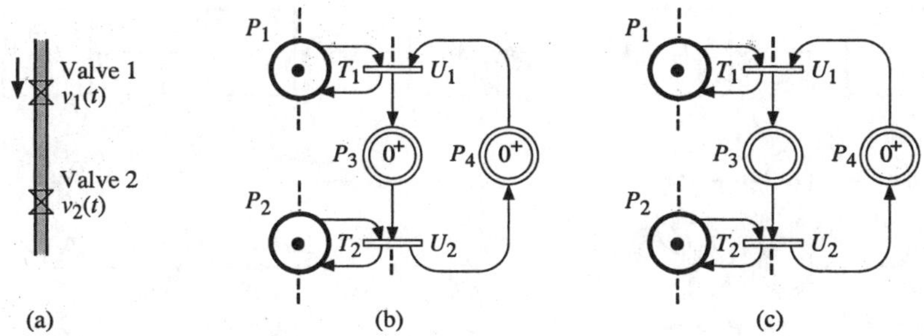

Figure 6.25 (a) Zero buffer. (b) Corresponding extended hybrid PN. (c) A possible initial marking.

If there is no buffer between valve 1 and valve 2, the system is as shown in Figure 6.25a. For this system, $v_2(t)$ depends on $v_1(t)$ as for the *empty buffer* in Figure 6.25, i.e. $v_2(t) \leq v_1(t)$ (for example, if valve 1 is closed then $v_2(t) = 0$). On the other hand, $v_1(t)$ depends on $v_2(t)$ as for the *full buffer* in Figure 6.25, i.e. $v_1(t) \leq v_2(t)$ (if valve 2 is closed then $v_1(t) = 0$). It follows that $v_1(t) = v_2(t)$.

Assume that $v_1(t) = v_2(t) > 0$ for the system in Figure 6.25a. The behavior corresponds to the hybrid PN in Figure 6.25b: both transitions are weakly enabled. Since $m_3(t) = 0$ (i.e., $\tilde{m}_3(t) \leq 0^+$), the balance of P_3 must be non-negative:

$$B_3(t) = v_1(t) - v_2(t) \geq 0. \tag{6.40}$$

Similarly, the balance of P_4 must be non-negative:

$$B_4(t) = v_2(t) - v_1(t) \geq 0. \tag{6.41}$$

Then, from (6.40) and (6.41) (given $v_1(t)$ and $v_2(t)$ are non-negative):

$$v_1(t) = v_2(t). \tag{6.42}$$

Because of the marking invariant $\tilde{m}_3 + \tilde{m}_4 = 0^+ + 0^+ = 0^+$, $\tilde{m}_3(t) \leq 0^+$ and $\tilde{m}_4(t) \leq 0^+$ at any time. It follows that (6.42) is verified at any time t.

Figure 6.25c presents a possible initial marking. Assume that $V_1(t) = U_1 \cdot m_1(t) > 0$, $V_2(t) = U_2 \cdot m_2(t) > 0$, and T_1 is enabled, for some time interval $t \in \left| 0, t_1 \right|$. From the marking in Figure 6.25c at $t = 0$, the marking in Figure 6.25b is obtained in this time interval (Remark 6.14 in Section 6.2.4).

Assume that $V_1(t) = 0$ for $t \in \left| t_1, t_2 \right|$, i.e. valve 1 is closed at t_1. Then, the marking in Figure 6.25c is reached (see Remark 6.14 in Section 6.2.4).

Assume now that $V_1(t) > 0$ but $V_2(t) = 0$ for $t \in \left| t_1, t_2 \right|$, i.e. valve 2 is closed at t_1 (but not valve 1). In this case, the marking $\tilde{m}_3 = 0^+$ and $\tilde{m}_4 = 0$ is reached.

Finally, any continuous marking $(\tilde{m}_3, \tilde{m}_4) = (0, 0^+)$ or $(0^+, 0)$ or $(0^+, 0^+)$ may be chosen for initial marking. The important feature is that $\tilde{m}_3 + \tilde{m}_4 = 0^+$. As a matter of fact, if the initial marking were such that $\tilde{m}_3 + \tilde{m}_4 = 0$, transitions T_1 and T_2 would never be enabled, then $v_1(t) = v_2(t) = 0$ for any $t \in \left| 0, \infty \right|$.

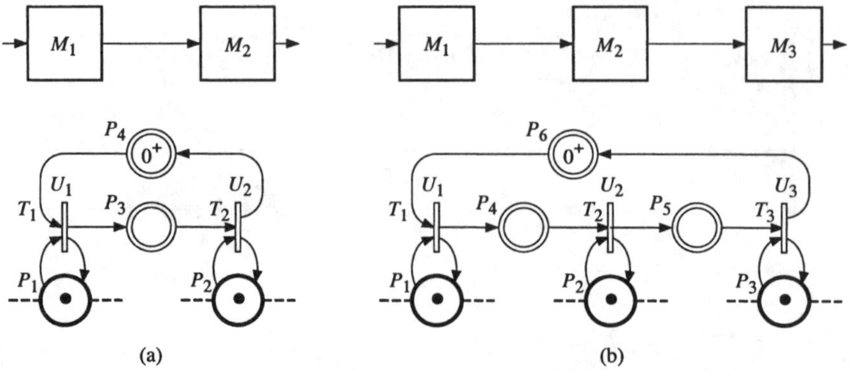

Figure 6.26 Transfer lines with zero buffers.

A hybrid model of transfer lines in which there are buffers between the machines was presented in Figure 6.21 (Section 6.3.2). If there are no buffers (i.e. the buffer capacities are zero), a two-machine and a three-machine line may be represented as in Figures 6.26a and b, respectively. Note that the PN model of the three-machine line could also have two continuous circuits as in Figure 6.21 (with an invariant marking 0^+ in each circuit).

Figure 6.27a illustrates a flow which is split into two flows. If $v_j(t)$ denotes a flow corresponding to a volume of liquid passing by time unit, then

$$v_1(t) = v_2(t) + v_3(t) \tag{6.43}$$

is true at any time t. Similarly, in Figure 6.27b, the continuous flows of parts on the machines ($v_j(t)$ for machine M_j) verify Equation (6.43) since there is neither a buffer between M_1 and M_2 nor a buffer between M_1 and M_3.

Both systems in Figures 6.27a and b can be modeled by the extended hybrid PN in Figure 6.27c. There is a marking invariant

$$\tilde{m}_4 + \tilde{m}_5 = 0^+ \tag{6.44}$$

ensuring that $\tilde{m}_4 \leq 0^+$ and $\tilde{m}_5 \leq 0^+$ at any time. Hence, the balances of the continuous places are such that

$$B_4(t) = v_1(t) - v_2(t) - v_3(t) \geq 0 \tag{6.45}$$

and $$B_5(t) = v_2(t) + v_3(t) - v_1(t) \geq 0, \tag{6.46}$$

from which (6.43) can be obtained.

Naturally, a reciprocal behavior can be modeled, i.e., convergence (gathering) of two flows without buffering. If all the arrows in Figures 6.27a and b are inverted, the model is obtained by inverting all the arrows in Figure 6.27c. Then (6.45) is replaced by $B_4(t) = v_2(t) + v_3(t) - v_1(t) \geq 0$, (6.46) is replaced by $B_5(t) = v_1(t) - v_2(t) - v_3(t) \geq 0$, whereas (6.43) and (6.44) remain unchanged.

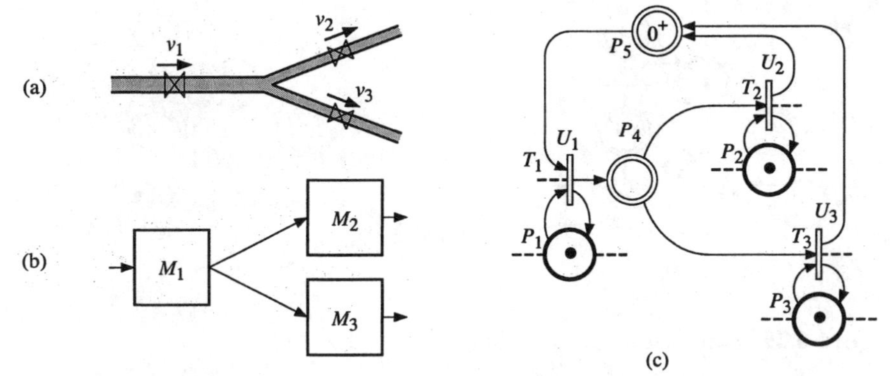

Figure 6.27 Divergence with zero buffers.

6.4.2 Arc Weight 0^+ for Testing if a C-Place is Empty

The use of arc weight 0^+ for zero test of a C-place was explained in Section 4.4.2 in the context of autonomous hybrid PNs. When time is involved (i.e., flow rates associated with C-transitions and timings associated with D-transitions), some behaviors must be specified.

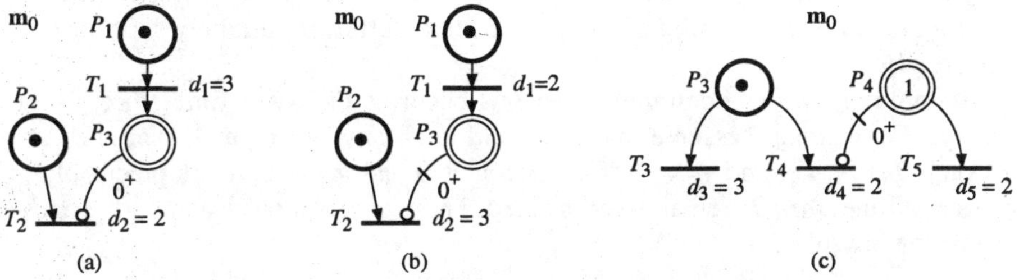

Figure 6.28 Examples of zero tests.

For the examples in Figure 6.28, the behavior is easy to understand.

In Figure 6.28a, T_1 is enabled and will be fired at $t = d_1 = 3$. Transition T_2 is enabled (since there is a token in P_2 and P_3 is empty); it will be fired at $t = 2$ since the markings of P_2 and P_3 do not change up to this time.

In Figure 6.28b, T_1 will be fired at $t = 2$. Transition T_2 is enabled from $t = 0$ to $t = 2$. Since P_3 becomes marked at $t = 2$, T_2 is no longer enabled: it will not be fired because $d_2 = 3 > 2$.

In Figure 6.28c, T_3 and T_5 are enabled whereas T_4 is not enabled because P_4 is marked. Transitions T_3 and T_5 are expected to be fired at $t = 3$ and $t = 2$ respectively. When T_5 is fired at $t = 2$, T_4 becomes enabled; it could be fired at $t = 2 + d_4 = 4$. However, when T_3 is fired, at $t = 3$, T_4 is disabled.

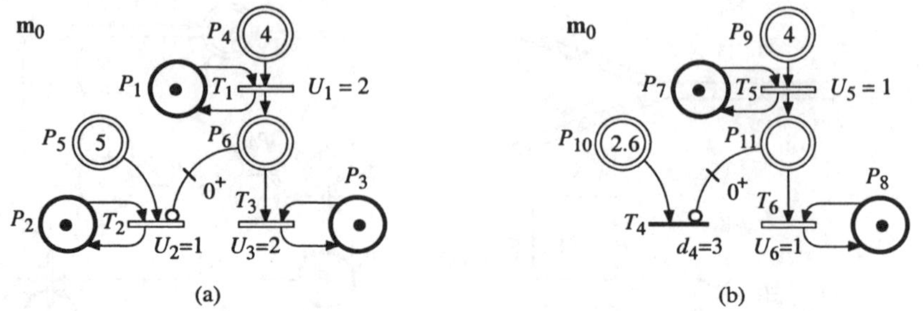

Figure 6.29 Zero test of places whose markings are 0^+.

For each example in Figure 6.28, the marking $\tilde{m}_i$ of C-place P_i whose zero marking was tested, was either 0 or a finite number. However, the marking of the C-place tested may be 0^+. For example, $\tilde{m}_6(t) = 0^+$ for $t \in |0, 2|$ in Figure 6.29a. In order to specify the semantics, let us point out some observations.

Let us consider first *non-immediate* transitions.

If a non-immediate D-transition is enabled, the duration of this enabling is finite: this enabling ends either when the transition is fired or when it is disabled by the firing of another transition. If follows that enabling of such a transition during an infinitely short time is not significant.

If a non-immediate C-transition is fired, the duration of this firing is finite or infinite. If follows that firing of such a transition during an infinitely short time is not significant.

Accordingly, if two contradictory events occur at the same time, there is no change. For example, assume that D-transition T_1 has two input D-places P_1 and P_2 such that $m_1 = 1$ and $m_2 = 0$. If m_1 becomes 0 and m_2 becomes 1 practically at the same time, then T_1 remains not enabled. These observations lead us to propose the following rule.

Rule 6.2

a) An interval during which a *non-immediate D-transition can be enabled* must be a finite interval such that $t \in |t_1, t_2|$, $t_1 < t_2 < \infty$.

b) An interval during which a *non-immediate C-transition can be fired* must be a finite or infinite interval such that $t \in |t_1, t_2|$, $t_1 < t_2$.

❑

This rule means that the conditions of enabling (for a D-transition) or firing (for a C-transition) are not taken into account at t_1 and at t_2 if they are different from the conditions in the interval (according to Notation 5.2 in Section 5.1.2).

Let us analyze the behavior of the hybrid PN in Figure 6.29a, using Rule 6.2. There is a continuous flow through T_1 and T_3 up to $t = 2$: $v_1(t) = v_3(t) = 2$ for $t \in |0, 2|$ and $v_1(t) = v_3(t) = 0$ for $t \in |2, \infty|$. Hence, $\tilde{m}_6(t) = 0^+$ for $t \in |0, 2|$ and $\tilde{m}_6(t) = 0$ for $t \in |2, \infty|$. Accordingly, T_2 becomes enabled at $t = 2$ and $v_2(t) = 1$ for $t \in |2, 7|$, whereas $v_2(t) = 0$ before and after this interval.

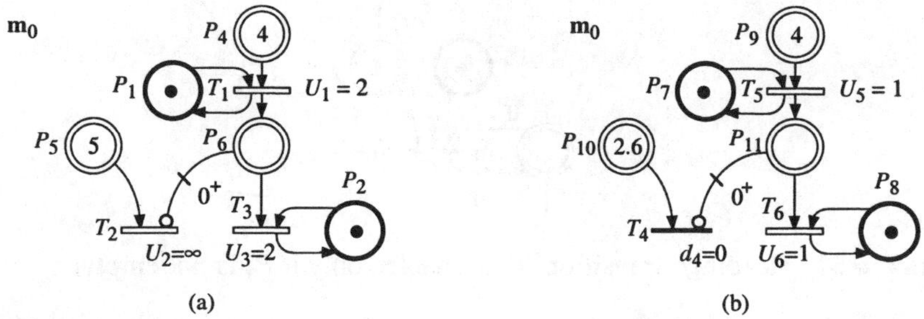

Figure 6.30 Immediate transitions.

For the PN in Figure 6.29b, there is a continuous flow $v_5(t) = v_6(t) = 1$ for $t \in \left| 0, 4 \right|$, then T_4 becomes 2-enabled at $t = 4$. There is a double firing $[T_4T_4]$ at $t = 4 + d_4 = 7$ (i.e., $m_{10} = 0.6$ after this double firing).

Consider now *immediate* transitions. Enabling of an *immediate D-transition* has no duration, and strong enabling of an *immediate C-transition* has no duration, by definition. Hence, Rule 6.2 cannot be applied in these cases.

Consider the example in Figure 6.30a. According to the analysis in Section 5.1.2, $\tilde{m}_6(t) = 0^+$ for $t \in]0, 2]$ but $\tilde{m}_6(0) = 0$. Since T_2 is an immediate transition, it is fired at $t = 0$: P_5 is instantaneously emptied.

Similarly, in Figure 6.30b, $\tilde{m}_{11}(0) = 0$. Since T_4 is an immediate transition, it is fired twice at $t = 0$.

From the beginning of this section, we were interested in the zero test by *inhibitor arcs*. The transition in question can be *fired only if the C-place concerned is empty*; then, obviously, the marking of this place is not changed by the firing of the transition. Let us consider now the *reading* case (see Figure 1.17b, Section 1.3, for the usual discrete case): a transition can be *fired only if the C-place concerned is* not *empty.*

Usually, for a reading, the same quantity is simultaneously drawn and deposited in the place concerned. Since the symbol 0^+ means "infinitely small but not nil" but does not really correspond to some "quantity", let us specify the semantics by the following rule.

Rule 6.3 If arcs $P_a \rightarrow T_j$ and $T_j \rightarrow P_b$ ($P_a \neq P_b$ or $P_a = P_b$) with weights 0^+ exist, then the *same* quantity is drawn from P_a and deposited in P_b when T_j is fired (even if this "quantity" cannot be measured, it is *the same*).

❑

Examples of models using readings with finite weights have already been presented (Figure 6.23 in Section 6.3.3, for example). Figure 6.31 illustrates the case of infinitely small weights. Transition T_1 is *enabled* only if P_3 is *not empty.* According to Rule 6.3, the firing of T_1 does not change the marking of P_3. Rule 6.2 is also relevant to this model.

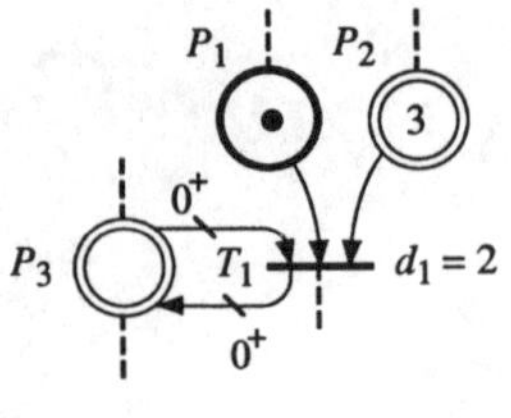

Figure 6.31 Reading: transition T_1 is enabled only if P_3 is not empty.

6.4.3 Pure Delay of a Continuous Flow

From a practical point of view, the *pure delay of a continuous flow* is an important phenomenon. A constant-speed *conveyor* corresponds to a pure delay: a part (or a flow of parts for a continuous model) entering a conveyor reaches the end of the conveyor after some constant time is elapsed. Similarly, a fluid entering a long *pipe* reaches the far end of the pipe after some constant time is elapsed (if the flow speed is constant).

It will be shown that a *pure delay of a continuous flow* from a C-place P_a to another P_b can be modeled by a *D-transition* T_{ab} between both, with *arc weights* 0^+ for arcs $P_a \rightarrow T_{ab}$ and $T_{ab} \rightarrow P_b$ (Rule 6.3). This behavior corresponds to the **continuous firing of a D-transition.**

The behavior of a simple conveyor is presented in Section 6.4.3.1. In Section 6.4.3.2, it is shown that the model is quite adapted for various behaviors of a conveyor. The pure delay in a simple pipe is illustrated in Section 6.4.3.3.

6.4.3.1 Simple Conveyor

Figure 6.32a represents a conveyor C with an input buffer B_1 and an output buffer B_2 (capacities not bounded). A batch of 40 parts is deposited in B_1 at $t = 0$. Conveyor length is $L_C = 10$ m and speed is $Q_C = 5$ m/min. Its maximal density is $D_C = 10$ parts/m (capacity 100 parts for the total length).

The behavior of this system is represented by the extended hybrid PN in Figure 6.32b where part flow is considered to be continuous. Markings of places P_3 and P_6 correspond to the parts in the input and output buffers. Transitions T_2 and T_3 correspond to parts entering and leaving the conveyor. The flow rates associated with these transitions correspond to:

$$U_2 = U_3 = Q_C \times D_C = 5 \times 10 = 50 \text{ parts/min.} \tag{6.47}$$

Place P_7 models the place available on the conveyor. The delay associated with transition T_1 corresponds to the time spent on the conveyor, i.e.,

$$d_1 = \frac{L_C}{Q_C} = \frac{10}{5} = 2 \text{ min.} \tag{6.48}$$

The weights 0^+ of arcs from P_4 to T_1 and from T_1 to P_5 mean that as soon as an infinitely small quantity of part is put on the conveyor (i.e., in P_4), transition T_1 is enabled and this quantity will be put at the end of the conveyor (i.e., in P_5) when the time d_1 has elapsed.

The dynamic behavior of the system modeled by Figure 6.32b is illustrated in Figure 6.32c and d.

At time $t = 0$, transition T_2 is strongly enabled since P_3 and P_7 are not empty and there is a token in P_1. Then $v_2(t) = U_2 \cdot m_1 = 50$. The marking in P_4 enables transition T_1, but its firing is delayed by $d_1 = 2$. Hence, m_4 increases (slope $v_2(t) = 50$ in Figure 6.32c). This behavior lasts as long as T_2 is enabled: the batch of 40 parts progressively enters the conveyor. This is represented by the left hand part of Figure 6.32d. There is a correspondence between the four parts of Figure 6.32d and Figure 6.32c, i.e., from left to right, the parts of Figure 6.32d correspond to the intervals $|0, 0.8|$, $|0.8, 2|$, $|2, 2.8|$, and $|2.8, \infty|$.

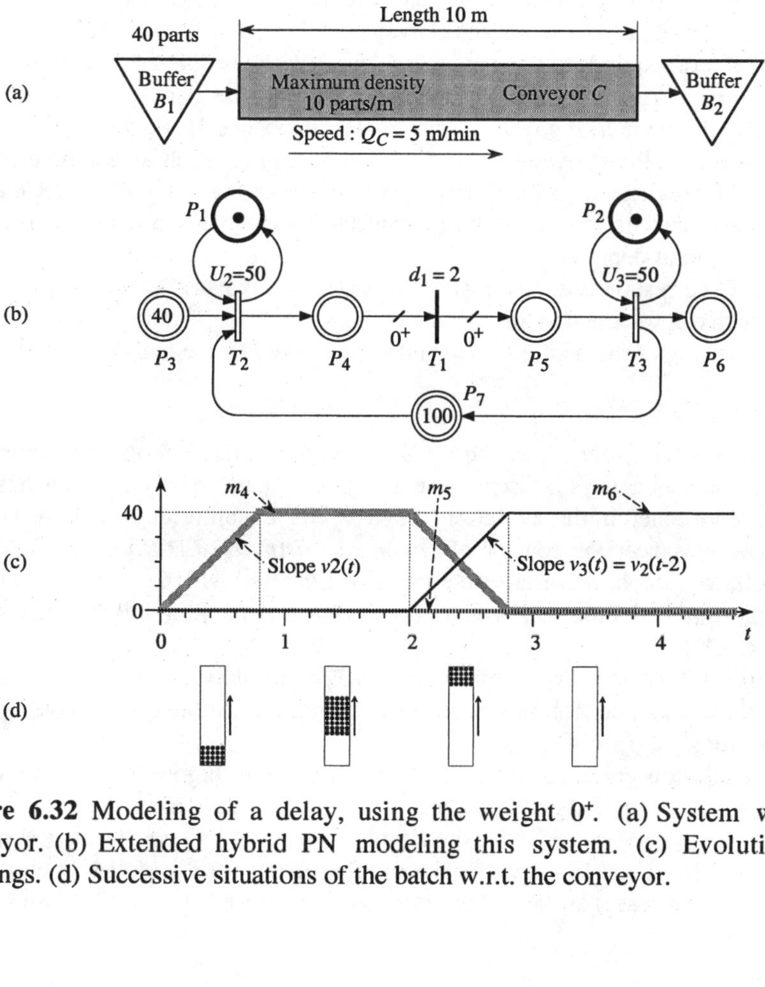

Figure 6.32 Modeling of a delay, using the weight 0^+. (a) System with a conveyor. (b) Extended hybrid PN modeling this system. (c) Evolution of markings. (d) Successive situations of the batch w.r.t. the conveyor.

At $t = 0.8$, P_3 becomes empty, then T_2 is no longer enabled. The marking in P_4 remains $m_4 = 40$, and no transition is fired up to $t = 2$. For $t \in |0.8, 2|$, the batch is completely on the conveyor (Figure 6.32d).

At time $t = 2$, the firing of T_1 starts since the delay $d_1 = 2$ is over. The weights 0^+ associated with the arcs $P_4 \rightarrow T_1$ and $T_1 \rightarrow P_5$ mean that every infinitely small marking deposited in P_4 between t and $t + dt$ is put into P_5, through T_1, between $t + 2$ and $t + dt + 2$. This modeling actually corresponds to a *continuous firing of a discrete transition.* Furthermore, each marking put into P_5 enables T_3 which is immediately fired. It follows that T_3 is fired at the same speed as T_2 with a delay of 2 minutes. The marking m_5 remains 0 (since T_1 and T_3 are fired at the same speed). The marking m_6 increases from 0 to 40 between $t = 2$ and $t = 2.8$. In the time interval $|2, 2.8|$, the batch leaves the conveyor.

At time $t > 2.8$, $m_3 = m_4 = m_5 = 0$ and the marking of P_6 will remain at $m_6 = 40$. The batch has left the conveyor as illustrated in Figure 6.32d.

Let us now specify the meanings of P_4, P_5, and P_7.

Place P_4 corresponds to the parts which are *on the conveyor but have not spent enough time to reach the end* of the conveyor.

Place P_5 corresponds to the parts which are still *on the conveyor but have spent enough time to reach the end* of the conveyor. For the conveyor represented in Figure 6.32b, the marking in P_5 is immediately deposited in P_6 through T_3, hence m_5 remains at value 0. However, if P_2 could be empty (such an example will be presented in Section 6.4.3.2), T_3 might not be enabled (even if P_5 is not empty). In this case, the marking in P_5 corresponds to parts which accumulated from the output (maximal density).

Place P_7 corresponds to space available on the conveyor, i.e. to the number of parts which can be added to the conveyor.

The marking of the places P_4, P_5, and P_7 are linked by the marking invariant

$$m_4 + m_5 + m_7 = L_C \times D_C = 10 \times 10 = 100 \text{ parts.} \tag{6.49}$$

Let us now consider the *evolution graph.* Before this section, the firing of a D-transition was always a discrete event, and the "state" of such a transition was always represented in the evolution graph by its enabling degree. However, we have now introduced the concept of *continuous firing of a D-transition*; hence, in the evolution graph, a D-transition continuously fired will be characterized by a *firing speed* (an additional information related to its enabling will be presented in Remark 6.21).

For the extended hybrid PN in Figure 6.32b, the only D-transition T_1 is fired continuously. Hence, for this example, v_1 is added to the speed vector of the C-transitions.

The evolution graph for the extended hybrid PN in Figure 6.32b is shown in Figure 6.33a. In the right hand part of IB-states, the firing speed v_1 of the D-transition T_1 has been added to the firing speeds of the C-transitions (because it is continuously fired like a C-transition). The firing speeds of the D-transitions (only one in this case) are separated from the firing speeds of the C-transitions by a semi-colon. The firing speed of the D-transition is shown in bold characters in Figure 6.33a. In this figure, the events associated with the changes between IB-States 2 and 3, and between IB-States 3 and 4 do not belong to the kinds of events presented in Property 6.2 (Section 6.1.5.3). They correspond to a fourth kind: the continuous firing of a D-transition either begins or ends.

Let us emphasize that *behavior can be calculated analytically* in every IB-state. For IB-state 3 in Figure 6.33, for example, $m_4(t)$ is given by:

$$m_4(t) = 40 + (v_2(t) - v_1(t))\,(t-2) = 40 - 50\,(t-2), \text{ for } t \in \left]2, 2.8\right[. \quad (6.50)$$

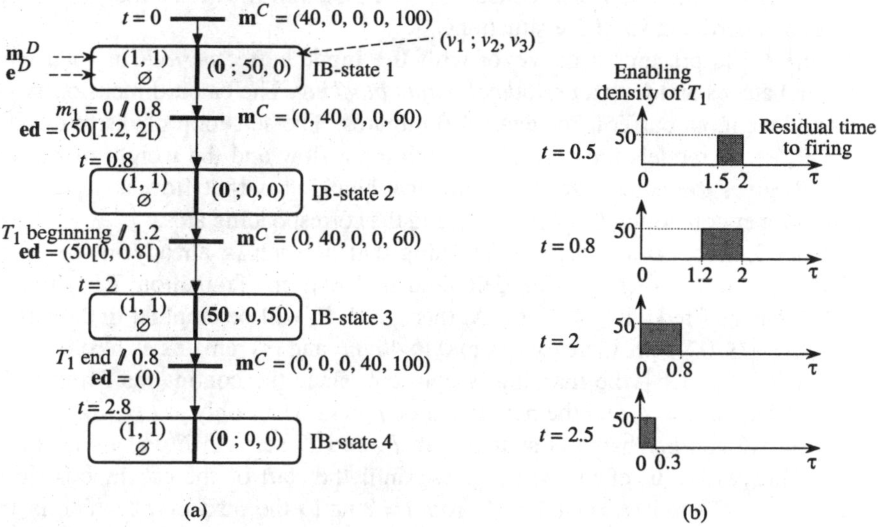

Figure 6.33 Evolution graph for the PN in Figure 6.32b.

Remark 6.21 Although it is not absolutely necessary for this simple example, an additional information related to the *enabling density* has been added in the left part of Figure 6.33a. This enabling density is illustrated in Figure 6.33b. At time t = 0.8 (beginning of IB-state 2), for example, there is some material on the conveyor (marking in P_4). The corresponding **enabling density**, denoted by 50[1.2, 2[, is 50 (corresponding to v_2) between the minimal *residual time*1.2 *to firing* (i.e. the head of batch will arrive at the end of the conveyor 1.2 minutes later, hence at t = 2) and the maximal *residual time* 2 *to firing*. This means that, for any τ in the range $1.2 \le \tau < 2$, the quantity of firing between $t = 0.8 + \tau$ and $t + dt$ will be 50 dt.

In the left part of each IB-state in Figure 6.33a, the enabling vector $\mathbf{e}^D$ corresponds to the "ordinary" D-transitions, i.e. not continuously fired (empty set in our example). The *enabling density vector* **ed** corresponds to the vector of D-transitions continuously fired (only T_1 in our example). If several batches enable the same D-transition, a concatenation of triples $A[B, C[$ represents its enabling density.

Even if this information is not necessary for this simple example, it is very useful for analyzing the behavior when there are *several batches* on the conveyor or if the *conveyor speed Q_C changes*. See Exercises 6.7 to 6.9.

6.4.3.2 Various Behaviors of a Conveyor

Besides the basic evolution described above, the functioning of a conveyor could involve some additional constraints. This section illustrates that the model proposed is pertinent for these situations.

Figure 6.34a presents a conveyor with the input buffer *periodically fed* with constant batches and with a *periodical output blocking*. The batch dimension is 40 parts and the input buffer is fed every 1.6 minutes. The self-loop of discrete nodes $T_1 \to P_1 \to T_1$ models the periodical batch input flow and the weight of the arc $T_1 \to P_5$ gives the batch size. The periodical blocking output flow is represented by the subnet containing T_3, T_4, P_3, P_4, and the corresponding arcs.

At $t = 0$, $m_5 = 40$ and $m_1 = 1$, meaning that a batch is already in the input buffer and another one is to be delivered in 1.6 min. Transition T_5, strongly enabled, can be fired with $v_5(t) = U_5$. At this speed, P_5 will be emptied in 0.8 min.

For $t \in |0, 0.8|$ m_6 increases from 0 to 40; m_7 and m_8 remains at zero.

For $t \in |0.8, 1.6|$ the marking is constant, since the continuous firing of T_2 will start only at $t = 2$, and the next firing of T_1 will occur only at $t = 1.6$.

At $t = 1.6$ another batch is available in P_5 and T_5 starts firing at its maximal speed. Thus, the value of m_6 will increase until the start of the continuous firing of T_2 at $t = 2$. Then $v_5(t) = v_2(t) = 50$ from $t = 2$ up to the next event which is the blocking of the conveyor output at $t = 2.2$ (firing of T_4: transition T_6 is no longer enabled).

At $t = 2.2$, $m_6 = 60$ and m_8 (which was increasing from zero to 40) has a value of 10. This means that the first batch was leaving the conveyor (10 parts of it being already in the output buffer and 30 still on the conveyor). The second batch is entering the conveyor and the distance between the batches is 4 m.

Because of the disabling of T_6, the value of $m_8 = 10$ will remain constant until $t = d_3 + d_4 = 12.2$. Parts from the first batch will accumulate at the end of the conveyor, i.e., m_7 will increase with the remnant of 30. The feeding speed for P_7 is equal to $v_2(t) = v_5(t - 2) = U_5 \cdot m_2 = 50$, so as $m_7 = 30$ at $t = 2.2 + 30 / 50 = 2.8$. At the same time $m_6 = 40$; this marking corresponds to the second batch which is completely on the conveyor. As illustrated in Figure 6.34d, the distance between the end of the first batch and the start of the second batch is decreasing and the second batch will also be blocked.

The next event to occur is at $t = 3.2$ when the third batch enters the conveyor. As in the previous section, the value of m_6 will increase until the time when a new firing of T_2 starts, i.e. at $t = 3.6$.

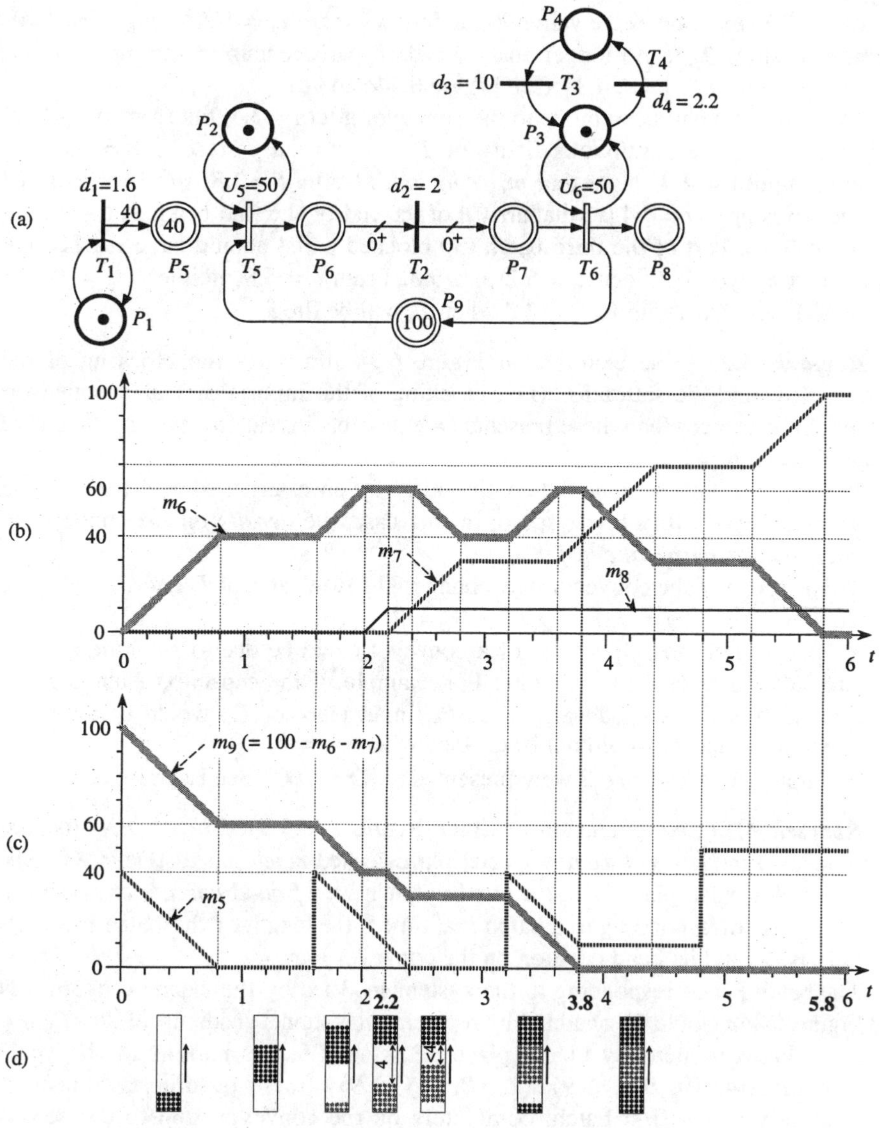

Figure 6.34 Batch accumulation and conveyor saturation.

With this example, we have seen that batch accumulation is correctly modeled; the end of the first batch is blocked and the second batch is blocked in turn when it reaches the first one. We shall now show that the whole conveyor can become saturated and that the model is still correct.

The behavior up to t = 3.6 has already been explained. At t = 3.6, m_6 = 60 is constant while m_7 increases from the value m_7 = 30.

At $t = 3.8$, m_7 reaches the value 40. It follows that $m_9 = 100 - m_6 - m_7 = 0$. Hence transition T_5 is no longer enabled and 10 parts remain in the input buffer, i.e., $m_5 = 10$ for $t \in |3.8, 4.8|$ (see Figure 6.34c and d).

Physically, no parts can move to the conveyor after $t = 3.8$, but from the model point of view, the continuous firing of T_2 can normally start at $t = 3.6$ and continue until $t = 4.4$. Note that $m_6 + m_7 = 100$ after $t = 3.8$; m_6 decreases and m_7 increases up to $t = 4.4$ (virtual arrival of the end of the 2nd batch at the end of the conveyor). Part of the third batch (30 blocked parts) should have reached the end of conveyor between $t = 5.2$ and 5.8. From $t = 5.8$, neither $m_6 = 0$ nor $m_7 = 100$ can change up to $t = 12.2$ when T_3 will be fired.

Remark 6.22 The example in Figure 6.34 illustrates the blocking of the output flow and the saturation (i.e., blocking of the input flow) of a conveyor. Other circumstances than those presented are possible in real functioning. We give examples below.

1) The input buffer can be fed continuously by an external source, for example another conveyor with a lower speed; in this case, the *density* on the conveyor is *smaller than its maximal density*.

2) The speed of the conveyor may change. The flow rates of T_2 and T_3 in Figure 6.32b would be $U_2(t) = U_3(t) = Q_C(t) \times D_C = 10\, Q_C(t)$.

3) Blocking of the output flow of a conveyor could be due to saturation of the output buffer if its capacity is finite. For example, if the capacity of buffer B_2 in Figure 6.32a is C_2, an additional place P_8 (input place of T_3) would be such that $m_6 + m_7 = C_2$, and T_3 would not be enabled when $\tilde{m}_7 = 0$.

Examples of cases 1 and 2 were presented in [DaCa 00]. See Exercise 6.8.

Remark 6.23 An extension of timed hybrid PNs called *batch PNs* (defined for conveyor modeling) contains special places called *batch places* [DePr 92]. The marking of a batch place is *not a number* but a *set of quadruples*, each quadruple $\langle L_i, D_i, x_i, t\rangle$ corresponding to a batch i: at time t, the length of the batch is L_i, its density is D_i, and its head position on the conveyor is x_i.

The batch PN corresponding to the system modeled by the extended hybrid PN in Figure 6.34a would be obtained by replacing the subnet made up of P_6, T_2, P_7, and arcs between them, by a batch place BP. At $t = 1.9$, the marking of BP would be the set $\{\langle 4, 10, 9.5, 1.9\rangle, \langle 1, 10, 1.5, 1.9\rangle\}$ (these quadruples correspond respectively to the first batch, completely on the conveyor, and to the second batch, partially on the conveyor).

6.4.3.3 *Fluid Example*

The example in Figure 6.35 illustrates a liquid flow in a pipe.

Figure 6.35a represents a pipe whose length is L_C. The flow in the pipe is controlled by two valves, R_1 and R_2. On the left hand side of R_1, there is always some liquid under pressure, such that, if R_1 is open, the liquid flows into the pipe at the linear speed Q_C. If R_2 is closed, it is assumed that the liquid flows up to R_2 (the air in the pipe is not taken into account).

An extended hybrid PN modeling this system is shown in Figure 6.35b. Transition T_6 represents the flow entering the pipe. It is enabled only if: 1) there is a token in P_1 (R_1 is open); 2) $m_7 > 0$ (the pipe is not full), or $\tilde{m}_7 = 0^+$ (the pipe is full but valve R_2 is open). Place P_5 corresponds to the liquid in the pipe which has not had time to reach the end of the pipe, and P_6 corresponds to the liquid in the pipe which has had time to reach the end of the pipe. Both places are separated by a discrete transition whose delay is the time necessary to flow from R_1 to R_2. The weights of arcs $P_5 \rightarrow T_5$ and $T_5 \rightarrow P_6$ are 0^+, modeling a continuous firing of the D-transition T_5. For simplicity, it is assumed that the flow rate U_7 of T_7 is the same when R_1 is closed and when it is open

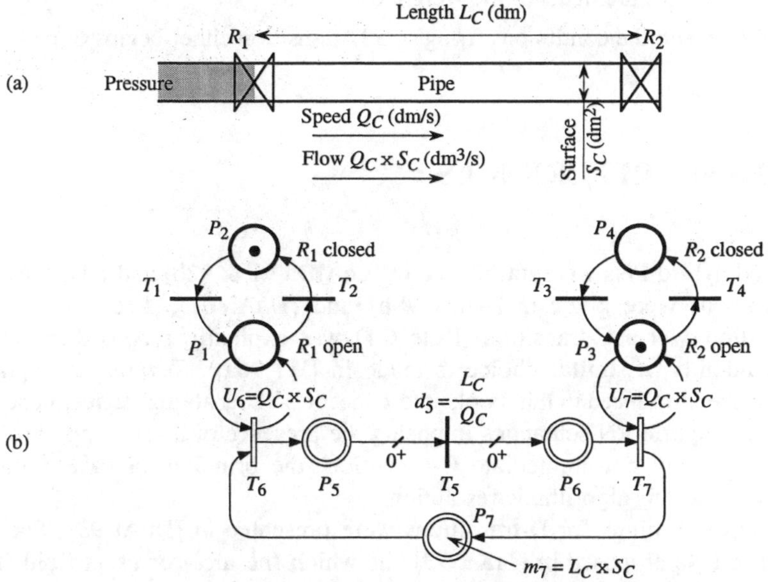

Figure 6.35 Flow in a pipe. (a) The pipe. (b) The corresponding extended hybrid PN.

6.4.4 Conclusion on Timed Extended Hybrid Petri Nets

Building an evolution graph for a timed *extended* hybrid PN requires consideration of *new kinds of events*, related to the additional modeling possibilities.

Initial marking 0^+. Even if an initial marking 0^+ is not allowed in a basic (not extended) timed hybrid PN, such a marking may be obtained at the end of a phase. Hence, the "tools" for building an evolution graph already exist.

Inhibitor arc. This possibility has been added. Hence, if the marking (increasing or decreasing) of a C-place, origin of an inhibitor arc, reaches the weight of this arc, this is an event which must be taken into account (called **C4-event** since related to a C-place).

Pure delay of a continuous flow. The concept of continuous firing of a D-transition has been added. Hence, the beginning and the end of such a continuous firing are new events that must be taken into account (**CD-event**).

Then, Property 6.4 is obtained.

Property 6.4 In a timed *extended* hybrid PN, a *change of IB-state* can occur only if an *event* belonging to one of the following types occurs.

C1-event, D1-event, D2-event, as in Property 6.2 (Section 6.1.5.3).

C4-event: the marking of a C-place (whose balance is not nil), origin of an inhibitor arc, reaches the weight of this arc.

CD-event: the continuous firing of a D-transition either begins or ends.

NOTES and REFERENCES

Timed hybrid PNs were introduced in [LeAlDa 91 & 92b] and all-compassing presentations were given in [AlDa 98b] and [DaAl 01]. The priority of a D-transition over a C-transition (Rule 6.1) was explicitly proposed in [Da 97] (independently, a similar choice is made in [Wi 96a]). Several concepts and methods are introduced in this book: the concept of D-enabling degree, hence, the new timed hybrid PN semantics imposing the presence of a self loop through a D-place for any non-immediate C-transition; the principle of case 4 conflict resolution and this algorithmic resolution.

Stochastic timings for D-transitions were presented in [DuAl 93]. The fluid stochastic PNs proposed in [TrKu 93], in which the arcs represent fluid flows, model the same kind of systems as the timed hybrid PN with a stochastic discrete part, while other simulation possibilities are added in [CiNiTr 97]. A numerical solution is proposed for the model in [Ho *et al.* 98].

Timed extended hybrid PNs were introduced in [DaAl 01]. Modeling of delays on continuous flows was developed in [DaCa 00], from which Figures 6.32 to 6.35 are drawn. Modeling of conveyors is important from a practical point of view. For this purpose, in [BrBl 90], timings have been added to places in continuous PNs (see also Notes & References in Chapter 5 and Appendix O). According to Section 3.4.2.1, in a P-timed PN, some tokens are unavailable and others are available. A continuous place in [BrBl 90] is equivalent to a subnet similar to P_4, T_1, P_5, and arcs among them, in Figure 6.32: m_4 and m_5 correspond respectively to the unavailable and available markings of the continuous place.

Batch PNs, mentioned in Remark 6.23 are presented in [DePr 92] [DeAuPr 97] [DeCaPr 98].

In [BaGiMe 98 & 00], a model called first-order hybrid Petri net is proposed. It is similar to our timed hybrid Petri net except that: 1) a firing speed has also a minimum value, i.e. $V_j' \leq v_j \leq V_j$ (it is possible that there is no solution for the speed vector **v**, for example if $V_2 = 2$ and $V_3' = 3$ in Figure 5.6); 2) all the C-transitions which are D-enabled, according to the definition in this book, are said to be enabled (this is equivalent to having a marking 0^+ in *all* the empty C-places: for the PN in Figure 5.15 (Section 5.3.1), for example, speeds $v_4 = v_6 = 1$ are obtained according to [BaGiMe 98 & 00], instead of $v_4 = v_6 = 0$ (another example, taken from [MuDaAl 04], is presented in Exercise 6.11). In [BaGiMe 00], the authors characterize all the possible speed vectors (also done in [Ha 03] as explained in Notes & References of Chapter 5).

Hybrid high-level PNs are proposed and used in [WiSo 95][Wi 96a][Wi 96b]. The proposed model is more detailed in the second of these references. In [Wi 96a], an adaptation of evolution graph (widely presented in this book) is proposed for hybrid high-level PNs, and object-oriented concepts (proposed in [ElCeOt 93]) are used.

The production system in Figure 6.19, the water supply system in Figure 6.22, the zero buffer production system in Figure 6.27b, and the conveyor system in Figure 6.32a, draw inspiration from [Al *et al.* 92] and [AlAl 98a], [Me 94], [BaGiMe 98], and [DePr 92], respectively. Other applications of timed hybrid PNs were recently proposed: modeling of a communication network under TCP/IP protocol [BiAl 03]; models of urban transportation networks [DiSa 03].

An important part of [SiRe 03] is devoted to structural analysis, control, and optimization of timed hybrid PNs (for example, choose initial marking or optimal firing speeds) by linear programming techniques.

The hybrid system model presented in [Al *et al.* 95] (called hybrid automaton in this book) is inspired by the phase transition systems of [MaMaPn 92] [NiSiYo 93], and can be viewed as a generalization of timed safety automata [AlDi 94][He *et al.* 94]. The main goal of this paper [Al *et al.* 95] is the algorithmic analysis of hybrid systems. Coupled linear hybrid automata are introduced and analyzed in [FaAlFl 99]. Let us note that three papers in [*SI* 01], namely [DaAl 01], [Ba *et al.* 01], and [TuChTr 01], discuss the connection of their respective models of hybrid systems with hybrid automata.

Batch processes operating on continuous flow materials is a typical class of hybrid systems. A verification of control procedures by means of predicate/transition nets (similar to grafcets) is proposed in [GeHaWö 94]. Models between hybrid automata and hybrid PNs may be used [Ch *et al.* 98a][Va 98]: a place in a PN may be associated with a phase in a recipe, and a set of differential equations may be associated with such a phase.

In [BrBoMi 94], the authors propose an unified framework for hybrid control, based on various models, namely [Ta 87][AnStLe 93][BaGuMy 93][Br 93] [NeKo 93].

7

Hybrid Petri Nets with Speeds Depending on the C-Marking

In the basic timed hybrid PN (called CHPN*), as well as in the variants, presented in Chapter 6, maximal speeds are independent from the C-marking. In this chapter, various models in which speeds depend on the C-marking are explained.*

A model called VHPN, in which the firing speed of a C-transition depends on the minimal marking of its input places, is presented in Section 7.1. This model usually provides an acceptable approximation for a discrete system (exactly modeled by a discrete T-timed PN). The instantaneous speeds are no longer piecewise constant.

An approximation of the previous model, called AHPN, is given in Section 7.2. It provides a satisfactory approximation with piecewise constant instantaneous speeds.

Various other models, adapted for modeling special systems, are presented in Section 7.3.

All these models have a common basis: *the autonomous hybrid Petri nets defined in Chapter 4.*

7.1 APPROXIMATION OF TIMED DISCRETE SYSTEMS BY VHPNs

As already expressed in Chapter 4, *the unit measuring a continuous value is arbitrary.* For example a volume of 9 m^3 of water can be measured as 9 000 liters or $9 \cdot 10^6$ cm^3 (usually, an engineer does not count the number of molecules!). On the other hand, *for discrete components, the unit is quite defined.* For example the number of parts is a buffer.

A real number (not integer) can be an *exact measurement* of a continuous value, but it can be only an approximate value for a number of discrete components. Assume that the number of 152.6 parts in a buffer results from a calculation: the exact value could be 152 or 153, the relative difference is small. Assume now that the number of 0.6 parts is found: the exact value could be 0 or 1 and the relative difference is large. This is the reason why the basic timed hybrid PN (or continuous PN) provides a good approximation for large numbers but not always for small numbers (Section 7.1.1). Another model, called VHPN (V standing for variable speed and H for hybrid), such that the firing speed depends on the C-marking, usually provides an acceptable approximation even for small numbers. Simple cases without conflict are given in Section 7.1.2, then the model is defined in Section 7.1.3, and application examples are presented in Section 7.1.4.

7.1.1 Weakness of Basic Timed Hybrid PNs for Small Numbers

As in Figure 5.33a in Section 5.4.1.1, the timed discrete PN R_T in Figure 7.1a may model the behavior of a production system made up of three single-server stations. Since $m_4 + m_5 + m_6 = 1$, there is a single pallet passing in turn on the three stations. A stationary behavior is immediately reached. According to Section 3.4.2.3, the firing frequencies of the transitions are $F(R_T) = F_1 = F_2 = F_3 = 1/1.5 = 0.67$ and the average number of tokens in P_4 to P_6 are $m_{av}(P_4) = m_{av}(P_5) = m_{av}(P_6) = 1/3 = 0.33$; this is illustrated in Figure 7.1c.

Let us assume that this discrete system is approximated by the basic timed hybrid PN (defined in Section 6.1 and called *CHPN*) R_{CH} in Figure 7.1b. According to Chapter 6, since $V_1 = U_1 \cdot m_1 = 2$, $V_2 = U_2 \cdot m_2 = 2$, and $V_3 = U_3 \cdot m_3 = 2$, the values $v_1(t) = v_2(t) = v_3(t) = 2 = v(R_{CH})$, $m_1(t) = 1$, and $m_5(t) = m_6(t) = 0$, are found for $t \in \left]0, \infty\right[$.

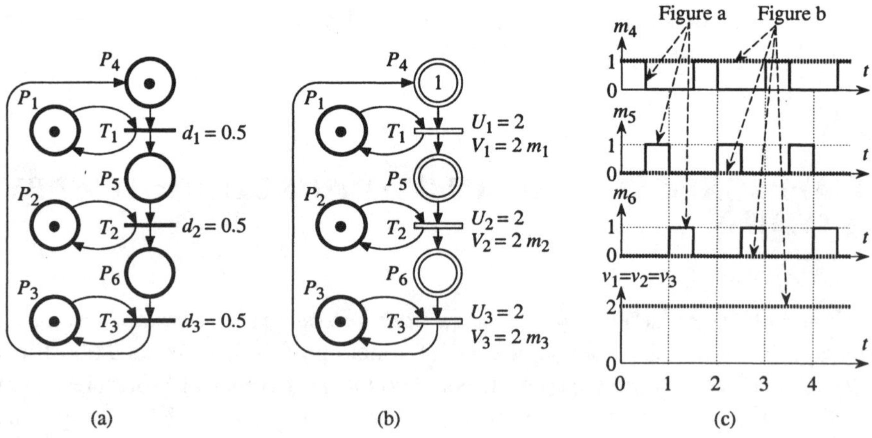

Figure 7.1 Unsatured circuit. (a) T-timed PN R_T. (b) Corresponding CHPN R_{CH}. (c) Behaviors.

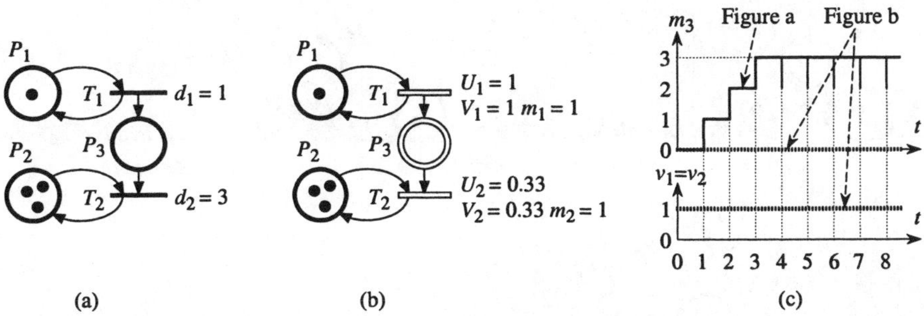

Figure 7.2 Transitory behaviors. (a) T-timed PN R_T. (b) Corresponding CHPN R_{CH}. (c) Behaviors.

Let us now return to the model in Figure 7.1a. For $m_4(0) + m_5(0) + m_6(0) = 1$, $F(R_T) = 0.67$ has been found. If the number of pallets in the 3-station circuit is greater, we have: for $m_4(0) + m_5(0) + m_6(0) = 2$, $F(R_T) = 1.33$, and for $m_4(0) + m_5(0) + m_6(0) \geq 3$, $F(R_T) = 2$. For the model in Figure 7.1b, $v(R_{CH}) = 2$ is obtained for any not nil initial marking in the circuit made up of places P_4, P_5, and P_6. It follows that $v(R_{CH})$ in Figure 7.1b is a correct approximation of $F(R_T)$ in Figure 7.1a if $m_4(0) + m_5(0) + m_6(0) \geq 3$. In other words, the firing speeds in the hybrid model in Figure 7.1b are equal to the firing frequencies of the model in Figure 7.1a if and only if there are enough pallets is the system for saturation of the circuit bottleneck (i.e. 3 pallets in our example).

For the T-timed discrete PN in Figure 7.2a, the firing frequencies are $F_1 = F_2 = 1$ when the stationary behavior is reached. For the timed hybrid PN in Figure 7.2b, the firing speeds are $v_1(t) = v_2(t) = 1$ for any $t > 0$. However, there is a transitory behavior for Figure a which does not appear for the model in Figure b: the first firing of T_2 in Figure a occurs only at $t = 4$. After $t = 3$, there are three tokens in P_3 in Figure a. As from then the three servers (tokens in P_2) are saturated, then v_2 in Figure b is equal to F_2 in Figure a.

It has been observed in both Figure 7.1 and 7.2 that a firing speed in a basic timed hybrid PN may be greater than the corresponding firing frequency of a timed discrete PN *when the marking of continuous places is too small for saturation of the considered transitions.* This observation will be taken into account in the model presented in the next section.

7.1.2 Simple Cases of Variable Speed Hybrid PN

Informally, the main idea of the model is as follows: the *firing speed of a C-transition is proportional to the minimum marking of its input places.* This model is called **variable speed hybrid PN, VHPN.**

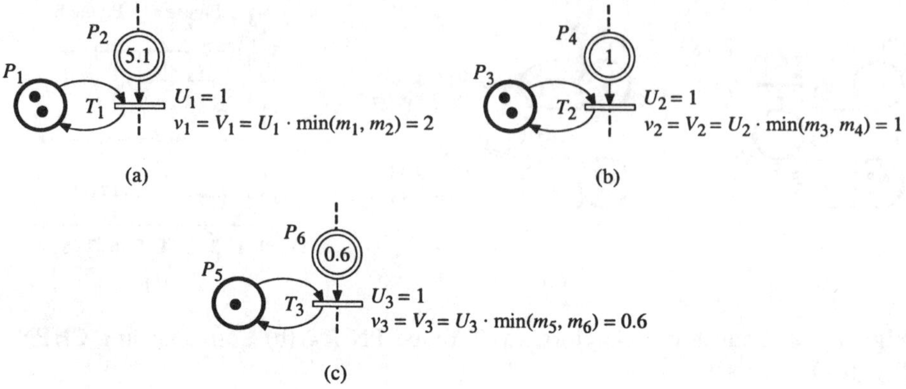

Figure 7.3 Simple examples of $v_j = V_j = U_j \cdot \min\limits_{P_i \in {}^\circ T_j} m_i$.

In order to explain simply the basic ideas, only simple examples are presented in this section: the weight of every arc is *one*, there is *no conflict* and *no immediate transition.*

Let us consider the example in Figure 7.3a. Transition T_1 corresponds to continuous processing by a double server (two tokens in P_1). If we assume that the marking of P_2 is an approximation of a discrete number of parts, it is clear that there are enough parts to occupy both servers: since $m_2 > m_1$, $\min(m_1, m_2) = m_1$. For this example, the instantaneous speed, which is equal to the maximal speed because there is no conflict, is

$$v_1 = V_1 = U_1 \cdot \min(m_1, m_2) = U_1 \cdot m_1 = 2; \tag{7.1}$$

this result would be similar with the basic model presented in Chapter 6 since the D-enabling degree of T_1 is $D(T_1, \mathbf{m}) = m_1 = 2$ (Equations (6.6) and (6.7) in Section 6.1.4).

Consider now the example in Figure 7.3b. The number of parts to be processed (i.e. $m_4 = 1$), is not sufficient to occupy both servers ($m_3 = 2$). The result

$$v_2 = V_2 = U_2 \cdot \min(m_3, m_4) = U_2 \cdot m_4 = 1 \tag{7.2}$$

expresses that the firing speed is equal to U_2 because only one server can be used.

In Figure 7.3c, there is a single server. But the number of parts to be processed is "estimated" by the number $m_6 = 0.6$ which is less than 1. The result

$$v_3 = V_3 = U_3 \cdot \min(m_5, m_6) = U_2 \cdot m_6 = 0.6 \tag{7.3}$$

is obtained. In the sequel, examples showing that a real number of "discrete" objects (like $m_6 = 0.6$ is Figure c) may correspond to an "average value".

Since instantaneous speeds depend on the C-marking, it is now necessary to express these speeds as functions of time.

If a maximal speed $V_j(t)$ is greater than 0, this implies that no input place of T_j is empty at time t (there is no immediate transition in this section). Then, there is no weakly enabled transition. It follows that $v_j(t) = V_j(t)$ if there is no conflict. Up to the end of this section, since only PNs without conflicts are concerned, only the instantaneous speeds $v_j(t)$ will be considered.

The discrete T-timed PN in Figure 7.1a is approximated by the VHPN in Figure 7.4a. The firing speed of T_1 is

$$v_1(t) = U_1 \cdot \min(m_1(t), m_4(t)). \tag{7.4}$$

The marking $m_1(t) = 1$ is constant. On the other hand, from the marking invariant $m_4(t) + m_5(t) + m_6(t) = 1$, it is known that $m_4(t) \leq 1$ for any t. It follows that $\min(m_1(t), m_4(t)) = m_4(t)$. Similar results are obtained for $v_2(t)$ and $v_3(t)$. Hence, given $U_1 = U_2 = U_3 = 2$:

$$v_1(t) = 2\, m_4(t),\ v_2(t) = 2\, m_5(t), \text{ and } v_3(t) = 2\, m_6(t). \tag{7.5}$$

The time derivative of $m_4(t)$ is the balance $B_4(t) = v_3(t) - v_1(t)$. Given (7.5),

$$\frac{\mathrm{d}m_4(t)}{\mathrm{d}t} = 2\,(m_6(t) - m_4(t)) \tag{7.6}$$

is obtained. Similar results are obtained for the time derivatives of m_5 and m_6:

$$\frac{\mathrm{d}m_5(t)}{\mathrm{d}t} = 2\,(m_4(t) - m_5(t)), \tag{7.7}$$

$$\frac{\mathrm{d}m_6(t)}{\mathrm{d}t} = 2\,(m_5(t) - m_6(t)). \tag{7.8}$$

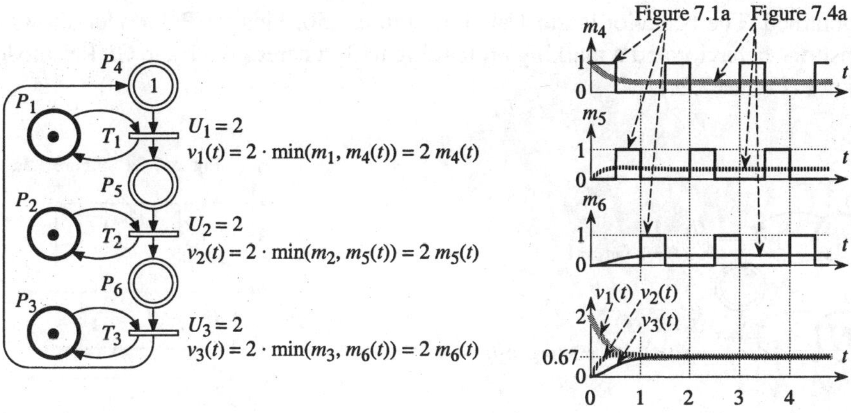

(a) (b)

Figure 7.4 VHPN approximating the discrete T-timed PN in Figure 7.1a.

From Equations (7.5) to (7.8) and given the initial marking $m_4(0) = 1$, $m_5(0) = m_6(0) = 0$, the following results are obtained:

$$m_4(t) = \frac{1}{3} + \frac{2}{3} \cdot e^{-3t} \cdot \cos\left(\sqrt{3}\,t\right), \tag{7.9}$$

$$m_5(t) = \frac{1}{3} + \frac{2}{3} \cdot e^{-3t} \cdot \sin\left(\sqrt{3}\,t - \frac{\pi}{6}\right), \tag{7.10}$$

$$m_6(t) = 1 - m_4(t) - m_5(t). \tag{7.11}$$

From these expressions, it is clear that the markings tend to asymptotic values $m_4(\infty) = m_5(\infty) = m_6(\infty) = 1/3 = 0.33$. Given (7.5) and (7.9) to (7.11), it is found that the firing speeds tend to asymptotic values $v_1(\infty) = v_2(\infty) = v_3(\infty) = 2/3 = 0.67$. These behaviors are illustrated in Figure 7.4b.

Let us note that *these asymptotic values correspond to the average markings and firing frequencies* presented at the beginning of Section 7.1.1 for the discrete T-timed PN: $m_i(\infty)$ in Figure 7.4a is equal to $m_{av}(P_i)$ in Figure 7.1a, for $i = 4, 5, 6$, and $v_j(\infty)$ in Figure 7.4a is equal to F_j in Figure 7.1a, for $j = 1, 2, 3$.

The discrete T-timed PN in Figure 7.2a is approximated by the VHPN in Figure 7.5a. The firing speeds of T_1 and T_2 are

$$v_1(t) = U_1 \cdot m_1 = 1, \text{ and} \tag{7.12}$$

$$v_2(t) = U_2 \cdot \min(m_2, m_3(t)) = 0.33\, m_3(t), \tag{7.13}$$

since, given the initial marking, $m_3(t) \leq 3 = m_2$ (because $v_2 = v_1$ for $m_3 = 3$). Given (7.12), (7.13), and the initial marking,

$$m_3(t) = 3 - 3 \cdot e^{-\frac{t}{3}}, \tag{7.14}$$

is obtained. The behavior is illustrated in Figure 7.5b. This VHPN model shows a transitory behavior and a marking m_3 tending to 3, whereas the basic CHPN model did not (Figure 7.2c).

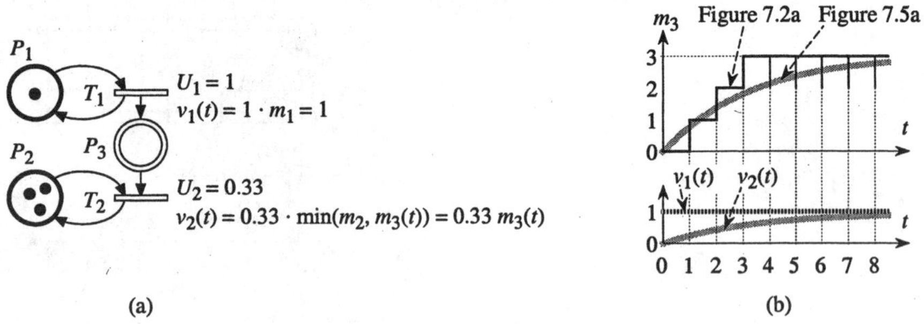

Figure 7.5 VHPN approximating the discrete T-timed PN in Figure 7.2a.

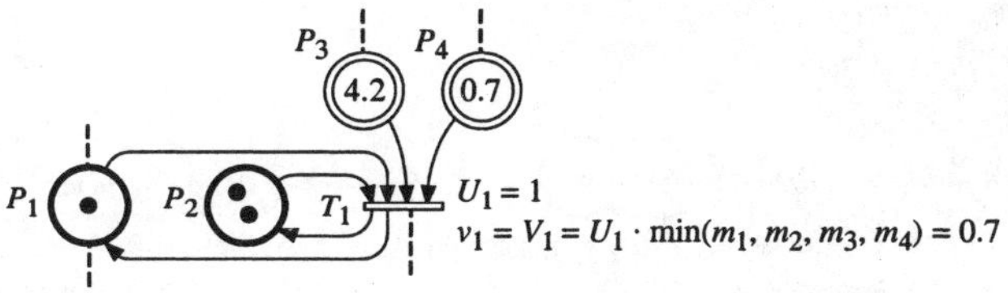

Figure 7.6 Example with several input D-places and several input C-places.

Let us end this section with a more general case illustrated in Figure 7.6. Transition T_1 in this figure may model a continuous assembly of parts represented by the markings of P_3 and P_4. This assembly requires two kinds of servers represented by the tokens in P_1 and P_2 (for example workmen and machines). For the marking in Figure 7.6, the D-enabling degree of T_1 is $\min(P_1, P_2) = 1$; for example, there is a workman available who can use one of the two machines. On the other hand, the minimal value of the input C-places (which will be called *C-enabling degree*) is $\min(P_3, P_4) = 0.7$. Since there is less than one part in the buffer represented by P_4 (this value may be an average number of discrete parts as illustrated by the behaviors of the PNs in Figure 7.1a and 7.4a), the processing speed is 0.7, which is obtained by

$$v_1 = U_1 \cdot \min(m_1, m_2, m_3, m_4) = 0.7. \qquad (7.15)$$

7.1.3 General Case of VHPN

The concept of *conflict* in a VHPN is explained in Section 7.1.3.1. Then, definitions and properties are given in Section 7.1.3.2, where *immediate transitions* and *arc weights* other than one are explicitly taken into account.

7.1.3.1 Conflicts

The concepts and resolutions for case 1, case 3, and case 4 conflicts are similar for a VHPN and for a basic timed hybrid PN (Section 6.1.3). However, behavior is different for a *case 2 conflict*, i.e., if the place and the transitions concerned are continuous.

The main idea is illustrated in Figure 7.7. The hybrid PNs in Figures a and b are equivalent as far as the markings in P_3 and P_4 are concerned. The only difference is that each server is explicitly modeled by a transition in Figure a whereas both servers are associated with the same transition in Figure b. For the marking in Figure b, $v_{12} = V_{12} = U_{12} \cdot \min(m_{12}, m_3) = U_{12} \cdot m_3 = 1.2$ is obtained. This result means that the marking of P_3 (i.e., $m_3 = 1..2$) is not sufficient for using both servers completely: only 1.2 server can be used. This is also true in Figure a, hence there is an actual conflict between T_1 and T_2 because each server is modeled by its own transition.

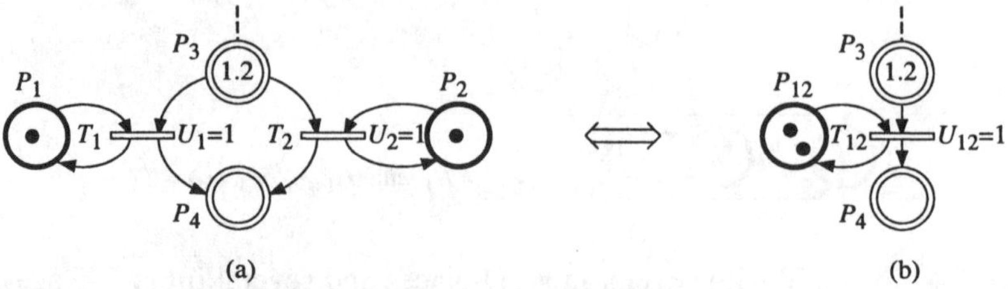

Figure 7.7 Illustration that there is a conflict between T_1 and T_2.

In Figure 7.7a, $V_1 = U_1 \cdot \min(m_1, m_3) = U_1 \cdot m_1 = 1$ and $V_2 = U_2 \cdot \min(m_2, m_3) = U_2 \cdot m_2 = 1$ are obtained. Then, $v_1 \leq V_1 = 1$ and $v_2 \leq V_2 = 1$; but $v_1 + v_2 = 1.2$ because of the conflict. As a matter of fact, the number 1.2 of parts (marking of P_3) is not sufficient to saturate both servers associated with T_1 and T_2. For example, if 0.8 part and 0.4 part are processed by the servers associated with T_1 and T_2, respectively, $v_1 = U_1 \cdot \min(1, 0.8) = 0.8$ and $v_2 = U_2 \cdot \min(1, 0.4) = 0.4$.

Let us note that there is an *actual conflict* for the VHPN in Figure 7.7a *if and only if* $0 < m_3 < 2$.

If $m_3 = 0$, then $V_1 = V_2 = 0$: no conflict.

If $0 < m_3 < 2$, then $V_1 = V_2 = \min(1, m_3)$ and $v_1 + v_2 = m_3 < V_1 + V_2$: conflict.

If $m_3 \geq 2$, then $v_1 = V_1 = 1$ and $v_2 = V_2 = 1$: no conflict.

Let us now consider the more general example of case 2 structural conflict in Figure 7.8.

In Figure 7.8a, $V_1 = U_1 \cdot \min(m_1, m_3) = 4 \cdot m_1 = 4$. Because of the arc weight $\mathrm{Pre}(P_3, T_2) = 3$, three marks in P_3 are necessary for firing T_2 at speed $V_2 = U_2$. Hence,

$$V_2 = U_2 \cdot \min\left(m_2, \frac{m_3}{\mathrm{Pre}(P_3,T_2)}\right) = 2 \cdot \min\left(1, \frac{5}{3}\right) = 2. \tag{7.16}$$

Firing of T_1 at speed $v_1 = V_1 = 4$ requires one mark in P_3 assigned to T_1 (because $\min(m_1, m_3) = 1$). Firing of T_2 at speed $v_2 = V_2 = 2$ requires three marks in P_3 assigned to T_2 (because $\min(m_2, a / 3) = 1$ for any $a \geq 3$). Hence, since $m_3 = 5 \geq 1 + 3$ (one mark for T_1 and three marks for T_2), there is no actual conflict.

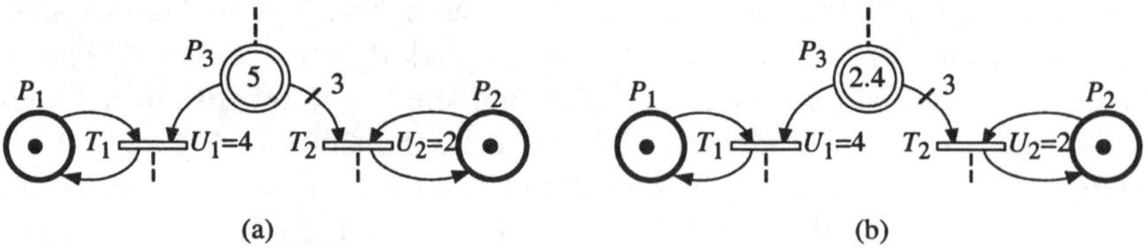

Figure 7.8 Structural case 2 conflict. (a) No actual conflict. (b) Actual conflict.

Consider now Figure 7.8b. As $\min(m_1, m_3) = 1$, $V_1 = U_1 = 4$ is obtained again, and $V_2 = U_2 \cdot \min(1, 2.4/3) = 2 \times 0.8 = 1.6$. But there is a conflict between both transitions. Several resolution rules are presented in the sequel.

Priority $T_1 < T_2$. One mark in P_3 is necessary for firing T_1 at speed $v_1 = V_1 = 4$. Hence, it remains $(2.4 - 1) = 1.4$ marks *available* in P_3 for firing of T_2. Then, $v_2 = U_2 \cdot \min(1, 1.4/3) = 2 \times 0.467 = 0.933$.

Priority $T_2 < T_1$. Firing T_2 at speed $v_2 = V_2 = 1.6$ requires 2.4 marks in P_3. It remains $(2.4 - 2.4) = 0$ mark *available* in P_3 for firing of T_1. Then, $v_1 = 0$.

Sharing. Let us assume that m_3 is shared in the following way: 0.6 mark allocated to T_1 and 1.8 marks allocated to T_2. Hence, $v_1 = U_1 \cdot \min(m_1, 0.6) = 2.4$ and $v_2 = U_2 \cdot \min(m_2, 1.8/3) = 1.2$.

For the part of VHPN in Figure 7.8, there is an *actual conflict if and only if*

$$0 < m_3 < m_1 + 3\,m_2 = 4. \tag{7.17}$$

7.1.3.2 Definition of the Model

At the beginning of this section, *only non-immediate transitions are concerned.* Immediate transitions will be tackled at the end of the section (Properties 7.2 and 7.3).

The definition of a timed hybrid PN (Definition 6.3 in Section 6.1.5.1) remains true for a VHPN. The D-enabling degree of a C-transition was defined in Section 6.1.4 (Definition 6.1). Let us now define the C-enabling degree.

Definition 7.1 The **C-enabling degree** of a non-immediate C-transition T_j for a marking $\mathbf{m}$, denoted by $C(T_j, \mathbf{m})$, is the enabling degree of T_j after *all the arcs from a D-place to a C-transition have been deleted.* Then,

$$C(T_j, \mathbf{m}) = \min_{P_i \in {}^\circ T_j \cap P^C} \frac{m_i}{\mathrm{Pre}(P_i, T_j)}. \tag{7.18}$$

Definition 7.2 In a VHPN, the **maximal firing speed** of a non-immediate transition T_j for a marking $\mathbf{m}$ is

$$V_j = U_j \cdot \min\big(D(T_j, \mathbf{m}), C(T_j, \mathbf{m})\big). \tag{7.19}$$

❑

Given (6.6) (expression of $D(T_j, \mathbf{m})$) and (7.18), (7.19) can be rewritten as:

$$V_j = U_j \cdot \min_{P_i \in {}^\circ T_j} \frac{m_i}{\mathrm{Pre}(P_i, T_j)}, \tag{7.20}$$

in which $\min_{P_i \in {}^\circ T_j} \frac{m_i}{\mathrm{Pre}(P_i, T_j)}$ is nothing but the **enabling degree** of a transition in a hybrid PN. Enabling degrees in discrete PNs (Definition 3.1 in Section 3.1) and in continuous PNs (Definition 4.2 in Section 4.1.2) are particular cases of this expression.

Let $K = \langle P_c, T(K)\rangle$ denote a structural case 2 conflict: P_c is a C-place and $T(K)$ is the subset of $P_c{}^\circ \cap T^C$ containing only *non-immediate* transitions.

Property 7.1 In a VHPN, there is an *actual conflict* $\langle P_c, T(K), \mathbf{m}\rangle$ for a marking $\mathbf{m}$ if and only if

$$0 < m_c < \sum_{T_j \in T(K)} \mathrm{Pre}(P_c, T_j) \cdot \min_{P_i \in {}^\circ T_j} \frac{m_i}{\mathrm{Pre}(P_i, T_j)}. \tag{7.21}$$

Proof This is a generalization of (7.17) in Section 7.1.3.1.
If $m_c = 0$, $V_j = 0$ for any T_j in $T(K)$, hence there is no actual conflict.
A marking $m_c > 0$ is possible only if no immediate transition in $P_c{}^\circ$ can be fired. In this case, there is an actual conflict if m_c is not sufficient to fire every transition T_j in $T(K)$ at its maximal speed. The number of marks in P_c necessary for firing T_j at $v_j = V_j$ is $\mathrm{Pre}(P_c, T_j) \cdot V_j / U_j$. Then, given (7.20), (7.21) is obtained.

Notation 7.1 If there is an actual conflict $\langle P_c, T(K), \mathbf{m}\rangle$, the *part of marking m_c assigned to T_j* in $T(K)$ will be denoted by $m_c^{(T_j)}$. Hence $\sum_{T_j \in T(K)} m_c^{(T_j)} = m_c$.

❑

Let us note that this notation can also be used if P_c is a D-place (i.e. for a case 4 conflict). Given Notation 7.1, the instantaneous firing speed of T_j is given by:

$$v_j = U_j \cdot \min_{P_i \in {}^\circ T_j} \frac{m_i^{(T_j)}}{\mathrm{Pre}(P_i, T_j)}. \tag{7.22}$$

In order to *simplify the presentation*, let us now assume that *all the conflicts are solved by priorities* (i.e., there are no sharings). This assumption will end with Remark 7.4.

Assume that the priorities among the transitions in $T(K) = \{T_1, T_2, \ldots, T_s\}$ are $T_1 < T_2 < \ldots < T_s$. Then,

$$m_c^{(T_1)} = \mathrm{Pre}(P_c, T_1) \cdot \min_{P_i \in {}^\circ T_1} \frac{m_i}{\mathrm{Pre}(P_i, T_1)}, \tag{7.23}$$

$$m_c^{(T_2)} = \min\left(\mathrm{Pre}(P_c, T_2) \cdot \min_{P_i \in {}^\circ T_2} \frac{m_i}{\mathrm{Pre}(P_i, T_2)}, \left(m_c - m_c^{(T_1)}\right)\right), \tag{7.24}$$

and so on. Equation (7.23) means that the marking $m_c^{(T_1)}$ assigned to T_1 allows $v_1 = V_1$. Equation (7.24) means that the marking $m_c^{(T_2)}$ assigned to T_2 depends on the available marking $(m_c - m_c^{(T_1)})$: if this available marking is sufficient for firing T_2 at $v_2 = V_2$, the corresponding marking is assigned to T_2; otherwise $m_c^{(T_2)} = (m_c - m_c^{(T_1)})$ and $v_2 < V_2$. And so on: the marking available for T_k is $(m_c - \sum_{h<k} m_c^{(T_h)})$.
For example, for the priority order $T_1 < T_2$ in Figure 7.8b, $m_c^{(T_1)} = 1$ and $m_c^{(T_2)} = 1.4$.

Definition 7.3 The *critical input place*, or **critical place** for short, of C-transition T_j at time t, denoted by $P(T_j)$ or $P(T_j, t)$, is a place P_k in $°T_j$ such that

$$\frac{m_k^{(T_j)}}{\text{Pre}(P_k,T_j)} = \min_{P_i \in °T_j} \frac{m_i^{(T_j)}}{\text{Pre}(P_i,T_j)}. \tag{7.25}$$

If several places satisfy (7.25), the place with the *smallest index k* is chosen.

Remark 7.1 Expression (7.25) in Definition 7.3 corresponds to a general case. Let us recall that, if there is no conflict involving place P_k, then $m_k^{(T_j)} = m_k$ (and similarly for any place P_i).

❑

Here are examples of critical places. At initial time, the *critical place* of T_1 in Figure 7.9a is $P(T_1, 0) = P_1$ since $m_1 = 1 = \min(m_1, m_2)$. For the PN in Figure 7.9b, $P(T_2, 0) = P_3$ and $P(T_3, 0) = P_5$ since $m_5 = 1 = \min(m_4, m_5, m_6/2)$.

The *critical place* of a transition is a *function of time*.

For T_1 in Figure 7.9a, $v_1(t) = U_1 \cdot \min(m_1(t), m_2(t)) = 5$ for $t \in |0, 0.4|$. As a matter of fact, for this time interval,

$$m_2(t) = m_2(0) - v_1(t) \cdot t = 3 - 5t, \tag{7.26}$$

whereas $m_1(t) = 1$ is constant. Hence, $m_2(t) > 1 = m_1(t)$ for $t < 0.4$, and $m_2(t) < 1 = m_1(t)$ for $t > 0.4$. It follows that the critical place of T_1 changes at $t = 0.4$:

$$P(T_1, t) = P_1 \text{ for } t \in |0, 0.4| \text{ and } P(T_1, t) = P_2 \text{ for } t \in |0.4, \infty|. \tag{7.27}$$

For the time interval $t \in |0.4, \infty|$,

$$v_1(t) = 5 \cdot m_2(t), \text{ and} \tag{7.28}$$

$$\frac{dm_2(t)}{dt} = -v_1(t), \text{ leading to } m_2(t) = 1 - \int_{0.4}^{t} v_1(u)\,du. \tag{7.29}$$

From (7.28) and (7.29), (7.30) is obtained:

$$m_2(t) = e^{-5(t-0.4)}. \tag{7.30}$$

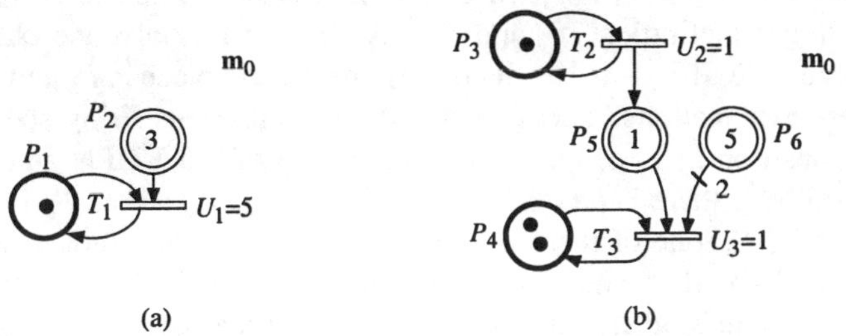

(a) (b)

Figure 7.9 Illustration for critical places and C5-events.

Consider now the VHPN in Figure 7.9b. From initial time, $m_6(t) = 5 - 2t$ decreases and will become less than m_5 from $t = 2$. Hence,

$$P(T_3, t) = P_5 \text{ for } t \in \left|0, 2\right| \text{ and } P(T_3, t) = P_6 \text{ for } t \in \left|2, \infty\right|. \tag{7.31}$$

It is clear that some critical places can change when a D1-event occurs (firing of a D-transition). The critical place of a transition can also change because of the continuous evolution of the C-marking; this change is defined as a C5-event (Definition 7.4). For example, C5-events occur at $t = 0.4$ and $t = 2$ for the VHPNs in Figures 7.9a and b, respectively.

Definition 7.4 A **C5-event** occurs at time t if there is a C-transition T_j such that the critical place of T_j changes at t, i.e. $P(T_j, t + dt) \neq P(T_j, t - dt)$, whereas this change is not *caused*[1] by a D1-event occurring at t.

Remark 7.2

a) For each C-transition, there is always a critical place, and this critical place is the same for a finite (or infinite) interval of time (otherwise the behavior would be unstable: if two places satisfy (7.25), their index implies the critical place).

b) According to (7.22) and (7.25), a *non-immediate transition* T_j may have a positive firing speed $v_j > 0$ if the critical place is a D-place or a C-place. On the other hand, the firing speed of an *immediate transition* T_k may be $v_k > 0$ only if the critical place is a C-place. As a matter of fact, at least one input place must be empty; if a D-place in $°T_k$ is empty, then $V_k = 0$ (whereas a C-place may be continuously fed).

❑

Let us now consider the behavior of immediate transitions (according to Remark 7.2b, immediate transitions without input D-place may be considered to simplify the presentation). In Figure 7.10a, the only input C-place of T_2 must remain empty; then, $v_2 = v_1$. In Figure 7.10b, given $m_3 > 0$, the critical place of T_5 is P_2 (the definition of a critical place, i.e. Definition 7.3, is also relevant to immediate transitions); then $v_5 = v_3$ since the critical place must remain empty (a formal proof is given below). If $v_3 < v_4$, there is no change of critical place; if $v_4 < v_3$, m_3 decreases and there is a change of critical place when m_3 becomes 0 (from this time m_2 increases). In Figure 7.10c, both places P_4 and P_5 are empty. This marking can last only if $v_6 = v_7$; in this case, the critical place is P_4 (smallest index according to Definition 7.3) and $v_8 = v_6$. If $v_6 \neq v_7$, only one place remains empty; for example, if $v_6 > v_7$, m_4 increases, the critical place is P_5 and $v_8 = v_7$.

From these explanations, it appears that the instantaneous firing speed v_k of an immediate transition T_k is obtained in a similar way for a VHPN and for a basic timed hybrid PN: if there is no conflict, $v_k = I_h / \text{Pre}(P_h, T_k)$ where P_h is the empty input C-place of T_k (one of them if there are several). What about the *maximal* firing speed V_k ? The flow rate of an *immediate transition* T_k is $U_k = \infty$ (necessary for this model whereas other values were possible for the basic model according to

---

[1] See Definition 6.6 in Section 6.2.2.

Section 6.1.5.2). By definition of such a transition, the marking of at least one of its input places, say P_h, is 0. What is the value of $V_k = U_k \cdot P_h$ (i.e., of $\infty \times 0$)? Property 7.2 answers this question.

Property 7.2 If T_k is an *immediate C-transition* and (7.20) is applied to it, then

$$V_k = 0 \tag{7.32}$$

if the critical place $P(T_j) \in P^D$. Otherwise:

$$V_k = \min_{h: P_h \in (^\circ T_k \cap P^C) \text{AND} m_h = 0} \frac{I_h}{\text{Pre}(P_h, T_k)}. \tag{7.33}$$

Proof The value in (7.32) was already explained in Remark 7.2b. If both an input D-place and an input C-place are empty, the critical place is the D-place (smallest index, see Definition 7.3).

To simplify the notation, the proof of (7.33) is given for the example in Figure 7.10b in which v_3 is assumed to be constant and $m_3 > 0$. The balance of P_2 is

$$\frac{dm_2}{dt} = v_3 - v_5. \tag{7.34}$$

Applying (7.20) to transition T_5, and given there is no conflict:

$$v_5 = V_5 = U_5 \cdot m_2. \tag{7.35}$$

Assuming that Figure 7.10b corresponds to time t_1, it follows from (7.34) and (7.35) that, for $t > t_1$ and as long as $P(T_5) = P_2$:

$$m_2(t) = \frac{v_3}{U_5} \cdot \left(1 - e^{-U_5(t-t_1)}\right). \tag{7.36}$$

Since m_3 cannot become zero immediately because $m_2 = 0$, a finite time interval $|t_1, t_2|$ exists, $t_1 < t_2 < \infty$, such that P_2 remains the critical place in this interval. During this time interval, given $U_5 = \infty$ and $(t - t_1)$ is finite, $e^{-U_5(t-t_1)} = 0$; hence, according to (7.36), $m_2(t) = v_3 / U_5 = v_3 / \infty = 0$, then $dm_2 / dt = 0$. It follows from (7.34) that

$$v_5 = v_3. \tag{7.37}$$

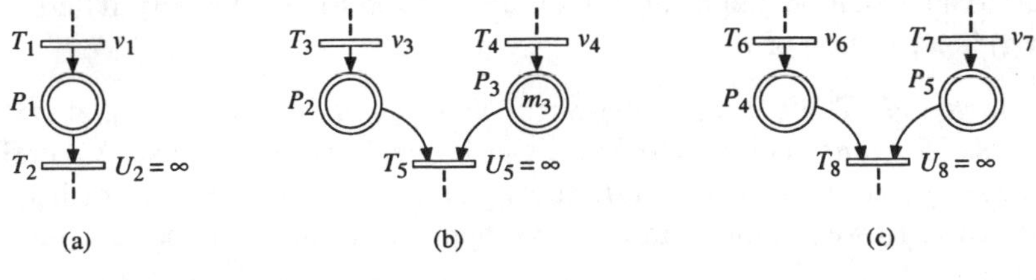

Figure 7.10 Immediate transitions.

Thus, from (7.35) and (7.37), $V_5 = v_3$ is obtained.

The generalization (several input transitions to P_2, several empty places like P_2, several non-empty places like P_3, arc weights, v_3 function of time, and conflict among T_5 and other immediate transitions) is straightforward.

Property 7.3 Let T_k be an immediate C-transition whose D-enabling degree is positive (which is obviously true if T_k has no input D-place). If $v_k(t_1) < \infty$ and P_h is the only empty input C-place in $°T_k$ and there is no firing of D-transition at t_1, then P_h will remain empty at least during a *finite time*. In other words, t_2 exists such that $t_2 > t_1$ and $m_h = 0$ in the interval $|t_1, t_2|$.

❑

The reason for this property was already explained in the proof of Property 7.2. The condition $v_k(t_1) < \infty$ excludes that an I-phase (Section 5.1.2.5) involving T_k is performed at t_1. Property 7.3 will appear to be useful (Section 7.1.4.1).

Remark 7.3 In a VHPN as in any timed hybrid PN, an immediate transition takes priority over a non-immediate transition. Hence, a conflict resolution may take place either among immediate transitions or among non-immediate transitions.

❑

Let us now define an *IB-phase* in a VHPN. Informally, an IB-phase corresponds to a period of time during which all the instantaneous firing speeds (then the markings) are given by an invariant system of equations. During an IB-phase, all the equations and the set of critical places remain unchanged. An IB-phase is similar to an *IB-state* in a basic hybrid PN (Definition 6.5, Section 6.1.5.3), except that the instantaneous speed vector **v** is constant in an IB-state whereas $\mathbf{v}(t)$ is a vector of functions of time in an IB-phase.

Definition 7.5 An **IB-phase** in a VHPN, is such that:

1) *the* marking $\mathbf{m}^D$ *of the D-places is* constant;
2) the *enabling vector* $\mathbf{e}^D$ of the D-transitions is *constant*;
3) every component $v_j(t)$ of $\mathbf{v}(t)$ is given by a time function which does not change during the IB-phase (critical places do not change);
4) when the IB-phase is reached, $\mathbf{m}^C$ always has the *same value*.

❑

The behavior of a VHPN can be represented by an *evolution graph* almost similar to an evolution graph for a basic timed hybrid PN. The only difference is that $\mathbf{v}(t)$ is a function of time.

Remark 7.4 The concepts of *part of marking m_c assigned* to T_j, of *critical place*, of *C5-event*, and of *IB-phase*, can be generalized to case 2 conflicts resolution by sharing (instead of priorities), but they are not easy to formalize in a general case. However, they can easily be applied in some particular cases. See Exercise 7.3.

7.1.3.3 Properties

Let us first observe that, according to the previous section, the behavior of an *immediate* C-transition is virtually similar in a VHPN and in a basic timed hybrid PN. However, *non-immediate* transitions behave differently.

Property 7.4 In a VHPN:

a) A *C1-event cannot occur* alone: it can only be caused by a D1-event (or initialization).

b) There is *no weak enabling* of non-immediate transitions (i.e., $v_j = V_j$ except possibly in case of conflict).

c) There is *no marking* 0^+.

Proof

a) Let P_k be the critical place of T_j, then $v_j \leq U_j \cdot m_k$ (and $v_j = U_j \cdot m_k$ if there is no conflict or if T_j takes priority over the other transitions). Since v_j tends to 0 when m_k tends to 0, m_k will never reach the value 0 by firing of T_j.

If a D-transition T_h in P_k° is fired when $\mathrm{Pre}(P_k, T_h) = m_k$, then m_k becomes 0. This is a C1-event caused by a D1-event.

b) and c): direct consequences of Property a.

❑

In a basic timed hybrid PN, a change of IB-state can occur if a C1 or D1 or D2-event occurs (Property 6.2, Section 6.1.5.3). In a VHPN, a C1-event cannot occur alone (Property 7.4a), and the speed vector can change when a C5-event occurs. Then, Property 7.5 is obtained.

Property 7.5 In a VHPN, a *change of IB-phase* can occur only if an *event* belonging to one of the following types occurs.

C5-event: change of critical place for a C-transition.

D1-event: *firing* of a D-transition.

D2-event: enabling degree of a D-transition changes because of the marking of a C-place.

Property 7.6 During an IB-phase of VHPN:

1) if $P(T_j)$ is a D-place, $v_j(t)$ is *constant*;

2) if $P(T_j)$ is a C-place, $v_j(t)$ is, *usually, a function of t* (in some particular cases $v_j(t)$ is constant).

Proof According to Definition 7.5 and Property 7.5, the critical place $P_k = P(T_j)$ does not change during an IB-phase. According to Definition 7.3 and (7.22), $v_j(t)$ is proportional to $m_k^{(T_j)}$ which is constant during the IB-phase if P_k is a D-place (in this case, $v_j(t)$ is the same for the VHPN model and for the basic timed hybrid PN), and which is, *usually*, a function of time if P_k is a C-place since $m_k(t)$ is a function of $v_j(t)$.

❑

For T_3 in Figure 7.9b, $P(T_3) = P_5$ for the IB-phase such that $t \in |0, 2|$. However, $v_3(t)$ is constant because $m_5(t)$ is constant: $I_5(t) = v_2(t) - v_3(t) = 0$.

Remark 7.5 In a basic timed hybrid PN, the maximal speed of a transition is proportional to its D-enabling (Definition 6.2, Section 6.1.4). It follows that each *non-immediate* transition must have at least one input D-place (Section 6.1.5.2).

On the other hand, if *a transition in a VHPN* has no input D-place, its maximal firing speed is proportional to its C-enabling (Definition 7.2, Section 7.1.3.2). It follows that, in this model, *no input D-place is required*, even for a non-immediate transition. This observation is also true for the model presented in Section 7.2 (called AHPN) which is based on the VHPN model.

Remark 7.6 An algorithm for obtaining equations of the behavior in the successive IB-phases is relatively simple. When all the critical places have been calculated (possibly taking into account priorities in case of conflict, as explained in Section 7.1.3.2), all the equations of markings and firing speeds can be calculated from equations usually of the form $dm_i(t) / dt = f(\mathbf{m})$. Occurrence time of the next C5, D1, or D2-event is then calculated (Property 7.5). And so on. Examples are presented in the next section.

7.1.4 Application Examples

7.1.4.1 Example 1

This is an example of modeling zero buffers with the VHPN model. In Section 6.4.1, a zero buffer was modeled using the marking 0^+ in an extended timed hybrid PN. The marking 0^+ does not exist in a VHPN (Property 7.4).

Let us consider the string of machines M_1M_2 in Figure 7.11a, whose processing times are $S_1 = 2$ and $S_2 = 1$, with no buffer between them. We assume that there are always unworked parts upstream M_1 and available space downstream M_2. The possible blocking of a machine is assumed to occur *after service* (blocking after service and blocking before service are discussed in Exercise 3.9).

A T-timed discrete PN modeling the specified behavior is given in Figure 7.11b. Tokens in P_1 and P_2 correspond to operational machines. These machines could become unavailable (failure or maintenance), but we assume here that both machines are operational. There is a marking invariant $m_3 + m_4 + m_7 = 1$ meaning that: either there is a processed part on M_1 (place P_3) or there is a part on M_1 whose processing is finished (place P_4) or there is no part on M_1 (place P_7). Similarly, there is a marking invariant $m_5 + m_6 + m_8 = 1$ for M_2. The timings associated with T_2 and T_4 are $d_2 = S_1 = 2$ and $d_4 = S_2 = 1$. The other transitions, namely T_1, T_3, and T_5, are immediate transitions. The initial marking is shown on the figure: no part on the machines. However, since T_1 is an immediate transition, it is fired at initialization (I-phase).

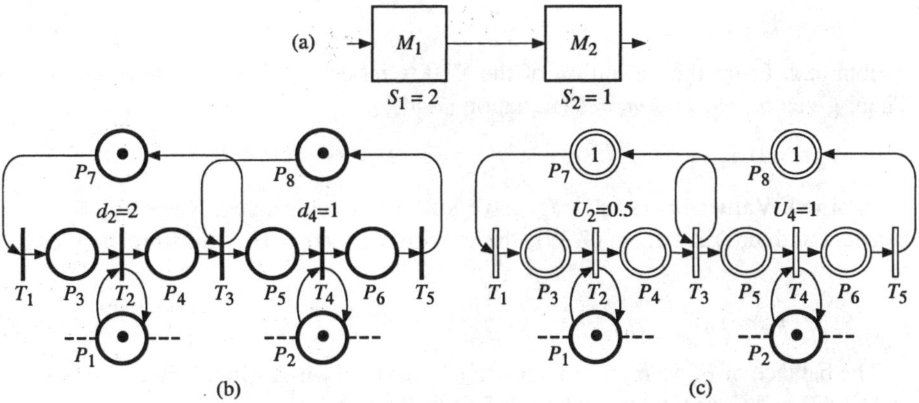

Figure 7.11 (a) Two-machine line without buffer. (b) T-timed discrete PN modeling the behavior (blocking after service). (c) Corresponding VHPN.

An approximation of this behavior can be obtained by the VHPN in Figure 7.11c. We still have the marking invariants:

$$m_3 + m_4 + m_7 = 1, \text{ and} \tag{7.38}$$

$$m_5 + m_6 + m_8 = 1. \tag{7.39}$$

Since T_1, T_3, and T_5, are immediate transitions, we have (after initialization):

$$m_7 = 0, \tag{7.40}$$

$$m_4 = 0 \text{ OR } m_8 = 0, \tag{7.41}$$

$$m_6 = 0. \tag{7.42}$$

According to Property 7.3 (Section 7.1.3.2), for a time interval from $t = 0$:

$$m_4 = 0. \tag{7.43}$$

From (7.38), (7.40), and (7.43),

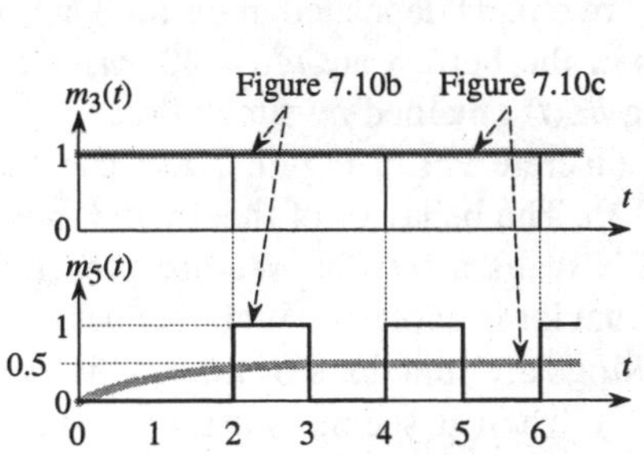

Figure 7.12 Behaviors of the T-timed discrete PN in Figure 7.11b and VHPN in Figure 7.11c.

$$m_3 = 1 \tag{7.44}$$

is obtained. From the definition of the VHPN model $v_2(t) = U_2 \cdot \min(m_1, m_3)$. Then, given $m_1 = 1$ and $m_3 = 1$ (Equation (7.44),

$$v_2(t) = 0.5 \tag{7.45}$$

is obtained. Values $m_5(t)$ and $v_4(t)$ have still to be calculated. Since $m_2 = 1$ is constant and $m_5(t) \leq 1$ from (7.39), the expression $v_4(t) = U_4 \cdot \min(m_2, m_5)$ leads to

$$v_4(t) = m_5(t). \tag{7.46}$$

The balance of P_5 is $B_5(t) = v_3(t) - v_4(t)$. Given T_3 is an immediate transition, and (7.43) and (7.45): $v_3(t) = v_2(t) = 0.5$. It follows that

$$\frac{dm_5(t)}{dt} = 0.5 - v_4(t). \tag{7.47}$$

The result in (7.48) is then obtained from (7.46) and (7.47):

$$m_5(t) = 0.5 - 0.5\, e^{-t}. \tag{7.48}$$

The value of $m_8(t)$ is obtained from (7.39), (7.42) and (7.48):

$$m_8(t) = 0.5 + 0.5\, e^{-t}. \tag{7.49}$$

The behaviors of the T-timed discrete PN in Figure 7.11b and VHPN in Figure 7.11c are illustrated in Figure 7.12. For both models, $m_3(t) = 1$ for any $t > 0$. The value $m_5(t)$ increases from 0 to the asymptotic value $m_5(\infty) = 0.5$.

7.1.4.2 Example 2

Figure 7.13a represents an assembly system. Every 0.2 time unit, a part in buffer 2 and a part in buffer 3 are assembled (firing of T_3) if these buffers are not empty. Buffer 2 is supplied by parts deposited every 0.5 time unit in buffer 1 (firing of T_1), then processed during 0.25 time unit (firing of T_2). Buffer 3 is supplied by firing of T_4: two parts deposited in buffer 3 every 4 / 7 time unit. The initial numbers of parts in the buffers are $m_5 = 40$, $m_6 = 5$, and $m_7 = 10$. Figure 7.14 shows the marking $m_6(t)$ obtained by simulation.

The behavior of the discrete PN in Figure 7.13a can be approximated by the hybrid PN in Figure 7.13b. The behavior of this hybrid model may be analyzed as a basic timed hybrid PN or as a VHPN. We are presently concerned with the VHPN model, whose behavior is explained in the sequel.

The speeds and markings are functions of time. This time dependence will be explicitly expressed only when it seems necessary. Since there is no conflict, $v_j = V_j$ for every C-transition. The firing instantaneous speeds of C-transitions and balances of C-places are:

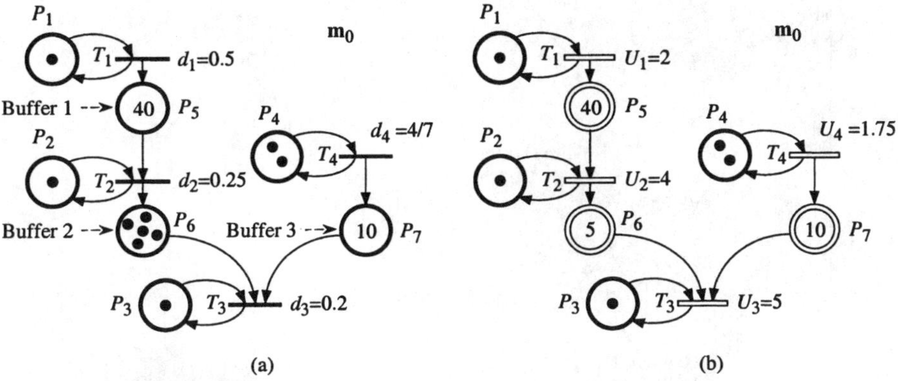

Figure 7.13 (a) T-timed discrete PN. (b) Corresponding VHPN.

$$v_1 = U_1 \cdot m_1 = 2, \tag{7.50}$$

$$v_2 = U_2 \cdot \min(m_2, m_5) = 4 \min(1, m_5), \tag{7.51}$$

$$v_3 = U_3 \cdot \min(m_3, m_6, m_7) = 5 \min(1, m_6, m_7), \tag{7.52}$$

$$v_4 = U_4 \cdot m_4 = 3.5, \tag{7.53}$$

$$B_5 = v_1 - v_2, \tag{7.54}$$

$$B_6 = v_2 - v_3, \tag{7.55}$$

$$B_7 = v_4 - v_3. \tag{7.56}$$

Let us note that $v_1 = 2$ and $v_4 = 3.5$ are constant. For each IB-phase, speeds v_2 and v_3, and markings m_5, m_6, and m_7 must be calculated.

IB-phase 1

Since, from $t = 0$, $m_5 > m_2$, and $m_6 > m_3$ and $m_7 > m_3$, the speeds

$$v_2 = 4 \text{ and } v_3 = 5 \tag{7.57}$$

are obtained from (7.51) and (7.52). Then, given $\mathbf{m}_0$ and (7.54) to (7.57):

$$m_5 = 40 - 2t,\ m_6 = 5 - t, \text{ and } m_7 = 10 - 1.5t. \tag{7.58}$$

The next C5-event will occur at $t = 4$. At this time, the *critical place* of T_3 changes: $P(T_3, t) = P_3$ for $t < 4$ and $P(T_3, t) = P_6$ for $t > 4$, because, according to (7.58), $m_6 > 1 = m_3$ for $t < 4$ and $m_6 < 1 = m_3$ for $t > 4$. Hence, IB-phase 1 corresponds to the time interval $t \in |0, 4|$ during which Equations (7.57) and (7.58) remain true.

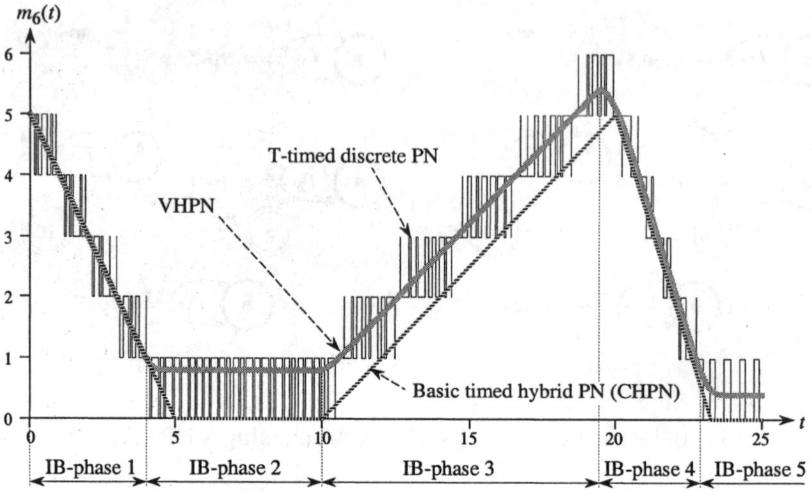

Figure 7.14 Behaviors of the T-timed discrete PN andVHPN in Figure 7.13.

IB-phase 2

This IB-phase begins at $t = 4$. At this time:

$$m_5(4) = 32,\ m_6(4) = 1, \text{ and } m_7(4) = 4. \tag{7.59}$$

Since $P(T_3, t) = P_6$, (7.52) becomes

$$v_3(t) = 5\, m_6(t). \tag{7.60}$$

Equations related to markings are then:

$$m_5(t) = 32 - 2(t - 4), \tag{7.61}$$

$$\frac{dm_6(t)}{dt} = 4 - v_3(t), \tag{7.62}$$

$$\frac{dm_7(t)}{dt} = 3.5 - v_3(t). \tag{7.63}$$

From (7.60) and (7.62),

$$m_6(t) = 0.8 + 0.2\, e^{-5(t-4)} \tag{7.64}$$

is obtained. The behavior of $m_6(t)$ is illustrated in Figure 7.14. The expression of $m_7(t)$ is obtained from (7.63), (7.60), and (7.64):

$$m_7(t) = 6 - 0.5\, t - 0.2\,(1 - e^{-5(t-4)}\,). \tag{7.65}$$

Since $m_7(t)$ decreases whereas $m_6(t) \approx 0.8$ is practically constant, there is a change of critical place at $t \approx 10$ (the value is not exactly 10 because m_6 is not exactly 0.8, but the difference is about 10^{-14}): $P(T_3, t) = P_7$ for $t > 10$. Hence IB-phase 2 corresponds to the time interval $t \in |4, 10|$.

IB-phase 3

The calculations are left as an exercise. Only some results are given: $m_5(10) = 20$, $m_6(10) = 0.8$, and $m_7(10) = 0.8$; $P(T_2) = P_2$, $P(T_3) = P_7$ (changed), $t \in |10, 19.5|$;

$$m_5(t) = 20 - 2(t - 10); \tag{7.66}$$

$$m_6(t) = -4.2 + 0.5t - 0.1(1 - e^{-5(t-10)}). \tag{7.67}$$

IB-phase 4

Results: $m_5(19.5) = 1$, $m_6(19.5) = 5.45$, and $m_7(19.5) = 0.7$; $P(T_2) = P_5$ (changed), $P(T_3) = P_7$, $t \in |19.5, 23|$;

$$m_5(t) = 0.5 + 0.5e^{-4(t-19.5)}; \tag{7.68}$$

$$m_6(t) = 35.2 - 1.5t - 0.5e^{-4(t-19.5)}. \tag{7.69}$$

IB-phase 5

Results: $m_5(23) = 0.5$, $m_6(23) = 0.7$, and $m_7(23) = 0.7$; $P(T_2) = P_5$, $P(T_3) = P_6$ (changed), $t \in |23, \infty|$ (hence this is the last IB-phase);

$$m_5(t) = 0.5; \tag{7.70}$$

$$m_6(t) = 0.4 + 0.3e^{-5(t-23)}; \tag{7.71}$$

$$m_7(t) = -34.1 + 1.5t + 0.3e^{-5(t-23)}. \tag{7.72}$$

***Remark** 7.7* The values of $m_6(t)$ in Figure 7.14 illustrate that the model provides an *analytical approximation* of the behavior of the discrete timed PN, even for the *transitory behavior*. For IB-phases 2 and 5, the *asymptotic values* (0.8 and 0.4, respectively) correspond *exactly* to the average number of tokens in the discrete model.

❑

The result obtained if the hybrid PN in Figure 7.13b is analyzed as a basic timed hybrid PN (CHPN) is also shown in Figure 7.14. The *relative* difference between both models is large when the marking is small (IB-phases 2 and 5).

7.2 ASYMPTOTIC HYBRID PETRI NETS (AHPNs)

The basic timed hybrid PN (CHPN) is easy to use because the firing speed vector is constant during an IB-state. However, the approximation of small numbers in a discrete system, by this model, is not always good (Section 7.1.1).

The VHPN model usually provides an acceptable approximation for a discrete system, even when the markings are small (Section 7.1.4). However, the model is not so easy to use because the markings and firing speeds can be functions of time in an IB-phase (exponential, sine shaped: see (7.10) in Section 7.1.2).

In this section, an approximation of VHPN behavior such that the *firing speed vector is constant* during an IB-state (the name IB-state is used again because the speed vector is constant) is presented. This model, based on the asymptotic behavior of a VHPN (see Section 7.1.4.2), is called AHPN, A standing for *asymptotic*.

In Section 7.2.1, the main idea is intuitively presented with examples in which a C-transition has a single input C-place. The case with several input C-places is considered in Section 7.2.2, a generalization is given in Section 7.2.3, and some properties of the model are given in Section 7.2.4.

Both VHPN and AHPN models will be considered in the sequel. In order to distinguish the corresponding values, let us introduce the following notation.

Notation 7.2 Values m_i (if P_i is a C-place) and v_j (if T_j is a C-transition), will be denoted by $m_{i(V)}$ and $v_{j(V)}$ for a VHPN interpretation, $m_{i(A)}$ and $v_{j(A)}$ for an AHPN interpretation.

7.2.1 A C-Transition Has a Single Input C-Place

7.2.1.1 Constant Feeding Speed of Input C-Place

Let us consider the part of timed hybrid PN in Figure 7.15a. It is assumed that the *instantaneous speed* $v_1 = 2$ *is constant*. Right now, this is an assumption, but it will appear that, by definition of the AHPN model, every instantaneous firing speed $v_{j(A)}$ will be constant during an IB-state.

Assume first that Figure 7.15a is interpreted as VHPN. Since $m_{2(V)} < m_1$,

$$v_{2(V)}(t) = U_2 \cdot m_{2(V)}(t). \tag{7.73}$$

The time derivative of marking $m_{2(V)}(t)$ is:

$$\frac{dm_{2(V)}(t)}{dt} = v_1 - v_{2(V)}(t). \tag{7.74}$$

Given $m_2(0)$, (7.75) can be obtained from (7.73) and (7.74):

$$m_{2(V)}(t) = \frac{v_1}{U_2} + \left(m_2(0) - \frac{v_1}{U_2}\right) \cdot e^{-U_2 t}. \tag{7.75}$$

For the example in Figure 7.15a, $m_{2(V)}(t) = 0.4 + 05e^{-5t}$ and $v_{2(V)}(t) = 2 + 2.5e^{-5t}$ are obtained from (7.75) and (7.73).

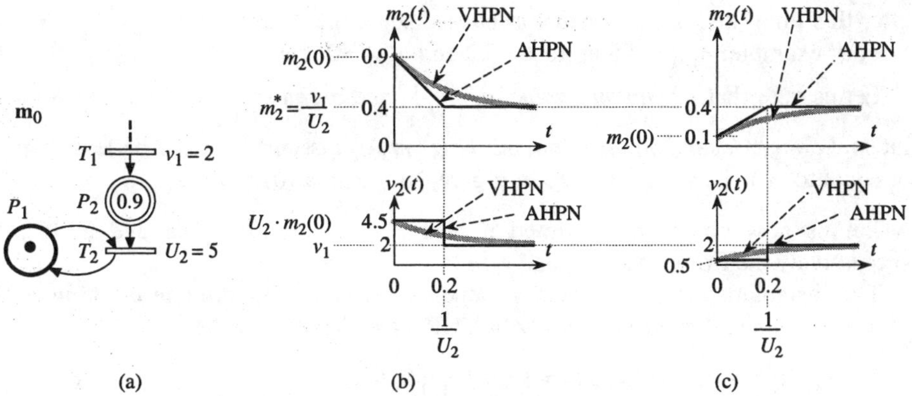

Figure 7.15 Basic idea leading to an AHPN model.

From (7.75), it can be observed that marking $m_{2(V)}(t)$ tends to asymptotic value v_1 / U_2 (which will be denoted by m_2^*). This is illustrated in Figure 7.15b for our example.

A classical approximation of this exponential evolution is presented in Figure 7.15b. This is the basic idea of an AHPN model: two time intervals may be considered.

1) For $t \in \left|0,\ 1/U_2\right|$, the exponential evolution is replaced by the segment of line tangent to the curve at $t = 0$:

$$m_{2(A)}(t) = m_2(0) + \left(\frac{v_1}{U_2} - m_2(0)\right) \cdot U_2\, t. \tag{7.76}$$

2) For $t \in \left|1/U_2,\ \infty\right|$, the exponential curve is replaced by its asymptote:

$$m_{2(A)}(t) = \frac{v_1}{U_2}. \tag{7.77}$$

For our example (Figures 7.15a and b), the values for the AHPN model are:

$$m_{2(A)}(t) = 0.9 - 2.5\, t \text{ for } t \in \left|0,\ 0.2\right|, \text{ and} \tag{7.78}$$

$$m_{2(A)}(t) = 0.4 \text{ for } t \in \left|0.2,\ \infty\right|. \tag{7.79}$$

Equations (7.76) and (7.77) correspond to *constant firing speeds* in each time interval:

$$v_{2(A)}(t) = U_2 \cdot m_2(0) \text{ for } t \in \left|0,\ 1/U_2\right|, \text{ and} \tag{7.80}$$

$$v_{2(A)}(t) = v_1 \text{ for } t \in \left|1/U_2,\ \infty\right|. \tag{7.81}$$

In other words, the speed $v_2(0) = U_2 \cdot m_2(0)$ (the same for both VHPN and AHPN) remains unchanged up to $t = 1/U_2$ for the AHPN model, then $v_{2(A)} = v_1$

after this time (i.e., $m_{2(A)}$ remains unchanged since the balance $B_2 = v_1 - v_2 = 0$). For our example, $v_{2(A)} = 4.5$ up to $t = 0.2$ and $v_{2(A)} = 2$ after.

Let us note that *asymptotic value* $m_2^* = \frac{v_1}{U_2}$ and *duration* $\frac{1}{U_2}$ of the first time interval *do not depend on the initial marking*[2] $m_2(0)$ (assumed here to be less than or equal to m_1). This is illustrated in Figure 7.15c for $m_2(0) = 0.1$.

Let us now consider the timed hybrid PN in Figure 7.16a and analyze successively the firing speeds v_1 and v_2.

The instantaneous firing speed v_1 depends only on the discrete marking of D-place P_1. Then it is constant for both VHPN and AHPN models:

$$v_{1(V)}(t) = v_{1(A)}(t) = U_1 \cdot m_1 = 3 \text{ for } t \in \left|0, \infty\right|. \tag{7.82}$$

Let us note that this value $v_1 = 3$ is also the firing speed obtained for the basic timed hybrid PN (CHPN).

Consider now the instantaneous firing speed v_2. During the first IB-phase, i.e. from $t = 0$, v_2 depends on D-place P_2 since $m_2 < m_3$. Then, it is also constant and the same for both models (as well as for the CHPN model):

$$v_{2(V)}(t) = v_{2(A)}(t) = U_2 \cdot \min(m_2, m_5) = U_2 \cdot m_2 = 5. \tag{7.83}$$

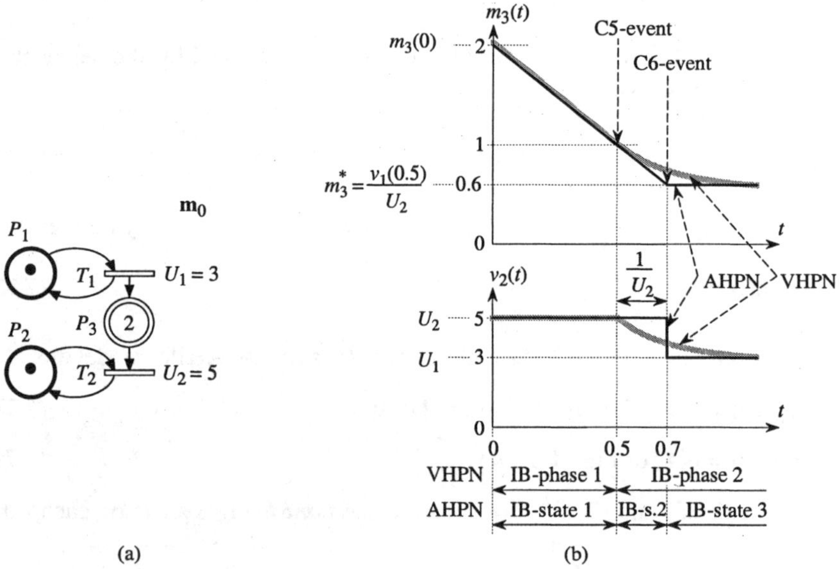

Figure 7.16 (a) Timed hybrid PN. (b) Behaviors of VHPN and AHPN models.

[2] An engineer may observe the similarity with the well known behavior of a resistance-capacitor circuit: the final voltage between boundaries of the capacitor depends only on the supply voltage (does not depend on initial voltage) and the time constant depends on the components but does not depend on the voltages (neither initial nor supply).

This equation remains true up to the next C5-event occurring at $t = 0.5$. At this time,

$$m_3(t) = 2 + (v_1 - v_2) \cdot t = 2 - 2\,t \tag{7.84}$$

reaches value $1 = m_2$ and P_3 becomes the new critical place of T_2. It follows that, from this time, given $v_{2(V)}(t) = 5\ m_{3(V)}(t)$ and $dm_{3(V)}(t) / dt = 3 - v_{2(V)}(t)$,

$$m_{3(V)}(t) = 0.6 + 0.4\ e^{-5(t-0.5)} \tag{7.85}$$

and $$v_{2(V)}(t) = 3 + 2\ e^{-5(t-0.5)} \tag{7.86}$$

are obtained for $t \in \left]0.5, \infty\right[$. This is illustrated in Figure 7.16b.

The values related to the AHPN model are obtained as for the example in Figure 7.15, i.e.,

$$v_{2(A)}(t) = 5 \text{ and } m_{3(A)}(t) = 1 - 2\,(t - 0.5) \text{ for } t \in \left]0.5, 0.7\right], \tag{7.87}$$

and $$v_{2(A)}(t) = 3 \text{ and } m_{3(A)}(t) = 0.6 \text{ for } t \in \left]0.7, \infty\right[. \tag{7.88}$$

This behavior of the AHPN model is also illustrated in Figure 7.16b. Let us note that neither the equation of $v_{2(A)}(t)$ nor the equation of $m_{3(A)}(t)$ change when the C5-event occurs. These equations change when the C6-event occurs, i.e. when $m_{3(A)}$ reaches its asymptotic value

$$m_3^* = \frac{v_1(0.5)}{U_2} = \frac{3}{5} = 0.6. \tag{7.89}$$

The C6-event occurs at $1 / U_2 = 0.2$ time units after the C5-event.

7.2.1.2 Change of Feeding Speed of the Input C-Place

In the previous section, it was observed that, in the AHPN model, the instantaneous firing speed of a transition could change from a constant value, say v_a, to another constant value, say v_b (see Figure 7.16b). For this reason, or because of a D1-event (firing of a D-transition), the feeding speed of the input C-place of the C-transition under consideration can change.

Let us analyze the behaviors of the PN in Figure 7.17a for the values $v_1(t)$ in Figure b, assuming a VHPN behavior, then an AHPN behavior. The feeding speed of P_1 is $v_1(t) = v_a = 0.9$ for t less than $t_1 = 0.7$ and $v_1(t) = v_b = 0.1$ for t greater than t_1.

Consider first the VHPN model. There are two IB-phases. From (7.75):

$$m_{1(V)}(t) = 0.45 + 0.75\ e^{-2t}, \text{ for } t \in \left]0, 0.7\right]. \tag{7.90}$$

For $t = t_1 = 0.7$, $m_{1(V)}(0.7) = 0.635$ is obtained from (7.90). Then,

$$m_{1(V)}(t) = 0.05 + 0.585\ e^{-2(t-0.7)}, \text{ for } t \in \left]0.7, \infty\right[\tag{7.91}$$

is obtained. This behavior is illustrated in Figure 7.17c.

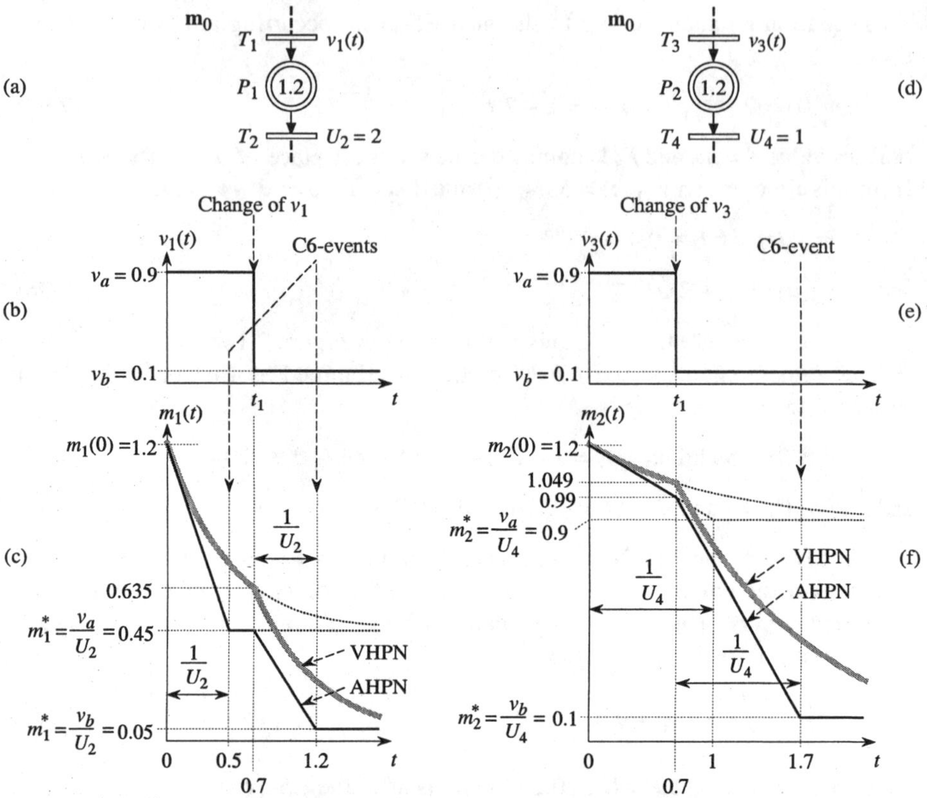

Figure 7.17 Change of feeding speed of input C-place. (a) $1 / U_2 < t_1$. (b) $1 / U_4 > t_1$.

In most cases, with an IB-phase during which a marking of C-place evolves exponentially in a VHPN model, two IB-states are associated in the corresponding AHPN model (see for example Figure 7.16b: with IB-phase 2 are associated IB-states 2 and 3). With the two IB-phases corresponding to (7.90) and (7.91), four IB-states are associated in the AHPN model, namely:

$$m_{1(A)}(t) = 1.2 - 1.5\,t \text{ and } v_{2(A)}(t) = 2.4 \text{ for } t \in \left|0, 0.5\right|, \tag{7.92}$$

$$m_{1(A)}(t) = 0.45 \text{ and } v_{2(A)}(t) = 0.9 \text{ for } t \in \left|0.5, 0.7\right|, \tag{7.93}$$

$$m_{1(A)}(t) = 0.45 - 0.8\,(t - 0.7) \text{ and } v_{2(A)}(t) = 0.9 \text{ for } t \in \left|0.7, 1.2\right|, \tag{7.94}$$

and $$m_{1(A)}(t) = 0.05 \text{ and } v_{2(A)}(t) = 0.1 \text{ for } t \in \left|1.2, \infty\right|. \tag{7.95}$$

These four IB-states exist because $1 / U_2 = 0.5$ (end of first IB-state) is less than $t_1 = 0.7$ (change of $v_1(t)$, beginning of second IB-phase). However, for the example presented in Figures 7.17d, e, and f, only three IB-states exist for the AHPN model. The speeds $v_3(t)$ and initial marking $m_2(0)$ in Figure d are similar to $v_1(t)$ and $m_1(0)$ in Figure a. However, $1 / U_4 > t_1$. In other words, the change of $v_3(t)$ from 0.9 to 0.1 occurs *before* the time $1 / U_4$ is reached, i.e. before the asymptotic value is reached in the ACPN model. The calculations leading to Figure 7.17f are left as an exercise (Exercise 7.4).

7.2.2 Several Input C-Places

If a C-transition has several input C-places, a change of critical place, among these C-places, can occur. In this section, the effect of this C5-event is analyzed.

Let us recall that the firing speed of a C-transition T_j depends only on the marking of its critical place (its flow rate U_j is assumed to be constant). It follows that the *vectors of firing speeds* $\mathbf{v}(t)$ at some time t *would be exactly* the same if all the non-critical places and related arcs were deleted from the PN. In addition, it was explained in the previous section that the speed of a C-transition is constant when its critical place is a D-place. Hence, only examples of C-transitions *with two C-places only will be considered,* without loss of generality.

Figure 7.18 presents a case where there is no change of critical place. During the first IB-state, $t \in \left| 0, 0.2 \right|$, the critical place $P(T_3, t) = P_1$, and $v_{3(A)}(t) = U_3 \cdot m_3(0) = 6$. After the C6-event occurring at $t = 0.2$, the critical place is the same since $m_{1(A)}(0.2) = 0.4 < m_{2(A)}(0.2) = 1.4$; the new firing speed is $v_{3(A)}(t) = v_1 = 2$. For $t > 0.2$, the marking $m_{2(A)}(t)$ increases because $v_2 - v_3$ is positive.

Two examples in which there is a change of critical place (C-places) are presented in Figure 7.19.

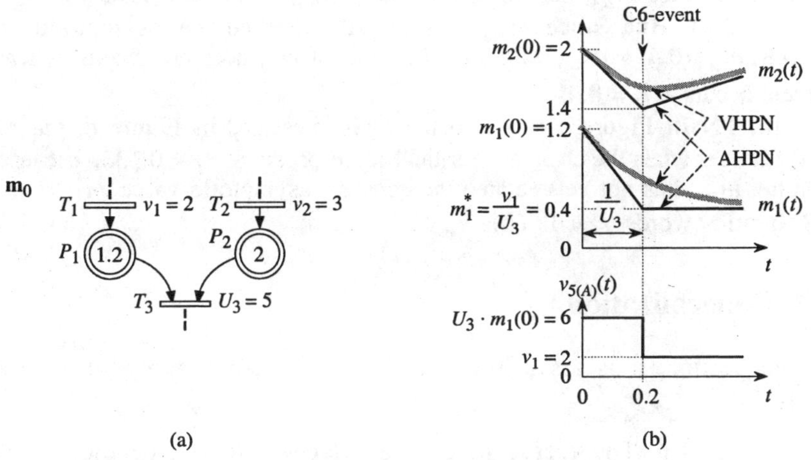

Figure 7.18 The critical place does not change.

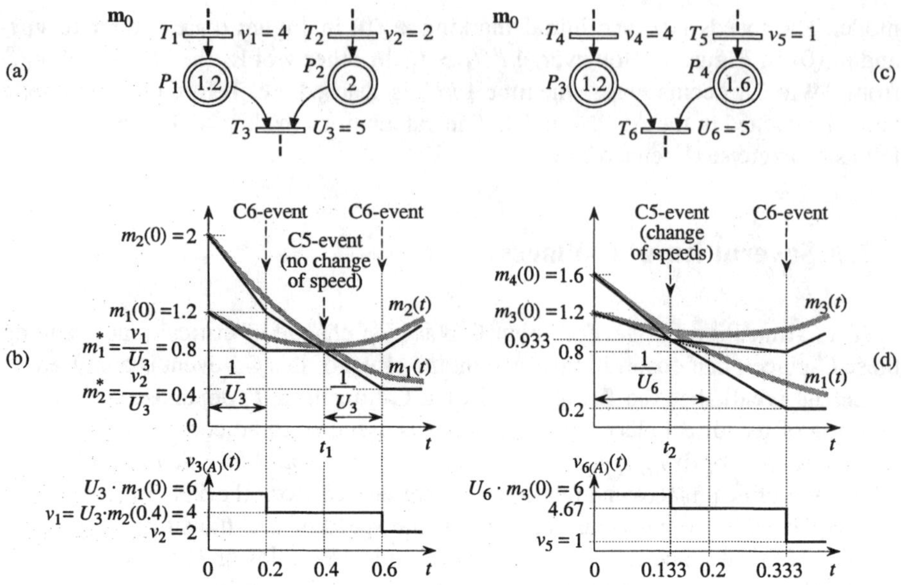

Figure 7.19 Change of critical place, i.e., C5-event. (a) Without change of speed. (b) With change of speeds.

For the PN in Figure a, whose behavior is shown in Figure b, there is a change of critical place at $t_1 = 0.4$ *without change of speed.* As a matter of fact, a C6-event occurred at $t = 0.2$ and the speed $v_{3(A)} = U_3 \cdot m_{1(A)}(0.2) = v_1 = 4$ was obtained for the IB-phase beginning at this time. At $t = 0.4$, there is a change of critical place since $m_{2(A)}$ becomes less[3] than $m_{1(A)}$. The new value of $v_{3(A)}$ is $U_3 \cdot m_{2(A)}(0.4)$. But, since $m_{1(A)}$ has already reached the asymptotic value $m_1^* = 0.8$, $m_{2(A)}(0.4) = m_{1(A)}(0.2)$; then the value of $v_{3(A)}$ does not change when the C5-event occurs at $t_1 = 0.4$.

For the PN in Figure c whose behavior is presented in Figure d, *the speed* $v_{6(A)}(t)$ *changes* when the change of critical place occurs at $t_2 = 0.133$, because, at this time, $m_{3(A)}$ has not yet reached the constant asymptotic value $m_3^* = v_4 / U_6 = 0.8$ (in other words, $t_2 < 1 / U_6$).

7.2.3 Generalization

In this section, it is assumed that there is no conflict, except at its end in Remark 7.9.

Definition 7.6 The **vector of critical places** at time t, denoted by $\mathbf{P}$ or $\mathbf{P}(t)$, is a $|T^C|$-dimensional vector containing all the critical places of

[3] Exactly at the same time, $m_{2(V)}$ becomes less than $m_{1(V)}$: see Property 7.11.

C-transitions at t. Since the index of D-transitions is lower than the index of C-transitions, the $(j - |T^D|)$th component of $\mathbf{P}(t)$ is critical place $P(T_j, t)$.

Remark 7.8 Let us emphasize a particular case which can occur in an AHPN. Assume transition T_j has two input places, D-place P_1 and C-place P_2, and the times $t_1 < t_2 < t_3$. If $m_2 < m_1$ in the time interval $|t_1, t_2|$, $m_2 = m_1$ at t_2, and $m_2 < m_1$ in $|t_2, t_3|$, there is *no change of critical place* at t_2: this is consistent with Definition 7.4 in Section 7.1.3.2. The property $m_1 \leq m_2$ defining P_1 as the critical place has a nul duration (then it is not taken into account).

A particular similar case can be observed at initialization. If $m_1(t) = \alpha$ and $m_2(t) = \alpha - \beta\, t$, α and $\beta > 0$, for $t \in |0, t_1|$, then P_2 is the critical place for this time interval.

An illustrative example will be given in Figure 7.23, Section 7.2.4.

❑

As specified in Definition 7.7, an IB-state for an AHPN is almost similar to an IB-state for a CHPN (basic timed hybrid PN). The only difference is that the vector of critical places is added.

Definition 7.7 An **IB-state** for an AHPN, is such that:

1) *the* marking $\mathbf{m}^D$ *of the D-places is* constant;

2) the *enabling vector* $\mathbf{e}^D$ of the D-transitions is *constant*;

3) the vector $\mathbf{v}$ of *instantaneous speeds* of the C-transitions is *constant.*

4) the vector of critical places $\mathbf{P}$ is constant;

5) when the IB-state is reached, $\mathbf{m}^C$ always has the *same value.*

❑

According to Section 7.2, a new type of event, C6-event, must be considered in addition to C5, D1, and D2-events already considered in VHPNs. This leads to Property 7.7. A C6-event is the reaching by critical place P_i of asymptotic value m_i^*. Let T_j be the output transition of P_i. In the general case m_i^* is equal to:

$$m_i^* = \frac{I_i}{U_j} = \frac{\sum_{T_k \in ({}^\circ P_i \cap T^C)} v_k \cdot \mathrm{Post}(P_i, T_k)}{U_j}. \tag{7.96}$$

Property 7.7 In an AHPN, a *change of IB-state* can occur only if an *event* belonging to one of the following types occurs.

C5-event: change of critical place for a C-transition.

C6-event: a critical place P_i reaches its asymptotic value m_i^*.

D1-event: *firing* of a D-transition.

D2-event: enabling degree of a D-transition changes because of the marking of a C-place.

❑

Since it is assumed that there is no conflict (i.e. $v_j = V_j$ for any C-transition), if v_j *is calculated at time* t_c, from Definitions 7.2 and 7.3 in Section 7.1.3.2:

$$v_j(t_c) = U_j \cdot \frac{m(P(T_j,t_c))}{\text{Pre}(P(T_j,t_c),T_j)} . \tag{7.97}$$

As the flow rate U_j and the weight Pre($P(T_j, t_c)$, T_j) are constant values given $P(T_j, t_c)$, Equation (7.97) shows clearly that the speed of T_j at t_c *depends only on the marking of its critical place at this time*. This observation leads to the following property.

Property 7.8 If the speed v_j is calculated at times t_c then t_c', the speed $v_j(t) = v_j(t_c)$ for $t \in \left| t_c, t_c' \right|$ is *independent from the feeding speed* of the critical place $P(T_j, t_c)$ during this time interval.

❑

This property is important. It ensures a decoupling among the firing speeds.

In Sections 7.2.1 and 7.2.2, the times when some firing speeds *must* be calculated were analyzed. However, if some speed v_j calculated at t_1 must be calcuted *at the latest* at t_3, there is no drawback in calculating it at some time t_2, $t_1 < t_2 < t_3$ (the effect on the quality of approximation will be explained in Section 7.2.4; see also Remark 7.10 in this section). From Property 7.7 (and given a D2-event cannot cause a change of speed) the following simple algorithm is obtained.

Algorithm 7.1 **Simple algorithm for AHPN without conflict**
Step 0. Enter data related to the timed hybrid PN, initial marking.
Step 1. Stabilization of marking (like *Step 3* in Algorithm 5.8, Section 5.3.4).
Step 2. Calculation of vector **P** of critical places.
Step 3. Calculation of vector **v** of speeds (in any order).
Step 4. Calculation of t_c, time of the next event, and $X(t_c)$, set of events occurring at this time. If $t_c = \infty$, then END.
Step 5. **If** $X(t_c)$ contains a D1-event **then** go to *Step 1* **else** go to *Step 2*.

❑

Algorithm 7.1 works, but it may carry out more speed calculations than really required. The cases where a new calculation of v_j is required are given in Property 7.9 (most cases were explained in Section 7.2).

Property 7.9 The calculation of speed v_j of C-transition T_j is required at time t_c if at least one of the following events occurs.

1) A D1-event occurs and either the critical place $P(T_j)$ changes OR $P(T_j)$ does not change but its marking $m(P(T_j))$ changes.

2) A C5-event related to T_j occurs (i.e. change of $P(T_j)$) AND i) the previous critical place was a C-place P_k, AND ii) m_k has not yet reached its asymptotic value m_k^*.

3) A C6-event related to T_j occurs.

4) A change in the feeding speed of $P(T_j)$ is caused by a D1, C5, or C6-event related to another transition.

Proof It is clear that a D2-event cannot imply a new speed calculation.

1) Obvious.

2) Let us show that all the cases of C5-event not corresponding to case 2 of the property do not require a new calculation, or are included in case 1.

I) If the previous critical place was a D-place and the next one a C-place, the speed does not change (Figure 7.16).

II) Changing from a critical D-place to another corresponds to a D1-event.

III) *If the previous critical place was a C-place P_k which has reached m_k^**: the next critical place cannot be a D-place without D1-event; the next critical place is a C-place and the speed does not change (Figure 7.19a and b).

3) Obvious (Figures 7.15 and 7.16).

4) Already explained with Figure 7.17.

❑

From Property 7.9, Algorithm 7.2 can be obtained. Only *Step 3* of Algorithm 7.1 is modified. In *Step 3.1*, the transitions satisfying one of the conditions 1, 2, or 3 of Property 7.9 are found. *Step 3.1* adds a set of transitions including all the transitions satisfying condition 4

Algorithm 7.2 **Algorithm for AHPN without conflict**

Similar to Algorithm 7.1, except *Step 3*.

Step 3. Calculation of speeds in vector **v** possibly changing.

Step 3.1. Find the set of transitions $T(X(t_c))$ satisfying one of the conditions 1, 2, or 3 of Property 7.9.

Step 3.2. **If** $T_k \in T(X(t_c))$ and $P(T_j) \in T_k°$ **then** $T(X(t_c)) = T(X(t_c)) \cup \{T_j\}$. Iterate *Step 3.2*.

Step 3.3. Calculate the speeds of transitions in $T(X(t_c))$.

❑

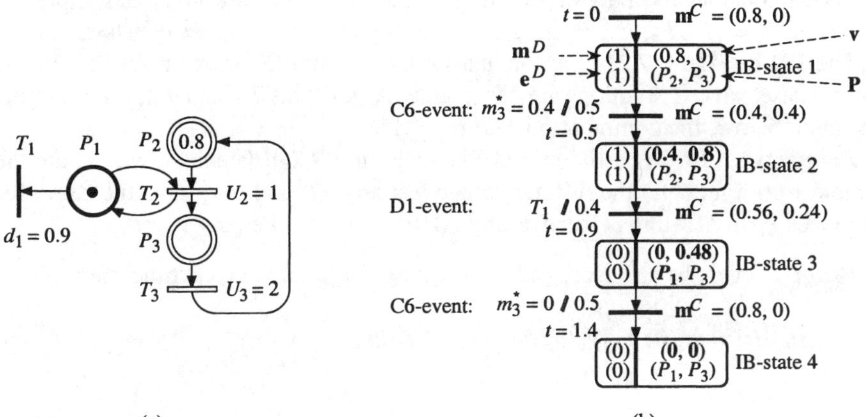

Figure 7.20 (a) Example of timed hybrid PN. (b) Evolution graph of the AHPN behavior.

This is illustrated with the example in Figure 7.20. The critical place of T_3 is always P_3 (single input place). At initial time, $P(T_2) = P_2$. The speeds $v_2 = U_2 \cdot m_2 = 0.8$ and $v_3 = U_3 \cdot m_3 = 0$ are obtained. Then, $m_2 = 0.8 - 0.8\,t$ and $m_3 = 0.8\,t$. This behavior corresponds to IB-state 1.

The next event is C6-event '*m_3 reaches its asymptotic value $m_3^* = 0.8/2 = 0.4$*' at $t = 0.5$. Then, in *Step 3.1* of Algorithm 7.2, $X(0.5) = \{m_3^*$ reached$\}$ and $T(X(0.5)) = \{T_3\}$. In *Step 3.2*, T_2 is added to $T(X(0.5))$ because a change of v_3 corresponds to a change of the feeding speed of $P(T_2)$: $P(T_2) = P_2 \in T_3°$. Hence $T(X(0.5)) = \{T_2, T_3\}$. It follows that both v_2 and v_3 are calculated in *Step 3.3*, using (7.97). These calculated values are written in bold characters in IB-state 2 (Figure 7.20b).

The next event is D1-event '*T_1 is fired*' at $t = 0.9$. The new critical place of T_2 is P_1 (in bold characters in IB-state 3). Both speeds must be calculated again.

The last event is C6-event '*m_3 reaches its new asymptotic value $m_3^* = 0$*' at $t = 1.4$. Both speeds are again calculated.

Remark 7.9 *Conflicts* in an AHPN model can be solved as in a VHPN model. See Section 7.1.3.2: Equations (7.21) to (7.25) presented for VHPNs can be used for AHPNs too. For a conflict resolution by *priorities*, *Step 2* in Algorithms 7.1 or 7.2 should be preceded by a calculation of markings assigned to transitions involved in a case 2 conflict (Equations (7.23), (7.24), etc.).

As for VHPNs models (Remark 7.4 in Section 7.1.3.2), resolution by *sharing* (instead of priorities) is not easy to formalize in a general case. However, such a resolution can easily be performed in some particular cases. See Exercise 7.3.

7.2.4 Differences Between VHPN and AHPN Behaviors

Let us consider the part of PN in Figure 7.21a. Transition T_j has input C-places in $°T_j = \{P_1, P_2, \ldots, P_a, \ldots\}$ such that the values $I_a = v_a$ are constant.

The PN in Figure 7.21a can be analyzed as a VHPN or as an AHPN. In the second case, several solutions are acceptable (Algorithms 7.1 and 7.2, for example. See also, below, the comment on Figure 7.21b).

Property 7.10 For the partial PN in Figure 7.21a (since v_1, v_2, ... are the same for both models) the difference between $m_{a(V)}(t)$ and $m_{a(A)}(t)$ is the same for all places P_a in $°T_j$ (this is true for any AHPN solution, i.e. any $v_{j(A)}(t)$).

Proof Let P_a be any place in $°T_j$. Markings $m_{a(V)}$ and $m_{a(A)}$ at time t are:

$$m_{a(V)}(t) = m_a(0) + \int_0^t v_a(u)\,du - \int_0^t v_{j(V)}(u)\,du\,, \tag{7.98}$$

and

$$m_{a(A)}(t) = m_a(0) + \int_0^t v_a(u)\,du - \int_0^t v_{j(A)}(u)\,du\,. \tag{7.99}$$

From (7.98) and (7.99), the result (7.100), *independent* from P_a, is obtained.

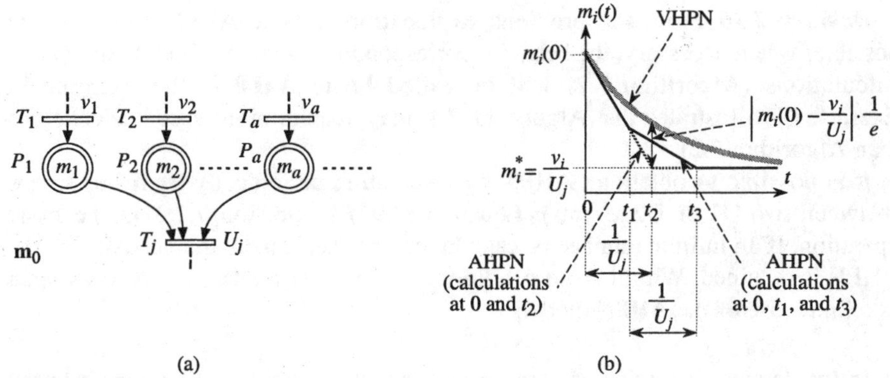

Figure 7.21 (a) PN with synchronization. (b) Effect of an additional calculation.

$$m_{a(V)}(t) - m_{a(A)}(t) = \int_0^t \left(v_{j(V)}(u) - v_{j(A)}(u)\right) du . \tag{7.100}$$

Property 7.11 In Figure 7.19b and d, changes of critical places occur at the same times in the VHPN and in the AHPN models, i.e. $m_{1(V)}(t_1) = m_{2(V)}(t_1)$ and $m_{1(V)}(t_2) = m_{2(V)}(t_2)$.

Proof This is a corollary of Property 7.10. For Figure 7.19a, for example: $m_{1(V)}(t_1) - m_{1(A)}(t_1) = m_{2(V)}(t_1) - m_{2(A)}(t_1)$ (according to Property 7.10); as $m_{1(A)}(t_1) = m_{2(A)}(t_1)$ by definition of time t_1, it follows that $m_{1(V)}(t_1) = m_{2(A)}(t_1)$.

❑

Let P_i be the critical place of T_j in Figure 7.21a. The difference $m_{i(V)} - m_{i(A)}$ is illustrated in Figure 7.21b.

According to Section 7.2.1.1 (and Algorithm 7.2), the firing speed $m_{i(A)}$ should be calculated at $t = 0$ then at $t = t_2 = 1 / U_j$. In this case the maximal value of $|m_{i(V)}(t) - m_{i(A)}(t)|$, reached for $t = t_2$, is:

$$|m_{i(V)}(t_2) - m_{i(A)}(t_2)| = \frac{1}{e} \cdot \left| m_i(0) - \frac{v_i}{U_j} \right| . \tag{7.101}$$

This result is formally shown in [Le 91].

Let us assume now that, after the speed calculation at $t = 0$, there is a new calculation of v_j at $t_1 < t_2$. This may happen, for example, if Algorithm 7.1 is used and there is another transition in the PN whose speed must be calculated at time t_1. The new speed is calculated from the marking $m_{i(A)}(t_1)$: $v_j = U_j \cdot m_{i(A)}(t_1)$. Then, at the latest, v_j will be calculated again at $t_3 = t_1 + 1 / U_j$. The corresponding behavior is illustrated in Figure 7.21b. It is clear that the maximal value of $|m_{i(V)}(t) - m_{i(A)}(t)|$, in this case, is less than in the previous case (calculations at times 0 and t_2). This result is formally shown in [Le 91].

Remark 7.10 From the previous explanation, several AHPN behaviors are possible. When necessary, the behavior corresponding to the minimum number of calculations (Algorithm 7.2) will be called **basic AHPN** (the example in Exercise 7.5 illustrates that Algorithm 7.1 may require more speed calculations than Algorithm 7.2).

It is possible to obtain an AHPN by calculating periodically its firing speeds (between two D1 or D2 events). Equation (7.97) in Section 7.2.3 is the basic operation. If an infinite number of calculations are performed, the behavior of the VHPN is obtained. Within these conditions, the VHPN model may be considered as a limit case of the AHPN model.

❑

In the sequel, examples are presented for illustrating the difference between VHPN and *basic* AHPN models.

For the hybrid PN in Figure 7.13b, the behavior of the VHPN model is illustrated in Figure 7.14. In Figure 7.22, $m_{6(V)}(t)$ is presented again as well as $m_{6(A)}(t)$ (basic AHPN). A good approximation of $m_{6(V)}(t)$ by $m_{6(A)}(t)$ may be observed. IB-state 1 corresponds to IB-phase 1 in Figure 7.14. Then, two IB-states in Figure 7.22 are associated with an IB-phase in Figure 7.14: IB-states 2 and 3 correspond to IB-phase 2, and so on. An evolution graph corresponding to this AHPN behavior is given in Solution 7.6.

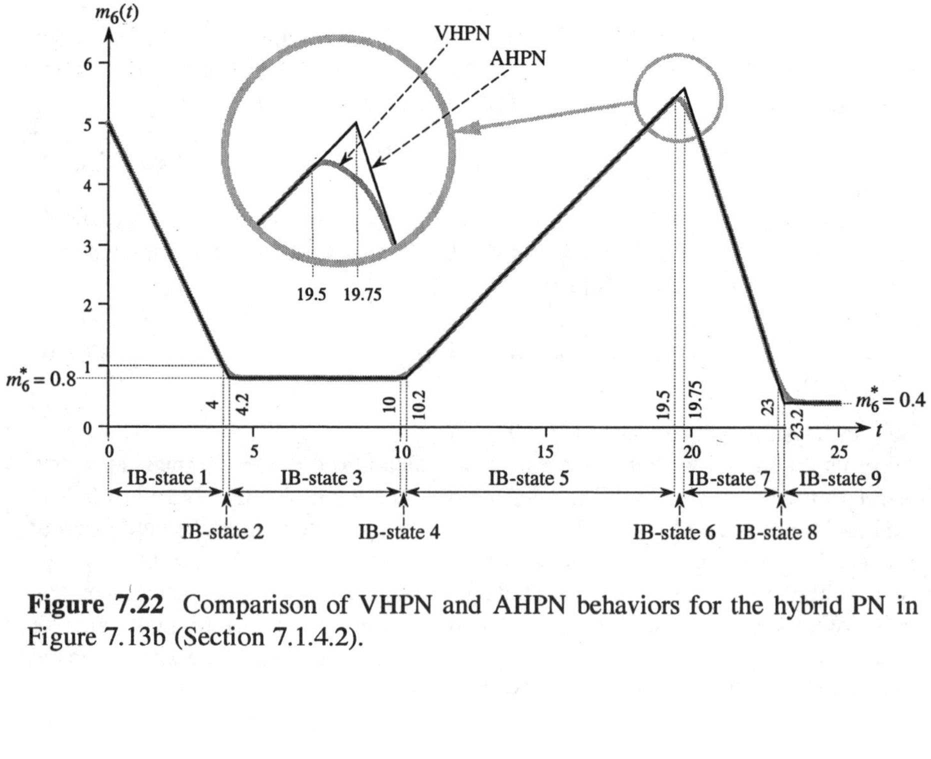

Figure 7.22 Comparison of VHPN and AHPN behaviors for the hybrid PN in Figure 7.13b (Section 7.1.4.2).

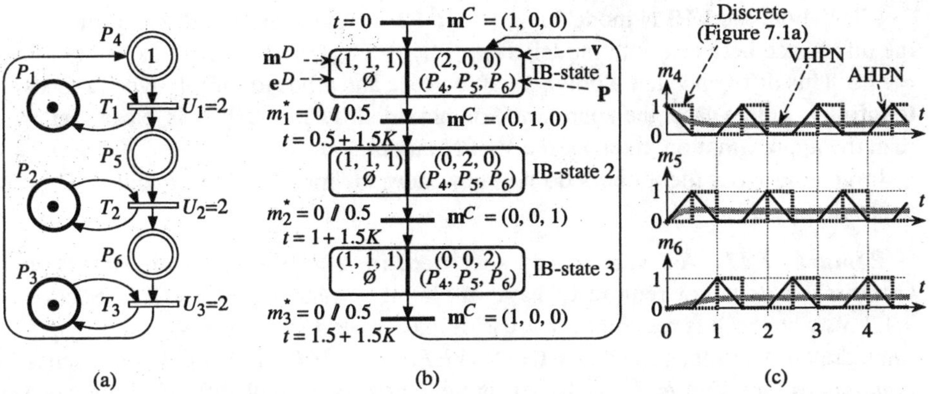

Figure 7.23 AHPN approximation of the discrete timed PN in Figure 7.1a.

Figure 7.23a is an hybrid approximation of the discrete PN in Figure 7.1a (Section 7.1.1). The VHPN approximation was presented in Figure 7.4 (Section 7.1.2). An evolution graph of the AHPN behavior is given in Figure 7.23b (this example is relevant to Remark 7.8): three IB-states and a feedback loop. Discrete, VHPN and AHPN behaviors are illustrated in Figure 7.23c. In this example, AHPN behavior, an approximation of the VHPN approximation of the discrete behavior, is closer to discrete behavior than VHPN behavior. For all cases, average markings are exact.

Figure 7.24a presents an artificial non-bounded example. Figure 7.24b is an approximation which can be considered as a VHPN or as an AHPN model. The three behaviors are illustrated in Figure 7.24c. For the discrete timed PN, $m_3(t) = \lfloor t \rfloor \cdot (\lfloor t \rfloor - 1) / 2$. The AHPN model is a very good approximation since $m_{3(A)}(t)$ is equal to the values $m_3(t)$ of the discrete model at the discrete times 1, 2, ..., N, ...

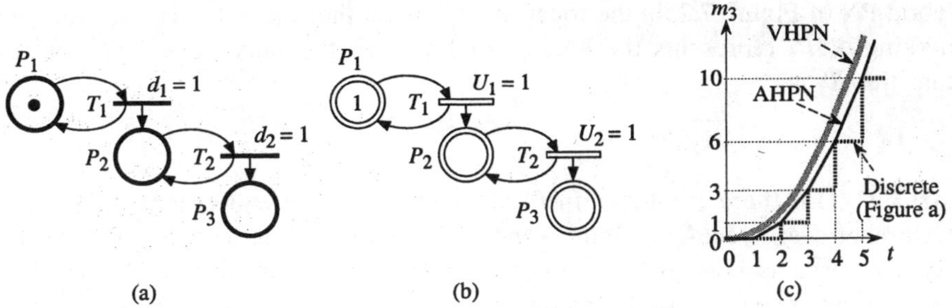

Figure 7.24 Example in which the difference betwen the ACPN and the VCPN may become infinite. The ACPN continues to be a good approximation.

What about the VHPN model? $m_{3(V)}(t) = t^2 / 2$ is found. At the discrete times t, the difference between both models is $m_{3(V)}(t) - m_{3(A)}(t) = t^2 / 2 - t(t-1) / 2 = t / 2$. This difference has a polynomial increase and tends to infinity when t tends to infinity. In this case, the approximation provided by the AHPN is clealy better than the approximation given by the VHPN model.

In general case, the authors do no know how to "measure" the quality of the approximations.

Remark 7.11 As was already specified, in a VHPN of in an AHPN, a C-transition does not require to have an input D-place (Remark 7.5, Section 7.1.3.3), whereas it is required for a basic timed hybrid PN (CHPN) for any non-immediate transition. It follows that *a VHPN or AHPN may have no discrete component*: see Figure 7.24b for example. There is no real difference between these models and the continuous corresponding models previously studied [LeAlDa 93].

7.3 OTHER MODELS

This section intends to illustrate that, from the same autonomous model (autonomous hybrid PN, in Section 4.2), *various systems and phenomena can be modeled using more complex functions* either for firing speeds of C-transitions or for firing rates of D-transitions (stochastic or constant timings).

7.3.1 Liquid Flow

The system presented in Figure 7.25a represents a tank filled with an unspecified input flow and an output flow v_2 which is proportional to the square root of H (height of the liquid in the tank). This system can be modeled by the hybrid PN of Figure 7.25b: the token in P_1 means that the valve is open and the marking of P_2 represents the volume of liquid in the tank. The flow rate of transition T_1 is

$$U_1(t) = k_2 \cdot \sqrt{m_2(t)}: \qquad (7.102)$$

$v_1(t) = U_1(t)$ if there is a token in P_1 at time t, and $v_1(t) = 0$ otherwise (k_2 is a function of coefficient k_1 in Figure a and of the surface of the section of the tank).

Let us assume now that the system is such that, at any time, $2 \leq m_2(t) \leq 4$ (i.e. feeding speed $I_2(t)$ and marking $m_1(t)$ are such that $m_2(t)$ remains in this range). In this case, $U_1(t)$ can be approximated by a linear function:

$$U_1(t) = k_2 \cdot (0.81 + 0.3\, m_2(t)). \qquad (7.103)$$

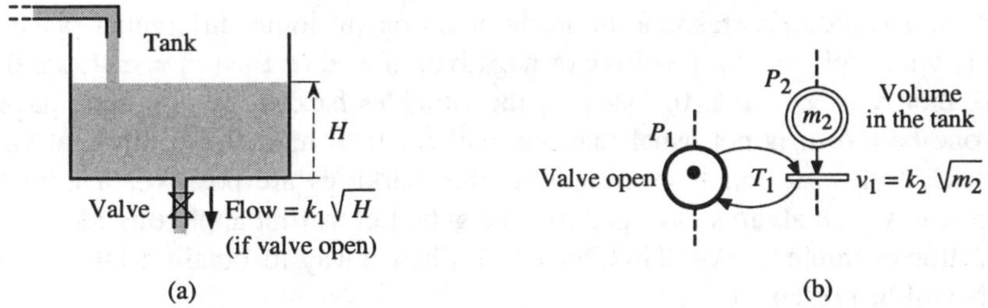

Figure 7.25 A liquid flow modeling.

Remark 7.12 Even if speed v_1 depends on m_2, there is no actual conflict in this model. Assume that there is another valve under the tank in Figure 5.25a. Then, in Figure b there is another C-transition T_2, an output transition of P_2, whose flow rate would be $U_2(t) = k_3 \cdot \sqrt{m_2(t)}$. There is a structural conflict $\langle P_2, \{T_1, T_2\}\rangle$ but *no actual conflict*: $m_2(t)$ can decrease more quickly, but equations of $v_1(t)$ and $v_2(t)$ are not modified.

Note that the meaning of C-markings are quite different in Figures 7.8 (Section 7.1.3.1) and in Figure 7.25b. In the first one, m_3 corresponds to parts which can be processed by a server 1 (firing of T_1) or a server 2 (firing of T_2) but not both: an actual conflict may exist is m_3 is not sufficient for saturating both servers. On the other hand, m_2 in Figure 7.25b, corresponding to height H, implies the pressure, hence the speed, but there is no "sharing" of height H between two or more possible valves. This example is an illustration that *new models of timed hybrid PNs could be defined by other researchers, for modeling other systems, from the same basis i.e. autonomous hybrid PNs* defined in Sections 4.2 and 4.3. Another example is given in the next section.

7.3.2 Differential Hybrid Petri Nets

In [DeKo 98], the authors defined a model called the *differential Petri net* (DPN). These authors define differential places and differential transitions, resembling our C-places and C-transitions, with some differences: the marking of a differential place may be negative and the weights of arcs to or from a differential place may be negative. Enabling of a differential transition is similar to D-enabling of a C-transition (Definition 6.1 in Section 6.14).

If all the markings and all the arc weights were non-negative, the behavior of a DPN could be modeled by a hybrid PN (i.e. based on the autonomous model presented in Section 4.2) with special speeds for C-transitions. Let us then propose a transformation of a DPN to a *model in which all these values are non-negative*, and call this model a **differential hybrid PN (DHPN)**.

If the marking of every place has *either an upper bound or a lower bound*, it can be replaced by a non-negative variable thanks to a simple change of variable

Let x_i be a variable corresponding to the marking of some differential place P_i. For any values of α and β (positive or negative): if $x_i \le \alpha$, then $y_i = \alpha - x_i \ge 0$; if $x_i \ge \beta$, then $y_i = x_i - \beta \ge 0$. Usually, the variables handled by engineers have at least one bound: it is not usual that the mathematical model of a physical value can reach both $-\infty$ and $+\infty$. When all the markings are positive, a model is obtained in which all arcs have positive weights. Let us first apply this idea to the application example proposed in [DeKo 98]. Then, a way to obtain a DHPN from a DPN will be presented.

Example drawn from [DeKo 98]: the behavior of a variable $x(t)$ switches between two models:

$$\frac{dx(t)}{dt} = -6 \text{ for } 5K \le t < 5K+2,\ K = 0, 1, 2, \ldots, \tag{7.104}$$

and $$\frac{dx(t)}{dt} = -x(t) \text{ for } 5K+2 \le t < 5(K+1),\ K = 0, 1, 2, \ldots. \tag{7.105}$$

In addition, a jump of magnitude 4 occurs at each switch from the first to the second model, i.e.:

$$x(t) := x(t) + 4 \text{ at } t = 5K+2,\ K = 0, 1, 2, \ldots, \tag{7.106}$$

and $$x(0) = 0. \tag{7.107}$$

A rough manual estimation of the behavior for one or two "periods" of 5 time units, shows that $x(t)$ will never be positive. Since $x(t) \le \alpha = 0$, $\alpha - x(t) \ge 0$; hence, the marking of a place (P_3 in Figure 7.26) may be $m_3(t) = -x(t) \ge 0$. Then, from (7.104) to (7.107), the timed hybrid PN in Figure 7.26 is obtained; all the markings and arc weights are non-negative; the flow rate associated with T_3 is a function of the C-marking: $U_3(t) = m_3(t)$.

A DHPN (Figure 7.26 for example) can be obtained from a DPN (model proposed in [DeKo 98]) in a systematic way, a C-place and a C-transition taking the places of a differential place and a differential transition, respectively.

Step 1. Change the variable of every C-place P_i in order that $m_i \ge 0$.

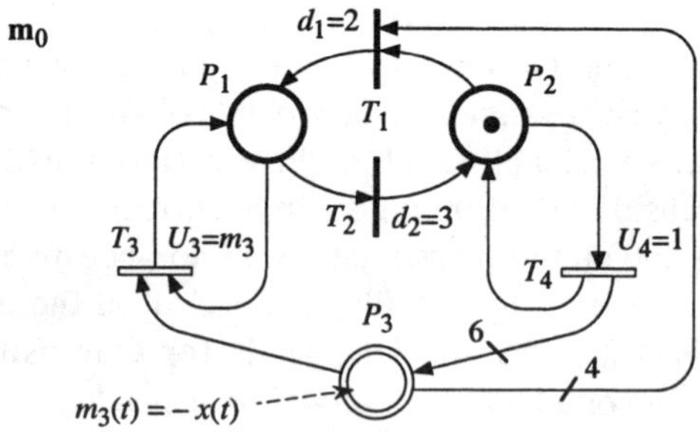

Figure 7.26 A timed hybrid PN (DHPN) satisfying (7.104) to (7.107).

Step 2. If $m_i = \alpha - x_i$ (where x_i is the marking of the initial differential place), change the weight sign of each arc $P_i \to T_j$ or $T_j \to P_i$.

Step 3. Replace each arc with a negative weight by an arc directed in the other direction and change the weight sign. For example, arc $P_i \to T_j$ with weight $\text{Pre}(P_i, T_j) = -5$ is replaced by arc $T_j \to P_i$ with weight $\text{Post}(P_i, T_j) = +5$.

Step 4. Adapt the flow rates of C-transitions according to variable changes in *Step 1*.

Step 5. If a flow rate is negative change its sign and the direction of all arcs linked to the corresponding transition. If a flow rate is sometimes positive and sometimes negative, split the transition into two transitions so that all the flow rates be positive. This step will be commented on in the sequel.

About Step 5. For initial value $x(0) = 0$ in (7.107), there is no difficulty: $U_3(t) = m_3(t)$ is positive and Figure 7.26 is obtained. On the other hand, assume now that the initial value is

$$x(0) = 10 \tag{7.108}$$

(instead of 0, i.e. (7.107) is replaced by(7.108)). In this case, it is easy to verify that $x(t) \leq 10$ for any $t \geq 0$. Then the marking of P_3 may be chosen as $m_3(t) = 10 - x(t)$ in order that $m_3(t) \geq 0$ for any t. It follows that the flow rate associated with T_3 should be

$$U_3(t) = m_3(t) - 10 \tag{7.109}$$

(this is the result obtained if 1) a DPN is built directly from the equations, then 2) *Steps 1* to *4* of the transformation explained above have been performed). A problem arises since $U_3(t) = m_3(t) - 10$ is sometimes negative and sometimes positive. The solution to this problem is to replace transition T_3 by two transitions T_3' and T_3'' as shown in Figure 7.27. Arcs linking transition T_3' to the places of the net are the same as for T_3 in Figure 7.26, and

$$U_3' = \max(U_3, 0) = \max(m_3 - 10, 0), \tag{7.110}$$

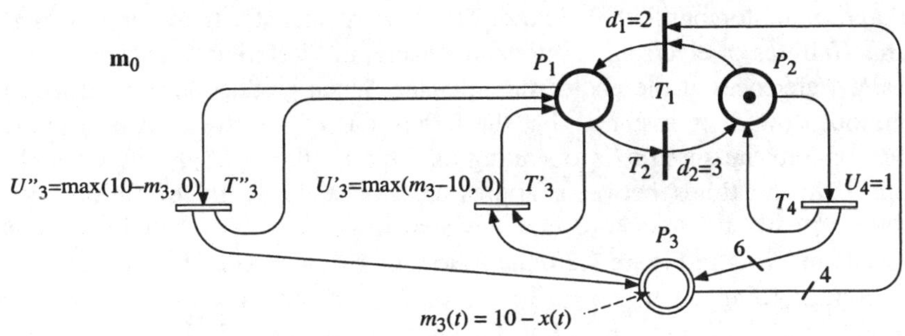

Figure 7.27 DHPN model if the initial marking of $x(t)$ is $x(0) = 10$.

never negative, is positive when $m_3 - 10$ is positive. Transition T_3'' is similar to T_3' with the following differences:

1) all arcs linking transition T_3'' to the places of the net are directed in the other direction (see Figure 7.27) and

2) $U_3'' = \max(-U_3, 0) = \max(10 - m_3, 0)$. (7.111)

Hence, U_3'', never negative, is positive when $m_3 - 10$ is negative.

Remark 7.13 An *abbreviation* replacing both transitions T_3' and T_3'' by a single transition T_3 could be used. The flow rate would be $U_3(t) = m_3(t) - 10$. The behavior of this transition is then understood as follows: if $U_3(t) \geq 0$, the behavior is "normal" (i.e. similar to T_3'); if $U_3(t) < 0$, transition T_3 behaves as if all the arcs linked to it were in the other direction and with a flow rate equal to $-U_3(t)$ (i.e. similar to T_3'').

Remark 7.14

a) If the upper bound (or lower bound) of some variable x_i is not easy to estimate, there is no drawback in using a non-reachable bound. For the example in Figure 7.27, $m_3(t) = 1000 - x(t)$ and $U_3(t) = m_3(t) - 1000$ could be used.

b) A flow rate $U_j(t)$ could be a non-linear function of the marking. See Exercise 7.8.

Remark 7.15 The VHPN model (Section 7.1) is a particular case of DHPN.

7.3.3 Transfer Line with Operation-Dependent Failures

In Section 6.3.2, a transfer line composed of three machines M_1, M_2 and M_3 in order, and two intermediate buffers B_1 and B_2 with respective finite capacities C_1 and C_2, was presented (Figure 6.21a). In the hybrid PN model of Figure 6.21b, the failure rate was *time dependent*, i.e. the machine can fail at any time when it is operational (it may be working or blocked or starved).

For a production system, the model of *operation dependent* failures (tool breakage or motor burnout for example) is more realistic than *time dependent* failures (failures of electronic systems for example). References will be given in Notes & References at the end of this chapter. When production is modeled by a continuous flow, it is assumed that the failure rate of a slowed down machine is proportional to the speed it is operating at. This is illustrated by the example in Figure 7.28: the times between failures are stochastic, λ_i is the failure rate of machine M_i when it works at its maximal speed, and λ_i' is the actual failure rate.

For the marking in Figure 7.28, the following firing speeds of C-transitions are obtained (according to Chapter 6). Transition T_7 is strongly enabled, then it is fired at speed $v_7 = V_7 = U_7 \cdot m_1 = 2$. For T_8, the maximal speed is $V_8 = U_8 = 4$ since there is a token in P_2. But, as T_8 is weakly enabled, $v_8 = I_9 = 2$ (i.e. machine M_2 is slowed down by machine M_1). Transition T_9 cannot be fired because there is no token in P_3 (M_3 is out of order).

It follows that the failure rates of the machines are $\lambda'_1 = \lambda_1$, $\lambda'_2 = \lambda_2 \cdot 2 / 4 = 0.5\ \lambda_2$, and $\lambda'_3 = 0$ (in fact not significant).

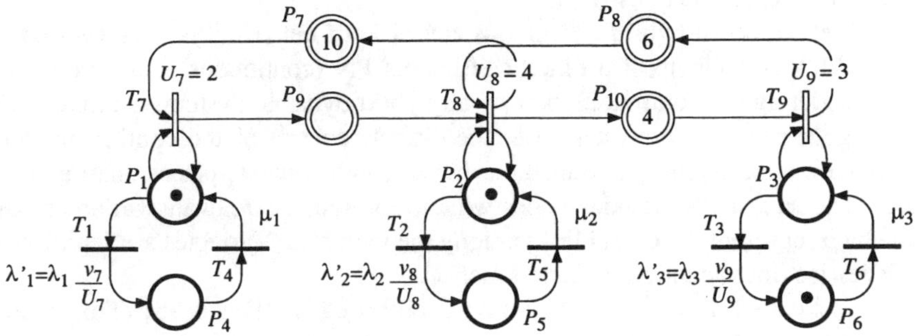

Figure 7.28 Model of transfer line with operation-dependent failures.

In this model, the firing rates λ'_1, λ'_2, and λ'_3, of the stochastic transitions T_1 to T_3, are functions of the instantaneous speeds v_j of the C-transitions, which are themselves functions of the marking $\mathbf{m}$. Thus, λ'_1 to λ'_3 are functions of $\mathbf{m}$.

NOTES and REFERENCES

Before introduction of timed hybrid PNs [LeAlDa 91 & 92b], the variable speed continuous PNs (called VCPN) were introduced in [DaAl 90] and used in [ZeAl 90]. The generalization of this model to the case of hybrid PNs is introduced in this book. It is the VHPN model presented in Section 7.1.

Similarly, the asymptotic continuous PN model (called ACPN), introduced in [Le 91] [LeAlDa 92a & 93] is generalized to hybrid PNs in this book. It is the AHPN model presented in Section 7.2.

In both cases, according to Remark 7.11 (Section 7.2.4), a VHPN or an AHPN may contain no discrete component (whereas this is not possible for a basic timed hybrid PN, studied in Chapter 6 and called CHPN).

In their revisiting of PN fluidification [ReSi 00], L. Recalde and M. Silva define a model called *Markovian P/T system*, or MPT. This model, although presented from a different point of view, corresponds to a VHPN behavior (without discrete component), as pointed out by the authors.

An interesting analysis of modeling possibilities of hybrid PNs is presented in [PeLe 95]. The authors suggest firing speeds depending on the marking for liquid flows (as in Section 7.3.1). They compare hybrid PNs with bound graphs.

In [Fl 96], a model in which the firing speed of a transition is proportional to the product of its input C-place markings is proposed. The examples given in this reference have a single input place. For this particular case, behavior can be modeled by a VHPN model too. (An example in which the firing speed is proportional to the product of C-place markings appears in the solution to Exercise 8.3, Figure S 8.3.F.)

A model called an hybrid continuous causal Petri net (HC^2PN) is proposed in [GoGe 96]. It is made up of a timed continuous PN (continuous causal Petri net, C^2PN) modeling the continuous part of an hybrid dynamic system, a discrete PN modeling the supervisory system, i.e. the discrete control of the continuous part, and an interface containing D-transitions having both input C-places and input D-places. The causal PN includes nodes which represent the relevant variables, and arcs which represent the causal links among the variables. Variables are represented by piecewise linear continuous time functions.

A model called differential PN (DPN) is proposed in [DeKo 98]. This model uses negative markings and arc weights. A model with the same modeling power and called differential hybrid PN (DHPN) is introduced in Section 7.3.2. This model is consistent with the autonomous hybrid PN presented in Section 4.2, i.e., all the markings and arc weights are non-negative. A VHPN is a particular case of DHPN.

Various models of transfer lines have been considered in the literature [DaGe 92]. The concepts of operation dependent failure and time dependent failure were specified in [BuHa 78]. The model assuming continuous production of machines and operation dependent failures (i.e. the model presented in Section 7.3.3, Figure 7.28) is now the most widely used (it was defined before its representation by an hybrid PN !). For this model, exact resolution (i.e. finding the average production and other parameters of the line) for a two-machine line was obtained, independently, in [GeSc 80] and [DuFo 82]. An approximate resolution for N-machine lines, $N > 2$, was given in [DaDaXi 89].

For the examples presented in this chapter, the speeds have a *constant coefficient* (U_i in Figure 7.1, k_2 in Figure 7.25, λ_j in Figure 7.28). However, as in Section 5.5, these coefficients can be *functions of time* and may be used as *control* variables. See [Le *et al.* 03] for example.

Postface

A variety of models have been presented, each one with its own particular feature. The *applications* of these models are summarized in Section Post.1. The relations between the *various models of hybrid PNs* are shown in Section Post.2.

Post.1 USE & ANALYSIS OF VARIOUS PN MODELS

Let us first summarize the application of the regular *discrete PNs.*

The basic model is the *autonomous PN* (ordinary or generalized). It enables discrete event systems of various kinds to be modeled. When a model has been established for a given system, it can be used to explain "how it works". This allows *qualitative validation* of a functioning process.

Non-autonomous PNs have two major application categories.

To begin with, *synchronized PNs* (of which *interpreted PNs* form part) enable the evolution of a *system subjected to external constraints* to be modeled. An important application is the description of *logic controlers* (as explained in the book, the theory of *Grafcet*, an international standard for logic controlers specification, is based on interpreted PNs) and *real time* systems.

Then, *timed and stochastic PNs* which take *time* into account (constant or stochastic timings) are best for *performance evaluation* (data processing systems, production systems, etc.).

Continuous PNs allow modeling of both continuous systems and discrete systems which can be approximated by continuous models. *Hybrid PNs* may be used when one part can be modeled by a continuous behavior while another part needs discrete modeling. As for discrete PNs, these models may be used for a *qualitative* validation as well as for *control* or *performance evaluation.*

In [Ce 79], one of the first methodologies for specifying and simulating hybrid systems is proposed. For more than a decade, hybrid systems became a specific research area for researchers in automatic control and computer science. See for example [Gr 93][QuGuBu 94][An *et al.* 95][*SI* 95][Ma 97][En 97][Za 01]. The hybrid Petri net is one of the models used for modeling these systems [Ch *et al.* 00].

Given a Petri net, various *analysis methods* can be used.

A structural analysis, based on invariants and other properties can be performed either for a regular discrete PN (Chapter 2) or for a continuous or hybrid PN (Section 4.3).

If time is involved, various parameters linked to performance evaluation can be obtained by analytical methods for discrete PNs with constant or stochastic timings (Section 3.4). For timed continuous and hybrid PNs (various models), evolution graphs can be used. Building an evolution graph is a kind of simulation, but the behavior between two events causing a change of IB-state (or IB-phase) can be modeled by continuous analytical variations. A variety of information can be obtained from a finite evolution graph (as from a reachability graph for a discrete PN). See for example [MüRa 00][OrAb 03].

A finite evolution graph cannot always be built (mix of constant and stochastic timings for example). For some cases (model presented in Section 7.3.3, for example), ad-hoc analytical methods are available. In any case, a Petri net model makes it possible to understand "how" a system works.

Petri nets, supporting various types of formalism, form a conceptual framework for modeling discrete event dynamic systems, called *the Petri net paradigm* in [SiTe 96]. Each model has its own specific character and special fields of application. Nevertheless, *Petri nets* form a common basis: they can be likened to a "*common language*" allowing dialog between people with very varied training backgrounds. Let us illustrate this with an anecdote.

In the late eighties, one of the authors and other researchers in the team got together around a board in order to understand and compare various kanban systems found in the literature. This was difficult because the figures in various papers were quite different, with unclear semantics, and the oral explanations of the participants were not always understood similarly by everybody... Then, to ensure that all participants understood the same behavior, it was suggested that a kanban system be represented by a Petri net: the semantics were clear (tokens in this place represent free kanbans, etc.). And this discussion led to a unified modeling of kanban systems using Petri nets [Di *et al.* 91].

Post.2 RELATIONS AMONG KINDS OF HYBRID PNs

Let us now present the relations between the *various models of hybrid PNs* presented in this book. See Figure Post.1 (for the moment do not take into account the grey arcs in this figure).

The first *hybrid PN* model is the *autonomous* one, presented in Sections 4.2 and 4.3. This model is obtained by mixing the discrete PNs (Chapters 1 and 2) and the continuous PNs (Sections 4.1 and 4.3). Hence, formally, these two models can be presented as particular cases of hybrid PNs: an hybrid PN without any continuous component is a discrete PN and an hybrid PN without any discrete component is a continuous PN (Remark 4.4 in Section 4.2.2). This is the meaning of the three lines of comments associated with the model shown at the top of Figure Post.1.

When *constant timings* are associated with transitions (constant delay for a D-transition and constant flow rate for a C-transition), the *basic timed hybrid PN*, called *CHPN*, is obtained.

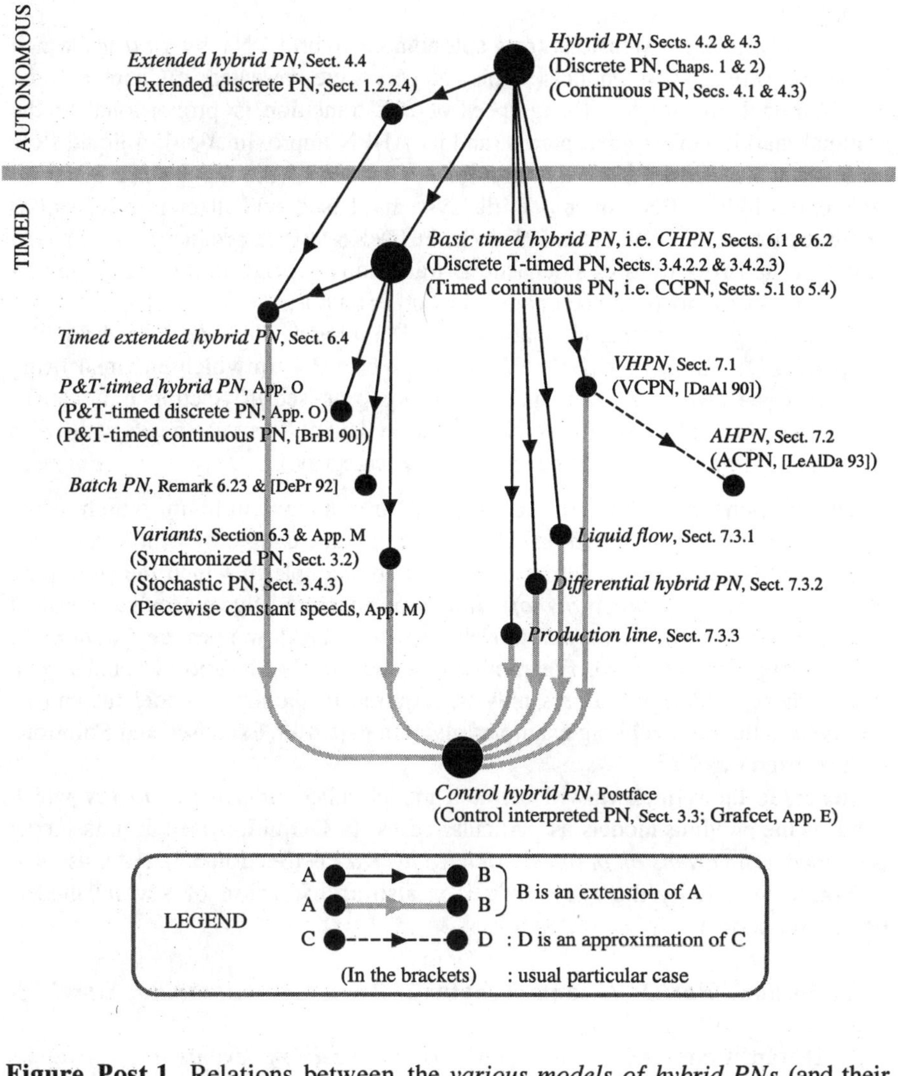

Figure Post.1 Relations between the *various models of hybrid PNs* (and their particular cases of discrete *or* continuous models).

These two models, represented by big points in the figure, constitute the basis of the hybrid models presented in the book: one without time and one with constant timings. All the other special cases of hybrid PNs can be presented as extensions of these two models. Informally, these extensions can be classed into *three categories.*

First category: extended hybrid PNs. This is an extension which does not involve time (an extended hybrid PN is autonomous). However, note that this model can be timed and that the CHPN model can also be extended to obtain the same model (which can be called either *extended timed* or *timed extended* hybrid PN).

Second category: extensions from autonomous hybrid PNs by *timings* which are *not constant.* Several models in this category were presented in Chapter 7: the VHPN model, in which a firing speed of a C-transition is proportional to the minimal marking of its input places (and its AHPN approximation); a liquid flow model in which firing speed is a non-linear function of the input place marking; differential hybrid PN for modeling systems based on differential equations (modification of the differential PN model in [DeKo 98]); a production line model in which the firing rates of stochastic D-transitions depend on the C-marking. A variety of other models can be imagined by other authors.

Third category: extensions from the CHPN model. *P&T-timed* hybrid PNs (Appendix O), *batch* PNs [DePr 92], timed hybrid PNs in which maximal firing speeds are *piecewise constant*, and the *variants* proposed in Section 6.3, can be classed in this category. (This classification is not formal: the variant with stochastic D-transitions might be in the second category.)

The variants presented in Section 6.3 are: an example in which some D-transitions are *synchronized* on external events, an example in which *stochastic timings* are associated with D-transitions, and an example in which the flow rates of C-transitions are *functions of time* (approximated by piecewise constant functions). In Chapter 7, several models in which the flow rates are *functions of the marking* are presented. For modeling a real-life big hybrid system, several among these modeling features may be required in the same model (as can be observed in the relatively big examples given in part 8 of Exercises and Solutions to these exercises).

Hence, in the sequel we will define a model called *control hybrid PN* which includes the previous models as particular cases. In Definition Post.1, this model is defined as an *extension of timed extended hybrid PN* (Section 6.4). As illustrated in Figure Post.1, by the *grey arcs*, it is also an extension of several models previously shown.

Definition Post.1 A **control hybrid PN** is a *timed extended hybrid PN* (Section 6.4) with, possibly:

1) D-transitions and C-transitions *synchronized on events or conditions* (Sections 3.2 and 3.3.3 for D-transitions, and Appendix G for C-transitions);

2) D-transitions with *timings* which may be stochastic, functions of time, of the marking, or of the firing speeds;
3) C-transitions whose *flow rates* may be functions of time, of the marking, or of the firing speeds.

❑

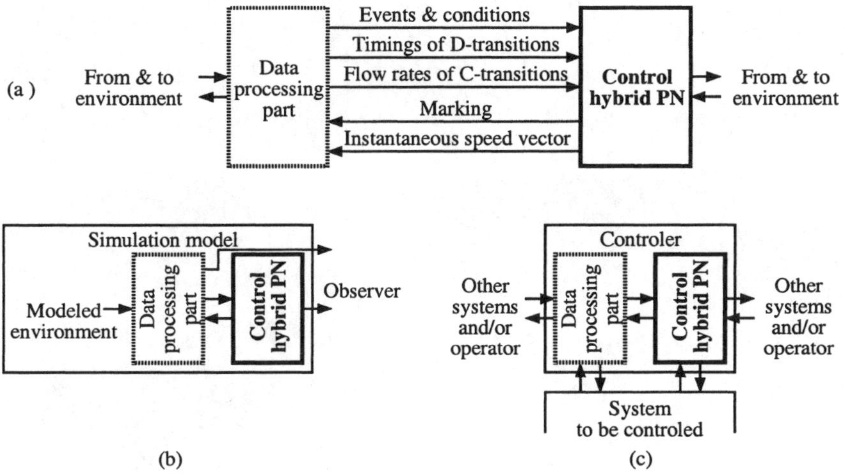

Figure Post.2 (a) Control hybrid PN and its data processing part. (b) Simulation. (c) Control.

According to this definition, hybrid PNs in Figures 6.19, 6.21, 6.23, 7.4, 7.5, 7.11, 7.13, and 7.25 to 7.28, are control hybrid PNs. The models presented in part 8 of Solutions to Exercises are also control hybrid PNs.

The concept of *control hybrid PN* is illustrated in Figure Post.2a. There is a linkage to a *data processing part* which calculates various data required by the control hybrid PN. Let us note that, according to Section 3.3.1 (see Figure 3.19), a control interpreted PN (also associated with a data processing part) is a particular case of control hybrid PN. The data processing part in Figure Post.2a receives informations from the environment and from the hybrid PN itself (marking and speed vector). It performs the necessary calculations in order to send the informations required by the hybrid PN: events and conditions, timings of D-transitions (stochastic value or function of speed for example), flow rates of C-transitions (functions of marking for example).

A control hybrid PN model, associated with its data processing part, will often be used for *simulation* purpose, as illustrated in Figure Post.2b. However, in some cases it can be used for *control* purpose, as illustrated in Figure Post.2c. The models are not exactly the same for both purposes: see end of Solution 8.1 for example.

Appendix A

Regular Expressions and Languages

A **language** is a set of **sequences** (also called **words** or **strings**) built from an **alphabet**. Let $D = \{a, b, ..., u\}$ be an alphabet; D may be the set T of transitions, a set of events, or any other set. A **regular language** can be represented by a **regular expression**. The three operations in regular expressions are union, concatenation, and iteration. Let $A = cabb$. This is a sequence built from the alphabet D. It corresponds to the successive occurrences of c, then a, then b twice. Let us say that A occurs if the sequence $cabb$ occurs.

Union is represented by +. If A and B are two sequences (or regular expressions), then $C = A + B$ means that C occurs if either A or B or both occur. For example $C = cabb + bd$ occurs if either the input sequence $cabb$ OR the input sequence bd occurs (the "OR" is inclusive). It is clear that $A + B = B + A$.

Concatenation is represented as a product[1]: $AB = (A)(B)$ means A followed by B (i.e. $AB \neq BA$). For example $E = (ab)(dc) = abdc$. If A and B represent several sequences, AB occurs if a sequence in A followed by a sequence in B occur. This corresponds to distributivity. For example $(a + dc)(b + c) = ab + ac + dcb + dcc$.

Iteration is represented by an asterix: B^* represents the repetition of B any finite number of times, including zero times. This can be written as $B^* = \varepsilon + B + BB + ... = \varepsilon + B + B^2 + ...$, where ε is the sequence of length 0. This sequence has the property: $\varepsilon A = A\varepsilon = A$. For example $b + ab = (\varepsilon + a)b$.

$D^* = (a + b + ...+ u)^*$ represents all the sequences that can be built from D. The set of sequences D^* is a monoid whose neutral element is ε.

Let α, β, γ, and δ be four sequences such that $\alpha = \beta\gamma\delta$. The sequences β, γ, and δ, are respectively called a **prefix** (proper prefix if $\gamma\delta \neq \varepsilon$), a **subsequence,** and a **suffix** of α. *The set of all the prefixes* of the regular expression B is denoted by $\overline{B}$. For example, $\overline{(ab)^*} = \varepsilon + a + ab + aba + abab + ...$

The sequences of transition firings in a PN represent a language. For the example in Figure A.1a, the longest sequences are in $\mathcal{L}_1 = T_1(T_1T_2 + T_2T_1)T_2$. Hence, the language generated from $\mathbf{m}_0$ is the set of prefixes of $\mathcal{L}_1$, i.e. $\overline{\mathcal{L}_1}$. Consider now Figure b: the initial marking $\mathbf{m}_0$ can be left then reached again by any firing sequence in $T_1(T_2T_3 + T_3T_2)T_4$. Hence, all the sequences starting from $\mathbf{m}_0$ and reaching $\mathbf{m}_0$ correspond to $\mathcal{L}_2 = (T_1(T_2T_3 + T_3T_2)T_4)^*$; the corresponding

[1] In this book, a concatenation is never represented by a dot: the *dot always represents a product* to avoid confusion (specially for events).

language is $\mathcal{L}_2$. For Figure c, one obtains $\mathcal{L}_3$ such that $\mathcal{L}_3 = (T_1(T_1T_2)^*T_2)^*$. The PN in Figure d is not bounded; no regular language can be associated with it.

Letters of any alphabet may be assigned to the transitions (possibly the same letter for several transitions). For Figure e, the language $\mathcal{L}_4$ such that $\mathcal{L}_4 = (a(ab + ba)c)^*$ is obtained. If a subset of reachable markings (usually called "marked" states) is considered, the firing sequences leading to one of these markings define a language. Example for Figure e: if the set of "marked" states is $\{\mathbf{m}_0, \mathbf{m}_1\}$, where $\mathbf{m}_1 = (0, 1, 1, 0, 0)$, then the language is $\mathcal{L}_5 = (a(ab + ba)c)^*(\varepsilon + a)$.

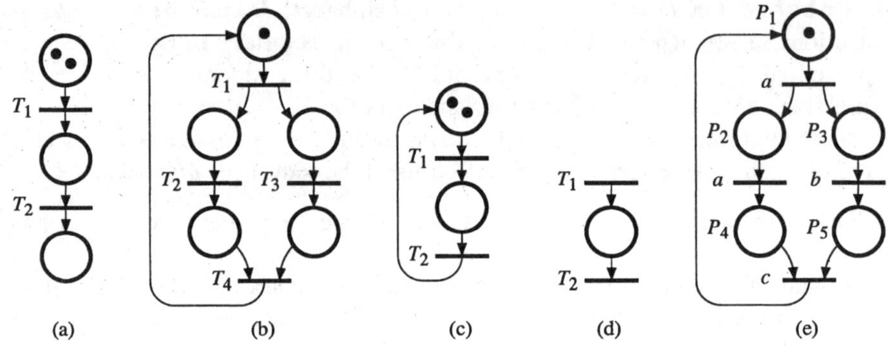

Figure A.1 Petri nets with labelled transitions ($\mathbf{m}_0$: initial marking for each).

Various types of languages correspond to various automata models. See [HoUl 79] or [Ca 89] for example. Relations among the sets of languages are given in Figure A.2. *Regular languages* correspond to finite state machines (a finite state automaton can be associated with any regular expression, and vice-versa). All the discrete event dynamic systems can be modeled by **Turing machines**. An ordinary Petri net can be constructed such that it corresponds to any regular language [Pe 81] since any *finite* state automaton can be modeled by such a PN. An example of context-free language modeled by a PN is given in [GiDi 94]. An example of PN language which is not context-free is given in [Pe 81]. Various results on PN languages are given in [GaGi 99].

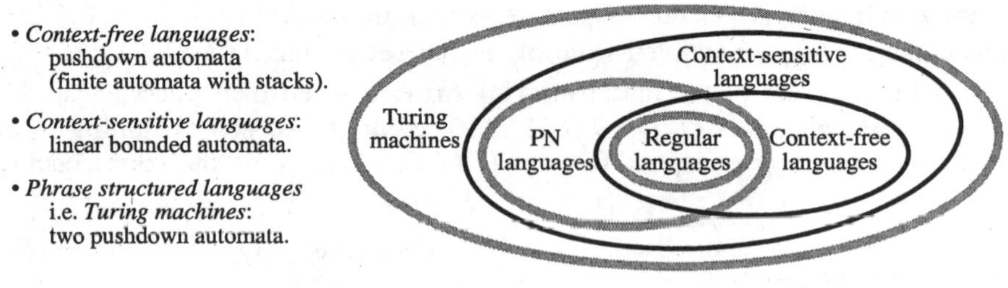

Figure A.2 Relations among the sets of formal languages.

Appendix B

Conflict Resolution

The PN in Figure B.1a contains the structural conflict $\langle P_5, \{T_2, T_5\}\rangle$. There is an effective conflict when P_1, P_3, and P_5 are marked (both T_2 and T_5 are enabled). This PN corresponds to a resource shared by two kinds of customers: the resource is available when there is a token in P_5; the tokens in P_1 and in P_3 model the customers, of types L and R respectively, waiting for a service. Allocating the resource to a type L or to a type R customer corresponds to firing T_2 or T_5. The **conflict resolution** consists of choosing the transition which is fired when both are enabled. The conflict resolution may be constrained in different ways. If the constraint is "rigid", the behavior is *deterministic*.

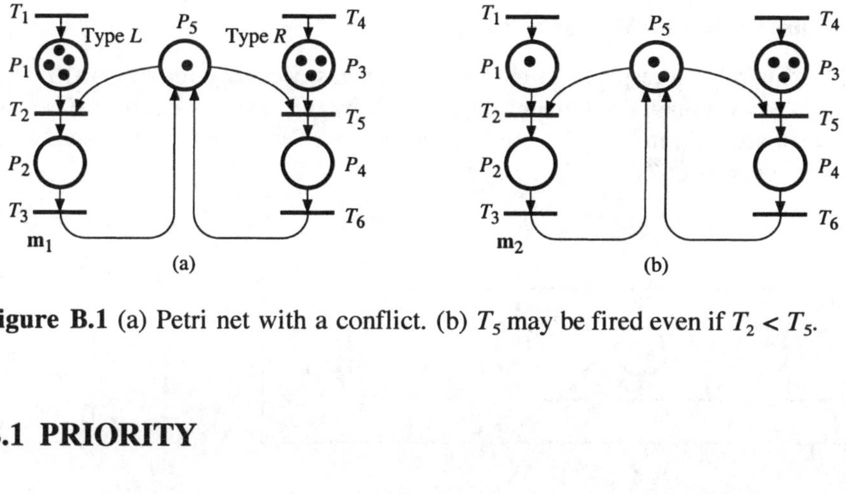

Figure B.1 (a) Petri net with a conflict. (b) T_5 may be fired even if $T_2 < T_5$.

B.1 PRIORITY

Resolution via priority is *deterministic*. In Figure B.1a, the priority may be given to T_2 over T_5, for example. This is denoted by $T_2 < T_5$ (Section 1.2.2.5). If T_2 is enabled, then it may be fired. In this example, as long as there is a token in P_1, T_5 cannot be fired. In Figure b, there are two resources (two tokens in P_5), T_2 is 1-enabled and T_5 is 2-enabled. If $T_2 < T_5$, a token in P_5 must be assigned to firing of T_2, because of the priority. However, since the enabling degree of T_2 is only one whereas there are two tokens in P_5, one of them may be used to fire T_5. According to Definition 2.10 (Section 2.1.4), there is a *general conflict*

$\langle P_5, \{T_2, T_5,\}, \mathbf{m}_2\rangle$ since $q(T_2, \mathbf{m}_2) + q(T_5, \mathbf{m}_2) < m_2(P_5)$ (see Definition 3.1 in Section 3.1). Let us now introduce the concept of *allowing degree.*

Definition B.1 The **allowing degree** of a transition T_j *in a priority PN*, for a marking **m**, is the number of firings of T_j which can be performed, given the marking and a partial order relation on the net transitions.

It is denoted by $r(T_j, \mathbf{m})$, $r(T_j, \mathbf{m}) \leq q(T_j, \mathbf{m})$, and T_j is said to be r-**allowed.** ❑

In Figure B.1b, both T_2 and T_5 are 1-allowed: we have $r(T_2, \mathbf{m}_2) = q(T_2, \mathbf{m}_2) = 1$ and $r(T_5, \mathbf{m}_2) = \min(m_2(P_3), (m_2(P_5) - r(T_2, \mathbf{m}_2))) = 1$.

Consider first a priority PN and a marking **m** such that there is a *single general conflict* $\langle P_i, \{T_1, \ldots, T_{j-1}, T_j, \ldots\}, \mathbf{m}\rangle$, and the priority order $T_{j-1} < T_j$ for all $j > 1$. The allowing degree of T_1 is equal to its enabling degree, i.e.,

$$r(T_1, \mathbf{m}) = q(T_1, \mathbf{m}). \tag{B.1}$$

Now, the allowing degree of another transition involved in the conflict depends on the allowing degrees of transitions with a higher priority. Let $\mathbf{s}^{(1,2,\ldots,j)}$ denote a m-component vector such that the kth component is $r(T_k, \mathbf{m})$ for $k = 1, \ldots, j$ and 0 for $k > j$. Let

$$\mathbf{m}^{(1,2,\ldots,j)} = \mathbf{m} - \mathbf{W}^- \cdot \mathbf{s}^{(1,2,\ldots,j)}. \tag{B.2}$$

Marking $\mathbf{m}^{(1)}$ corresponds to **m** minus the tokens necessary for $r(T_1, \mathbf{m})$ firings of T_1. We now obtain the allowing degree of T_2 by $r(T_2, \mathbf{m}) = q(T_2, \mathbf{m}^{(1)})$. Then we can calculate $r(T_3, \mathbf{m}) = q(T_3, \mathbf{m}^{(1,2)})$ and so on. We obtain successively the allowing degrees of $T_2, \ldots, T_j, \ldots$ by iteration of

$$r(T_j, \mathbf{m}) = q(T_j, \mathbf{m}^{(1,2,\ldots,j-1)}). \tag{B.3}$$

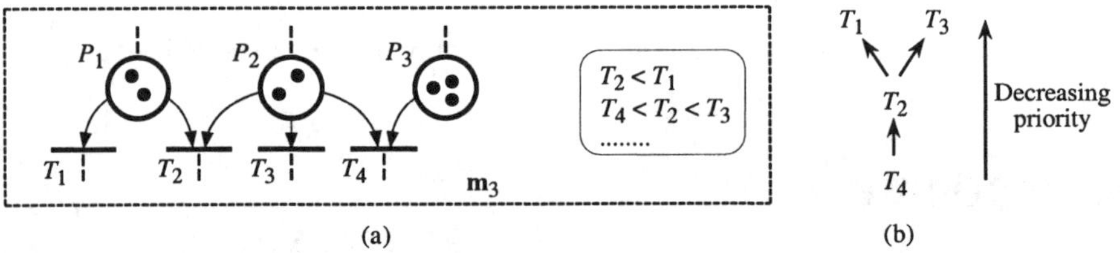

Figure B.2 (a) General conflicts in a not simple PN. (b) Priority graph.

The case must now be considered where there are *several general conflicts* for **m**. If the PN is simple (Section 1.2.1.5), no transition is involved in two conflicts hence the previous algorithm using iteratively (B.2) and (B.3) may be used for each conflict (in any order). If the PN is not simple, a priority graph is drawn as illustrated in Figure B.2. From the partial priority orders in Figure a, the graph in Figure b is obtained. In a general case this graph may contain cycles but not any circuit (see Appendix C). The allowing degree of a transition is calculated only when allowing degrees of all the transitions with a higher priority are known, using (B.1) to (B.3). For our example, we calculate the enabling degrees in the following order[1]: T_4, then T_2, then T_1 and T_3 in any order. Allowing degrees 2, 0, 2, and 0 are obtained respectively for T_4, T_2, T_1, and T_3.

B.2 PROBABILISTIC CHOICE

A firing probability may be assigned to each transition[2], for example 0.75 for T_2 and 0.25 for T_5 in Figure B.1a. When both transitions are enabled, a random drawing is performed such that: Pr[firing T_2 | T_2 and T_5 enabled] = 0.75. This random drawing is such that T_2 will be fired roughly 3 times more often than T_5. In this case, the resolution is *not deterministic.* If probability 0.5 is assigned to both transitions, it corresponds practically to the absence of constraint. The probabilities could be functions of other variables (enabling degrees for example).

B.3 ALTERNATE FIRING

In Figure B.3a, two places P_6 and P_7 have been added to the PN in Figure B.1a such that T_2 and T_5 will be fired in turn[3]. There is a token in P_6; after firing of T_2, there will be no token in P_6 but a token in P_7, allowing T_5 to be fired when P_5 and P_3 are marked, and so on. There is no longer any effective conflict in the PN on Figure B.3a, hence the resolution is *deterministic*.

If places P_1 and P_3 are never empty, the resource never remains idle. However, assume there is no token in P_1 when there are tokens in P_5 and P_6; transition T_2 is not enabled since $m(P_1) = 0$, but the resource cannot be allocated to type R customers because the "allocating token" is not in P_7 but in P_6. This problem may be solved by adding the transition T_5' as illustrated in Figure B.3b; this transition is fired if $m(P_1) = 0$ and $m(P_i) > 0$ for i = 3, 5, 6; T_5' has the same output places as T_5. Transition T_2' plays a symmetric role. Note that if the initial PN were bounded, a solution without inhibitor arc could be obtained.

A firing sequence for the PN in Figure B.3a is in $(T_1 + ... + T_6)^*$ (see Appendix A). The projection of this sequence in $(T_2+T_5)^*$ consists of replacing all the T_j except T_2 and T_5 by ε. Hence, the language generated by the PN in figure B.3a, projected in $(T_2+T_5)^*$ is the set of prefixes of $(T_2T_5)^*$. It is possible to obtain other ratios between the firings. For figures B.3c and d, the languages obtained are

[1] The indices of transitions must obviously be adapted: for our example, Equation (B.1) becomes $r(T_4, \mathbf{m}_3) = q(T_4, \mathbf{m}_3)$.

[2] The type of randomness is different from a stochastic PN (Section 3.4.3).

[3] The same behavior could be obtained with a simpler PN (as well as for Figure B.3b), but it is the authors' will to keep the basic structure of Figure B.1a in all examples.

the prefixes of $(T_2T_2T_2T_5)^*$ and $(T_2T_5T_2T_2T_5)^*$. In a general case, a is the weight of $P_6 \rightarrow T_2$ and $T_2 \rightarrow P_7$, b is the weight of $P_7 \rightarrow T_5$ and $T_5 \rightarrow P_6$, and the initial marking is such that $m(P_6)+m(P_7)=a+b-1$; we then obtain b firings of T_2 and a firings of T_5 in a period of $a + b$ firings; behavior is *deterministic*.

If $m(P_6) + m(P_7) > a + b - 1$, the ratio is approximately the same for a long sequence, but behavior is *not deterministic* (sometimes, T_2 and T_5 are enabled simultaneously). These behaviors have been studied in [KlLa 82].

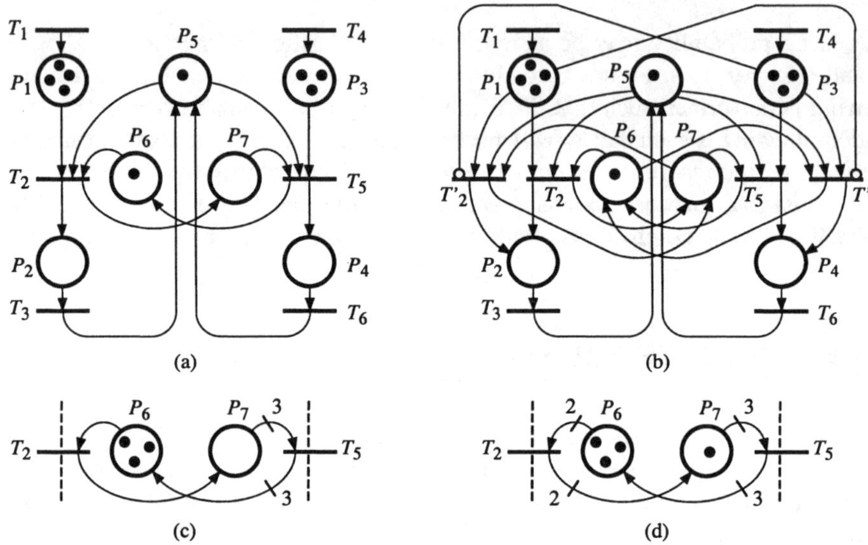

Figure B.3 (a) Alternate firing of T_2 and T_5. (b) Improvement of the PN in Figure a. (c) and (d) Different firing ratios for T_2 and T_5.

B.4 CONCLUSION

In a general case, resolutions presented in Sections B.1, B.2, and B.3, could be mixed in the same PN. In fact, if solutions similar to Figures B.3a to d are used, the corresponding conflicts disappear. Hence, for the remaining conflicts, priorities and probalistic choices can be used. Both solutions can be mixed; for example, for the structural conflict $\langle P_8, \{T_1, T_2, T_3\}\rangle$, the priorities $T_1 < T_2$ and $T_1 < T_3$ can be chosen, with a probalistic choice between T_2 and T_3.

Appendix C

Elements of Graph Theory

A graph (see [Be 70] for example) is made up of points and arrows, each arrow connecting two points. A point is called a **vertex** (or *node*) and an arrow is called an **arc** (or *directed edge*, the word *edge* being used if the direction of the arc is not taken into account).

Consider the graph in Figure C.1a. The set of vertices is $X = \{a, b, c, d, e, f, g, h, i\}$ and the set of arcs is $U = \{1, 2, 3, 4, 5, 6, 7, 8, 9, 10, 11, 12\}$. The arc *1*, for example, may be denoted by $a \rightarrow c$, where a and c are called the *initial extremity* and the *terminal extremity*, respectively; the vertex a is a *predecessor* of c, and c is a *successor* of a. The graph G is denoted by the pair $G = (X, U)$. The sets of successors and predecessors of a vertex x are denoted by $\Gamma_G^+(x)$ and $\Gamma_G^-(x)$; for example, $\Gamma_G^+(h) = \{c, g, i\}$ and $\Gamma_G^-(h) = \{e, g\}$.

Two vertices x and y are *adjacent* if they are the extremities of an arc, i.e., there is an arc $x \rightarrow y$ or (inclusive) $y \rightarrow x$. Two arcs (or edges) are *adjacent* if they have a common extremity. For example, in Figure C.1a, vertices b and c are adjacent; arcs *2* and *3* are adjacent, and arcs *6* and *7* are adjacent.

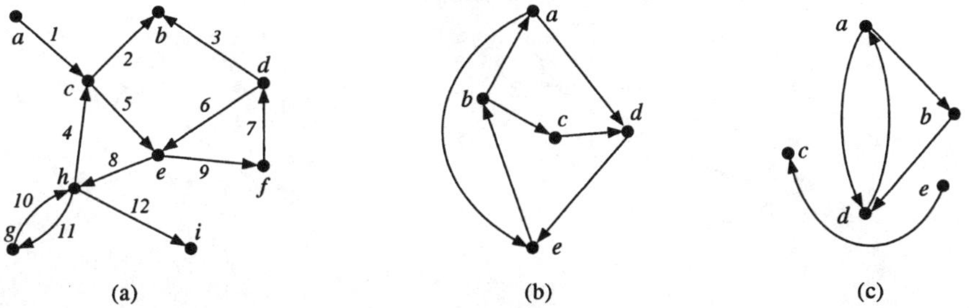

Figure C.1 Graphs (a) Connected. (b) Strongly connected. (c) Non-connected.

A **chain** of length $L > 0$ is a sequence of L arcs such that each arc in the sequence has an extremity common with the preceding arc and the *other* extremity common with the following arc. For example, in Figure C.1a, the sequence of arcs *2 3 7* is a chain of length 3. A **path** of length $L > 0$ is a particular chain such that the terminal extremity of the ith arc in the sequence is the initial extremity of the (i+1)th arc in the sequence (where $i < L$). For example, the chain *2 3 7* is not a path, but the sequence *10 4 5 9* is a path of length 4.

In this book, only graphs such that there is *at most one arc* $x \to y$ are considered; then, *this feature is implicit.* For such a graph, a chain or a path is completely defined by the sequence of vertices encountered. For example the chain *2 3 7* corresponds to *cbdf*, and the path *10 4 5 9* to *ghcef.* The vertices g and f are the initial extremity and the terminal extremity of the path *ghcef.* A **cycle** is a *chain* such that the initial extremity is also the terminal extremity; for example *cbdec* is a cycle. A **circuit** is a *path* such that the initial extremity is also the terminal extremity: the cycle *cbdec* is not a circuit but *hceh* is a circuit (hence, it is also a cycle). A chain, path, cycle, or circuit, is qualified *elementary* if it does not encounter the same vertex twice: *hceh* and *defd* are **elementary circuits**, while the circuit *hcefdeh* is not elementary.

A graph is **connected** if, for any pair of vertices (x, y), there is a *chain* between x and y; the graphs in Figures C.1a and b are connected, while the graph in Figure c is not connected. A graph is **strongly connected** if, for any pair of vertices (x, y), there is a *path* from x to y; the graph in Figure b is strongly connected, while the graph in Figure a is not (no path from b to h, for example). In a strongly connected graph, since there is a path from x to y and a path from y to x, there is at least one circuit passing by x and y. In fact, a graph is strongly connected if and only if there is a circuit passing by all the vertices: *adebaebcdeba* in Figure b. The number of elementary circuits is equal to (number of arcs – number of vertices + 1); these circuits form a base from which the other circuits (i.e., non-elementary) can be obtained. In Figure b, there are (7 – 5 + 1) = 3 elementary circuits, namely *adeba*, *bcdeb*, and *aeba*.

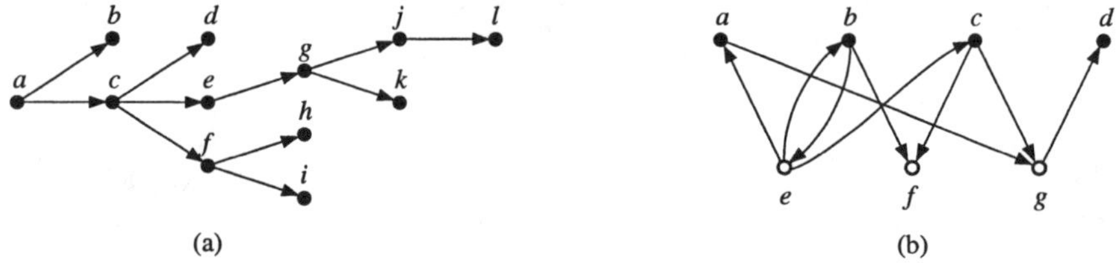

Figure C.2 (a) Rooted tree. (b) Bipartite graph.

A *tree* is a connected graph without a cycle. A *root* is a vertex x such that all the other vertices can be reached by a path starting at x. A **rooted tree** is a tree with a root; the graph in Figure C.2a is a rooted tree the root of which is a.

A graph is **bipartite** if the set of vertices can be partitioned in two classes X_1 and X_2 such that two vertices in the same class are never adjacent. The graph in Figure C.2b is bipartite: $X_1 = \{a, b, c, d\}$ and $X_2 = \{e, f, g\}$. *A Petri net is a bipartite graph* in which X_1 and X_2 are the set of places and the set of transitions.

Appendix D

Algebra of Events

The behavior of a discrete event dynamic system may depend on two kinds of information stemming from its environment: *conditions* and *events*.

We may think of a logic controller for example (its environment is made up of the process to be controlled, an operator, other systems), but the concepts of conditions and events may be applied to other kinds of systems. The state of a logic controller may change if a condition is true (a condition is expressed by a Boolean variable), when an event occurs. These notions, specified in Section 3.3.1, have been clearly defined by M. Moalla [Mo 85], and form a basis for Grafcet interpretation [Da 95]. Similar notions are presented in [SrKr 91].

The first kind of information relates to *the state* of the environment of the discrete event dynamic system modeled. For example[1] [*the number of parts in the buffer is greater than or equal to 10*], or [*the level in the tank is high*].

Observation D.1 The state of a discrete event system (assumed finite state) can always be defined by Boolean values.

❑

This is illustrated by the example in Figure D.1. The level in the tank (Figure D.1a) is a continuous variable. If this tank is considered to have three states (low, middle and high), its state can be modeled as in Figure D.1b. Obviously, three discrete states need at least two Boolean variables to be encoded. One of the possible solutions is presented in Figure D.1c: the Boolean variable $x_1 = 0$ if the level is low, and $x_1 = 1$ otherwise, and the Boolean variable $x_2 = 0$ if the level is low or middle, and $x_2 = 1$ otherwise.

A **predicate** is a proposal which may be either true or false, and which then can be modeled by a Boolean variable[1]. For example let x_3 denote the Boolean variable corresponding to the predicate [*the number of parts in the buffer is greater than or equal to 10*], i.e., $x_3 = 1$ if this proposal is true. If we assume that the number of parts in the buffer is encoded as a 4-digit binary number $y_3\, y_2\, y_1\, y_0$ (maximum number of parts = 15), then $x_3 = y_3\,(y_1 + y_2)$.

The second kind of information relates to a *change in the state* of the environment. Such a change is called an event[1]. An event has no duration, whereas the value of a Boolean variable lasts some time. For example, the event *'the level*

[1] In this book, a predicate is written in square brackets and an event between inverted commas (when they are specified by a text): [*predicate*] and *'event'*.

in the tank changes from low to middle' occurs at times t_1 and t_2 illustrated in Figure D.1d. This event is denoted $\uparrow x_1$ since it corresponds to the rising edge of the Boolean value x_1.

Observation D.2 An event can always be defined as a rising (or falling) edge of a Boolean value (variable or function).

❑

This is a direct consequence of Observation D.1, since an event is a change of discrete variable state. It follows that all the inputs of a logic controller may be Boolean variables, since the events can be defined from their changes.

In order to have a clear understanding of the relations between conditions and events, let us give some formal definitions and properties which were introduced in [DaAl 89 & 92], then improved in [Da 95].

The following definition is illustrated in Figure D.2a.

Definition D.1 Let $f(a_1, a_2, ..., a_m)$ be a Boolean function whose value is defined from the initial time 0, and such that, for $t_1 < t_2 < t_3 < ... < t_n < t_{n+1}$ ...:

1) $f = 0$ in the time intervals $[0, t_1[, [t_2, t_3[, ..., [t_{2p}, t_{2p+1}[, ...$;
2) $f = 1$ in the time intervals $[t_1, t_2[, [t_3, t_4[, ..., [t_{2p+1}, t_{2p+2}[, ...$

If $t_1 > 0$, the event $\uparrow f = \uparrow f(a_1, a_2, ..., a_m)$ occurs at times $t_1, t_3, ..., t_{2p+1}, ...$ and the event $\downarrow f$ $(= \uparrow f')$ occurs at times $t_2, t_4, ..., t_{2p}, ...$

If $t_1 = 0$, i.e., if the initial value of f is 1, the event $\uparrow f = \uparrow f(a_1, a_2, ..., a_m)$ occurs at times $t_3, ..., t_{2p+1}, ...$ and the event $\downarrow f$ occurs at times $t_2, t_4, ..., t_{2p}, ...$

❑

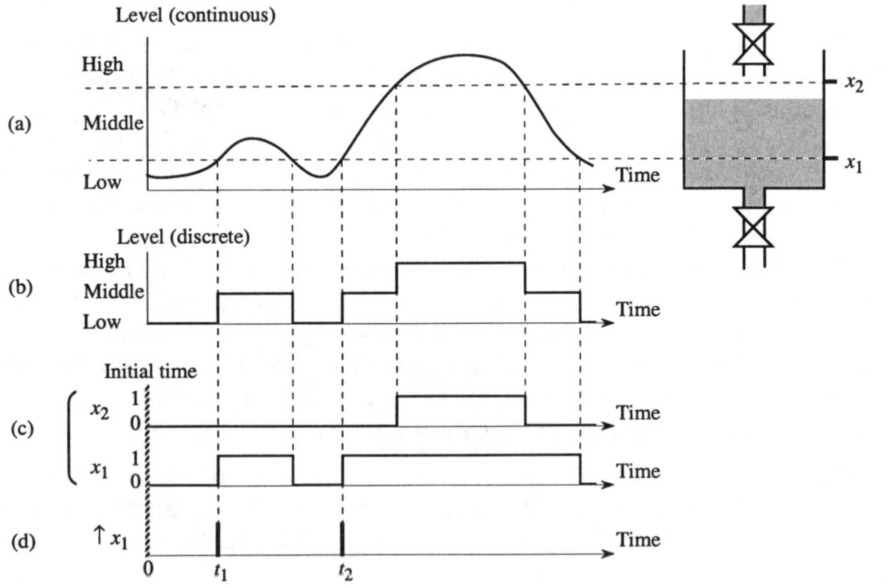

Figure D.1 Illustration of some concepts. (a) Continuous state. (b) Discrete state. (c) Boolean encoding of discrete states. (d) Occurrences of an event.

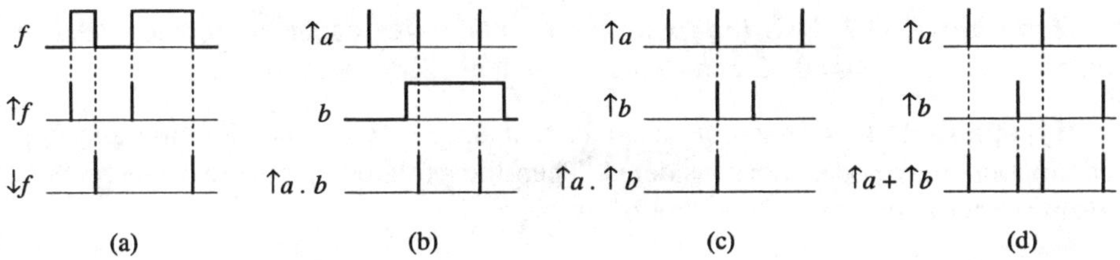

Figure D.2 Illustration of Definitions D.1 and D.2.

From now on, a, b, ... will denote either a variable or function (Boolean).

Notation D.1 The symbol $\uparrow$ takes precedence over the logic symbols $\cdot$ and +; in other words, $\uparrow a \cdot b = (\uparrow a) \cdot b$ and $\uparrow a + b = (\uparrow a) + b$.

❑

Definitions D.2a, b and c are illustrated in Figure D.2b, c and d respectively.

Definition D.2

a) The product $\uparrow a \cdot b$ is an event occurring at the same time as $\uparrow a$, each time that $b = 1$ at the corresponding time.

b) The product $\uparrow a \cdot \uparrow b$ is an event occurring at the times when $\uparrow a$ and $\uparrow b$ occur simultaneously (only possible if a and b are not independent, as will be seen).

c) The sum $\uparrow a + \uparrow b$ is an event occurring whenever either $\uparrow a$ or $\uparrow b$ occurs.

d) Let S be a system whose behavior depends on the set of events E. Let $E(t)$ be the event signal associated with S at time t: $E(t) \in E \cup \{\varepsilon\}$ where ε is the absence of an event in E at time t.

Notation D.2

a) Let E_1 and E_2 be two events. We write: $E_1 = E_2$ if E_1 and E_2 are equivalent, i.e., if they always occur at the same time; $E_1 > E_2$ if $E_1 \cdot E_2 = E_2$ and $E_1 \neq E_2$.

b) We write $E_1 = 0$, if E_1 is an event which never occurs.

c) We denote $e = \sum_{E_i \in E} E_i + \varepsilon$. We call e the "*always occurring*" event.

Definition D.3

Two events E_1 and E_2 are *independent* if there is no event E_i such that

$$E_1 = X + A \cdot E_i \quad \text{and} \quad E_2 = Y + B \cdot E_i,$$

where X and Y are events, A and B are Boolean functions or events such that

$$X + Y + A \cdot B \cdot E_i > X + Y.$$

❑

For example $E_1 = \uparrow f$ and $E_2 = \uparrow g$ may be not independent if both depend on the same Boolean variable a, e.g. $f = a \cdot b$ and $g = a$. In this example the reader may verify (using Property D.1d below) that $E_i = \uparrow a$, $X = \uparrow b \cdot a$, $A = b$, $Y = 0$ and $B = 1$.

Hypothesis D.1 Two *independent* events never occur simultaneously. In other words, $\uparrow a \cdot \uparrow b = 0$, if a and b are two independent variables.

❑

Hypothesis D.1 is based on the fact that an event has no duration and that continuous-time models are considered. Then the probability that two independent events occur at the same time is zero.

Remark D.1 It appears that we have the following features:
a) The product of two events is an event (Definition D.2b).
b) The sum of two events is an event (Definition D.2c).
c) The product of an event and a condition is an event (Definition D.2a).
But, what about the sum of an event and a condition? It is also an event: this is commented on in Section 3.3.3.

Property D.1

a) $\uparrow a = \downarrow a'$.

b) $\uparrow a \cdot a = \uparrow a$, $\quad \uparrow a \cdot a' = 0$, $\quad \downarrow a \cdot a' = \downarrow a$, $\quad \downarrow a \cdot a = 0$.

c) $\uparrow a \cdot \uparrow a = \uparrow a$, $\quad \uparrow a \cdot \uparrow a' = 0$, $\quad \uparrow a \cdot e = \uparrow a$.

d) If a and b are two independent variables, then:

$$\uparrow(a \cdot b) = \uparrow a \cdot b + \uparrow b \cdot a \quad \text{and} \quad \uparrow(a + b) = \uparrow a \cdot b' + \uparrow b \cdot a'.$$

e) If a, b and c are 3 independent variables, then:

$$\uparrow(a \cdot b) \cdot \uparrow(a \cdot c) = \uparrow a \cdot b \cdot c.$$

❑

Properties D.1a, b and c are obvious from Definitions D.1 and D.2. The proofs of Properties D.4d and e, based on Hypothesis D.1, can be found in [DaAl 92]. Property D.1e shows that the events $\uparrow(a \cdot b)$ and $\uparrow(a \cdot c)$ can occur simultaneously since they are not independent.

Remark D.2 In Figure D.1, the event $\uparrow x_1$ is defined *from* the Boolean variable x_1. What about an event defined "from scratch"? We can then introduce, more or less artificially, a Boolean value such that Observation D.2 is always verified. Here are some examples.

Event: *'failure of a machine'*; the predicate [*the machine is out of order*] is false before the failure and true after. Event: *'end of a timing'*; the predicate [*the time is elapsed*] is false before the end of timing and true after. Event: *'clap of thunder'*; the number of claps since some initial time changes from N to $N + 1$ (this number may be encoded by Boolean variables).

Appendix E

About Grafcet

Let us first present the main resemblances and differences between the control interpreted Petri net (Section 3.3) and the Grafcet[1] model (references in Notes & References at the end of Chapter 3).

Each model possesses two types of nodes, namely the *steps* and the *transitions* for the Grafcet, and the *places* and *transitions* for the interpreted PN (the original graphical representation of the Grafcet was similar to that of the interpreted PNs). In both models, the evolutions of the markings are *synchronized* on the occurrences of *external events*.

The *marking* is Boolean for a grafcet step and numerical for an interpreted PN place. However, this difference is not significant if control interpreted PNs are considered since this model is safe (Definition 3.7, Section 3.3.1).

In fact, *if a control interpreted PN is interpreted as a grafcet, it has exactly the same behavior*. On the other hand, given some usual minor restrictions (explained in Section E.3), *if a grafcet is interpreted as a control interpreted PN, it has exactly the same behavior*. Thus, *Appendix D* (algebra of events) *together with Section 3.3 make up a theoretical basis of Grafcet.* Additional information is given in this appendix: the normalized representation is briefly presented in Section E.1, various concepts leading to abbreviations (some of them allow a structured and hierarchical specification) are given in Section E.2, followed by some comments in Section E.3.

E.1 NORMALIZED REPRESENTATION

The normalized representation [IE 02] of the main elements is presented in the sequel. The reader, now familiar with interpreted PN, can easily compare both representations; an *arc* in a PN is called a **directed link** in a grafcet.

Figure E.1 presents the basic elements: steps, transitions and directed links.

A **step** (corresponding to a *place* in a PN) is represented by a square, or a double square if it has to be active when the system is started.

[1] We shall write Grafcet (with a capital G) when speaking of the tool in general, and grafcet (with a small g) when referring to a particular logic controller model.

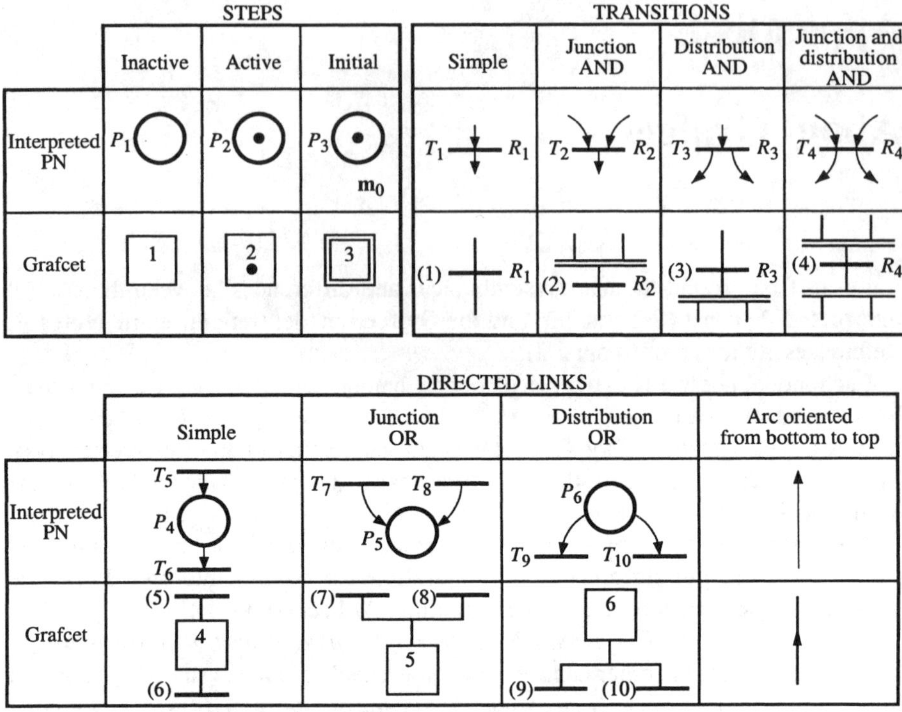

Figure E.1 Representation of basic components.

A *vertical link* without arrow is running from top to bottom. The *transition* symbol is a bar, but the latter must be preceded or/and followed by a double bar, when two or more arcs join this transition or/and when two or more arcs leave it.

When two or more *directed links* join the same step they are grouped together, like (7) → 5 and (8) → 5 in Figure E.1. When two or more *directed links* leave the same step they have a common departure point (see 6 → (9) and 6 → (10)).

Notation E.1 Here are some normalized symbols.

a) The complement of Boolean value a is denoted by $\bar{a}$ (whereas it is denoted by a' in the rest of the book).

b) Xi: Boolean variable whose value is 1 when step i is active.

c) $t1/a$: Boolean variable whose value is 1 when a time at least equal to $t1$ has elapsed since the *last time* Boolean variable a became true (i.e. since last event $\uparrow a$, or possibly $t = 0$ if $a = 1$ at initialization). For example, the normalized notation for the predicate $[t/P_3/2\text{ min}]$ in Figure 3.20 (Section 3.3.1) is $2\text{min}/X3$.

Notation $t1/a/t2$ is also normalized ($t1 < t2$). It is equivalent to $(t1/a)(\overline{t2/a})$.

d) A predicate is represented with square brackets: $[N > 5] = 1$ if $N > 5$ and $[N > 5] = 0$ if $N \le 5$ (similar in Section 3.3).

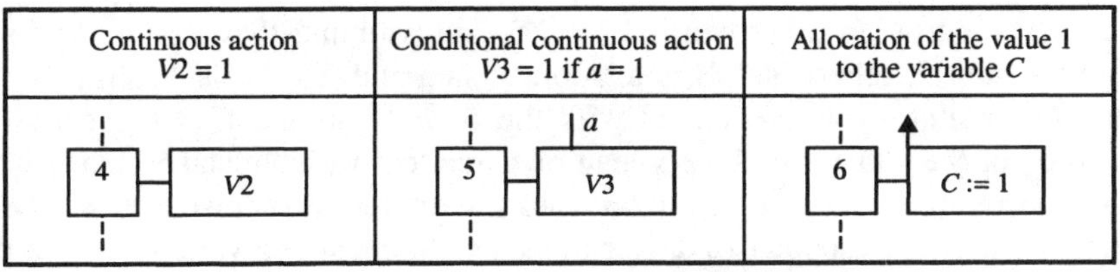

Figure E.2 Normalized representation of actions.

e) An *impulse action* (which was presented in Section 3.3.1) corresponds to **allocation of a value** to a variable (Boolean or numerical). This allocation is denoted by :=, for example B := 1 (similar in Section 3.3).

Receptivities As in an interpreted PN, a receptivity may be either a *condition* (i.e. a Boolean function), or an *event*, or the *product of an event and a condition*. The normalized representation of the always true receptivity (corresponding to an *immediate transition*) is $\underline{1}$.

Boolean variables Xi are allowed in the receptivities. If step i and transition (j) are in the same connected[2] component of the grafcet, the designer must be *very careful* when using Xi in R_j.

Evolution rules The set of active steps is called the **situation** (*marking* in a control interpreted PN). The behavior of a grafcet is similar to the behavior of a control interpreted PN (Section 3.3), except for the following rule:

In a grafcet, if several transitions can be simultaneously fired, they are *simultaneously fired*.

Remark E.1 Let us comment on this last rule. Assume that transitions (1) and (2) are simultaneously firable. If they have no common input place, the behavior is similar in a grafcet and in a control interpreted PN. On the other hand, if transitions (1) and (2) have a common input place, both are fired in a grafcet whereas, since there is an actual conflict, only one of them can be fired in a control interpreted PN. Fortunately, [IE 02] suggests this case be avoided: "*The designer should ensure that the timing, logical or mechanical aspects of the transition conditions are mutually exclusive*".

Actions The outputs of a grafcet are obtained as in a control interpreted PN. The normalized representation is illustrated in Figure E.2: an action is written in a rectangle linked to the corresponding step by a horizontal short line segment.

Level action $V2$ associated with step 4 is a **continuous action**: $V2 = 1$ as long as step 4 is active. Level action $V3$ associated with step 5 is a **conditional continuous action**: $V3 = 1$ as long as step 5 is active *and* Boolean variable (or

[2] The structure of a grafcet corresponds to a bipartite graph in which the steps and the transitions are vertices and the oriented links are arcs. The definition of a connected grafcet follows (see Appendix C).

function) a is true; in other words $V3 = X5 \cdot a$ (assuming that $V3$ is associated only with step 5; otherwise $V3$ is a Boolean sum and $X5 \cdot a$ is a term of this sum). The rising arrow associated with the *impulse action* $C := 1$ means that allocation of the value 1 to C (Boolean or numerical variable) takes place when step 6 becomes active (i.e. simultaneously with the **internal event** $\uparrow X6$). Variable C may correspond to a device in the environment or a variable in the data processing part (see Figure 3.19 in Section 3.3.1).

Other possibilities proposed in [IE 02] are commented on in Section E.3.

E.2 MACROSTEPS AND MACROACTIONS

Various abbreviations were proposed to simplify the logic controller specification and to obtain a structured and hierarchical specification. *Macrosteps* were introduced in [MoDa 81]. The concept of *macroaction* was introduced in [De 83] who defined the macroactions *inactivate*, *initialize*, and *mask*. The notions of *forcing* (which include and generalize those of inactivation and initialization) and *freezing* are specified in [GR 85]. The concept of *enclosure* is proposed in [IE 02]. All these concepts correspond to *abbreviations*. This was shown for some of them in [DaAl 89 &92]: it is always possible to build a grafcet equivalent (same input/output behavior) without these abbreviations.

In Figure E.3, grafcet 1 contains two **macrosteps**, namely macrosteps 3 and 5. The complete definition of the grafcet requires knowledge of the macrosteps *expansions*. In our example both macrosteps have the same expansion called M20, but they could be different. A macrostep **expansion** contains exactly *one entry step* and *one exit step* (which could be the same) respectively denoted by E20 and S20 in our example.

In most cases, an expansion Mk corresponds to a single macrostep i: the entry step Ek becomes active when an input transition of i is fired, and the output transitions of i may be enabled when the exit step Sk is active (may depend also on other steps or macrosteps). If two or more macrosteps with the same expansion can be simultaneously active (as in our example), the implementation requires several copies of the steps in the expansion.

When describing complex systems, the *size of the grafcets* may increase so that they become difficult to work out and thus to understand, correct, update, etc. Consideration of safety, in particular, is an important reason for increasing complexity. The concept of **hierarchy** naturally springs to mind. A (global) grafcet may be made up of several connected grafcets. A **partial grafcet** is formed by *one or more connected grafcets.*

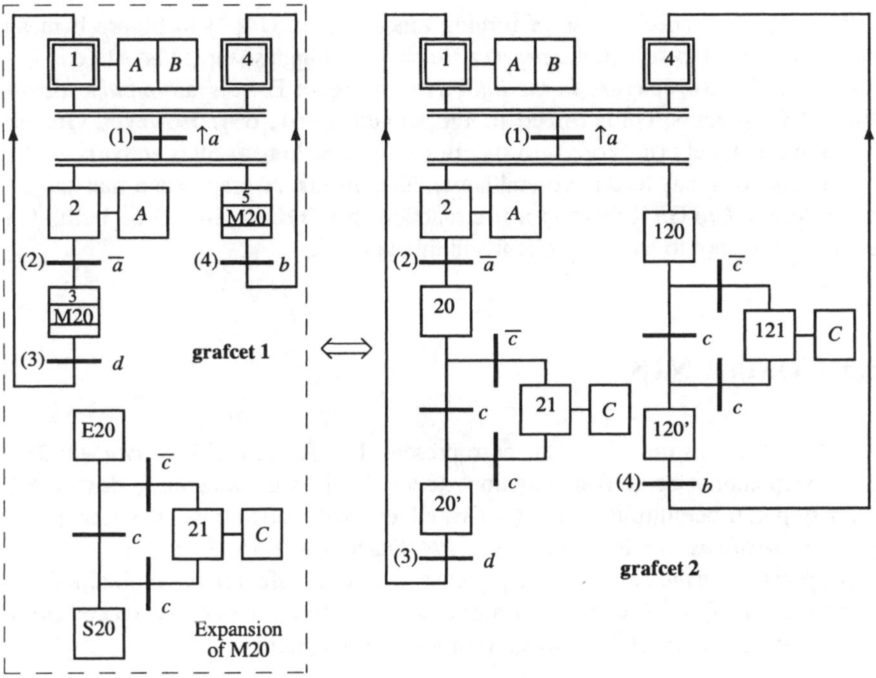

Figure E.3 Grafcet 1 contains two macrosteps (same expansion). The behavior of grafcet 2, without macrostep, is equivalent.

Let Gk and Gh be two *partial grafcets*. Grafcet Gk may modify the behavior of Gh by producing a **macroaction**: a macroaction is *homogeneous with an action* from the point of vue of Gk (level action or impulse action); it will be represented in a double rectangle instead of a simple one.

Figure E.4a represents a level macroaction **forcing**: when step 17 in grafcet G1 is active, G2 is forced to be in the situation {22, 24}. This means that steps 22 and 24 in G2 remain active (and only these) as long as step 17 is active. Afterwards, G2 can evolve as usual. Note that G2{} deactives grafcet G2.

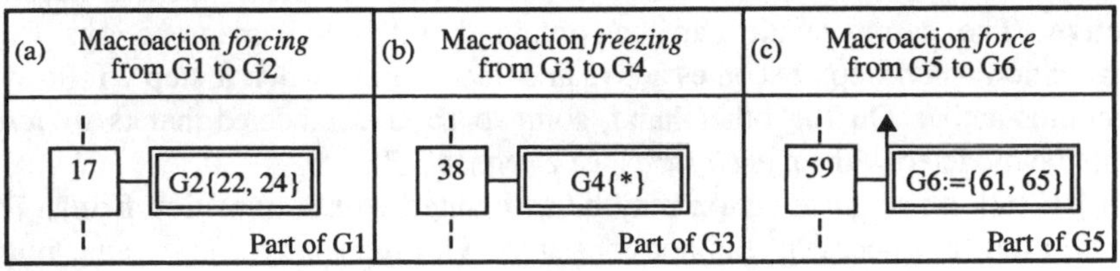

Figure E.4 Macroactions.

Freezing is a special case of forcing macroaction: G4{*} in Figure b means that G4 is forced to stay in the *present situation* as long as step 38 is active.

Macroaction **force** (*verb in the infinitive*) in Figure E.4c is an *impulse action.* When $\uparrow X59$ occurs, G6 is forced to the situation {61, 65}. However, G6 may evolve immediately (not forced to stay in {61, 65} as long as 59 is active).

Macroactions may lead to several hierarchical levels. A graph such that there is an arc from G*k* to G*h* if there is a macroaction from G*k* to G*h* can be built. It is recommended not to have any circuits in this graph.

E.3 COMMENTS

a) A grafcet is **sound** if: 1) the case presented in Remark E.1 is avoided; 2) an active step cannot be activated again (except if it is simultaneously deactivated); 3) no step can be simultaneously activated by two transitions. *A sound grafcet behaves exactly as a control interpreted PN* [DaDe 83a][DaAl 89 & 92].

Grafcets are usually sound. In any case, a sound grafcet may be obtained by a minor modification. Note that, in order to satisfy point 2 above, the receptivity of a source transition must be an event (not a Boolean condition).

b) A concluding remark in [IE 02] is that "At the present time, there is no rule to translate a GRAFCET specification into a SFC[3] program". Algorithm 3.3 in Section 3.3.2 could be a basis for an interpretation program. However, this algorithm is adapted to a *basic grafcet* (or control interpreted PN), i.e. *without abbreviations.* Hence, it should be modified to take the abbreviations into account (this suggests *not having too many authorized abbreviations*!).

c) The concept of *macrostep*, as well as the word, is now generally accepted [EI 02]. In this reference, the macroactions *forcing* and its special case *freezing* are also proposed (the word macroaction is not used). The authors believe that the *impulse* macroaction *force*, not in [IE 02], may be very useful: it allows initialization or deactivation of a partial grafcet (without making it necessary to stay in the corresponding situation for a certain duration, as *forcing* does).

The abbreviation *enclosure* in [IE 02] initializes one or more partial grafcets when *enclosing step i* becomes active and deactivates them when *i* becomes inactive. The same result can be obtained with two macroactions *force*: initialization when step *i* becomes active and deactivation when a step *j* following *i* becomes active. On the other hand, some behaviors modeled thanks to *force*, cannot be modeled with an *enclosure* (see Exercise 3.7).

In [IE 02], an impulse action may be associated with a transition firing. This abbreviation (not necessary) is not consistent with the *robust dichotomy*: *inputs associated with transitions* (in receptivities) and *actions associated with steps.*

[3] Sequential Function Chart.

Appendix F

Modeling Power of Synchronized PNs

The aim of this appendix is to state that synchronized PNs with single-server semantics [MoPuSi 78] and with infinite-server semantics ([Le 92] and this book) have the same modeling power. In other words, the behavior modeled by one of these models can be modeled by the other. For this purpose, let us specify the meaning of two equivalent behaviors.

Let us first introduce the concept of *significant marking*. In Figure F.1a, the markings of P_1, P_2, P_3, and P_4, may change, while P_5 contains always one token. The **significant marking** is the marking restricted to the places which *do not have the same marking in all the stable states*. In both Figures F.1a and b, the significant marking is the 4-component vector $\mathbf{m} = (m(P_1), m(P_2), m(P_3), m(P_4))$. Let us say that two models are *equivalent* if they have the same initial significant marking and, for any reachable *stable* marking, *any occurrence of an external event* (or set of events) leads to the *same stable significant marking*.

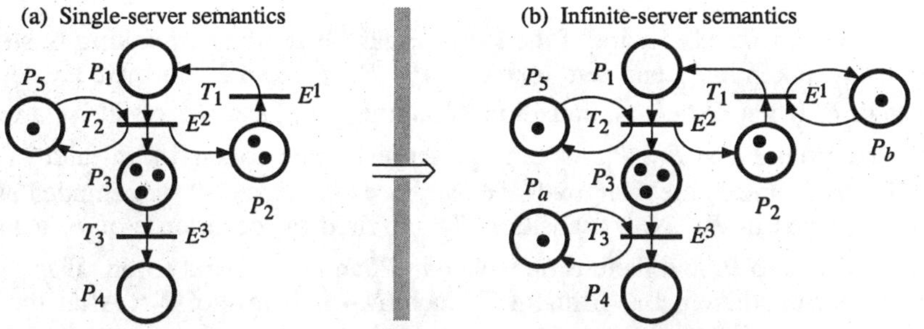

Figure F.1 From single-server semantics to infinite-server semantics.

According to Figure 3.1 in Section 3.1, the behavior of the PN in Figure F.1a with a single-server semantics is modeled by the PN in Figure b with an infinite-server semantics: a self loop with one token is associated with every transition.

Consider now the PN in Figure F.2b (*similar* to Figure F.1a in which P_5 was redundant). Assume an infinite-server semantics, and that events E^2 and E^3 are compatible. How can we model this behavior with a single-server semantics? A solution is given in Figure F.2a, where event e is associated with all the transitions without explicit synchronization. The enabling degrees of T_1 and T_2 cannot be more than 2 (invariant $m_1 + m_2 = 2$) and 1 ($m_5 = 1$), respectively. These transitions, corresponding to a 2-server and an 1-server are easily modeled (Figure F.2a). But the enabling degree of T_3 is *unbounded*: it corresponds to an *infinite-server* since place P_3 must be emptied if E^3 occurs.

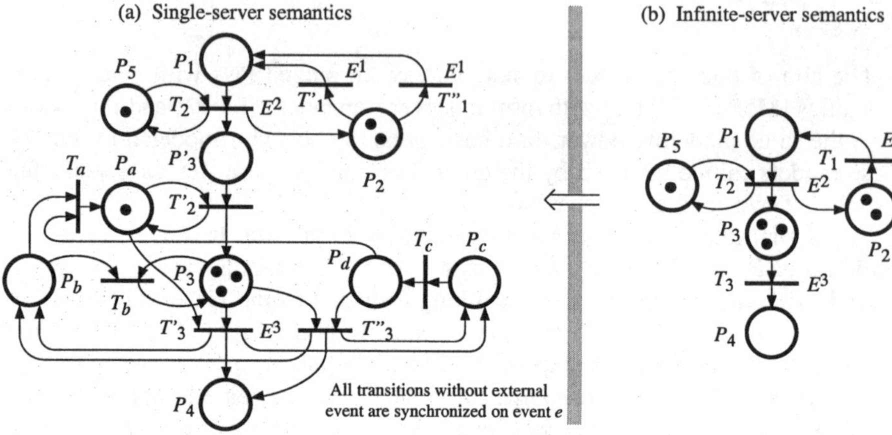

Figure F.2 From infinite-server semantics to single-server semantics.

Let us explain the behavior of the PN in Figure F.2a. If the marking is stable, there is one token in P_a and zero token in P_3', P_b, P_c, and P_d. If there are several tokens in P_3 when E^3 occurs, transition T_3' is fired once, and the result is: a token is passed from P_3 to P_4; P_a is empty; there is one token in P_b and in P_c. Transitions T_b and T_c are then fired on occurrence of *e*: P_b and P_c are emptied and a token is placed in P_d. Now transition T_3'' is fired on occurrence of *e*: a token passes from P_3 to P_4 and there is one token in P_b and in P_c. And so on. If $m_3 = N$ when E^3 occurs, there is one firing of T_3' then $(N-1)$ firings of T_3''. After the last firing of T_3'', there is one token in P_b and in P_c. Transition T_b is not fired because P_3 is empy but T_c is fired. There is now a token in P_b and in P_d: transition T_a is fired and a stable marking is reached.

Places P_3' and P_a and transition T_2' are necessary only because E^2 and E^3 are compatible (ensuring that a token placed in P_3 on firing of E^2 is not immediately taken out). If E^2 and E^3 are not compatible, there is an arrow from T_2 to P_3, and T_a is a sink transition. If there were another transition in actual conflict with T_3, another solution would be required in order to preserve all possible resolutions [Mo 01]. The idea, proposed by M. Moalla, is always the same: there is a kind of priority of T_b over T_a because the token arriving in P_d is delayed by T_c.

Appendix G

Timed Petri Nets Are Special Cases of Synchronized Petri Nets

Sections G.1 and G.2 of this appendix explain that a timed PN can be considered as a special case of synchronized PN. This is true for P-timed and T-timed PNs at *maximal speed*, and for stochastic PNs (any distribution). The aim is not to propose a new way for modeling timed PNs (it would be more complicated!) but to allow the timed PNs to inherit all the properties of synchronized PNs presented in Section 3.2.

In Section G.3, it is shown that a synchronized PN can be considered as a special case of timed PN. This observation leads to the new concept of synchronized C-transition, hence of synchronized continuous PN.

G.1 T-TIMED PETRI NETS

Let us first consider three cases, illustrated by Figure G.1, for a transition with a single input place[1].

Case 1: assume that the *maximal enabling degree* of T_2 (Figure a) is *one* (i.e., $m(P_1) \leq 1$ for any reachable marking). The synchronized PN in Figure b has the same behavior. Note that T_2 is synchronized by a *condition*: [*a token has been present in* P_1 *for at least* d_2]. According to Section 3.3.3, a transition may be synchronized by a condition instead of an event: formally the firing occurs on event *e* if the condition is true (in this case, according to Remark 3.9b, T_2 could be synchronized by the event '*a token has been present in* P_1 *for exactly* d_2').

Case 2: the *enabling degree may be greater than one* (Figure c) and the tokens are *deposited in* P_3 at t_1 *and* t_2, $t_2 < t_1 + d_4$. Firings of T_4 occur at $t_1 + d_4$ and at $t_2 + d_4$. In Figure d, the behavior is similar. Condition C_4 becomes true at $t_1 + d_4$, hence T_4 is fired; the first token leaves P_3 but P_3 is not emptied since T_4 was only 1-enabled because of place P_4'. After the first firing, condition C_4 = [*a token has been present in* P_3 *for at least* d_4] is no longer true since the remaining token has been present for only $t_2 - t_1$. Condition C_4 will become true again at $t_2 + d_4$.

[1] According to Section 3.4.2.1, similar behaviors could be shown for P-timed PNs. For the cases in Figure G.1, this is quite obvious.

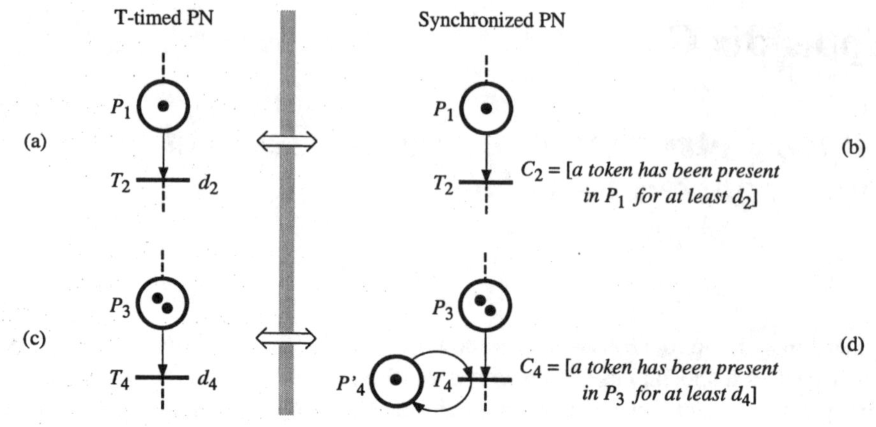

Figure G.1 (a) and (b) The maximal enabling degree of T_2 is one. (c) and (d) The enabling degree of T_4 is not bounded.

Case 3: the *enabling degree may be greater than one* (Figure c) and the tokens are *deposited simultaneously* at t_1. Condition C_4 becomes true at $t_1 + d_4$: a single firing of T_4 is performed, then a second firing is performed since C_4 continues to be true after the first one. This behavior corresponds to an iterated firing T_4T_4 (similar to firing from (0, 2) to (2, 0) in Figure 3.22, Section 3.3.3). Note that in this case, condition C_4 cannot be replaced by an event (whereas it is possible for Cases 1 and 2) because an event implies a *single* firing of the 1-enabled transition.

Remark G.1 If both tokens are deposited at the same time t_1 in Figure c, the T-timed model specifies that transition T_4 is fired twice at $t_1 + d_4$. However, it does not specify whether they are simultaneously fired ($S_1 = [T_4T_4]$) or iteratively fired ($S_2 = T_4T_4$); hence, both cases are admissible. For the synchronized PN in Figure d (which works for any number of tokens, arriving at any time), $S_2 = T_4T_4$ is performed. If we were sure that the tokens in P_3 would *always* be deposited at the same time (i.e., P_3 is always empty when one or several tokens are deposited in it), additional self-loop place P_4' would not be necessary; in this case, the double firing $S_1 = [T_4T_4]$ would be performed at $t_1 + d_4$.

❑

Figure G.2 illustrates the general case where the considered transition may have more than one input place.

The condition C_7 = [*a token has been present in* P_5 *and a token has been present in* P_6 *for at least* d_7] is simply a generalization of condition C_4 in Figure G.1d. When this condition becomes true, a firing of T_7 is performed; then, either a second firing of T_7 is performed if the condition remains true or one waits for the time necessary for the condition to become true again.

Naturally, if the weights of arcs $P_5 \rightarrow T_7$ and $P_6 \rightarrow T_7$ were respectively r and s, the condition would be C_7 = [*r tokens have been present in* P_5 *and s tokens have been present in* P_6 *for at least* d_7].

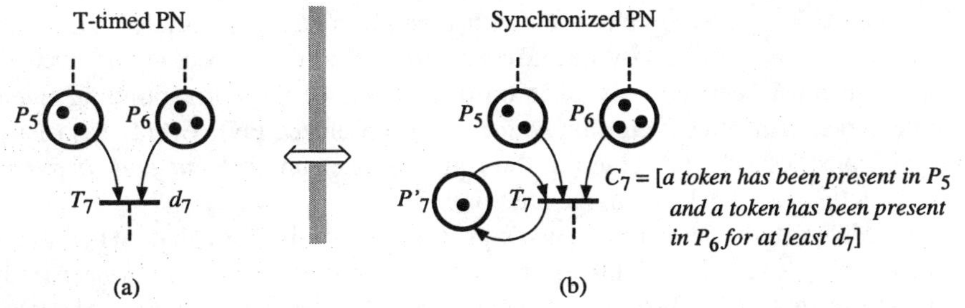

Figure G.2 Several input places for the transition under consideration.

G.2 STOCHASTIC PETRI NETS

We shall consider in turn stochastic PNs and generalized stochastic PNs.

A stochastic PN exhibits an interesting property: since continuous time is considered, the probability that two firings occur at the same time is zero (Section 3.4.3.1); then, *all the transitions may be synchronized on events* (while synchronization by a condition was necessary for a T-timed PN). In order to avoid a complicated abstract expression of the event, it is sufficient to consider simply the "physical event" corresponding to the system modeled. Let us illustrate this idea with the system specified in Section 3.4.3.1 and modeled by the stochastic PN in Figure 3.30a. The corresponding synchronized PN is given in Figure G.3.

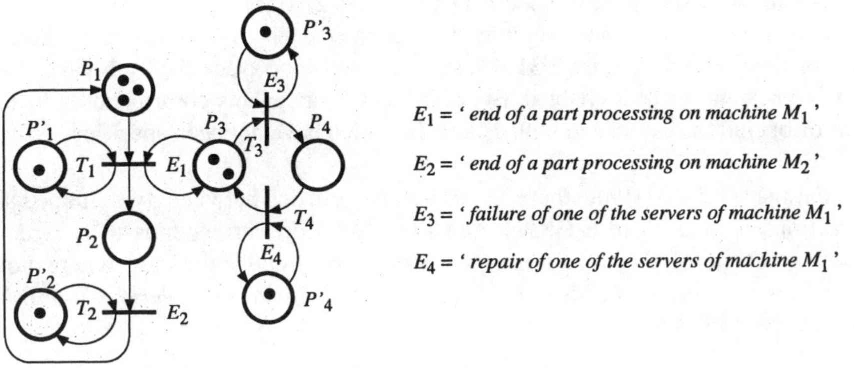

Figure G.3 Synchronized PN corresponding to the stochastic PN in Figure 3.30.

In Figure G.3, a self-loop place P_j' with a single token is associated with each transition T_j. This is necessary because every transition may become at least 2-enabled but must be fired only once on occurrence of the corresponding event (infinite-server semantics is considered for the synchronized PN). Events associated with the transitions are presented in the figure. For example E_1 = *'end of a part processing on machine M_1'* is associated with T_1.

Be careful with the semantics. The single token in P_1' implies that, when event E_1 occurs, only one token is taken out of P_1 (the processing of *only one part* is completed at this time). It does not mean that only one part is being processed when there are three tokens in P_1 and two tokens in P_3; this information is not signified by the PN in Figure G.3 which is *synchronized but not timed.*

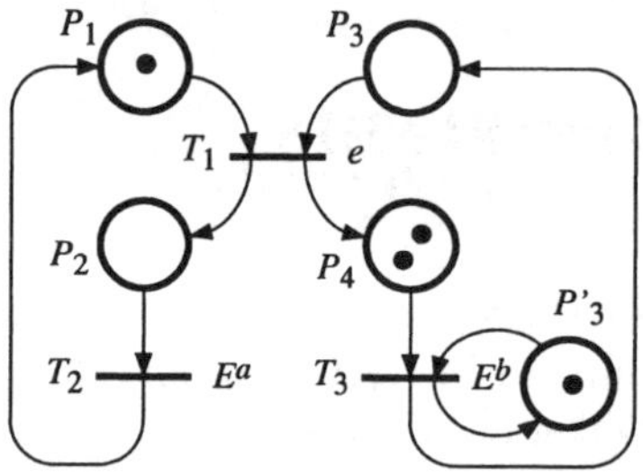

Figure G.4 Synchronized PN corresponding to the generalized stochastic PN in Figure 3.31.

For a generalized stochastic PN, the corresponding synchronized PN is obtained as for a basic stochastic PN, except that an *immediate transition*: 1) is synchronized by the "always occurring event" e; 2) does not require a self-loop place P_j' (even if this transition may be q-enabled). This is illustrated in Figure G.4 for the GSPN in Figure 3.31. There is no self-loop place for T_2 because this transition can never be 2-enabled. Events E^a and E^b are events corresponding to the ends of operations associated with T_2 and T_3, which have not been specified.

Remark G.2 Assume there is an actual conflict between two immediate transitions T_1 and T_2 in a GSPN, and the resolution corresponds to a random drawing of one of the transitions (we assume implicitely the case where both transitions are 1-enabled). Then, $E_1 = e$, $E_2 = e$, C_1 = [*firing of T_1 has been drawn*], C_2 = [*firing of T_2 has been drawn*].

Property G.1 A (basic) stochastic PN is totally synchronized and each external *event* is associated with a *single transition.* Hence, from Section 3.2.3.3, it has all the properties (bounded, live, etc) of the corresponding autonomous PN.

G.3 SYNCHRONIZED PNs ARE SPECIAL CASES OF TIMED PNs

A T-timed PN is usually a PN in which a constant timing is associated with every transition (Section 3.4.2.2). However, according to Remark 3.15, the timing associated with a transition may change with time or with a function of time. Accordingly, a synchronized transition (Sections 3.2 and 3.3.3) behaves as a timed transition, as illustrated in Figure G.5.

Transition T_2 in Figure G.5a is synchronized on condition C_2. If T_2 is *q*-enabled (it is 2-enabled in the figure), it will be fired *q* times as soon as C_2 becomes 1. When $C_2 = 1$, T_2 cannot remain enabled since it is immediately fired. This behavior corresponds to the timed PN in Figure b: the timing is $d_2 = 0$ when $C_2 = 1$ (immediate firing) and $d_2 = \infty$ when $C_2 = 0$ (no firing).

Transition T_4 in Figure c is synchronized on event E_4. If T_4 is *q*-enabled when E_4 occurs, it is fired *q* times at this instant. This behavior corresponds to the timed PN in Figure d: the timing is $d_4 = 0$ when event E_4 occurs (immediate firing) and $d_4 = \infty$ otherwise (no firing as long as E_4 does not occur).

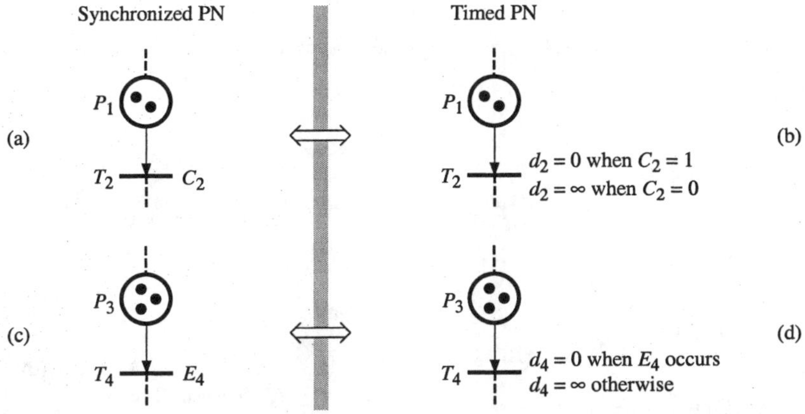

Figure G.5 A synchronized PN is a special case of timed PN.

The concept of *synchronized continuous PN* was evoked in Remark 5.6 (Section 5.1.2.5). The relation between a T-timed discrete PN and its continuous counterpart is clearly established in Section 5.1: with a D-transition with a zero timing is associated a C-transition with an infinite speed, and vice-versa. It follows that, on the basis of the correspondence between synchronization and timing illustrated in Figure G.5, the concept of **synchronized C-transition**, illustrated in Figure G.6, is obtained.

Transition T_6 in Figure G.6a is synchronized on condition C_6. If T_6 is q-enabled (it is 9.2-enabled in the figure), it will be fired as soon as C_6 becomes 1, quantity q of firing. When $C_6 = 1$, T_6 cannot remain enabled since it is immediately fired. This behavior corresponds to the timed continuous PN in Figure b: the maximal speed is $V_6 = \infty$ when $C_6 = 1$ (immediate transition) and $V_6 = 0$ when $C_6 = 0$ (no firing).

Transition T_8 in Figure c is synchronized on event E_8. If T_8 is q-enabled when E_8 occurs, its quantity of firing is q at this instant. This behavior corresponds to the timed PN in Figure d: the maximal speed is $V_6 = \infty$ when event E_8 occurs (immediate transition) and $V_6 = 0$ otherwise (no firing as long as E_8 does not occur).

From this new concept of synchronized C-transition, a **continuous synchronized PN** can be defined. If all the transitions are synchronized, the firings are not continuous: quantities of firing (not integer in the general case) occur at discrete times. It is also possible to have synchronized and timed transitions (i.e. with maximal speeds) in the same model. See Exercise 5.4.

In the Solution to Exercise 8.3, synchronized C-transitions will be used. They allow to empty and/or to assign some real-number markings to C-places (instantaneously).

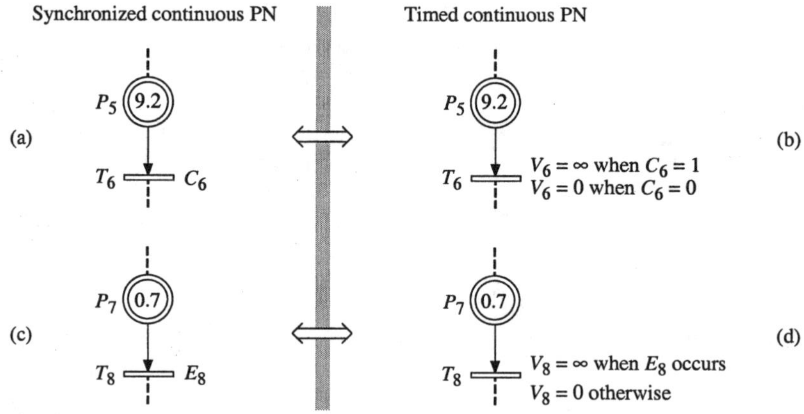

Figure G.6 Illustration of the concept of synchronized continuous PN.

Appendix H

Time Petri Nets

While in a timed PN, a duration, either deterministic or random, is associated with each transition or each place, in a **time PN**, a *time interval* $[a, b]$ is associated with each transition or each place. The models will be called T-time PN and P-time PN, respectively.

H.1 T-TIME PETRI NETS

T-time PNs were introduced by P. M. Merlin [Me 76][MeFa 76]. This model have been used by various authors. The definition given here is based on [BeDi 91]. All these authors call this model time PN or TPN.

As expressed by the title of the first papers, this model was defined in order to model the behavior of communication protocols. It is adapted to analysis of the *time behavior of existing systems.*

Definition H.1 A **T-time PN** is a pair $\langle R, \text{Inter}\rangle$ such that:

R is a marked PN;

Inter is a function from the set T of transitions to the set of pairs (a, b), where[1] $a \in Q_+$, $b \in Q_+ \cup \{\infty\}$, with $a \leq b$. Inter$(T_j) = [a, b]$ = firing time interval associated with T_j.

Operating of a T-time PN. Let Inter$(T_1) = [a_1, b_1]$. Assume first that the maximal enabling degree of T_1 is one. If T_1 becomes enabled at t, then T_1 will be *certainly fired* in the interval $[t + a_1, t + b_1]$, except if T_1 is disabled by the firing of another transition during this time interval. This is the meaning of the model: it describes an *existing system* such that, if T_1 remains enabled, it will certainly be fired between the *earliest firing time* $t + a_1$ and is the *latest firing time* $t + b_1$.

Assume now that T_1 becomes 2-enabled at t_i, $t \leq t_i < t + b_1$; it will be fired once more in the interval $[t_i + a_1, t_i + b_1]$. And so on. Now, if the enabling degree of T_1 decreases, because of the firing of another transition, the most recent enablings are suppressed while the oldest enabling are kept (other interpretations seem possible concerning the firing rules [BeDi 91]).

[1] The set of positive or zero rational numbers is denoted by Q_+.

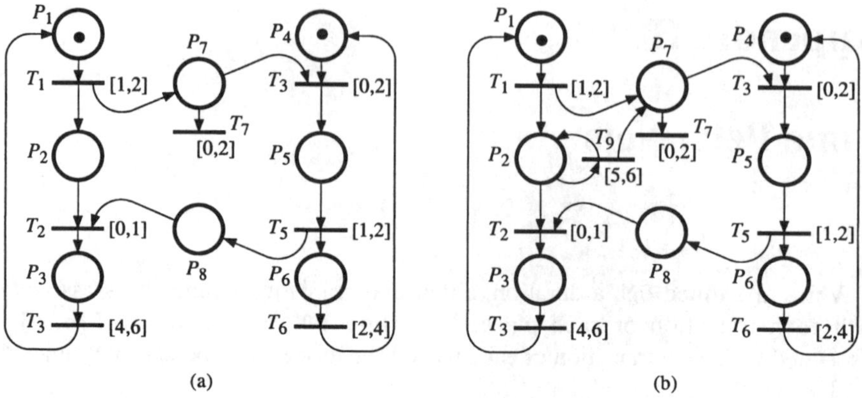

Figure H.1 T-time PNs. (a) Communication protocol. (b) Recoverable system.

Let us now discuss the conflict concept, restricting our attention to the case of enabling degree not greater than one. If the structural conflict $K = \langle P_i, \{T_1, T_2\}\rangle$ exists, a marking **m** such that the effective conflict $K^E = \langle P_i, \{T_1, T_2\}, \mathbf{m}\rangle$ exists may occur. Then, existence of an actual conflict depends on the times when T_1 and T_2 become enabled and on their firing intervals $[a_1, b_1]$ and $[a_2, b_2]$. Assume both transitions become enabled at t when a token is deposited in P_i (and no other transition is enabled). If $a_1, a_2 \leq b_1 < b_2$, transition T_1 can be fired in $[t + a_1, t + b_1]$ and T_2 can be fired in $[t + a_2, t + b_1]$ (since T_2 would be disabled by the firing of T_1 at $t + b_1$ at the latest); then an actual conflict exists in the interval $[t + \max(a_1, a_2), t + b_1]$. Now, there is no actual conflict if $b_1 < a_2$ since the earliest firing time of T_2 is greater than the latest firing time of T_1 (but T_2 is not necessarily redundant if it may happen that T_2 is enabled while T_1 is not).

Let us illustrate this model with the example presented in Figure H.1a (drawn from [MeFa 76]). System A (whose states are modeled by places P_1, P_2, and P_3) sends messages to system B (whose states are modeled by places P_4, P_5, and P_6). P_1: system A is ready to send a message. P_2: system A is waiting for acknowledge the message. P_3: system A prepares a new message. P_4: system B is ready to receive a message. P_5: system B has received the message. P_6: system B prepares for receiving a new message. P_7: system A has sent a message, which has not yet been read. P_8: system B has sent acknowledge the message.

A time interval is associated with every transition. For example, interval $[1, 2]$ is associated with T_1. This means that this transition will be fired between one and two time units after becoming enabled. It is assumed that a sent message may be lost; this is represented by T_7 whose firing empties P_7. Let t denote the firing time of T_1. There is an actual conflict between T_3 and T_7 during interval $[t, t + 2]$. If T_3 is fired (message received), a token will be deposited in P_8 at the latest at $t + 4$ (after firings of T_3 and T_5). Hence, if T_2 is not yet enabled at $t + 4$, it means that the message was lost.

The model may be modified as illustrated in Figure b for having a recoverable system. If there is a token in P_2 after $t+4$, we know that the message was lost. Hence, the message may be sent again. This is represented by T_9: its firing deposits a token in P_7, meaning that a new message have been sent. Note that an actual conflict between T_2 and T_9 cannot occur.

H.2 P-TIME PETRI NETS

P-time PNs were introduced by J.-P. Denat and co-workers [KhDeCo 96] [Kh 97]. This model was defined in order to specify *time constraint systems*: the sojourn in some state *must have a value between a minimum value and a maximum value*. Hoist scheduling problems [PhHu 76][ShNu 89] appear in various domains such as electroplating, thermic treatements, dispatching of freshs goods.

Definition H.2 A **P-time PN** is defined as a T-time PN, except that Inter(P_i) = $[a, b]$ = sojourn time interval associated with place P_i.

Operating of a P-time PN. Let Inter(P_1) = $[a_1, b_1]$. It means that a token deposited in P_1 at t, *must be taken out* (by firing an output transition) between the *earliest leaving time* $t + a_1$ and the *latest leaving time* $t + b_1$. Before $t + a_1$, the token is unavailable; after $t + b_1$, the *token is dead*. A dead token corresponds to a wasted object. The model specifies a *scheduling problem* such that dead tokens should be avoided.

The process in Figure H.2a is as follows. Parts are transfered from the input buffer (never empty) to Vat 1, then to Vat 2, then to the output buffer (never full). All the transfer operations are performed by a single robot; each transfer lasts 2 time units (no additional time required between two transfers). The parts must stay between 5 and 8 time units in Vat 1, and between 4 and 5 time units in Vat 2. Each vat cannot contain more than 4 parts. This behavior is specified by the P-time PN in Figure b (the state represented is such that there are 3 parts in Vat 1 and 2 parts in Vat 2). If no time constraint is assigned to some place, the corresponding sojourn time interval is $[0, \infty]$; for our example, Inter(P_6) = Inter(P_7) = Inter(P_8) = $[0, \infty]$.

Here is an example of problem. Given the initial marking $\mathbf{m}_0 = (0, 0, 0, 0, 0, 1, 4, 4)$ find a schedule such that no part stays in a vat more than its maximum allowed time (maximizing if possible the throughput). A solution may start as follows: transfer 3 parts in vat 1 (6 time units), wait 1 time unit, transfer the three parts from vat 1 to vat 2 (6 time units), etc. After $t = 13$, several choices are possible. The reader may verify that if the beginning were the transfer of 4 parts in vat 1 (possible according the maximun capacity of the vat), there is no way to avoid a dead token at $t = 17$.

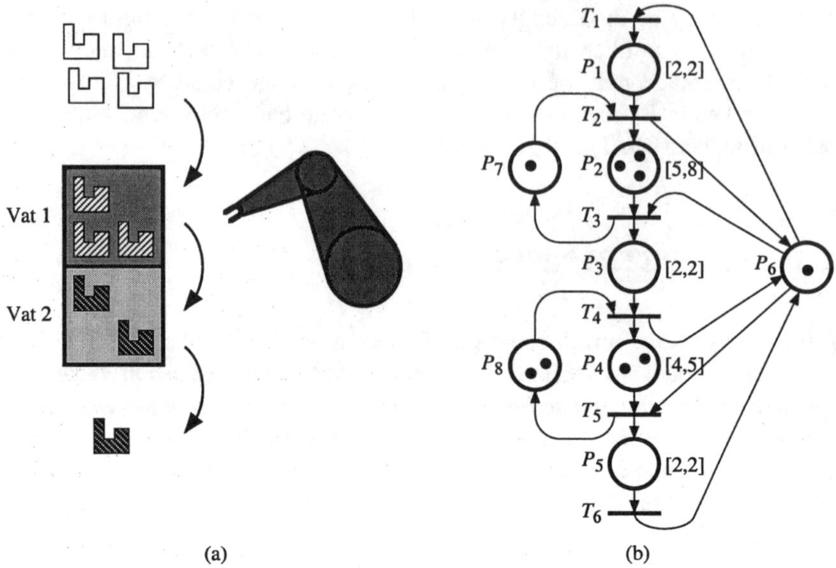

Figure H.2 (a) Process. (b) P-time PN.

Remark H.1 **Comparison of both models**

In general, it is not possible to change-over from a T-time PN to a P-time PN, and vice-versa [Kh 97]. Changing over is only possible for *special cases or using extensions* (priority, inhibitor arc, etc). A *P-time state graph* is easily changed-over to a T-time *state graph* [Kh 97]. A *T-time state graph* may be changed-over to a P-time PN [De 00]. A *P-time event graph* may be almost changed-over to a T-time PN (not exactly equivalent but may be used in practice) [De 00]. A *T-time event graph* may be changed-over to a P-time event graph with the *additional concept* of immediate transition [Kh 97].

A system modeled by a P-timed PN (respectively T-timed PN) is included in the class of systems which can be modeled by P-time PNs (respectively T-time PNs): timing d_i may be replaced by interval $[d_i, \infty]$.

❑

In [Ha 93], a time interval $[d_{ij1}, d_{ij2}]$ is associated with every arc $P_i \rightarrow T_j$.

Appendix I

Linearity of the Fundamental Equation for Continuous Petri Nets

Let us first define two operations related to firing sequences in a continuous PN. These operations are called *splitting* of a firing sequence and *interleaving* between two firing sequences. Then, on the basis of these operations, a function whose linearity is shown is presented.

I.1 Splitting and Interleaving

Consider first the **splitting** operation. Let $[U_1]$ be an OG-firing whose characteristic vector is $\mathbf{u}_1$. *Splitting* of $[U_1]$ consits of replacing $[U_1]$ by a sequence of OG-firings, denoted by $S^\#(U_1)$, having the same characteristic vector, i.e., $\mathbf{s}^\#(U_1) = \mathbf{u}_1$:

$$S^\#(U_1) = [U_{11}][U_{12}]\ldots[U_{1a}] \text{ such that } \mathbf{u}_{11} + \mathbf{u}_{12} + \ldots + \mathbf{u}_{1a} = \mathbf{u}_1,\ a \geq 1. \qquad \text{(I.1)}$$

It is clear that, if $[U_1]$ can be fired from a marking $\mathbf{m}$, $S^\#(U_1)$ can also be fired from this marking because $\mathbf{m}$ is sufficient to enable all the transitions according to their respective components in $\mathbf{u}_1$. In other words, $\mathbf{m} \xrightarrow{[U_1]}$ implies $\mathbf{m} \xrightarrow{[U_{11}][U_{12}]\ldots[U_{1a}]}$ and furthermore the marking reached is the same (same characteristic vector).

An OG-firing can be split into an infinite number of different sequences. For example $[U_i] = [(T_1)^{1.5}(T_2)^{0.6}]$ may be split into $S_1^\#(U_i) = [(T_1)^{1.5}][(T_2)^{0.6}]$ or $S_2^\#(U_i) = [(T_2)^{0.2}][(T_1)^{1.5}][(T_2)^{0.4}]$ or $S_3^\#(U_i) = [(T_1)^{0.5}(T_2)^{0.6}][(T_1)^1]$.

Splitting of a sequence $S = [U_1][U_2]\ldots[U_k]$ consists of splitting every OG-firing in S according to (I.1):

$$S^\#(S) = S^\#([U_1][U_2]\ldots[U_k]) = S^\#(U_1)S^\#(U_2)\ldots S^\#(U_k). \qquad \text{(I.2)}$$

As for an OG-firing, a firing sequence can be split into an infinite number of different sequences, and all of them allow to reach the same marking, i.e.,

$$\text{if } \mathbf{m} \xrightarrow{S} \mathbf{m}', \text{ then } \mathbf{m} \xrightarrow{S^\#(S)} \mathbf{m}' \text{ for any } S^\#(S). \qquad \text{(I.3)}$$

Consider now the **interleaving** operation between two firing sequences. Let $S_1 = [U_{1,1}][U_{1,2}]...[U_{1,k}]$ and $S_2 = [U_{2,1}][U_{2,2}]...[U_{2,l}]$ be two firing sequences such that $\mathbf{m} \xrightarrow{S_1}$ and $\mathbf{m} \xrightarrow{S_2}$. An interleaving between S_1 and S_2 consists of:

1) cutting both S_1 and S_2 into N subsequences: $S_1 = S_{11}S_{12}...S_{1N}$ and $S_2 = S_{21}S_{22}...S_{2N}$, such that $1 \leq N \leq \min(k, l) + 1$ and all the subsequences are different from the null sequence ε except, possibly, S_{11} and S_{2N};

2) interleaving the subsequences for obtaining the sequence

$$S_{11}S_{21}S_{12}S_{22}...S_{1N}S_{2N}. \tag{I.4}$$

Note that the order of the subsequences is preserved. For example, in (I.4), S_{12} appears after S_{11}, S_{13} after S_{12}, and so on.

Here are some examples of interleaving between $S_1 = [U_{1,1}][U_{1,2}]$ and $S_2 = [U_{2,1}][U_{2,2}][U_{2,3}]$ (hence $1 \leq N \leq 3$ since $\min(k, l) + 1 = 3$).

Example 1: $N = 1$; $S_{11} = [U_{1,1}][U_{1,2}]$, $S_{21} = [U_{2,1}][U_{2,2}][U_{2,3}]$, the sequence obtained is $S_a = [U_{1,1}][U_{1,2}][U_{2,1}][U_{2,2}][U_{2,3}]$.

Example 2: $N = 2$; $S_{11} = [U_{1,1}]$, $S_{12} = [U_{1,2}]$, $S_{21} = [U_{2,1}][U_{2,2}]$, $S_{22} = [U_{2,3}]$, the sequence obtained is $S_b = [U_{1,1}][U_{2,1}][U_{2,2}][U_{1,2}][U_{2,3}]$.

Example 3: $N = 2$; $S_{11} = [U_{1,1}]$, $S_{12} = [U_{1,2}]$, $S_{21} = [U_{2,1}][U_{2,2}][U_{2,3}]$, $S_{22} = \varepsilon$, the sequence obtained is $S_c = [U_{1,1}][U_{2,1}][U_{2,2}][U_{2,3}][U_{1,2}]$.

Example 4: $N = 3$; $S_{11} = \varepsilon$, $S_{12} = [U_{1,1}]$, $S_{13} = [U_{1,2}]$, $S_{21} = [U_{2,1}]$, $S_{22} = [U_{2,2}]$ $S_{23} = [U_{2,3}]$, the sequence obtained is $S_d = [U_{2,1}][U_{1,1}][U_{2,2}][U_{1,2}][U_{2,3}]$.

Remark I.1 The *concatenation* is a particular case of interleaving: S_1S_2 is obtained for $N = 1$ (Example 1); S_2S_1 is obtained for $N = 2$ with $S_{11} = S_{22} = \varepsilon$.

Remark I.2 Even if S_1 and S_2 are firing sequences, the sequence (I.4) obtained by interleaving of S_1 and S_2 is *not always a firing sequence*. This appendix analyses somes properties related to this possibility.

I.2 Linearity of the Fundamental Equation

Let $\mathbf{m}_0$ denote the initial marking of a continuous PN. According to Remark 4.1 in Section 4.1.3.2, $\mathcal{L}(\mathbf{m}_0)$ denotes the set of firing sequences which are possible from $\mathbf{m}_0$. In other words, $\mathcal{L}(\mathbf{m}_0)$ is the set of sequences S such that $\mathbf{m}_0 \xrightarrow{S}$. According to the fundamental equation (4.5) in Section 4.1.3.2, for any firing sequence S in $\mathcal{L}(\mathbf{m}_0)$, the marking $\mathbf{m}$ reached by firing S from $\mathbf{m}_0$ is given by:

$$\mathbf{m} = \mathbf{m}_0 + \mathbf{W} \cdot \mathbf{s}, \tag{I.5}$$

where $\mathbf{s}$ is the characteristic vector of S.

Notation I.1 Let $\Delta\mathbf{m} = \mathbf{m} - \mathbf{m}_0$. (I.6)

❑

Hence, *given the PN* (which implies $\mathbf{W}$) *and the initial marking* ($\mathbf{m}_0$), the reached marking may be characterized by $\Delta\mathbf{m} = \mathbf{W} \cdot \mathbf{s}$ which is a *function* of the firing sequence S:

$$f(S) = \Delta\mathbf{m}. \quad (I.7)$$

If αS (see Notation 4.2b in Section 4.1.3.2) is in $\mathcal{L}(\mathbf{m}_0)$, then

$$f(\alpha S) = \alpha\Delta\mathbf{m} \quad (I.8)$$

is obtained from (I.5), (I.6) and (I.7) since the characteristic vector of αS is $\alpha\,\mathbf{s}$. According to Property 4.3c in Section 4.1.3.2:

$$\text{if } 0 \leq \alpha \leq 1, \text{ then } \alpha S \in \mathcal{L}(\mathbf{m}_0). \quad (I.9)$$

Let us now define a "sum" of firing sequences, using the notation $\oplus$ in order to avoid a confusion with an union of sequences.

Notation I.2 Let S_1 and S_2 be two sequences in $\mathcal{L}(\mathbf{m}_0)$. The sum $S_1 \oplus S_2$ denotes a sequence S_i obtained by the following process.

1) The sequences S_1 and S_2 are *split* into sequences $S^{\#}(S_1)$ and $S^{\#}(S_2)$.

2) The sequence S_i is obtained by *interleaving* of $S^{\#}(S_1)$ and $S^{\#}(S_2)$.

Property I.1 Given $f(S_1) = \Delta\mathbf{m}_1$, $f(S_2) = \Delta\mathbf{m}_2$, and $\alpha,\ \beta \geq 0$, the following properties are true.

a) If $\alpha S_1 \oplus \beta S_2 \in \mathcal{L}(\mathbf{m}_0)$, then

$$f(\alpha S_1 \oplus \beta S_2) = \alpha\Delta\mathbf{m}_1 + \beta\Delta\mathbf{m}_2. \quad (I.10)$$

b) If $0 \leq \alpha + \beta \leq 1$, then

$$\alpha S_1 \oplus \beta S_2 \in \mathcal{L}(\mathbf{m}_0). \quad (I.11)$$

Proof

a) From the definitions, the characteristic vector of $S_i = \alpha S_1 \oplus \beta S_2$ is $\alpha\,\mathbf{s}_1 + \beta\mathbf{s}_2$. The marking reached by the firing of S_i is $\mathbf{m} = \mathbf{m}_0 + \mathbf{W} \cdot (\alpha\,\mathbf{s}_1 + \beta\mathbf{s}_2)$. The property follows.

b) Let us first consider the case $\alpha + \beta = 1$. From (I.2), (I.4) and Notation I.2, the sequence $S_i = \alpha S_1 \oplus \beta S_2$ may be writen as

$$S_i = \alpha S_{11}\beta S_{21}\alpha S_{12}\beta S_{22}\ldots\,\alpha S_{1N}\beta S_{2N}, \quad (I.12)$$

such that $S_{11}S_{12}\ldots S_{1N} = S^{\#}(S_1)$ and $S_{21}S_{22}\ldots S_{2N} = S^{\#}(S_2)$. Let $\mathbf{m}_{ab}$, $a \in \{1, 2\}$ and $b \in \{1, 2, ..., N\}$, denote the marking reached by the firing sequence $S_{a1}S_{a2}\ldots S_{ab}$ from $\mathbf{m}_0$.

According to Property 4.3b (and given S_1 and S_2 can be fired from $\mathbf{m}_0$), $\alpha\mathbf{m}_0 \xrightarrow{\alpha S_{11}\alpha S_{12}\ldots\alpha S_{1N}}$ and $\beta\mathbf{m}_0 \xrightarrow{\beta S_{21}\beta S_{22}\ldots\beta S_{2N}}$. Hence,

$$\alpha\mathbf{m}_0 \xrightarrow{\alpha S_{11}} \alpha\mathbf{m}_{11} \xrightarrow{\alpha S_{12}} \alpha\mathbf{m}_{12} \ldots \xrightarrow{\alpha S_{1N}} \alpha\mathbf{m}_{1N}, \qquad (I.13)$$

and similarly for the firing of S_2 from $\beta\mathbf{m}_0$.

Let us divide $\mathbf{m}_0$ into two parts: $\mathbf{m}^{(1)} = \alpha\,\mathbf{m}_0$ and $\mathbf{m}^{(2)} = \beta\mathbf{m}_0$ (then $\mathbf{m}^{(1)} + \mathbf{m}^{(2)} = \mathbf{m}_0$).

Assume first that marks of $\mathbf{m}^{(2)}$ are "*frozen*", i.e., *these marks do not move.* According to (I.13), the firing sequence αS_{11} can be performed from the marking $\mathbf{m}^{(1)} = \alpha\mathbf{m}_0$, i.e., $\alpha\mathbf{m}_0 \xrightarrow{\alpha S_{11}} \alpha\mathbf{m}_{11}$ (unfrozen marking).

Assume now that, after this firing of αS_{11}, the marks of $\alpha\mathbf{m}_{11}$ are frozen whereas the marks in $\mathbf{m}^{(2)} = \beta\mathbf{m}_0$ are no longer frozen. According to Property 4.3b, the firing sequence βS_{21} can be performed from the unfrozen marking $\mathbf{m}^{(2)} = \beta\mathbf{m}_0$, i.e., $\beta\mathbf{m}_0 \xrightarrow{\beta S_{21}} \beta\mathbf{m}_{21}$.

Now, after firing of $\alpha S_{11}\beta S_{21}$, the marks of $\alpha\mathbf{m}_{11}$ are unfrozen while the marks of $\beta\mathbf{m}_{21}$ are frozen. The subsequence αS_{12} can then be fired according to (I.13). And so on.

Finally, after firing of the last subsequence βS_{2N}, all the marks are unfrozen: the reached marking is then $\mathbf{m} = \alpha\mathbf{m}_{1N} + \beta\mathbf{m}_{2N}$. Note that $\mathbf{m}_{1N} = \mathbf{m}_1$ such that $\mathbf{m}_0 \xrightarrow{S_1} \mathbf{m}_1$ and $\mathbf{m}_{2N} = \mathbf{m}_2$ such that $\mathbf{m}_0 \xrightarrow{S_2} \mathbf{m}_2$. Hence,

$$\mathbf{m}_0 \xrightarrow{\alpha S_1 \oplus \beta S_2} \alpha\mathbf{m}_1 + \beta\mathbf{m}_2. \qquad (I.14)$$

Consider now the case where $\alpha + \beta < 1$. From the previous result and Property 4.3c, $\mathbf{m}_0 \xrightarrow{\alpha S_1 \oplus \beta S_2}$.

Remark I.3 Property I.1a is also true for a discrete PN if α and β are positive or zero integers. This is not very useful. On the other hand, for a continuous PN, α and β may be positive or zero real numbers and Property I.1b is powerful.

Remark I.4 The results presented for a pair on firing sequences may be easily generalized to a larger number of firing sequences. For example,

$$(S_1 \oplus S_2) \oplus S_3 = S_1 \oplus S_2 \oplus S_3$$

since the operation may be performed in any order (associativity).

Property I.1b may also be generalized. For example, for three sequences:

if $S_1, S_2, S_3 \in \mathcal{L}(\mathbf{m}_0)$ and $0 \le \alpha + \beta + \gamma \le 1$,

then $\alpha S_1 \oplus \beta S_2 \oplus \gamma S_3 \in \mathcal{L}(\mathbf{m}_0)$.

Appendix J

Notation 0^+ and Non-Standard Analysis

After an intuitive presentation of the notation 0^+, it will be shown that the corresponding concept is supported by the non-standard analysis [Ro 66].

J.1 INTUITIVE PRESENTATION

According to Notation 4.5 in Section 4.4.2, the ratio

$$\frac{\alpha}{k}, \text{ where } \alpha \text{ is a finite positive number and } k \to \infty, \text{ is denoted by } 0^+. \quad \text{(J.1)}$$

From this notation, we may write that $0 < 0^+$ and there is no positive real number x such that $x < 0^+$.

Assume *β is a finite positive number*. Since the product $\alpha\beta$ is a finite positive number too,

$$\frac{\alpha\beta}{k} \text{ is also denoted by } 0^+ \text{ for } k \to \infty. \quad \text{(J.2)}$$

From (J.1) and (J.2), one can write

$$\beta \cdot 0^+ = 0^+, \quad \text{(J.3)}$$

hence from (J.3):

$$0^+ + 0^+ = 0^+. \quad \text{(J.4)}$$

The quantity 0^+ is a *first order* infinitely small quantity. For practical reasons, a *second order* infinitely small quantity may be ignored and written simply as 0.

Then, if $0 < \beta < 1$,

$$(\beta)^h \cdot 0^+ = 0 \text{ for } h \to \infty. \quad \text{(J.5)}$$

If $\beta > 1$, the product

$$(\beta)^h \cdot 0^+ \text{ for } h \to \infty \text{ is undetermined.} \quad \text{(J.6)}$$

The following notations may be used:

$$\beta - 0^+ = \beta^-, \text{ and } \beta + 0^+ = \beta^+. \tag{J.7}$$

J.2 NON-STANDARD ANALYSIS

The non-standard analysis was introduced by A. Robinson in the early sixties [Ro 66]. Since then, various authors have worked on this topic [Sa 99].

Robinson defines the building of an *enlargement* of a mathematical structure. Such a widening introduces some new elements qualified *ideal*. His theory, when applied to the set $\mathcal{R}$ of real numbers, leads to the set $^*\mathcal{R}$ of *hyperreal* numbers. The set $^*\mathcal{R}$ contains all the *standard* numbers in $\mathcal{R}$ plus *non-standard* numbers.

Let us limit our attention to non-negative numbers, i.e. to the sets $\mathcal{R}_+$ and $^*\mathcal{R}_+$. "Ideal" numbers in $^*\mathcal{R}_+$, usually denoted by ε, are *infinitely small*: they are smaller than any positive real number in $\mathcal{R}_+$, whereas greater than 0. They correspond to our notation[1] 0^+. The set $^*\mathcal{R}_+$ also contains infinitely large numbers (larger than any real number), which are not useful for our purpose.

In the widening $^*\mathcal{R}$, it is possible to have an infinitesimal approach different from the usual one: the *quantification* of an infinitely small value is not necessary. For example, the sum of two "infinitely small values" is an "infinitely small value"; with our notation, this property corresponds[2] to (J.4).

If β is a standard number (i.e., in $\mathcal{R}$), then β^- and β^+, according to (J.7), are non-standard numbers (in $^*\mathcal{R}$). According to A. Robinson, for any bounded number x in $^*\mathcal{R}$, there is a standard number infinitely close to x. This number, denoted by $st(x)$, is called the *standard part* of the *hyperreal* x considered. For example:

$$st(\beta^-) = st(\beta) = st(\beta^+) = \beta. \tag{J.8}$$

[1] The notation 0^+ was proposed to the authors by an anonymous reviewer of [DaAl 01]. Notation ε is often used as a parameter to which some value in $\mathcal{R}$ can be assigned. In this book, ε has this meaning or denotes a sequence of zero length.

[2] Whereas, in the *standard* infinitesimal approach a quantification exists, for example $\mathrm{d}x + \mathrm{d}x = 2\mathrm{d}x$.

Appendix K

Sharing Between Two Transitions

Let us first recall Notation 5.5: given the structural conflict $K = \langle P_c, \{T_a, T_b\}\rangle$, the speeds of T_a and T_b, the speed vector, and the balance obtained are denoted by $v_a^{(a)}$, $v_b^{(a)}$, $\mathbf{v}^{(a)}$, and $B_c^{(a)}$, for $T_a < T_b$, and by $v_a^{(b)}$, $v_b^{(b)}$, $\mathbf{v}^{(b)}$, and $B_c^{(b)}$, for $T_b < T_a$.

This appendix explains various cases of sharing between two transitions, in order to understand the meaning of *Algorithm 5.5* in Section 5.3.3.1. Part of this algorithm is recalled here in the simplified case where $\alpha_a = \alpha_b$, and splitting the first two conditions in *Step 4* into five conditions (K.3) to (K.7).

Step 3.
If $v_b^{(a)} > v_b^{(b)}$ **then** $\mathbf{v} = \mathbf{v}^{(a)}, \ldots$ (K.1)
else if $v_a^{(b)} > v_a^{(a)}$ **then** $\mathbf{v} = \mathbf{v}^{(b)}, \ldots$ (K.2)
Step 4.
If $v_a^{(a)} \le v_b^{(a)}$ **then** $\mathbf{v} = \mathbf{v}^{(a)}, \ldots$ (K.3)
else if $v_b^{(b)} = v_b^{(a)}$ and $v_a^{(b)} = v_a^{(a)}$ **then** $\mathbf{v} = \mathbf{v}^{(a)}, \ldots$ (K.4)
else if $v_b^{(b)} = v_b^{(a)}$ and $B_c^{(a)} > 0$ **then** $\mathbf{v} = \mathbf{v}^{(a)}, \ldots$ (K.5)
else if $v_b^{(b)} \le v_a^{(b)}$ **then** $\mathbf{v} = \mathbf{v}^{(b)}, \ldots$ (K.6)
else if $v_a^{(a)} = v_a^{(b)}$ and $B_c^{(b)} > 0$ **then** $\mathbf{v} = \mathbf{v}^{(b)}, \ldots$ (K.7)
else if $v_a^{(a)} = v_a^{(b)}$ **then** $v_a = v_a^{(a)}$ and $v_b = v_a, \ldots$ (K.8)
else if $v_b^{(b)} = v_b^{(a)}$ **then** $v_b = v_b^{(b)}$ and $v_a = v_b, \ldots$ (K.9)
Step 5. .... **else** $v_a = v_b, \ldots$ (K.10)

The graphical illustration used for explaining Algorithm 5.5 is presented in Figure K.1. The Y-axis represents firing speeds and the X-axis is not significant. The firing speeds $v_a^{(a)}$ and $v_a^{(b)}$ are represented, respectively, by a big white dot and a small black dot, and all the possible firing speeds v_a correspond to the line segment between $v_a^{(b)}$ and $v_a^{(a)}$ (and similarly for v_b).

Figure K.1a represents the conflict resolution $T_a < T_b$: a *dotted line* connects the values corresponding to this solution, i.e. $v_a = v_a^{(a)}$ and $v_b = v_b^{(a)}$. Figure K.1b represents the resolution $T_b < T_a$: a *dotted line* connects the values obtained, i.e. $v_a = v_a^{(b)}$ and $v_b = v_b^{(b)}$. These solutions are obviously *acceptable* (Section 5.2.2) by definition of the values $v_j^{(k)}$.

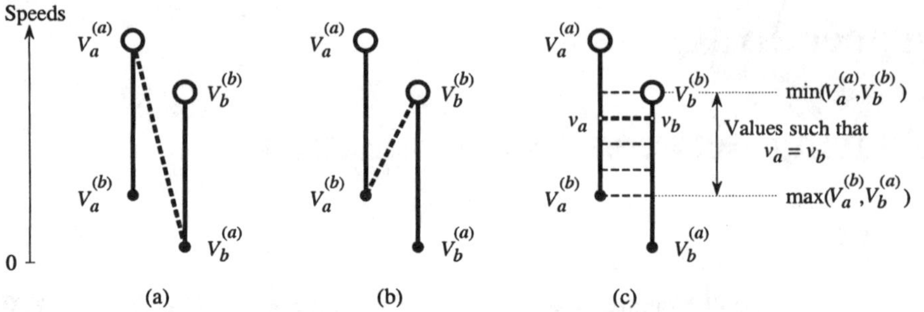

Figure K.1 Graphical representation. (a) Resolution $T_a < T_b$. (b) Resolution $T_b < T_a$. (c) Resolution $[T_a, T_b]$.

There are other solutions such that $v_a^{(b)} \le v_a \le v_a^{(a)}$ and $v_b^{(a)} \le v_b \le v_b^{(b)}$. The sharing resolution $[T_a, T_b]$ leads to *one* of them. In the case where

$$\max(v_a^{(b)}, v_b^{(a)}) \le \min(v_a^{(a)}, v_b^{(b)}), \tag{K.11}$$

a solution such that

$$\max(v_a^{(b)}, v_b^{(a)}) \le v_a = v_b = \beta \le \min(v_a^{(a)}, v_b^{(b)}) \tag{K.12}$$

is obtained, as illustrated in Figure K.1c. This solution is such that $B_c = 0$. A solution such that $v_a = v_b > \beta$ is *impossible* because B_c would be negative; a solution such that $v_a = v_b < \beta$ is *not acceptable* since it corresponds to $B_c > 0$.

A solution such as (K.12) (corresponding to (K.10) in the algorithm) cannot always be obtained. In the sequel, various cases are presented. Each one is illustrated by a graphical representation similar to Figure K.1, near the corresponding PN. For each example, the solution is presented in a relatively intuitive manner, then a formal reference to the (part of) algorithm above is given (for simplicity, all the examples are such that $\alpha_a = \alpha_b$).

Figure K.2 shows three examples without feedback, i.e. I_c does not depend on v_a and v_b. In Figures a, b, and c, T_1 is strongly enabled, then $I_c = v_1 = V_1$.

In Figure K.2a, the flow $I_c = v_1$ feeding P_c is divided into two equal flows. The solution $v_a = v_b = v_1 / 2 = 1$ satisfies $v_a \le V_a$, $v_b \le V_b$, and $v_a + v_b = v_1$. According to Notation 5.5, $v_a^{(a)} = 2$, $v_b^{(a)} = 0$, $v_a^{(b)} = 0$, and $v_b^{(b)} = 2$. This example corresponds to case (K.10): it is an application of Figure K.1c.

In Figure b, the solution $v_a = 1$ is not acceptable since $V_a < 1$. Then $v_a = \min(V_a, v_1 / 2) = 0.5$. It follows that $v_b = v_1 - 0.5 = 1.5$. Formally: $v_a^{(a)} = 0.5$, $v_b^{(a)} = 1.5$, $v_a^{(b)} = 0$, and $v_b^{(b)} = 2$, corresponds to case (K.3); since Equation (K.11) is not satisfied, the acceptable solution minimizing the absolute value $|v_a - v_b|$ is obtained.

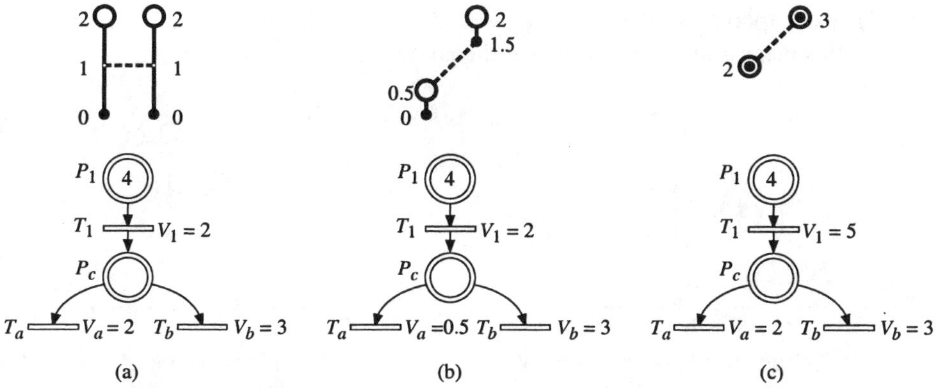

Figure K.2 Sharing without feedback. (a) Case (K.10). (b) and (c) Case (K.3).

In Figure K.2c, the solution is obviously $v_a = V_a = 2$ and $v_b = V_b = 3$ since $v_1 = V_a + V_b$. Formally: $v_a^{(a)} = 2$, $v_b^{(a)} = 3$, $v_a^{(b)} = 2$, and $v_b^{(b)} = 3$, i.e. also case (K.3). This example is a limit case in which there is no actual conflict whereas $B_c = 0$: if $v_1 = V_1 < 5$, there is an actual conflict; if $v_1 = V_1 > 5$, then $B_c > 0$, hence there is no actual conflict.

Figure K.3 presents two examples with feedback. The flow to be shared between T_a and T_b is $I_c = v_1 + v_2$. In both Figures a and b, T_1 is strongly enabled, then $v_1 = V_1 = 1$. It follows that v_a has no influence on v_1. On the other hand, v_2 depends on v_b.

What are the speeds v_a and v_b in Figure a? Intuitively, the following behavior may be analyzed. At the very beginning, $I_c = v_1 + v_2 = 1 + 0 = 1$ is divided into $v_a = v_b = 0.5$. But $v_b = 0.5 \Rightarrow v_2 = 0.5 \Rightarrow v_1 + v_2 = 1.5 \Rightarrow v_a = v_b = 0.75 \Rightarrow v_2 = 0.75$ $\Rightarrow v_1 + v_2 = 1.75 \Rightarrow v_a = v_b = 0.875$ etc. In fact, the solution satisfying all the equations including $v_a = v_b$ is $v_a = v_b = 1$. This value corresponds to

$$v_a = v_b = 0.5 + \frac{1}{2}0.5 + \left(\frac{1}{2}\right)^2 0.5 + \left(\frac{1}{2}\right)^3 0.5 + \ldots = 1.$$

After this intuitive explanation, let us specify the set of equations leading to this result. In addition to $v_1 = 1$, $v_2 = \min(1.5, v_b)$, and $v_a + v_b = v_1 + v_2$ (since it is clear that $v_1 + v_2 < V_a + V_b$), Algorithm 5.5 is used. If $T_a < T_b$, $v_a^{(a)} = 1$ and $v_b^{(a)} = 0$ are obtained. If $T_b < T_a$, $v_1 = 1 \Rightarrow v_b = 1 \Rightarrow v_2 = 1 \Rightarrow v_1 + v_2 = 2$ $\Rightarrow v_b = 2 \Rightarrow v_2 = 1.5 \Rightarrow v_1 + v_2 = 2.5 \Rightarrow v_b^{(b)} = 2$ and $v_a^{(b)} = 0.5$. It follows that the equation $v_a = v_b$ is added (case (K.10)).

For the PN in Figure K.3b, the behavior is completely different. At the very beginning, $v_a = v_b = 0.5$. But $v_b = 0.5 \Rightarrow v_2 = 0.5 \Rightarrow I_c = v_1 + 2v_2 = 2 \Rightarrow v_a = v_b = 1 \Rightarrow$ $v_2 = 1 \Rightarrow I_c = 3 \Rightarrow v_a = v_b = 1.5 \Rightarrow v_2 = 1.5 \Rightarrow I_c = 4 \Rightarrow v_a = v_b = 2 \Rightarrow v_2 = 2 \Rightarrow I_c = 5$. The balance B_c is positive (hence there is no actual conflict) because $I_c = 1 + 2v_b$:

the feedback loop $P_c \rightarrow T_b \rightarrow P_2 \rightarrow T_2 \rightarrow P_c$ feeds P_c with the speed $2\,v_b$ whereas it drains this place with the speed v_b, i.e. the gain of the feedback is 2.

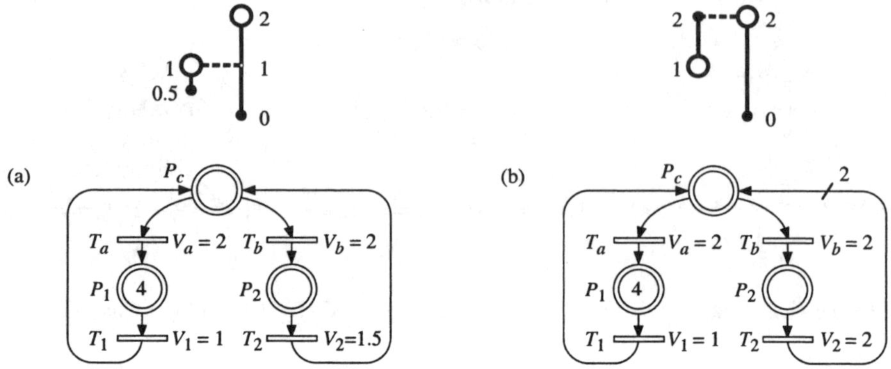

Figure K.3 Sharing with feedback. (a) Case (K.10). (b) Case (K.2).

Let us now apply Algorithm 5.5 to this example. If $T_a < T_b$, $v_a^{(a)} = 1$ and $v_b^{(a)} = 0$ are obtained. If $T_b < T_a$, $v_b^{(b)} = 2$ and $v_a^{(b)} = 2$ are obtained. This is the case (K.2) in which $v_a^{(b)} > v_a^{(a)}$ (the speed v_a is greater when T_b takes priority over T_a than when T_a takes priority over T_b !). Because of the gain greater than 1 in the feedback loop $P_c \rightarrow T_b \ldots \rightarrow P_c$, $B_c > 0$, hence there is no actual conflict (thus, the maximal values are obtained for both v_a and v_b).

Let us now consider two examples for which $v_a^{(a)} = v_a^{(b)}$ (Figure K.4). In these cases, the value of $B_c^{(b)}$ is important.

For the example in Figure K.4a, the intuitive analysis leads to the same result as in Figure K.3a, i.e. $v_a = v_b = 1$.

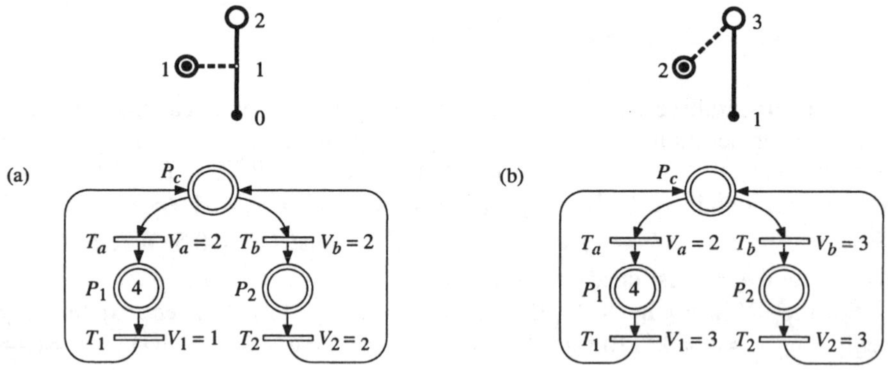

Figure K.4 Examples with $v_a^{(a)} = v_a^{(b)}$. (a) $B_c^{(b)} = 0$: case (K.8). (b) $B_c^{(b)} > 0$: case (K.7).

Formally: $v_a^{(a)} = 1$, $v_b^{(a)} = 0$, $v_a^{(b)} = 1$, $v_b^{(b)} = 2$, and $B_c^{(b)} = 0$ are obtained. Value $v_a^{(a)} = 1$ is obtained when T_a takes priority over T_b. On the other hand, $v_a^{(b)} = 1$ is obtained because the quantity $I_c - v_b = 3 - 2 = 1$ is "left" to T_a when T_b takes priority over T_a. For this priority case, however, the balance is $B_c^{(b)} = 0$: there is an actual conflict. The solution is given by (K.8), i.e. $v_a = 1$, then $v_b = v_a = 1$.

For the example in Figure K.4b, if both V_a and V_b were greater than or equal to 3, the limit value of both v_a and v_b would be $1.5 + 0.5 \times 1.5 + 0.5^2 \times 1.5 + \ldots = 3$. But $V_a < 3$, then $v_a = \min(V_a, 3) = 2$ and $v_b = \min(V_b, 3 + 3 - v_a) = 3$. There is no actual conflict.

Formally: $v_a^{(a)} = 2$, $v_b^{(a)} = 1$, $v_a^{(b)} = 2$, $v_b^{(b)} = 3$, and $B_c^{(b)} = 1$ are obtained. The balance $B_c^{(b)} > 0$ implies that there is no actual conflict. The values $v_a = 2$ and $v_b = 3$ are given by (K.7).

Taking into account that (K.1) is symmetrical to (K.2), and that (K.5), (K.6), and (K.9) are symmetrical to (K.7), (K.3), and (K.8), respectively, all the cases in Algorithm 5.5 have been illustrated in the previous examples.

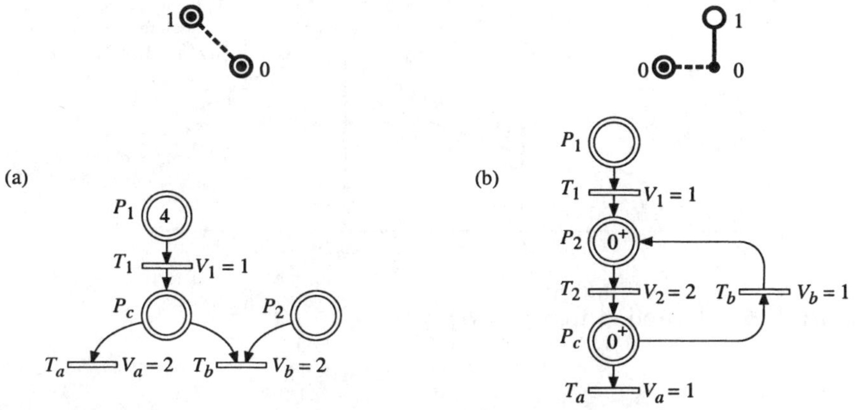

Figure K.5 (a) Illustration of (K.4). (b) Illustration of (K.3).

Let us now present two additional examples presented in Figure K.5.

In Figure K.2c, $v_a^{(a)} = v_a^{(b)}$ and $v_b^{(a)} = v_b^{(b)}$ because T_a and T_b can be fired at their maximal speeds. Figure K.5a illustrates another case where $v_a^{(a)} = v_a^{(b)}$ and $v_b^{(a)} = v_b^{(b)}$: v_a is limited by v_1, but its value is the same for $T_a < T_b$ and $T_b < T_a$ because T_b is not enabled. The values $v_a^{(a)} = 1$, $v_b^{(a)} = 0$, $v_a^{(b)} = 1$, and $v_b^{(b)} = 0$, are obtained (case (K.4)).

The example in Figure K.5b may correspond to a marking obtained at the end of a phase (Figure 5.32 in Section 5.4.1.2). Then, the speeds for the next phase must be calculated from this marking. If $T_a < T_b$, $v_a^{(a)} = 0$ and $v_b^{(a)} = 0$ are obtained, because places P_c and P_2 are immediately emptied through T_a (Property 5.4 in Section 5.1.3.3). If $T_b < T_a$, $v_b^{(b)} = 1$ and $v_a^{(b)} = 0$ are obtained (Algorithm 5.7, Section 5.3.2.3). For the sharing $[T_a, T_b]$, $v_a = v_b = 0$ are obtained according to (K.3). Note that this result is consistent with Equation (J.5) in Appendix J, with $\beta = 0.5$.

Remark K.1 The examples given above correspond to the simple case $\alpha_a = \alpha_b$. Consider now the general case where $\alpha_a \neq \alpha_b$ (general case in Algorithm 5.5). There is a comparison between v_a / α_a and v_b / α_b (instead of a comparison between v_a and v_b). This is illustrated in Figure K.6. The X-axis is now significant since it corresponds to the values α. For $\alpha_a < \alpha_b$, the possible pair of cases such that $v_a / \alpha_a = v_b / \alpha_b$ is illustrated in Figure K.6. There is no solution $v_a / \alpha_a = v_b / \alpha_b$ if $v_a^{(a)} / \alpha_a < v_b^{(a)} / \alpha_b$ or $v_b^{(b)} / \alpha_b < v_a^{(b)} / \alpha_a$: correspond respectively to cases (K.3) and (K.4).

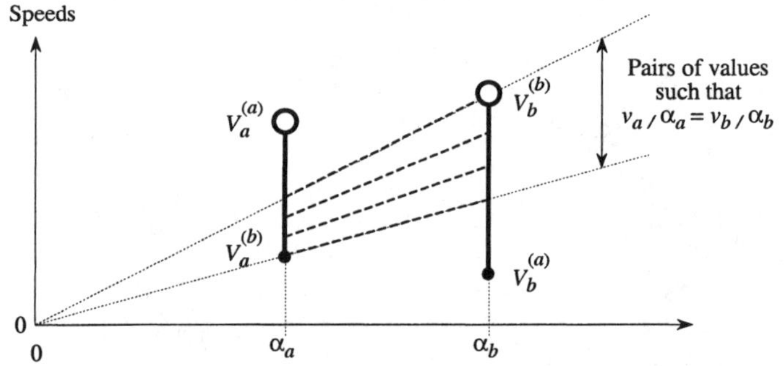

Figure K.6 Illustration for $\alpha_a \neq \alpha_b$.

Appendix L

Graph of Relations Among Conflicts

Let us present a way for verifying whether or not Hypothesis 5.2a is satisfied. The process is illustrated by the example in Figure L.1. Figure a shows the structure of a continuous PN, and Figure b is the corresponding **graph**[1] **of relations among conflicts**. This graph is built as follows: it contains two kinds of nodes, illustrated in Figure b: the first kind is associated with every transition not involved in a conflict (T_1, T_4, and T_{12}, in the example); the second kind is associated with every set (maximal) of transitions involved in a structural conflict ($\{T_2, T_3\}$, $\{T_5, T_6, T_7\}$, etc. in the example). There is an arc from node n_1 to node n_2 in Figure b if there is at least one path of length 2 in Figure a from a transition in n_1 to a transition in n_2. For example: the path $T_1 \rightarrow P_1 \rightarrow T_2$ in Figure a implies the arc $T_1 \rightarrow \{T_2, T_3\}$ in Figure b; $T_2 \rightarrow P_2 \rightarrow T_1$ in Figure a implies $\{T_2, T_3\} \rightarrow T_1$ in Figure b; $T_3 \rightarrow P_3 \rightarrow T_6$ in Figure a implies $\{T_2, T_3\} \rightarrow \{T_5, T_6, T_7\}$ in Figure b. For the time being, *let us assume that the PN is simple* (Section 1.2.1.5); the case where a transition may be involved in two or more conflicts will be explained in Remark L.1.

Let a **group** G_h of nodes in the *graph of relations among conflicts* be defined as follows: the subgraph made of the nodes in G_h and the arcs among them is a strongly connected component, and it is maximal (i.e., there is no $G_l \supset G_h$ such that G_l is a strongly connected component). In Figure b, the groups G_1 to G_5 are found (all these groups are disjoints, this is a well known property). For example $G_1 = \{T_1, \{T_2, T_3\}\}$. Let us denote by $T(G_h)$ the set of transitions corresponding to the transitions in G_h; for example, $T(G_1) = \{T_1, T_2, T_3\}$ (by abuse of language we may say that T_2 is in G_1). We have, then, the following property.

Property L.1 A *graph of relations among conflicts* is such that:

a) A transition of the PN is *in exactly one group.*

b) The structural conflict $K = \langle P_i, T(K)\rangle$ satisfies *Hypothesis 5.2a if and only if there is a group* G_h such that: 1) $T(G_h) = T(K)$; 2) there is no self-loop from the node associated with $T(K)$ to itself.

Proof This proof is given here for a *simple PN*, which is the present hypothesis. However, it will be shown in Remark L.1 that the property may be adapted if the PN is not simple.

---

[1] For the concepts and vocabulary related to graphs, see Appendix C.

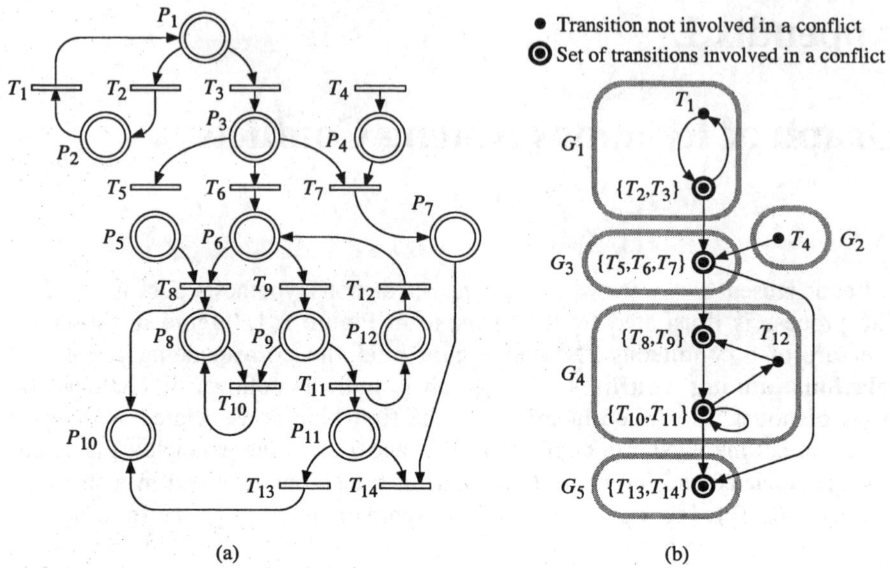

Figure L.1 (a) Structure of a PN. (b) Graph of relations among conflicts.

a) By definition, in a simple PN, a transition may be involved at most in one structural conflict. Then, by construction, a transition is associated with exactly one node of the graph. Since the groups are disjoints, the property is true.

b) *Necessary condition.* 1) Assume there is a group containing both the node n_K associated with $T(K)$ and another node n_j. If the transitions associated with n_j were placed in $UP(K)$, an arc from a component in $T(K)$ to a component of $UP(K)$ would exist since there is a path from n_K to n_j. If the transitions in n_j were placed in $DOWN(K)$, an arc from a component in $DOWN(K)$ to a component of $T(K)$ would exist since there is a path from n_j to n_K. 2) If there is $T(G_h) = T(K)$ and a self-loop from n_K to itself, then there is a an arc from a transition in $T(K)$ to P_i.

Sufficient condition. If $T(G_h) = T(K)$ and there is no self-loop from n_K to itself, the conditions in Hypothesis 5.2a are obviously satisfied.

❑

According to Property L.1b and Figure L.1b, the structural conflicts $K_3 = \langle P_3, \{T_5, T_6, T_7\}\rangle$ and $K_{11} = \langle P_{11}, \{T_{13}, T_{14}\}\rangle$ verify Hypothesis 5.2, whereas the other conflicts do not.

Remark L.1 Let us consider the case where *the PN is not simple*, i.e. a transition may be involved in two or more conflicts. An example is given in Figure L.2 where both conflicts $K_1 = \langle P_1, \{T_3, T_4\}\rangle$ and $K_2 = \langle P_2, \{T_4, T_5\}\rangle$ involve T_4. In the *graph of relations among conflicts*, two nodes are associated with $T(K_1)$ and $T(K_2)$ and there is an edge between these nodes, meaning that $T(K_1) \cap T(K_2) \neq \varnothing$ (see Figure b). Such an edge may be seen as an arc which may be crossed in both directions. Then both nodes must be in the same strongly connected component, i.e. in the same group: in Figure b, G_3 contains both $T(K_1)$ and $T(K_2)$, i.e., $T(G_3) = \{T_3, T_4, T_5\}$.

What about Property L.1 in this case? Property L.1a is still verified. Let us now consider Property L.1b.

Let $GK = \{K_1, K_2, \ldots\}$ denote a maximal set of conflicts such that for any K_a in GK, there is at least another K_b in GK such that $T(K_a) \cap T(K_b) \neq \varnothing$. Property L.1b can then be replaced by: every structural conflict in GK satisfies *Hypothesis 5.2a if and only if there is a group* G_h such that: 1) $T(G_h) = \bigcup_{K_a \in GK} K_a$; 2) for every $K_a \in GK$, there is no self-loop from the node associated with $T(K_a)$ to itself.

Then, according to Hypothesis 5.2b, a sharing among the transitions of at most one K_a in GK is accepted for Algorithm 5.5. For the example in Figure L.2, a sharing $[\alpha_3 T_3, \alpha_4 T_4]$ or $[\alpha_4 T_4, \alpha_5 T_5]$ is accepted but not both (i.e. the other conflict is solved by a priority among the transitions).

❑

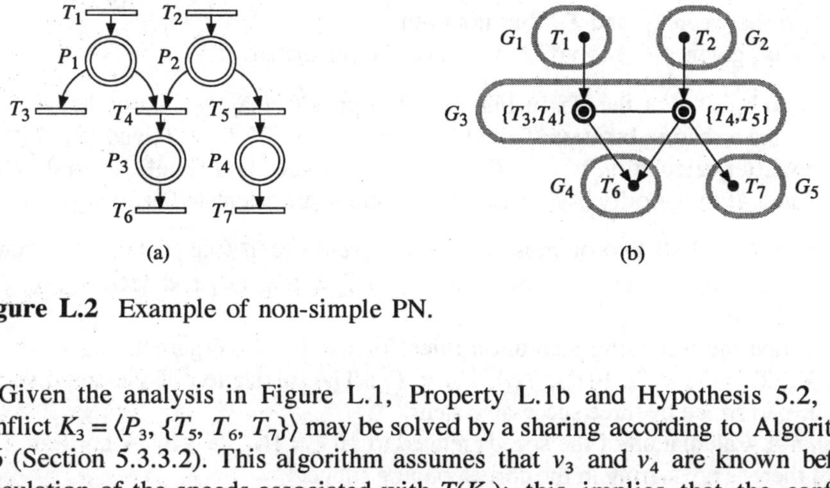

Figure L.2 Example of non-simple PN.

Given the analysis in Figure L.1, Property L.1b and Hypothesis 5.2, the conflict $K_3 = \langle P_3, \{T_5, T_6, T_7\}\rangle$ may be solved by a sharing according to Algorithm 5.6 (Section 5.3.3.2). This algorithm assumes that v_3 and v_4 are known before calculation of the speeds associated with $T(K_3)$: this implies that the conflict $K_1 = \langle P_1, \{T_2, T_3\}\rangle$ has already been solved, hence that v_1 is also known since v_2 and v_3 depend on v_1. The result of this analysis may be summarized as follows: since there are flows $G_1 \to G_3$ and $G_2 \to G_3$ (Figure L.1b), the instantaneous firing speeds of transitions in $T(G_1)$ and $T(G_2)$ must be known before the resolution of K_3.

The groups illustrated in Figure L.1b can be ordered in a way such that: if there is a flow $G_h \to G_k$, then G_h is before (not necessarily just before) G_k. Let $\mathcal{G}$ denote this ordered set. For our example

$$\mathcal{G} = (G_1, G_2, G_3, G_4, G_5) \tag{L.1}$$

may be obtained (a permutation between G_1 and G_2 is also possible). The meaning of this ordered set is that: the speeds of $T(G_h)$ do not depend on the speeds of $T(G_k)$

if G_k is after G_h. It follows that, according to (L.1), the firing speeds may be calculated in the following order: speeds of $T(G_1)$, then speeds of $T(G_2)$, ..., then speeds of $T(G_5)$. Let $\mathbf{v}(G_k)$ denote the speed vector related to the set of transitions $T(G_k)$. For example in Figure L.1, $\mathbf{v}(G_1) = (v_1, v_2, v_3)$, $\mathbf{v}(G_2) = (v_4)$, $\mathbf{v}(G_3) = (v_5, v_6, v_7)$, etc. If the transitions are ordered in a way such that the index of transitions in G_k are smaller than the index of transitions in G_{k+1}, then:

$$\mathbf{v} = (\mathbf{v}(G_1), \mathbf{v}(G_2), \mathbf{v}(G_3), \mathbf{v}(G_4), \mathbf{v}(G_5)). \quad \text{(L.2)}$$

Assume the priority $T_2 < T_3$ is chosen for K_1. From Algorithm 5.4 in Section 5.3.2, $\mathbf{v}(G_1)$ can be obtained.

Assume now the sharing $[T_5, T_6, 2T_7]$ is chosen for K_3. This structural conflict K_3 satisfies Hypothesis 5.2, then Algorithm 5.6 can be used. After the speeds in $UP(K_3)$, i.e. v_1, v_2, v_3, and v_4, the speeds of v_5, v_6, and v_7 can be obtained by this algorithm. Once the speed vectors $\mathbf{v}(G_1)$, $\mathbf{v}(G_2)$, and $\mathbf{v}(G_3)$ are known, they are input data for G_4. Hence, $\mathbf{v}(G_4)$ can be calculated by Algorithm 5.4, if there is no sharing but only priorities, or by Algorithm 5.5, if there is a sharing between T_8 and T_9 *or* between T_{10} and T_{11} (but not both).

Finally, given $\mathbf{v}(G_1)$ to $\mathbf{v}(G_4)$, $\mathbf{v}(G_5)$ can be calculated .

Remark L.2 For the PN in Figure L.2, a priority may be chosen between T_3 and T_4, and a sharing between T_4 and T_5, for example. If $T_3 < T_4$ and $[T_4, T_5]$ are chosen: after calculation of v_1 and v_2, v_3 is calculated and finally v_4 and v_5 are calculated. If the priority is $T_4 < T_3$, then v_4 and v_5 are calculated before v_3.

Remark L.3 If two or more *successive groups* in $\mathcal{G}$ (see (L.1)) are gathered, the result is a **macrogroup**. For example, $MG_a = \{G_1, G_2\}$ and $MG_b = \{G_4, G_5\}$ are macrogroups.

Assume the following resolution rules for the PN in Figure L.1a: $T_2 < T_3$, $[T_5, T_6, 2T_7]$, $T_8 < T_9$, $[4T_{10}, T_{11}]$, $T_{14} < T_{13}$. The calculation of the speed vector can be performed by three successive steps.

Step A. Calculation of the speeds related to $MG_a = \{G_1, G_2\}$ by Algorithm 5.4, since there is no sharing in the corresponding subnet.

Step B. Calculation of the speeds related to G_3 by Algorithm 5.6, since the structural conflict K_3 satisfies Hypothesis 5.2 and MG_a corresponds to $UP(K_3)$.

Step C. Calculation of the speeds related to $MG_b = \{G_4, G_5\}$ by Algorithm 5.5, since the speeds of transitions feeding MG_b, namely v_6 and v_7, are already known and there is a single sharing between two transitions in MG_b.

Algorithm 5.7 in Section 5.3.3.3 is a generalization of this example.

Appendix M

Piecewise Constant Maximal Speeds

M.1 INTUITIVE PRESENTATION

Consider the timed continuous PN in Figure M.1a, whose maximal firing speeds are specified in Figure c: $V_1(t) = 3$ for $t \in [0, 4[$, $V_1(t) = 0$ for $t \in [4, 11[$, $V_1(t) = 6$ for $t \in [11, \infty[$; $V_2(t) = 2$ for $t \in [0, 11[$, $V_2(t) = 1$ for $t \in [11, \infty[$. Both $V_1(t)$ and $V_2(t)$ are piecewise constant speeds.

During the interval $t \in [0, 4[$, the maximal speeds are constant: $\mathbf{V}(t) = (3, 2)$. The behavior corresponds to the CCPN (constant speed continuous PN) R_0 in Figure b. It can be analyzed by the methods presented in Section 5.3:

$$v_1(t) = 3,\ v_2(t) = 2,\ m_1(t) = 8 - t,\ m_2(t) = 2 + t, \text{ for } t \in \left]0, 4\right[. \tag{M.1}$$

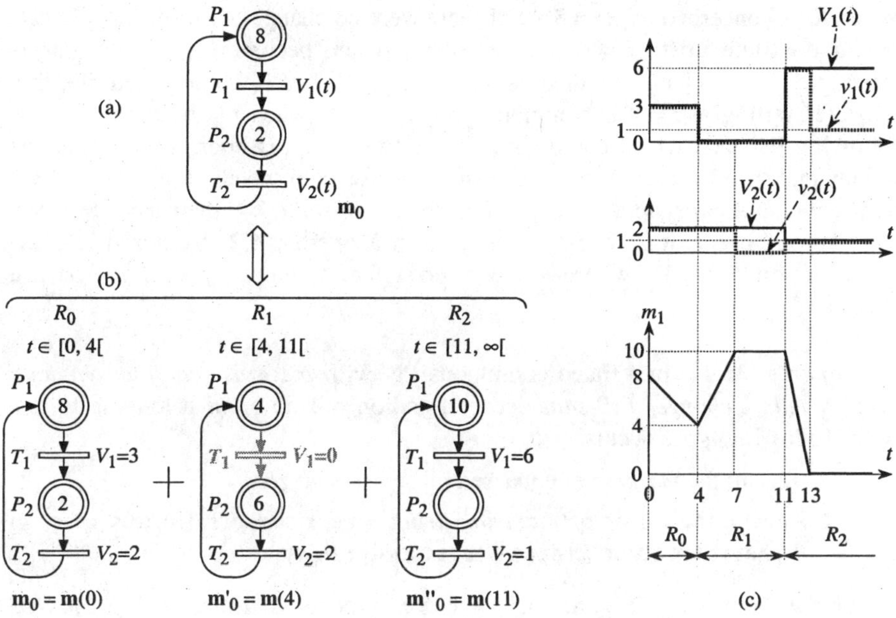

Figure M.1 (a) Timed continuous PN with piecewise constant maximal speeds. (b) Equivalent set of CCPNs. (c) Maximal speeds and behavior.

At $t = 4$, the marking is $\mathbf{m}(4) = (4, 6)$. At this time, the maximal speed vector changes and becomes $\mathbf{V}(t) = (0, 2)$ for $t \in [4, 11[$. During this interval, the behavior corresponds to the CCPN R_1 in Figure b. The "initial" marking, at time 4, is $\mathbf{m}_0' = \mathbf{m}(4) = (4, 6)$. As before, the behavior can be analyzed by the methods presented in Section 5.3 (m_2 becomes 0 at $t = 7$):

$$v_1(t) = 0,\ v_2(t) = 2,$$
$$m_1(t) = 4 + 2(t-4),\ m_2(t) = 6 - 2(t-4), \text{ for } t \in |4, 7|; \qquad \text{(M.2)}$$

$$v_1(t) = 0,\ v_2(t) = 0,\ m_1(t) = 10,\ m_2(t) = 0, \text{ for } t \in |7, 11|. \qquad \text{(M.3)}$$

At $t = 11$, the marking is $\mathbf{m}(11) = (10, 0)$; the maximal speed vector becomes $\mathbf{V}(t) = (6, 1)$ for $t \in [11, \infty[$. The behavior then corresponds to R_2 in Figure b. The "initial" marking is $\mathbf{m}_0'' = \mathbf{m}(11) = (10, 0)$. The behavior is given by (M.4) and (M.5) (m_1 becomes 0 at $t = 13$):

$$v_1(t) = 6,\ v_2(t) = 1,$$
$$m_1(t) = 10 - 5(t-11),\ m_2(t) = 5(t-11), \text{ for } t \in |11, 13|; \qquad \text{(M.4)}$$

$$v_1(t) = 1,\ v_2(t) = 1,\ m_1(t) = 0,\ m_2(t) = 10, \text{ for } t \in |13, \infty|. \qquad \text{(M.5)}$$

Hence, the behavior of the PN in Figure M.1a, corresponds to the concatenation of behaviors of the successive CCPNs R_0, R_1, and R_2, in Figure b. When R_0 is concerned, $m_1(t) = 8 - t$. If there were no change of model, the IB-state would continue up to $t = 8$ (i.e., when m_1 would become 0); this IB-state is interrupted by the change of maximal speed vector. When R_1 is concerned, the first IB-state starting at $t = 4$ is complete (ends at $t = 7$, when m_2 becomes 0). The following IB-state, which could continue up to infinity, is interrupted by change of maximal speed vector. And so on. An **IB-state** corresponds to a time interval during which the *instantaneous speed vector is constant*; we then have Property M.1. The *C1-events* were already considered in Properties 4.2 (Section 4.1.2) and 5.5 (Section 5.3.2.3), whereas **C3-events**, i.e. *changes of vector* $\mathbf{V}(t)$, are introduced here.

Property M.1 In a timed continuous PN with *piecewise constant maximal firing speeds*, a *change of IB-state* occurs if and only if an *event* belonging to one of the following types occurs.

C1-event: the *marking* of a marked place *becomes zero*.

C3-event: the *vector of maximal firing speeds changes*. (In this case, an I-phase may occur between two successive IB-states.)

❑

For the example in Figure M.1, C1-events occur at times 7 and 13, and C3-events occur at times 4 and 11.

Let us now specify the notations which will be used.

The time when the kth *change of speed vector* occurs is denoted by τ_k, and the pair $(\tau_k, \mathbf{V}_k)$ specifies that the *maximal speed vector* is $\mathbf{V}_k$ for $t \in [\tau_k, \tau_{k+1}[$ (if the

kth change is the last one, then $\tau_{k+1} = \infty$). The maximal speed vector $\mathbf{V}(t)$, for $t \in [0, \infty[$, is then specified by the string (in which $\tau_0 = 0$):

$$\mathbf{V}(t) = (\tau_0, \mathbf{V}_0)(\tau_1, \mathbf{V}_1)\ldots(\tau_k, \mathbf{V}_k) \ldots \tag{M.6}$$

For the PN in Figure M.1, $\mathbf{V}(t) = (0, (3, 2))(4, (0, 2))(11, (6, 1))$.

Let $Q = \langle P, T, \text{Pre}, \text{Post}\rangle$ denote an unmarked PN. The triple

$$R_k = \langle Q, \mathbf{V}_k, \tilde{\mathbf{m}}(\tau_k)\rangle \tag{M.7}$$

specifies the CCPN R_k (this is the general case; if there is no conflict, $\mathbf{m}(\tau_k)$ may be used instead of $\tilde{\mathbf{m}}(\tau_k)$). If $R(t)$ denotes the *timed continuous PN with maximal firing speeds piecewise constant*, $R(t)$ is completely defined by the successive R_k:

$$\begin{aligned} &R_0 = \langle Q, \mathbf{V}_0, \mathbf{m}(0)\rangle \text{ for } t \in [0, \tau_1[, \\ &R_1 = \langle Q, \mathbf{V}_1, \tilde{\mathbf{m}}(\tau_1)\rangle \text{ for } t \in [\tau_1, \tau_2[, \\ &\ldots\ldots\ldots\ldots\ldots\ldots\ldots\ldots\ldots, \\ &R_k = \langle Q, \mathbf{V}_k, \tilde{\mathbf{m}}(\tau_k)\rangle \text{ for } t \in [\tau_k, \tau_{k+1}[, \\ &\ldots\ldots\ldots\ldots\ldots\ldots\ldots\ldots\ldots \end{aligned} \tag{M.8}$$

For the example in Figure M.1, if Q_1 denotes the unmarked PN in this figure: $R_0 = \langle Q_1, (3, 2), (8, 2)\rangle$ for $t \in [0, 4[$, $R_1 = \langle Q_1, (0, 2), (4, 6)\rangle$ for $t \in [4, 11[$, and $R_2 = \langle Q_1, (6, 1), (10, 0)\rangle$ for $t \in [11, \infty[$.

M.2 CALCULATION ALGORITHMS

The calculation of successive instantaneous firing speed vectors and markings is a kind of simulation leading to an evolution graph. Two algorithms will be presented. Both assume that there is no interruption process in a real time evolution. The vector of maximal speeds $\mathbf{V}(t) = (V_1(t), V_2(t), \ldots, V_m(t))$ can be represented by the string (M.6). The first algorithm assumes that $\mathbf{V}(t)$ *is known progressively* during the simulation, and the second algorithm that $\mathbf{V}(t)$ *is completely known* at the beginning of the simulation.

M.2.1 The Vector V(*t*) is Progressively Known

Let us observe that, given $R_{k-1} = \langle Q, \mathbf{V}_{k-1}, \tilde{\mathbf{m}}(\tau_{k-1})\rangle$ and τ_k, $\tilde{\mathbf{m}}(\tau_k)$ can be calculated since R_{k-1} is a CCPN. This calculation may be carried out automatically by Algorithm 5.8 in Section 5.3.4 (with a slight adaptation for stopping the calculation at τ_k). It follows that the behavior of the PN $R(t)$ can be completely calculated given

$$R_0 = \langle Q, \mathbf{V}_0, \mathbf{m}(0)\rangle \quad \text{AND} \quad \mathbf{V}(t) = (0, \mathbf{V}_0)(\tau_1, \mathbf{V}_1)\ldots(\tau_k, \mathbf{V}_k) \ldots \tag{M.9}$$

Hypothesis M.1 When a C3-event (i.e. change of vector $\mathbf{V}_k$) is treated, the *time* of the next C3-event is *known*.

❑

The simulation consists in calculating the evolution of the PN for the intervals $[\tau_0, \tau_1[$, then $[\tau_1, \tau_2[$ and so on. When the calculation due to the change $(\tau_k, \mathbf{V}_k)$ is performed, time τ_k and vector $\mathbf{V}_k$ are necessarily known; Hypothesis M.1 assumes that the time τ_{k+1} is also known (but not necessarily $\mathbf{V}_{k+1}$). This assumption, which allows the evolution in the interval $[\tau_k, \tau_{k+1}[$ to be calculated may be applied to a wide range of systems. For example, in a real time environment, the vector $\mathbf{V}(t)$ may change only at specified times (synchronous behavior by periodical reading of external data).

The corresponding algorithm is based on Algorithm 5.8 (*Steps 3* to *5* of Algorithm M.1 correspond to *Steps 2* to *4* of Algorithm 5.8). Algorithm 5.8 treats the C1-events; Algorithm M.1 adds the treatment of every C3-event occurring between two C1-events (or simultaneously with one of them).

Algorithm M.1 **$\mathbf{V}(t)$ progressively known: Hypothesis M.1**

Step 1. Initialization of:

Step 1.1. Structure Q of the PN, initial marking $\mathbf{m}(0)$, $\tilde{\mathbf{m}} = \mathbf{m}(0)$, local conflict resolutions, ordered set G of groups (or macrogroups). Let[1] $s = 1$.

Step 1.2. Time-ordered sequence of pairs: $TOS = (0, \mathbf{V}_0)(\tau_1, \mathbf{V}_1)\ldots$ Let $\tau_k = 0$.

Step 2.

Step 2.1. Updating of the time-ordered sequence TOS (new known values).

Step 2.2. Let $\mathbf{V} = \mathbf{V}(\tau_k)$. Then delete the first pair in TOS.

Step 2.3. Let τ_k = the first time in TOS (if TOS is empty then $\tau_k = \infty$).

Step 3. Setting the balances (literal expressions).

Step 4. Calculations related to I-phase, if necessary (*Step 3* in Algorithm 5.8).

Step 5. Calculations for IB-state s: Algorithm 5.7 (or 5.3 or 5.4) without *Step 0*.

Step 6. **If** $t_s = \tau_k = \infty$ **then** END

else **if** $t_s < \tau_k$: let $\tilde{\mathbf{m}} = \tilde{\mathbf{m}}(t_s)$ and $s = s + 1$, and go to *Step 5*.

Step 7. $\tau_k \le t_s$: calculate $\tilde{\mathbf{m}} = \tilde{\mathbf{m}}(\tau_k)$, let $s = s + 1$, and go to *Step 2*.

❑

When the time-ordered sequence is initialized (*Step 1.2*), it must contain at least $\tau_0 = 0$, $\mathbf{V}_0$, and τ_1 (except if $\tau_1 = \infty$). When new values are known, they are added in *Step 2.1*. The feedback to *Step 5* is performed in *Step 6* if the next event is a C1-event; if it is a C3-event, the feedback to *Step 2* is performed in *Step 7*.

M.2.2 The Vector V(t) is Completely Known

It may be interesting to identify the influence of maximal speeds on the behavior of a system. It is then necessary to know $\mathbf{V}(t)$ for all the positive values of t. Since it is not possible to store an infinite number of pairs $(\tau_k, \mathbf{V}_k)$, the following hypothesis is made.

[1] Let us recall that s and t_s denote an IB-state and its end (in Algorithms 5.3, 5.4, 5.7, and 5.8).

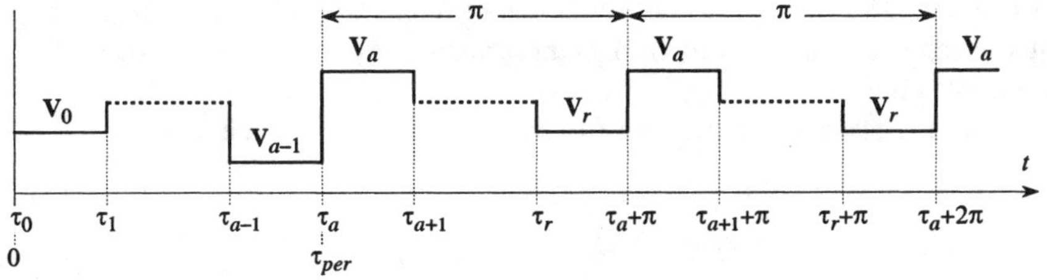

Figure M.2 Illustration of the notation related to Hypothesis M.2.

Hypothesis M.2 After a *finite transient behavior*, the maximal speed firing vector $\mathbf{V}(t)$ has a *stationary behavior which is either constant or periodical.* ❑

A wide range of systems can be modeled, exactly or approximately, from this hypothesis. Its interest is to express a behavior, unbounded in time, with a *finite number of symbols.*

The notation used is illustrated in Figure M.2.

Let $[\pi, (\tau_a, \mathbf{V}_a)(\tau_{a+1}, \mathbf{V}_{a+1})\ldots(\tau_r, \mathbf{V}_r)]$ denote a periodical behavior: the period is π; the periodical maximal speed vector begins at τ_a, $0 \le \tau_a$; and $0 < (\tau_{a+1} - \tau_a) < \ldots < (\tau_r - \tau_a) < \pi$. It means that $\mathbf{V}(t) = \mathbf{V}_a$ in the intervals $[\tau_a, \tau_{a+1}[$, $[\tau_a + \pi, \tau_{a+1} + \pi[$, and more generally $[\tau_a + K\pi, \tau_{a+1} + K\pi[$, where $K = 0, 1, 2 \ldots$ (and similarly for $\mathbf{V}_{a+1},\ldots, \mathbf{V}_r$). Then, given Hypothesis M.2 is verified, the general expression of $\mathbf{V}(t)$ will be:

$$\mathbf{V}(t) = (\tau_0, \mathbf{V}_0)(\tau_1, \mathbf{V}_1)\ldots(\tau_{a-1}, \mathbf{V}_{a-1})[\pi, (\tau_a, \mathbf{V}_a)(\tau_{a+1}, \mathbf{V}_{a+1})\ldots(\tau_r, \mathbf{V}_r)] \quad \text{(M.10)}$$

or $$\mathbf{V}(t) = (\tau_0, \mathbf{V}_0)(\tau_1, \mathbf{V}_1)\ldots(\tau_{a-1}, \mathbf{V}_{a-1})$$

if the stationary behavior of $\mathbf{V}(t)$ is not periodical. Note that the expression of $\mathbf{V}(t)$ is not unique if its stationary behavior is periodical; for example $(0, \alpha)[6, (5, \beta)(8, \alpha)]$ and $(0, \alpha)$ $[6, (2, \alpha)(5, \beta)]$ represent the same behavior.

Algorithm M.2 will show how to calculate the successive IB-states if $\mathbf{V}(t)$ is completely known. Since the number of changes of $\mathbf{V}(t)$ is unbounded, the number of successive IB-states may also be unbounded. Fortunately, because of the stationary behavior of $\mathbf{V}(t)$, a stationary behavior of $\mathbf{v}(t)$ will be found. Hence, the calculation of the successive IB-states will end and the evolution graph will contain a feedback to an IB-state already encountered.

Even if the stationary behavior of $\mathbf{V}(t)$ is periodical as in (M.10), the stationary behavior of $\mathbf{v}(t)$ may be constant. For example, consider a PN consisting of a loop $P_1T_1P_2T_2P_1$, with an initial marking (3, 0), and $V_1(t) = 1$ and $V_2(t) = [10, (0, 2)(5, 3)]$. It is clear that $\mathbf{v}(t) = (1, 1)$ is constant. If the stationary behavior of $\mathbf{v}(t)$ is not constant, Algorithm M.2 can end as a result of the following property.

Property M.2 Let $R(t)$ be a timed continuous PN without conflict or such that the conflicts are solved in a deterministic way, and $\mathbf{V}(t)$ be such that its stationary behavior has the period π.

If there is $t \geq \tau_{per} + \pi$ (t different from a limit of IB-state) such that

$$\mathbf{v}(t) = \mathbf{v}(t-\pi) \text{ AND } \tilde{\mathbf{m}}(t) \geq \tilde{\mathbf{m}}(t-\pi), \tag{M.11}$$

then $\mathbf{v}(t+\pi) = \mathbf{v}(t)$ AND $\tilde{\mathbf{m}}(t+\pi) = \tilde{\mathbf{m}}(t) + (\tilde{\mathbf{m}}(t) - \tilde{\mathbf{m}}(t-\pi))$. (M.12)

❑

Proof Let P_i be such that $\tilde{m}_i(t) > \tilde{m}_i(t-\pi)$ and $P_i \in {}^{\circ}T_j$. Given (M.11), $v_j(t)$ does not depend on $\tilde{m}_i(t)$. If all the places with this property and their adjacent arcs are deleted, the vector $\mathbf{v}(t)$ is the same, i.e. it depends only on the places P_h such that $\tilde{m}_h(t) = \tilde{m}_h(t-\pi)$. Given the periodicity of $\mathbf{V}(t)$, the proof follows.

❑

In the algorithm, this property will be used as follows. When the features of IB-state s are calculated (i.e. at t_{s-1}), $\mathbf{v}_s$ is calculated from $\tilde{\mathbf{m}}(t_{s-1})$. Since $\mathbf{v}_s$ is definite in $\left| t_{s-1}, t_s \right|$, its value in this interval is considered (but not $\mathbf{v}(t_{s-1})$ which is not specified). See *Step 7*.

Algorithm M.2 **$\mathbf{V}(t)$ completely known: Hypothesis M.2**

This algorithm is similar to Algorithm M.1, except *Steps 1.2*, *2.1*, and *6*.

Step 1.2. The initial time-ordered sequence is: $TOS = (0, \mathbf{V}_0)\ldots(\tau_{a-1}, \mathbf{V}_{a-1})$ $(\tau_a, \mathbf{V}_a)\ldots(\tau_r, \mathbf{V}_r)$. The values π and $\tau_{per} = \tau_a$ are let[2], in addition to $\tau_k = 0$.

Step 2.1. Let $(\tau_{(1)}, \mathbf{V}_{(1)})$ denote the first pair in TOS. **If** $\tau_{(1)} \geq \tau_{per}$, add the pair $(\tau_{(1)} + \pi, \mathbf{V}_{(1)})$ in TOS.

Step 6. **If** $t_s = \tau_k = \infty$ **then** END
else **if** $t_{s-1} \geq \tau_{per} + \pi$ AND $\mathbf{v}(t_{s-1}^+) = \mathbf{v}(t_{s-1}^+ - \pi)$
AND $\tilde{\mathbf{m}}(t_{s-1}) \geq \tilde{\mathbf{m}}(t_{s-1} - \pi)$ **then** END
else if $t_s < \tau_k$: let $\tilde{\mathbf{m}} = \tilde{\mathbf{m}}(t_s)$ and $s = s + 1$, and go to *Step 5*.

❑

The second and third lines in *Step 6 stop the calculation when the IB-state reached was previously found* (feedback in the evolution graph).

The following example is illustrated in Figure M.3. The time unit is one month, the marking unit 1000 parts, and thus the speed unit 1000 parts / month. A company intends to manufacture parts C, as from next January, for several years. Each part C needs one part A and two parts B which are produced externally. Parts A may be provided as from January, at 3 speed units, continuously except during August and September: speed $V_1(t)$ (integer time t is the end of tth month). Parts B can be provided at 5 speed units even during the holidays, but not before next April: $V_2(t)$. However, there is an inventory: 12 units immediately available. Parts C can be produced at 4 speed units by a machine which will be available only during the first nine months of each year: $V_3(t)$. The maximum shipping of product C will be 2 speed units during the first three months and 4 units subsequently. The buffers for parts A, B, and C are assumed to be unbounded.

---

[2] If the stationary behavior of $\mathbf{V}(t)$ is not periodical: the last term in TOS is $(\tau_{a-1}, \mathbf{V}_{a-1})$, $t_{per} = \infty$, and $\pi = \infty$.

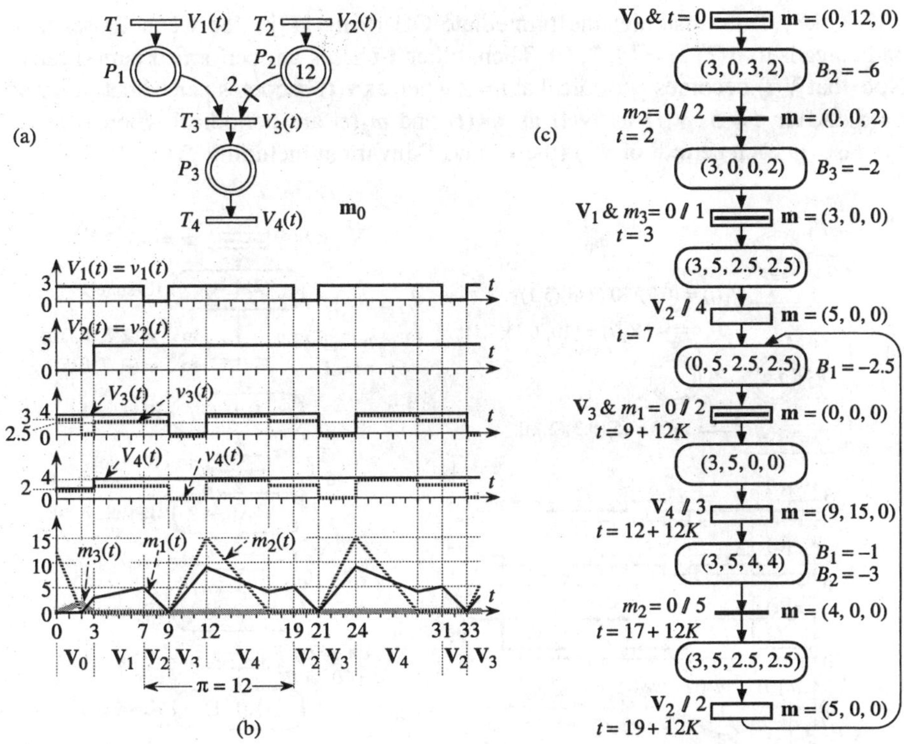

Figure M.3 (a) PN. (b) Maximal speeds and behavior. (c) Evolution graph.

The corresponding PN is given in Figure M.3a, the detailed behavior in Figure b, and the evolution graph in Figure c: when the *IB-state change is due to a C3-event, this is represented by a white rectangle* (the rectangle for $\mathbf{V}_0$ is redundant since $t = 0 \Rightarrow \mathbf{V}_0$). All the markings become periodical because the data are such that there is a solution giving $\int v_1(t)\cdot dt = 1/2\int v_2(t)\cdot dt = \int v_3(t)\cdot dt = \int v_4(t)\cdot dt = 30$ during a period (T-invariant $\mathbf{y} = (1, 2, 1, 1)$).

Remark M.1 The vector $\mathbf{V}(t)$ of functions may be given directly as (M.10) or given as $\mathbf{V}(t) = (V_1(t), \ldots, V_m(t))$. In the second case, a string (M.10) can be built. Each $V_j(t)$ ends with a constant value or a periodical value whose period is π_j. If all the times in the vectors $V_j(t)$ are rational numbers, π is the lowest common multiple of the values π_j. Example in Figure M.3: $V_1(t) = (0, 3)[12, (7, 0)(9, 3)]$, $V_2(t) = (0, 0)(3, 5)$, etc. lead to $\mathbf{V}(t) = (0, \mathbf{V}_0)(3, \mathbf{V}_1)(7, \mathbf{V}_2)[12, (9, \mathbf{V}_3)(12, \mathbf{V}_4)(19, \mathbf{V}_5)]$. Another example is presented in Figure M.4, in which $\pi_1 = 2\pi_2 = \pi_3$. ❑

Figure M.4 illustrates a case where I-phases are performed (*Step 4*, Algorithm M.2). At $t = 2$, $m_1 = 2.7$ when V_1 becomes infinite. The marking $\mathbf{m}^{bef}(2)$

= (2.7, 4.3, 0) is unstable: the immediate OG-firing $[T_1]^{2.7}$, between IB-states 1 and 2, leads to $\mathbf{m}^{aft}(2) = (0, 7, 0)$. Then, other I-phases are performed periodically. Note that $\mathbf{V}(t)$ becomes periodical at $t = 2$ whereas $\mathbf{v}(t)$ becomes periodical later, at $t = 5$. After $t = 5$, $\mathbf{v}(t)$ as well as $m_1(t)$ and $m_2(t)$ are periodical whereas $m_3(t)$ increases at each period[3] of $\mathbf{v}(t)$ (there is no T-invariant including T_3).

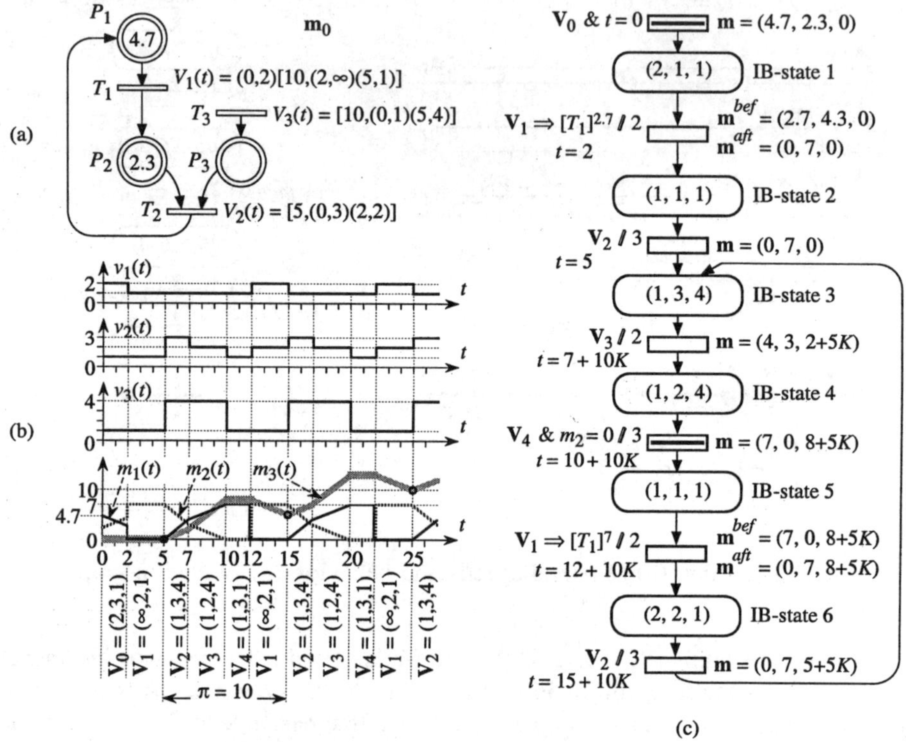

Figure M.4 (a) Example with I-phases and increasing marking.

[3] The system is said to be *quasi-periodical* (see Appendix N).

Appendix N

From Hybrid PNs to Hybrid Automata

In Chapters 4 to 6, it is shown that a hybrid PN inherits all the advantages of PN models. It provides models designed in an intuitive way. Meanwhile, except for the properties related to invariants, a quantitative analysis can be performed only via the construction of the evolution graph, which is a kind of simulation.

Hybrid automata are another tool for the modeling of hybrid dynamic systems. Several procedures have been proposed to analyze systems modeled by linear hybrid automata [Al *et al.* 95]. Although this analysis is oriented mainly towards the verification of system specifications, these procedures may also be used for other analysis goals. Hybrid automata are difficult to use for the modeling of complex systems. The goal of this Appendix is to associate the *modeling power of hybrid PNs* with the *analysis power of linear hybrid automata* by an automatic transformation from a hybrid PN into a hybrid automaton [AlAl 98b][DaAl 01]. This approach is similar to the approach combining stochastic PNs with Markov chains (automatic transformation from a stochastic PN into a Markov chain for which powerful analysis methods exist: see Section 3.4.3).

N.1 HYBRID AUTOMATA

Let us introduce intuitively the hybrid automaton model. Here is an example taken from [Al *et al.* 95]. The water level in a tank is controled through a monitor, which continuously senses water level and turns a pump *on* or *off*. Water level is represented by h. When the pump is *off*, the water level falls by 2 dm/min. When it is *on*, the level rises by 1 dm/min. Initially the level is 6 dm and the pump is turned *on*. We wish to keep the level of water between 1 and 12 dm. Moreover, there is a delay of 2 minutes from the time when the monitor signals the status of the pump until the change becomes effective. An extended *hybrid PN* describing this *hybrid system* is shown in Figure N.1a. Places P_1 and P_2 represent respectively the *on* and *off* status of the pump and P_3 the level in the tank. Transition T_3 is continuously fired at speed $v_3 = V_3 = U_3 \cdot m_1 = 1$. As soon as $m_3 = 10$, transition T_2 is enabled; it is fired 2 minutes later: T_3 is no longer enabled and T_4 becomes enabled. Transition T_1 will be enabled as soon as $m_3 < 5$ ...

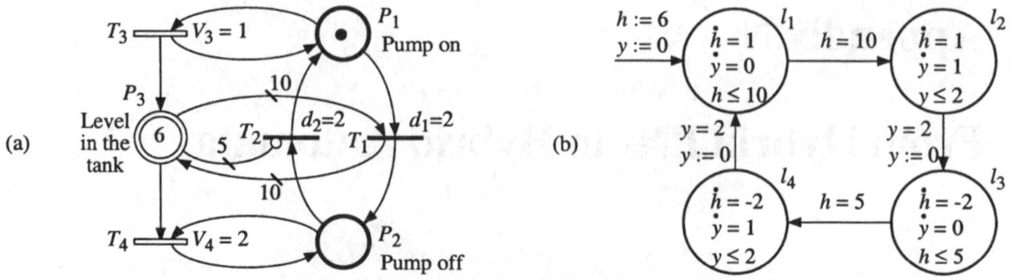

Figure N.1 Two Models. (a) (Extended[1]) Hybrid PN, (b) Hybrid automaton.

It is easy to see that this model is natural since each node represents a physical entity. The corresponding *hybrid automaton* is given in Figure N.1b. The construction of this model (taken from [Al *et al.* 95]) is less intuitive. It has four locations: in locations l_1 and l_2, the pump is turned *on*, and in l_3 and l_4, the pump is *off*. The variable y models time delays, $\dot{y}$ is the time derivative of y. The change from l_1 to l_2 corresponds to enabling of T_2 (condition $h = 10$); the change from l_2 to l_3 corresponds to firing of T_2 (condition $y = 2$, and y is reset); and so on.

Adopting the terminology of [Al *et al.* 95], a hybrid automaton[2] is defined as follows (the examples refer to Figure N.1b):

Definition N.1 A **hybrid automaton** is a six-tuple $H = \langle Loc, Var, Lab, Edg, Act, Inv \rangle$:

Loc is a finite set of vertices called *locations* (e.g.: $Loc = \{l_1, l_2, l_3, l_4\}$).

Var is a finite set of real-valued *variables* $\{x\}$ (e.g.: $Var = \{h, y\}$). A valuation u for the variables is a function that assigns a real-value $u(x)$ to each variable x. The set of valuations is denoted by U.

A *state* is a pair (l, u) consisting in a location l in *Loc* and a valuation u in U.

Lab is a finite set of *synchronization labels* that contains the *stutter label* τ (e.g.: $Lab = \{\tau, h = 10, h = 5, y = 2\}$).

Edg is a finite set of edges called *transitions*. Each transition is defined by a quadruple (l, a, μ, l'): $l \in Loc$ is a source location and $l' \in Loc$ a target location, $a \in Lab$ is a synchronization label, and $\mu \in U^2$ is a transition relation. It is required that, for each location l, there is a stutter transition of the form (l, τ, Id, l) where Id denotes the identity function (valuation u remains unchanged). E.g.: $e_1 = (l_1, h = 10, Id, l_2)$, $e_2 = (l_2, y = 2, y := 0, l_3)$.

Act is a function that assigns to each location $l \in Loc$ a set of activities. Each activity is a *time-invariant function* from the non-negative real numbers,

[1] An inhibitor arc is not necessary. Assume that the heigh of the tank is 14 dm. A C-place P_4 with arcs $T_4 \rightarrow P_4$ and $P_4 \rightarrow T_3$ may be added, such that $m_4 = 14 - m_3$. Inhibitor arc $P_3 \rightarrow T_1$ whose weight is 5 *can be replaced by an ordinary arc* $P_4 \rightarrow T_1$ whose weight is $14 - 5 = 9$.

[2] The model called hybrid *automaton* in this book, is called a hybrid *system* in [Al *et al.* 95].

representing the time since this location was reached, to U (e.g. in l_1: $\dot{h} = 1$ and $\dot{y} = 0$ }).

Inv is a function that assigns to each location $l \in Loc$ an invariant $Inv(l) \in U$ (e.g. in l_1: $h \leq 10$).

❑

A particular class of hybrid automata is represented by *linear hybrid automata.* A *linear term* over the set *Var* of variables is a linear combination of the variables in *Var* with integer coefficients. A *linear formula* over *Var* is a Boolean combination of inequalities between linear terms over Var. A *hybrid automaton* is **linear** if its activities, invariants and transition relations can be defined by linear expressions over the set *Var* of variables.

N.2 FROM HYBRID PNs TO HYBRID AUTOMATA

An algorithm was proposed in [AlAl 98b] for the construction of the hybrid automaton associated with a hybrid PN. The hybrid PN functioning corresponding to a location in a hybrid automaton may be characterized by the *evolution vector*, made up of the *enabling vector* of D-transitions (Section 3.4.2.2) and the *balance* of every C-place (Section 5.1.3.3).

One can notice that *an IB-state* (Definition 6.5, Section 6.1.5.3) *implies an evolution vector*. Hence, a linear hybrid automaton can be obtained from the evolution graph [DaAl 01]: 1) each IB-state is transformed into a location; then 2) several locations corresponding to the same evolution vector may be merged. This transformation is explained in Section N.2.1. It works only for bounded hybrid PNs (as well as the algorithm in [AlAl 98b]). However, its use may be extended to unbounded hybrid PNs thanks to the concept of macro-IB-state, as shown in Section N.2.2.

N.2.1 An IB-State Implies a Location

In a hybrid PN (hence in an *IB-state*), the continuous variables are: 1) the *residual time* to firing for every enabling of a timed D-transition; 2) the *marking* of every C-place. Similarly, in every *location*, the continuous variables correspond to: 1) the time elapsed since enabling for every enabling of D-transition (the *'delay not yet elapsed'* corresponds to the invariant of the location which is still satisfied, and *'delay elapsed'* is a synchronization label leading to a transition to another location); 2) the *marking* of every C-place.

Let us illustrate this transformation from the hybrid PN in Figure N.1a. Its evolution graph is presented in Figure N.2a. To build the hybrid automaton, the following continuous variables are considered. 1) A variable y_j for every timed

D-transition T_j. When T_j is enabled but not yet fired: $\dot{y}_j = 1$ (in case of multiple enabling, new variables will be added when necessary: $y_{j,1}$, $y_{j,2}$, ...). 2) A variable m_i for every *C-place* P_i; at any time $\dot{m}_i = B_i$ (Section 5.1.3.3).

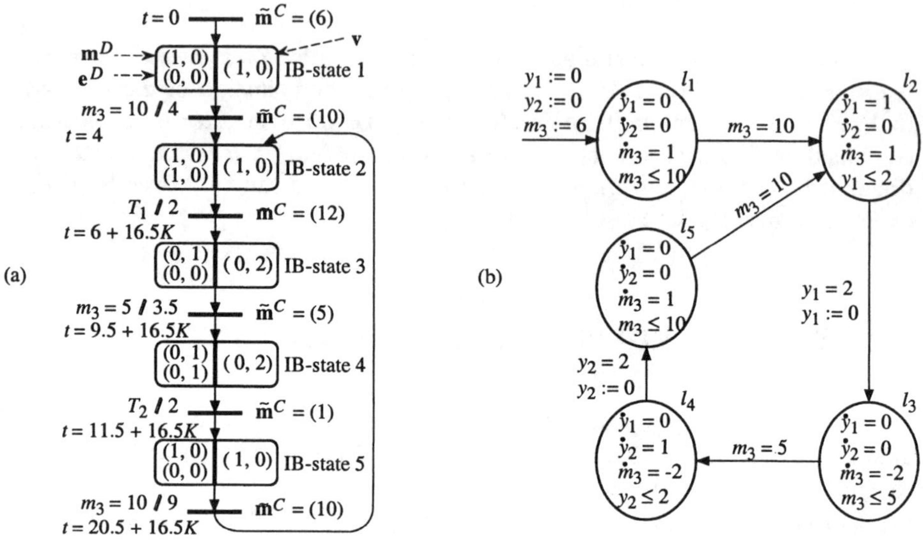

Figure N.2 (a) Evolution graph for the hybrid PN in Figure N.1a. (b) Corresponding hybrid automaton.

For our example, the evolution vector is $(\dot{y}_1, \dot{y}_2, \dot{m}_3)$ and the hybrid automaton obtained is shown in Figure N.2b. At initial time, $(y_1, y_2, m_3) = (0, 0, 6)$.

Initial location l_1 is associated with initial IB-state 1. Neither T_1 nor T_2 is enabled, hence $\dot{y}_1 = \dot{y}_2 = 0$. Since $v_3 = 1$ and $v_4 = 0$, $B_3 = \dot{m}_3 = 1$. According to Figure N.2a, the next event causing a change of IB-state is 'm_3 *reaches the value* 10'. In Figure N.2b, the invariant associated with l_1 is $m_3 \le 10$ and the synchronization label associated with the transition from l_1 to l_2 is $m_3 = 10$ ($m_3 = h$ in Figure N.1).

Location l_2 is associated with IB-state 2. In this location $\dot{y}_1 = 1$ since T_1 is enabled[3] and $\dot{m}_3$ is still 1. The next transition, from l_2 to l_3, is performed when '$y_1 = 2$' (delay 2 for transition T_1 elapsed); the value of y_1 is reset. And so on.

Figure N.2b shows that $(\dot{y}_1, \dot{y}_2, \dot{m}_3) = (0, 0, 1)$ in both l_1 and l_5. In addition, the invariant is the same, and the synchronization label and the transition relation associated with the transitions to l_2 are the same. These locations may be merged.

[3] For $t \in |4, 6|$, the system is in IB-state 2 and location l_2. In IB-state 2, a continuous value is the *residual time* to firing of T_1 i.e. $6 - t$ (which becomes 0 at $t = 6$ since T_1 is enabled at time 4). In location l_2, the corresponding continuous variable is $y_1 = t - 4$ (because this location is reached at time 4), and the next firing occurs when $y_1 = 2$, i.e. at t $t = 6$.

It can be noticed that, after merging, this automaton is isomorphic to the automaton of Figure N.1b. The only remaining difference is: one clock y in Figure N.1b and two clocks y_1 and y_2 in Figure N.2b (y_1 and y_2 are never activated simultaneously).

N.2.2 Macro-IB-States

According to Definition 6.5 in Section 6.1.5.3, an IB-state is characterized by four features. Two of them, i.e. the *enabling vector is constant* and the *vector* **v** *of speeds is constant*, correspond to the features of a location in a linear hybrid automaton. The other two features, (i.e. the discrete marking is constant and the continuous marking has always the same value when the IB-state is reached) do not correspond to required constraints for a location in a hybrid automaton. For example, in Figure N.2a the difference between IB-states 1 and 5 is that $m_3 = 6$ when the first one is reached whereas $m_3 = 1$ when IB-state 5 is reached. However, the corresponding locations, l_1 and l_5, can be merged into a single one.

Let us define a **macro-IB-state** as a *set of IB-states having the same enabling vector and the same speed vector*. In other words, IB-states corresponding to different markings (discrete or continuous) may be in the same macro-IB-state.

Let us consider the timed discrete PN in Figure N.3a. It is a particular case of hybrid PN without C-place or C-transition. However, it represents a hybrid system since the residual times to firings are continuous variables [AlDi 94].

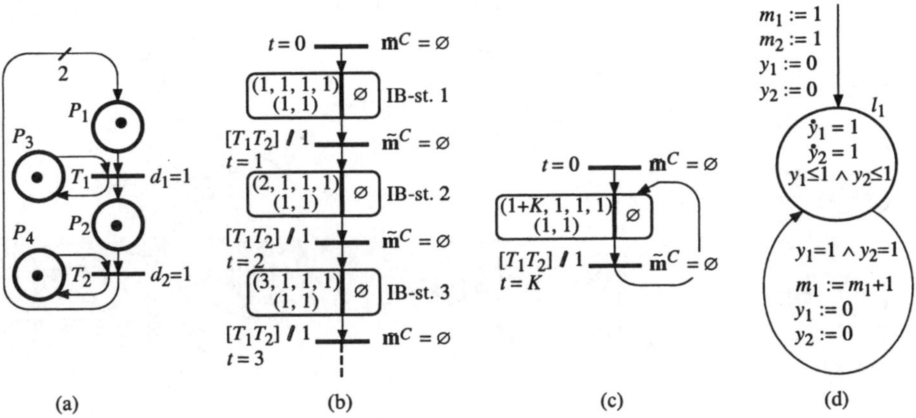

Figure N.3 (a) Unbounded PN. (b) Beginning of the evolution graph. (c) Representation with a macro-IB-state. (d) Hybrid automaton.

This PN is *unbounded*. The beginning of the evolution graph is shown in Figure N.3b. The marking is (1, 1, 1, 1) in IB-state 1, (2, 1, 1, 1) in IB-state 2, etc. For each time unit there is a double firing $[T_1T_2]$ and m_1 increases by one unit.

According to Section 2.2.1.2, the "marking" (ω, 1, 1, 1), where ω denotes an arbitrarily large number, can be reached. For any value of ω, the enabling vector is always (1, 1). It follows that all the reachable IB-states can be grouped in a single *macro-IB-state* illustrated in Figure N.3c: at $t = K$ (non-negative integer), the marking $\mathbf{m} = (1 + K, 1, 1, 1)$ is reached and the enabling vector is (1, 1). Let us say that the behavior is **quasi-periodical**. From this Figure N.3c, the hybrid automaton in Figure N.3d can be obtained.

The behavior of the timed PN in Figure N.4a (unbounded) is also quasi-periodical. The *quasi-period* is the *lowest common multiple* of $d_1 = d_2 = 1$ and $d_3 = 1.5$, i.e. 3. The behavior may be represented by the quasi-periodical evolution graph in Figure b, made up of four macro-IB-states. According to the definition of a macro-IB-state, the four macro-IB-states in Figure b can be *merged into a single one*. In this case, there is no representation of the behavior by an evolution graph. However, the hybrid automaton in Figure c can be obtained: location l_1 corresponds to the global macro-IB-state mentioned above.

If $d_3 = 1.4$ instead of 1.5, the quasi-period is 14, the evolution graph replacing Figure b contains 23 macro-IB-states, and the hybrid automaton is similar to Figure c except that the value 1.4 replaces 1.5. If $d_3 = \sqrt{2}$ (out of our usual hypotheses), Figure b would contain an infinite number of macro-IB-states, but the hybrid automaton would be similar to Figure c, $\sqrt{2}$ replacing 1.5.

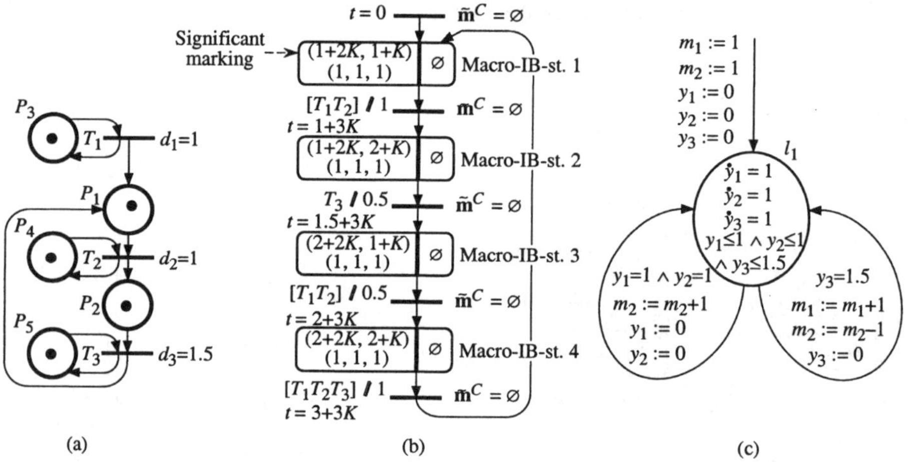

Figure N.4 (a) Timed PN. (b) Quasi-periodical behavior. (c) Hybrid automaton.

Appendix O

P&T-Timed Petri Nets and Modeling Power

In this appendix some comments will be made on PNs with both timed places and timed transitions. Let us call these nets **P&T-timed PNs**. Usually, a timed PN is either P-timed *or* T-timed but not both. An important reason is that it is easier to obtain *analysis methods* for a timed PN if its timings are homogeneous (i.e. associated with only places or only transitions). For example: [Si 77] [Ch 84] [Ba *et al.* 92]. In addition, this is not an important restriction as far as discrete PNs are concerned, as specified in Section O.1.

Notation O.1 In this appendix, the timings in a P&T-timed PN will be denoted d_j for transition T_j (as usual) and D_i for place P_i.

O.1 DISCRETE PETRI NETS

Is it possible to model by a P-timed or by a T-timed PN any system which can be modeled by a P&T-timed PN? Answers are given in Properties O.1 and O.2. Let us now specify what exactly is meant by "modeling the same[1] system".

Definition O.1 Models PN1 and PN2 **model the same system** if: 1) to each *reachable stable marking* of PN1 corresponds a non-empty set of reachable stable markings in PN2 (and vice-versa), and 2) to each *simultaneous firing of a set of transitions* in PN1, corresponds a simultaneous firing of a set of transitions in PN2, possibly empty (and vice-versa). Hence, informally, both PN1 and PN2 have the "same" reachability graph.

Property O.1 P&T-timed PNs and T-timed PNs have the same modeling power.

Proof A T-timed is a (particular case of) P&T-timed PN. On the other hand, a P&T-timed PN can be changed-over to a T-timed PN as illustrated in Figure O.1. Figure a presents a place P_i, whose timing is D_i, and the transitions linked to it in a P&T-timed PN. In Figure b, P_i has been split into two places P_i' and P_i''

[1] If outputs were associated with places or/and transitions, there would be no difficulty in obtaining the same outputs at any time for both models if they satisfy the conditions in Definition O.1.

separated by transition T_i' with which the timing D_i is associated. Tokens in P_i' and P_i'' correspond respectively to *unavailable* and *available* tokens in P_i.

Properties O.1 and O.2 are illustrated in Figure O.2. Let *S1* denote the set of systems which can be modeled by a P&T-timed PN. According to Property O.1, the set of systems which can be modeled by a T-timed PN is the same, i.e. *S1*. Let *S2* and *S3* denote respectively the set of systems which can be modeled by a P-timed PN and the subset of *S1* corresponding to P&T-timed PNs without conflict or bounded. According to Property O.2 below, $S3 \subset S2 \subseteq S1$.

Property O.2

a) The set of systems modeled by P-timed PNs is included in the set of systems modeled by P&T-timed PNs.

b) The set of systems modeled by P&T-timed PNs which are *without conflict or* (inclusive) *bounded* is a *proper subset* of systems modeled by P-timed PNs.

Proof

a) Obvious since a P-timed is a particular case of P&T-timed PN.

b) The proof contains three parts, namely i), ii), and iii).

i) If there is *no conflict* in the P&T-timed PN, a transition T_j with timing d_j can be replaced by a sequence $T_j' \rightarrow P_j' \rightarrow T_j''$; timing d_j is associated with P_j'.

ii) If a P&T-timed PN is *bounded*, its *reachability graph* can be built (as in Section 3.4.2.2 with additional information on residual times of unavailability for tokens in timed places). Then, a P-timed PN can be built in which a timed place is associated with each node of the reachability graph. This process, explained below, is illustrated by Figure O.3. From i) and ii), $S3 \subseteq S2$.

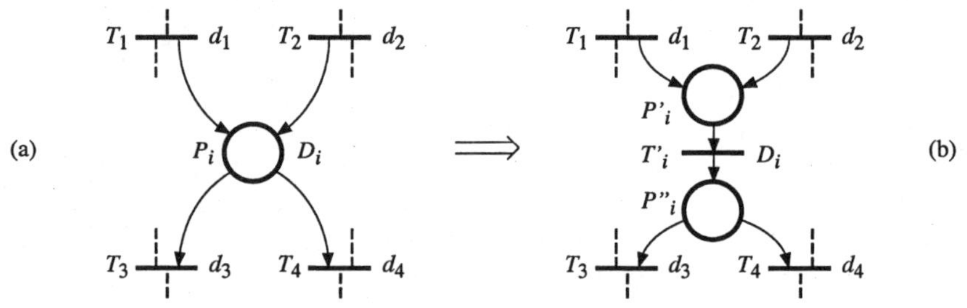

Figure O.1 Change-over from a P&T-timed PN to a T-timed PN.

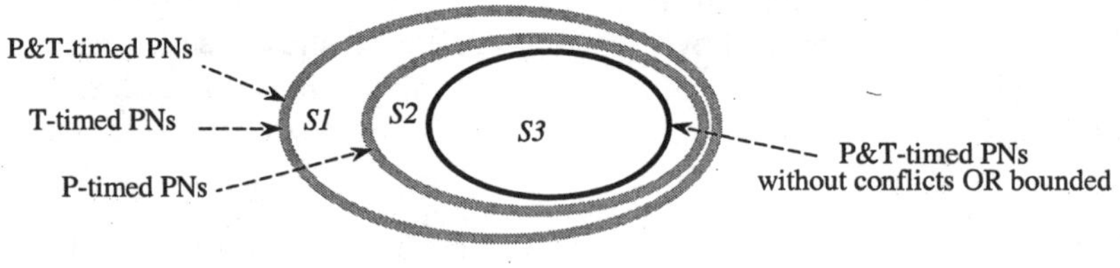

Figure O.2 Illustration of Properties O.1 and O.2.

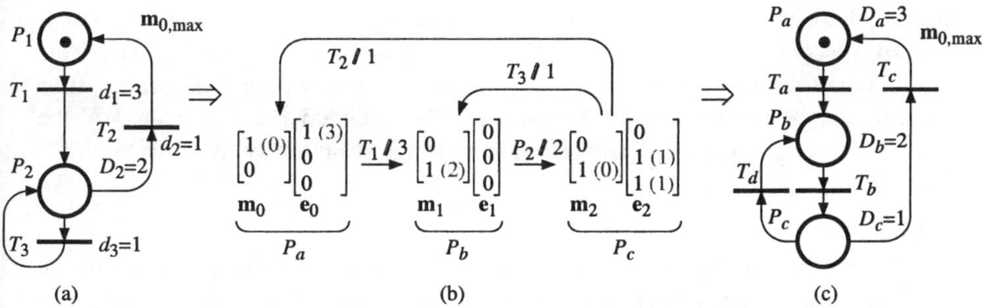

Figure O.3 Change-over from a bounded P&T-timed PN to a P-timed PN.

iii) An example of system in *S2* but not in *S3* will be given in Remark O.1a. ❑

Figure O.3a represents a P&T-timed PN with an initial marking[2] $\mathbf{m}_{0,\max}$; According to Remarks3.12 and3.13 in Section 3.4.2, this notation means that, at initial time the *residual times* (to availability for a token and to firing for an enabling) *are maximal.* Only place P_1 is not timed ($D_1 = 0$ is not written in the figure). From this PN, the reachability graph in Figure b can be obtained. The building of a reachability graph is explained in Section 3.4.2.2 for a T-timed PN: each node N_k of this graph consists of a pair ($\mathbf{m}_k$, $\mathbf{e}_k$) in which $\mathbf{m}_k$ is the marking and $\mathbf{e}_k$ the enabling vector; in addition, the residual times (to availability for a token, to firing for a transition enabling) are specified.

In Figure O.3b, the marking $\mathbf{m}_0$ = (1, 0) is written with the additional information in brackets that the residual time to availability is 0 (i.e. the token is available). The corresponding enabling vector is $\mathbf{e}_0$ = (1, 0, 0), i.e. transition T_1 is 1-enabled, and the residual time to firing is 3 (corresponding to d_1 = 3). The next event will be the firing of T_1, occurring three time units later and leading to $\mathbf{m}_1$ = (0, 1). When this marking is reached, the residual time to availability of the token is 2 (corresponding to D_2 = 2) and no transition is enabled. Next event: the token in P_2 becomes available and the node N_2 = ($\mathbf{m}_2$, $\mathbf{e}_2$) is reached. The transition is denoted by P_2/ 2, meaning that the token in P_2 becomes available 2 time units after N_1 was reached. For N_2 = ($\mathbf{m}_2$, $\mathbf{e}_2$), both T_2 and T_3 are enabled; since both can be fired one time unit later, an actual conflict will appear at this time. From the reachability graph in Figure b, the P-timed PN in Figure c can be obtained. Each node N_k = ($\mathbf{m}_k$, $\mathbf{e}_k$) in Figure b, is associated with a place P_h in Figure c and the timing D_h is equal to the delay corresponding to arcs outgoing from N_k. As a matter of fact, by construction (according to *Operating of a T-timed PN* in Section 3.4.2.2), if there are several arcs outgoing from N_k, they can be fired at the same time (actual conflict).

According to Definition O.1, the PNs in Figures O.3a and c model the same system. Reachable markings: $\mathbf{m}_0$ = (1, 0) and $\mathbf{m}_1$ = (0, 1) for Figure a,

[2] Another initialization could be chosen.

$\mathbf{m}_a = (1,0,0)$, $\mathbf{m}_b = (0,1,0)$, and $\mathbf{m}_c = (0,0,1)$ for Figure c. Hence, $\{\mathbf{m}_a\} \Leftrightarrow \{\mathbf{m}_0\}$, $\{\mathbf{m}_b, \mathbf{m}_c\} \Leftrightarrow \{\mathbf{m}_1\}$, $\{T_a\} \Leftrightarrow \{T_1\}$, $\{T_b\} \Rightarrow \varnothing$, $\{T_c\} \Leftrightarrow \{T_2\}$, $\{T_d\} \Leftrightarrow \{T_3\}$. For this simple example, the PN in Figure c could be intuitively found. However, in a general case, the structure of the P-timed PN obtained (i.e., structure of the reachability graph) may be quite different from the structure of the initial P&T-timed PN.

Remark O.1
a) Let us consider a new PN defined as follows: similar to Figure O.3a with an additional place P_3 and an additional arc $T_3 \rightarrow P_3$. The system modeled by this new PN is not in *S3* (it is *unbounded* and contains a conflict) but it is in *S2*. As a matter of fact the system can be modeled thanks to the following modification of Figure O.3c: add a place P_d and an arc $T_d \rightarrow P_d$. Hence, $S3 \subset S2$.

b) Property O.2a showed that $S2 \subseteq S1$. Is the exact property $S2 \subset S1$ or $S2 = S1$? The authors do not know the answer to this question. They are familar with examples of P&T-timed PNs for which they have not been able to find a P-timed PN modeling the same system. But they have no proof that such a solution does not exist.

Remark O.2 *T-timed PNs with reserved tokens*, not used in this book but briefly presented in Section 3.4.1, have the same modeling power as P-timed PNs [DaAl 89 & 92]. (They are nothing but an abbreviation of P-timed PNs.)

O.2 HYBRID PETRI NETS

Figure O.4 presents a production system made up of two machines, a conveyor linking them and four buffers. It is assumed that the capacities of the buffers and of the conveyor are unbounded. Each machine can process at most one part at a time. Processing times are $S_1 = 3$ for machine M_1, $S_2 = 2$ for machine M_2, and the time to cover the length of conveyor C is $S_C = 10$. At initial time, there are twelve parts in B_1, to be processed by M_1 then M_2.

P-timed and T-timed models of this system, taking into account the processing times in order to have a *minimal* representation *in place and transition numbers*, are given in Figure S 3.10.

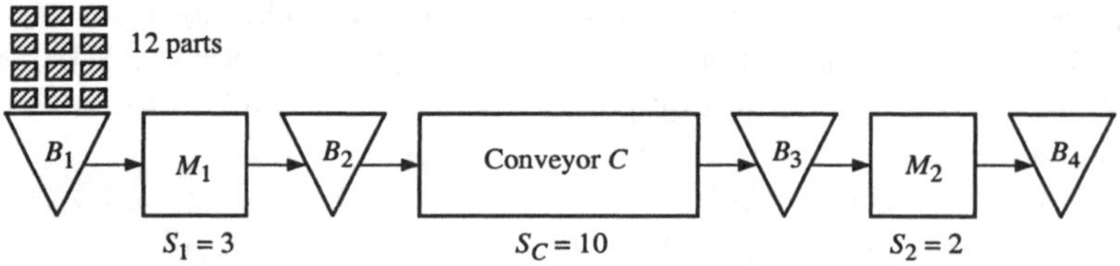

Figure O.4 Production system with 12 parts to be processed.

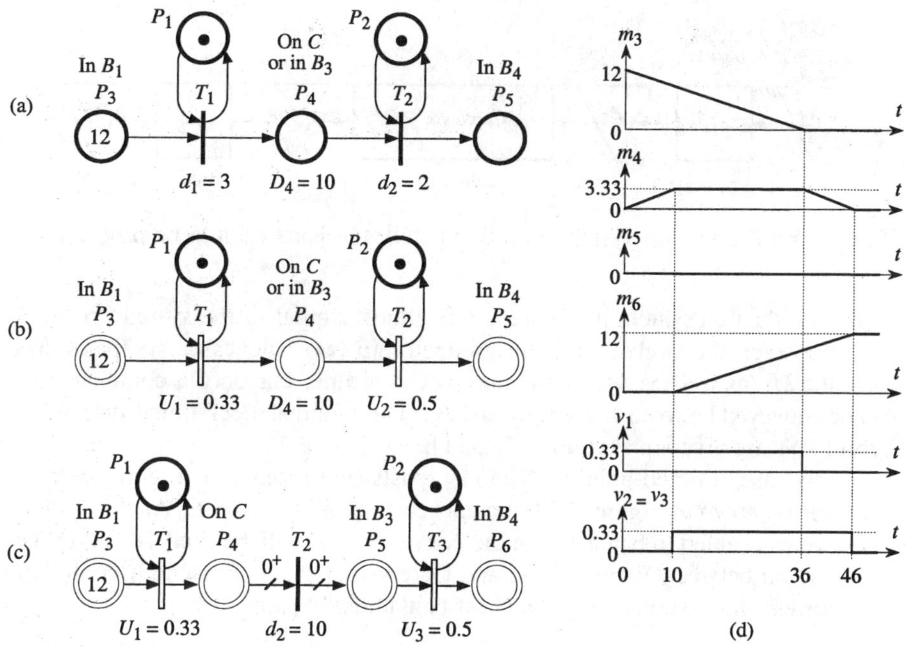

Figure O.5 Modelings for system in Figure O.4. (a) P&T-timed discrete PN. (b) P&T-timed hybrid PN. (c) Extended timed hybrid PN. (d) Behavior of Figure c.

A P&T-timed discrete model and a P&T-timed hybrid model (approximation by continuous flows) are given in Figures O.5a and b (in a general case, timings could also be associated with D-transitions and D-places in a P&T-timed hybrid PN). These are the *minimal representations* for both models. Some quantity of marking deposited in P_4 becomes available 10 time units later for enabling T_2, in Figure O.5b too. This means that *unavailable* marking in P_4 corresponds to parts on the *conveyor* and that *available* marking in P_4 models parts in *buffer* B_3.

It was observed in the Solution to Exercise 3.11 that there are two kinds of time, varying in nature: *processing times* (for machines M_1 and M_2) and *transportation time* (for conveyor C). It is interesting to note that, for both P&T-timed discrete and hybrid PNs, in the *minimal representations*, the *processing times* are associated with *transitions* and the *transportation times* with *places*.

Now, if a basic timed hybrid PN is the target (i.e. with *only timed transitions*, according to Chapter 6), there is no solution without an *extended* timed hybrid model (Section 6.4.3). The model and its behavior are presented in Figures O.5c and d, respectively. In Figure O.5c, the parts on the conveyor (m_4) and the parts in buffer B_3 (m_5) are associated with two different places linked by a timed discrete transition T_2 whose timing is $d_2 = S_C = 10$.

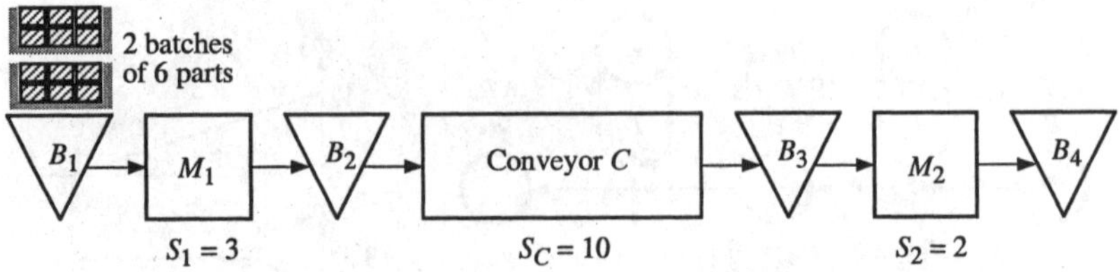

Figure O.6 Production system with two batches of parts each to be processed.

The production system in Figure O.6 is almost similar to the system in Figure O.4. However, the twelve parts are grouped into two batches of six parts each. Machine M_1 (as well as M_2) treats one part at a time, but only a complete batch can be conveyed between buffers B_2 and B_3. The minimal P&T-timed discrete and hybrid PNs are given in Figures O.7a and b.

In this case, a timed hybrid PN model exists (*not* extended): Figure O.7c. Its behavior is shown in Figure O.7d: the dots correspond to $m_4 = 6$. At these times, T_2 becomes enabled (a batch put on the conveyor) and will be fired 10 units later. Comparison between Figures O.5c and O.7c shows that conveying a continuous flow corresponds to conveying "batches" containing 0^+ part.

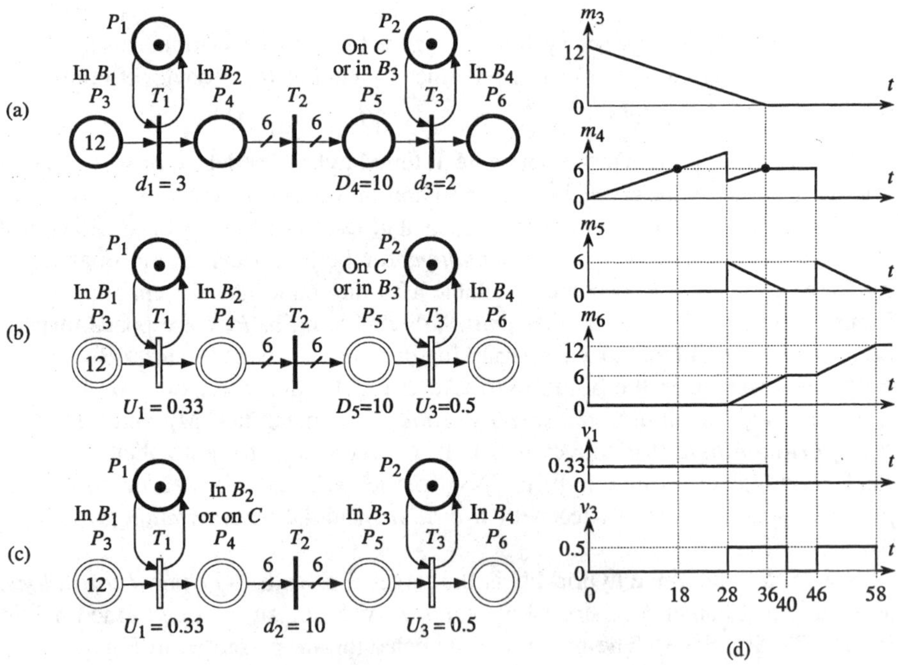

Figure O.7 Modelings for system in Figure O.5. (a) P&T-timed discrete PN. (b) P&T-timed hybrid PN. (c) Timed hybrid PN. (d) Behavior of Figure c.

Exercises

The proposed exercises are divided into eight parts. The first seven parts correspond to exercises related to Chapters 1 to 7, respectively. The 8th part contains three case studies whose resolution uses the results of various chapters; the general model including these various possibilities is the control hybrid PN *defined in the Postface.*

Notation ES If there are several figures in the same exercise (or solution), these will be distinguished by a bold capital letter. In this exercise (or solution), the figure may be referred to only by this letter. Example: the first figure in Exercise E 8.1 is referred to as Fig. E 8.1.**A** or simply Fig. **A**.

Chapter 1

The authors recommend that you concretize the tokens using small objects. This is very useful, in order to avoid errors.

1.1 a) Are the nets in Figure E.1.1 PNs? For those that are Petri nets, indicate:
b) the transitions which are enabled;
c) the markings after firing;
d) the transitions which are still enabled after firing.

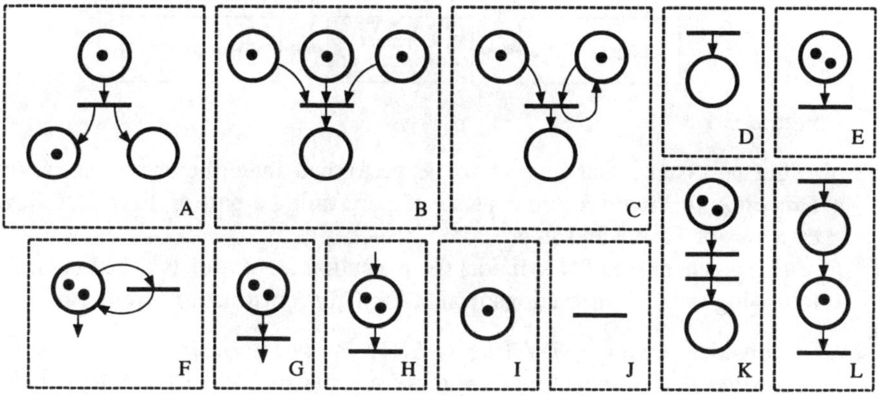

Figure E 1.1

1.2 a) Is the PN in Fig. 1.1 a state graph? An event graph? Without conflict? Free choice? Simple? Pure?

b) Same questions for the PN in Fig. 1.8a.

1.3 Consider the example given in Fig. 1.11a (Section 1.2.2.4) with an additional constraint: the administration only lets in one customer at a time.

a) Describe the functioning by an inhibitor arc PN.

b) Describe the functioning by an ordinary PN (thus without inhibitor arc)

c) Now consider that 4 customers can be let in (at the most) before closing the door. Describe the functioning by a generalized PN, without inhibitor arc.

1.4 See Figure E 1.4. When its "production" is complete (production of one part at a time), a *producer* deposits the part produced in a buffer, if there is any room in this buffer whose capacity is 3 units. As soon as he has deposited it, he starts to produce another part. A *consumer,* as soon as he has finished consumption (one part at a time) takes out a part from the buffer if this is not empty. Represent the functioning of this system by a PN with an initial marking corresponding to the figure.

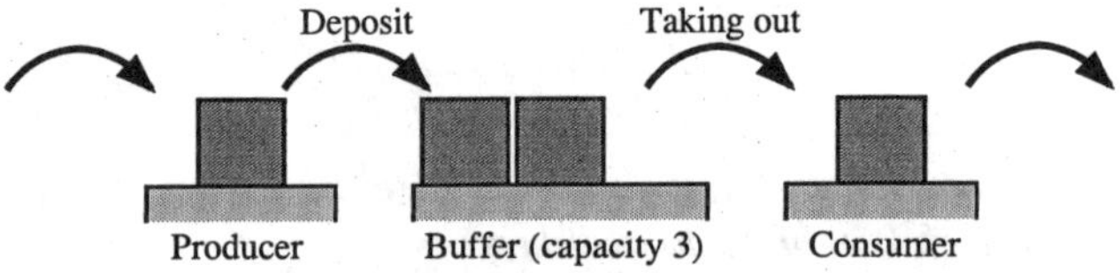

Figure E 1.4

1.5 *Two computers use a common memory* (Figure E 1.5). We assume that each computer may have three states: either it does not need the memory, or it requests it but does not yet use it, or it uses it. Model the functioning of this system using a PN.

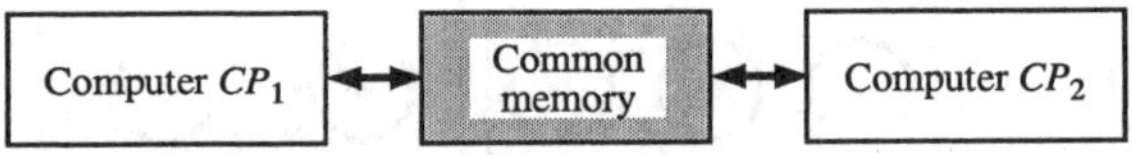

Figure E 1.5

1.6 Let us assume that 2 tasks to be performed share the same central unit. *Execution in Round-Robin* consists of performing a part of Task 1, followed by a part of Task 2 and so on.

Use a generalized PN to model the execution in turn of two tasks with the following rates: 3 instructions from Task 1 for 5 instructions from Task 2.

1.7 Consider a *task scheduling (PERT: Program Evaluation and Review Technique)* with recycling. Six tasks are defined, execution of which is determined by the following rules:

At the start, Task 1 can be performed. Tasks 2 and 3 can only be performed when Task 1 is complete (warning, this does not mean that the execution of these tasks begins immediately after Task 1 is complete, nor even that they begin simultaneously). Task 4 can only be performed after Task 3, Task 5 after Tasks 2 and 4, and Task 6 after Tasks 4 and 5.

Finally, Task 1 can only be performed again after Task 2 is complete, Task 3 after Tasks 1 and 6, and the cycle starts again indefinitely.

Remark. If a Task j can only be performed after Task i is complete, and if Task i is performed several times in a row without Task j being performed, then Task j can be performed one or more times in a row. For example, if Task 1 has been performed 5 times and Task 3 has not been performed at all, then Task 3 could be performed up to 5 times before Task 1 is next complete.

Represent the linking up of these tasks by a PN such that their executions are associated with places. Can this PN be simplified? Check that this PN is an event graph and explain why.

1.8 Consider the *flow shop* symbolized in Fig. E 1.8. Parts enter the buffer B_1, pass through machine M_1, then enter buffer B_2, finally pass through machine M_2, before leaving the flow shop. The buffers have an unlimited capacity. There can only be one single part on each machine.

There are two types of parts, p_1 and p_2. The parts arrive in no particular order (thus they are mixed in buffer B_1), but they pass through the machines in a well defined order, i.e. with an alternance p_1 then p_2 then p_1 etc.

Use a PN to describe this functioning.

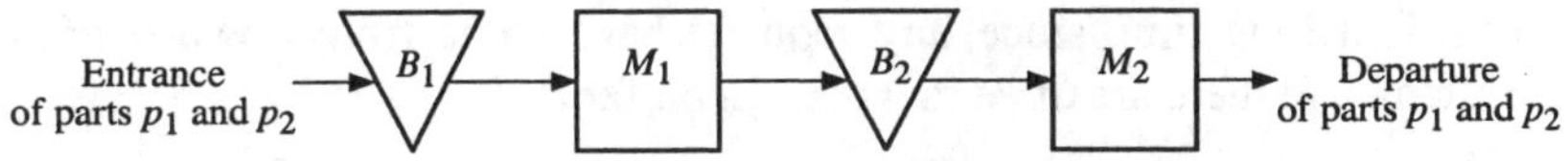

Figure E 1.8

1.9 See Fig. E 1.9. Machine M_1 receives a part, treats it and then deposits it in buffer B_1 if this buffer is not full. It can then receive another part. Machine M_2 functions in an identical manner with buffer B_2. Machine M_3 carries out the assembly: it takes up a part in B_1 and a part in B_2 which it assembles before depositing the assembled part in a downsteam buffer which is not represented and which is assumed never to be full. It can then start up again. Buffers B_1 and B_2 have capacities of 3 and 4 units respectively. We assume in addition that a machine can break down while a part is being produced. After breakdown, a repair will be carried out so that production can be continued.

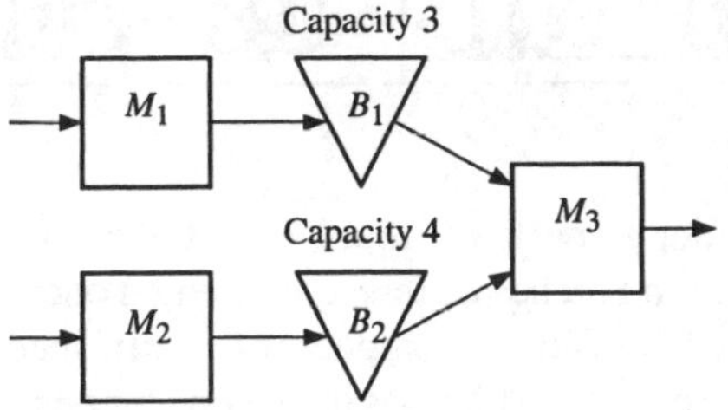

Figure E 1.9

Use a PN to represent the functioning of this system. Do not represent either the upstream of machines M_1 and M_2 or the downstream of machine M_3.

1.10 Figure E 1.10a represents the *control of a production system by kanbans* (*kanban* is a Japanese word which means "label"). This system is made up of two in series production meshes. Mesh i is made up of the production system PS_i and its buffer of finished products B_i (the parts in a buffer are not arranged: they may be considered as being loose, since they are all identical). The raw parts are in buffer B_o. In order for a part which is in buffer B_{i-1} to enter production system PS_i, it must carry a kanban i (i = 1, 2). When it is finished, it is deposited in buffer B_i with its *kanban* which remains attached to it. When a part is removed from B_i in order to satisfy a downstream request (request from an external customer for B_2, or request from mesh 2 by the arrival of a *kanban* 2 for B_1), it is separated from its kanban i and it is given a kanban i + 1 (except if mesh i is the last one). *Kanban i* is then brought back to the entrance of production system PS_i to be allocated to another part.

a) Use an ordinary PN to represent this system by taking the following hypotheses: a) for each mesh i, there is a place corresponding to the parts in production system PS_i and a place corresponding to the parts in buffer B_i, the changeover from one to the other being effected by the firing of a transition which corresponds to the execution of the part, and only one part can be treated at a time; b) there are two *kanbans* for mesh 1 and three for mesh 2; c) it is a long time since any requests have come from downstream in the system, and there are three raw parts in buffer B_o.

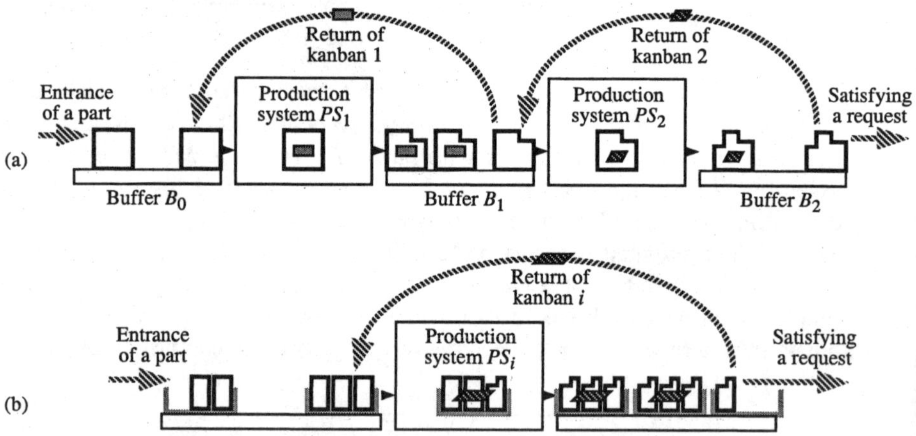

Figure E 1.10

b) Now consider a mesh i (Figure E 1.10b) in which the parts enter the production system in batches of three, are treated one by one and are deposited in the downstream buffer by batches of 3 finished parts. One *kanban* is associated with one batch. The parts arrive one by one and are requested one by one. As soon as the last part is removed from a batch present in B_i, treatment of a fresh batch is authorized (return of the *kanban*). Use a generalized PN to represent this system assuming that there are two *kanbans* and that the initial marking is as follows: it is a long time since there have been any requests from downsteam; b) there are 4 parts in B_{i-1} and 4 parts in B_i (i.e. a complete batch plus one part).

Chapter 2

2.1 For each of the PNs, A, B, C, D and E of Figure E 2.1, answer the questions: is it bounded? live? deadlock free?

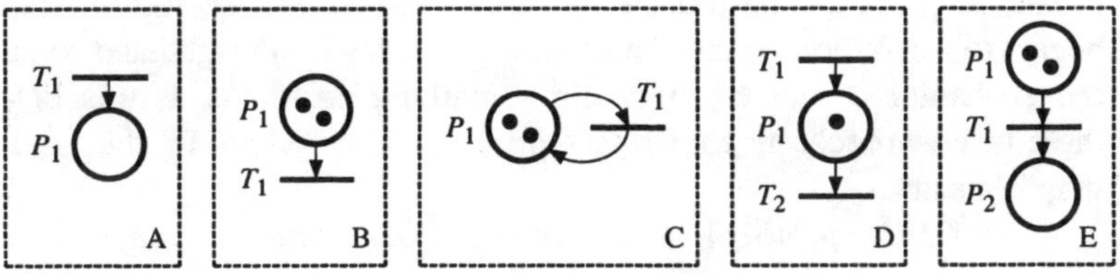

Figure E 2.1

2.2 For each of the PNs A to F of Fig. E 2.2, certain of which are generalized PNs, answer the following questions: is it bounded? live? deadlock free?

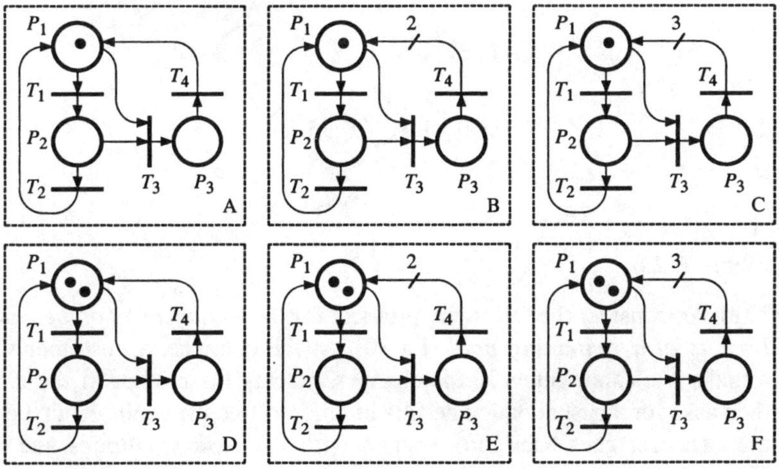

Figure E 2.2

2.3 Are the PNs in Figure E 2.1 repetitive or increasing repetitive?

2.4 For both PNs in Figs. S 1.4 and S 1.6, answer the following questions:

a) Is it a strongly connected event graph?

b) Give the P-invariants and T-invariants, the marking invariants, and one or several repetitive sequences.

c) Show that it is live (using Properties 2.8 and 2.10).

d) Is it consistent?

2.5 For the PN in Fig. S 1.8, answer the following questions:
a) Give the P-invariants.
b) Show that it is live using Property 2.11.

2.6 As shown in Figure E 2.6, four *philosophers, $Phil_1$* to *$Phil_4$* are seated round a table, laying out rods B_1 to B_4 arranged among themselves. A philosopher may essentially have two states: he thinks or he eats. In order to eat, he needs the two rods which are on either side of him. At the initial state, all the philosophers are thinking and the rods are placed on the table.

a) Using a PN describe the following protocol: when a philosopher wishes to eat, he picks up the rod on his right, followed by the rod on his left and begins to eat. When he has finished, he sets down the righthand rod and then the left-hand rod. Indicate the minimal marking invariants. Is this PN live? If there is a deadlock, give a firing sequence which results in this and explain why it occurs.

b) Define a protocol such that deadlock cannot occur and give the corresponding PN.

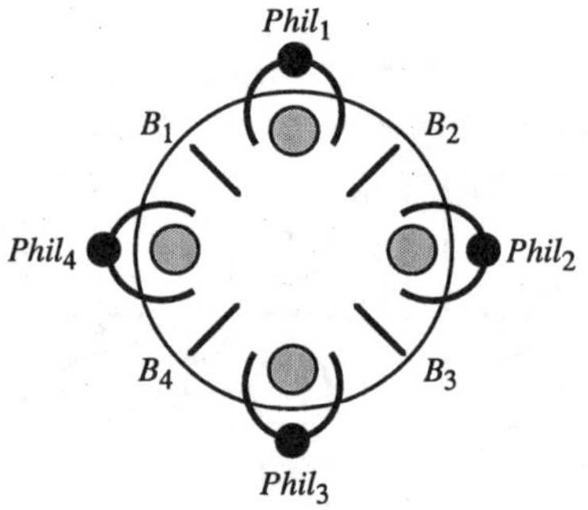

Figure E 2.6

2.7 a) We consider the following protocol for the *management of the cabins* and *baskets of a swimming pool* [La 80]. At the entrance, a customer who has found a free cabin enters it and changes, placing his clothes in the cabin. He then asks for a basket which he fills in order to free the cabin. After his swim, the customer goes back into a cabin with his basket, empties and frees it. Then he gets dressed and frees the cabin.

Let N_c be the number of cabins and N_p the number of baskets. Model this protocol with $N_c = 3$ and $N_p = 5$. Is the number of customers swimming bounded (i.e. after undressing and before redressing)? Is the PN bounded? Show that there is a state of deadlock. Is there deadlock for all values of N_c and N_p?

b) Define a protocol such that there is no deadlock, and give the corresponding PN.

c) Modify the PN of b) in order to model the number of customers who are waiting for a cabin to enter the pool.

2.8 Consider the PN in Fig. 2.16a.

a) Give the incidence matrix of this PN

b) Assume the initial marking $\mathbf{m}'_0 = (1, 0, 1, 0, 0, 0)$. Build the graph of markings and give the set of possible firing sequences.

2.9 Construct the coverability root tree for the PN of Figure E 2.9. Work out from this the places which are not bounded.

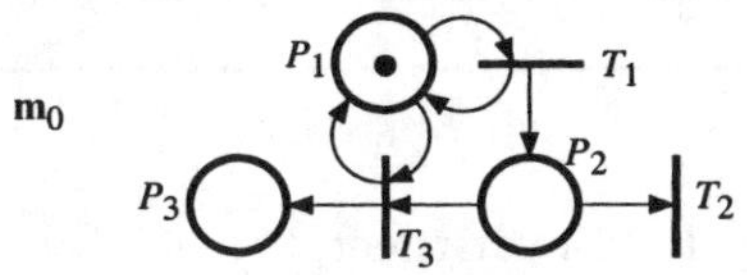

Figure E 2.9

2.10 a) Consider the unmarked PN corresponding to Fig. 2.4a. Intuitively look for the P-invariants and T-invariants of this net and then prove this result.

b) Give the dual PN of Fig. 2.4a. What are the P-invariants and T-invariants of this net?

2.11 For both PNs in Fig. E 2.11, find the minimal P-invariants using Algorithm 2.2. Then give the marking invariants.

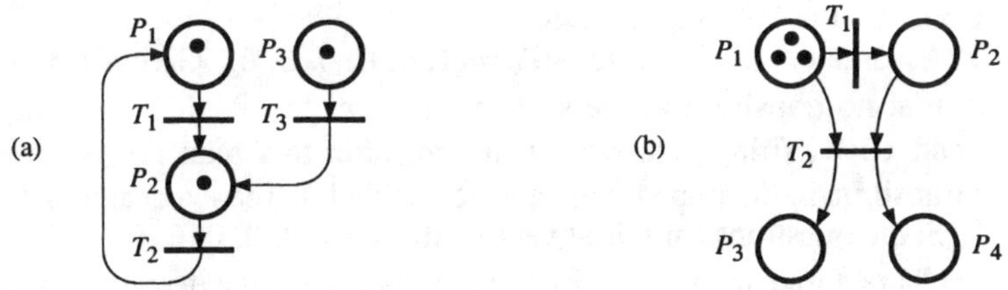

Figure E 2.11

2.12 For the marking **m** of the (part of) PN presented in Fig. E 2.12:

a) What are the enabling degrees of transitions T_1, T_2, and T_3?

b) What are the maximal concurrent firings possible?

c) Is $\langle P_2, \{T_1, T_2, T_3\}, \mathbf{m}\rangle$ an effective conflict? Is it a general conflict?

d) Are $\langle P_2, \{T_1, T_2\}, \mathbf{m}\rangle$, $\langle P_2, \{T_1, T_3\}, \mathbf{m}\rangle$, and $\langle P_2, \{T_2, T_3\}, \mathbf{m}\rangle$, general conflicts?

e) Are $\langle P_1, \{T_1, T_2\}, \mathbf{m}\rangle$ and $\langle P_3, \{T_2, T_3\}, \mathbf{m}\rangle$ general conflicts?

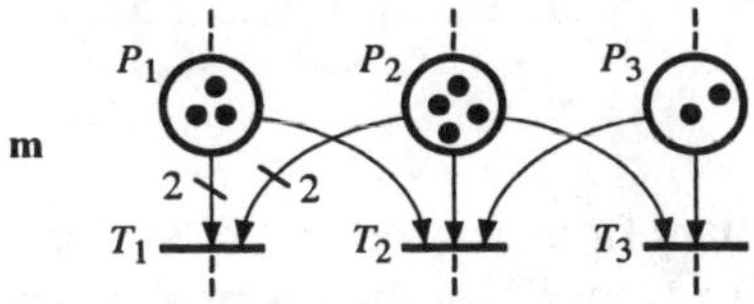

Figure E 2.12

2.13 Consider the priority PN presented in Fig. E 2.13 and its marking **m**.

a) What are the enabling degrees of transitions T_1 to T_5?

b) For the priorities given in the figure, give the priority graph and the allowing degrees (Definition B.1 in Appendix B) of transitions T_1 to T_5.

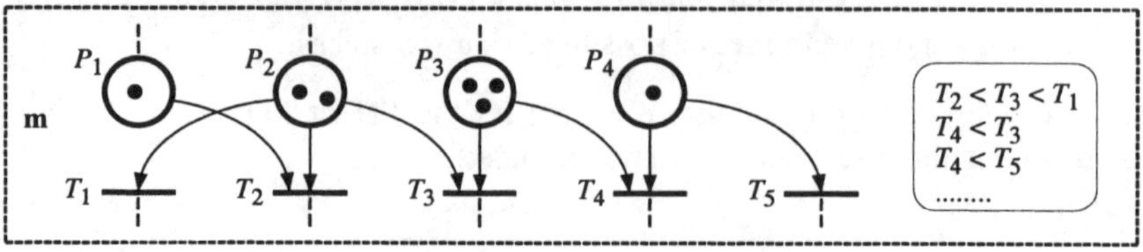

Figure E 2.13

2.14 Is the PN in Fig. E 2.14 persistent?

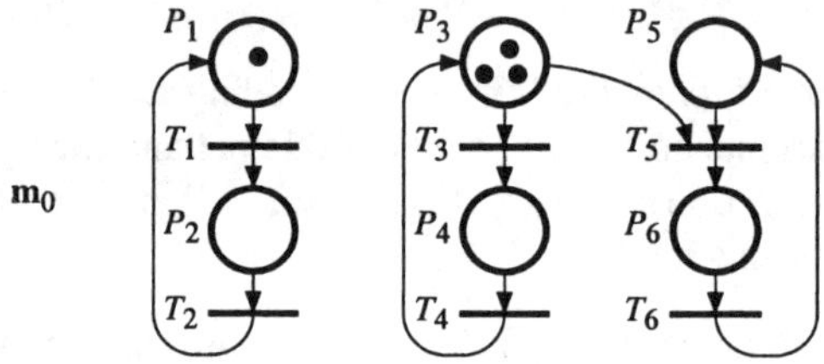

Figure E 2.14

2.15 Consider the PN in Fig. 2.24c.

a) Assume the initial vector $\mathbf{m}'_0 = (1, 0, 0, 0, 0, 0)$. Give a firing sequence such that no transition in the siphon will ever be enabled from the marking reached. Give a firing sequence such that, after this sequence is fired, at least one transition in the trap will always be enabled. Is-there a deadlock?

b) Same questions for initial vector $\mathbf{m}''_0 = (3, 0, 0, 0, 0, 0)$.

c) Characterize the set of initial markings such that a deadlock exists.

Chapter 3

3.1 Construct the graph of stable markings for the synchronized PNs in Fig. E 3.1.

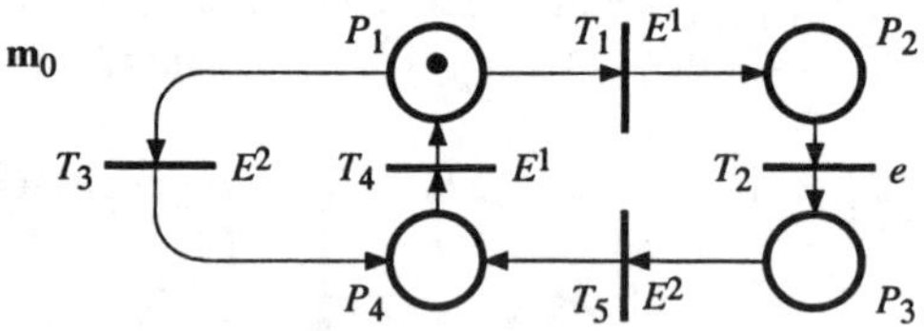

Figure E 3.1

3.2 Give the elementary firing sequences (EFSs) for the following cases.

a) With respect to event E^2 for the PN of Fig. 3.6a (Section 3.2.1).

b) With respect to event E^1 for the PN of Fig. 3.6b.

c) With respect to event E^1 for the PN of Fig. E 3.2a.

d) With respect to event E^1 for the PN of Fig. E 3.2b.

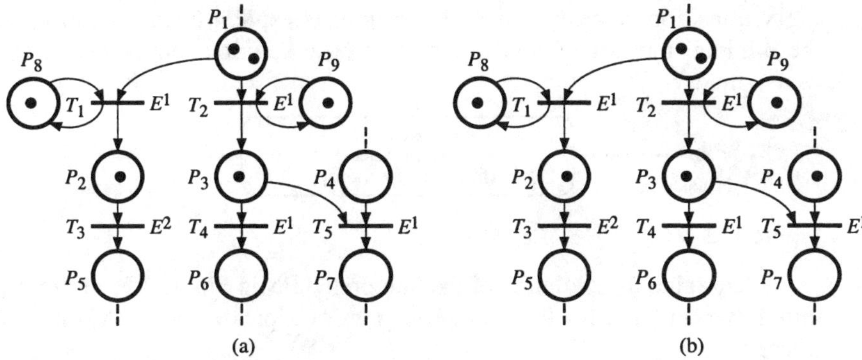

Figure E 3.2

3.3 We consider a buffer able to contain an unlimited number of parts (Fig. E 3.3). Its functioning is synchronized on the following two external events: event E^1, arrival of a part, and event E^2, arrival of a part request. When a part is requested, this request is immediately satisfied if there is a part in the buffer.

Use a synchronized PN to model the behavior of the buffer and construct the coverability root tree of reachable markings in the following two cases:

a) We assume that an unsatisfied request (because there are no parts in the buffer) is "lost" (the user will thus be forced to renew his request).

b) We assume that an unsatisfied request is memorized, and will be satisfied as soon as a new part arrives.

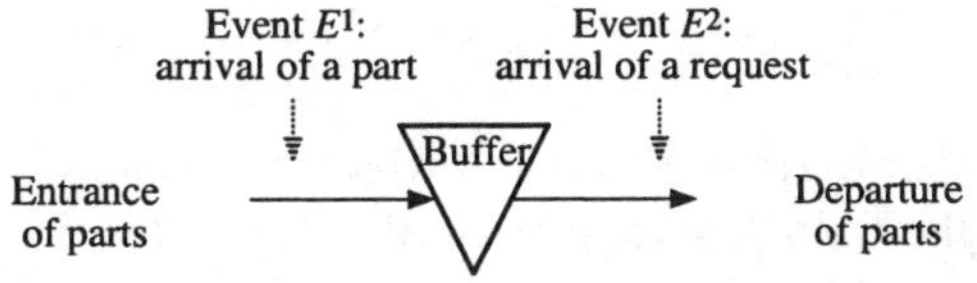

Figure E 3.3

3.4 Consider the system presented in Fig. E 3.4. It is placed between a production and a consumption whose continuous flows are respectively d_p and d_c. The system to be modeled has four possible configurations. If $d_p < a$, the system is in configuration 4 (a is a small value which may be ignored, i.e., produced flow d_p is considered as nul if it is less than a). Otherwise, the system is in configuration 1 if $-a < d_p - d_c < a$, in configuration 2 if $d_p - d_c > a$, and in configuration 3 if $d_c - d_p > a$.

The following Boolean variables will be used:

$$D_0 = [d_p < a],\ D_{pc} = [-a < d_p - d_c < a],\ D_p = [d_p - d_c > a],\ D_c = [d_c - d_p > a].$$

a) Model the transitions among the configurations by an interpreted PN without output (Sect. 3.3.3), assuming that $d_p(t)$ and $d_c(t)$ are continuous functions of time, and that $d_c(t) > 2a$ for any $t \geq 0$.

b) Same question, assuming that $d_p(t)$ and $d_c(t)$ are piecewise constant functions of time, and that $d_c(t) \geq 0$ for any $t \geq 0$. Give a solution in which every transition between two configurations is explicit, then a solution with the minimal number of transitions. Are these solutions equivalent if there were outputs?

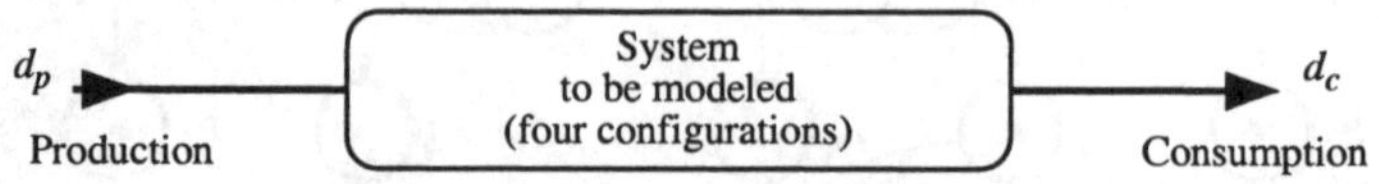

Figure E 3.4

3.5 a) Determine the evolution of the interpreted PN in Fig. E 3.5, as from the initial marking and for the timing diagram of inputs x and y, shown in the figure.

b) Show that the interpreted PN obtained is a control interpreted PN (Definition 3.7, Sect. 3.3.1). Give the corresponding grafcet.

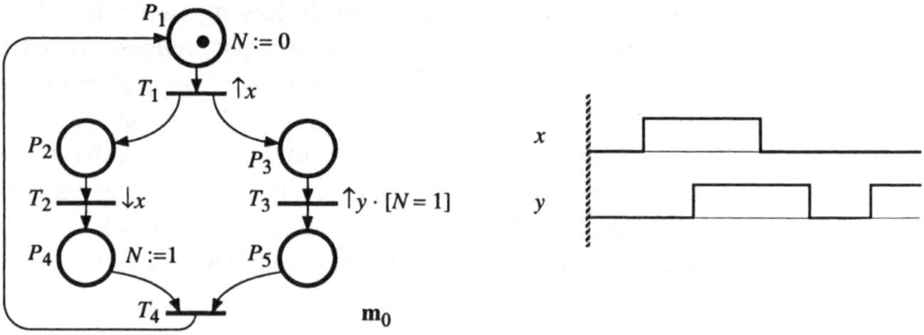

Figure E 3.5

3.6 A train passes in front of points B which are equidistant. A signal $b = 1$ is given when the train passes in front of a point. The number of periods of a clock h is counted between two points, and the *speed* of the *train* is calculated from this number.

a) Using a control interpreted PN, describe the functioning which is specified by the timing diagram in Fig. E 3.6.

b) Comment on the cases where the rising front of h occurs while $b = 1$, and where the rising front of b occurs while $h = 1$.

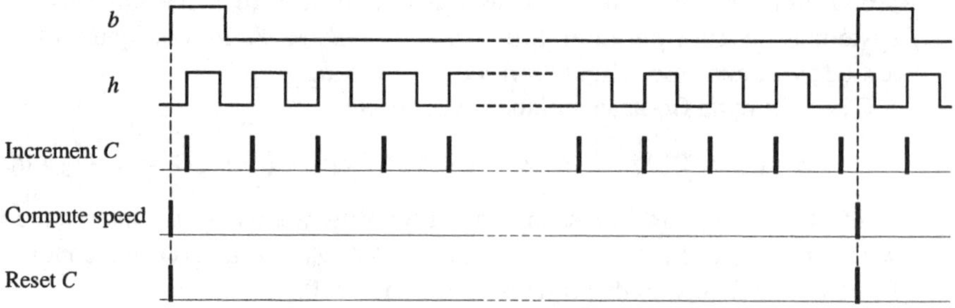

Figure E 3.6

c) Comment on the case of "apparent simultaneity" of $\uparrow h$ and $\uparrow b$.
d) Give the corresponding grafcet.

3.7 The system to be described has 6 Boolean inputs, as shown in Fig. E 3.7, and produces 6 actions including one impulse action. Model it by a grafcet, using the macroactions presented in Appendix E (Sect. E.2).

1) As from the initial state, as soon as event $\uparrow a$ occurs, the level action A is set to state 1, together with action B if $b = 0$ or C if $b = 1$. As soon as variable b assumes value 1, action B is stopped and action C started. Then, when $a\,b' = 1$, actions A and C are stopped and the impulse action D^* is performed. Finally, on the rising edge of c, there is a return to the initial state.

2) The operation of this system may very well be disturbed by the appearance of *two faults* assumed not to be able to occur simultaneously. The Boolean variables x and y indicate the presence of the first and second faults, respectively. When one of the faults appears, all the actions previously defined are reset to zero ($A = B = C = 0$ and action D^* prohibited), and an acoustic alarm is set to 1 (Boolean variable S). When the fault has cleared, operation starts up again, but not necessarily at the initial state. In the case of the *first fault*, there is restart at the initial state. In the case of the *second*, there is restart in the first state after the initial state (i.e. the state reached as soon as event $\uparrow a$ occurs). In both cases, the acoustic alarm stops when the fault has cleared.

3) If the first fault occurs twice in less than a minute, the fact that this fault clears will not be enough for operation to resume as indicated in 2). An additional alarm, M, requiring manual intervention is set. A new initialization is performed at the end of this intervention: changeover from 0 to 1 of the Boolean variable *restart*.

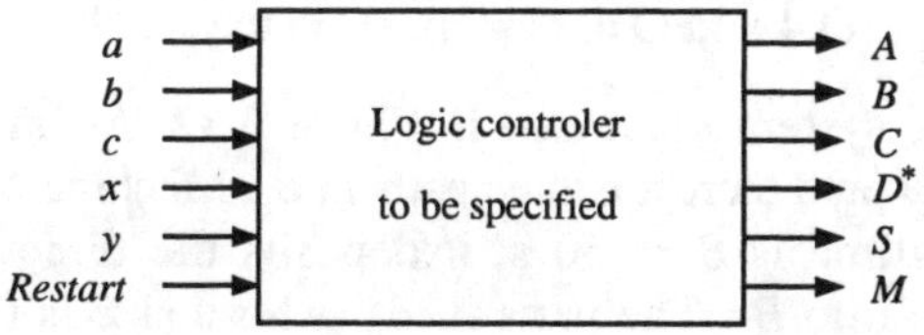

Figure E 3.7

3.8 Consider the *PERT with recycling* of Exercise 1.7. Let D_i be the duration associated with the execution of Task i.

a) How should the PN of Figure S 1.7 be timed, in order to model these durations?

b) If the initial marking corresponds to the time $t = 0$, what can we say about the instant at which the first execution of Task 5 will begin?

3.9 When a machine is followed by a *finite capacity* buffer, this machine may be blocked (i.e. it cannot process a new part) when the buffer is full. *Blocking before service*: processing a new part may begin only if there is a place in the buffer for depositing it when it is finished (the machine may be a furnace for example, the time in this furnace must not exceed a fixed value). *Blocking after service*: processing a new part may begin if the machine is free, and the part remains on the machine, after its processing, if the buffer is full.

Consider the 3-machine, 2-buffer line in Fig. E 3.9. The service time of M_i is denoted by S_i and the capacity of B_j is denoted by C_j. It is assumed that the buffer after M_3 is unbounded and that the machines do not fail.

a) *Blocking before service* is assumed. Model the behavior of this system by a T-timed PN in the two following cases: i) C_1 and $C_2 > 0$; ii) $C_1 = C_2 = 0$.

b) *Blocking after service* is assumed. Same questions.

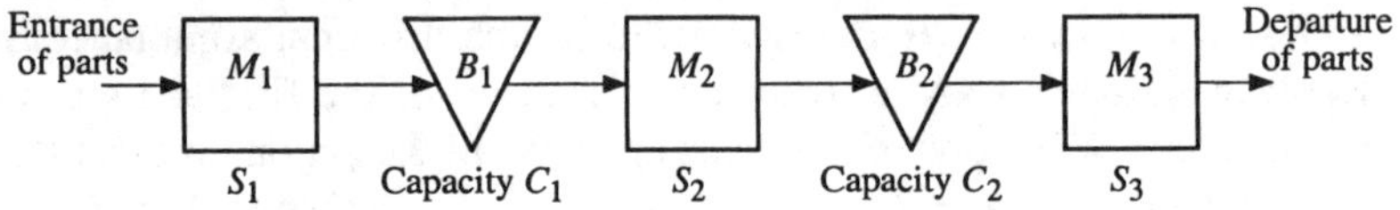

Figure E 3.9

3.10 Consider the production system presented in Fig. O.4 (Appendix O). Model this system, taking into account the processing times in order to have a representation minimal in place and transition numbers:

a) with a P-timed PN:

b) with a T-timed PN.

3.11 Consider the production system presented in Fig. O.6 (Appendix O).

a) Model the system with a P-timed PN.

b) Model the system with a T-timed PN.

c) Compare the arrival times in B_4 of the first part and of the 12th part for the cases in Figs. O.4 and O.6. Comment on the results.

3.12 Consider the system illustrated in Fig. E 3.12. At initial time, there is a 3-part palet in B_1 and there are three parts in B_2. Machine M_1 is an unpalletizer whose service time is $S_1 = 50$ s; it deposits the three parts on the pallet (initially in B_1) into B_2. The parts in B_2 (whose size is 0.2 m) enter one by one on the conveyor. The other features are given in the figure.

Model the behavior of the system by a T-timed PN and give the corresponding reachability graph.

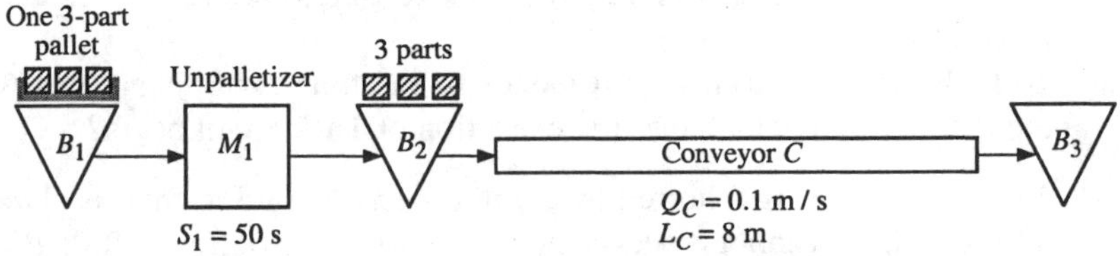

Figure E 3.12

3.13 We consider that there are two pallets (each one carrying a part) which pass in turn through machines M_1 and M_2 (Fig. E 3.13). Machine M_1 can only treat one part at a time and its service time is $S_1 = 2$. Machine M_2 can treat two parts at a time and its service time for a part is $S_2 = 3$.

a) Use a P-timed PN to represent the functioning of this system. The initial state is such that both pallets are in buffer B_1.

b) Represent the graph of reachability for functioning at maximal speed (indicating the residual times to availability for each token).

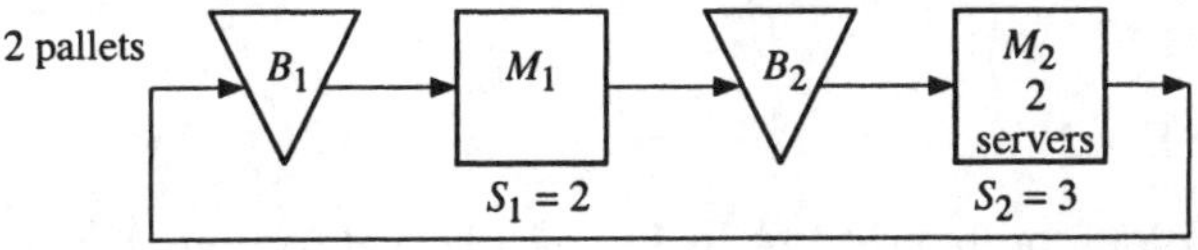

Figure E 3.13

3.14 For the T-timed PN in Fig. 3.26 (Sect. 3.4.2.3) calculate the firing frequencies of the stationary behavior.

3.15 Consider a production station made up of a machine and a buffer upstream. The capacity of the station (buffer + machine) is limited to 2 (Fig. E 3.15). We call μ_1 the arrival rate of the parts in the station when this station is not saturated (we assume that the arrival of the parts forms a Poissonian process and that this process is interrupted when the station is full) and μ_2 the station service rate (exponential distribution).

The variables involved in the following questions are defined in Section 3.4.3, but the calculation methods were not presented (they can be found, for example, in [KeSn 76] for Markov chains and [FlNa 85] for stochastic PNs). However, interested readers may find some solutions by themselves, or at least understand the solution given for this simple example.

a) Construct the stochastic PN corresponding to this system (initial marking: station empty). Give its graph of reachable markings and use it to deduce the associated Markov chain.

b) Calculate the probabilities of the states in stationary behavior in the particular case $\mu_1 = 2$ and $\mu_2 = 1$.

c) Calculate the mean markings, the mean firing frequencies and the mean dwelling times in stationary behavior.

e) Show that the mean markings and frequencies verify the preservation laws by calculating the marking and firing invariants.

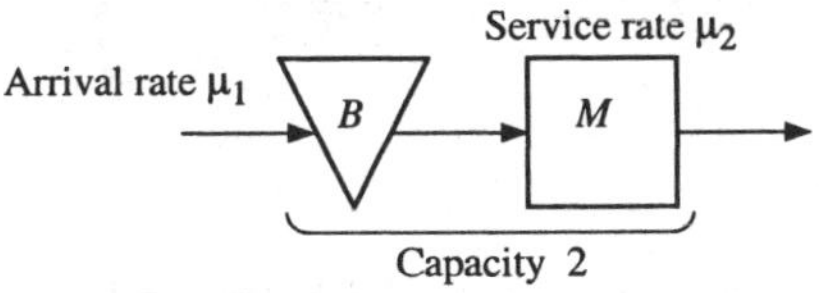

Figure E 3.15

Chapter 4

4.1 Give the reachability graph for the continuous PN in Fig. 4.3a.

4.2 Give the possible OG-firings related to the partial hybrid PNs in Figs. E 4.2a and b.

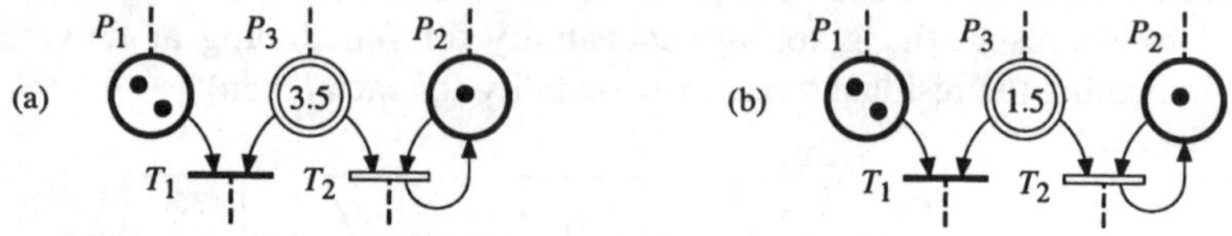

Figure E 4.2

4.3 Consider the hybrid PN in Fig. 4.13a and its initial marking $\mathbf{m}_0$. Are the following firing sequences possible from this initial state?

a) $S_1 = [T_4]^{1.8}T_1[T_4]^{0.9}T_1[T_3]^{0.6}$.
b) $S_2 = [T_4]^{1.8}T_1[(T_4)^{0.9}T_1][T_3]^{0.6}$.
c) $S_3 = [T_4]^{1.8}T_1[T_4]^{1.2}][T_1]^2$.
d) $S_4 = [T_4]^{1.8}T_1[T_4]^{1.2}][T_3]^{2.2}$.
e) $\mathcal{L}_1 = ([T_4]^{1.8}T_1[T_3]^{0.8}T_2)^*$.

4.4 Give the reachability graph for the hybrid PN in Fig. E 4.4.

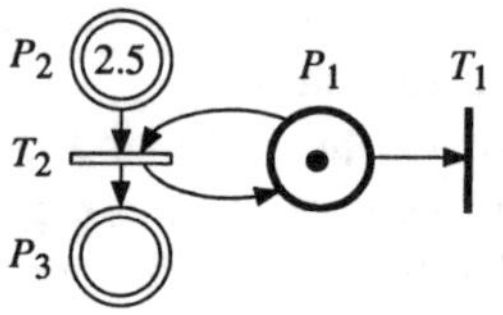

Figure E 4.4

4.5 Consider the continuous PN in Fig. E 4.5 (inspired from [ReTeSi 99]).

a) Using Property 4.10 (and Definition 4.9) in Sect. 4.3.2, show that the marking $\mathbf{m}_1 = \mathbf{m}_0 + \mathbf{W} \cdot [0\ \ 0.3\ \ 0.3\ \ 0\ \ 0]^T$ is reachable.

b) Give one or more firing sequences leading from $\mathbf{m}_0$ to $\mathbf{m}_1$.

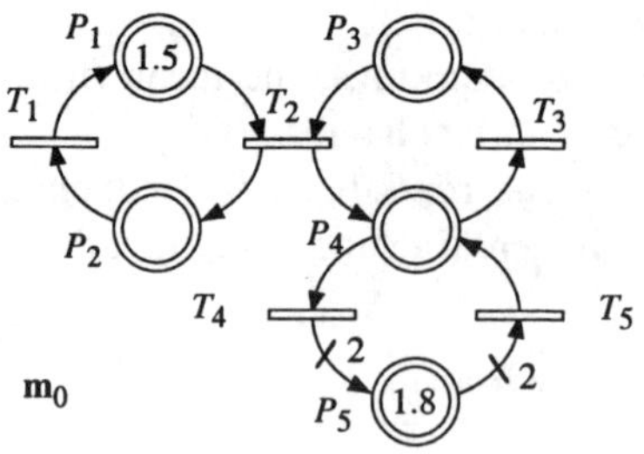

Figure E 4.5

4.6 Are the continuous PN in Fig. E 4.6, live, ε-live, lim-live?

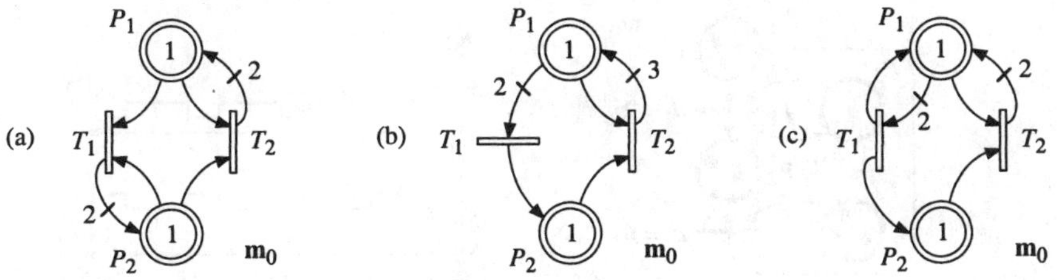

Figure E 4.6

4.7 What about liveness of the discrete counterparts of the PNs in Fig. E 4.6.

4.8 Consider the following chemical process. A solution contains 100 g of soda and 50 g of hydrochloric acid. The soda is decomposed: for 40 g of NaOH, 23 g of ions Na^+ and 17 g of ions OH^- are obtained. Similarly, 36.5 g of HCl are decomposed into 35.5 g of ions Cl^- and 1 g of ions H^+. Now, ions OH^- plus ions H^+ produce water H_2O, while ions Cl^- plus ions Na^+ produce sodium chloride, NaCl.

Model this process by a continuous PN. The reaction speeds and the function of water (solution, then evaporation to obtain NaCl) are not represented.

4.9 Which transitions are enabled in Fig. E 4.9?

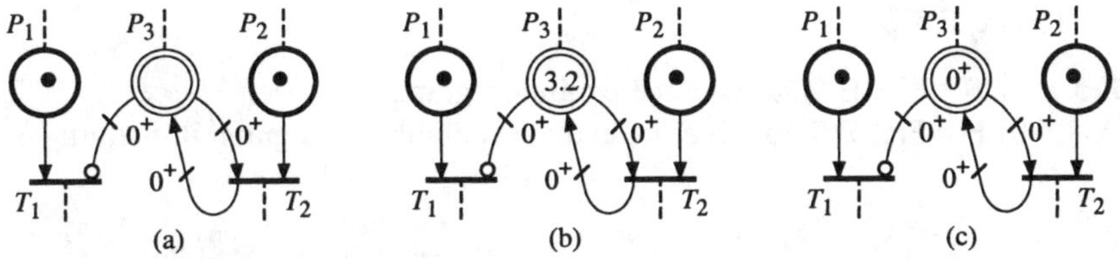

Figure E 4.9

4.10 Consider a stadium containing 10 000 places. The doors are usually closed. When a show is planned, the 12 doors are open and the spectators may enter (they can also leave when they want). Twelve people can enter at once. After the show, when all the spectators have left, the doors are closed again.

Model this system by a hybrid PN, in which entrance and departure of spectators are modeled by continuous firings. Is there any conflict? what does it mean?

Chapter 5

5.1 Consider the T-timed discrete PN in Fig. E 5.1. Give a continuous timed PN approximating the behavior of this model.

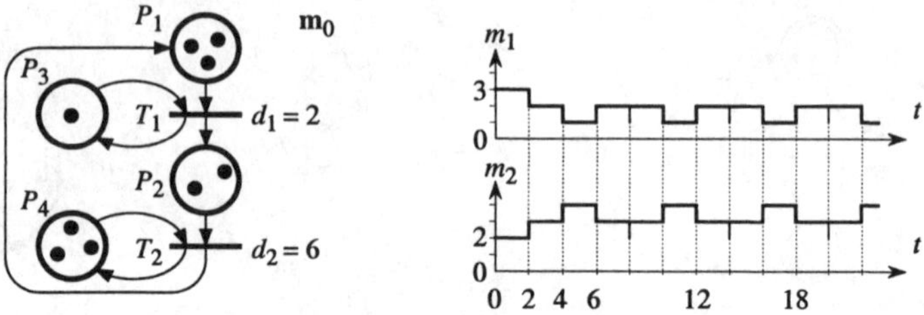

Figure E 5.1

5.2 Consider the assembly system modeled by the continuous PN in Fig. E 5.2 and the periodical maximal speeds $V_1(t)$ and $V_2(t)$ illustrated in the figure.

a) Express $V_1(t)$, $V_2(t)$, and $\mathbf{V}(t)$, according to Sect. M.2 in Appendix M (Remark M.1 and (M.10)).

b) Give the graphs of $m_1(t)$, $m_2(t)$, and $v_3(t)$.

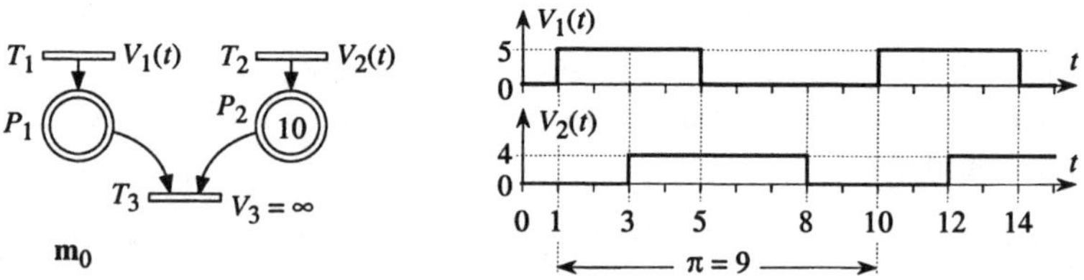

Figure E 5.2

5.3 a) For Fig. E 5.3a, give the evolution graph.

b) For Fig. E 5.3b, give the evolution graph and a graph illustrating $m_1(t)$.

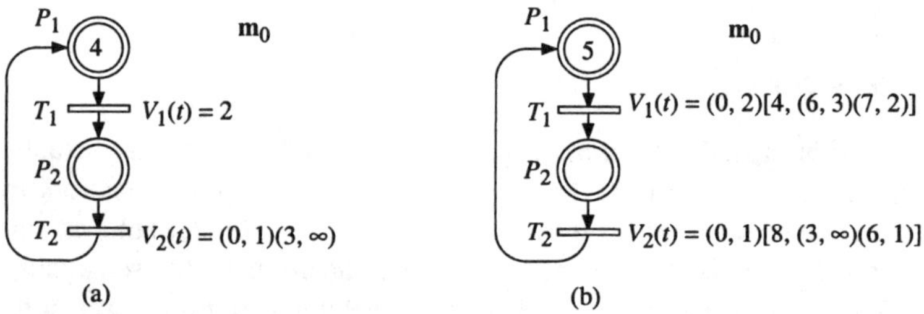

Figure E 5.3

5.4 According to Section G.3 in Appendix G, transitions in a continuous PN can be synchronized. Figure E 5.4a presents a synchronized continuous PN: T_1 and T_2 are synchronized on event $\uparrow x$ and on condition y, respectively. The continuous PNs in Fig. E 5.4b and c are partially synchronized and partially timed. Give the timing diagrams illustrating the behaviors of these PNs for the timing diagram of Boolean variables x and y in the figure.

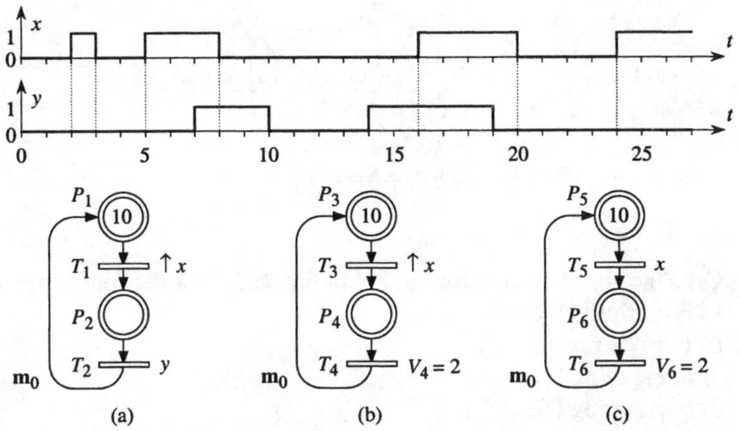

Figure E 5.4

5.5 Figure E 5.5 presents a timed continuous PN and its priority graph corresponding to the local resolution rules: $T_1 < T_2$, $T_3 < T_4$, and $T_5 < T_4$. From $\mathbf{m}_0$, it is intuitively clear that the instantaneous speed vector will be $\mathbf{v} = (1, 1, 2, 0, 2)$: since T_5 is strongly enabled, $v_5 = 2$; then, since $I_2 = 2$, $v_1 = 1$ and $v_2 = 1$ are obtained; finally, $v_3 = 2$ and $v_4 = 0$ because $T_3 < T_4$.

At the end of the first passage (end of *Step 6* in Algorithm 5.4) the results are as follows: $T_{SF} = \{T_1, T_3, T_5\}$, $T_{pf} = \varnothing$; maximization of $J_1 = v_1 + v_3 + v_5$ has given $v_1 = 1$, $v_3 = 1$, $v_5 = 2$; $C_3(1) = \{v_2 = 0, v_4 = 0\}$, $C_4(1) = \{v_1 = 1, v_3 \geq 1, v_5 = 2\}$, $T_{ndc} = \{T_2, T_3, T_4\}$. Then, in *Step 7* in Algorithm 5.4, Algorithm 5.3 will be performed.

a) Observe that T_4 will be deleted from J_2 in *Step 6* of Algorithm 5.3.

b) Explain the effect of this deletion.

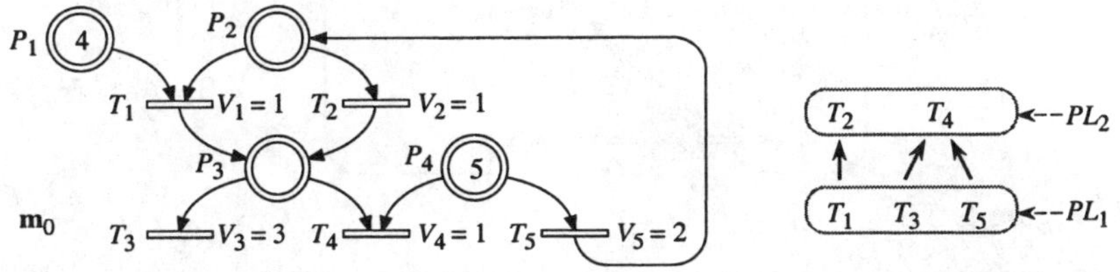

Figure E 5.5

5.6 Figure E 5.6 presents a timed continuous PN (with initial marking and at $t = 3$) and its priority graph corresponding to $T_2 < T_3$. For the first IB-state, $\mathbf{v} = (2, 1, 1)$ is obtained, then $m_1(t) = 3 - t$ and $m_2(t) = 0$ (in fact, $\tilde{m}_2(t) = 0^+$ because $B_2(t) = 0$ and $I_2(t) > 0$, i.e. P_2 has a zero balance and is fed). This IB-state ends at $t = 3$ and $\tilde{\mathbf{m}}(3) = (0^+, 0^+)$ (Property 5.3, Sect. 5.1.3.3).

What happens in the second IB-state? It is suggested to look particularly at *Steps 1.1*, *2*, *3*, *4*, and *5.1* of Algorithm 5.4.

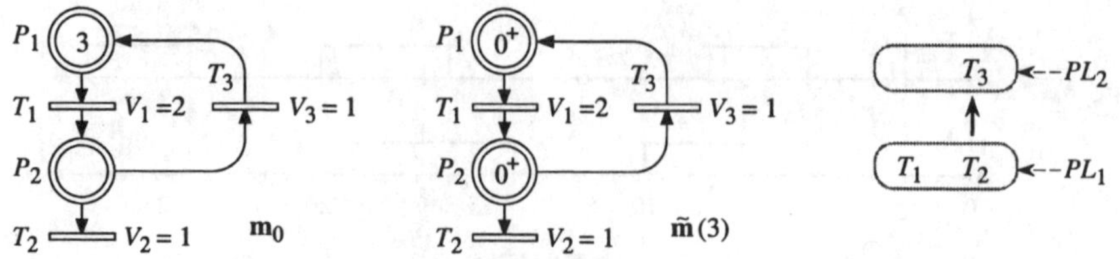

Figure E 5.6

5.7 Give the speed vectors for the PN in Fig. E 5.7 for the following cases.

a) For priority $T_2 < T_3$.
b) For priority $T_3 < T_2$.
c) For sharing $[T_2, T_3]$.
d) For sharing $[T_2, 3\,T_3]$.

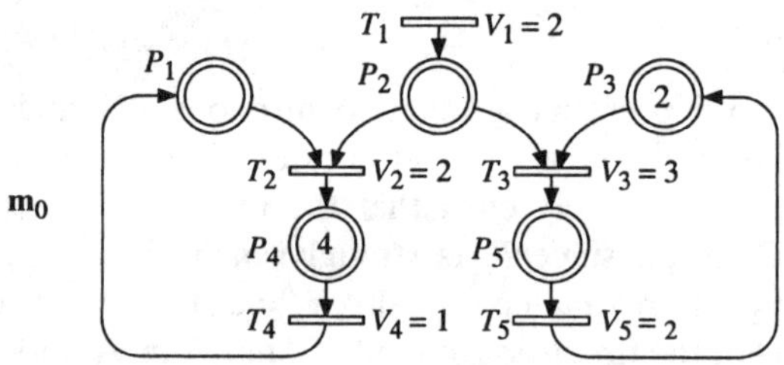

Figure E 5.7

5.8 Consider the timed continuous PN in Fig. E 5.8 and the resolution rule $T_1 < [T_2, T_3]$. What is the firing speed vector for the first IB-state?

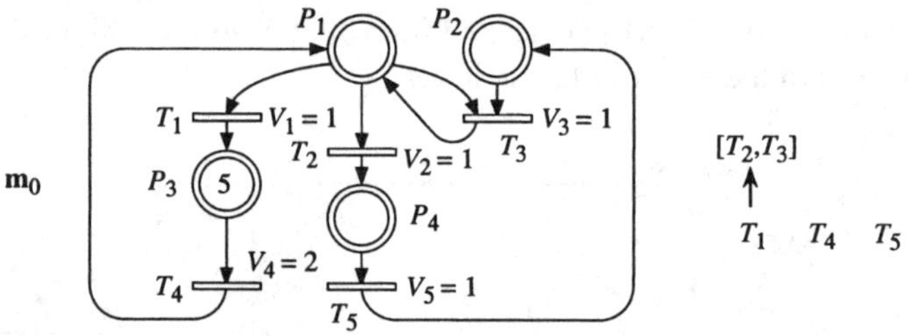

Figure E 5.8

5.9 Consider the timed continuous PN in Fig. E 5.9. Comment on the relations between both resolutions of conflicts. What is the firing speed vector for the first IB-state in the following cases?

a) Priorities $T_1 < T_2$ and $T_3 < T_4$.
b) Priorities $T_1 < T_2$ and $T_4 < T_3$.
c) Priorities $T_2 < T_1$ and $T_3 < T_4$.
d) Priorities $T_2 < T_1$ and $T_4 < T_3$.
e) Sharings $[T_1, T_2]$ and $[T_3, T_4]$.

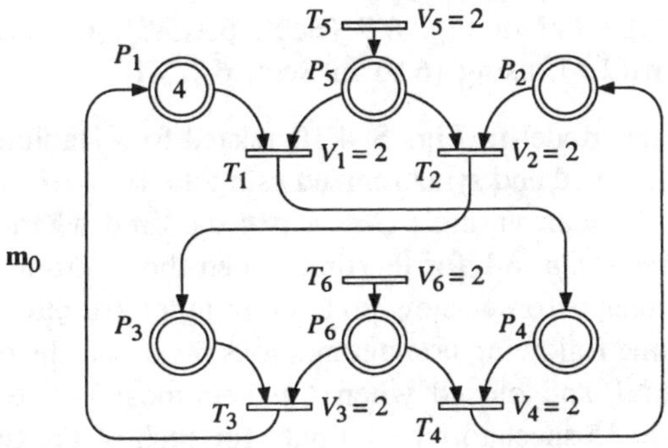

Figure E 5.9

5.10 Consider the continuous PN in Fig. E 5.10a and the priority graph in Fig. b, corresponding to $T_4 < T_6 < T_5$ and $T_7 < T_8$. For the first IB-state, Algorithm 5.4 gives the speed vector $\mathbf{v} = (2, 4, 1, 2, 0.75, 1.25, 0.5, 0.25)$.

a) Show that this result does not correspond exactly to the priority graph.

b) What should be the exact result?

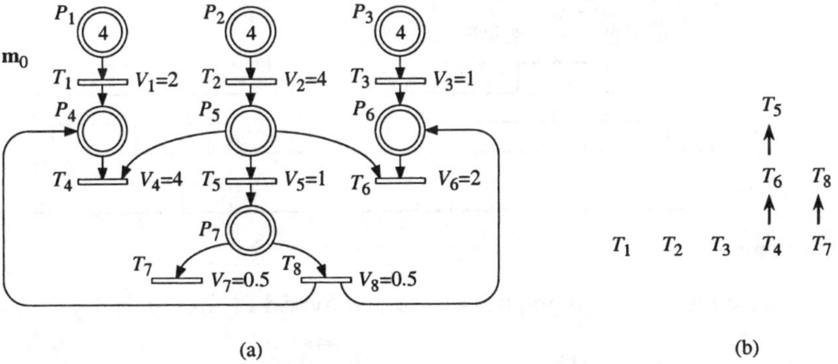

Figure E 5.10

Chapter 6

6.1 Consider the (partial) PN in Fig. 6.1 and its present marking $\mathbf{m}$.

a) What are the D-enabling degrees of C-transitions T_1, T_2, and T_3?

b) What are the maximal speeds for these transitions?

c) For which values of v_4 is there an actual conflict?

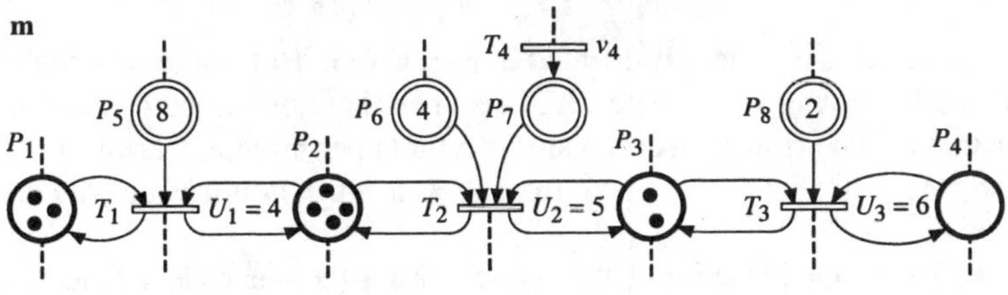

Figure E 6.1

6.2 Consider the PN in Fig. 6.9 (Sect. 6.1.5.3). Calculate the marking at $t = 2.5$, i.e. $\mathbf{m}(2.5)$, using (6.9) in Sect. 6.1.5.1.

6.3 Consider the model in Fig. S 4.10 related to a stadium. This autonomous model will be timed and synchronized using the Boolean variables presented in Fig. E 6.3. Boolean variable $Sh = 1$ for the time when a show is planned. Boolean variable $Op = 1$ for the time when the stadium should be open, i.e. from 90 minutes before a show up to 60 minutes after the end of the show.

a) Model the following non-autonomous behavior: the doors are open when Op becomes 1 and closed when Op becomes 0. The entrance flow rate corresponds to 15 spectators per minute for each of the twelve doors, and the departure flow rates are 25 spectators per minute for every door. It is assumed that the spectators enter as soon as the doors are open and leave as soon as the show is over.

b) Same question, except that the spectators arrive randomly during the 90 minutes preceding the show, i.e. when the Boolean variable $Ar = 1$. The time between two successive spectators follows an *exponential law of rate* $\mu_0 = 100$ (i.e. average of 100 spectators per minute).

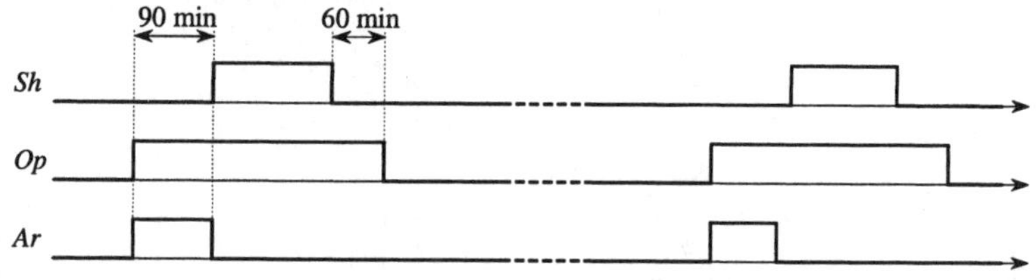

Figure E 6.3

6.4 Give the evolution graph for the timed hybrid PN in Fig. E 6.4.

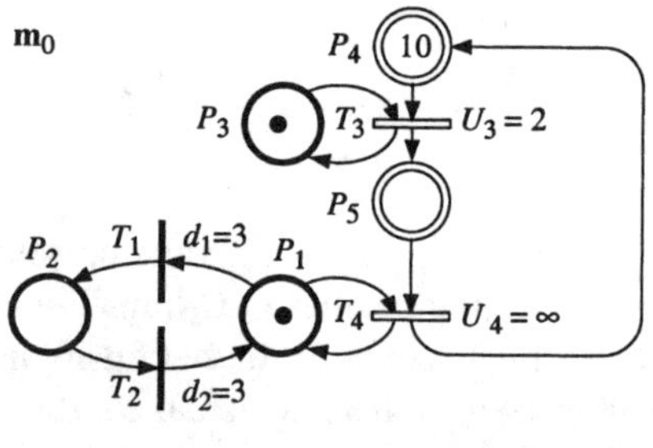

Figure E 6.4

6.5 Consider the system illustrated in Fig. E 6.5. Hot water and cold water are mixed for giving warm water. It is assumed that hot and cold water are always available. The flow of warm water is 5 liters per minute. Various ratios of hot and cold water can be obtained (from 0 % to 100 % hot water) by rotation of part A.

a) Model the behavior of this system by a (non-extended) timed hybrid PN in which the ratio is specified by a sharing between two C-transitions.

b) Model the behavior of this system by an *extended* timed continuous PN.

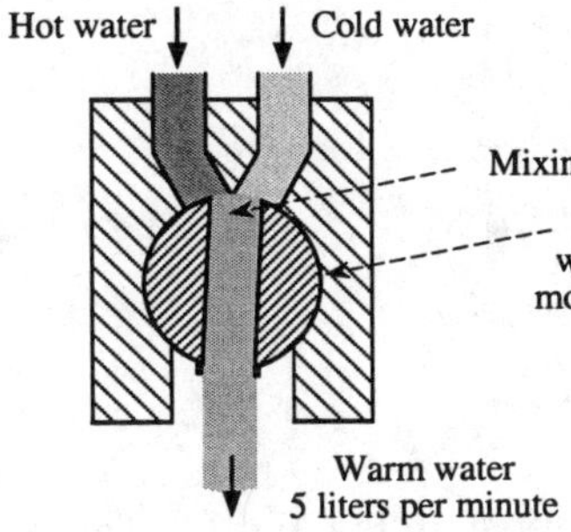

Figure E 6.5

6.6 Give a timed extended hybrid PN modeling the behavior of the system presented in Fig. E 6.6.

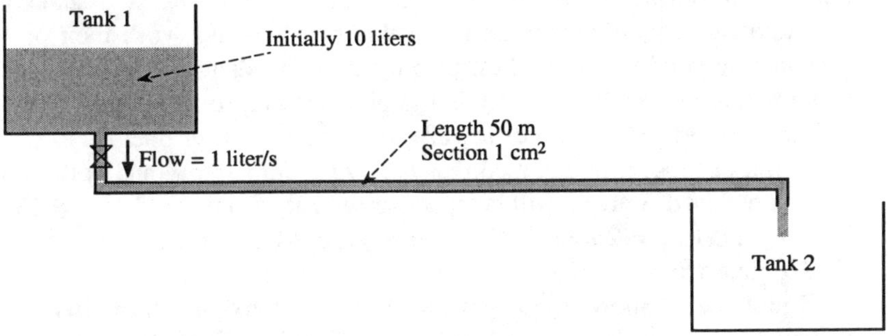

Figure E 6.6

6.7 Consider again the system presented in Exercise 3.12.
Model the behavior of the system by an extended hybrid PN, assuming a discrete modeling of the pallet and a continuous flow of parts. Give the corresponding evolution graph. Compare this graph with the reachability graph in Fig. S 3.12**B**.

6.8 Consider the conveyor system presented in Fig. 6.32a in Sect. 6.4.3.1. It is assumed now that the conveyor speed is $Q_C(t) = 5$ for $t < 0.4$ and $Q_C(t) = 10$ for $t \geq 0.4$. Model the behavior by a hybrid PN (some parameters are functions of time t) and give the corresponding evolution graph.

6.9 Same question as in Exercise 6.8, assuming the following values for $Q_C(t)$: $Q_C(t) = 5$ for $t < 0.4$, $Q_C(t) = 0$ for $0.4 \leq t < 1.4$, and $Q_C(t) = 5$ for $t \geq 1.4$. I.e., the conveyor is completely stopped for one time unit.

6.10 Compare the behaviors of hybrid PNs in Fig. E 6.10a and b: the firings of T_1 and T_3 will be compared.
a) If $m_3 = 0$ when the token is deposited in P_1 (time t_1) and P_3 is not fed.
b) If $m_3 = 12$ when the token is deposited in P_1, and P_3 is not fed.

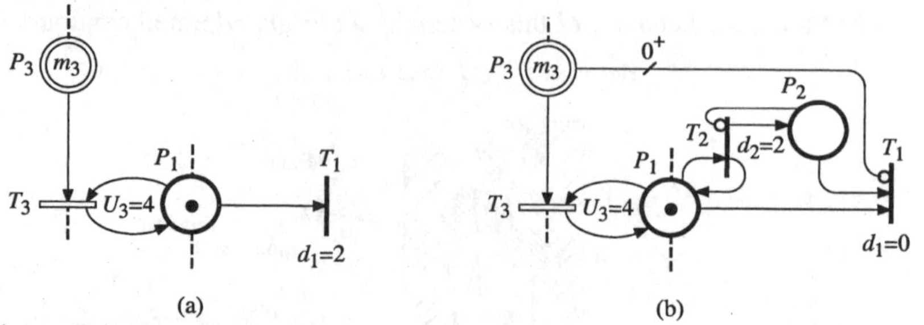

Figure E 6.10

6.11 Let us consider the system in Fig. E 6.11. This system is made up of two tanks (numbers 1 and 2), four valves (numbers 5, 6, 7, and 9, corresponding to the index of speeds associated with them) which could be open or closed, and a pump. There is a permanent supply of tank 1, 1 liter/s, when valve 5 is open. Liquid flows from tank 1 to tank 2 via valve 6 (2 liter/s, assumed to be always open). Liquid in tank 2 can leave it in three ways, with their priorities obtained by gravity. The highest priority corresponds to valve 7 (1 liter/s, if it is open). The second priority is the pipe leading to the pump (1 liter/s). The third priority corresponds to valve 9 (1 liter/s, always open).

At initial time, it is assumed that *both tanks are empty* and both valves 5 and 7 are closed. Valve 5 will be open periodically from $t = 7$ to $t = 13$, then from $t = 20$ to $t = 26$ etc. Valve 7 will be open periodically from $t = 16$ to $t = 26$, then from $t = 42$ to $t = 52$ etc.

Figure E 6.11 shows the behavior when both tanks are empty while valve 5 is open and valve 7 is closed (between $t = 7$ and $t = 13$ for example).

Model the behavior of this system with a timed hybrid PN and analyze this behavior for $t < 20$.

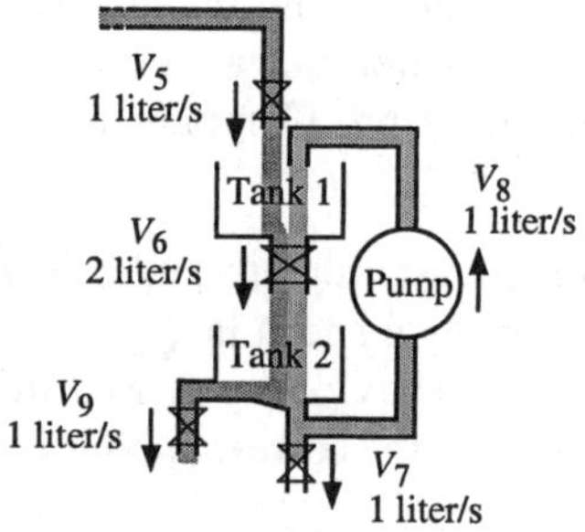

Figure E 6.11

Chapter 7

7.1 Consider again the production system in Fig. E 3.9.

a) Model the behavior of this system by a hybrid PN. A continuous flow on the machines and blocking before service are assumed. Numerical values: $C_1 = C_2 = 10$, $S_1 = 0.25$, $S_2 = 0.5$, $S_3 = 0.2$.

b) Assuming that the behavior is modeled by a VHPN, give the marking corresponding to the stationary behavior.

7.2 Consider the system in Fig. E 7.2. Variables VO_i (m^3), H_i (m), and S_i (m^2), denote the volume of liquid, its height, and the section of tank i ($i = 1, 2$). Liquid flows $u_0(t)$ (input flow), $u_1(t)$, and $u_2(t)$ (output flow), are expressed in m^3/s. A coefficient R_i (s/m^2) is associated with valve i such that:

$$u_i(t) = \frac{H_i(t)}{R_i}, \; i = 1, 2. \qquad \text{(E 7.1)}$$

a) Model the behavior of this system by a VHPN, using the following notations:

$$\tau_1 = R_1 \cdot S_1 \text{ and } \tau_2 = R_2 \cdot S_2. \qquad \text{(E 7.2)}$$

b) From the differential equations expressing the dynamics of the volumes VO_1 and VO_2, deduce the transfer function between the input flow $u_0(t)$ and the output flow $u_2(t)$. At initial time both tanks are empty.

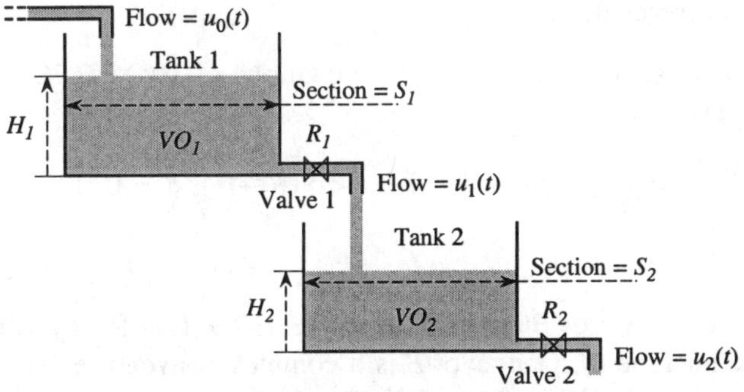

Figure E 7.2

7.3 a) Analyze the behavior of the hybrid PN in Fig. E 7.3, assuming the following hypothesis: it is a VHPN and there is a *sharing aiming at* $m_3^{(T_1)} = m_3^{(T_2)}$ (Notation 7.1 in Sect. 7.1.3.2). In other words, in case of conflict, half of marking m_3 is assigned to each transition.

b) Same question except that an AHPN is considered instead of a VHPN.

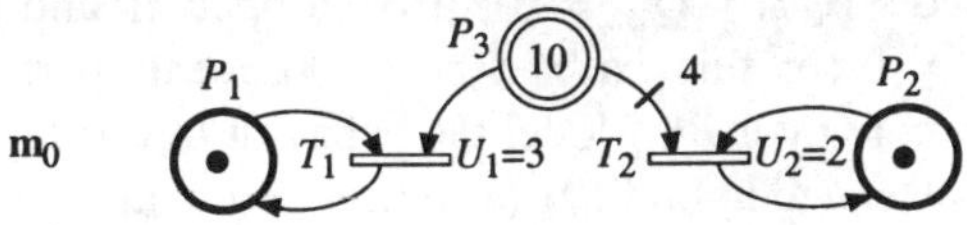

Figure E 7.3

7.4 Give the equations and numerical results allowing the graph to be drawn in Fig. 7.17b (Sect. 7.2.1.2).

a) For the VHPN model.

b) For the AHPN model.

7.5 Consider the timed hybrid PN in Fig. E 7.5.

a) Give the evolution graph for an AHPN interpretation of this hybrid PN.

b) Specify the instantaneous speeds which must be calculated if Algorithm 7.1 or 7.2 is used (Sect. 7.2.3).

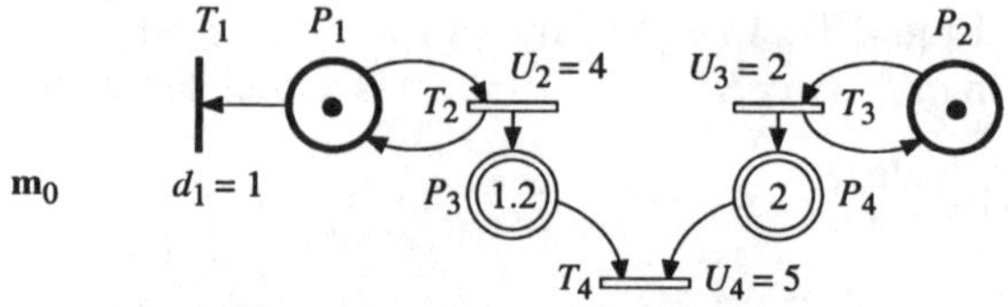

Figure E 7.5

7.6 Consider the timed hybrid PN in Fig. 7.13b (Sect. 7.1.4.2). Give the evolution graph for an AHPN interpretation of this hybrid PN. (This AHPN interpretation is illustrated by the graph in Fig. 7.22 in Sect. 7.4.2.)

7.7 Give a DHPN (differential hybrid PN, Sect. 7.3.2) producing the function of time $\sin t$. A place whose marking is $\sin t + 1$ will be used (the value $\sin t + 1$ is never negative).

7.8 Give a DHPN model satisfying Equations (7.104), (7.106) and the two following equations:

$$\frac{dx(t)}{dt} = -x^3(t) \text{ for } 5K + 2 \le t < 5(K+1),\ K = 0, 1, 2, \ldots. \qquad \text{(E 7.3)}$$

$$x(0) = 10. \qquad \text{(E 7.4)}$$

7.9 Consider conveyor B represented in Fig. E 7.9. It is fed by conveyor A and supplies machine M. Conveyor B is a complex conveyor essentially made up of two components: the width of the first component is such that the maximal density is 30 parts / m (3-strip conveyor) and the maximal density of the second component is 10 parts / m (1-strip conveyor); both components have different speeds. The move from the first to the second component takes 0.4 m as shown in the figure.

a) It is assumed that conveyor B may have two working states: either it works with the speeds on the figure ($Q_{B1} = 3$ m / min and $Q_{B2} = 10$ m / min), or it is stopped ($Q_{B1} = Q_{B2} = 0$); it is stopped if and only if the Boolean variable $ST_B = 1$ (in this case, a barrier prevents parts from passing from conveyor A to conveyor B). Model the behavior of conveyor B by a timed and synchronized extended hybrid PN (a timing function of a maximal speed may be used in the model). The marking represented will assume that the three parts close to machine M have been on the conveyor for more than 0.56 min.

b) How should the model be modified if $Q_{B2} = 5$ m / min instead of 10 m / min? Comment on the results.

c) How should the model obtained in a) be modified if the speeds Q_{B1} and Q_{B2} could change (not only become 0)? Comment on the results.

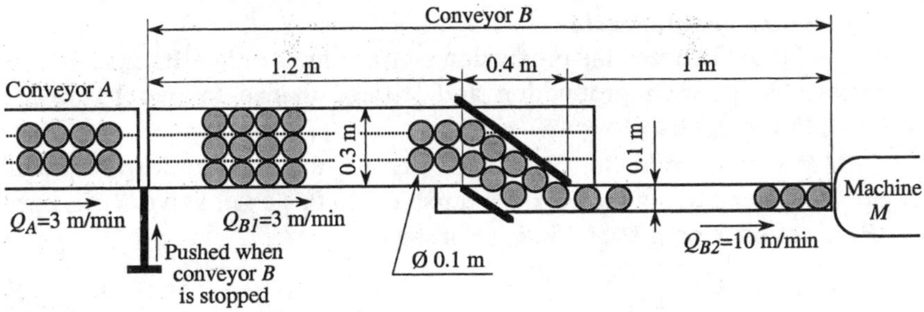

Figure E 7.9

8 *Application Examples*

8.1 GAS STORAGE

This example is a benchmark for hybrid system modeling proposed in [Ch *et al.* 98b]. It is a simplified version of a case study presented in [LaTa 97]. This example was chosen by the authors of [Ch *et al.* 98b] because "the continuous view is sufficiently complex to illustrate some hybrid issues". It involves differential and non-linear algebraic equations.

Most of the notations used here are similar to those in [Ch *et al.* 98b]. However, the index of a pressure (P_p for example) is always a letter (possibly Greek) whereas the index of a place (P_1 for example) is always a number, and notation V_j is reserved for a maximal speed of transition.

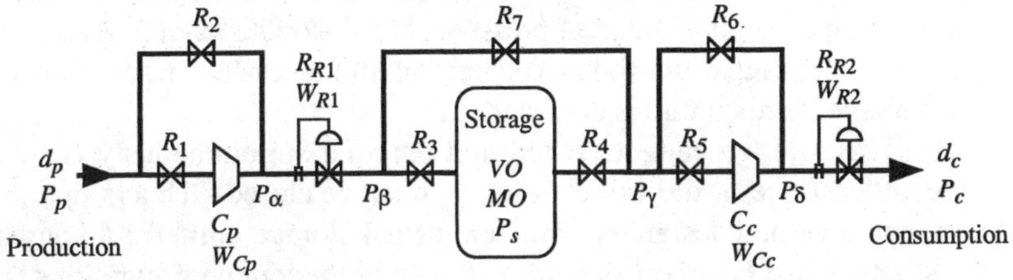

Figure E 8.1.**A** Gas storage unit.

The general structure of the gas storage unit is presented in Fig. **A**. It is made up of the actual storage (characterized by its volume VO, its number MO of moles, and its pressure P_s), a compressor C_p, and a compressor C_c. The gas storage is placed between a production unit and a consumption unit. Its goal is to introduce a buffer in order to facilitate the balance between production and demand.

The storage is fed by the production unit at pressure P_p and flow rate d_p. It is drawn by the consumer at pressure P_c and flow rate d_c (d_p and d_c are expressed in number of moles per second). Temperature is assumed to be constant and identical throughout the system, i.e. $T = 298°$ K.

Upstream compression
Thi is placed between the production unit and the storage. Its goal is to adapt pressures between production and storage thanks to two basic devices: compressor C_p and valve R_{R_1}.

If $P_p < P_\beta$, compressor C_p is used to ensure that gas flows from production to storage and not in the opposite direction. In this case, valve R_1 is open and R_2 is closed. Compression rate is constant:

$$P_\alpha = 5 \cdot P_p \tag{E 8.1}$$

when the compressor is used. If $P_p > P_\beta$, compressor C_p is not required. In this case, valve R_2 is open and R_1 is closed.

In both cases (C_p used or not), valve R_{R_1} is required to connect compressor output (pressure P_α) with the storage input (pressure P_β) because $P_\alpha > P_\beta$. The energy loss resulting from flow rate d_p through R_{R_1} compensates the pressure difference $P_\alpha - P_\beta$.

Downstream compression
Its behavior is similar to that upstream. Its goal is to ensure that P_δ is greater than P_c. When the compressor is used, compression rate is constant:

$$P_\delta = 2 \cdot P_\gamma \tag{E 8.2}$$

Storage
This is a *hydraulic subterranean storage* (see Fig. **B**) which means that storage capacity depends on its internal pressure. The material stored is a *compressible gas*. This means that the molar capacity of storage, *MO*, depends on pressure and volume (temperature is constant).

The gas storage cannot be fed and emptied simultaneously (i.e., at any time, at least one of the valves R_3 or R_4 must be closed). This is due to safety considerations and design issues of the actual storage unit. As a consequence, the storage must be driven according to one of the four configurations below.

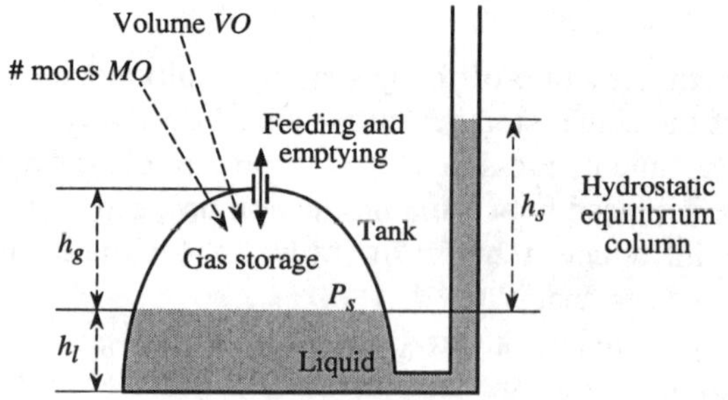

Figure E 8.1.**B** Simplified view of the gas storage.

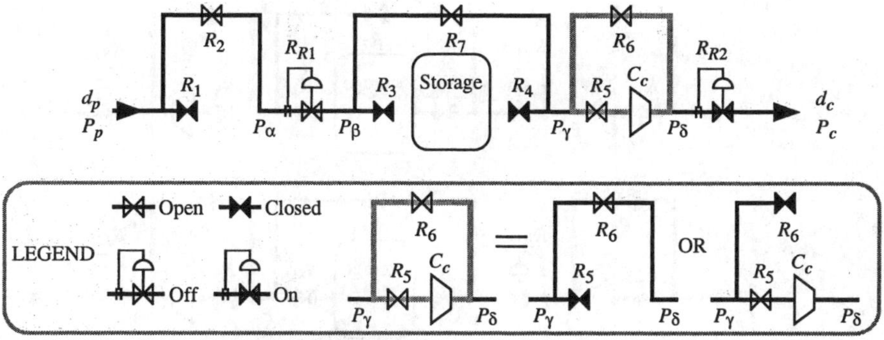

Figure E 8.1.**C** Configuration 1.

Configuration 1
Production flow rate and consumer flow rate are identical (Fig. **C**). The storage is by-passed and input compressor and dissipator are off. (Note that another solution could be obtained by setting the output compressor and dissipator to off and operating with the input ones, but this is not the specification in [Ch *et al.* 98b]). Hence,

$$P_\alpha = P_\beta = P_\gamma = P_p. \tag{E 8.3}$$

As illustrated in the legend in Fig. **C**, either the output compressor is by-passed or used (depending on ratio P_p / P_c).

Configuration 2
This configuration is active if production flow is greater than consumer flow. Part of the gas flows from the production unit to the consumption unit, and the surplus is filling the storage (Fig. **D**a):

$$P_\beta = P_\gamma = P_s. \tag{E 8.4}$$

Configuration 3
This configuration is active if production flow is lower than consumer flow. The produced gas flows to the consumption unit, and the lack of gas is compensated by emptying the storage (Fig. **D**b):

$$P_\beta = P_\gamma = P_s. \tag{E 8.5}$$

Configuration 4
This configuration is active if the consumed gas comes only from the storage (Fig. **D**c):

$$P_\gamma = P_s. \tag{E 8.6}$$

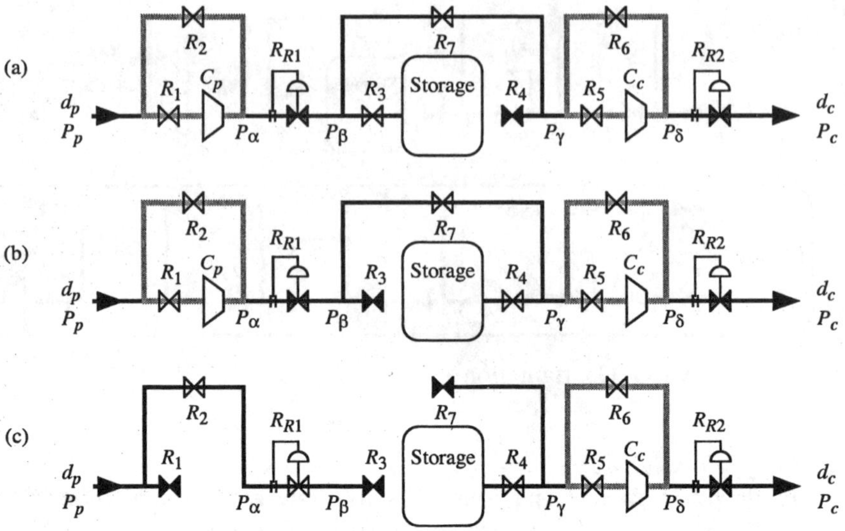

Figure E 8.1.**D** (a) Configuration 2. (b) Configuration 3. (c) Configuration 4.

Gas storage equations

Storage volume depends on its pressure. As the quantity of gas increases, the liquid surface in the storage is pushed down, following the hydrostatic law. Figure **B** shows the physical principle of storage volume variation. The sum $h_g + h_l$ remains constant and the gas volume is determined by the height h_g. The pressure at the liquid/gas interface in the tank is determined by the hydrostatic equilibrium column h_s. The following equation can be derived:

$$P_s = P_o + \rho_l \cdot g \cdot h_s. \quad \text{(E 8.7)}$$

where P_o is a constant (reference pressure) and ρ_l is the volumic mass of the liquid. If pressure P_s increases, height h_s also increases. This implies that h_l decreases. Then gas volume increases with its pressure.

Since the interior shape of the tank is irregular, the following expression[1] of the gas storage volume can be obtained [Ch *et al.* 98b]:

$$VO = 3.536 \cdot 10^{-3} \cdot P_s \cdot (10^{-4}(P_s - P_o))^{2.5}. \quad \text{(E 8.8)}$$

Storage height and volume are such that $0 \le h_g \le 130$ m and $0 \le VO \le 4.6 \cdot 10^9$ m^3.

The gas is assumed to be perfect. So, the perfect gas law can be applied:

$$P_s \cdot VO = MO \cdot R \cdot T, \quad \text{(E 8.9)}$$

where R is the constant of perfect gas ($R = 8.314$ J / mol · K).

[1] All units are those of the international system (unless otherwise specified). Hence, for (E 8.8), VO is in m^3 and the pressures is Pa.

Compressor equations
The input compressor C_p has two possible states: on an off. When C_p is *off*, it does not use any energy, i.e. $W_{C_p} = 0$, whereas, when C_p is *on*, the power produced is:

$$W_{C_p} = \frac{\gamma \cdot R \cdot T}{\eta(\gamma - 1)} \left(\left(\frac{P_\alpha}{P_p} \right)^{\frac{\gamma-1}{\gamma}} - 1 \right), \tag{E 8.10}$$

with $\eta = 0.8$ and $\gamma = 1.31$. Similarly, for compressor C_c, the power is given by $W_{C_c} = 0$ if it is *off* and by (E 8.11) if it is *on*:

$$W_{C_c} = \frac{\gamma \cdot R \cdot T}{\eta(\gamma - 1)} \left(\left(\frac{P_\delta}{P_\gamma} \right)^{\frac{\gamma-1}{\gamma}} - 1 \right). \tag{E 8.11}$$

Dissipation valve equations
Valves R_{R_1} and R_{R_2} act as dissipators. Using the thermodynamic theory, a behavioral model is used under the assumption of isenthalpic process. So, since the upstream and downstream pressures are known, power is calculated using the following formulae (when these valves are *on*):

$$W_{R_1} = -R \cdot T \cdot \ln\left(\frac{P_\beta}{P_\alpha} \right); \tag{E 8.12}$$

$$W_{R_2} = -R \cdot T \cdot \ln\left(\frac{P_c}{P_\delta} \right). \tag{E 8.13}$$

1) **Simulation purpose**
The first goal of this exercise is to propose to simulate process behavior.

Given that reference pressure P_o and temperature T are constant values, the variables can be divided into three sets, as illustrated in Fig. **E**. By *abuse of notation*, the names of compressors and valves are also used as Boolean values which are true when the corresponding devices are *on* or *open* (e.g., $C_p = 1$ if compressor C_p is *on*, $R_{R_1} = 1$ if dissipation valve R_{R_1} is *on*, $R_1 = 1$ if valve R_1 is *open*).

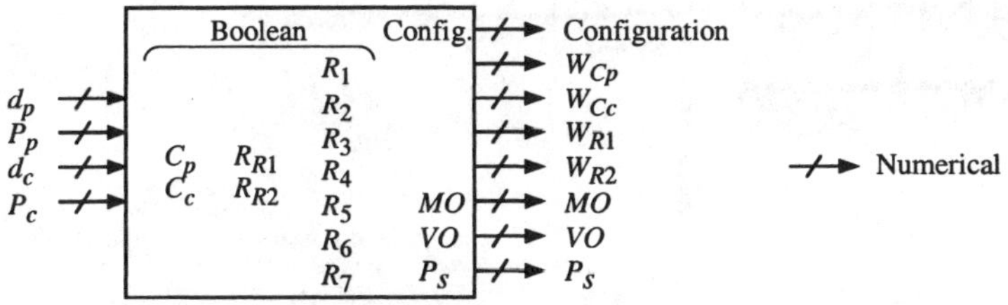

Figure E 8.1.**E** Input, internal, and output variables for simulation purpose.

Input variables: $\{d_p, P_p, d_c, P_c\}$;
Internal variables: {Config., C_p, C_c, R_{R_1}, R_{R_2}, R_1, …, R_7, MO, VO, P_s};
Output variables: {Config., W_{C_p}, W_{C_c}, W_{R_1}, W_{R_2}, MO, VO, P_s}.

The input variables are given in the specifications. The internal variables are the other variables whose values have an influence on behavior. The output variables are those that an observer would like to know. Their choice is relatively arbitrary; in Fig. **E**, it corresponds to the current configuration, the power consumed, and the features of the gas stored. Note that $W_{C_p} > 0$ implies $C_p = 1$ and that $W_{R_1} > 0$ implies $R_{R_1} = 1$, for example).

A model is sought for simulation of system behavior, assuming that the input vector $\mathbf{I} = (d_p, P_p, d_c, P_c)$ is piecewise constant. Take into account the failures cases which can be determined from the above specifications.

2) **Control**
Propose modifications to the simulation model obtained in 1) in order to obtain control of the system modeled.

8.2 BOTTLING CHAIN

A bottling chain for Perrier mineral water was studied both in [De 94] and [Au 95]. This chain is represented in Fig. **A**. The bottling chain is a quasi-linear structure made up of machines in series. The machines are linked by unidirectional multi-strip conveyors being used to buffer. In order, the machines are an unpalletizer (placing simultaneously a number of empty boxes on the first conveyor), a washing machine, a rotary filler (in which each box is filled then closed), a checking machine, and a packaging machine (ending by a furnace as explained further on).

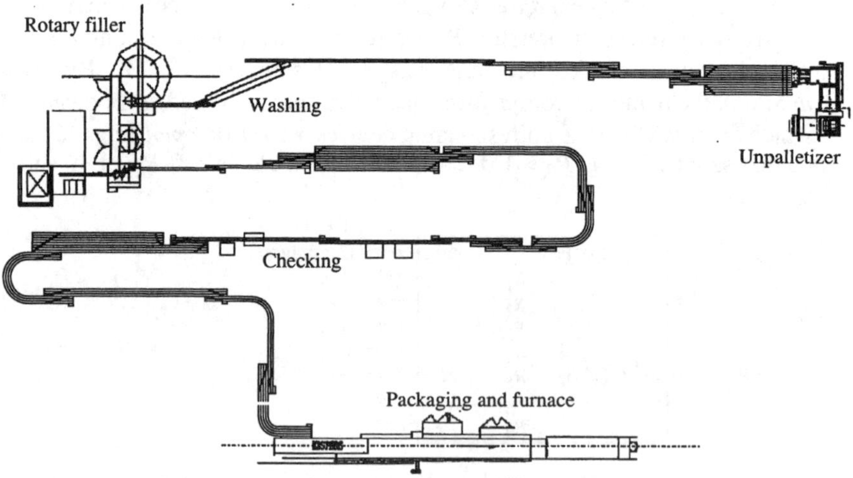

Figure E 8.2.A Perrier mineral water bottling chain.

Modeled bottling chain

A simplified chain is modeled in [Au 95] (according to the author, its modeling is quite significant of the modeling problems in the true chain). We propose to modeling this simplified chain because all the data are available in this reference. Hence, the bottling chain presented below corresponds to the specifications in [Au 95], except that more details on the conveyors are given in this exercise.

The structure of the bottling chain to be modeled is presented in Fig. **B**. All the components of this chain are presented below.

There is a *controler* for the bottling chain (not specified in the reference above). The controler inputs are given by various sensors detecting accumulations at some places in the chain and measuring all the conveyor speeds (including internal conveyors in the rotary filler, packaging and furnace). The controler outputs (see Sect. 3.3.1) are an impulse action (i.e. event) ordering unpalettization and level actions (i.e. Boolean variables) acting on the various components of the chain. For each component in Fig. **B**, except the output buffer, there are two Boolean variables: a *stop* signal (denoted by ST with an index) and a failure *signal* (denoted by FA with an index).

Give a hybrid PN modeling of the behavior of the bottling chain using the controler outputs as inputs of this model, according to the specifications given in the sequel.

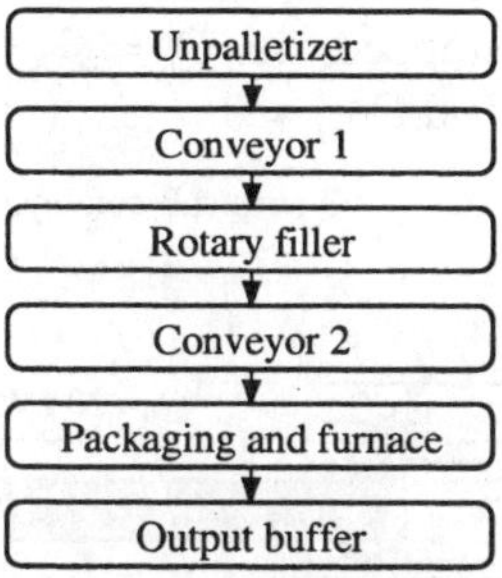

Figure E 8.2.**B** Bottling chain to be modeled.

Unpalletizer

The unpalletizer is illustrated in Fig. **C**. On a pallet, there are 360 boxes and unpalletization consists of pushing all these boxes on the head of conveyor 1 (this operation is assumed to have no duration). The order "unpalletize" is given by the event E_u which can be produced by the controler *only* if there is enough place at the head of the conveyor[2].

If $ST_u = 1$ (i.e. stop unpalletizer), no unpalletizing is allowed even if event E_u occurs. Unpalletizing will be allowed again when $ST_u = 0$.

[2] The mechanical part of the unpalletizer is able to unpalletize 1800 boxes/min (12 s between two unpalletizings), which is more than required by the bottling chain.

If $FA_u = 1$ (i.e. failure of unpalletizer), no unpalletizing is allowed even if E_u occurs. Unpalletizing will be allowed again when $FA_u = 0$.

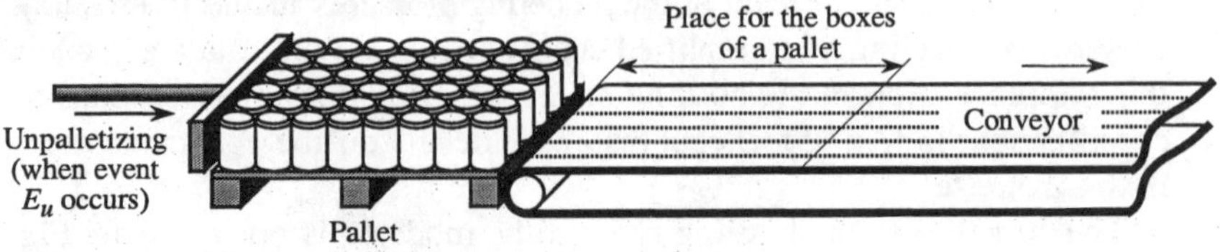

Figure E 8.2.**C** Unpalletizer.

Conveyor 1

Conveyor 1 is illustrated in Fig. **D**. Informally it is made up of a 12-strip[3], a 3-strip, and a 1-strip sub-conveyors (in series), respectively called sub-conveyors *a*, *b*, and *c*.

The 12-strip sub-conveyor is divided into two parts: the head corresponding to the place required for unpalletizing (Figure **D** represents it just after unpalletizing) and a conveying part.

The behavior of the conveyor will be approximated by a string of three sub-conveyors (performing calculations similar to Exercise 7.9): *a*, 12 strips, Q_a = 5 m / min, length 0.6 m; *b*, 3 strips, Q_b = 20 m / min, length 2m; *c*, 1 strip, Q_c = 60 m / min, length 3 m.

If $ST_{c1} = 1$ (i.e. stop conveyor 1) or $FA_{c1} = 1$ (i.e. failure of conveyor 1), the whole conveyor is stopped, i.e. $Q_a = Q_b = Q_c = 0$. Normal behavior is resumed when $ST_{c1} = 0$ and $FA_{c1} = 0$.

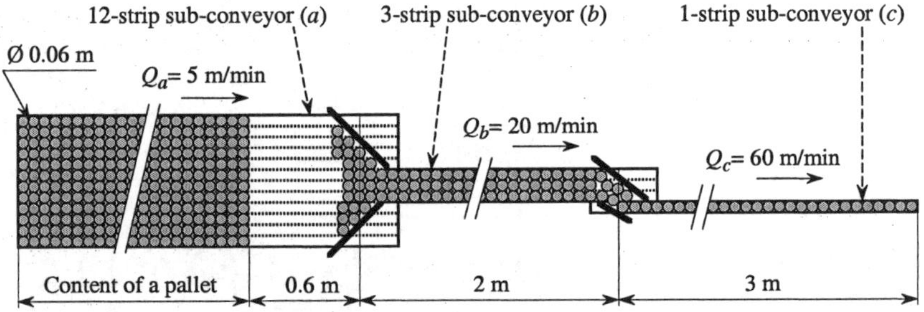

Figure E 8.2.**D** Conveyor 1.

Rotary filler

In the rotary filler, the boxes are filled then closed as shown in Fig. **E**. These boxes are conveyed on a 1-strip *conveyor without accumulation*. This means that a box is always at the same place on the conveyor from when it enters to when it leaves the rotary filler. This is illustrated in Fig. **E**: if some spots are

[3] Conveyor strips could be larger than the boxes to be conveyed. In this exercise, the width of all the conveyor strips is 0.06 m, i.e. the width of a box. Conveyor capacities are easier to calculate.

not occupied by a box, they will not be occupied from the beginning to the end of the rotary filler.

Conveyor speed Q_{rf} may change if necessary. Its nominal value is $Q_{rf} = 52.8$ m / min. However, it may slow down if the downstream component (i.e. conveyor 2 in our example) is not able to absorb the corresponding output flow. Slowing down is required because the boxes cannot slide on the rotary filler conveyor.

The length of the rotary filler conveyor is 8.778 m and the distance between two spots is 0.066 m, hence there are 133 spots in the rotary filler.

If $ST_{rf} = 1$ (i.e. stop rotary filler), no box can enter the rotary filler but the conveyor continues working in order to evacuate all the boxes in the rotary filler, then is stopped.

If $FA_{rf} = 1$ (i.e. failure of rotary filler), the rotary filler conveyor is immediately stopped, i.e. all the boxes stay where they are.

In order to satisfy these constraints, Q_{rf} is calculated by the controler. Specify when and how these calculations must be performed.

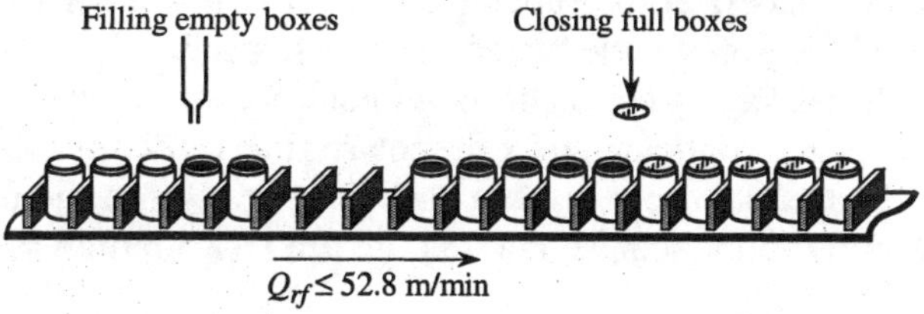

Figure E 8.2.**E** Rotary filler.

Conveyor 2

Conveyor 2 is illustrated in Fig. **F**. Its speed is $Q_{c2} = 160$ m / min and its length is 60 m: 3 strips for 20 m and one strip for the other 40 m.

If $ST_{c2} = 1$ or $FA_{c2} = 1$, the whole conveyor is stopped, i.e. $Q_{c2} = 0$. Normal behavior is resumed when $ST_{c2} = 0$ and $FA_{c2} = 0$.

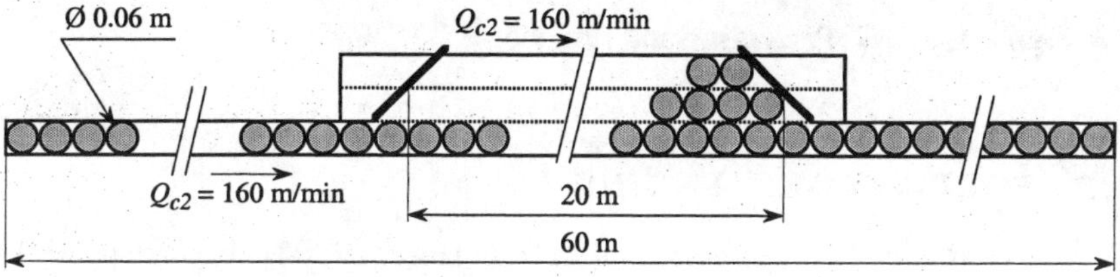

Figure E 8.2.**F** Conveyor 2.

Packaging and furnace, and output buffer

The packaging/furnace component produces packs of 24 boxes. Its behavior is illustrated in Fig. **G**. At the end of conveyor 2, boxes are gathered. When 24 boxes have been gathered, and if there is enough room on the head of the packaging conveyor, the 24 boxes are pushed (operation assumed to be without duration) on the conveyor.

The *packaging conveyor is without accumulation.* Its length is 5 m (after the depositing place), corresponding to 20 packs (0.25 m for a pack). Along this conveyor, the pack is surrounded by a cardboard box then by a plastic film.

After packaging, packs pass through a furnace on an ordinary conveyor (i.e. with possible sliding): the plastic film is tightened around the pack. When leaving the furnace, 2.5 m long, a pack reaches an output buffer whose capacity is 10 000 packs. This buffer may be emptied at maximal speed $V_D(t)$ (external demand in packs / min) which is piecewise constant.

The maximal speed Q_p of the packaging conveyor is the same as the speed Q_f of the furnace conveyor: $Q_p \leq 12.5$ m / min $= Q_f$.

In case of a *stop* order ($ST_{pf} = 1$) or *failure* ($FA_{pf} = 1$) of the packaging/furnace component, behavior is the same: 1) the packaging conveyor is stopped immediately ($Q_p = 0$, all the packs on this conveyor stay at the same place); 2) the furnace conveyor continues working in order to evacuate all the packs in the furnace, then is stopped.

Since the packaging conveyor is without accumulation, its speed must be calculated by the controler (as for the rotary filler). Specify when and how. In the model, take into account that a pack must not stay in the furnace for more than 0.2 min (normal conveying time: 2.5 m / 12.5 m/min = 0.2 min).

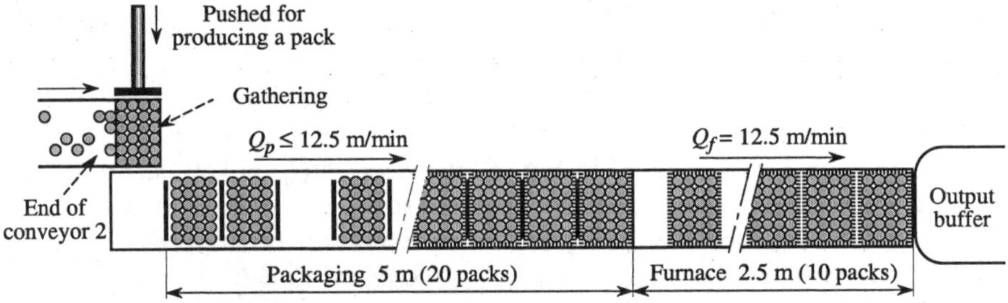

Figure E 8.2.G Packaging and furnace.

8.3 FOUR-STROKE ENGINE

The example presented here is taken from [Ba *et al.* 03]. It corresponds to idle speed control of a 4-cylinder and 4-stroke engine. As illustrated in Fig. **A**, the four stokes are: intake, compression, expansion, and exhaust. In [Ba *et al.* 03], it is implicitly *assumed that, at any time, the four cylinders are in four different strokes*. This document takes into account non-linearities in the intake manifold and models both the torque generation mechanism and the effect of clutch switching on the power-train dynamics.

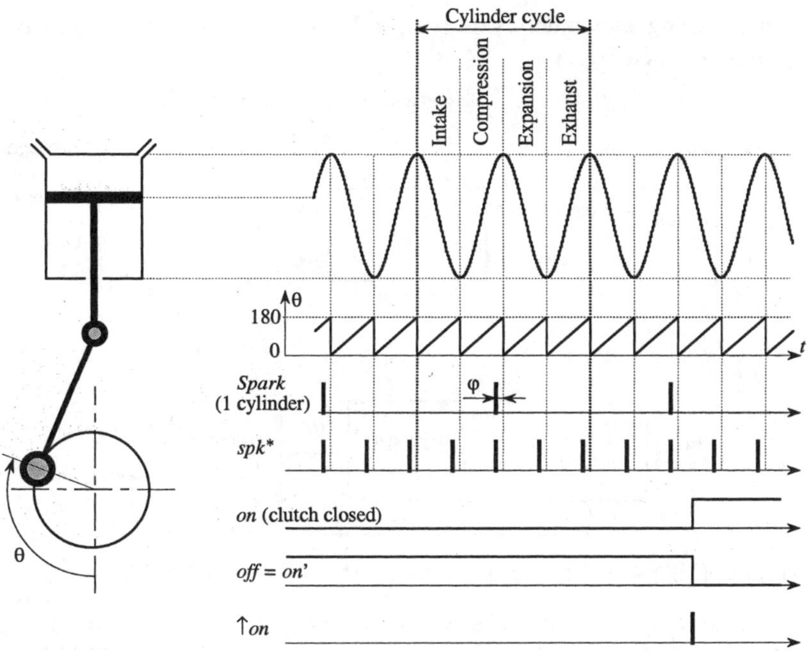

Figure E 8.3.**A** Four strokes, spark, and clutch switching.

The control strategy for an internal combustion engine in idle operating conditions aims at maintaining engine speed as close as possible to a reference constant engine speed despite disturbances. As illustrated in Fig. **B**a, disturbances are *load torque disturbance*, denoted by T_l (i.e. the air conditioning system, the steering wheel servo-mechanism), and engagements and disengagements of the transmission occurring when the driver operates the *clutch*. The Boolean variable *on* = 1 when the clutch is closed. As illustrated in Fig. **A**, *off* = *on*' represents the complementary Boolean variable and ↑*on* is the event[4] corresponding to engagement (the disengagement event ↓*on* is not required in the model).

For indirect injection engines, assuming a stoichiometric air-to-fuel ratio, stability of engine speed is achieved by controling the amount of air supplied to the engine (*throttle valve* whose *angle* is denoted by α) and the *spark* ignition times (control by the amount of air can provide a large potential of control but is relatively slow, whereas spark control is much faster but has limited potential). According to Fig. **B**a, both control inputs are *angle* α and event[5] spk^*. As illustrated in Fig. **A**, the spark related to a cylinder occurs φ degrees *before* the end of compression stroke ($20 \geq \varphi \geq -15$, i.e. the spark occurs between 20° before the end of compression and 15° afterwards). Since there is a spark for each cylinder, for the engine the event spk^* occurs four

[4] For notation ↑, see Appendix D.

[5] The asterisk ending spk^* shows that it corresponds to an *event* (according to Sect. 3.3.1).

times during a cylinder cycle (i.e., *spk** is the event *'spark for one of the cylinders'*; see Fig. **A**).

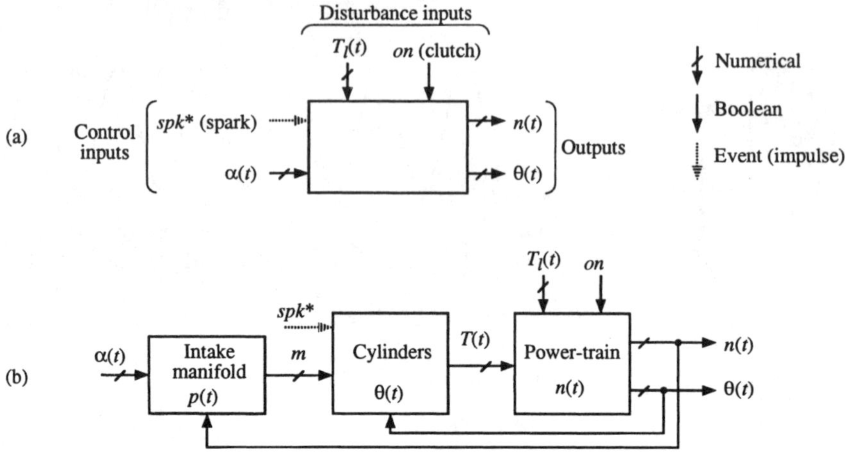

Figure E 8.3.**B** (a) Inputs and outputs. (b) The engine blocks.

According to Fig. **B**a, the two system outputs are the revolution speed of the crankshaft, denoted by n, and its angular position modulo 180°, denoted by θ (see also Fig. **A**).

More details are given in Fig. **B**b. Pressure p of the intake manifold depends on the throttle valve angle α and the crankshaft revolution speed n. Manifold pressure determines the mass of air mixture m loaded in the cylinders. In each stroke of the 4-stroke engine cycle, cylinder evolution is described in terms of the crankshaft angle θ, which determines the position of the piston within the given stroke. The torque T generated by the cylinders depends on both the mass m and the spark ignition time. Finally, the power-train dynamics, controled by the generated torque T, is subject to the load torque T_l and depends on clutch position. The gear is assumed to be in idle position.

Engine dynamic behavior is characterized by two sets of variables called *CTS* and *DES*. The set *CTS* of continuous variables, is subject to *continuous-time dynamics*: $CTS = \{p, n, n_g, \theta\}$ contains the revolution speed n_g of the segment of the driveline from the clutch to the gear, in addition to manifold pressure p, crankshaft revolution speed n, and piston position θ already presented. The set *DES* contains five analogical variables with a *discrete-time evolution*: $DES = \{T_C, T_E, m_C, m_E, \varphi\}$ contains the torque T_C absorbed by the cylinder in the compression stroke, the torque T_E generated by the cylinder in the expansion stroke (T_C and T_E are assumed to be constant during a stroke, i.e. they are average values), the mass of mixture m_C loaded in the cylinder in the compression stroke, the mass of mixture m_E in the cylinder in the expansion stroke, and the applied spark advance φ.

Aim of this exercise

This exercise consists of *modeling the dynamic evolution of the variables in CTS* and *calculating the values of variables in DES* when required. Such dynamic modeling and calculations are based on the equations given in the sequel and the numerical data in Figs. **C** and **D**. All these values are those found in [Ba *et al.* 03]. However, the sign of all the negative constant values in [Ba *et al.* 03] have been changed (hence, all the values in our Fig. **C** are positive) as have obviously the corresponding signs in the equations. In addition, revolution speeds are expressed in radians per second instead of revolutions per minute.

Crankshaft angle gain	K_c	$180/\pi = 57.3$ deg rad^{-1}
Crankshaft momentum of inertia (clutch open)	J_0	0.9 kg m^2
Crankshaft momentum of inertia (clutch closed)	J_c	1.04 kg m^2
Viscosity coefficient	B	0.02 kg m^2 s^{-1}
Pumping and friction torque	T_p	20 N m
Expansion torque gain	G^e	520 000 N m kg^{-1}
Expansion torque offset	T_0^e	18.93 N m
Compression torque gain	G^c	130 000 N m kg^{-1}
Compression torque offset	T_0^c	4.7325 N m
Spark efficiency function	$\eta(\varphi)$	$\eta = 0.73 + \sqrt{0.083 + 0.005\,\varphi}$
Manifold dynamic parameters	a_p	$1.935 \cdot 10^7$ Pa kg^{-1}
	b_p	$4.515 \cdot 10^9$ Pa s^{-1} m^{-2}
Air flow rate generation parameters	O_{nn}	$5.55 \cdot 10^{-4}$ kg s^{-1}
	G_{nn}	$1.32 \cdot 10^{-5}$ kg rd^{-1}
	O_{np}	$7.78 \cdot 10^{-9}$ kg s^{-1} Pa^{-1}
	G_{np}	$1.43 \cdot 10^{-9}$ kg Pa^{-1} rd^{-1}
Throttle angle / surface conversion parameters	a_s	$1.87 \cdot 10^{-7}$ m^2 deg^{-2}
	b_s	$1.92 \cdot 10^{-7}$ m^2 deg^{-1}
	c_s	$6.143 \cdot 10^{-6}$ m^2

Figure E 8.3.**C** Parameters of the engine's nonlinear model.

Intake manifold pressure

The intake manifold is subject to the following dynamics:

$$\frac{\mathrm{d}p(t)}{\mathrm{d}t} = a_p\, f_{out}(p(t), n(t)) + b_p\, S(\alpha(t)), \qquad \text{(E 8.14)}$$

where

$$f_{out}(p(t), n(t)) = O_{np}\, p(t) + G_{nn}\, n(t) - O_{nn} - G_{np}\, n(t) \cdot p(t) \qquad \text{(E 8.15)}$$

represents air flow rate, and

$$S(\alpha) = a_s\, \alpha^2 + b_s\, \alpha + c_s \qquad \text{(E 8.16)}$$

stands for the equivalent throttle area.

Power-train dynamics when the clutch is open

When the clutch is in the *open* position (Boolean variable *on* = 0), the two segments of the driveline are disconnected, and the power train dynamics is given by:

$$\frac{\mathrm{d}n(t)}{\mathrm{d}t} = -\frac{B}{J_o}n(t) + \frac{1}{J_o}(T_g(t) - T_l(t)), \tag{E 8.17}$$

where $T_g(t)$ is the *effective torque* given in (E 8.27) and $T_l(t)$ is the *load torque disturbance* (see Fig. **B**), with J_o denoting the inertial momentum of the segment of the power-train from the crankshaft to the clutch, and B denoting the viscous friction coefficient (assumed to be the same from the crankshaft to the gear);

$$\frac{\mathrm{d}n_g(t)}{\mathrm{d}t} = -\frac{B}{J_c - J_o}n_g(t), \tag{E 8.18}$$

with J_c denoting the inertial momentum of the power-train from the crankshaft to the gear; and

$$\frac{\mathrm{d}\theta(t)}{\mathrm{d}t} = k_c\, n(t). \tag{E 8.19}$$

Power-train dynamics when the clutch is closed

When the clutch is in the *close* position (Boolean variable *on* = 1), the two segments of the driveline are connected and thus have the same speed $n_g(t) = n(t)$.

At engagement (i.e. when event $\uparrow on$ occurs), the values of $n(t)$ and $n_g(t)$ change instantaneously. Their common value "after" is an average value of both values "before", weighted by the respective inertial momenta:

$$n_g := n := \frac{J_o\, n + (J_c - J_o)n_g}{J_c}. \tag{E 8.20}$$

In this case, (E 8.19) is unchanged whereas (E 8.17) and (E 8.18) are replaced by:

$$\frac{\mathrm{d}n_g(t)}{\mathrm{d}t} = \frac{\mathrm{d}n(t)}{\mathrm{d}t} = -\frac{B}{J_c}n(t) + \frac{1}{J_c}(T_g(t) - T_l(t)). \tag{E 8.21}$$

Discrete-time evolution variables

The variables in the set $DES = \{T_C, T_E, m_C, m_E, \varphi\}$ are now considered. The torque T_C absorbed by the cylinder in the compression stroke is given by:

$$T_C = G^c\, m_C - T_0^c, \tag{E 8.22}$$

and the torque T_E generated by the cylinder in the expansion stroke is

$$T_E = \left(G^e\, m_E - T_0^e\right)\cdot\eta(\varphi)\,, \qquad \text{(E 8.23)}$$

where T_0^c and T_0^e are the offset values and $\eta(\varphi)$ the spark efficiency function given in Fig. **C**.

The masses of mixture m_C loaded in the cylinder in the compression stroke and m_E in the cylinder in the expansion stroke, are obtained from (E 8.15) by

$$m_C = m_E = -\frac{30}{n(t)} f_{out}(p(t), n(t))\,. \qquad \text{(E 8.24)}$$

The advance angle φ is obtained from the crankshaft angle $\theta(spk^*)$ corresponding to the value θ when the even spk^* occurs:

If $\theta(spk^*) \geq 160°$ **then** $\varphi = 180 - \theta(spk^*)$ **else** $\varphi = -\,\theta(spk^*)$. (E 8.25)

Note that, as previously specified, $\theta(spk^*) \leq 15°$ or $\theta(spk^*) \geq 160°$.

The torque T applied on the power train is obtained from (E 8.22) and (E 8.23) by:

$$T = T_E - T_C. \qquad \text{(E 8.26)}$$

Given the loss of generated torque T_p due both to pumping and friction (assumed to be constant, $T_p = 20$ according to Fig. **C**), the *effective torque* T_g generated by the engine is:

$$T_g = T - T_p = T - 20. \qquad \text{(E 8.27)}$$

Manifold pressure	p	24 768 Pa
Torque	T	21.34 N m
Compression mass	m_C	$0.9 \cdot 10^{-4}$ Kg
Expansion mass	m_E	0 Kg
Spark advance	φ	20°
Crankshaft speed	n	83.8 rd s^{-1}
Throttle angle	α	4.29°

Figure E 8.3.**D** Set-point values of state and input variables.

Solutions to Exercises

Chapter 1

1.1 a) Nets A to E are PNs. Nets F to H are not PNs because each one has a directed arc for which a node is missing (place or transition) at one of its ends. Net I is not a PN because there are no transitions and J is not a PN because there are no places. Net K is not a PN because there is an arc between two transitions, and net L is not a PN because there is an arc between two places.

b) All the transitions of nets A to E are enabled. In example D, there are no places upstream of the transition. The property that "each place" upstream should contain at least one token is trivially verified. This is a *source transition.*

c) The markings after firing are indicated in Fig. S 1.1. In example C, there is a place which is both upstream of and downstream from the same transition. When the transition is fired, a token is removed from and a token is added to this place at the same time. Its marking therefore does not change.

d) Only the transitions of cases D and E are still enabled. The transition of example D is a source transition and will *always be enabled.* The transition of example E is a *sink transition.* It is still enabled because there were 2 tokens at the start. If it is fired again, it will no longer be enabled.

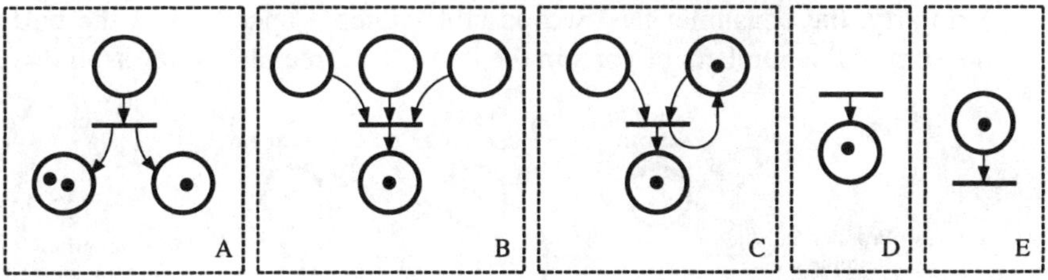

Figure S 1.1

1.2 a) The PN of Fig. 1.1: i) is not a state graph (transition T_2 has 2 output places and transitions T_5 and T_6 have 2 input places); ii) is not an event graph (place P_3 has two output transitions); iii) is not conflict free (place P_3 has two output transitions); iv) is not free-choice (transition T_6 which takes part in conflict $\langle P_3, \{T_3, T_6\}\rangle$ has another input place, P_7); v) is simple (there is only one conflict and thus no transition concerned by 2 conflicts); vi) is pure.

b) The PN of Fig. 1.8d: i) is not a state graph (T_1 has 2 input places); ii) is not an event graph (P_1 has no input transition); iii) is conflict free (hence, is free-choice and simple); iv) is pure.

1.3 a) Figure S 1.3a answers the first question. Place P'_2 is complementary of P_2: we always have $m(P_2) + m(P'_2) = 1$. With the exception of this place P'_2 and its input and output arcs, the figure is identical to Fig. 1.11a.

b) Since place P'_2 is marked when and only when place P_2 is empty, the inhibitor arc can be replaced by reading of place P'_2 (Fig. S 1.3b).

c) The solution is presented in Fig. S 1.3c. Arcs $P'_2 \rightarrow T_4$ and $T_4 \rightarrow P'_2$ have a weight 4. Transition T_4 is only enabled if there are 4 tokens in P'_2, therefore 0 token in P_2. When T_4 is fired, marking of P'_2 is not changed. The initial marking (for the 3 cases) corresponds to the case when the door is open and no customer has come in since it was opened.

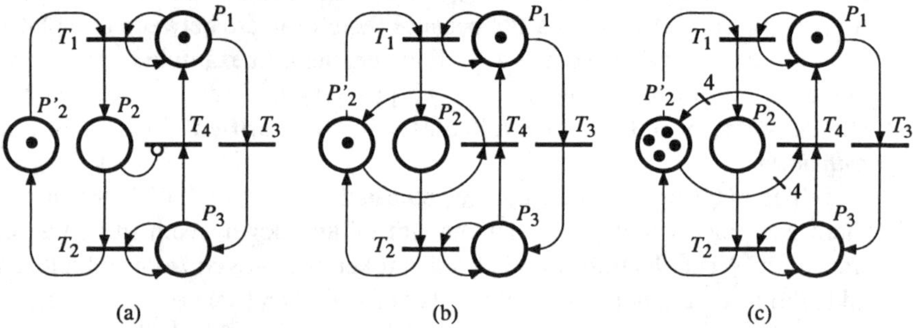

Figure S 1.3

1.4 The PN is indicated in Fig. S 1.4. The producer has two possible states: a "production" state on the one hand and a "ready to deposit" state on the other. Similarly, the consumer has two possible states. Capacity 3 of the buffer is the sum of the numbers of parts in the buffer and free places: $m_3 + m_4 = 3$.

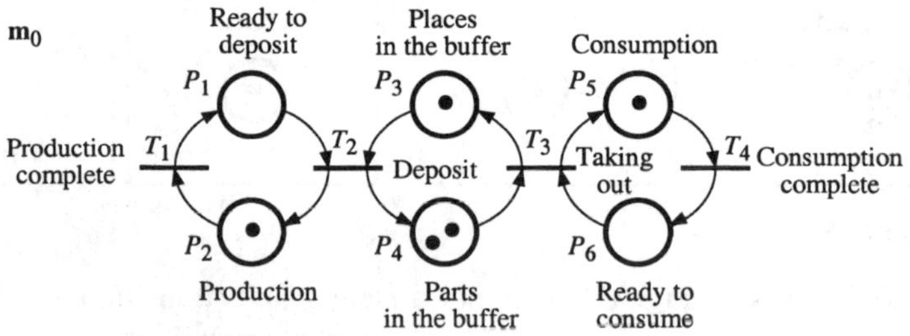

Figure S 1.4

1.5 Figure S 1.5 presents the PN, with an initial marking. The three states of computer CP_1 corresponds to P_1, P_2, and P_3 (and similarly for CP_2). The memory has also three states: idle (P_7), used by CP_1 (P_2), and used by CP_2 (P_6).

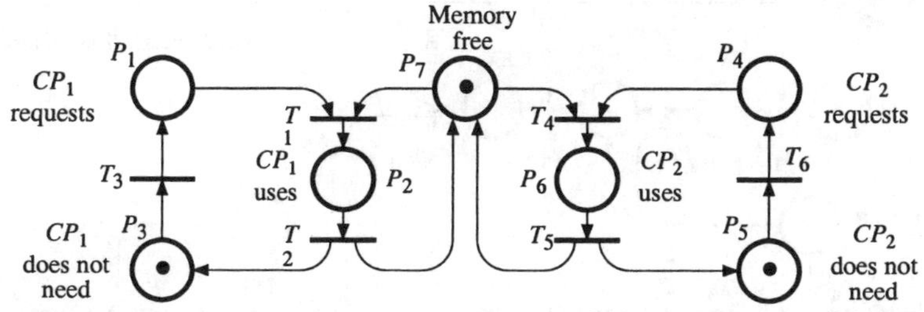

Figure S 1.5

1.6 The generalized PN is presented in Fig. S 1.6. For the initial marking indicated, only transition T_1 is enabled. When it is fired, a token is deposited in place P_2: execution of an instruction from task 1. Then (firing of T_2), task 1 is placed in stand-by (token in place P_7) and a second execution of an instruction from task 1 is authorized, etc. The initial marking is such that 3 instructions from task 1 can be performed (3 tokens in place P_1). The firing sequence T_1T_2 must be carried out 3 times in a row before another transition becomes enabled. In fact, after the firing sequence $T_1T_2T_1T_2T_1T_2$, place P_1 is empty and 3 tokens have been deposited in P_3. Transition T_3 is then enabled. Its firing consists of removing the 3 tokens from P_3 and of adding 5 tokens to P_4. We shall thus be able to perform 5 instructions from task 2, etc.

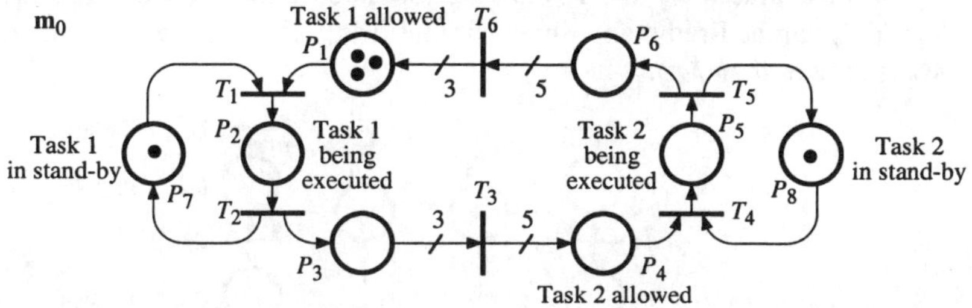

Figure S 1.6

1.7 The PN is represented in Fig. S 1.7. It corresponds to the specifications which have been given, but place P_7 is redundant: it is *implicit*, i.e. a token in P_9 implies a token in P_7. Place P_7 can be deleted together with the two corresponding arcs.

This PN is an event graph since each place has exactly one input and one output transition. In this PN, there are two types of place, namely those corresponding to the execution of tasks (EX_1, EX_2, ...) and standby places corresponding to conditions necessary for the execution of tasks (P_1, P_2...). A place such as EX_1 has exactly one input and one output transition (corresponding to the start and finish of execution, respectively). A standby place such as P_3 corresponds to a condition necessary to perform Task 3. It memorizes that Task 1 has been performed. This place is thus located between the transitions corresponding to the end of Task 1 and the start of Task 3.

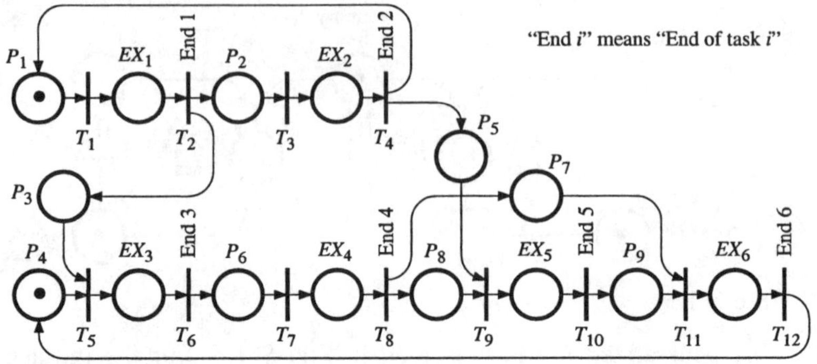

Figure S 1.7

1.8 The PN is the one in Fig. S 1.8. The initial marking indicated corresponds to empty buffers, with both machines available to treat a part p_1. As soon as there is a token in P_1 (firing of T_1), T_2 can be fired (the part p_1 is then processed by M_1). When this processing is finished (firing of T_3), T_4 becomes enabled (part p_1 can then be processed by M_2) and a token is deposited in P_6. From this instant, a part p_2 can be processed by M_1 if there is a token in P_9. And so on.

Note that P_1 and P_9 are not bounded because T_1 and T_6 are source transitions. Places P_3 and P_{11} are not bounded either because the sequence $T_2T_3T_7T_8$ can be fired many times without firing of T_4 (m_3 and m_{11} increase if M_2 is slower than M_1).

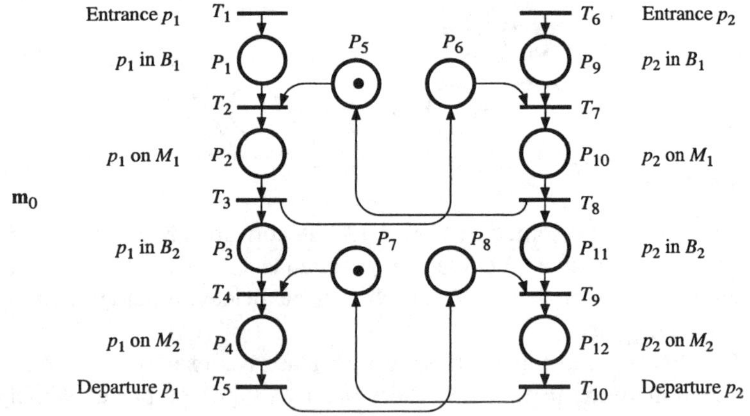

Figure S 1.8

1.9 The PN of Fig. S 1.9 represents the system proposed. We shall observe that when a machine (M_1 or M_2) is occupied by a finished part that has not yet been deposited in the downstream buffer, this machine cannot break down. Places P_5 and P'_5 on the one hand, and P_{10} and P'_{10} on the other are complementary (number of parts and places available in buffers B_1 and B_2).

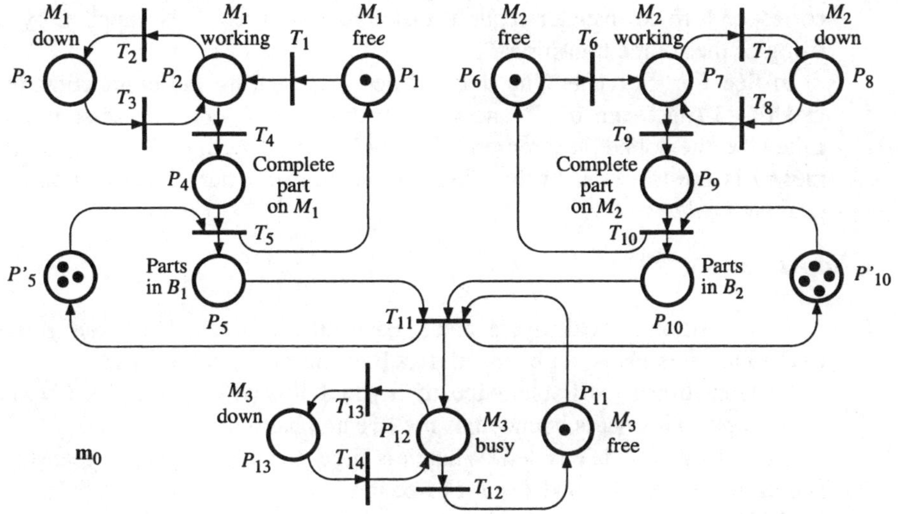

Figure S 1.9

1.10 a) The PN is the one in Fig. S 1.10a.

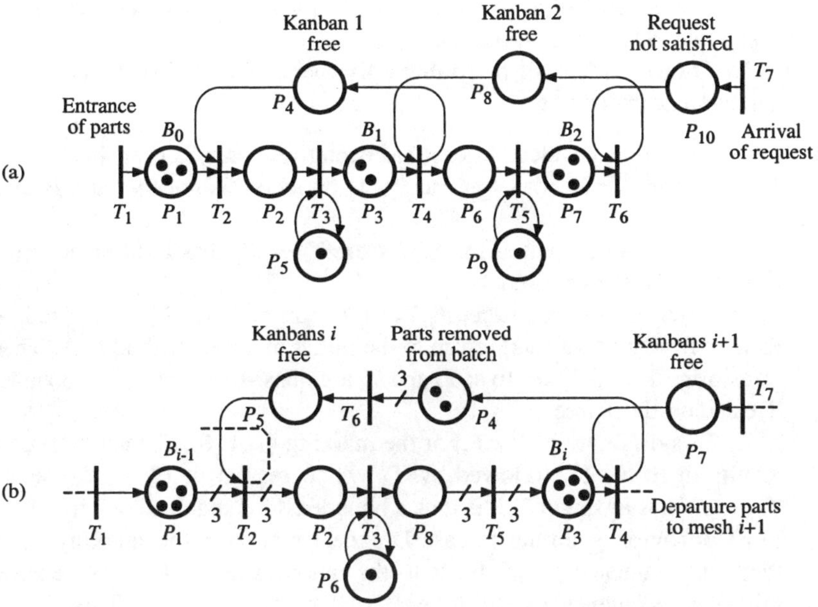

Figure S 1.10

For mesh 1, place P_2 corresponds to the parts in the production system PS_1, place P_3 to buffer B_1 and place P_4 to the *kanbans* possibly free waiting for parts in B_0. Place P_5 indicates that only one part is treated at a time. Mesh 2 has a similar behavior. Place P_1 corresponds to the raw parts at the system entrance. It is supplied by the firing of the source transition T_1. Place P_{10}

corresponds to downstream requests that are not satisfied. It is supplied by the firing of the source transition T_7.

b) See Fig. S 1.10b. The dotted lines representing the connection with mesh $i-1$ (upstream of T_1 and downstream from T_2) do not exist if $i=1$. Likewise, the dotted lines representing the link with mesh $i+1$ do not exist if mesh i is the last one. In this case, the unsatisfied external requests take the place of *kanbans* $i+1$.

Chapter 2

2.1 A) Transition T_1 is always enabled. Each time T_1 is fired, a token is added to P_1. Thus this PN is not bounded. It is live and thus deadlock free.

B) Transition T_1 is firable twice, after which it is no longer enabled because P_1 is empty. This net is bounded, is not live and has a deadlock state.

C) Transition T_1 is enabled. When it is fired, the marking is not changed. It is thus a bounded, live and deadlock free PN.

D) Transition T_1 is and will always be enabled. If we have the firing sequence $S = T_1T_1T_1$ ..., tokens are accumulated in P_1. The PN is thus unbounded. Transition T_2 is enabled. If it is fired, it will no longer be enabled, but it will become so again as soon as the firing of T_1 has added a token to P_1. It is a live PN and thus deadlock free.

E) Transition T_1 will be firable only twice. This PN is bounded, not live and has a deadlock state.

2.2 A) Only T_1 is enabled. After it has been fired, only T_2 is enabled and so on. Thus T_3 and T_4 are not live. The PN is bounded, is not live and is deadlock free.

B and C) Same result as for A. The arc $T_4 \rightarrow P_1$ has a different weight but transition T_4 is never fired.

D) After T_1 is fired, either T_1,T_2 or T_3 can be fired. If T_3 is fired, T_4 can then be fired and we shall return to the initial marking of the PN A. Therefore, transitions T_1 and T_2 are live; T_3 and T_4 are quasi-live. The PN is bounded, not live and deadlock free.

E) T_1 and T_2 may be fired. For the marking (1, 1, 0), T_3 may be fired which results in (0, 0, 1), followed by T_4 which results in (2, 0, 0) because arc $T_4 \rightarrow P_1$ has a weight 2. It is thus a bounded, live and deadlock free PN.

F) Behavior is similar to case D except that from the marking (2, 0, 0), the firing sequence $T_1T_3T_4$ leads to the marking (3, 0, 0). The repetition of this firing sequence causes tokens to accumulate in P_1. This PN is not bounded. It is live and deadlock free.

2.3 A) Yes. The sequence $S = T_1$ is increasing repetitive, thus the PN is increasing repetitive.

B) No. The sequence $S = T_1$ is not repetitive.

C) Yes. The sequence $S = T_1$ is repetitive.

D) Yes. The sequence $S = T_1T_2$ is repetitive. It contains all the transitions (T_2T_1 is also repetitive).

E) No. The sequence $S = T_1$ is not repetitive.

2.4 A) For the PN in Fig. S 1.4.

a) As each place has exactly one input transition and one output transition, and all the arc weights are 1, it is an event graph. Since the path $P_1T_2P_4T_3P_5T_4P_6T_3P_3T_2P_2T_1P_1$ passes through all the nodes, the event graph is strongly connected.

b) According to Property 2.8, there is a P-invariant for each elementary circuit, namely $\mathbf{x}_1 = (1, 1, 0, 0, 0, 0)$, $\mathbf{x}_2 = (0, 0, 1, 1, 0, 0)$, and $\mathbf{x}_3 = (0, 0, 0, 0, 1, 1)$, and a single T-invariant $\mathbf{y}_1 = (1, 1, 1, 1)$.

Marking invariants: $m_1 + m_2 = 1$ (states of the producer), $m_3 + m_4 = 3$ (number of places in the buffer), and $m_5 + m_6 = 1$ (states of the consumer).

Examples of repetitive sequences: $T_1T_2T_4T_3$ or $T_1T_4T_3T_4T_3T_2T_1T_2$ (same $\mathbf{y}_1$).

c) According to Property 2.8c, the PN is live since there is at least a token in every elementary circuit.

d) Firing sequence $T_1T_2T_4T_3$ is complete and repetitive from $\mathbf{m}_0$. Hence, the PN is consistent.

B) For the PN in Fig. S 1.6.

a) It has the structure of an event graph, but it is not an event graph because all the arc weights are not 1.

b) Building the graph of markings shows that, for each reachable marking there is only one transition enabled (no concurrency). The firing sequence $S = (T_1T_2)^3T_3(T_4T_5)^5T_6$ is such that $\mathbf{m}_0 \xrightarrow{S} \mathbf{m}_0$. It corresponds to the T-invariant $\mathbf{y}_2 = (3, 3, 1, 5, 5, 1)$. There is no other minimal repetitive firing sequence.

There is a marking invariant for each elementary circuit: $m_2 + m_7 = 1$, $m_5 + m_8 = 1$, and $5\,m_1 + 5\,m_2 + 5\,m_3 + 3\,m_4 + 3\,m_5 + 3\,m_6 = 15$, corresponding to P-invariants $\mathbf{x}_4 = (0, 1, 0, 0, 0, 0, 1, 0)$, $\mathbf{x}_5 = (0, 0, 0, 0, 1, 0, 0, 1)$, and $\mathbf{x}_6 = (5, 5, 5, 3, 3, 3, 0, 0)$.

c) According to b) above, $\mathbf{m}_0$ is a home state. Since S is complete, the PN is live according to Property 2.10. (Property 2.11 is also verified).

d) The PN is consistent since S is complete and repetitive from $\mathbf{m}_0$.

2.5 a) P-invariants: $\mathbf{x}_1 = (0, 1, 0, 0, 1, 1, 0, 0, 0, 1, 0, 0)$ and $\mathbf{x}_2 = (0, 0, 0, 1, 0, 0, 1, 1, 0, 0, 0, 1)$. The PN is not conservative since P_1, P_3, P_9, and P_{11}, are not in the support of a P-invariant. These places are unbounded.

b) Since there is no conflict, the PN is persistent. From $\mathbf{m}_0$ the firing sequence $T_1T_2T_3T_4T_5T_6T_7T_8T_9T_{10}$ is possible. This is a complete repetitive sequence and thus the PN is live according to Property 2.11.

2.6 a) The PN is presented in Figure S 2.6a. The initial state corresponds to the case where all the philosophers are thinking. Places P_{17} to P_{20} correspond to the availability of rods B_1 to B_4. There are four marking invariants corresponding to each of the philosophers. For example, $m_1 + m_2 + m_3 + m_4 = 1$ for $Phil_1$. There are equally four marking invariants each corresponding to a rod, e.g. $m_{18} + m_1 + m_2 + m_5 + m_8 = 1$. This corresponds to the fact that rod B_2 is either available (P_{18}) or picked up by $Phil_1$ (P_1 and P_2) or picked up by $Phil_2$ (P_8 and P_5).

The net is not live. Indeed, the firing sequence $T_3T_7T_{11}T_{15}$ results in a deadlock: there is a token in each of places P_4, P_8, P_{12} and P_{16} and only in these places. This comes about because each philosopher has picked up the rod on his right-hand side and is waiting for the one on his left to be freed. This will never occur since each of them needs two rods in order to eat. In order to avoid this unfortunate situation in which the philosophers would inevitably die of hunger, we shall propose another protocol.

b) We note that philosopher $Phil_i$ needs two resources in order to eat: rod B_i and rod $B_{i+1(\text{mod. } 4)}$. Since *the possession of just one of these is of no use, both resources must be taken up simultaneously,* in order to avoid taking up one which might have been useful to someone else. The PN of Figure S 2.6b is obtained, if each philosopher simultaneously takes up the two rods he needs and simultaneously puts them back.

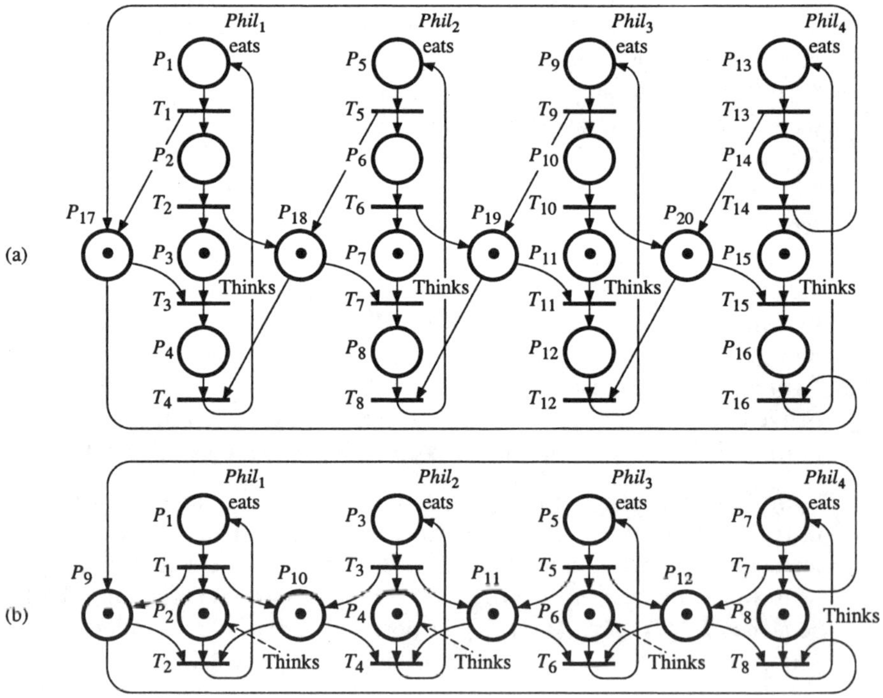

Figure S 2.6

2.7 a) The PN is the one in Fig. S 2.7a. The number of customers swimming corresponds to the number m_3 of tokens in P_3. There is a marking invariant corresponding to the number of baskets: $m_7 + m_2 + m_3 + m_4 = 5$, thus $m_3 \leq 5$. There is also a marking invariant corresponding to the number of cabins: $m_6 + m_1 + m_2 + m_4 + m_5 = 3$. Each place belongs to at least one marking invariant, thus the PN is bounded. As from the initial marking of the figure, the firing sequence $(T_1T_2T_3)^5\ (T_1)^3$ results in the marking (3, 0, 5, 0, 0, 0, 0) which is a deadlock. Five people are swimming and 3 are in the cabins waiting for baskets which will never be freed. Whatever the numbers N_c and N_p, the sequence $(T_1T_2T_3)^{N_p}(T_1)^{N_c}$ results in a deadlock.

b) The customer who enters needs *two resources,* namely a cabin and a basket. If he *takes these simultaneously,* when both are free, deadlock will not occur. See Figure S 2.7b (without the grey part).

c) The modification is the grey part of Figure S 1.7b. The firing of T_0 (arrival of a customer) adds a token to P_0 (number of customers waiting).

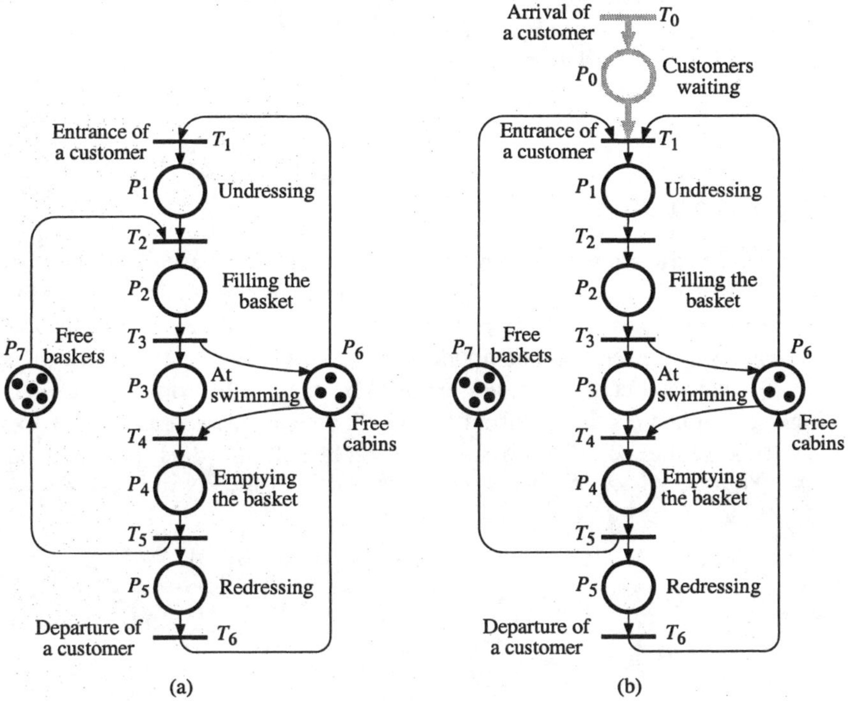

Figure S 2.7

2.8 a) The incidence matrix is given in Fig. S 2.8a.

b) The graph of markings is given in Fig. 2.8b. The set of possible firing sequences is the set of prefixes $\overline{\mathcal{L}_1}$ of $\mathcal{L}_1 = (T_1(T_2T_4 + T_3T_5))^*T_1(T_2T_3 + T_3T_2)$. In other words, from $\mathbf{m}'_0$, $T_1T_2T_4$ and $T_1T_3T_5$ are repetive firing sequences, whereas $T_1T_2T_3$ and $T_1T_3T_2$ lead to a deadlock.

$$\begin{array}{c} \begin{matrix} T_1 & T_2 & T_3 & T_4 & T_5 \end{matrix} \\ \begin{bmatrix} -1 & 0 & 1 & 0 & 1 \\ 1 & -1 & 0 & 0 & -1 \\ 1 & -1 & 0 & -1 & 0 \\ 1 & 0 & -1 & -1 & 0 \\ 0 & 1 & -1 & 0 & 0 \\ 0 & 0 & 0 & 1 & -1 \end{bmatrix} \begin{matrix} P_1 \\ P_2 \\ P_3 \\ P_4 \\ P_5 \\ P_6 \end{matrix} \end{array}$$

(a)

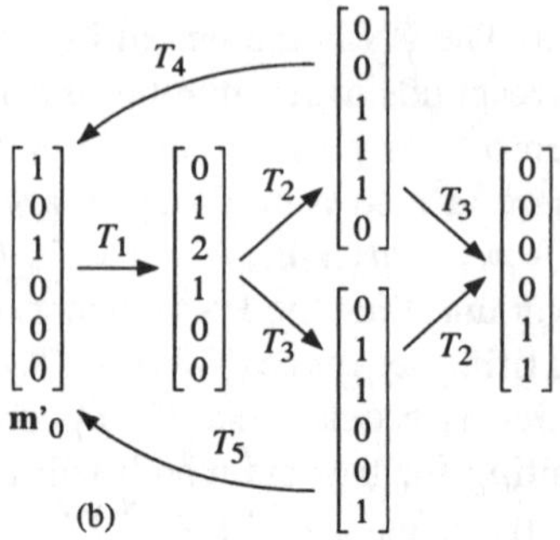

(b)

Figure S 2.8

2.9 Figure S 2.9 represents the coverability root tree. Places P_2 and P_3 are not bounded.

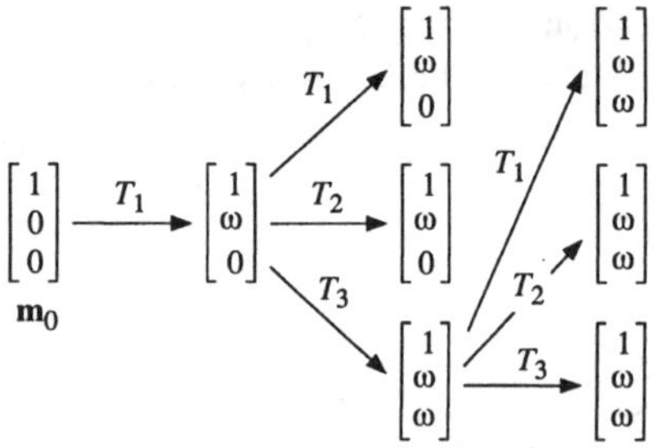

Figure S 2.9

2.10 a) We disregard the initial marking. If T_3 and T_4 are never fired, the number of tokens initially in $\{P_1, P_2\}$ is maintained, but if T_3 is fired, then this number decreases and if T_4 is fired, this number increases. Thus $\{P_1, P_2\}$ cannot be a conservative component (for all $\mathbf{m}_0$). When the firing sequence T_3T_4 is carried out, 2 tokens are removed from P_1 and 1 token is put into P_3, then this token is removed from P_3 while 2 tokens are restored to P_2 at the end of this sequence. A P-invariant $\mathbf{x} = (1, 1, 2)$ is thus obtained. Indeed $\mathbf{x}^T \cdot \mathbf{W}_1 = 0$, where $\mathbf{W}_1$ is the incidence matrix of the PN in Fig. 2.4a,

$$\text{i.e., } \begin{bmatrix} 1 & 1 & 2 \end{bmatrix} \cdot \begin{bmatrix} -1 & +1 & -2 & 0 \\ +1 & -1 & 0 & +2 \\ 0 & 0 & +1 & -1 \end{bmatrix} = \begin{bmatrix} 0 & 0 & 0 & 0 \end{bmatrix}.$$

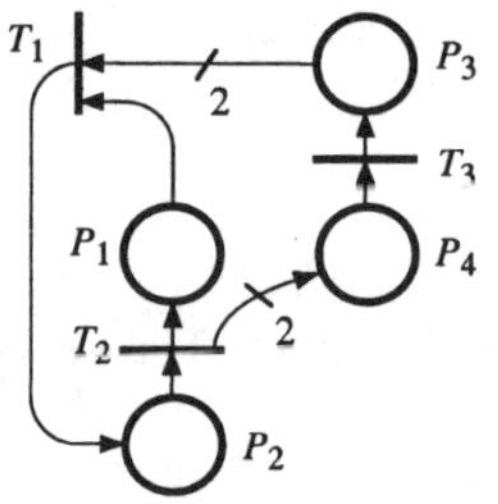

$$\mathbf{W}_2 = \begin{array}{c} \begin{matrix} T_1 & T_2 & T_3 \end{matrix} \\ \begin{bmatrix} -1 & +1 & 0 \\ +1 & -1 & 0 \\ -2 & 0 & +1 \\ 0 & +2 & -1 \end{bmatrix} \begin{matrix} P_1 \\ P_2 \\ P_3 \\ P_4 \end{matrix} \end{array}$$

Figure S 2.10

Assume that the initial marking is $\mathbf{m}_0 = (2, 0, 0)$. The firing sequence T_1T_2 is repetitive. It corresponds to the T-invariant $\mathbf{y}_1 = (1, 1, 0, 0)$. Now the firing sequence $T_3T_4T_2T_2$ is also repetitive. It corresponds to the *T*-invariant $\mathbf{y}_2 = (0, 2, 1, 1)$. We may verify that $\mathbf{W}_1 \cdot \mathbf{y}_1 = 0$ and $\mathbf{W}_1 \cdot \mathbf{y}_2 = 0$.

b) The dual PN and its incidence matrix $\mathbf{W}_2$ are given in Fig. S 2.10 (place P_i is the dual of transition T_i, and vice-versa). According to the end of Sect. 2.2.2.5, this PN has two P-invariants, $\mathbf{x}_1 = (1, 1, 0, 0)$ and $\mathbf{x}_2 = (0, 2, 1, 1)$, and one T-invariant, $\mathbf{y} = (1, 1, 2)$. This result may be observed in Fig. S 2.10, and the proof follows from $\mathbf{W}_2 = \mathbf{W}_1^{\mathrm{T}}$.

2.11 Application of Algorithm 2.2 to the PN in Fig. E 2.11a and b are shown in Figs. S 2.11a and b, respectively.

a) Figure S 2.12a shows that there is a single minimal support P-invariant: $\mathbf{x} = (1, 1, 1)$. The corresponding marking invariant is $m_1 + m_2 + m_3 = 3$.

b) According to Fig. S 2.12b, there are two minimal support P-invariants, namely $\mathbf{x}_1 = (1, 1, 2, 0)$ and $\mathbf{x}_2 = (1, 1, 0, 2)$. According to Property 2.5 in Sect. 2.2.2.3, $\mathbf{x}_1 + \mathbf{x}_2 = (2, 2, 2, 2)$ is a P-invariant. This P-invariant is not minimal since all its components can be divided by 2. But $\mathbf{x}_3 = (\mathbf{x}_1 + \mathbf{x}_2) / 2 = (1, 1, 1, 1)$ is a minimal P-invariant. Hence, P-invariant $\mathbf{x}_3$ is *minimal*, whereas it is *not minimal support* (Remark 2.12).

Corresponding to $\mathbf{x}_1$, $\mathbf{x}_2$, and $\mathbf{x}_3$, the marking invariants are $m_1 + m_2 + 2m_3 = 3$, $m_1 + m_2 + 2m_4 = 3$, and $m_1 + m_2 + m_3 + m_4 = 3$.

A			B			
P_1	P_2	P_3	T_1	T_2	T_3	
1	0	0	−1	+1	0	P_1
0	1	0	+1	−1	+1	P_2
0	0	1	0	0	−1	P_3
1	1	0	0	0	+1	$P_1 + P_2$
1	1	1	0	0	0	$P_1+P_2+P_3$

(a)

A				B		
P_1	P_2	P_3	P_4	T_1	T_2	
1	0	0	0	−1	−1	P_1
0	1	0	0	+1	−1	P_2
0	0	1	0	0	+1	P_3
0	0	0	1	0	+1	P_4
1	1	0	0	0	−2	P_1+P_2
1	1	2	0	0	0	$P_1+P_2+2P_3$
1	1	0	2	0	0	$P_1+P_2+2P_4$

(b)

Figure S 2.11

2.12 a) The enabling degrees of T_1, T_2, and T_3 are 1, 2, and 2, respectively.

b) Maximal concurrent firings: $\{T_1T_2T_3\}$, $\{T_1(T_3)^2\}$, and $\{(T_2)^2\}$.

c) $\langle P_2, \{T_1, T_2, T_3\}, \mathbf{m}\rangle$ is not an effective conflict since the concurrent firing $\{T_1T_2T_3\}$ is possible. It is a general conflict since there are not enough tokens in P_2 for firing simultaneously *once* T_1, *twice* T_2, and *twice* T_3.

d) Neither $\langle P_2, \{T_1, T_2\}, \mathbf{m}\rangle$ nor $\langle P_2, \{T_1, T_3\}, \mathbf{m}\rangle$ nor $\langle P_2, \{T_2, T_3\}, \mathbf{m}\rangle$, are general conflicts. As a matter of fact, the *four tokens in* P_2 *are sufficient* for firing the pairs of transitions according to their enabling degrees, i.e., T_1 *once* and T_2 *twice*, or T_1 *once* and T_3 *twice*, T_2 *twice* and T_3 *twice*.

e) The *three tokens in* P_1 *are not sufficient* for firing T_1 *once* and T_2 *twice*: hence, $\langle P_1, \{T_1, T_2\}, \mathbf{m}\rangle$ is a general conflict. The *two tokens in* P_3 *are not sufficient* for firing T_2 *twice* and T_3 *twice*: hence, $\langle P_3, \{T_2, T_3\}, \mathbf{m}\rangle$ is a general conflict.

2.13 a) The enabling degrees are $q(T_1, \mathbf{m}) = 2$, $q(T_2, \mathbf{m}) = 1$, $q(T_3, \mathbf{m}) = 2$, $q(T_4, \mathbf{m}) = 1$, and $q(T_5, \mathbf{m}) = 1$.

b) The priority graph is given in Fig. S 2.13. Transitions T_2 and T_4 have the *highest priority*, in which case their allowing degree r equals their enabling degree q: for $\mathbf{m} = (1, 2, 3, 1)$, $r(T_2, \mathbf{m}) = 1$ and $r(T_4, \mathbf{m}) = 1$. Deleting the tokens necessary for firing T_2 and T_4, $\mathbf{m}^{(2,4)} = (0, 1, 2, 0)$ is obtained (notation in Section B.1, Appendix B). *Second priority level*, $r(T_3, \mathbf{m}) = q(T_3, \mathbf{m}^{(2,4)}) = 1$ and $r(T_5, \mathbf{m}) = q(T_5, \mathbf{m}^{(2,4)}) = 0$. Then, $\mathbf{m}^{(2,4,3,5)} = (0, 0, 1, 0)$ is obtained. *Third priority level*, $r(T_1, \mathbf{m}) = q(T_1, \mathbf{m}^{(2,4,3,5)}) = 0$.

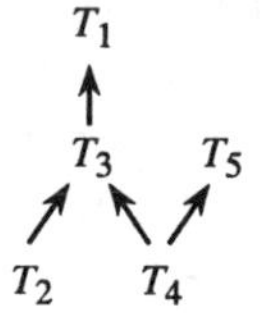

Figure S 2.13

2.14 There is no token in the circuit $P_5T_5P_6T_6P_5$. Hence, T_5 and T_6 will never be enabled (the PN is not quasi-live). For any reachable marking in $\mathcal{L}(\mathbf{m}_0)$ only transitions in $\{T_1, T_2, T_3, T_4\}$ can be enabled. Since there is no conflict between these transitions, for any reachable marking, all the enabled transitions can be fired concurrently according to their enabling degrees (for example, from $\mathbf{m}_0$, the concurrent firing $\{T_1(T_3)^3\}$ is possible). This implies that the PN in Fig. E 2.14 is *persistent.*

Note that the unmarked PN of Fig. E 2.14 (i.e. considered without marking) is *not structurally persistent.* For an initial marking $\mathbf{m}_0'$ such that there is at least one token in the circuit $P_3T_3P_4T_4P_3$ and one token in the circuit $P_5T_5P_6T_6P_5$, a conflict between T_3 and T_5 can occur: the PN is not persistent for this marking.

2.15 a) Initial vector $\mathbf{m}_0' = (1, 0, 0, 0, 0, 0)$. After $S_1 = T_2T_4$, no transition in the siphon will ever be enabled. After S_1 was fired, always one transition in $\{T_5, T_6, T_7\}$ is enabled. There is no deadlock because T_3 can not be enabled from $\mathbf{m}_0'$ (it is not possible to have a token in both P_1 and P_2).

b) Initial vector $\mathbf{m}_0'' = (3, 0, 0, 0, 0, 0)$. After $S_2 = T_2T_4T_2T_4T_2T_4$, no transition in the siphon will ever be enabled. After S_1 was fired, always one or several transitions in $\{T_5, T_6, T_7\}$ are enabled. There is no deadlock: as a matter of fact, T_3 can be fired once, but a token remains in one of the places in $\{P_1, P_2, P_4, P_5, P_6\}$.

c) A deadlock can exist if and only if the initial marking is such that the number of tokens in the siphon is even and the number of tokens in the trap is nil: if there are $2N$ tokens in the siphon and zero token in the trap, then T_3 might be fired N times; the marking reached is a deadlock.

Chapter 3

3.1 Figure S 3.1 represents the graph of stable markings. As from marking (1, 0, 0, 0), the occurrence of E^1 causes transition T_1 and immediately afterwards T_2 to be fired, thereby resulting in a stable marking (0, 0, 1, 0). The other firings correspond to a single transition.

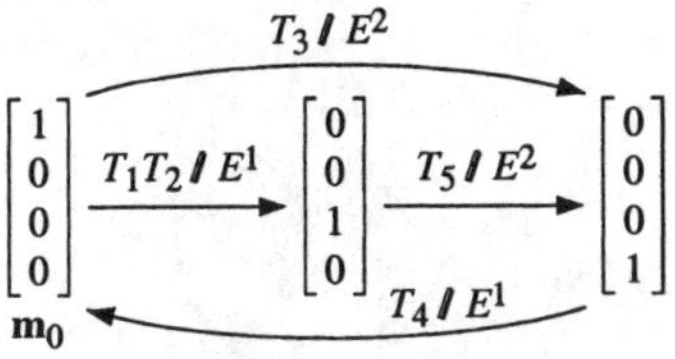

Figure S 3.1

3.2 a) The two enabled transitions are receptive to E^2. Since there is no conflict, we find $S_2 = [T_2T_5]$, which is a maximal EFS.

b) The two enabled transitions are receptive to E^1. Since they are not in conflict, we find $S_2 = [T_1T_3]$, which is maximal.

c) There are four enabled transitions, three of which are receptive to E^1. These are T_1, T_2 and T_4. Transitions T_1 and T_2 are in structural conflict, but there are two tokens in P_1 and thus no effective conflict. We find a maximal EFS which is $S_3 = [T_1T_2T_4]$.

d) There are four enabled transitions which are receptive to E^1. These are T_1, T_2, T_4 and T_5. There is an effective conflict between T_4 and T_5 and thus an actual conflict if event E^1 occurs. We find two EFSs, namely $S_4 = [T_1T_2T_4]$ and $S_5 = [T_1T_2T_5]$. There is no maximal EFS: not deterministic behavior.

3.3 a) See Fig. S 3.3a. There is a source transition T_1 synchronized on E^1 and a sink transition T_2, synchronized on E^2. Place P_1 is not bounded. The coverability root tree of reachable markings is shown in the figure.

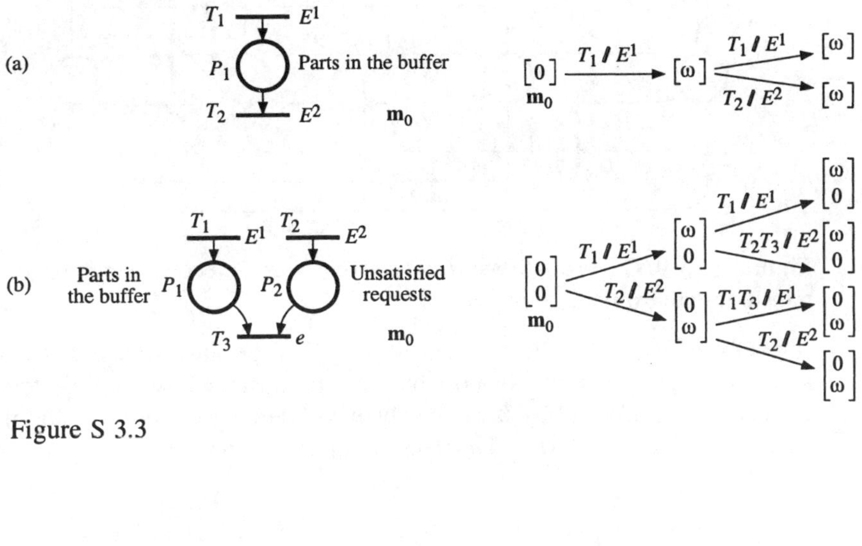

Figure S 3.3

b) See Fig. S 3.3b. There are now two source transitions which are synchronized on E^1 and E^2, and a sink transition synchronized on e (i.e. immediate transition). All stable markings are thus such that at least one of the two places P_1 or P_2 is empty. Neither place is bounded.

3.4 The configurations of the system and the Boolean variables presented in Exercise 3.4 are illustrated in Fig. **A**. Configurations 2 and 4 correspond respectively to $D_p = 1$ and $D_o = 1$. Configurations 1 and 3 correspond respectively to $D_{pc}D_o' = 1$ and $D_cD_o' = 1$.

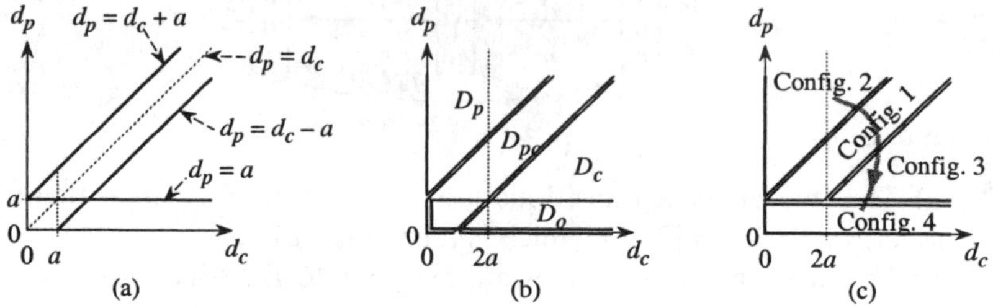

Figure S 3.4.**A** Illustration of configurations and Boolean variables.

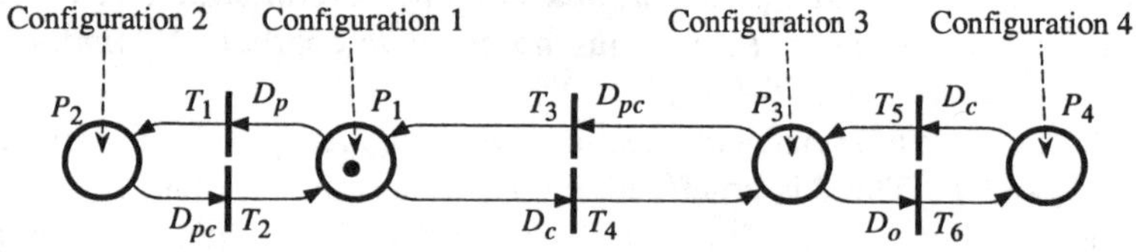

Figure S 3.4.**B** Flows $d_p(t)$ and $d_c(t)$ are continuous functions of time.

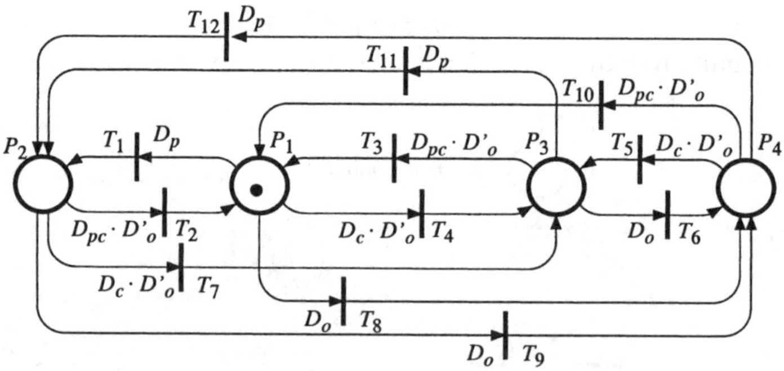

Figure S 3.4.**C** Functions $d_p(t)$ and $d_c(t)$ are piecewise constant. All transitions are explicitly represented.

a) Given $d_c(t) > 2a$, the problem is simplified because all the transitions among configurations are not possible; this is illustrated by the trajectory in Fig. **A**c. The model in Fig. **B** can be obtained (there is no other transition).As a matter of fact, since $d_p(t)$ and $d_c(t)$ are continuous functions of time, between a time when $d_p - d_c > a$, and a time when $d_c - d_p > a$, there is necessarily a time interval where $-a < d_p - d_c < a$; in other words, a direct transition from P_2 to P_3 is impossible (and vice-versa). Similarly, no direct transition between P_1 to P_4 is possible.

b) We consider now that $d_p(t)$ and $d_c(t)$ are piecewise constant functions of time. Let t_1 denote a time when the vector $\mathbf{d}(t) = (d_p(t), d_c(t))$ changes. From any configuration before t_1, any configuration can be reached by this change. As in Fig. **B**, in Fig. **C** there is a token in P_i when the system is in configuration i, $i = 1, 2, 3, 4$. Hence, in this figure, all transitions between places exist. For example, from P_1, there are transitions to P_2 (T_1, condition $D_p = 1$), to P_3 (T_4, condition $D_c D'_o = 1$), and to P_4 (T_8, condition $D_o = 1$).

Now, the structure in Fig. **B** can also be obtained if $d_p(t)$ and $d_c(t)$ are piecewise constant. This is illustrated in Fig. **D**. The PN in Fig. **C** is 1-stable (each marking reached is stable) whereas the PN in Fig. **D** is 3-stable (up to 2 unstable markings can be reached before a stable one). For example, transitions T_4 and T_8 in Fig. **C** are replaced by T_4 in Fig. **D**, with a condition which is the sum of both: $D_c D'_o + D_o = D_c + D_o$. From P_1, if $D_c + D_o = 1$, T_4 is fired; if $D_c D'_o = 1$, then the token is deposited in P_3, and this marking is stable; otherwise T_6 is fired and the marking corresponding to a token in P_4 is stable. The firing of T_8 in Fig. **C**, corresponds to the iterated firing of the sequence $T_4 T_6$ in Fig. **D**. The conditions associated with transitions T_1 to T_6 in Fig. **D**e are illustrated in Fig. **D**a–d.

Let us consider another example of behavior. Assume there is a token in P_2 and there a change such that D_o becomes 1. Figures **D**a and e show that the condition associated with T_2 is $D_{pc} + D_c$. According to Fig. **A**b, $D_o = 1$ implies $D_{pc} + D_c = 1$. Hence, all the conditions associated with T_2, T_4, and T_6, are true. It follows that there is a firing sequence $T_2 T_4 T_6$, ending with a token in P_4.

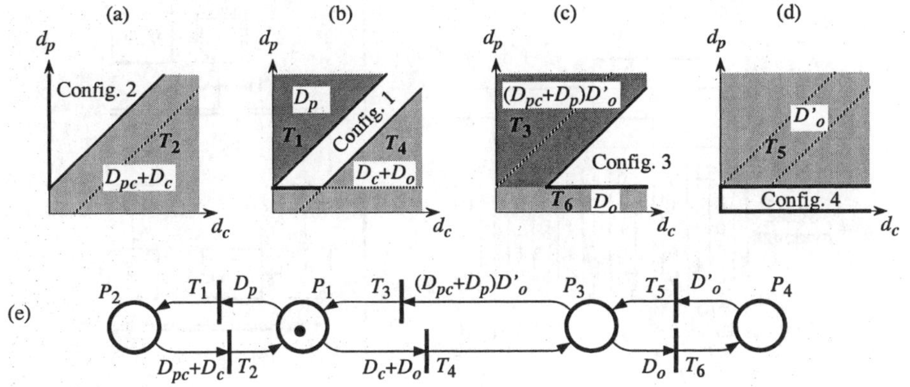

Figure S 3.4.**D** Functions $d_p(t)$ and $d_c(t)$ are piecewise constant. Minimal representation.

Let us now answer the following question: if outputs were associated with places P_1 to P_4, would the solutions in Figs. **C** and **D** be equivalent (i.e., same input/output behavior)?

1) If all outputs are *level outputs*, the interpreted PNs obtained from Figs. **C** and **D** are equivalent: from Algorithm 3.3 in Sect. 3.3.2 and the subsequent comments, the level outputs are significant only in the stable markings.

2) If some *impulse outputs* are associated with places whose marking may be unstable (namely P_1 and P_3), the interpreted PNs obtained from Figs. **C** and **D** are *not* equivalent: from Algorithm 3.3 in Sect. 3.3.2 and the subsequent comments, the impulse outputs are performed even in the unstable markings.

3.5 a) The stable markings and the states of the variable N are indicated in Fig. S 3.5a.

b) In order to show that it is a *control interpreted PN*, we have to show that the first three conditions in Definition 3.7 (Sect. 3.3.1) are verified.

1) The synchronized PN in Fig. E 3.5 has two elementary circuits, namely $P_1T_1P_2T_2P_4T_4P_1$ and $P_1T_1P_3T_3P_5T_4P_1$. Since T_1, common to both circuits, is synchronized by an event ($\uparrow x$), this is sufficient for the PN to be *stable*, according to Property 3.2 in Sect. 3.2.3.1.

2) This PN has two marking invariants: $m_1 + m_2 + m_4 = 1$ and $m_1 + m_3 + m_5 = 1$. Since each place belongs to one of these invariants, the PN is *safe*.

3) Verification that it is *deterministic* (Property 3.11). i) There is no structural conflict, hence *no actual conflict.* ii) Two incompatible operations are performed: $N := 0$ (place P_1) and $N := 1$ (place P_4). Because of the marking invariant $m_1 + m_2 + m_4 = 1$, these *incompatible operations cannot be performed simultaneously.*

According to Appendix E, a *control interpreted PN* corresponds to a *grafcet,* which is given in Fig. S 3.5b for our example.

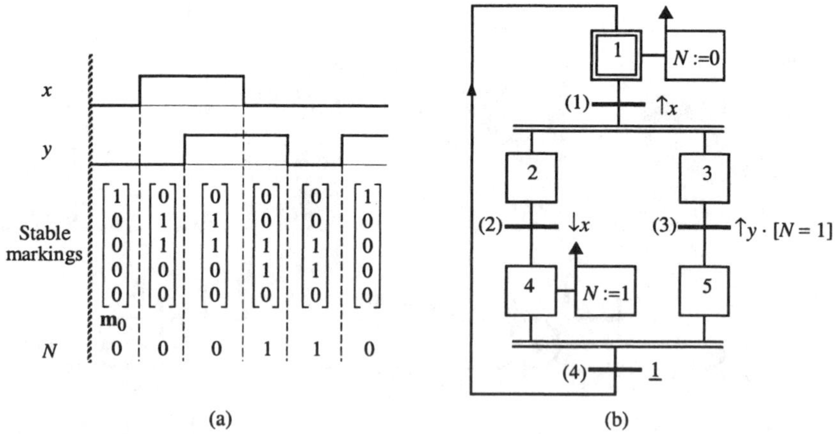

Figure S 3.5

3.6 a) See Fig. S 3.6a. Once T_1 has been fired, the token in the cycle $P_2 \rightarrow P_3 \rightarrow P_2 \rightarrow ...$ counts the edges $\uparrow h$. When the train passes in front of a point B, we have $\uparrow b$ and thus changeover to marking (0, 0, 0, 1), where the speed is calculated and the impulse counter reset (note that the speed is only calculated from the 2nd point B onwards). It should be observed that, after the firing of T_1, the only stable marking is (0, 1, 0, 0) since T_4 and T_5 are immediate transitions. This implies that neither of the events $\uparrow h$ or $\uparrow b$ is "lost".

b) If we had $R_5 = b'$ instead of $R_5 = e$, we could "lose" an impulse $\uparrow h$ which would occur while $b = 1$. This would create a small error in the speed calculated. If we had $R_4 = h'$ instead of $R_4 = e$, we could "lose" an impulse $\uparrow b$ which would occur while $h = 1$. This would create a *large error* in the speed calculated (twice its real value).

c) In the event of "apparent simultaneity" of $\uparrow h$ and $\uparrow b$, *priority must* on all accounts *be given* to $\uparrow b$ in order to prevent a serious error being made. For this we can use the redundant receptivity $R_2 = \uparrow h \cdot b'$ instead of $R_2 = \uparrow h$.

d) The corresponding grafcet is given in Fig. 3.6b.

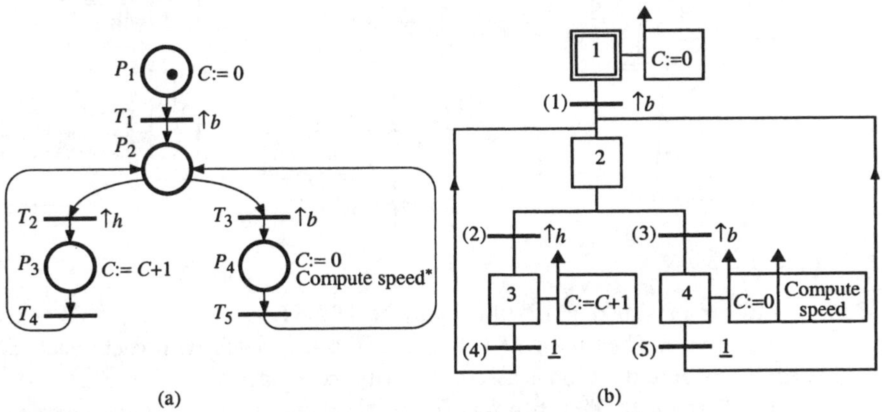

Figure S 3.6

3.7 The set of 3 grafcets {G1, G2, G3 } of Fig. S 3.7 meets the specifications.

1) Grafcet G1 corresponds to functioning without faults.

2) Grafcet G2 modifies the behavior of G1. When a fault appears, step 22 becomes active, thereby inactivating the entire grafcet G1: *level macroaction forcing* G1{} (G1 is forced into a situation where there are no active steps), and giving rise to action S (alarm). When the fault clears, grafcet G1 is reinitialized either in situation {11} for an x fault (*impulse macroaction force*[1] G1{11} associated with step 23), or in situation {12, 13}, for an y fault (*impulse action force* G1{12,13} for step 24).

3) Grafcet G3 could prevent grafcet G2 from evolving if the event $\uparrow x$ occurred twice with less than a minute's interval, by the *level macroaction*

---

[1] The *impulse* macroaction *force* is distinguished from the *level* macroaction *forcing* by the vertical arrow (top-left of the corresponding rectangle, see Appendix E).

freezing denoted by G2{*}: grafcet G2 must remain in the present situation (i.e. {22} just after the second $\uparrow x$) as long as step 33 is active. When the operator ends its manual intervention, *restart* allows a new initialization of G1 and G2 (*impulse actions force* G1{11} and G2{21}).

A set of three grafcets is thus obtained which **interact** with one another: G2 acts on G1, while G3 acts on both G1 and G2. A hierarchy is thus established between grafcets.

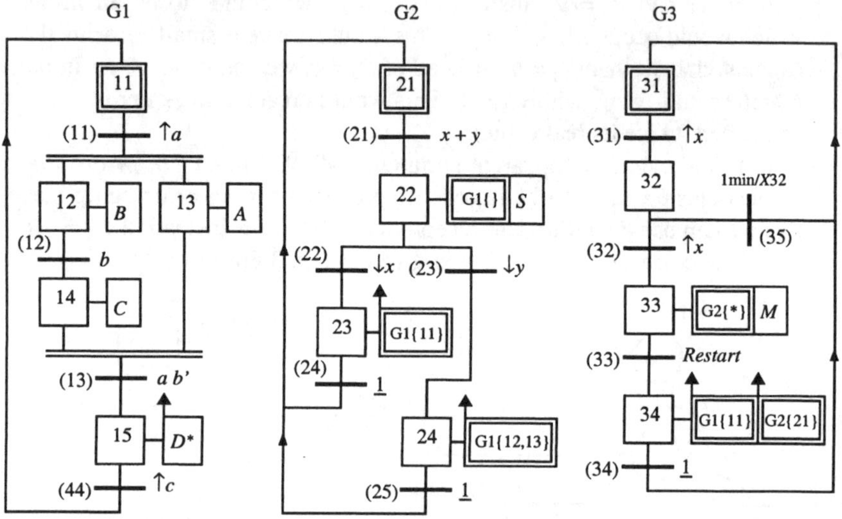

Figure S 3.7

3.8 a) Either a P-timed or a T-timed may be defined.

i) *P-timed.* The timing D_i $(i = 1, ..., 6)$ is associated with each place EX_i. A timing of zero duration is associated with each place $P_1, P_2, ..., P_9$.

ii) *T-timed.* The timing D_i is associated with each transition corresponding to "end of Task *i*". A timing of zero duration is associated with the other transitions, namely T_1, T_3, T_5, T_7, T_9, and T_{11}.

b) The P-timed PN is considered. Task 5 must begin after Task 2 is complete, which requires at least Tempo(EX_1) + Tempo(P_2) + Tempo(EX_2) + Tempo(P_5) = $D_1 + 0 + D_2 + 0$. Task 5 must also begin after Task 4 is complete, which requires at least Tempo(EX_1) + Tempo(P_3) + Tempo(EX_3) + Tempo(P_6) + Tempo(EX_4) + Tempo(P_8) = $D_1 + 0 + D_3 + 0 + D_4 + 0$. The minimum duration before starting to perform Task 5 is thus $\max((D_1 + D_2), (D_1 + D_3 + D_4))$. The execution of this task will therefore begin at $t \geq D_1 + \max(D_2, (D_3 + D_4))$.

3.9 a) *Blocking before service.*

i) The case where C_1 and $C_2 > 0$ is given in Fig. **A**a. The invariant $m_1 + m_2 + m_6 = C_1$ ensures that the number of parts in buffer B_1 (i.e. m_2) is less than C_1 if there is a part on M_1 ($m_1 = 1$). The same applies to B_2.

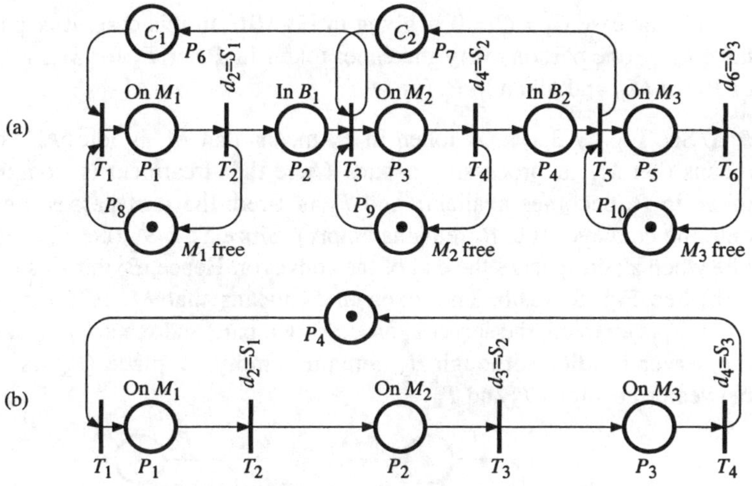

Figure S 3.9.**A** Blocking before service. (a) C_1, $C_2 > 0$. (b) $C_1 = C_2 = 0$.

ii) The case $C_1 = C_2 = 0$ is given in Fig. **A**b. Since there are no buffers between the machines, a part may enter M_1 only if machines M_2 and M_3 are free. This is ensured by the marking invariant $m_1 + m_2 + m_3 + m_4 = 1$.

b) *Blocking after service.*

i) The case where C_1 and $C_2 > 0$ is given in Fig. **B**a. When a part has been completed by M_1, a token is deposited in P_2. Then, if B_1 is not full (i.e. $m_8 \geq 1$), transition T_3 can be fired (the part is deposited in the buffer). Otherwise the token remains in P_2 up to the next firing of T_4, immediately followed by a firing of T_3.

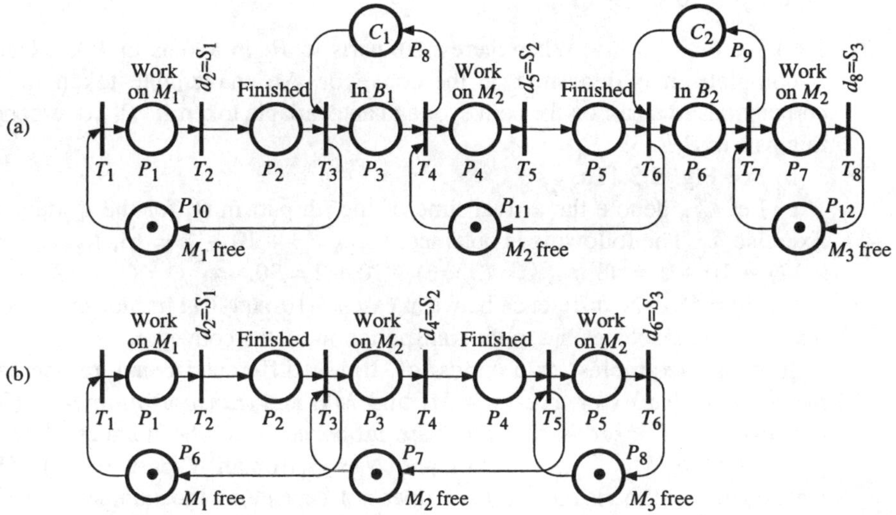

Figure S 3.9.**B** Blocking after service. (a) C_1, $C_2 > 0$. (b) $C_1 = C_2 = 0$.

ii) The case $C_1 = C_2 = 0$ is given in Fig. **B**b. In this case, it is possible to have up to one part on every machine: token in P_1 or P_2 for M_1, token in P_3 or P_4 for M_3, and token in P_5 for M_3.

3.10 a) See Fig. S 3.10a. A token in P_1 means that M_1 is idle. A token in P_4 means that M_1 is processing a part. Once this treatment is completed, the token in P_4 becomes available and T_2 is fired: the part passes immediately onto the conveyor (i.e. B_2 remains empty). Since $S_2 < S_1$ (i.e. $d_6 < d_4$), M_2 is idle when a part reaches the end of the conveyor. Hence, B_3 remains empty.

b) See Fig. S 3.10b. The token in P_1 means that M_1 is a single server: when T_1 is enabled, the server is processing a part, and when T_1 is not enabled the server is idle. Although B_3 remains empty, a place (P_5) is necessary between transitions T_2 and T_3.

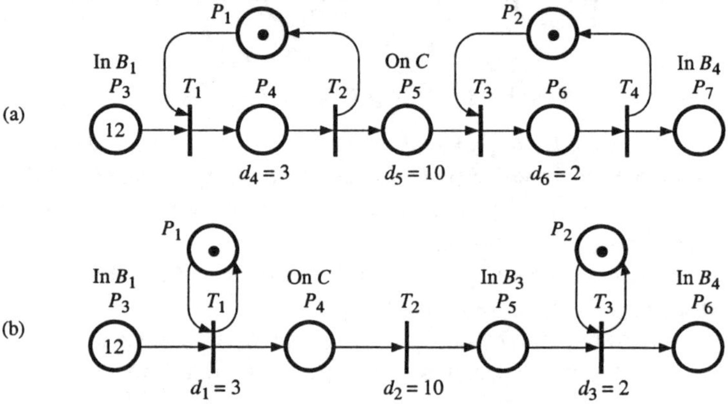

Figure S 3.10

3.11 a) See Fig. S 3.11a. When there are 6 parts in B_2 (6 tokens in P_5), a batch is complete: it is deposited on the conveyor. An unavailable token in P_6 corresponds to a part on the conveyor and an available token in P_6 corresponds to a part in B_3.

b) See Fig. S 3.11b.

c) Let $t_{i(3.j)}$ denote the arrival time of the *i*th part in B_4 for the system in Exercise 3.*j*. The following is obtained: $t_{1(3.10)} = 3 + 10 + 2 = 15$, $t_{12(3.10)} = (3 \times 12) + 10 + 2 = 48$, $t_{1(3.11)} = (3 \times 6) + 10 + 2 = 30$, $t_{12(3.11)} = (3 \times 12) + 10 + (2 \times 6) = 58$. The difference between cases 3.10 and 3.11 is not due to the treatment on machines but to the transportation on the conveyor.

In these examples, two kinds of time, different in nature, appear: *processing times* (for machines M_1 and M_2) and *transportation time* (for conveyor *C*). The *processing* times are *proportional* to the number of parts processed (for M_1: S_1 for a part and $N \cdot S_1$ for an *N*-part batch). The *transportation* time is *constant* and does not depend on the number of parts transported (for *C*: S_C for a part or for an *N*-part batch); it corresponds to a pure delay.

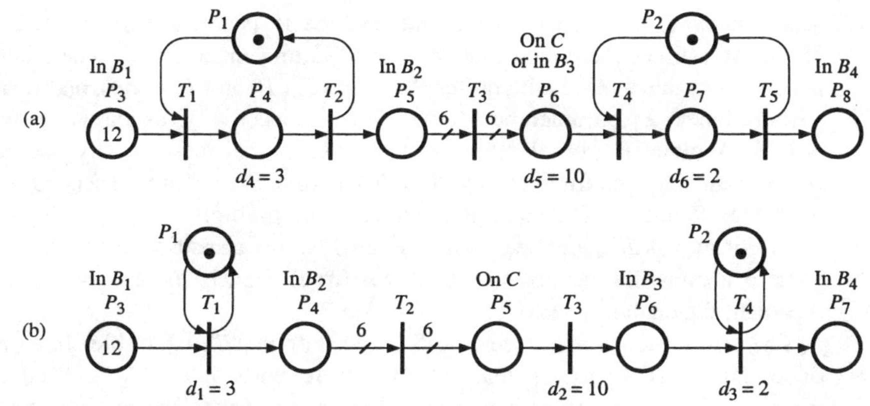

Figure S 3.11

3.12 The T-timed PN is given in Fig. **A**. When P_3 is not empty, a part enters the conveyor every 2 s (size 0.2 m, speed 0.1 m / s). A part on the conveyor reaches B_3 in 80 s.

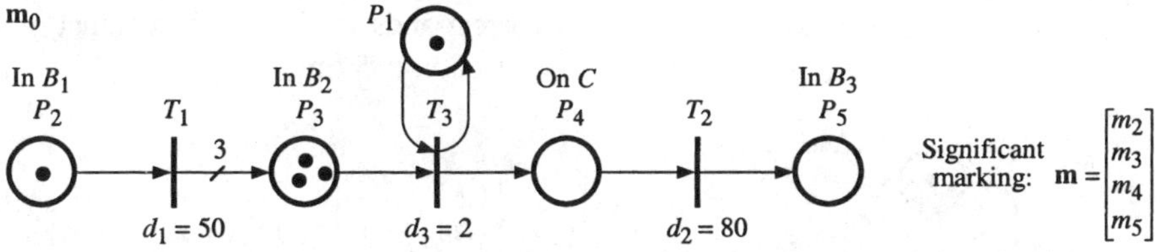

Figure S 3.12.**A** T-timed PN.

The corresponding reachability graph is given in Fig. **B**. Transition T_3 is fired three times at $t = 2$, 4, and 6. Then, T_1 is fired 44 s later. Transition T_3 is fired again three times 2 s, 4 s, and 6 s later. Finally, T_2 is fired six times.

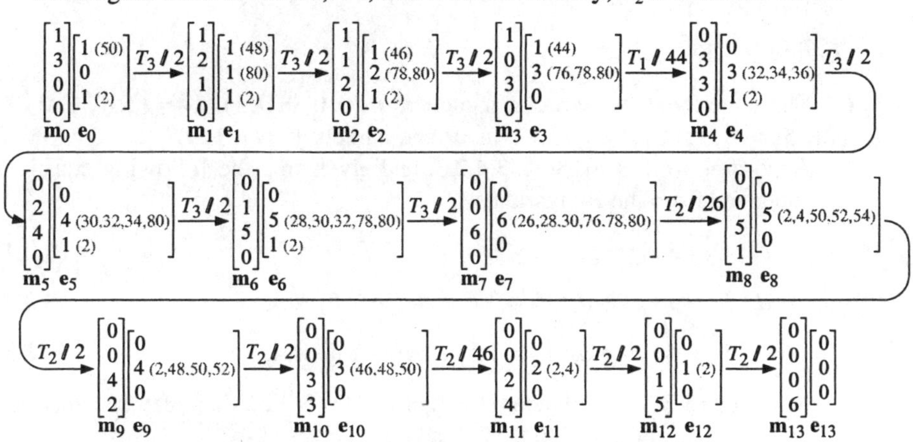

Figurc S 3.12.**B** Reachability graph.

3.13 a) Since there are only two parts and machine M_2 is able to serve them both at the same time, there are never any parts waiting in buffer B_2. There is no need to associate a place with buffer B_2. Places P_1, P_2 and P_4 correspond to the parts in buffer B_1 on machine M_1 and on M_2 respectively, as shown in Fig. S 3.13. A place P_3 complementary of P_2 ensures that there is only one part on machine M_1. The timings $d_2 = S_1 = 2$ and $d_3 = S_2 = 3$ are associated with the places P_2 and P_4. The other places have a zero timing.

b) For any *stable marking* of a P-timed PN, no transition is enabled (as soon as a transition becomes enabled it is fired). Hence, there is no need to represent the enabling vector as in a T-timed PN.

The initial state is $\mathbf{m}_0$ (see Fig. S 3.13). For $\mathbf{m}_0$, T_1 is firable, but only once. The marking $\mathbf{m}_1$ is reached. Two time units later, T_2 is fired and transition T_1, which becomes enabled again, is also fired. The marking $\mathbf{m}_2$ is obtained. Two time units later, T_2 is fired again. The marking $\mathbf{m}_3$ is obtained. One unit later the first token reaching P_3 becomes available and T_3 is fired. Since T_1 is immediately firable, $\mathbf{m}_4$ is obtained. Two time units later, transitions T_2 and T_3 become simultaneously enabled. They are simultaneously fired, then transition T_1 is immediately firable and we return to $\mathbf{m}_2$. The stationary behavior thus corresponds to the periodic functioning $\mathbf{m}_2 \rightarrow \mathbf{m}_3 \rightarrow \mathbf{m}_4 \rightarrow \mathbf{m}_2$ whose cycle duration is 5 time units.

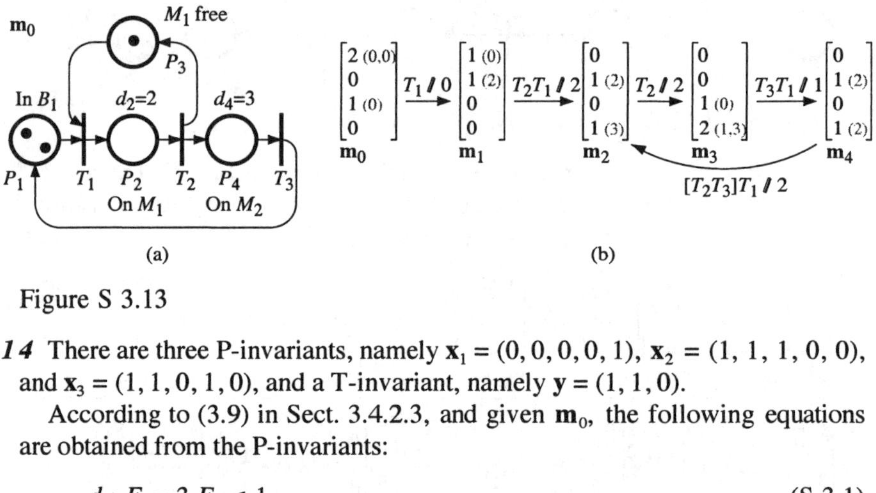

Figure S 3.13

3.14 There are three P-invariants, namely $\mathbf{x}_1 = (0, 0, 0, 0, 1)$, $\mathbf{x}_2 = (1, 1, 1, 0, 0)$, and $\mathbf{x}_3 = (1, 1, 0, 1, 0)$, and a T-invariant, namely $\mathbf{y} = (1, 1, 0)$.

According to (3.9) in Sect. 3.4.2.3, and given $\mathbf{m}_0$, the following equations are obtained from the P-invariants:

$$d_1 \cdot F_1 = 3\,F_1 \leq 1, \tag{S 3.1}$$

$$d_1 \cdot F_1 + d_2 \cdot F_2 + d_3 \cdot F_3 = 3\,F_1 + 4\,F_2 + 2\,F_3 \leq 3, \tag{S 3.2}$$

$$d_1 \cdot F_1 + d_2 \cdot F_2 + d_3 \cdot F_3 = 3\,F_1 + 4\,F_2 + 2\,F_3 \leq 4. \tag{S 3.3}$$

Note that (S 3.3) is redundant (compare with (S 3.2), then may be canceled. From T-invariant $\mathbf{y}$, (S 3.4) is obtained.

$$F_1 = F_2. \tag{S 3.4}$$

Since T_3 is not in a support of T-invariant, there is no average firing frequency F_3. Given (S 3.4) and $F_3 = 0$, $F_1 = F_2 \leq 3/7$ is obtained from (S 3.2). From (S 3.1), $F_1 \leq 1/3$ is obtained. Then, the stationary behavior (maximal speed) corresponds to:

$$F_1 = F_2 = \min\left(\frac{3}{7}, \frac{1}{3}\right) = 0.33. \quad \text{(S 3.5)}$$

3.15 a) In Figure S 3.15a, place P_1 corresponds to the parts which are in the station (in buffer or on the machine), and place P_2 to the places available in the station. Place P_3 performs the following constraint: the mean time separating two part arrivals is equal to $1/\mu_1$ (if this place were not present, we could have an arrival rate greater than μ_1). Likewise, P_4 models the fact that the machine can only serve one part at a time. The markings of P_3 and P_4 are always equal to one; they are not represented in the significant marking.

The reachability graph and the Markov process are given in Figs. S 3.15b and c, respectively. The system has 3 states.

b) In the particular case considered, we have $\mu_1 = 2$ and $\mu_2 = 1$. Let $p(i)$ denote the probability to be in state i, $i = 1, 2, 3$. The following equations are obtained (at equilibrium, i.e. when the $p(i)$ are constant):

$$\mu_2 \cdot p(2) = \mu_1 \cdot p(1), \text{ i.e., } p(2) = 2\,p(1), \text{ because } p(1) \text{ is constant;} \quad \text{(S 3.6)}$$

$$\mu_2 \cdot p(3) = \mu_1 \cdot p(2), \text{ i.e., } p(3) = 2\,p(2), \text{ because } p(3) \text{ is constant;} \quad \text{(S 3.7)}$$

$$p(1) + p(2) + p(3) = 1. \quad \text{(S 3.8)}$$

From (S 3.6) to (S 3.8), (S 3.9) is obtained.

$$p(1) = \frac{1}{7},\ p(2) = \frac{2}{7},\ p(3) = \frac{4}{7}. \quad \text{(S 3.9)}$$

c) Mean markings of places P_1 and P_2. From the markings in Fig. S 3.15b:

$$m^*(P_1) = 1 \cdot p(2) + 2 \cdot p(3) = \frac{10}{7}, \quad \text{(S 3.10)}$$

$$m^*(P_2) = 2 \cdot p(1) + 1 \cdot p(2) = \frac{4}{7}. \quad \text{(S 3.11)}$$

Mean firing frequencies of transitions T_1 and T_2. From the transitions in Fig. S 3.15b:

$$F_1^* = \mu_1 \cdot p(1) + \mu_1 \cdot p(2) = \frac{6}{7}, \quad \text{(S 3.12)}$$

$$F_2^* = \mu_2 \cdot p(2) + \mu_2 \cdot p(3) = \frac{6}{7}. \quad \text{(S 3.13)}$$

Mean dwelling times in places P_1 and P_2:

$$D^*(P_1) = \frac{m^*(P_1)}{F_1^*} = \frac{5}{3}, \tag{S 3.14}$$

$$D^*(P_2) = \frac{m^*(P_2)}{F_2^*} = \frac{2}{3}. \tag{S 3.15}$$

d) There are three P-invariants, namely $\mathbf{x}_1 = (1, 1, 0, 0)$, $\mathbf{x}_2 = (0, 0, 1, 0)$, and $\mathbf{x}_3 = (0, 0, 0, 1)$, and a T-invariant, namely $\mathbf{y} = (1, 1)$. The marking invariant $m_1 + m_2 = 2$ is deduced from $\mathbf{x}_1$. This invariant is verified for the mean markings in stationary behavior, according to (S 3.10) and (S 3.11). From $\mathbf{y}$ we deduce that the two mean firing frequencies of transitions T_1 and T'_1 are equal. This is verified by (S 3.12) and (S 3.13).

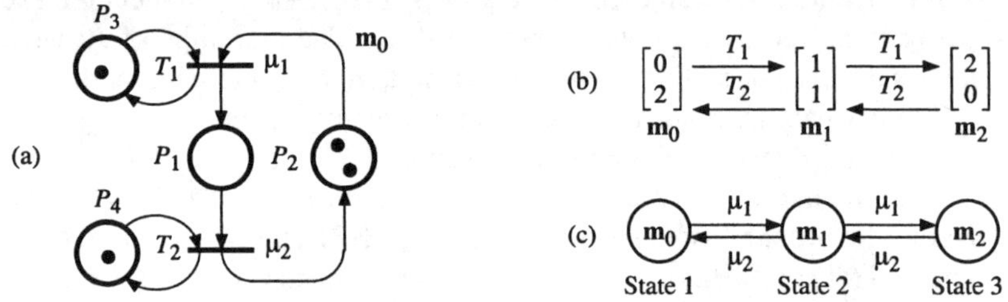

Figure S 3.15

Chapter 4

4.1 The reachability graph for the PN in Fig. 4.3a is given in Fig. S 4.1.

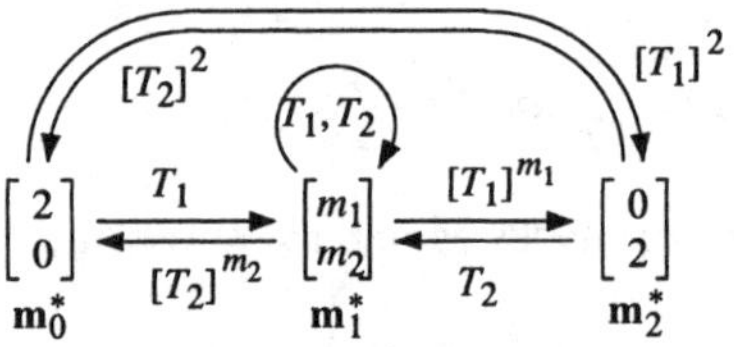

Figure S 4.1

4.2 For Fig. E 4.2a, there is no general conflict: T_1 is 2-enabled and T_2 is 1-enabled, whereas m_3 is greater than the sum of these enabling degrees. The possible OG-firings are: $[(T_1)^\alpha(T_2)^\beta]$ such that $\alpha \in \{0, 1, 2\}$ and $\beta \in [0, 1]$.

For Fig. E 4.2b, there is a general conflict: both T_1 and T_2 are 1-enabled, whereas m_3 is less than the sum of these enabling degrees. The possible OG-firings are: $[T_1(T_2)^\beta]$, $\beta \in [0, 0.5]$ and $[(T_2)^\beta]$, $\beta \in [0, 1]$.

4.3 a) Yes: $(2, 0, 0, 3) \xrightarrow{[T_4]^{1.8}} (2, 0, 1.8, 1.2) \xrightarrow{T_1} (1, 1, 0.8, 2.2) \xrightarrow{[T_4]^{0.9}} (1, 1, 1.7, 1.3) \xrightarrow{T_1} (0, 2, 0.7, 2.3) \xrightarrow{[T_3]^{0.6}} (0, 2, 0.1, 2.9)$.

b) No: $(2, 0, 0, 3) \xrightarrow{[T_4]^{1.8} T_1} (1, 1, 0.8, 2.2)$; then, m_3 is not sufficient for firing T_1 (simultanously with $[T_4]^{0.9}$).

c) No: $(2, 0, 0, 3) \xrightarrow{[T_4]^{1.8}T_1[T_4]^{1.2}} (1, 1, 2, 1)$; then, m_1 is not sufficient for firing $[T_1]^2$.

d) No: $(2, 0, 0, 3) \xrightarrow{[T_4]^{1.8}T_1[T_4]^{1.2}} (1, 1, 2, 1)$; then, m_3 is not sufficient for firing $[T_3]^{2.2}$.

e) Yes: $(2, 0, 0, 3) \xrightarrow{[T_4]^{1.8}T_1[T_3]^{0.8}T_2} (2, 0, 0, 3)$. The firing sequence $[T_4]^{1.8}T_1[T_3]^{0.8}T_2$ is repetitive. Hence, all the sequences in $\mathcal{L}_1$ are possible.

4.4 The reachability graph for the PN in Fig. E 4.4 is given in Fig. S 4.4. Let us recall that $(T_j)^\alpha$ does not denote a OG-firing but a sequence of OG-firings related to the same transition T_j (Notation 4.4, Section 4.2.3.1).

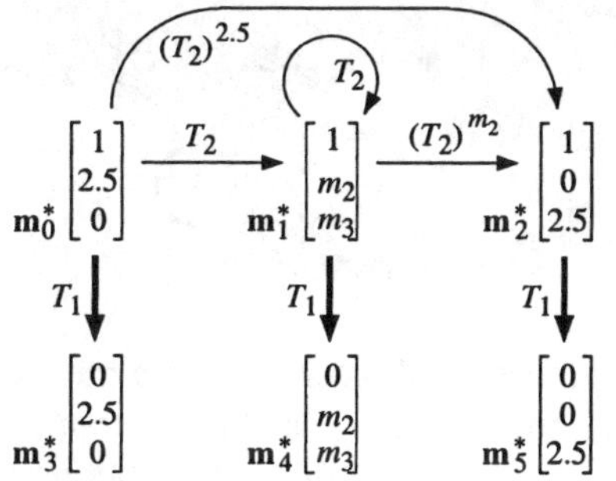

Figure S 4.4

4.5 a) From $\mathbf{m}_0$, the firing sequence $S_1 = [T_5]^{0.9}[T_3]^{0.9}[T_2]^{0.9}[T_1]^{0.9}[T_4]^{0.9}$ is possible and $\mathbf{m}_0 \xrightarrow{S_1} \mathbf{m}_0$. Hence, since S_1 contains all the transitions, the continuous PN is quasi-live and consistent. From Property 4.10, it follows that $\mathbf{m}_1 = (1.2, 0.3, 0, 0, 1.8)$ is reachable.

b) Marking $\mathbf{m}_1$ is obtained from $\mathbf{m}_1 = \mathbf{m}_0 + \mathbf{W} \cdot \mathbf{y}$, where $\mathbf{y} = (0, 0.3, 0.3, 0, 0)$. However, there is no firing sequence from $\mathbf{m}_0$ with this characteristic vector.

Firing sequences $S_2 = [T_5]^{0.3}[T_3]^{0.3}[T_2]^{0.3}[T_4]^{0.3}$, $S_3 = [T_5]^{0.9}[T_3]^{0.3}[T_2]^{0.3}[T_4]^{0.9}$, and an infinity of others lead from $\mathbf{m}_0$ to $\mathbf{m}_1$.

4.6 a) Figure E 4.6a. $S_1 = [T_2]^1$ leads to $\mathbf{m}_1 = (2, 0)$ which is a dealock. Hence, the PN is not live (implying that it is neither ε-live nor lim-live).

b) Figure E 4.6b. $S_2 = [T_1]^{0.5}$ leads to $\mathbf{m}_2 = (0, 2)$ which is a deadlock.

c) Figure E 4.6c. The continuous PN is ε-live (hence live): the enabling degree of T_1 is 0.5; after firing of $[T_1]^{0.5}$, the enabling degree of T_1 is 0.25, and so on. If successive firings of T_1 take place, the marking tends towards $\mathbf{m}_2 = (0, 2)$ *without reaching it*. Now, the PN is reversible since successive firings of T_2 may lead to $\mathbf{m}_0$ from $\mathbf{m} = (\alpha, 2 - \alpha)$ where α is any small positive value.

The PN is not lim-live since $\mathbf{m}_2 = (0, 2)$ is a deadlock.

4.7 The discrete conterparts of the PNs in Fig. E 4.6a are given in Fig. S 4.7.
a) Figure S 4.7a. $S_1 = T_2$ leads to a deadlock (similar to the continuous PN).
b) Figure S 4.7b. From $\mathbf{m}_0$, only T_2 is enabled and $\mathbf{m}_0 \xrightarrow{T_2} \mathbf{m}_3 = (3, 0)$. From $\mathbf{m}_3$, only T_1 is enabled and $\mathbf{m}_3 \xrightarrow{T_1} \mathbf{m}_0$. It follows that the discrete PN is live whereas its continuous counterpart is not live.
c) Figure S 4.7c. From $\mathbf{m}_0$, only T_2 is enabled and $\mathbf{m}_0 \xrightarrow{T_2} \mathbf{m}_1 = (2, 0)$. From $\mathbf{m}_1$, only T_1 is enabled and $\mathbf{m}_1 \xrightarrow{T_1} \mathbf{m}_0$. It follows that the discrete PN is live. Its continuous counterpart is also live (ε-live).

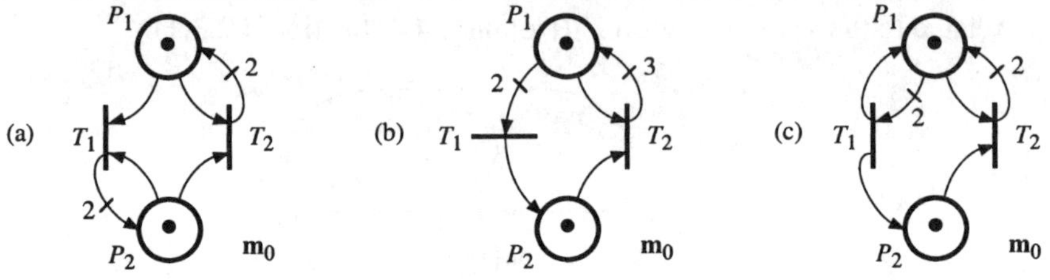

Figure S 4.7

4.8 The chemical process and its initial state are shown in Fig. S 4.8. If we assume that transitions T_1 and T_2 are fired according to their enabling degrees before firing of T_3 and T_4, the firing sequence[2] $S = \{(T_1)^{2.5}(T_2)^{1.37}\}\{(T_3)^{1.37}(T_4)^{1.37}\}$ leads to $\mathbf{m} = (0, 0, 26, 19.2, 0, 0, 80.1, 24.7)$.

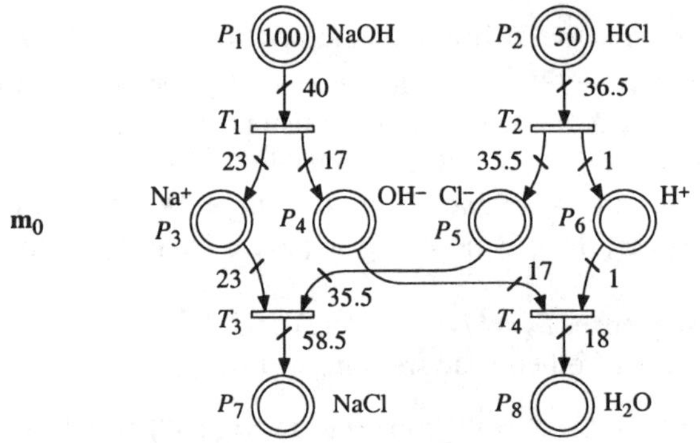

Figure S 4.8

4.9 a) In Fig. E 4.9a, T_1 is enabled and T_2 is not enabled.
b) In Fig. E 4.9b, T_1 is not enabled and T_2 is enabled.
c) In Fig. E 4.9c, T_1 is not enabled and T_2 is enabled.

4.10 The system is modeled by the hybrid PN in Fig. S 4.10. The marking m_4 corresponds to the free places in the stadium (10 000 at initial state) and m_3 corresponds to the spectators in the stadium. Firings of T_3 and T_4 correspond to entrance and departure of the spectators, respectively. These transitions are

[2] According to Sect. 2.1.4, $\{X, Y\}$ models the concurrency of X and Y.

not enabled when the doors are closed ($m_2 = 0$). They can be enabled when the doors are open (firing of T_1 leading to $m_2 = 12$). After the show, the doors are closed (firing of T_2) only if the stadium is empty (zero test by inhibitor arc $P_3 \rightarrow T_2$).

There is a structural conflict: $\langle P_2, \{T_3, T_4\} \rangle$. The doors have a limited flow capacity: when many people enter, the crowd makes it impossible to leave!

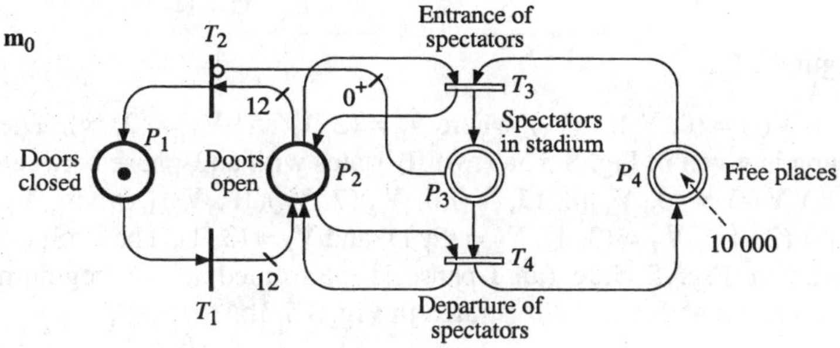

Figure S 4.10

Chapter 5

5.1 The continuous timed PN is given in Fig. S 5.1.

Transition T_1 corresponds to a single server. Hence, the timing $d_1 = 2$ is replaced by the maximal firing speed $V_1 = 1 / d_1 = 0.5$.

Transition T_2 corresponds to a triple server. Hence, according to the explanations illustrated by Fig. 5.2, the timing $d_2 = 6$ is replaced by the maximal speed $V_2 = 3 / d_2 = 0.5$.

The markings are not the same as for the discrete counterpart (they remain constant) but the firing speeds correspond exactly to the firing frequencies.

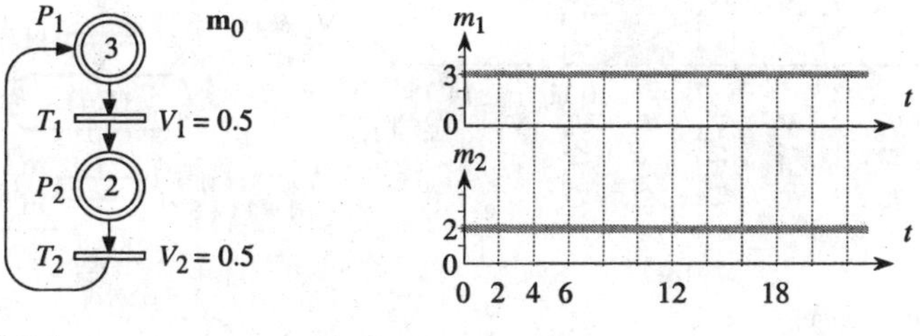

Figure S 5.1

5.2 a) $V_1(t) = (0,0)[9, (1,5)(5,0)]$, $V_2(t) = (0,0)[9, (3,4)(8,0)]$, and $\mathbf{V}(t) = (0,(0,0))[9, (1,(5,0))(3,(5,4))(5,(0,4))(8,(0,0))]$,

b) The graphs of $m_1(t)$, $m_2(t)$, and $v_3(t)$ are given in Fig. S 5.2. Since the firing quantity is the same for both T_1 and T_2 (i.e. 20 units per period) neither $m_1(t)$ nor $m_2(t)$ increase with time.

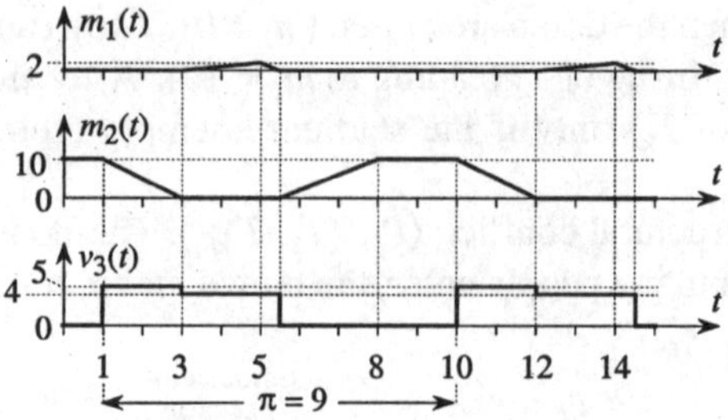

Figure S 5.2

5.3 a) $\mathbf{V}(t) = (0, \mathbf{V}_0)(3, \mathbf{V}_1)$, where $\mathbf{V}_0 = (2, 1)$ and $\mathbf{V}_1 = (2, \infty)$. The evolution graph is given in Fig. S 5.3a (two IB-states with an I-phase between them).

b) $\mathbf{V}(t) = (0, \mathbf{V}_0)[8, (3, \mathbf{V}_1)(6, \mathbf{V}_2)(7, \mathbf{V}_3)(10, \mathbf{V}_4)]$, where $\mathbf{V}_0 = (2, 1)$, $\mathbf{V}_1 = (2, \infty)$, $\mathbf{V}_2 = (3, 1)$, $\mathbf{V}_3 = (2, 1)$, and $\mathbf{V}_4 = (3, 1)$. The evolution graph is given in Fig. S 5.3c (an I-phase is performed at the beginning of each $\mathbf{V}_1 = (2, \infty)$) and $m_1(t)$ is illustrated in Fig. S 5.3b.

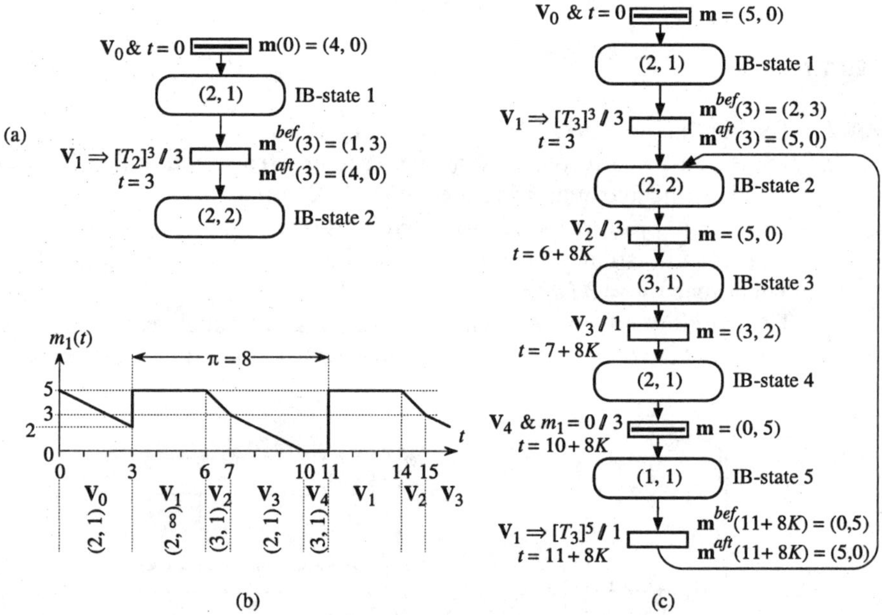

Figure S 5.3

5.4 The behaviors are illustrated in Fig. S 5.4. Because of the marking invariants, $m_2(t) = 10 - m_1(t)$, $m_4(t) = 10 - m_3(t)$, and $m_6(t) = 10 - m_5(t)$. Hence, only the markings $m_1(t)$, $m_3(t)$, and $m_5(t)$, are represented.

Note that the variable y is significant only for the PN in Fig. E 5.4a.

a) *Synchonized continuous PN in Fig. E 5.4a.* Event $\uparrow x$ occurs at $t = 2$. At this instant T_1 is an immediate transition and the marking changes instantaneously from $\mathbf{m}^{bef}(2) = (10, 0)$ to $\mathbf{m}^{aft}(2) = (0, 10)$. From this time, only T_2 is enabled and is receptive to y. When y becomes 1, at $t=7$, T_2 is fired (according to its enabling degree: 10) and $\mathbf{m}_0 = (10, 0)$ is reached again. From this time, only T_1 is enabled and is receptive to $\uparrow x$. When this event occurs, at $t=16$, T_1 is fired to reach $\mathbf{m}_1 = (0, 10)$; but this marking is unstable because $y = 1$ at this time, and T_2 is fired immediately. Hence, at $t=16$, we have $(10, 0) \xrightarrow{T_1T_2} (10, 0)$; in other words, $\mathbf{m}^{bef}(16) = \mathbf{m}^{aft}(16) = (10, 0)$. The next firing of T_1 will occur at $t=24$. Note that, since all the transitions are synchronized, there is no continuous firing.

b) *PN in Fig. E 5.4b*. When $\uparrow x$ occurs, at $t=2$, we have a change from $\mathbf{m}^{bef}(2) = (10, 0)$ to $\mathbf{m}^{aft}(2) = (0, 10)$. From this time, T_4 is strongly enabled, hence fired at speed $v_4(t) = V_4 = 2$ up to $t=5$. At this instant, $\uparrow x$ occurs again, resulting in $\mathbf{m}^{bef}(5) = (6, 4) \rightarrow \mathbf{m}^{aft}(5) = (0, 10)$. Then $v_4(t) = V_4 = 2$ for $t \in \left| 5, 10 \right|$. At $t = 10$, the marking is $\mathbf{m} = (10, 0)$ and T_4 is no longer enabled. There is no change up to the next event $\uparrow x$. And so on.

c) *PN in Fig. E 5.4c*. When x becomes 1, at $t=2$, we have a change from $\mathbf{m}^{bef}(2) = (10, 0)$ to $\mathbf{m}^{aft}(2) = (0, 10)$. From this time, T_6 is strongly enabled, hence fired at speed $v_6(t) = V_6 = 2$. However, T_5 remains an immediate transition as long as $x = 1$. Hence, $v_5(t) = v_6(t) = 2$ as long as $x = 1$, i.e. up to $t=3$. From this time, $v_5(t) = 0$ since $x = 0$, whereas $v_6(t) = 2$. This lasts up to $t=5$, when x becomes again 1. And so on. See part c of Fig. S 5.4.

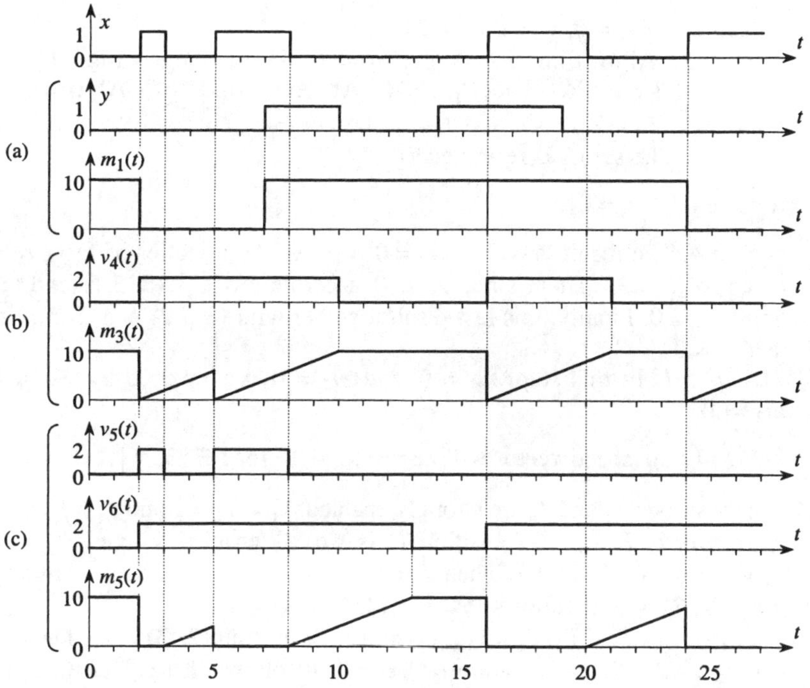

Figure S 5.4

5.5 a) At the end of *Step 5* of Algorithm 5.3, $T_{SF} = T$, $T_{pf} = \varnothing$, and $T_{ndc} = \{T_2, T_3, T_4\}$ (in addition $T_{wait} = \varnothing$, from the beginning of the calculation process).

Step 6 of Algorithm 5.3. Initially, $T(J_2) = T_{SF} \cap T_{ndc} = \{T_2, T_3, T_4\}$. Since T_3 and T_4 are involved in the same structural conflict, T_4 is deleted because $T_3 < T_4$. Hence, $T(J_2) = \{T_2, T_3\}$ and $J_2 = v_2 + v_3$.

We then return to Algorithm 5.4.

Steps 7.2, 7.3, 3, then *4*, are performed.

Step 4 of Algorithm 5.4. Maximization of $J_2 = v_2 + v_3$ leads to $v_2 = 1$ and $v_3 = 2$.

Algorithm 5.4 will end without any other change of firing speed.

b) If T_4 were not deleted from $T(J_2)$, criterion J_2 would be $J_2 = v_2 + v_3 + v_4$. The result of *Step 4* of Algorithm 5.4 would give values of v_3 and v_4 satisfying: $v_3 \geq 1$ (first passage), and $v_3 + v_4 = 2$.

There is an infinity of solutions. Among them: $v_3 = 2$ and $v_4 = 0$, which is the right one, or $v_3 = 1$ and $v_4 = 1$, which is not consistent with $T_3 < T_4$.

5.6 Here are some steps of Algorithm 5.4, performed from $\tilde{\mathbf{m}}(3) = (0^+, 0^+)$.

Step 1.1. The set C_2 of constraints related to the balances includes

$$v_3 - v_1 \geq 0, \text{ and} \quad \text{(S 5.1)}$$

$$v_1 - v_2 - v_3 \geq 0. \quad \text{(S 5.2)}$$

And $C_3 = \{v_1 = 0, v_2 = 0, v_3 = 0\}$.

Step 2. Algorithm 5.2 is performed. At the beginning, $T_{SF} = \varnothing$, $T_{pf} = \{T_1, T_2\}$, $P_{ne}^{(1)} = \{P_1, P_2\}$. At the end of Algorithm 5.2, $T_{SF} = \{T_1, T_2\}$, $T_{pf} = \varnothing$, and $J_1 = v_1 + v_2$.

Step 3. The set C_3 is restricted to

$$C_3 = \{ v_3 = 0\}. \quad \text{(S 5.3)}$$

Step 4. The result is: $v_1 = 0$, $v_2 = 0$, $v_3 = 0$. As a matter of fact, $v_3 = 0$ is given in (S 5.3). Then, since $v_1 \geq 0$ according to C_1, (S 5.1) and (S 5.3) imply $v_1 = 0$. Finally, this last result together with (S 5.2) and (S 5.3) imply that $v_2 = 0$.

Step 5.1. In this step, $\tilde{m}_1 = 0^+$ and $\tilde{m}_2 = 0^+$ are replaced by $\tilde{m}_1 = 0$ and $\tilde{m}_2 = 0$.

Finally the speed vector will be $\mathbf{v} = (0, 0, 0)$ for $t \in |3, \infty|$.

5.7 Transitions T_1 and T_4 are strongly enabled: $v_1 = V_1 = 2$ and $v_4 = V_4 = 1$.

a) *Priority* $T_2 < T_3$. Transition T_2 is weakly enabled: $v_2 = v_{2,\max} = \min(I_1, I_2, V_2) = \min(1, 2, 2) = 1$. Then $v_3 = I_1 - v_2 = 2 - 1 = 1$, and $v_5 = \min(I_5, V_5) = \min(1, 2) = 1$. It follows that $\mathbf{v} = (2, 1, 1, 1, 1)$.

b) *Priority* $T_3 < T_2$: $v_3 = v_{3,\max} = \min(I_1, V_3) = \min(2, 3) = 2$. Then $v_2 = I_1 - v_3 = 2 - 2 = 0$, and $v_5 = \min(I_5, V_5) = 2$. It follows that $\mathbf{v} = (2, 0, 2, 1, 2)$.

c) *Sharing* $[T_2, T_3]$. The aim $v_2 = v_3$ is possible: $\mathbf{v} = (2, 1, 1, 1, 1)$.

d) *Sharing* $[T_2, 3T_3]$. The aim $v_2 = v_3 / 3$ is possible: $\mathbf{v} = (2, 0.5, 1.5, 1, 1.5)$.

5.8 Transition T_4 is strongly enabled: $v_4 = V_4 = 2$. Then, $I_1 \geq v_4 = 2$. Since $I_1 > V_1$, and T_1 takes priority over the other transitions in structural conflict, $v_1 = V_1 = 1$.

Transition T_3 is not yet enabled. However, since $I_1 > v_1$, it is certain that $v_2 = I_4 = v_5 = I_2 > 0$. Then, T_3 will be enabled. Equation to be satisfied:

$$I_1 = 2 + v_3 \leq 3; \tag{S 5.4}$$

$$v_2 + v_3 = I_1 - v_1 = I_1 - 1. \tag{S 5.5}$$

From (S 5.4) and (S 5.5), $v_2 = 1$ is obtained. Then, the maximum value $v_3 = V_3 = 1$ can be obtained. The speed vector is $\mathbf{v} = (1, 1, 1, 2, 1)$.

5.9 All the maximal firing speeds have the same value 2. The feeding speed of P_5, $I_5 = v_5 = 2$, is to be shared between T_1 and T_2, and the feeding speed of P_6, $I_6 = v_6 = 2$, is to be shared between T_3 and T_4. Because of the structure of the PN, if priority is given to T_1 over T_2, this is favorable to transition T_4 (over T_3), whose firing is favorable to T_2 (over T_1), itself favorable to T_3 (over T_4). Hence, the expected result is not really obvious! All the results intuitively presented in the sequel can be obtained by Algorithm 5.7.

a) Right at the beginning, T_1 is enabled (P_1 marked and P_5 fed by T_5) whereas T_2 is not enabled (P_2 empty and not yet fed). Since $T_1 < T_2$, it is clear that $v_1 = 2$ and $v_2 = 0$. It follows that $I_3 = 0$ and $I_4 = 2$, thus $v_3 = 0$ and $v_4 = 2$ (even if $T_3 < T_4$). Then, $I_2 = 2$, but this does not modify $v_2 = 0$ because $T_1 < T_2$. The result is $\mathbf{v} = (2, 0, 0, 2, 2, 2)$.

b) Same result: $\mathbf{v} = (2, 0, 0, 2, 2, 2)$.

c) As in case a, T_1 is fired, then T_3 is fired, then P_2 is fed and T_2 is enabled. But in the present case, $T_2 < T_1$. Should we have $v_2 = 2$? No because $v_2 = 2$ implies $v_1 = 0$, and $v_2 \leq v_1$ because of the path $T_1P_4T_4P_2T_2$. It follows that the equibrium is obtained for $v_1 = v_2 = 1$ (greatest possible value for v_2, given it depends on v_1). The result is $\mathbf{v} = (1, 1, 1, 1, 2, 2)$.

d) Same result as in c: $\mathbf{v} = (1, 1, 1, 1, 2, 2)$. Note that in all cases a to d, and *for this initial marking*, the priority between T_3 and T_4 has no influence.

e) The speed vector $\mathbf{v} = (1, 1, 1, 1, 2, 2)$ is also obtained in this case since it satisfies both $v_1 = v_2$ and $v_3 = v_4$.

The speed vector $\mathbf{v} = (1, 1, 1, 1, 2, 2)$ lasts up to infinity (i.e. single IB-state for cases c, d, and e). The speed vector $\mathbf{v} = (2, 0, 0, 2, 2, 2)$ lasts 2 time units; at t = 2, $m_1 = 0$ and $m_2 = 4$. From this time, $\mathbf{v} = (1, 1, 1, 1, 2, 2)$ is obtained in cases a and b. Hence, for all cases the *final* speed vector is such that $v_1 = v_2 = v_3 = v_4$.

5.10 a) Since T_4 has the highest priority (among $\{T_4, T_5, T_6\}$), $v_4 = \min(I_4, I_5)$. As $I_4 = v_1 + v_8 = 2 + v_8 \leq 2.5$ and $I_5 = v_2 = 4$, we can conclude that $v_4 = 2 + v_8$. However, the vector $\mathbf{v} = (2, 4, 1, 2, 0.75, 1.25, 0.5, 0.25)$ does not satisfy this equation since $v_4 = 2$ and $v_8 = 0.25$.

Let us note that, in this example, the speed of *high priority* transition T_4 increases if the speeds of *low priority* transitions T_5 then T_8 increase. The inaccuracy of the result is due to this kind of model "inconsistency".

b) The following set of equations can be established.

$$v_1 = 2;\ v_2 = 4;\ v_3 = 1; \tag{S 5.6}$$

$$v_4 = 2 + v_8; \tag{S 5.7}$$

$$v_5 = 4 - v_4 - v_6; \tag{S 5.8}$$

$$v_6 = \min((4 - 2 - v_8), (1 + v_8)) = 1 + v_8 \text{ (since } v_8 \le 0.5); \tag{S 5.9}$$

$$v_7 = \min(v_5, 0.5); \tag{S 5.10}$$

$$v_8 = \max((v_5 - 0.5), 0). \tag{S 5.11}$$

The solution $\mathbf{v}$ = (2, 4, 1, 2.167, 0.667, 1.167, 0.5, 0.167) is obtained by resolution of the following LPP: maximize $J = v_4 + v_6$ given the constraints (S 5.6) to (S 5.11).

Let us observe that the result given by Algorithm 5.4, even if it does not satisfy the conflict resolution rules exactly (small differences on v_8, hence on v_4 to v_6), is quite consistent with the basic rules of the timed continuous PN model, then "acceptable" from this point of view.

Chapter 6

6.1 a) D-enablings: $D(T_1, \mathbf{m}) = 3$, $D(T_2, \mathbf{m}) = 2$, and $D(T_3, \mathbf{m}) = 0$.

b) Maximal speeds: $V_1 = 12$, $V_2 = 10$, and $V_3 = 0$.

c) Since T_3 is not enabled, there is no actual conflict related to P_3.

There is no actual conflict related to P_2 if $v_1 = V_1 = 12$ and $v_2 = \min(v_4, V_2)$. For $v_1 = 12$, three tokens in P_2 are required. The fourth token in P_2 allows $v_2 \le U_2 \times 1 = 5$. Hence, there is an actual conflict only if $v_4 > 5$.

6.2 Between $t = 0$ and $t = 2.5$, T_1 is fired once at $t = 2$ and T_2 is fired continuously at speed $v_2 = 1$. Hence, the characteristic vector of the firing sequence is (1, 2.5). It follows that:

$$\mathbf{m}(2.5) = \begin{bmatrix} 1 \\ 3.2 \\ 0 \end{bmatrix} + \begin{bmatrix} 0 & 0 \\ 0 & -1 \\ -2 & 1 \end{bmatrix} \cdot \begin{bmatrix} 1 \\ 2.5 \end{bmatrix} = \begin{bmatrix} 1 \\ 0.7 \\ 0.5 \end{bmatrix}. \tag{S 5.12}$$

6.3 a) A solution is given in Fig. S 6.3a. As soon as T_1 is fired (event $\uparrow Op$ or condition Op), 12 tokens are in T_1 (12 doors) and T_3 is fired at speed $v_3 = V_3 = 12\, U_3 = 180$ spectators per minute. When $\downarrow Sh$ occurs (end of show), all the spectators become ready to leave (T_4' is a synchronized C-transition: all the marks in P_3 move immediately to P_3'), then $v_4 = V_4 = 12\, U_4 = 300$. One hour after the show, Op' becomes 1 and T_2 is fired. Given the hypothesis concerning the departure of the spectators, P_3' is empty at this time (i.e. the inhibitor arc is redundant).

b) See Fig. S 6.3b. The marking of place P_5 represents the queue of spectators waiting to enter the stadium.

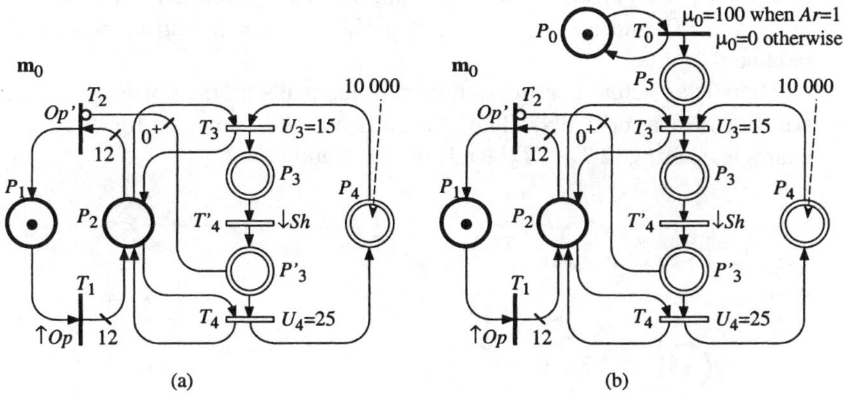

Figure S 6.3

6.4 The evolution graph is presented in Fig. S 6.4.

During the first IB-state, both T_3 and T_4 are D-enabled. Since $V_4 = \infty > V_3 = 2$, both are fired at the same speed $v_3 = v_4 = 2$, i.e. $\mathbf{v} = (2, 2)$.

During the second IB-state (i.e. after firing of T_1), the D-enabling of T_4 is 0, hence $V_4 = 0$ (even if $U_4 = \infty$ because T_4 is *not enabled*) while $v_3 = 2$, i.e. $\mathbf{v} = (2, 0)$.

When T_2 is fired, 3 time units later, the C-marking is $\mathbf{m}^{C,bef} = (4, 6)$ which is unstable because immediate transition T_4 becomes enabled again. Hence, an I-phase is performed: instantaneous firing (quantity of firing: 6) of T_4 to reach the stable marking $\mathbf{m}^{C,aft} = (10, 0)$.

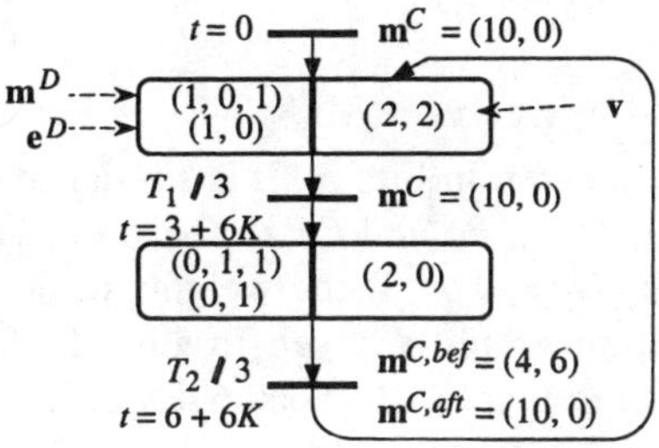

Figure S 6.4

6.5 Models are given in Fig. S 6.5.

a) See Fig. S 6.5a. Since $U_1 = U_2 = 5$, the token in P_1 models that $v_1 + v_2 = 5$. Since $v_3 = v_1 + v_2 = 5$, any value $U_3 \geq 5$ would give the same result. Hence, there is some redundancy in this model: $U_3 = \infty$ may be used and P_2 may be deleted.

There is a case 4 actual conflict. If there is a priority $T_1 < T_2$, then $v_1 = 5$ and $v_2 = 0$. If $T_2 < T_1$, then $v_1 = 0$ and $v_2 = 5$. For a sharing [2 T_1, 3 T_2], for example, $v_1 = 2$ and $v_2 = 3$ are obtained.

b) See Fig. S 6.5b. This is an *extended continuous PN* and $V_1 = V_2 = \infty$ since they are not specified (Remark 6.5b, Section 6.1.5.2). Initial marking 0^+ in either P_1 or P_4 ensures that T_1, T_2, and T_3, can be fired. As a matter of fact, $m_1 + m_4 = 0^+$ and $v_1 + v_2 = v_3$. Since $V_3 = 5$, $v_3 = 5$ and $v_1 + v_2 = 5$ are obtained.

There is an actual conflict in this continuous PN (corresponding to a case 2 conflict in a hybrid PN). The solutions are the same as for Fig. a. For example, sharing [2 T_1, 3 T_2] leads to $v_1 = 2$ and $v_2 = 3$.

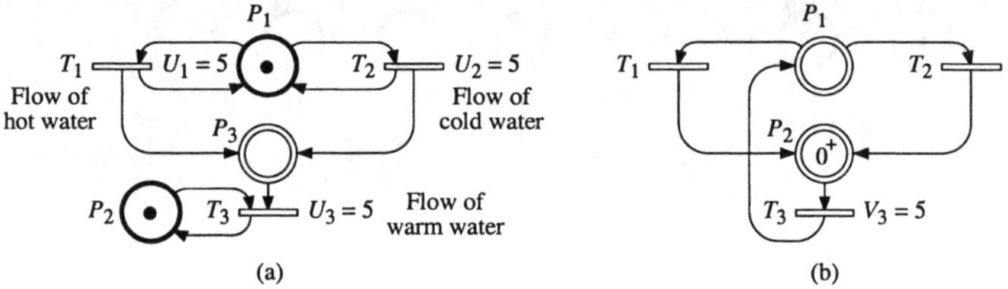

Figure S 6.5

6.6 The model is given in Fig. S 6.6. C-transition T_2 models the flow: $v_2 = V_2 = U_2 \cdot m_1 = 1$ liter/s as long as P_2 is not empty. D-transition T_1 models the delay in the pipe. Given the section is 1 cm^2, one liter corresponds to a length of 1000 cm = 10 m. Hence, the delay is equal to 50 m / (10 m / s) = 5 s.

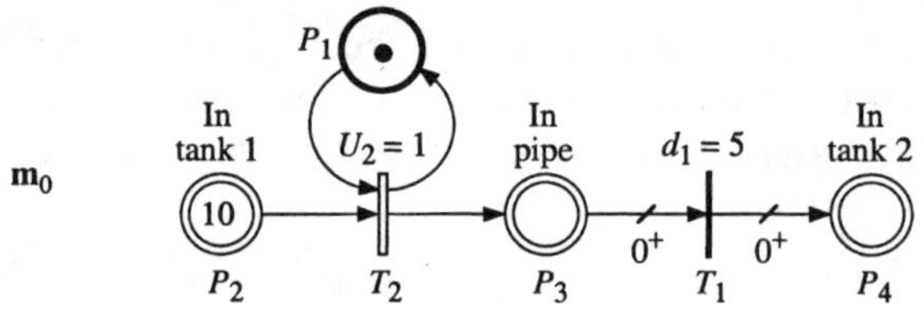

Figure S 6.6

6.7 The hybrid PN is given in Fig. **A**. It is similar to the discrete PN in Fig. S 3.12**A**, except that: 1) P_3, P_4, and P_5 are C-places; 2) T_3 is a C-transition and $d_3 = 2$ is replaced by $U_3 = 0.5$; 3) the weights of arcs $P_4 \to T_2$ and $T_2 \to P_5$ are 0^+. The evolution graph is given in Fig. **B**. The notation of enabling density presented in Remark 6.19 (Sect. 6.4.3.1) is used.

Let us compare this figure with Fig. S 3.12**B** (discrete modeling). For marking $\mathbf{m}_7$, there are 6 parts on the conveyor, corresponding to two bundles: according to $\mathbf{e}_7$, three parts will leave the conveyor at 26, 28, and 30 seconds from now, then three other parts will leave at 76, 78, and 80 seconds from now. When $\mathbf{m}_7$ is reached, at $t = 56$, IB-state 4 is also reached in the hybrid model. At this time, there are two batches of three parts each on the conveyor:

the enabling density is 0.5 between τ = 24 to 30 s from now (then, corresponding to 0.5 × 6 = 3 parts) and is also 0.5 for τ = 74 to 80 s from now. The main difference is that a bundle leaves the conveyor by three firings of T_2 (three successive markings) in the discrete model whereas it corresponds to a single IB-state in the hybrid model. The reachability marking in Fig. S 3.12.**B** contains 14 markings and this number would be 202 if each batch contains 50 parts. The evolution graph in Fig. **B** contains 8 IB-states and this number would be the same if each batch contains 50 parts.

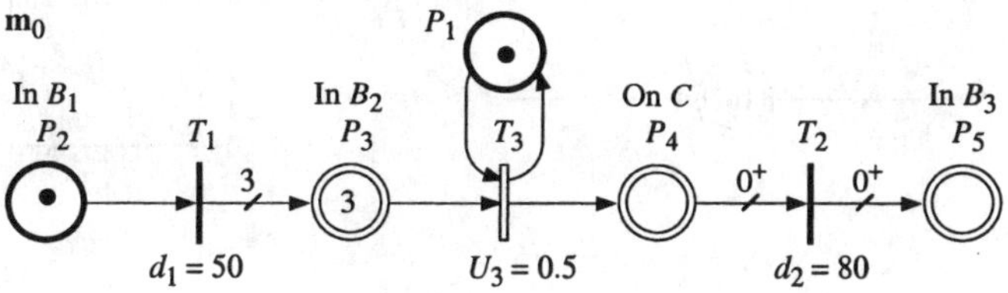

Figure S 6.7.**A**

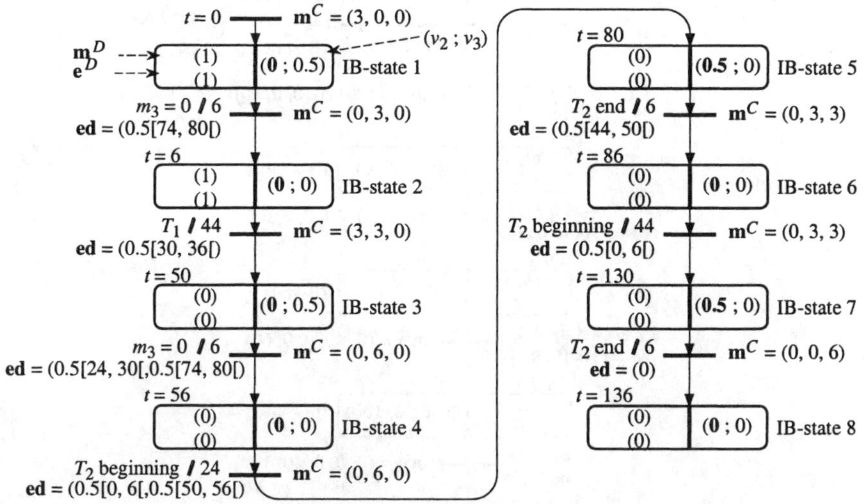

Figure S 6.7.**B**

6.8 As illustrated in Fig. **A**a, U_2, U_3, and d_1, are functions of Q_C. Since Q_C is a function of time, these variables are also functions of time as shown in Fig. **A**b (since $m_1 = 1$ and $m_2 = 1$, always, $V_2(t) = U_2(t)$ and $V_3(t) = U_3(t)$).

The evolution graph is in Fig. **B**. Just before Q_C changes of value, the enabling density of T_1 is 50 from τ = 1.6 to τ = 2 (input flow v_2 = 50 from t = 0 to t = 0.4). When Q_C changes from 5 to 10, the enabling density is multiplied by two whereas the residual times to firing are divided by two. Hence the enabling density is denoted by 100[0.8, 1[when IB-state 2 is reached. The other transitions between IB-states are similar to Fig. 6.33 in Sect. 6.4.3.1 (with different numerical values).

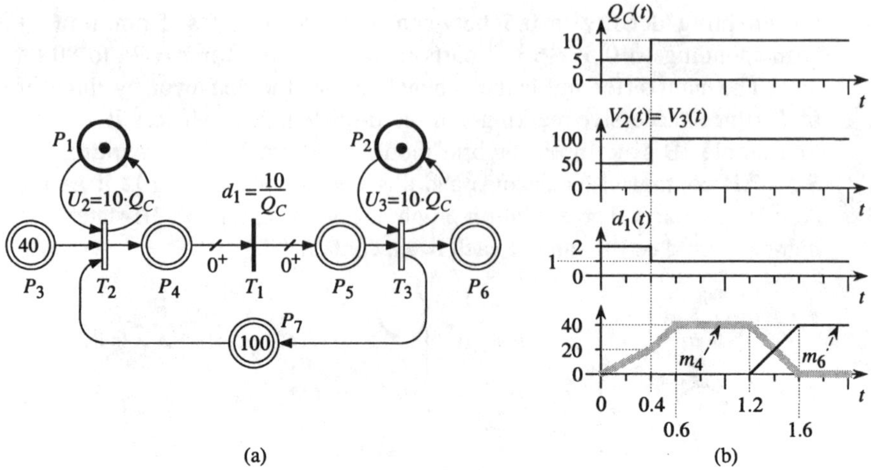

Figure S 6.8.**A**

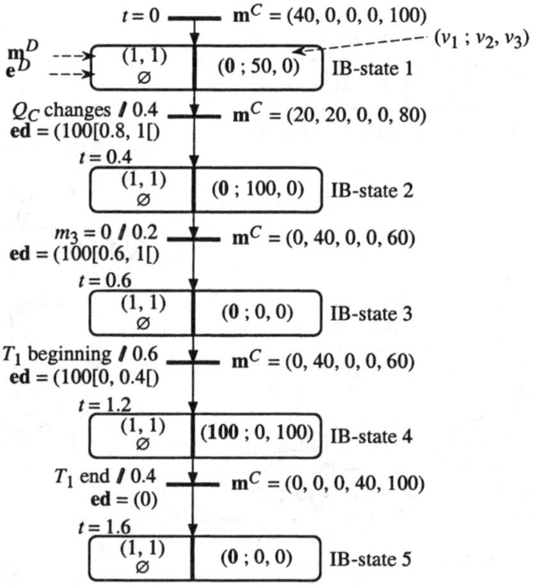

Figure S 6.8.**B**

6.9 The hybrid PN is the same as in Fig. S 6.8.**A**a. But in this case, $Q_C = 0$ for $0.4 \leq t < 1.4$. Then, during this time interval, $V_2 = V_3 = 0$ and $d_1 = \infty$. The behavior is illustrated in Fig. S 6.9a and the evolution graph is in Fig.b.

Just before the first change of Q_C, at the end of IB-state 1, the enabling density of T_1 is 50 from $\tau = 1.6$ to $\tau = 2$. Just after the change, the value 50 must be multiplied by the ratio $\frac{Q_C \text{ after}}{Q_C \text{ before}}$ whereas the values 1.6 and 2 are divided by this ratio. It follows that **ed** = (0[∞, ∞[). In order to keep some information on the values before the change of Q_C, we write **ed** = (50/∞[1.6∞, 2∞[) (see Fig.b). In that way, when Q_C changes again and becomes positive, we have kept the information to calculate the new enabling density. For our example, when Q_C re-assumes the value 5, **ed** = (50[1.6, 2[) at the beginning of IB-state 3. After this time, the behavior is the same as in Fig. 6.33 in Sect. 6.4.3.1 (with a shift of one time unit).

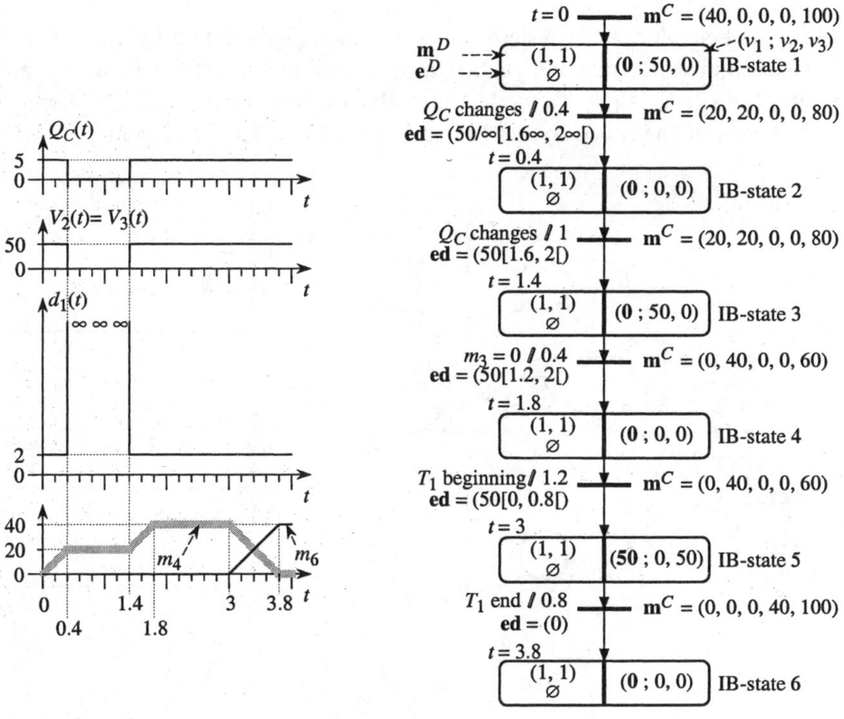

(a) (b)

Figure S 6.9

6.10 a) Assume that $m_3 = 0$ when the token is deposited in P_1, at time t_1.

In Fig. E 6.10a, only T_1 is enabled: it will be fired at $t = t_1 + d_1 = t_1 + 2$.

In Fig. b, only T_2 is enabled: it will be fired at $t = t_1 + d_2 = t_1 + 2$. Hence, immediate transition T_1 becomes enabled and is also fired at $t = t_1 + 2$.

b) Assume that $m_3 = 12$ when the token is deposited in P_1.

In Fig. a, both T_1 and T_3 are enabled. D-transition T_1 takes priority over T_3 (Rule 6.1, Section 6.1.3). Hence, it will be fired at $t = t_1 + 2$. C-transition T_3 is continuously fired at $v_3 = 4$ from t_1 to $t_1 + 2$, and is no longer enabled from $t_1 + 2$. Then $m_3 = 4$ from this time.

In Fig. b, both T_2 and T_3 are enabled. D-transition T_2 takes priority over T_3. Hence, it will be fired at $t = t_1 + 2$. At this time there is a token in P_1 and in P_2, but T_1 is not enabled because P_3 is not empty ($m_3 = 4$). Hence,

C-transition T_3 is continuously fired at $v_3 = 4$ from t_1 to $t_1 + 3$ (m_3 becomes 0 at $t_1 + 3$). At this time $t_1 + 3$, T_1 can be fired.

From the analysis above, it appears that T_1 takes priority over T_3 in Fig. a (basic rule) whereas T_3 takes priority over T_1 in Fig. b. In other words, although Rule 6.1 is respected, the priority of a C-transition over a D-transition can be modeled thanks to an extended hybrid PN.

6.11 The behavior of the system can be modeled by the timed hybrid PN in Fig. S 6.11a in which places P_8 and P_9 model the liquid quantities in tanks 1 and 2, and transitions T_5 to T_9 model the valves (same index) and the pump. The behavior of the timed hybrid PN up to $t = 20$ is illustrated in Fig. b.

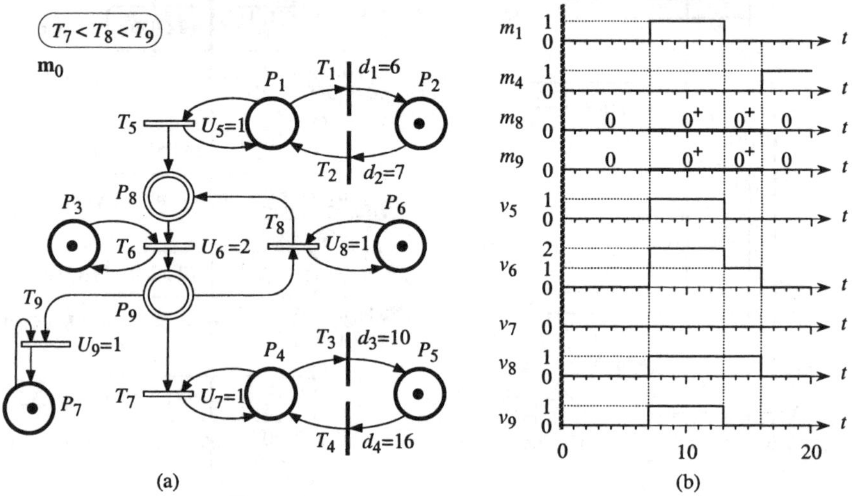

Figure S 6.11

During the first IB-state, $v_5 = v_7 = 0$ since $m_1 = 0$ and $m_4 = 0$. Hence, since $v_5 = 0$, places P_8 and P_9 remain empty, i.e. $\tilde{m}_8 = 0$ and $\tilde{m}_9 = 0$. In Fig. E 6.11, the tanks and pipes under valve 5 are empty.

At $t = 7$, P_1 becomes marked (valve 5 is open). From this time, $v_5 = 1$. Because of the priority $T_8 < T_9$ (T_7 is not enabled), $v_8 = 1$ and the speed vector is $\mathbf{v} = (v_5, v_6, v_7, v_8, v_9) = (1, 2, 0, 1, 1)$. This behavior is the sum of two flows: a flow whose speed is $v = 1$ from T_5 to T_9 and another flow whose speed is also $v = 1$ in the loop $P_8T_6P_9T_8P_8$, as illustrated in Fig. E 6.11.

When valve 5 is closed, at $t = 13$, the flow from T_5 to T_9 disappears whereas the flow in the loop continues (due to priority $T_8 < T_9$). In this case, the speed vector $\mathbf{v} = (0, 1, 0, 1, 0)$ is expected for the third IB-state. During the second IB-state, $\tilde{m}_8 = \tilde{m}_9 = 0^+$ because P_8 and P_9 are simultaneously fed and drained. It follows from Property 5.3 in Sect. 5.1.3.3, that $\tilde{m}_8(13) = \tilde{m}_9(13) = 0^+$; from these values, Algorithm 5.4 (used in *Step 3* of Algorithm 5.7, inside Algorithm 5.8), gives the vector $\mathbf{v} = (0, 1, 0, 1, 0)$.

(The marking 0^+ in P_8 and P_9 corresponds to infinitely small values remaining in the places, illustrated by the liquid in the pipes in the loop in Fig. E 6.11.)

At $t = 16$, T_4 is fired and T_7 becomes enabled. This high priority transition empties the 0^+ marking in the loop (*Step 5* in Algorithm 5.4: valve 7 open, the loop is emptied). The speed vector is then $\mathbf{v} = (0, 0, 0, 0, 0)$.

Let us observe that the algorithm in [AlDa 98a] finds $v_6 = v_8 = 0$ instead of $v_6 = v_8 = 1$ for $t \in |13, 16|$, because, in this algorithm, the speeds were calculated from $(m_8(13), m_9(13)) = (0, 0)$. On the other hand, the algorithm in [BaGiMe 00] gives a correct value for the second and third IB-states, but it gives $v_6 = v_8 = 1$, instead of $v_6 = v_8 = 0$, for $t \in |0, 7|$ and $t \in |16, 20|$, because the model used is equivalent to assuming 0^+ in all the empty places when a speed calculation is performed.

Chapter 7

7.1 a) The hybrid PN is shown in Fig. S 7.1. The structure is similar to the structure of the discrete PN in Fig. S 3.9.**A**a, except that the place corresponding to a part on a machine and its input and output transitions are replaced by a single C-transition (transition T_1 in Fig. S 7.1 take the place of the subnet made up of T_1, P_1, T_2, and the arcs between them, in Fig. S 3.9.**A**a). Flow rates: $U_1 = 1 / S_1 = 4$, $U_2 = 1 / S_2 = 2$, $U_3 = 1 / S_3 = 5$. Buffer capacities: $m_4 + m_5 = C_1 = 10$ and $m_6 + m_7 = C_2 = 10$.

b) The behavior could be analyzed as a CHPN, a VHPN, or an AHPN. In all cases, it is clear that the bottleneck corresponding to T_2 imposes its lower speed on the other transitions. Hence, the stationary behavior is such that: $v_1 = v_2 = v_3 = V_2 = 2$ (as a matter of fact, $V_2 = U_2 \cdot \min(m_2, m_4, m_7) = 2$ because the buffer capacities are large enough to have $m_4, m_7 \geq m_2 = 1$).

Consider now the VHPN interpretation: $v_1 = U_1 \cdot \min(1, m_5) = 2$ is obtained for $m_5 = 0.5$, and $v_3 = U_3 \cdot \min(1, m_6) = 2$ is obtained for $m_6 = 0.4$. It follows that $m_4 = 10 - m_5 = 9.5$, and $m_7 = 10 - m_6 = 9.6$.

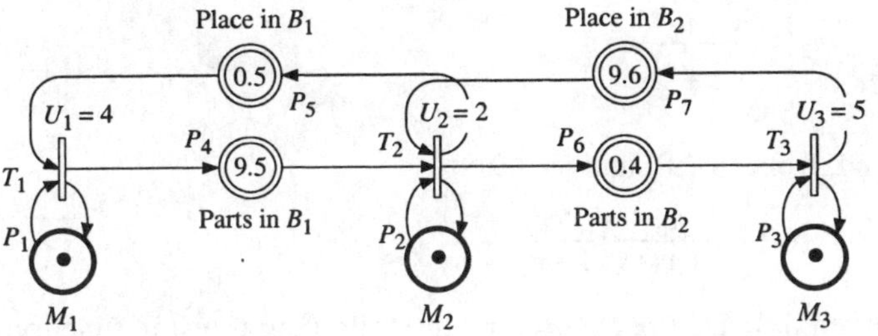

Figure S 7.1

7.2 The VHPN model is given in Fig. S 7.2. The markings of P_2 and P_3 correspond to the liquid volumes in the tanks: $m_2 = VO_1$ and $m_3 = VO_2$.

Transition T_1: $v_1(t) = u_0(t)$ is obtained by $U_1(t) = u_0(t)$ and $m_1 = 1$, since $v_1(t) = U_1(t) \cdot m_1$.

Transition T_2: $v_2(t) = u_1(t)$ is sought. From (E 7.1), $u_1(t) = H_1(t) / R_1$, and given $H_1(t) = VO_1(t) / S_1$,

$$u_1(t) = VO_1(t) / (S_1 \cdot R_1) \tag{S 7.1}$$

is obtained. From (S 7.1) and (E 7.2), and given $VO_1(t) = m_2(t)$,

$$u_1(t) = m_2(t) / \tau_1 \tag{S 7.2}$$

is obtained. Hence, since $v_2(t) = U_2 \cdot m_2(t)$ (according to the VHPN model), $v_2(t) = u_1(t)$ is obtained for (see Fig. S 7.2):

$$U_2 = 1 / \tau_1. \tag{S 7.3}$$

Transition T_3: similarly,

$$U_3 = 1 / \tau_2. \tag{S 7.4}$$

b) The balances of P_2 and P_3 correspond to:

$$\frac{dm_2}{dt} = v_1 - v_2 = u_0 - \frac{m_2}{\tau_1}; \tag{S 7.5}$$

$$\frac{dm_3}{dt} = v_2 - v_3 = \frac{m_2}{\tau_1} - \frac{m_3}{\tau_2}. \tag{S 7.6}$$

From (S 7.5) and (S 7.6), respectively, and given the nil initial conditions, the following Laplace transforms are obtained:

$$s \cdot m_2 = u_0 - \frac{m_2}{\tau_1}; \tag{S 7.7}$$

$$s \cdot m_3 = \frac{m_2}{\tau_1} - \frac{m_3}{\tau_2}. \tag{S 7.8}$$

Hence, from (S 7.7):

$$m_2(s) = \frac{u_0}{s + \dfrac{1}{\tau_1}}. \tag{S 7.9}$$

And, from (S 7.8) and (S 7.9):

$$m_3(s) = \frac{u_0}{1 + \tau_1 \cdot s} \cdot \frac{\tau_2}{1 + \tau_2 \cdot s}. \tag{S 7.10}$$

Given $u_2 = U_3 \cdot m_3 = m_3 / \tau_2$, the following transfer function is obtained from (S 7.9) and (S 7.10):

$$G(s) = \frac{u_2(s)}{u_0(s)} = \frac{1}{(1 + \tau_1 \cdot s)(1 + \tau_2 \cdot s)}. \tag{S 7.11}$$

This is a second order transfer function with two time constants and a static gain equal to 1.

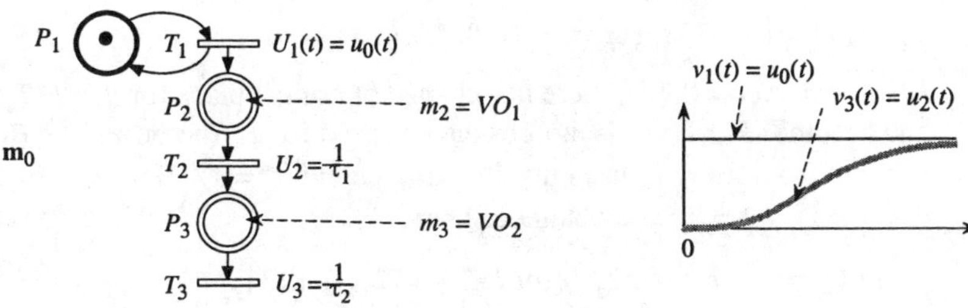

Figure S 7.2

7.3 The VHPN and AHPN behaviors are illustrated in Fig. S 7.3. In both cases, $v_1 \leq U_1 \cdot m_1 = 3$ and $v_2 \leq U_2 \cdot m_2 = 2$. The values $v_1 = 3$ and $v_2 = 2$ are obtained, respectively, for $m_3^{(T_1)} = 1$ and $m_3^{(T_2)} = 4$. Hence, there is an actual conflict if and only if $m_3 < 5$.

a) *VHPN.*

IB-phase 1. There is no actual conflict: the critical places (Definition 7.3, Sect. 7.1.3.2) are $P(T_1) = P_1$ and $P(T_2) = P_2$, then $v_1 = 3$ and $v_2 = 2$. Hence, $B_3 = \mathrm{d}m_3 / \mathrm{d}t = -v_1 - 4v_2 = -11$. It follows, as long as $m_3 \geq 5$,

$$m_3(t) = 10 - 11\,t, \text{ for } t \in |0, 0.455|. \tag{S 7.12}$$

IB-phase 2. For $t > 0.455$, $m_3 < 5$, then there is an actual conflict. The parts of m_3 assigned to the transitions could be $m_3 / 2$, but T_2 cannot use more than $m_3^{(T_1)} = 1$. Hence, $m_3^{(T_2)} = m_3 - 1$. The critical places are $P(T_1) = P_1$ and $P(T_2) = P_3$, It follows that $v_1 = 3$ and $v_2 = 2 \times (m_3 - 1) / 4$, then $B_3 = \mathrm{d}m_3 / \mathrm{d}t = -v_1 - 4v_2 = 1 - 2m_3$. Thus, as long as $m_3 \geq 2$,

$$m_3(t) = -\frac{1}{2} + \frac{11}{2} e^{-2(t-0.455)}, \text{ for } t \in |0.455, 0.849|. \tag{S 7.13}$$

IB-phase 3. For $t > 0.849$, $m_3 < 2$, then the speed v_1 is also dependent on the actual conflict because $P(T_1) = P(T_2) = P_3$. The parts of m_3 assigned to the transitions are $m_3^{(T_1)} = m_3^{(T_2)} = m_3 / 2$. It follows that $v_1 = 3 \times m_3 / 2 = 1.5\,m_3$ and $v_2 = 2 \times (m_3 / 2) / 4 = 0.25\,m_3$, then $B_3 = \mathrm{d}m_3 / \mathrm{d}t = -\frac{5}{2}\,m_3$. Hence,

$$m_3(t) = 2 \cdot e^{-\frac{5}{2}(t-0.849)}, \text{ for } t \in |0.849, \infty|. \tag{S 7.14}$$

For any t, m_3 and $\mathrm{d}m_3 / \mathrm{d}t$ are continuous functions of time.

b) *AHPN.*

IB-state 1. No actual conflict. The behavior is similar for both AHPN and VHPN models.

IB-state 2. At $t = 0.455$, there is a change of critical place for T_2: $P(T_2)$ was P_2 and becomes P_3. The speed vector, calculated with $m_3^{(T_1)} = 1$ (because the part $m_3^{(T_1)} = 2.5$ cannot be used by T_2) and $m_3^{(T_2)} = 4$, does not change. Marking m_3 is given by the same equation as long as $m_3 \geq 2$, i.e. up to $t = 0.727$:

$$m_3(t) = 10 - 11\,t, \text{ for } t \in |0, 0.727|. \qquad \text{(S 7.15)}$$

IB-state 3. At $t = 0.727$, there is a change of critical place for T_1: $P(T_1)$ was P_1 and becomes P_3. There is no change of speed for T_1 (because $m_3^{(T_1)}(0.727) = m_1$), but from now, T_1 has only its part, i.e. $m_3^{(T_2)} = 1$). Hence, $v_1 = 3$ and $v_2 = 2 \times m_3^{(T_2)} / 4 = 0.5$ are obtained. Then,

$$m_3(t) = 2 - 5\,(t - 0.727) \text{ for } t \in |0.727, 1.127|, \qquad \text{(S 7.16)}$$

since $m_3(t)$ reaches its asymtotic value $m_3^* = 0$ at $t = 1.127$.

IB-state 4. From $t = 1.127$, $v_1 = v_2 = 0$, and

$$m_3(t) = 0 \text{ for } t \in |1.127, \infty|, \qquad \text{(S 7.17)}$$

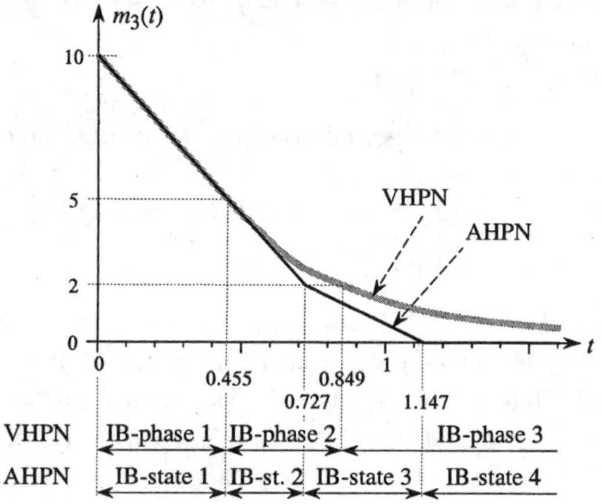

Figure S 7.3

7.4 a) *VHPN*. Since $U_4 = 1$, $v_4 = U_4 \cdot m_2 = m_2$.
For $t \in |0, 0.7|$, $m_2(t) = 0.9 + 0.3\,e^{-t}$. Then, $m_2(0.7) = 1.049$.
For $t \in |0.7, \infty|$, $m_2(t) = 0.1 + (1.049 - 0.1)\,e^{-(t-0.7)} = 0.1 + 0.949\,e^{-(t-0.7)}$.

b) *AHPN*.
For $t \in |0, 0.7|$, $v_3(t) = 0.9$, $v_4(t) = 1.2$. Then, $m_2(t) = 1.2 + (v_3(t) - v_4(t))\,t = 1.2 - 0.3\,t$. Then, $m_2(0.7) = 0.99$.
For $t \in |0.7, 1.7|$, $v_3(t) = 0.1$, $v_4(t) = 0.99$. Then, $m_2(t) = 0.99 + (v_3(t) - v_4(t))\,(t - 0.7) = 0.99 - 0.89\,(t - 0.7)$. Then, $m_2(1.7) = 0.1$.
For $t \in |1.7, \infty|$, $v_3(t) = 0.1$, $v_4(t) = 0.1$. Then, $m_2(t) = 0.1 + (v_3(t) - v_4(t))\,(t - 1.7) = 0.1$.

7.5 a) The evolution graph is given in Fig. S 7.5. Numerical values of IB-states 1 to 4 correspond to the graph in Fig. 7.19b (Sect. 7.2.2). The critical places of T_2 and T_3 are always P_1 and P_2 (single input place for each); for T_4, the critical place is P_3, then P_4 from IB-state 3, then P_3 again from IB-state 6.

b) If Algorithm 7.1 is used, all the instantaneous firing speeds are calculated for each IB-state (i.e. 21 calculations). If Algorithm 7.2 is used, the only instantaneous speeds calculated are those presented in bold fonts in the evolution graph, namely v_2, v_3, and v_4 at initial time, then v_4 at $t = 0.2$, v_4 at $t = 0.6$, v_2 at $t = 1$, and v_4 at $t = 1.8$. According to Property 7.9 in Sect. 7.2.3, the other instantaneous speeds do not change when a new IB-state is reached.

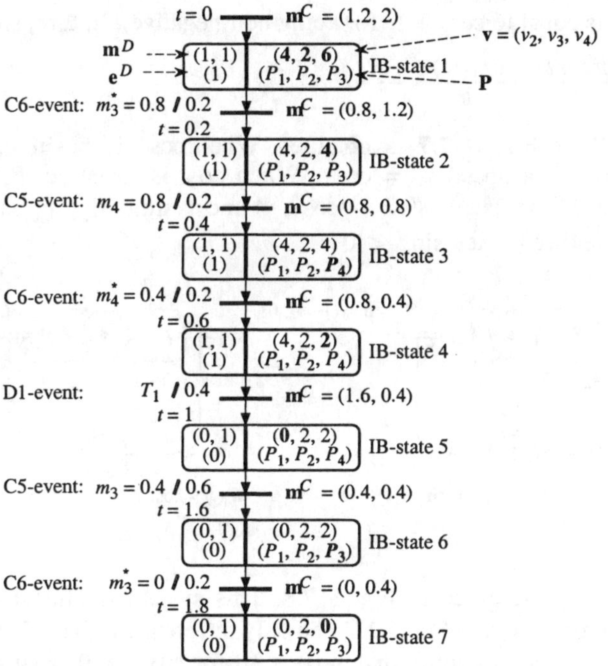

Figure S 7.5

7.6 The evolution graph is presented in Fig. S 7.6 (see also Fig. 7.22).

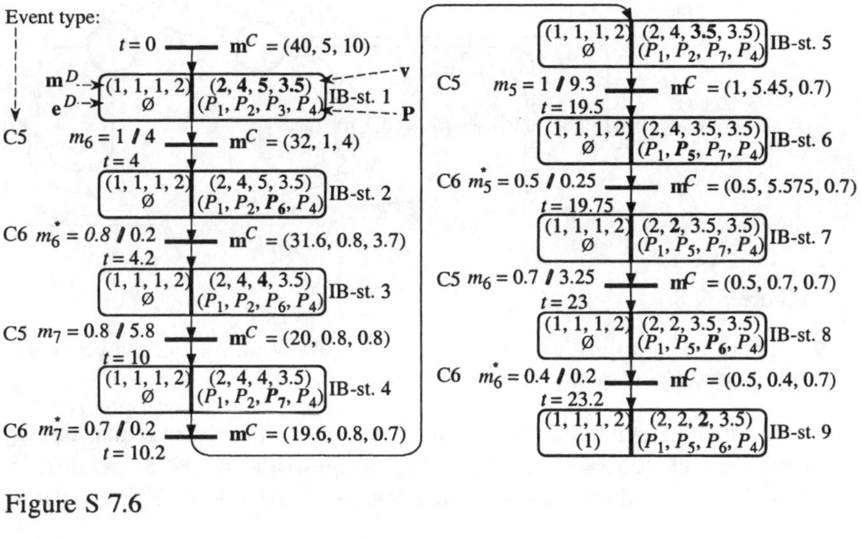

Figure S 7.6

7.7 A solution is given in Fig. S 7.7. Let $m_3 = \sin t + 1$. Since

$$\frac{d(\sin t)}{dt} = \frac{dm_3(t)}{dt} = \cos t, \tag{S 7.18}$$

P_3 must be fed by a speed equal to $\cos t$. This results in a place P_4 whose marking is equal to $\cos t + 1$ (for being non-negative). In turn, since

$$\frac{d(\cos t)}{dt} = \frac{dm_4(t)}{dt} = -\sin t, \tag{S 7.19}$$

the DHPN in Fig. S 7.7 is obtained. When $\cos t$ is positive, P_3 is fed by transition T_1' at speed $v_1' = \cos t$. When $\cos t$ is negative, P_3 is drained by transition T_1'' at speed $v_1'' = -\cos t$. Almost similarly, P_4 is drained when $\sin t > 0$ and fed when $\sin t < 0$.

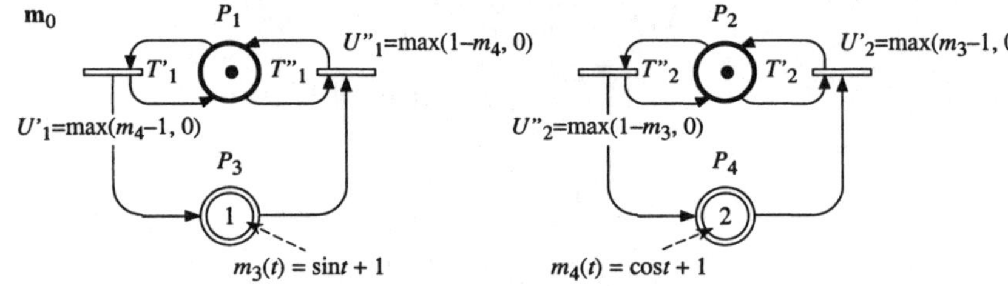

Figure S 7.7

7.8 A solution is given in Fig. S 7.8. This system is almost similar to the system represented in Fig. 7.27. The only difference is that (7.105) is replaced by (E 7.3). One can verify that $m_3(t) \le 10$ for any $t \ge 0$. Hence, the solution is similar to Fig. 7.27, except that $(10 - m_3)^3$ take the place of $(10 - m_3)$.

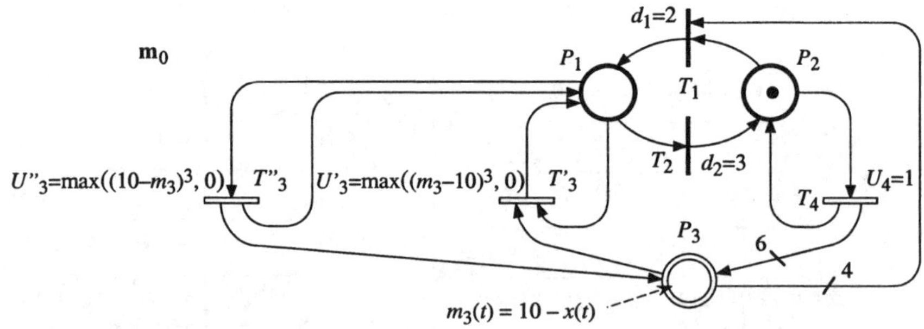

Figure S 7.8

7.9 There are similarities with the model for a simple conveyor given in Fig. 6.32b (Sect. 6.4.3.1).

a) The maximal entering flow rate is the product of the maximal density by the speed of subconveyor B_1: $V(B_1)$ = 30 parts / m × 3 m / min = 90 parts / min. Similarly, the maximal leaving flow rate is $V(B_2)$ = 10 × 10

= 100 parts / min. Let us observe that $V(B_1) < V(B_2)$. This means that the entering flow cannot be slowed down by the downstream part of the conveyor (but only by a component downstream from the conveyor under consideration, machine M in our example). It follows that both sub-conveyors in series can be modeled as a single conveyor as shown in the sequel. Two cases will successively be considered: the conveyor works or is stopped.

α) *Conveyor B works.* In Fig. S 7.9, $V_4 = U_4 \cdot m_2 = V(B_1) = 90$ and $V_5 = U_5 \cdot m_3 = V(B_2) = 100$. For some calculations, the conveyor may be divided into three subconveyors, namely B_1 (length 1.2 m), B_2 (length 1 m), and between them B_{12} (length 0.4 m). For B_{12}, speed and density are approximated by average speed 6.5 m / min, and average density 20 parts / m [De 94][Au 95]. From these values, the time spent by a part on the conveyor is $1.2/3 + 0.4/6.5 + 1/10 = 0.56$ min (if not slowed down by other parts), hence $d_1 = 0.56$ min, and the capacity of the conveyor is $1.2 \times 30 + 0.4 \times 20 + 1 \times 10 = 54$ parts. As the three parts close to M have spent more than 0.56 min on the conveyor, $m_6 = 3$; the other parts on the conveyor correspond to $m_5 = 26$; finally, $m_7 = 54 - m_5 - m_6 = 25$.

β) *Conveyor B is stopped.* If $ST_B = 1$, T_1 is fired and the D-enabling of both T_4 and T_5 is 0. Hence, $V_4 = V_5 = 0$. The delay d_3 becomes infinite. We have chosen to express this delay as $d_3 = 0.56 \times 90 / V_4$ (when the conveyor works, $d_3 = 0.56$ because $V_4 = 90$, and when it is stopped, $d_3 = \infty$ because $V_4 = 0$). The expression $d_3 = 0.56 \times 100 / V_5$ may also be chosen[3]. Note that a solution disabling T_3 (like T_4 is disabled when $ST_B = 1$) is not correct because the residual time to firing is d_3 when T_3 *becomes* enabled (Sect. 3.4.2.2) whereas a memory of the enabling density is required, as shown in Solution 6.9 (i.e., the parts on the conveyor do not come back at to the beginning of the the conveyor when it is stopped!).

Remark S 7.1 When the conveyor is working, C-transitions T_4 and T_5 are D-enabled by two different D-places, namely P_2 and P_3. Let us note that both transitions cannot be enabled by the same D-place because this structure would correspond to a case 4 conflict (Sect. 6.1.3).

❑

b) If $Q_{B2} = 5$ instead of 10, $U_5 = 50$ instead of 100 in Fig. S 7.9. It is possible that entry on sub-conveyor B_1 will be slowed down by B_2. Assume that conveyor B is empty (hence, $m_7 = 54$) and a flow of 90 parts / min arrives on conveyor A. From a model like Fig. S 7.9 (in which d_3 and U_5 would have different values), transition T_4 may be fired at speed $v_4 = 90$ during $m_7 / v_4 = 54 / 90 = 0.6$ min. However, this is not possible because it corresponds to a density of 30 parts / m for a length of 1.8 m (whereas the length of sub-conveyor B_1 is only 1.2 m). Thus, in this case, the complete conveyor would have to be modeled by a string of three conveyors (the maximal capacity of the first conveyor would be 36 parts for a length of 1.2 m).

---

[3] Another solution: $d_3 = (0.56$ if $m_1 = 0$ or ∞ otherwise). This solution is less interesting for part c.

c) Let us consider two cases weither or not the ratio Q_{B1} / Q_{B2} is constant.

α) *The ratio Q_{B1} / Q_{B2} is constant*. The model in Fig. S 7.9 remains correct with a parametrization of U_4 and U_5: $U_4 = 30 \times Q_{B1}$, $U_5 = 10 \times Q_{B2}$. For example if $Q_{B1} = 6$ and $Q_{B2} = 20$, $U_4 = 180$ and $U_5 = 200$. The delay $d_3 = 0.56 \times 90 / V_4$ is still correct: for $V_4 = U_4 = 180$, $d_3 = 0.28$ is obtained.

Remark S 7.2 It follows from the previous comments that: if a string of sub-conveyors is such that 1) the *maximal flow rate of the first sub-conveyor* is at most equal to the other maximal flow rates, and 2) the *ratios among the speeds* of the sub-conveyors are constant, then this series of sub-conveyors (with different widths and speeds) may be *modeled as a single conveyor* (i.e. with four parameters like U_4, U_5, d_3, and $m_5 + m_6 + m_7$ in the figure). See Fig. E 8.2.**D** and **F** for example.

❑

β) *The ratio Q_{B1} / Q_{B2} may change*. In this case, the complete conveyor must be represented by a string of three conveyors [De 94][Au 95], namely B_1, B_{12}, and B_2, each with four parameters.

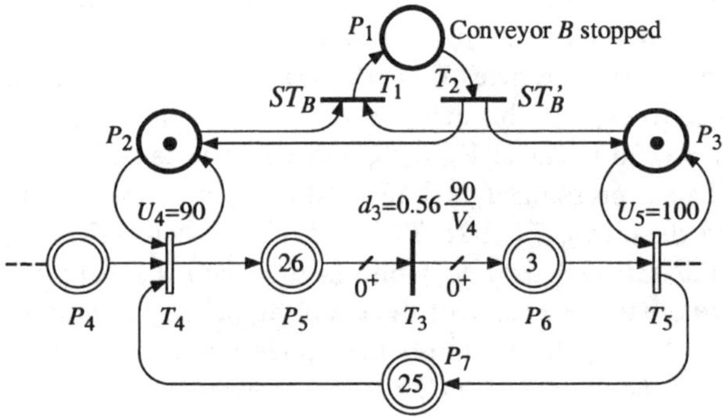

Figure S 7.9

8 Application Examples

8.1 GAS STORAGE

1) Simulation

The hybrid PN model will be progressively built. In order, the *configurations*, the use of the *compressors*, the *continuous gas flows*, and the possible *failures* will be analyzed. Then the *complete model* and a *simulation algorithm* will be given.

Configurations

According to the specifications, configuration 1 is used when the produced flow d_p equals the consumed flow d_c. From a mathematical point of view, the probability that $d_p = d_c$ is zero since these are continuous and independent values. From an engineer point of view, the values may be considered as equal when the difference between both may be ignored. Let us then define the Boolean value[4]

$$D_{pc} = [-a < d_p - d_c < a], \tag{S 8.1}$$

where a is a small value which may be ignored. Similarly,

$$D_p = [d_p - d_c > a],\ D_c = [d_c - d_p > a], \text{ and } D_0 = [d_p < a], \tag{S 8.2}$$

can be defined (the produced flow d_p is considered as nul if it is less than a).

From these notations, the transitions between configurations can be modeled as shown in Fig. **A**. The method for obtaining the interpreted PN in Fig. **A** was explained in Exercise 3.4, part b (without outputs). A token in P_1 means that the system is in configuration 1 (i.e., $-a < d_p - d_c < a$). If $d_p - d_c$ increases, D_p becomes 1 and the token is deposited in P_2 (configuration 2); if $d_p - d_c$ decreases, D_c becomes 1, or (inclusive) possibly D_o becomes 1, and the token is deposited in P_3: this marking is stable if $D_o = 0$ and unstable (then P_4 is reached) if $D_o = 1$. And so on.

Near every place, the variables in $\{R_3, R_4, R_7, R_{R_1}, R_{R_2}\}$ whose Boolean values are 1 for the corresponding configuration are written. Note that $R_{R_2} = 1$ for all configurations. For example, for configuration 1, $R_7 = 1$ (i.e. valve R_7 is open, whereas R_3 and R_4 are closed and R_{R_1} is off).

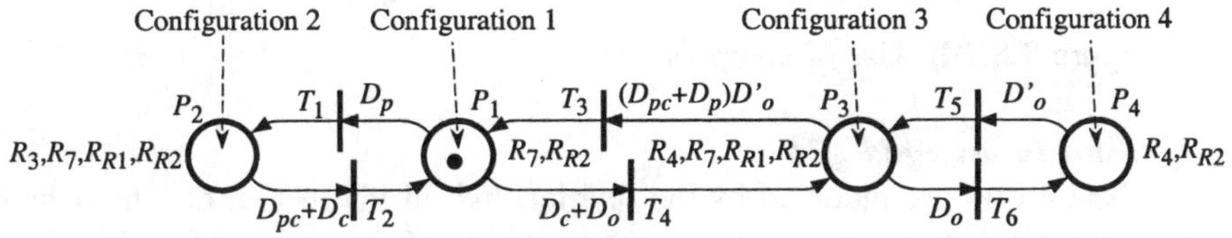

Figure S 8.1.**A** Transitions among configurations for discrete time model.

Compressors

The use of compressor C_p depends on both the configuration of the storage unit (since it is never used for configurations 1 and 4) and the Boolean value

$$Q_{ps} = [P_p > P_s]. \tag{S 8.3}$$

This is illustrated in Fig. **B**, in which the variables in $\{C_p, R_1, R_2\}$ are written near the places where their Boolean values are 1. For the current marking C_p is not used (token in P_6, valve R_2 open). It will be used (token in P_5, valve R_1 open) when T_7 or T_8 is fired. Firing of T_7 occurs if P_2 is marked (configuration 2) and $Q'_{ps} = 1$ (i.e., $P_p < P_s$) and firing of T_8 occurs if P_3 is marked and $Q'_{ps} = 1$. When C_p is used, it may become not used if T_9, T_{10}, T_{11}, or T_{12} is fired. Transition T_{10} or T_{11} is fired if the system is in configuration 2 or 3,

[4] According to the predicate notation introduced in Sect. 3.3.1, $[X] = 1$ if X is true.

respectively, and $Q_{ps} = 1$. Transition T_9 or T_{12} is fired if the system is in configuration 1 or 4, respectively, without any condition[5].

The behavior concerning the use of compressor C_c is almost the same. The difference is that it may be used for any configuration and that the Boolean variables concerned are

$$Q_{pc} = [P_p > P_c] \text{ and } Q_{sc} = [P_s > P_c], \tag{S 8.4}$$

for configuration 1 and configuration 2, 3, or 4, respectively.

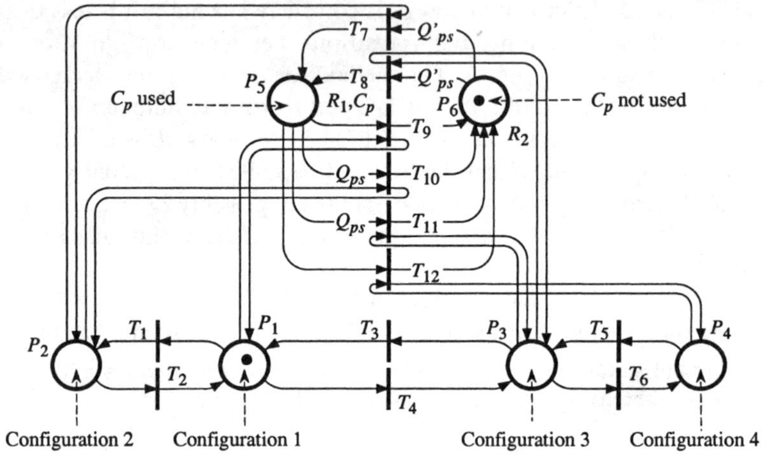

Figure S 8.1.**B** Use of compressor C_p.

Continuous gas flows

These flows are modeled by the (partial) hybrid PN in Fig. **C**. The quantity stored (*MO*, number of moles) is the marking of C-place P_{12}. This place is fed by continuous firing of T_{25} when P_2 is marked (configuration 2). It is emptied by firing of T_{26} or T_{27} when P_3 or P_4 is marked (configuration 3 or 4).

Informally, C-places P_{13} and P_{14} model the contents of the pipes, from production ($V_{28} = U_{28} = d_p$, since the D-enabling of T_{28} is always 1) to storage and from storage to consumption ($V_{30} = U_{30} = d_c$), respectively (informally however, the speed v_{29} may correspond to the flow through valve R_7). According to Sect. 6.4.1, the absence of buffer in these pipes is modeled by markings 0^+ in places P_{15} and P_{16} at initialization. Let us explain the behavior of this part of PN for configurations 1 and 2.

Configuration 1. There is a token in P_1 and no token in P_2, P_3, and P_4. Hence, T_{25}, T_{26}, and T_{27} are not enabled: only T_{28}, T_{29}, and T_{30} are enabled. The sum $m_{13} + m_{15} = 0^+$ implies that $v_{28} = v_{29}$. Similarly, $v_{29} = v_{30}$. Thus,

$$v_{28} = v_{29} = v_{30} = \min(V_{28}, V_{29}, V_{30}) = \min(d_p, d_c). \tag{S 8.5}$$

---

[5] No event or condition is written near transitions T_9 and T_{12}. They are immediate transitions, according to Notation 3.3 in Sect. 3.3.1.

This behavior is consistent with the use of configuration 1: $d_p \approx d_c$.

Configuration 2. There is a token in P_2 and no token in P_1, P_3, and P_4. Hence, T_{25}, T_{28}, T_{29}, and T_{30} are enabled. The flow v_{28} feeds P_{13} and is split between v_{29} and v_{25}: $v_{28} = v_{29} + v_{25}$. Since T_{29} is an immediate transition ($U_{29} = \infty$), it takes priority over T_{25} (Remark 3.2 in Sect. 3.2.1); this behavior corresponds to the fact that the production must satisfy the demand, and only the surplus is stored. Hence, since $v_{29} = v_{30}$ (due to $m_{14} + m_{16} = 0^+$), and given $d_p > d_c$ in this configuration, (S 8.6) to (S 8.8) are obtained:

$$v_{28} = V_{28} = d_p, \tag{S 8.6}$$

$$v_{29} = v_{30} = V_{30} = d_c, \tag{S 8.7}$$

$$v_{25} = v_{28} - v_{29} = d_p - d_c. \tag{S 8.8}$$

The behaviors for configurations 3 and 4 can be analyzed in a similar way.

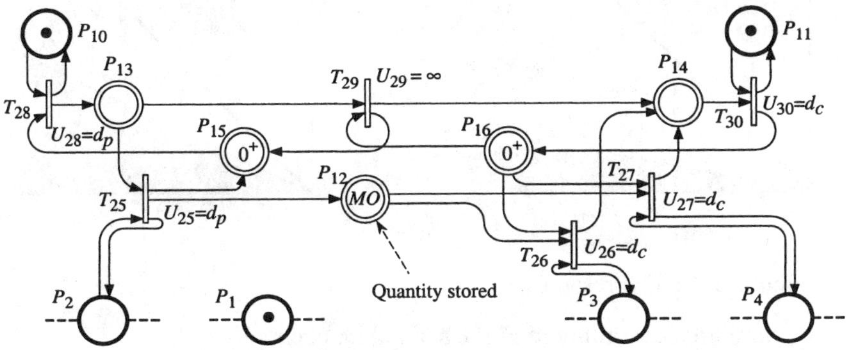

Figure S 8.1.**C** Continuous gas flows.

Failures

Only failures which can be derived from the specifications are analyzed here. Let us first observe that the storage volume has a maximal value $VO_{\max}$ (i.e. the volume of the tank). Hence, from (E 8.8), $VO_{\max}$ implies $P_{s,\max}$, and from (E 8.9), $MO_{\max}$ is obtained from $VO_{\max}$ and $P_{s,\max}$. Thus the first case of failure is as follows: the tank is full and the production flow is greater than the consumption. This is illustrated by the partial PN in Fig. **D**a (configuration 2 and $MO_{\max}$ is reached).

Figure S 8.1.**E**a presents the plane (P_s, P_p). When the system is in configuration 2 or 3, C_p is used if $P_p < P_s$. However, if $P_p < P_s / 5$, i.e.

$$Q_p = [P_p > P_s / 5] = 0, \tag{S 8.9}$$

P_p is too low and there is a failure. According to Fig. **B**, if C_p is used (token in P_5), the system is certainly in configuration 2 or 3 (T_9 and T_{12} are immediate transitions). It follows that this second case of failure can be modeled as in Fig. **D**b.

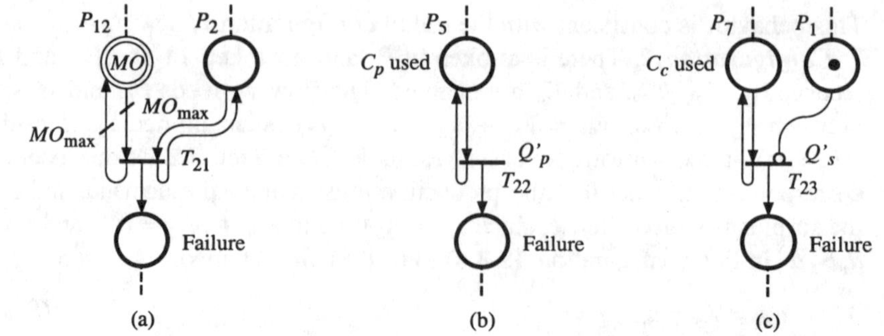

Figure S 8.1.**D** Failures

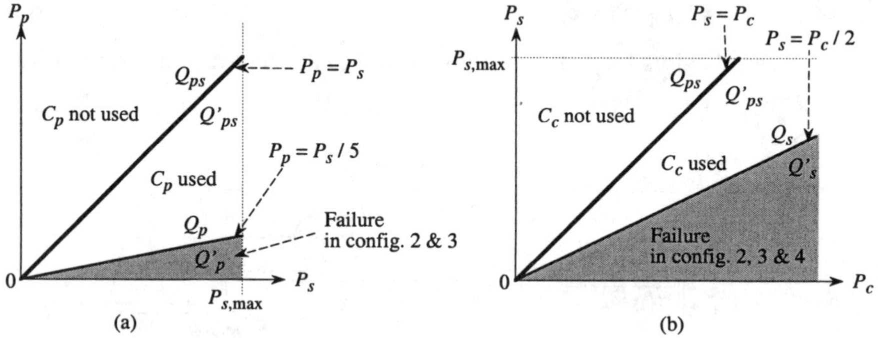

Figure S 8.1.**E** Pressures.

Similarly, according to Fig. **E**b, a failure occurs if

$$Q_s = [P_s > P_c / 2] = 0, \qquad \text{(S 8.10)}$$

and the system is in configuration 2, 3 or 4. Since the system is in configuration 2, 3 or 4 if and only if it is not in configuration 1, this case of failure can be modeled as in Fig. **D**c (without inhibitor arc, three transitions would be required, for configurations 2, 3 and 4, respectively).

Complete model

Partial PNs in Figs. **A–D** plus the part modeling the use of compressor C_c can be merged as in Fig. **F**. In this PN, the three cases of failure presented in Fig. **D** are merged in place P_9. It is assumed that there is an alarm and that manual behavior by an operator is required.

Simulation algorithm

Let $\mathbf{I}(t) = (d_p(t), P_p(t), d_c(t), P_c(t))$ be the input vector at time t. This time dependent vector can be given by the string $\mathbf{I}(t) = (\tau_0, \mathbf{I}_0)(\tau_1, \mathbf{I}_1)\ldots(\tau_k, \mathbf{I}_k)$ ... which denotes the successive values of $\mathbf{I}(\tau_k) = (d_p(\tau_k), P_p(\tau_k), d_c(\tau_k), P_c(\tau_k))$. The vector $\mathbf{I}(t)$ changes at times τ_1, τ_2, etc, i.e. $\mathbf{I}(t)$ is constant for $t \in [\tau_k, \tau_{k+1}[$. This notation is quite similar to the notation of $\mathbf{V}(t)$ in

Appendix M; after a *finite transient behavior*, $\mathbf{I}(t)$ may have a *periodical behavior* (the notation is similar to (M.10), $\mathbf{I}(t)$ taking the place of $\mathbf{V}(t)$).

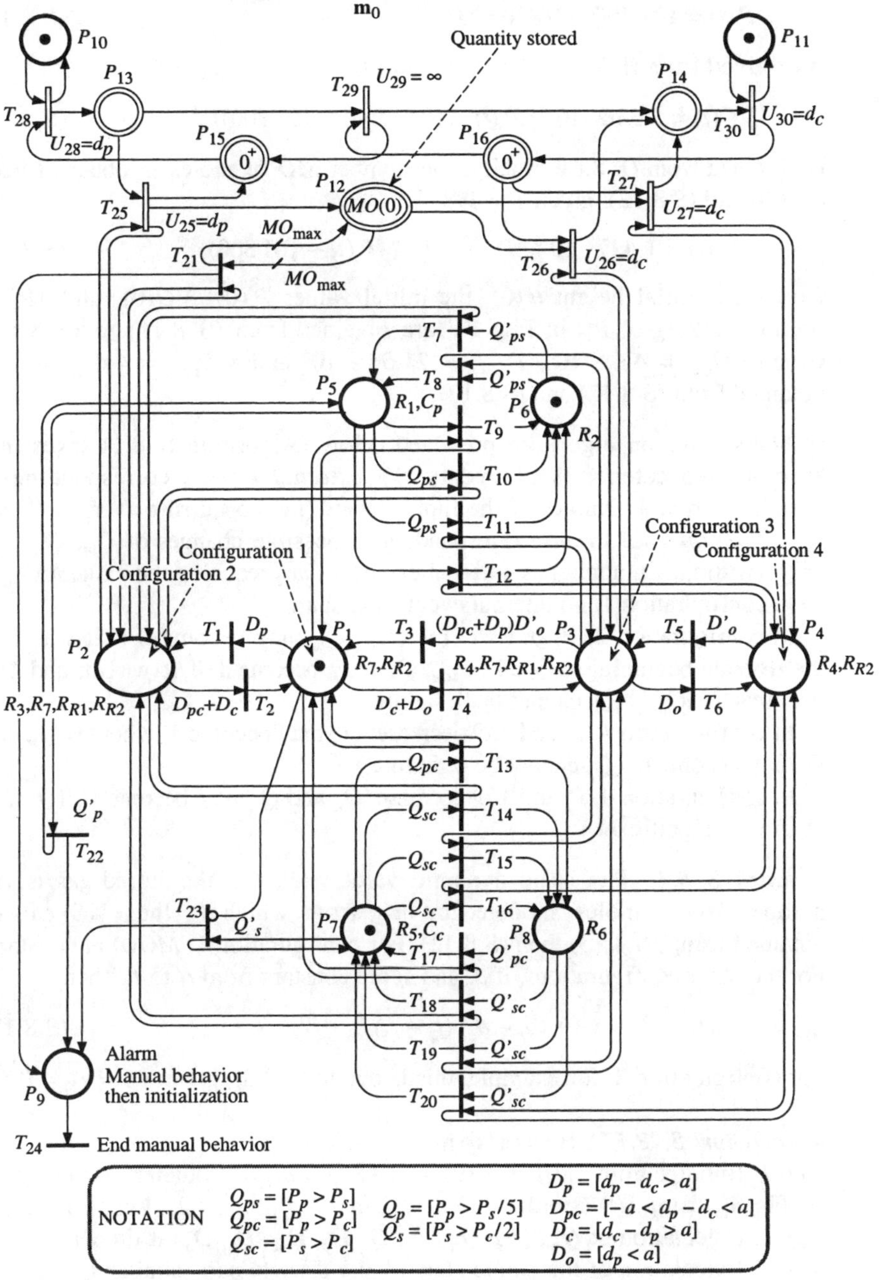

Figure S 8.1.**F** Hybrid PN model for the gas storage unit.

In addition to $\mathbf{I}(t)$, the knowledge of $h_s(0)$ is required (Fig. E 8.1.**B**). Given P_o= 101 300 Pa (i.e., 1.013 bar), ρ_l = 1000 kg/m^3 and g = 9.81 m/s^2,

$$P_s(t) = 101\,300 + 9\,810 \cdot h_s(t) \tag{S 8.11}$$

is obtained from (E 8.7). Then, the volume

$$VO(t) = 3.536 \cdot 10^{-3} \cdot P_s(t) \cdot (10^{-4}\,(P_s(t) - 101\,300))^{2.5} \tag{S 8.12}$$

is obtained from (E 8.8). Finally, the number MO of moles is obtained from (E 8.9) and (S 8.12), given $T = 298°$ K:

$$MO(t) = 1.427 \cdot 10^{-6} \cdot P_s(t)^2 \cdot (10^{-4}\,(P_s(t) - 101\,300))^{2.5}. \tag{S 8.13}$$

Given the initial height $h_s(0)$, the initial values $P_s(0)$, $VO(0)$, and $MO(0)$ (initial marking of P_{12} in Fig. **F**), are obtained from (S 8.11) to (S 8.13). Given $VO_{\max} = 4.6 \cdot 10^9$, $P_{s,\max} = 21.54 \cdot 10^5$ and $MO_{\max} = 4.0 \cdot 10^{12}$ are obtained from (S 8.12) and (S 8.13).

The simulation algorithm presented below (Algorithm S 8.1) takes into account two categories of events: 1) *external events* corresponding to initialization and changes of the input vector (i.e., occurring at $\tau_0 = 0$, τ_1, τ_2, ...); 2) *internal events* corresponding to possible changes of Q_{ps}, Q_{sc}, Q_p, or Q_s (without simultaneous external event), or m_{12} reaching the value $MO_{\max}$.

In configuration 1, no internal event can occur.

In configuration 2, P_s and MO increase: Q_{ps} may become 0 if it was 1 at the IB-state beginning, as well as Q_p; Q_{sc} may become 1 if it was 0; and MO may reach $MO_{\max}$. (Q_s cannot be 0: failure.)

In configuration 3, P_s and MO decrease: Q_{ps} may become 1, whereas Q_{sc} and Q_s may become 0. (Q_p cannot be 0: failure.)

In configuration 4, P_s and MO decrease: Q_{sc} and Q_s may become 0. (Q_{ps} and Q_p are not significant.)

Remark S 8.1 The basic dynamic value related to the stored gas is the number MO of moles, modeled by m_{12}, from which P_s then VO can be obtained using (S 8.13) then (S 8.12). For configuration 1, $MO(t)$ is constant. For the other configurations, if d_p and d_c are constant from t_1 to t_2, then:

$$MO(t_2) = MO(t_1) + (d_p - d_c)\,(t_2 - t_1). \tag{S 8.14}$$

For configuration 2, for example, this is due to (S 8.8), given $v_{26} = v_{27} = 0$.

***Algorithm S 8.1* Simulation**

Step 1. Initialization: $h_s(0) \Rightarrow P_s(0) \Rightarrow MO(0) = m_{12}(0)$; other markings as in Fig. **F**; string $\mathbf{I}(t)$ (similar to *Step 1.2* in Algorithm M.2). Let $\mathbf{I}(\tau) = \mathbf{I}(0)$.

Step 2. Calculation of Q_{ps}, Q_{pc}, Q_{sc}, Q_p, Q_s, D_p, D_{pc}, D_c, D_o, at time τ.

Step 3. Calculation of the stable state of the discrete part (similar to *Steps 3* and *4* in Algorithm 3.1, Sect. 3.2.2.2).

Step 4. Calculation of the speed vector **v** for the continuous part (similar to Algorithm 5.4, Sect. 5.3.2.3).

Step 5. Calculation of the outputs W_{C_p}, W_{C_c}, W_{R_1}, W_{R_2}, *MO*, *VO*, P_s, at τ.

Step 6. Calculation of the time of the next event (next change in $\mathbf{I}(t)$ or internal event). Let τ be this time. **If** $\tau <$ "end of simulation" **then** go to *Step 2* **else** END.

❑

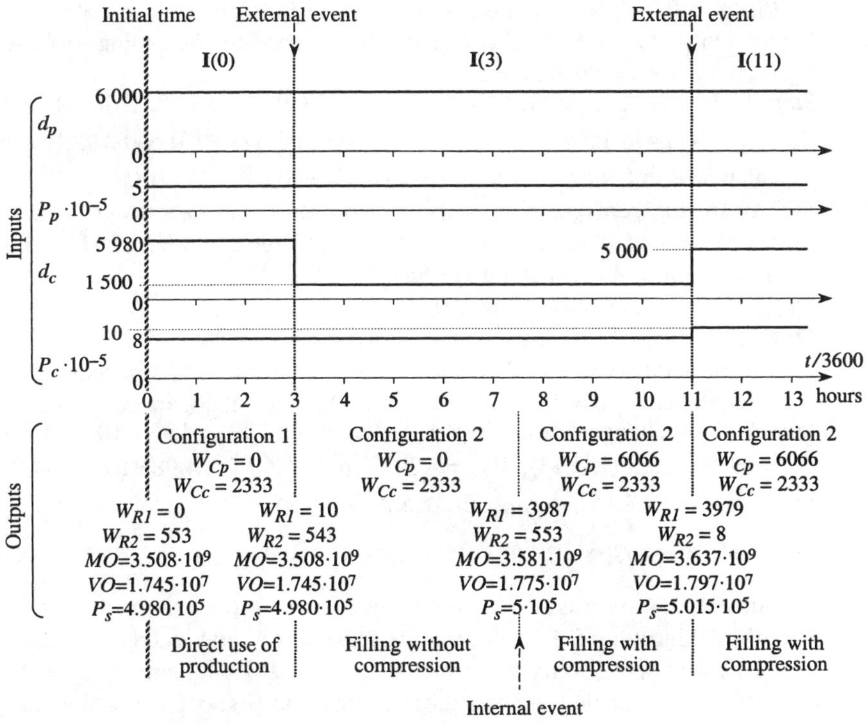

Figure S 8.1.**G** Simulation example.

An example of application of this algorithm is given in the sequel. The corresponding inputs and outputs are illustrated in Fig. **G**. Let H denote the value 3600; hence, the time 3 H, for example, corresponds to 10 800 s, i.e. 3 hours. Assume the value $a = 30$ (a is presented in (S 8.1) and (S 8.2)).

The configuration, as well as some output variables (namely W_{C_p} and W_{C_c}) does not change during an IB-state. The other output variables may change during an IB-state. They are calculated only at the beginning of every IB-state (the possibility to calculate them periodically could be added, but they change very slowly during an IB-state).

Application of Algorithm S 8.1

Step 1. $h_s(0) = 40.44$ m $\Rightarrow P_s(0) = 4.980 \cdot 10^5 \Rightarrow MO(0) = m_{12}(0) = 3.508 \cdot 10^9$; $\mathbf{I}(t) = (0, (6000, 5 \cdot 10^5, 5980, 8 \cdot 10^5))(3\,H, (6000, 5 \cdot 10^5, 1500, 8 \cdot 10^5))(11\,H, (6000, 5 \cdot 10^5, 5000, 10 \cdot 10^5))$. End of simulation at $t = 13.5\,H$. Let $\tau = 0$, $\mathbf{I}(0) = (6000, 5 \cdot 10^5, 5980, 8 \cdot 10^5)$.

Step 2. $Q_{ps}=1$, $Q_{pc}=0$, $Q_{sc}=0$, $Q_p=1$, $Q_s=1$, $D_p=0$, $D_{pc}=1$, $D_c=0$, $D_o=0$.

Step 3. Token in P_1 (configuration 1) since $d_p - d_c = 20 < a = 30$ (i.e. $D_{pc} = 1$); token in P_6 (C_p not used) is stable; token in P_7 (C_c used) also stable.

Step 4. Only C-transitions T_{28}, T_{29}, and T_{30} are enabled. According to (S 8.5), $v_{28} = v_{29} = v_{30} = 5980$ is obtained.

Step 5. $W_{C_p} = 0$; from (E 8.11), $W_{C_c} = 2333$ W; $W_{R_1} = 0$; from (E 8.15), $W_{R_2} = -\ 2478 \ln(8 \cdot 10^5 / 2 \times 5 \cdot 10^5) = 553$ W; $P_s(0)$ and $MO(0)$ where calculated in *Step 1*, and from (S 8.12), $VO(0) = 1.745 \cdot 10^7$.

Step 6. In this configuration, only an external event can occur. The next change (end of IB-state 1) occurs at $t = 3\,H$: Let $\tau = 3\,H$, $\mathbf{I}(3\,H) = (6000, 5 \cdot 10^5, 1500, 8 \cdot 10^5)$ and go to *Step 2*.

Step 2. Only changes: $D_p = 1$ and $D_{pc} = 0$.

Step 3. Firing of T_1, then a token in P_2 (configuration 2).

Step 4. In addition to T_{28}, T_{29}, and T_{30}, T_{25} is enabled. The speeds $v_{28} = d_p = 6000$, $v_{29} = v_{30} = d_c = 1500$, and $v_{25} = d_p - d_c = 4500$, are obtained.

Step 5. Only changes: from (E 8.12), $W_{R_1} = -\ 2478 \ln(4.98 \cdot 10^5 / 5 \cdot 10^5) = 10$ W; from (E 8.13), $W_{R_2} = -2478 \ln(8 \cdot 10^5 / 2 \times 4.98 \cdot 10^5) = 543$ W.

Step 6. From (S 8.14), during IB-state 2:

$$m_{12}(t) = MO(t) = 3.508 \cdot 10^9 + 4500\,(t - 3\,H). \qquad \text{(S 8.15)}$$

Since $MO(t)$ increases, $P_s(t)$ increases too. According to the analysis just before Algorithm S 8.1, Q_{ps} may become 0, Q_{sc} may become 1, and MO may reach $MO_{\max}$. Given $P_s = 4.98$, $P_p = 5$, $P_c = 8$, and $P_{s,\max} = 21.54 \cdot 10^5$, it is clear that, among these events, the first occurring will be Q_{ps} becoming 0. This event will occur when $P_p = P_s$, i.e. when $P_s = 5 \cdot 10^5$. From (S 8.13), this event will occur when $MO = 3.581 \cdot 10^9$, thus, from (S 8.13), this event occurs at $t_1 = 27\,022 \approx 7.5\,H$.

As $t_1 = 7.5\,H < \tau_2 = 11\,H$, IB-state 2 ends when the internal event '*Q_{ps} becomes* 0' occurs. Let $\tau = 7.5\,H$, $\mathbf{I}(7.5\,H) = \mathbf{I}(3\,H)$ and go to *Step 2*.

Step 2. Only changes: $Q_{ps} = 0$.

Step 3. Firing of T_7, then a token in P_5 (C_p will be used).

Step 4. No change.

Step 5. Only changes: $W_{R_1} = 3987$ W; $W_{R_2} = 553$ W; $MO = 3.581 \cdot 10^9$ and $P_s = 5 \cdot 10^5$ were calculated in *Step 6*, and it follows that $VO = 1.775 \cdot 10^7$.

Step 6. The next event is the change of $\mathbf{I}(t)$ occurring at $t = 11\,H$: Let $\tau = 11\,H$, $\mathbf{I}(11\,H) = (6000, 5 \cdot 10^5, 5000, 10 \cdot 10^5)$ and go to *Step 2*.

Step 2. No change.

Step 3. No change.

Step 4. Changes: $v_{29} = v_{30} = d_c = 5000$, and $v_{25} = d_p - d_c = 1000$.
Step 5. Only changes: $W_{R_1} = 3979$ W; $W_{R_2} = 8$ W; from (S 8.14):

$$m_{12}(11) = MO(11) = 3.581 \cdot 10^9 + 4500\,(11 - 7.5)) = 3.637 \cdot 10^9;$$

$VO = 1.797 \cdot 10^7$; $P_s = 5.015 \cdot 10^5$.
Step 6. The next event is the end of simulation at 13.5 H. Then END.

❑

Remark S 8.2 For higher values of P_s (about $15 \cdot 10^5$ for example), P_s changes very slowly: $MO(t)$ is linear in time and, from (S 8.13), P_s is approximately proportional to $MO^{-4.5}$. It follows that, practically, there are few internal events. Most of the events are external.

2) Control

For the control purpose, Algorithm S 8.1 should be modified (the inputs/outputs of the corresponding controler are shown in Fig. **H**):

1) Since the inputs $\mathbf{I}(t)$ are progressively known, these inputs are periodically read (as in Algorithm M.1 in Appendix M).

2) The Boolean values of C_p, C_c, R_{R_1}, R_{R_2}, R_1 to R_7, are outputs acting on the process.

3) In order to avoid a slow drift of $MO(t)$ in the model, $h_s(t)$ is added in the set of inputs. Thus, the value of $MO(t)$ can be updated.

Finally, the treatment of failures could be developed.

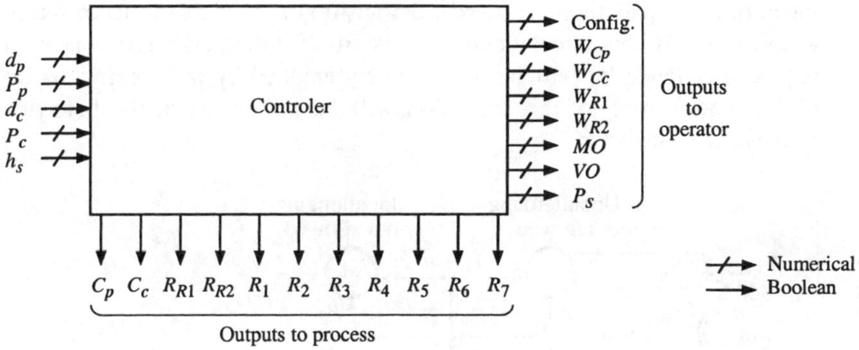

Figure S 8.1.**H** Inputs and outputs for control purpose.

According to Fig. Post.2c in the postface, the controler in Fig. **H** can be divided into a *control hybrid PN* and its data processing part. For this example, the diagram in Fig. **I** can be obtained from the information in Figs. **F** and **H**.

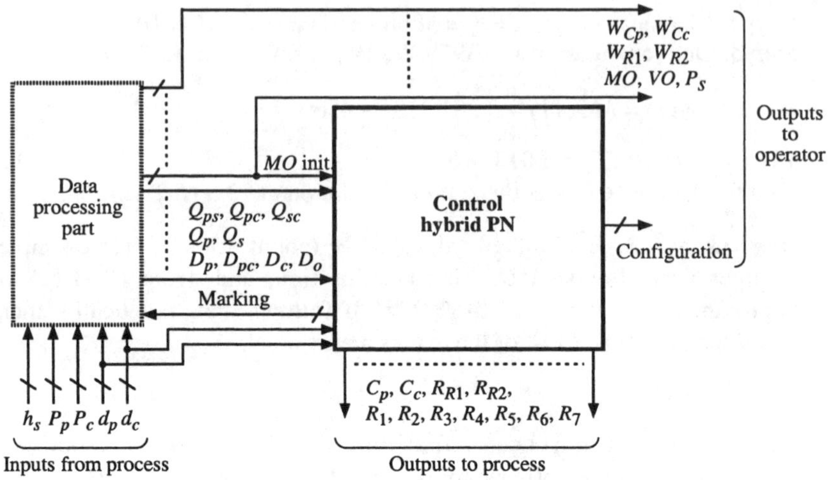

Figure S 8.1.I Control hybrid PN and its data processing part.

8.2 BOTTLING CHAIN

Models for every component in isolation (i.e. without upstream or downstream constraint) are presented in the sequel. Then, a model of the complete chain is given.

Unpalletizer

From the specifications, the model in Fig. **A** can be obtained. If there is a token in P_1 (unpalletizing allowed), D-transition T_1 is enabled, hence receptive to event E_u. If this event occurs, T_1 is fired: 360 marks are deposited in C-place P_{16}. Place P_{16} will be progressively emptied by the continuous firing of T_{18}. The features of this transition will be known when the behavior of conveyor 1 is analyzed.

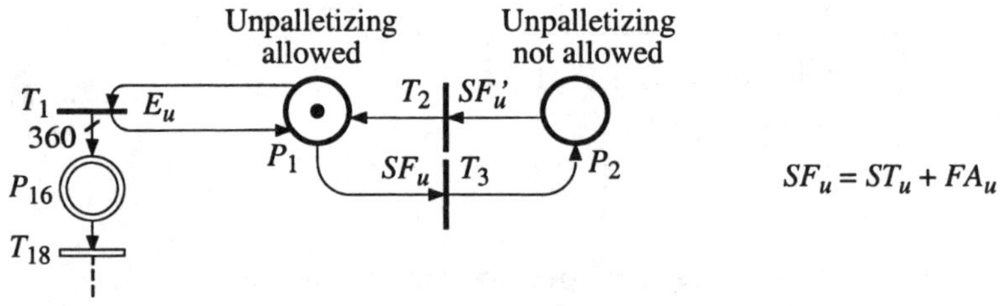

Figure S 8.2.A Unpalletizer model in isolation.

Since the behavior is similar when the decision is made to stop the unpalletizer or when there is a failure, the Boolean variable

$$SF_u = ST_u + FA_u \tag{S 8.16}$$

can be used: $SF_u = 1$ if $ST_u = 1$ (*stop* unpalletizer) OR $FA_u = 1$ (*failure* of unpalletizer). If $SF_u = 1$, T_3 is fired and T_1 is no longer enabled. When the complement value $SF_u' = ST_u' \cdot FA_u' = 1$, T_2 is fired and T_1 is again enabled.

Conveyor 1

The model for conveyor 1 in isolation is given in Fig. **B**. Its structure is similar to the hybrid PN in Fig. S 7.9 (because conditions in Remark 7.2 are satisfied: the flow rate of conveyor a is not greater than the others). Let us calculate the various parameters:

$$m_{19}(0) = \text{capacity} = 12\frac{0.6\,\text{m}}{0.06\,\text{m}} + 3\frac{2\,\text{m}}{0.06\,\text{m}} + 1\frac{3\,\text{m}}{0.06\,\text{m}} = 270;$$

$$U_{18} = \text{flow rate sub-conveyor } a = 12\,\text{boxes}\text{x}\frac{5\,\text{m/min}}{0.06\,\text{m}} = 1000\ \text{boxes/min};$$

$$\text{flow rate sub-conveyor } b = 3\,\text{boxes}\text{x}\frac{20\,\text{m/min}}{0.06\,\text{m}} = 1000\ \text{boxes/min};$$

$$U_{19} = \text{flow rate sub-conveyor } c = 1\,\text{box}\text{x}\frac{60\,\text{m/min}}{0.06\,\text{m}} = 1000\ \text{boxes/min};$$

$$d_4\ (\text{if conveyor works}) = \frac{0.6\,\text{m}}{5\,\text{m/min}} + \frac{2\,\text{m}}{20\,\text{m/min}} + \frac{3\,\text{m}}{60\,\text{m/min}} = 0.27\ \text{min}.$$

The timing associated with T_4 is $d_4 = 0.27\,\frac{1000}{V_{18}}$ (i.e. $d_4 = 0.27$ min when the conveyor works and $d_4 = \infty$ when it is stopped).

When the conveyor works, two places (P_3 and P_4) must contain a token according to Remark S 7.1 in Solution 7.9. The move from the state where the conveyor works to stop (token in P_5), occurs if $SF_{cl} = ST_{cl} + FA_{cl} = 1$ (similarly to the unpalletizer).

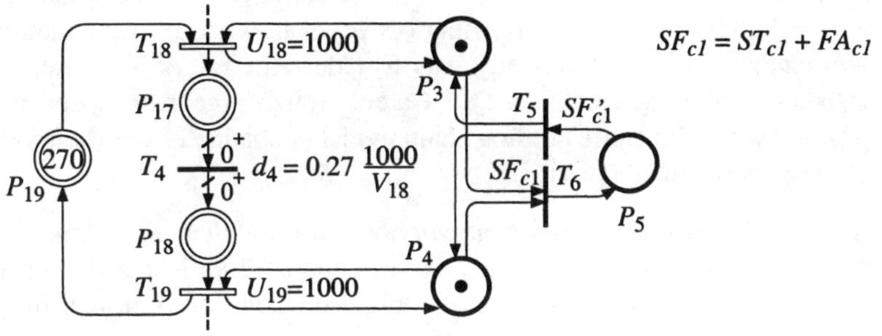

Figure S 8.2.**B** Conveyor 1 model in isolation.

Rotary filler

The model in isolation is shown in Fig. **C**. When the rotary filler is working at its maximal speed ($Q_{rf} = 52.8$ m / min), the input and output flow rates are

$$U_{19} = U_{20} = \frac{1\,\text{box}}{0.066\,\text{m}} Q_{rf} = 15.15\, Q_{rf} = 15.15 \times 52.8 = 800 \text{ boxes/min},$$

and the time spent by a box in this machine is

$$d_9 = \frac{8.778\,\text{m}}{Q_{rf}} = \frac{8.778}{52.8} = 0.166 \text{ min.}$$

Let us now consider the cases when Q_{rf} must be less than 52.8 m / min.

a) *Failure*, i.e. $FA_{rf} = 1$. The conveyor speed is modified when the event $\uparrow FA_{rf}$ occurs (for the notation, see Appendix D): $Q_{rf} = 0$, i.e. the conveyor is stopped. The conveyor speed is modified again when the event $\downarrow FA_{rf}$ occurs: $Q_{rf} = 52.8$ if the downstream conveyor is not full, or another value less than 52.8 if it is full (Q_{rf} depends on the output flow of the conveyor as explained below).

b) *Stop*, i.e. $ST_{rf} = 1$. When $\uparrow ST_{rf}$ occurs, transition T_8 is fired, hence V_{19} becomes 0 since the D-enabling of T_{19} becomes 0 (no boxes enter the rotary filler). However, the value of Q_{rf} is not immediately changed. When the time $d_9 = 8.778 / Q_{rf}$ is spent (after $\uparrow ST_{rf}$), all the boxes which were in the rotary filler are evacuated, then $Q_{rf} = 0$ from this time.

When the event $\downarrow ST_{rf}$ occurs, transition T_7 is fired, i.e. a token is deposited in P_6, and the new value of Q_{rf} is calculated as after a failure.

c) *The downstream component cannot absorb the output flow*. If the output flow of the rotary filler (i.e. v_{20}) is greater than the output flow of the downstream component (conveyor 2 in our example), the number of boxes on this conveyor increases. If this number reaches the conveyor capacity, it will be necessary to decrease the input flow of the conveyor (i.e. the output flow of the rotary filler). Since the rotary filler conveyor is without accumulation (the boxes cannot slide on it), the only way to reduce the rotary filler output flow consists of reducing the speed Q_{rf}. Calculation of the new speed will be explained when the entire bottling chain model is obtained from the models of the components in isolation.

d) *The downstream component can absorb more than the present output flow*. If the present value of Q_{rf} is less than 52.8, it can be increased. Calculation of the new speed will be explained when the entire bottling chain model is obtained from the models of the components in isolation.

Note that, since the boxes cannot slide on this conveyor, a feedback loop ensuring that the capacity is not passed beyond (like place P_{19} in Fig. **B**) is not required.

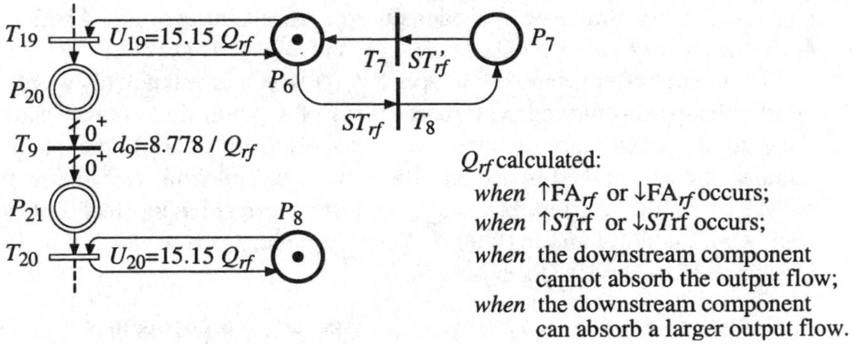

Figure S 8.2.**C** Rotary filler model in isolation.

Conveyor 2

The model for conveyor 2 in isolation is given in Fig. **D**. It is similar to Fig. **B**, except for the parameters:

$$m_{24}(0) = \text{capacity} = 3\,\frac{20\,\text{m}}{0.06\,\text{m}} + 1\,\frac{40\,\text{m}}{0.06\,\text{m}} = 1667 \text{ boxes;}$$

$$U_{20} = U_{21} = 1\,\text{box} \times \frac{160\,\text{m/min}}{0.06\,\text{m}} = 2667 \text{ boxes/min;}$$

$$d_{10}\,(\text{if conveyor works}) = \frac{60\,\text{m}}{160\,\text{m/min}} = 0.375 \text{ min.}$$

The move from the state where the conveyor works (tokens in P_8 and P_9) to stop (token in P_{10}), occurs if $SF_{c2} = ST_{c2} + FA_{c2} = 1$.

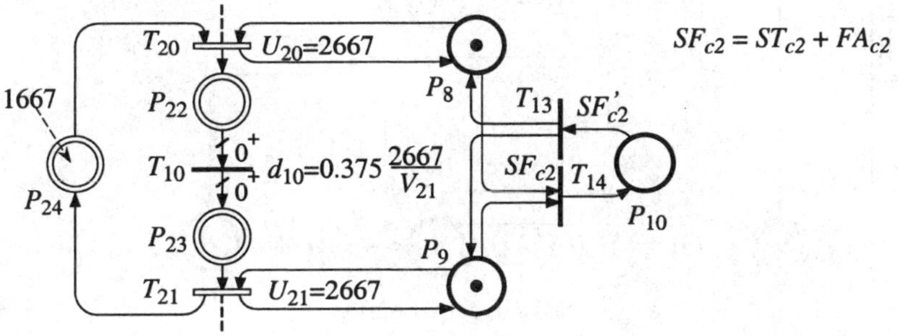

Figure S 8.2.**D** Conveyor 2 model in isolation.

Packaging and furnace

The model is presented in Fig. **E**. Instantaneous firing speed v_{21} is only limited by the upstream. When 24 boxes have been gathered ($m_{25} = 24$) and if there is enough place on the conveyor head ($m_{28} = 1$), a pack is produced (T_{13}

is fired). Since the maximal density on the conveyor is 4 packs / m, $U_{22} = U_{23} = 4\,Q_p$, $d_{14} = 5\,/\,Q_p$, $U_{24} = 4\,Q_f$, and $d_{15} = 2.5\,/\,Q_f$.

The output buffer is emptied at speed $V_D(t)$ which is unknown. In order to be sure that packs entering the furnace (firing of T_{23}) will find enough place in the output buffer when leaving the furnace (firing of T_{24}), the maximal number of packs in the furnace plus the output buffer is limited to the number of places in this buffer: place P_{34} and the corresponding feedback loop ($m_{31} + m_{32} + m_{33} + m_{34} = 10\,000$). This precautionary behavior is known as blocking before service (Exercise 3.9).

a) *Stop or failure.* When the event $\uparrow SF_{pf}$ occurs, the packaging conveyor is immediately stopped ($Q_p = 0$) whereas the furnace conveyor is stopped when it is empty, i.e. after firing of T_{20} requiring $m_{31} = m_{32} = 0$ ($Q_f = 0$ when $m_{14} = 1$).

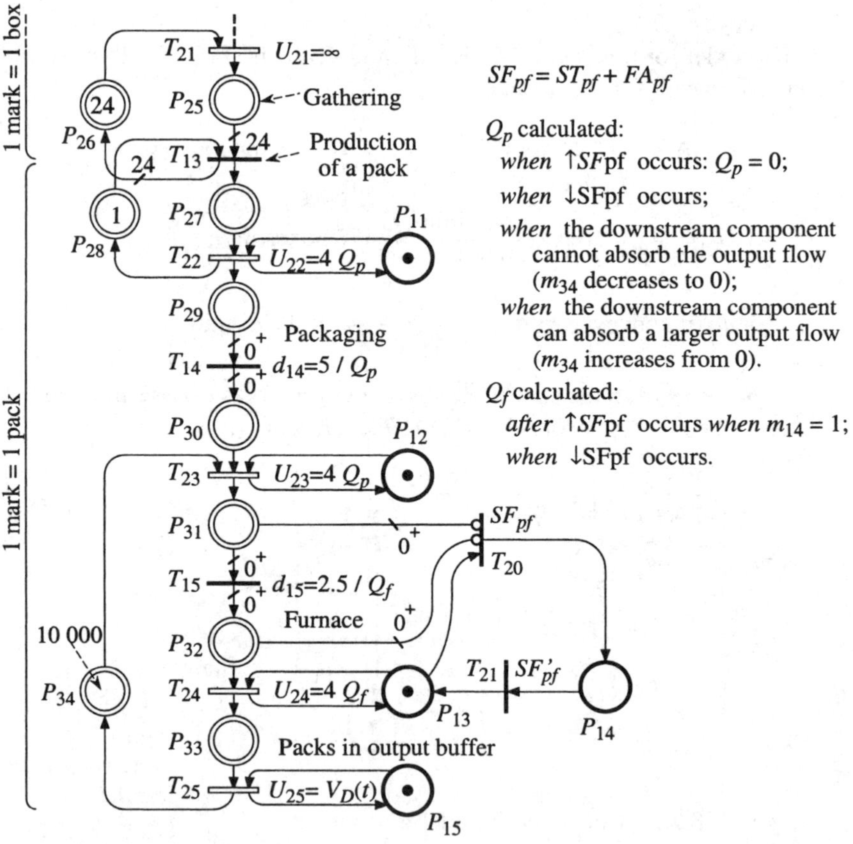

Figure S 8.2.E Packaging and furnace model in isolation.

b) *Required change of* Q_p. If $v_{23} > v_{25}$, m_{34} decreases. If m_{34} reaches the value 0 at time t_1, then, as T_{23} becomes weakly enabled, v_{23} decreases immediately to the present value of v_{25}, i.e. $v_{23} = v_{25}$ from t_1. But, since the packaging conveyor is without accumulation (without sliding), v_{23} must be reduced by reducing the conveyor speed Q_p, i.e.

$$Q_p := Q_p \times \frac{v_{25}}{v_{23}}. \qquad \text{(S 8.17)}$$

In (S 8.17), v_{23} is the value just before t_1 and the value v_{25} is the same before and after t_1 since $V_D(t)$ is piecewise constant.

Assumes that $v_{23} = v_{25}$ and that $m_{34} = 0$ for some time and that m_{34} increases from t_2. This means that either $v_{23}(t_2 + dt) < v_{23}(t_2 - dt)$ (the density on the rotary filler is lower, i.e. the proportion of spots reaching the output which are occupied is less than before) or $v_{25}(t_2 + dt) > v_{25}(t_2 - dt)$ (the outside demand increases). Hence, speed v_{23} may be increased by increasing Q_p:

$$Q_p := \min(Q_p \times \frac{v_{25}}{v_{23}}, 12.5) \qquad \text{(S 8.18)}$$

because the maximal value of Q_p is 12.5 m / min. Note that (S 8.18) may also be used instead of (S 8.17) since, when (S 8.17) is used, the resulting Q_p value is always less than the maximum value.

a) *End of stop or failure*. When the event $\downarrow SF_{pf}$ occurs, values of Q_p and Q_f are calculated:

If $m_{34} > 0$ **then** $Q_p = 12.5$ **else** $Q_p = \min(v_{25} / 4, 12.5)$; $Q_f = 12.5$. (S 8.19)

(If $m_{34} = 0$, $v_{23} \le v_{25}$ is obtained.)

Entire bottling chain

The model of the entire bottling chain is shown in Fig. **F**. It is obtained by merging the models is isolation and adjusting some parameters. For example, the output flow rate of conveyor 1 in isolation, $U_{19} = 1000$ according to Fig. **B**, is greater that the input flow rate of the rotary filler, $U_{19} = 15.15\, Q_{rf}$ according to Fig. **C** (since $Q_{rf} \le 52.8$, $15.15\, Q_{rf} \le 800$). Hence, in Fig. **F**, the minimum of these values is kept: $U_{19} = 15.15\, Q_{rf}$.

According to part c in the section devoted to the rotary filler, Q_{rf} must be reduced if the downstream conveyor cannot absorb the output flow of the rotary filler, i.e. if $m_{24} = 0$ and $v_{20} > v_{21}$. In a way similar to the calculations for the packaging conveyor, the following results are obtained. When m_{24} decreases to 0 or increases from 0

$$Q_{rf} := \min(Q_{rf} \times \frac{v_{21}}{v_{20}}, 52.8). \qquad \text{(S 8.20)}$$

When the event $\uparrow ST_{rf}$ occurs: $Q_{rf} := 0$. When the event $\downarrow ST_{rf}$ occurs:

If $m_{24} > 0$ **then** $Q_{rf} = 52.8$ **else** $Q_{rf} = \min(v_{21} / 15.15, 52.8)$. (S 8.21)

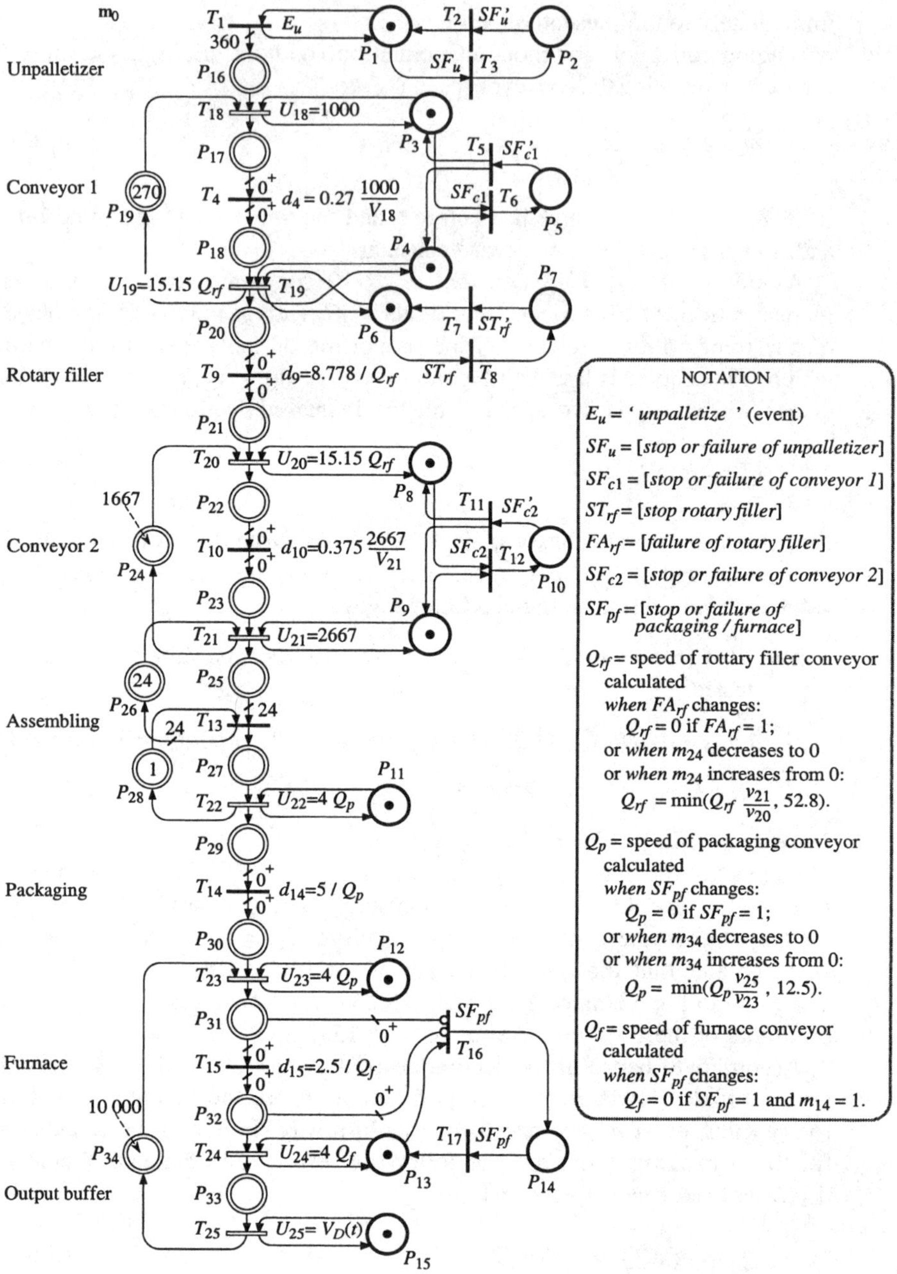

Figure S 8.2.F Entire bottling chain.

Remark S 8.3 There are two examples of "stop with emptying" and two different possible solutions have been chosen (relatively arbitrarily). When

$\uparrow ST_{rf}$ occurs, the rotary filler conveyor works sufficiently long enough after this event to empty the conveyor in any case. If $\uparrow ST_{pf}$ occurs, the furnace conveyor is stopped as soon as it is empty (even if the conveying time O.2 min is not elapsed). The second solution is less simple but quicker.

Remark S 8.4 **Simulation and control**
The model presented in Fig. **F** is used to similate the bottling chain. As explained at the end of Solution 8.1, some slight modifications of the model, based on practical considerations, can be made in order to perform system control. For example, since every calculation and mechanical change requires some time, the calculation of Q_{rf}, theoretically performed when m_{24} reaches the value 0, could be performed as soon as m_{24} is lower than some threshold, for example 10 or 20. Similarly, an increased value of Q_{rf} can be calculated if some threshold is exceeded (same threshold or another).

8.3 FOUR-STROKE ENGINE

Partial models are given from Figs. **A** to **E**, then the complete model of continuous variables subject to *continuous-time dynamics* is given in Fig. **F**.

Intake manifold pressure
From (E 8.14) to (E 8.16), given the values in Figs. E 8.3.**C** and **D**, the model in Fig. **A** can be obtained:

$$p(t) = m_{14}(t) \text{ and } \frac{dp(t)}{dt} = B_{14}(t) = v_{10}(t) + v_{11}(t) - v_{12}(t) - v_{13}(t). \qquad \text{(S 8.22)}$$

Transitions T_{10}, T_{12}, and T_{13}, correspond to the first term in (E 8.14) whereas T_{11} correspond to the second term. Transition T_{10} is associated with the two first terms in (E 8.15): in the flow rate U_{10}, value p is denoted by m_{14} since $p = m_{14}$ whereas n will be specified below (it is also possible to replace T_{10} by two transitions, one for each term). The third and fourth terms in (E 8.15) correspond respectively to T_{12} and T_{13}.

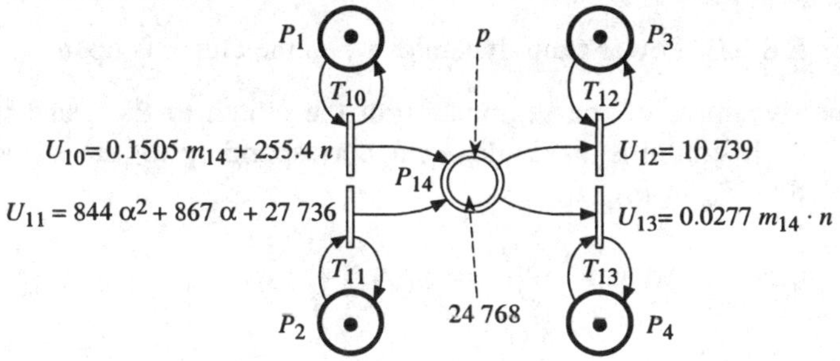

Figure S 8.3.**A** Intake manifold pressure.

Power-train dynamics when the clutch is open

When the clutch is in the *open* position (Boolean variable *on* = 0), the two segments of the driveline are disconnected.

The dynamics of the segment from the crankshaft to the clutch, given by (E 8.17) may be rewritten as:

$$\frac{dn(t)}{dt} = \frac{1}{J_o} T_g(t) - \frac{B}{J_o} n(t) - \frac{1}{J_o} T_l(t). \tag{S 8.23}$$

It is modeled in Fig. **B**:

$$n(t) = m_{15}(t) \text{ and } \frac{dn(t)}{dt} = B_{15}(t) = v_{15}(t) - v_{18}(t) - v_{20}(t). \tag{S 8.24}$$

The first term in (S 8.23) corresponds to $v_{15}(t) = V_{15}(t) = m_6 \cdot U_{15}(t)$, which is the effect of the effective torque $T_g = T - 20$ given in (E 8.27). The second term, corresponding to viscous friction, is modeled by T_{18}. The third term, corresponding to load torque T_l, is modeled by T_{20}.

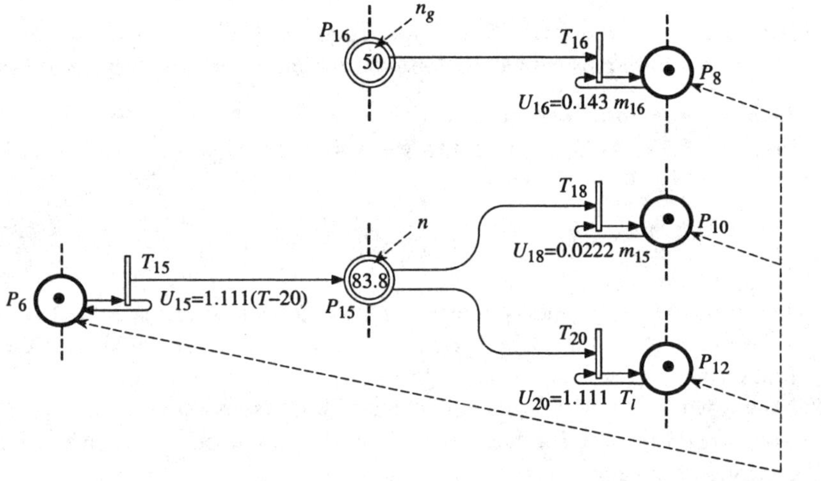

Figure S 8.3.**B** Power-train dynamics when the clutch is open.

The dynamics of the segment from the clutch to the gear[6] is given by (E 8.18). It contains a single term corresponding to the viscous friction modeled by T_{16} in Fig. **B**:

$$n_g(t) = m_{16}(t) \text{ and } \frac{dn_g(t)}{dt} = B_{16}(t) = -\, v_{16}(t). \tag{S 8.25}$$

[6] The value n_g = 50 is arbitrary because the time since the last disengagement is not specified. However, $n_g < n$ = 83.8.

Remark S 8.5 This remark is similar to Remark S 7.1: C-transitions T_{15}, T_{16}, T_{18}, and T_{20} are D-enabled by different D-places, namely P_6, P_8, P_{10}, and P_{12}. Let us note that these transitions cannot be enabled by the same D-place because this structure would correspond to a case 4 conflict (Sect. 6.1.3).

Power-train dynamics when the clutch is closed
When the clutch is in the *closed* position (Boolean variable $on = 1$), the two segments of the driveline are connected.

At the engagement, i.e. when event $\uparrow on$ occurs, both revolution speeds n and n_g reach the same common value given by (E 8.20). This is modeled in Fig. **C**. According to Section G.3 in Appendix G, C-transition T_{21} in Fig. **C** behaves as an immediate transition (infinite speed) when $\uparrow on$ occurs (but only at this time). Hence, when $\uparrow on$ occurs, firing of T_{21} empties P_{15} and, simultaneously, deposits 0.865 of the previous value in both P_{15} and P_{16} ($J_o / J_c = 0.865$). At the same time, firing of T_{22} empties P_{16} and, simultaneously, deposits 0.135 of the previous value in both P_{15} and P_{16} ($(J_c - J_o) / J_c = 0.135$). Thus, just after event $\uparrow on$, n and n_g have the value given in (E 8.20).

The dynamics of the segment from the clutch to the gear is given by (E 8.21) and may be rewritten as:

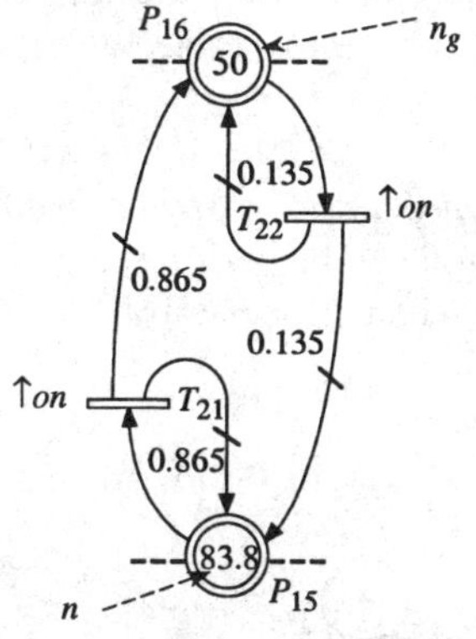

Figure S 8.3.**C** At the engagement, n and n_g reach the same common value.

$$\frac{dn_g(t)}{dt} = \frac{dn(t)}{dt} = \frac{1}{J_c} T_g(t) - \frac{B}{J_c} n(t) - \frac{1}{J_c} T_l(t). \tag{S 8.26}$$

It is modeled in Fig. **D** in which $n_g(t) = m_{16}(t) = n(t) = m_{15}(t)$. The dynamics is the same for both $n_g(t)$ and $n(t)$:

$$\frac{dn_g(t)}{dt} = \frac{dn(t)}{dt} = B_{16}(t) = B_{15}(t) = v_{14}(t) - v_{17}(t) - v_{19}(t). \tag{S 8.27}$$

The three terms in (S 8.27) correspond, in order, to the effect of the effective torque, to the viscous friction, and to the load torque.

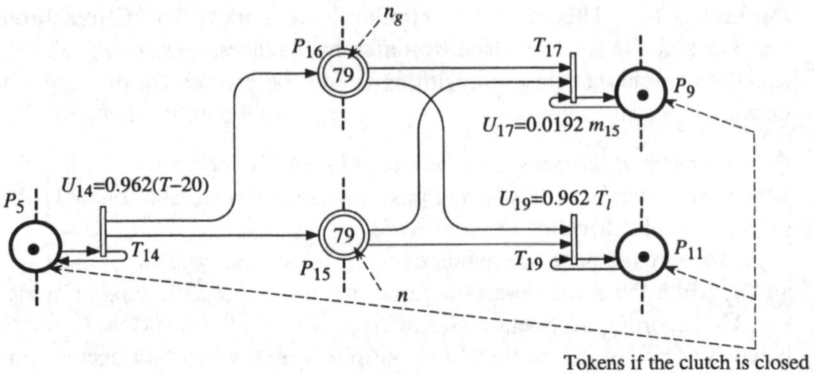

Figure S 8.3.**D** Power-train dynamics when the clutch is closed.

Piston position and discrete-time evolution variables
The piston position is given by the crankshaft angle θ ($0° \leq \theta < 180°$, according to Fig. E 8.3.**A**). Value θ is obtained by integration of $n(t)$ according to ((E 8.19), which is modeled by the part of Fig. **E** containing T_{23}, P_{13}, P_{17}, and the arcs between them. When θ reaches the value 180, it returns to value 0. This is modeled by T_9 and the arc $P_{17} \rightarrow T_9$: when m_{17} reaches the value 180, transition T_9 becomes enabled and is immediately fired (hence, P_{17} is emptied).

Let us now consider the variables in the set $DES = \{T_C, T_E, m_C, m_E, \varphi\}$. They are calculated *once in each stroke* thanks to (E 8.22), (E 8.23) and (E 8.24). This calculation can be performed in the interval during which event spk^* cannot occur, i.e. when $15 < \theta < 160$.

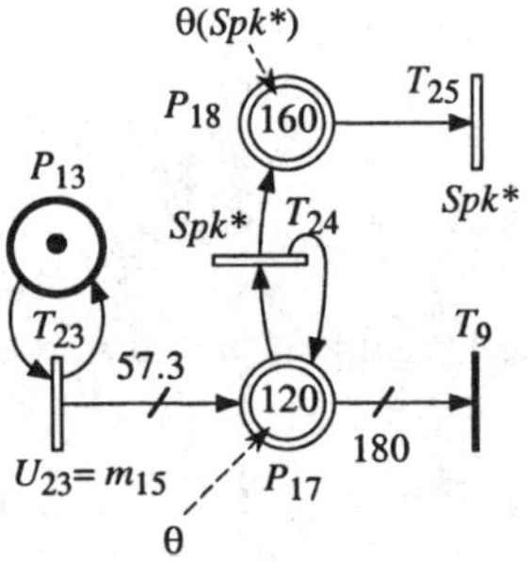

Figure S 8.3.**E** Model for θ and $\theta(spk^*)$.

Equations (E 8.22) to (E 8.24) require knowledge of p (i.e. m_{14}), n (i.e. m_{15}), and φ obtained from $\theta(spk^*)$ by (E 8.25). Hence, the value $\theta(spk^*)$ has still to be obtained, i.e. the angle θ when event spk^* occurs. This is modeled in Fig. **E** by C-place P_{18}, C-transitions T_{24} and T_{25}, and related arcs. When event spk^* occurs, T_{24} and T_{25} are fired: firing of T_{25} empties P_{18} whereas firing of T_{24} deposits in P_{18} the value in m_{17} at this time (which is $\theta(spk^*)$, by definition). The marking m_{18} remains constant up to the next occurrence of spk^*.

Complete dynamic evolution of the variables in CTS
From Figs. **A** to **E**, the complete model of continuous variables subject to *continuous-time dynamics* (in $CTS = \{p, n, n_g, \theta\}$), can be obtained as shown in Fig. **F**. From these variables and $\theta(spk^*)$, also modeled in Fig. **F**, the variables in $DES = \{T_C, T_E, m_C, m_E, \varphi\}$, with a *discrete-time evolution*, can be calculated periodically as explained above.

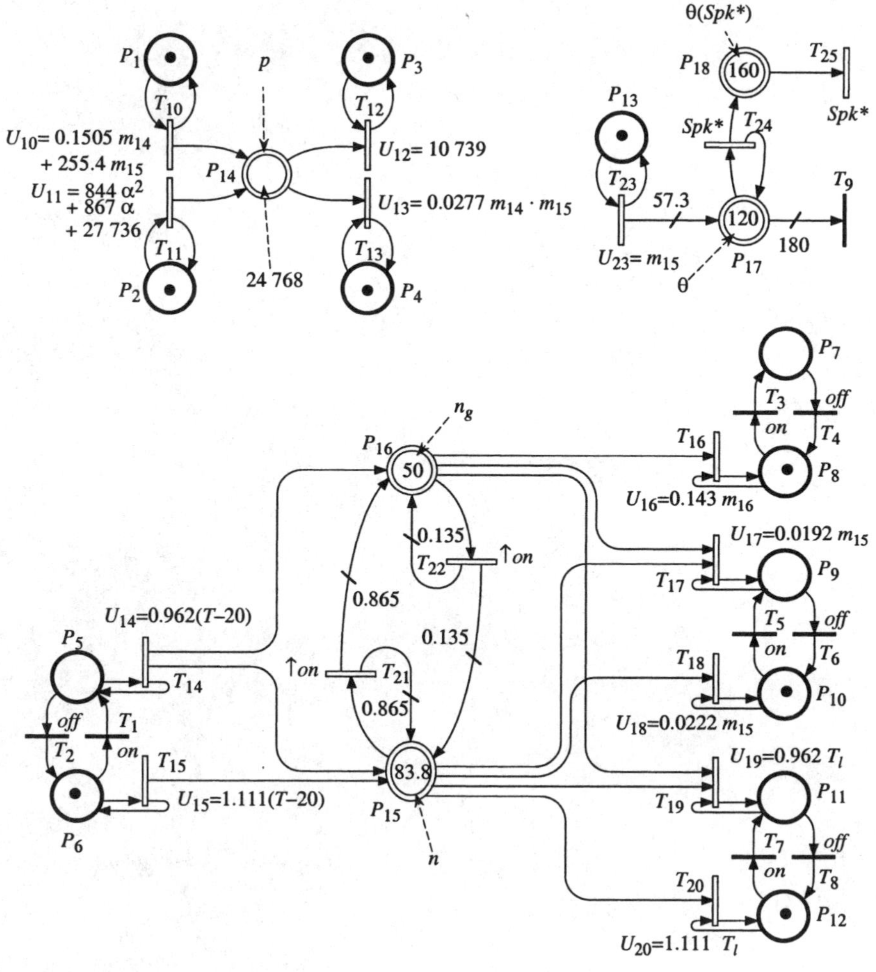

Figure S 8.3.F Complete model for the variables in $CTS = \{p, n, n_g, \theta\}$.

Variable n in Fig. **A** corresponds to m_{15} from Fig. **B** then in Fig. **F**.

As shown in Fig. **F**, a token in P_6 (Fig. **B**) moves to P_5 (Fig. **C**) by firing of T_1, i.e. when the Boolean variable *on* takes the value 1. The opposite change is obtained by firing T_2, when the complement variable *off* takes the value 1. Similar switchings occur between P_8 and P_7, between P_{10} and P_9, and between P_{12} and P_{11}.

References

[AF77] AFCET, "Normalisation de la représentation du cahier des charges d'un automatisme logique", *Final report of AFCET Commission*, August 1977, published in the *J. Automatique et Informatique Industrielles* (61–62), November–December 1977.

[AgFl 73] T. Agerwala and M. Flynn, "Comments on Capabilities, Limitations and Correctness of Petri Nets", *First Annual Symposium on Computer Architecture*, Florida (US), pp. 81–86, 1973.

[Aj *et al.* 89] M. Ajmone-Marsan, G. Balbo, A. Bobbio, G. Chiola, G. Conte, and A. Cumati, "The Effect of Execution Policies on the Semantics and Analysis of Stochastic Petri Nets for the Performance of Multiprocessor Systems", *IEEE Trans. on Software Engineering*, vol. 15, n° 7, pp. 832–846, 1989.

[AjBaCo 84] M. Ajmone-Marsan, G. Balbo, and G. Conte, "A Class of Generalized Stochastic Petri Nets", *ACM Trans. Computer Syst.*, vol. 2, n° 1, May 1984.

[AjCh 84] M. Ajmone-Marsan and G. Chiola, "On Petri Nets with Deterministic and Exponential Transition Firing Times", *Proc. 7th European Workshop on Theory and Applications of Petri Nets*, Oxford (GB), December 1984.

[Al 87] H. Alla, *Réseaux de Petri colorés et réseaux de Petri continus : Application à l'étude des sytèmes à événements discrets*, Thèse d'Etat, Grenoble University (INPG), June 1987.

[Al *et al.* 92] H. Alla, J.-B. Cavaillé, J. Le Bail, and G. Bel, "Les systèmes de production par lots : une approche discret-continu utilisant les réseaux de Petri hybrides", *Proc. Int. Conf. on Automation of Mixed Processes: Hybrid Dynamic Systems* (ADPM), Paris, January 1992.

[AlDa 98a] H. Alla and R. David, “A Modelling and Analysis Tool for Discrete Event Systems: Continuous Petri Net”, *Performance Evaluation*, vol. 33, pp. 175–199, 1998.

[AlDa 98b] H. Alla and R. David, “Continuous and Hybrid Petri Nets”, *Journal of Circuits, Systems and Computers*, Special Issue on Petri Nets, Vol. 8, N° 1, pp. 159–188, 1998.

[AlAl 98a] M. Allam and H. Alla, "Modeling and Simulation of an Electronic Component Manufacturing System Using Hybrid Petri Nets", *IEEE Trans. on Semiconductor Manufacturing*, vol. 11, n° 3, pp. 374–383, 1998.

[AlAl 98b] M. Allam and H. Alla, "From Hybrid Petri Nets to Automata", *Journal Européen des Systèmes Automatisés*, vol. 32, n°9–10, pp. 1165–1185, 1998.

[Al *et al.*95] R. Alur, C. Courcoubetis, N. Halwachs, T.A. Henzinger, P.H. Ho, X. Nicollin, A. Olivero, J. Sifakis, and S. Yovine, "The Algorithmic Analysis of Hybrid Systems", *Theoretical Computer Science* 138, pp. 3–34, 1995.

[AlDi 94] R. Alur and D. L. Dill, "A Theory of Timed Automata", *Theoretical Computer Science*, vol. 126, pp. 183–235, 1994.

[An *et al.* 95] P. Antsalkis, W. Khone, A. Nerode, and S. Sastry (Coordinators), *Hybrid Systems II*, Lecture notes in Computer Science 999, Springer-Verlag, 1995.

[AnStLe 93] P. J. Antsalkis, J. A. Stiver, and M. D. Lemmon, "Hybrid System Modeling and Autonomous Control Systems", *Workshop on Theory of Hybrid Systems*, Lecture notes in Computer Science 736, Springer-Verlag, pp. 336–392, 1993.

[Au 95] N. Audry, *Modélisation et simulation des systèmes de production à haute cadence avec convoyage: réseaux de Petri lots commandés génériques,* Doctorat thesis, Montpellier University, 1995.

[Ba *et al.* 92] F. Baccelli, G. Cohen, G. J. Olsder, and J.-P. Quadrat, *Synchronization and Linearity,* Wiley, 1992.

[BaGuMy 93] A. Back, J. Gukenheimer, and M. Myers, "A Dynamical Simulation Facility for Hybrid Systems", *Workshop on Theory of Hybrid Systems*, Lecture notes in Computer Science 736, Springer-Verlag, pp. 255–267, 1993.

[BaGiMe 98] F. Balduzzi, A. Giua, and G. Menga, "Hybrid Stochastic Petri Nets: Firing Speed Computation and FMS Modelling", *WODES '98*, Gagliari (IT), 1998.

[BaGiMe 00] F. Balduzzi, A. Giua, and G. Menga, "First-Order Hybrid Petri Nets: A Model for Optimization and Control", *IEEE Trans. on Robotics and Automation*, vol. 16, n° 4, pp. 382–399, 2000.

[Ba *et al.* 03] A. Balluchi, M. Zonca, T. Villa, and A. L. Sangiovanni-Vincentelli, *A Non-Linear Hybrid Model of a 4-Cylinder Engine for Idle Speed Control*, Report European Community Projects IST-2001-33520 CC (Control & Computation), `http://www.dii.unisi.it/~hybrid/cc/`, Parades, Rome, 2003.

[BeBe 91] A. Benveniste and G. Berry, "The Synchronous Approach to Reactive and Real-Time Systems", *IEEE Proceedings*, vol. 79, n° 9, September 1991.

[Be 70] C. Berge, *Graphes et hypergraphes*, Dunod, Paris, 1970.

[BeCoGo 87] G. Berry, P. Couronne, and G. Gonthier, "Programmation synchrone des systèmes réactifs : le langage ESTEREL", *Technique et Science Informatiques*, vol. 6, n°4, 1987.

[Be 83] G. Berthelot, *Transformations et analyse de réseaux de Petri: application aux protocoles*, Thèse d'Etat, Paris VI University, 1983.

[Be 85] G. Berthelot, "Transformations des réseaux de Petri", *Technique et Science Informatiques*, vol. 14, n° 1 , pp. 91–101, 1985.

[BeDi 91] B. Berthomieu and M. Diaz, "Modeling and Verification of Time Dependent Systems Using Time Petri Nets", *IEEE Trans. on Software Engineering,* vol. 17, n° 3, pp. 259–273, 1991.

[BiAl 03] M. Bitam and H. Alla, "Modeling of a Communication Network under TCP/IP Protocol Using Hybrid Petri Nets", *IFAC Conf. on Analysis and Design of Hybrid Systems* (ADHS 03), Saint-Malo (FR), pp. 105–110, 2003.

[Bo 78] J. Boussin, "Synthesis and Analysis of Logic Automation Systems", *7th Triennal World Congress,* Helsinki, Pergamon Press, 1978.

[Br 83] G. W. Brams, *Réseaux de Petri : théorie et pratique*, Masson Pub., Paris, 1983.

[BrBoMi 94] M. S. Branicky, V. S. Borkar, and S. K. Mitter, "A Unified Framework for Hybrid Control: Background, Model, and Theory", *Proc. 33rd IEEE Conf. on Decision and Control,* Lake Buena Vista (US), pp. 4228–4234, 1994.

[BrBl 90] P. L. Brinkman and W. A. Blaauboer, "Timed Continuous Petri Nets :A tool for Analysis and Simulation of Discrete Event Systems", *Proc. 1990 European Simulation Symposium*, Ghent (BE), pp.164–168, November 1990.

[Br 93] R. W. Brockett, "Hybrid Models for Motion Control Systems", in H. L. Trentelman and J. C. Willems eds., *Essays in Control: Perspectives in the Theory and its Applications*, Birkhäuser, Boston, pp. 29–53, 1993.

[BuHa 78] J.A Buzacott and L.E. Hanifin, "Models of Automatic Transfer Lines with Inventory Banks— A Review and Comparison", *IIE Transactions*, vol. 10, pp. 197–207, 1978.

[By75] W. H. Byrn, *Sequential Processes, Deadlocks, and Semaphores Primitives*, Tech. Report 7-75, Harvard Univ., Cambridge, MA, 1975.

[Ca *et al.* 89] J. Campos, G. Chiola, J. M. Colom, and M. Silva, "Tight Polynomial Bounds for Steady-State Performance of Marked Graphs", in *Proc. 3rd Int. Workshop on Petri Nets and Performance Models (PNPM '89)*, Kyoto (JP), December 1989.

[CaSi 92] J. Campos and M. Silva, "Structural Techniques and Performance Bounds of Stochastic Petri Net Models. Design Methods Based on Nets (DEMON)", *Advances in PN '92*, G. Rosenberg Ed., *LNCS 607*, pp. 352–391, Springer-Verlag, Berlin, 1992.

[CaVaDu 96] J. Cardoso, R. Valette, and D. Dubois, "Fuzzy Petri Nets: An Overview", in *Proc. 13th IFAC World Congress*, San Francisco (US), July 1996.

[Ca 89] J. Carroll, *Theory of Finite Automata*, Prentice Hall, Engelwood Cliffs, NJ, 1989.

[Ca 87] P. Caspi, Personal communication, 1987.

[Ca *et al.* 87] P. Caspi, D. Pilaud, N. Halbwachs, and J. Plaice, "LUSTRE, a Declarative Language for Real-Time Programming", *Proc. Conf. on Principles of Programming Languages*, Munich (DE), 1987.

[Ce 79] F. E. Cellier, *Combined Continuous/Discrete System Simulation by Use of Digital Computer,* Dissertation, Diss ETH 6483, Swiss Federal Institute of Technology Zurich (CH), 1979.

[Ch*etal.*98a] R. Champagnat, P. Esteban, H. Pingaud, and R. Valette, " Petri Net Based Modeling of Hybrid Systems", *Computers in Industry*, vol. 36, pp. 139–146, 1998.

[Ch*etal.*98b] R. Champagnat, H. Pingaud, H. Alla, C. Valentin-Roubinet, J.-M. Flauss, and R. Valette, "A Gas Storage Example as a Benchmark for Hybrid Modelling", *European Journal of Automation*, vol. 32, n° 9–10, 1998.

[ChZhWa93] D. T. Chao, M. Zhou, and D.T. Wang, "Multiple-Weighted Marked Graphs," in *Proc. IFAC 12th Triennal World Congress*, Sydney (AU), vol. 1, pp. 259–262, 1993.

[Ch 93a] F. Charbonnier, *Etude des conflicts dans les réseaux de Petri hybrides*, Student Report, Laboratoire d'Automatique, INP Grenoble, September 1993.

[Ch 93b] G. Chiola, "On the Structural and Behavioural Characterization of P/T Nets", in *Proc. 5th Int. Workshop on Petri Nets and Performance Models (PNPM '93)*, Toulouse (FR), pp.66–75, 1993.

[Ch *et al.* 00] M. Chouikha, G. Decknatel, R. Drath, G. Frey, C. Müller, C. Simon, J. Thieme, and K. Wolter, "Petri Net-Based Descriptions for Discrete-Continuous Systems", *Automatisierungstechnik*, vol. 48, n° 9, pp. 415–425, 2000.

[Ch 84] P. Chrétienne, "Exécutions contrôlées des réseaux de Petri temporisés", *Technique et Science Informatiques*, vol. 3, n°1, pp.23–31, 1984.

[ChCa 83] P. Chrétienne and J. Carlier, "Modelling Scheduling Problems with Timed Petri Nets", *Proc. 4th European Workshop on Theory and Applications of Petri Nets*, Toulouse (FR), September 1983.

[CiNiTr 97] G. Ciardo, D. Nicol, and K. S. Trivedi, "Discrete-event Simulation of Fluid Stochastic Petri Nets", *Petri Nets & Performance Models (PNPM '97)*, Saint Malo (FR), pp. 217–225, June 1997.

[CoGaQu95] G. Cohen, S. Gaubert, and J.-P. Quadrat, "Asymptotic Throughput of Continuous Timed Petri Nets", *Conference on Decision and Control,* New Orleans (US), 1995.

[CoGaQu 98] G. Cohen, S. Gaubert, and J.-P. Quadrat, "Algebraic System Analysis of Timed Petri Nets", *Idempotency*, Editor J. Gunawardena, Cambridge University Press, Cambridge (GB), 1998.

[CoSi 89] J. M. Colom and M. Silva, "Improving the Linearly Based Characterization of P/T Nets", *10th Int. Conference on Application and Theory of Petri Nets*, Bonn (DE), June 1989.

[Co 72] F. Commoner, *Deadlocks in Petri Nets*, Applied Data Research Inc., Wakefields Mass., CA 7206-2311, 1972.

[Da 76] L. Dadda, "The Synthesis of Petri Nets for Controlling Purposes and the Reduction of their Complexity", *Euromicro,* North-Holland Publ., Amsterdam, 1976.

[DaDaXi 89] Y. Dallery, R. David, and X.-L. Xie, "Approximate Analysis of Transfer Lines with Unreliable Machines and Finite Buffers", *IEEE Transactions on Automatic Control*, vol. 39, n° 4, pp. 818–823, 1994.

[DaGe 92] Y. Dallery and S. B. Gershwin, "Manufacturing Flow Line Systems: A Review of Models and Analytical Results", *Queueing Systems, Theory and Applications*, vol. 12, pp. 3–94, 1992.

[Da95] R. David, "Grafcet: A Powerful Tool for Specification of Logic Controllers", *IEEE Trans. on Control Systems Technology,* vol. 3, n° 3, pp.253–268, 1995.

[Da97] R. David, "Modeling of Hybrid Systems Using Continuous and Hybrid Petri Nets", *Petri Nets & Performance Models (PNPM'97),* Saint Malo (FR), pp. 47–58, June 1997.

[DaAl 87] R. David and H. Alla, "Continuous Petri Nets", *8th European Workshop on Application and Theory of Petri Nets,* Zaragoza (ES), pp. 275–294, June 1987.

[DaAl 89] R. David and H. Alla, *Du Grafcet aux réseaux de Petri,* Hermès Pub., Paris, 1989; Second Edition 1992.

[DaAl 90] R. David and H. Alla, "Autonomous and Timed Continuous Petri Nets", *11th International Conference on Application and Theory of Petri Nets,* Paris, pp. 367–386, June 1990. Then Lecture notes in Computer Science, Springer-Verlag, pp. 71–90, 1993.

[DaAl92] R. David and H. Alla, *Petri Nets and Grafcet: Tools for Modelling Discrete Event Systems,* Prentice Hall Int., London, 1992.

[DaAl01] R. David and H. Alla, "On Hybrid Petri Nets", *J. of Discrete Event Dynamic Systems: Theory and Applications,* vol.11, pp. 9–40, 2001.

[DaAl 03] R. David and H. Alla, "Reachability Graph for Autonomous Continuous Petri Nets", in *Proc. 1st Int. Symp. of Positive Systems* (POSTA 03), Rome, Lecture Notes in Control and Information Sciences, vol. 294, pp. 63–70, Springer, 2003.

[DaCa 00] R. David and S. Caramihai, "Modeling of Delays on Continuous Flows Thanks to Extended Hybrid Petri Nets", *Proc. Int. Conf. on Automation of Mixed Processes: Hybrid Dynamic Systems* (ADPM). Dortmund (DE) , pp.343–350, September 2000.

[DaDe83a] R. David and H. Deneux, "Deterministic Safe Interpreted Petri Nets - Relation with the Grafcet Model", *4th European Workshop on Theory and Applications of Petri Nets,* Toulouse (FR), September 1983.

[DaDe83b] R. David and H. Deneux, "Sur le déterminisme du Grafcet". *RAIRO Automatique,* n° 3, pp. 223–240, 1983.

[DaXiDa 90] R. David, X.L. Xie, and Y. Dallery, "Properties of Continuous Models of Transfer Lines with Unreliable Machines and Finite Buffers". *IMA J. of Mathematics Applied in Business and Industry,* pp. 281–308, n°6, 1990.

[De 94] I. Demongodin, *Les réseaux de Petri lots : Modélisation des systèmes de production à haute cadence en régime transitoire,* Doctorat thesis, Montpellier University, 1994.

[DeAuPr97] I. Demongodin, N. Audry, and F. Prunet, "Batches Petri Nets", *Proc. of Int. IEEE Conf. on Systems, Man, and Cybernetics,* Le Touquet (FR), pp.607–617, October 1997.

[DeCaPr98] I. Demongodin, M. Caradec, and F. Prunet., "Fundamental Concepts of Analysis in Batches Petri Nets", *Proc. of Int. IEEE Conf. on Systems, Man, and Cybernetics,* San Diego (US), pp.845–850, Oct. 1998.

[DeKo98] I. Demongodin and N. T. Koussoulas, "Differential Petri Nets: Representing Continuous Systems in a Discrete Event World", *IEEE Transactions on Automatic Control.* Vol. 38, n°4, pp. 573–579, 1998.

[DePr92] I. Demongodin and F. Prunet, "Extension of Hybrid Petri Nets to Accumulation Systems", in *Proc. IMACS Int. Symp. on Mathematical Modelling ans Scientific Computing*, Bangalore (IN), 1992.

[De 00] J.-P. Denat, *Conduite robuste d'ateliers à contraintes de temps de séjour : application à la galvanoplastie*, Lecture Notes, transmitted personally in 2000.

[De83] H. Deneux, *Contribution à l'étude des réseaux d' automates interconnectés,* Doctorat thesis, Grenoble University (INPG), February 1983.

[DiSa 03] A. Di Febbraro and N. Sacco, "Experimental Validation of a Hybrid Petri-Net Based Model of Urban Transportation Networks", *IFAC Conf. on Analysis and Design of Hybrid Systems* (ADHS 03), Saint-Malo (FR), pp. 111–116, 2003.

[Di *et al.* 91] M. Di Mascolo, Y. Frein, Y. Dallery, and R. David, "A Unified Modeling of Kanban Systems using Petri Nets", *International Journal of Flexible Manufacturing Systems,*vol.3, n° 3/4, pp. 275–307, 1991.

[DuFo82] D. Dubois and J.-P. Forestier, "Productivité et en-cours moyens d'un ensemble de deux machines séparées par un stock", *RAIRO Automatique*, vol. 16, n°2, pp. 105–132, 1982.

[Du95] E. Dubois, *Sur les réseaux de Petri continus à vitesses maximales fonction du temps, constantes par paliers*, Doctorat thesis, Grenoble University (INPG), June 1995.

[DuAl93] E. Dubois and H. Alla, "Hybrid Petri Nets with a Stochastic Discrete Part", *European Control Conference*, Groningen (NL), June 1993.

[DuAlDa94a] E. Dubois, H. Alla, and R. David, "Les réseaux de Petri à vitesses fonction du temps", *Revue d'Automatique, Productique et Informatique Industrielle*, vol. 28, n° 5, pp.425–443, June 1994.

[DuAlDa94b] E. Dubois, H. Alla, and R. David, "Continuous Petri Net with Maximal Speeds Depending on Time", *Proc. 4th Int. Conf. on CIM & Automation Technology*, Troy (US), pp 32–39, October 1994.

[Du*et al.*84] J. B. Dugan, K. S.Trivedi, R. M. Geist, and V. F. Nicola, "Extended Stochastic Petri Nets: Applications and Analysis", *Proc. Performance '84*, Paris, December 1984.

[ElCeOt93] H. Elmqvist, F. E. Cellier, and M. Otter, "Object-Oriented Modeling of Hybrid Systems", *Proc. of the European Simulation Symposium*, Delft (NL), pp.31–41, 1993.

[En97] S. Engell, "Modeling and Analysis of Hybrid Systems", *Proc. 2nd IMACS MATHMOD Conference*, Vienna (AT), pp. 17–31, 1997.

[FaAlFl99] A. Favela, H. Alla, and J.-M. Flauss, "Analysis of the Coupled Linear Hybrid Automata Dynamics", *European Control Conference*, Karlsruhe (DE), September 1999.

[Fl 96] J.-M. Flauss, "Modélisation hybride de procédés batch", *Journées d'étude sur l'analyse et la supervision des systèmes dynamiques hybrides*, Lyon (FR), February 1996.

[FlNa 85] G. Florin and S. Natkin, "Les réseaux de Petri stochastiques", *Technique et Science Informatiques*, Vol. 4, n° 1, pp. 143–160, 1985.

[GaGi 99] S. Gaubert and A. Giua, "Petri Net Languages and Infinite Subsets of N^m", *Journal of Computer and System Sciences*, vol. 59, pp. 373–391, 1999.

[GeHaWö94] H. J. Genrich, H.-M. Hanisch, and K Wöllhaf, "Verification of Recipe-Based Control Procedures by Means of Predicate/Transition Nets", *Int. Conf. on Application and Theory of Petri Nets*, Zaragoza (ES), Lecture notes in Computer Science 815, Springer-Verlag, pp. 278–297, 1994.

[GeLa79] H. J. Genrich and K. Lautenbach, "The Analysis of Distributed Systems by Means of Predicate/Transition Nets, Semantics of Concurrent Computation", in *Lecture Notes in Computer Sciences*, vol. 70, G. Khan Ed., pp. 123–146, Springer Verlag 1979,.

[GeSc80] S. B. Gershwin and I. C. Schick, *Continuous Model of an Unreliable Two-Stage Material Flow System With a Finite Interstage Buffer*, Technical report MIT LIDS-R-1032, September 1980.

[GiDi 94] A. Giua and F. DiCesare, "Blocking and Controllability in Supervisory Control", *IEEE Transactions on Automatic Control*, vol. 39, n° 4, pp. 818–823, 1994.

[GoGe96] M. M. Gomaa and S. Gentil, "Hybrid Industrial Dynamical System Supervision Via Hybrid Continuous Causal Petri Nets (HC^2PNs)", *Proc. IMACS/IEEE SMC, CESA '96*, Lille (FR), pp. 285–290, 1996.

[Gr *et al.* 93] R. L. Grossman, A. Nerode, A.P. Ravn, and H. Rischel (Coordinators), *Hybrid Systems*, Lecture notes in Computer Science 736, Springer-Verlag, 1993.

[GR 85] GREPA, *Le Grafcet, de nouveaux concepts,* Editions Cepadues, Toulouse 1985, GREPA: Groupe Equipement de Production automatisée, leaded by J.-M. Toulotte.

[Ha72] M. Hack, *Analysis of Production Schemataby Petri Nets*, M. S. thesis, Dep. Electrical Engineering, MIT, Cambridge, MA, 1972.

[Ha74] M. Hack, *Extended State-Machine Allocatable Nets, an Extension of Free-Choice Petri Nets Results*, Comput. Struct. Gr. Memo 78-1, MIT Project MAC, Cambridge, MA, 1974.

[Ha75] M. Hack, "Petri Nets Languages" *Extended State-Machine Allocatable Nets, an Extension of Free-Choice Petri Nets Results*, Comput. Struct. Gr. Memo 124, MIT Project MAC, Cambridge, MA, 1975.

[Ha 93] H.-M. Hanisch, "Analysis of Place/Transition Nets with Timed Arcs and its Application to Batch Process Control", *Int. Conf. on Application and Theory of Petri Nets*, Chicago (US), Lecture notes in Computer Science 691, Springer-Verlag, pp. 282–299, 1993.

[Ha03] Z. Hanzalek, "Continuous Petri Nets and Polytopes", *Proc. of Int. IEEE Conf. on Systems, Man, and Cybernetics*, Washington, October 2003.

[He *et al.* 94] T. A. Henzinger, X. Nicollin, J. Sifakis, and S. Yovine, "Symbolic Model Checking for Real-Time Systems", *Information and Computation*, vol. 111, n° 2, pp. 193–244, 1994.

[HoVe 87] M. A. Holliday and M. K. Vernon, "A Generalized Timed Petri Net Model for Performance Analysis", *IEEE Trans. on Software Engineering,* vol. 13, n° 12, pp.1297–1310, 1987.

[HoKrGi97] L. E. Holloway, B. H. Krogh, and A. Giua, "A survey of Petri Net Methods for Controlled Discrete Event Systems", *J. of Discrete Event Dynamic Systems: Theory and Applications,* vol.7, pp. 151–190, 1997.

[HoCo 70] A. W. Holt and F. Commoner, *Events and Conditions*, Applied Data Research Inc., New York, 1970.

[HoUl 79] J. E. Hopcroft and J. D. Ullman, *Introduction to Automata Theory, Languages, and Computation*, Addison-Wesley, MenloPark, CA, 1979.

[Ho *et al.* 98] G. Horton, V. Kulkani, D. Nicol, and K. S. Trivedi, "Fluid Stochastic Petri Nets: Theory, Application and Solution", *European Journal of Operational Research*, vol. 105, pp. 184–201, 1998.

[HuJeSh 89] P. Huber, K. Jensen, and R. M. Shapiro, "Hierarchies in Coloured Petri Nets," *10th International Conference on Theory and Applications of Petri Nets*, Bonn (DE), June 1989.

[IE 88] IEC, International Electrotechnic Commission, *Preparation of Function Charts for Control Systems*, publication 848, 1988.

[IE92] IEC, International Electrotechnic Commission, *Programmable Controllers, Part 3: Programming Languages*, publication 1131-1, 1992.

[IE 02] IEC, International Electrotechnic Commission, *Grafcet Specification Language for Sequential Function Charts*, publication 60848, 2002.

[Je 81] K. Jensen, "Coloured Petri Nets and the Invariant Method", *Theorical Computer Sciences*, vol. 14, pp. 317–336, 1981.

[JeRo 91] K. Jensen and G. Rosenberg, *High-level Petri Nets – Theory and Application*, Springer Verlag, Berlin, 1991.

[KaMi 79] T. Kaisai and R. Miller, *Homorphisms Between Models of Parallel Computation*, IBM Research Report RC 7796 (33742), 1979.

[KaMi 69] R. Karp and R. Miller, "Parallel Program Schemata", *Journal of Computer and System Sciences*, vol. 3, n°4, pp. 167–195, 1969.

[KeSn 76] J. G. Kemeny and J. L. Snell, *Finite Markov Chains*, Springer Verlag, Berlin, 1976.

[Kh 97] W. Khansa, *Réseaux de Petri P-temporels : contribution à l'étude des systèmes à événements discrets*, Doctorat thesis, University of Savoie, Annecy (FR), March 1997.

[KhDeCo 96] W. Khansa, J.-P. Denat, and S. Collart-Dutilleul, "P-Time Petri Nets for Manufacturing Systems", *Int. Workshop on Discrete Event Systems* (*Wodes '97*), Edinburg (GB), pp.94–102, August 1996.

[KlLa 82] W. E. Kluge and K. Lautenbach, "The Orderly Resolution of Memory Access Conflicts Among Competing Channel Processes", *IEEE Trans. on Computers*, vol. C-31, n° 3, pp. 194–207, March 1982.

[Ko 82] S. R. Kosaraju, "Decidability of Reachability in Vector Addition System", in *Proc. 14th Annual Symposium Theory Computing*, San Francisco (US), pp. 267–281, 1982.

[LaTa 97] C. Lansade and P. Tayrac, *Simulation mixte*, ANVAR Project TLS 0089, Pau (FR), January 1997.

[La 80] M. Latteux, "Synchronisation de processus", *Journal RAIRO Computer Science Series*, vol. 14, n° 2, 1980.

[LaSc 74] K. Lautenbach, H. A. Schmid, "Use of Petri Nets for Proving Correctness of Concurrent Process Systems", *Proc. IFIP Congress 74*, North-Holland Publ., Amsterdam, pp. 187–191, 1974.

[Le 91] J. Le Bail, *Un nouveau modèle continu : les réseaux de Petri continus asymptotiques*, Research Report 91-22, Laboratoire d'Automatique, INP Grenoble, April 1991.

[Le92] J. Le Bail, *Notion de séquence de franchissement élémentaire : Application à l'étude des réseaux de Petri avec conflicts*, Research Report 92-019, Laboratoire d'Automatique, INP Grenoble, February 1992.

[LeAlDa 91] J. Le Bail, H. Alla, and R. David, "Hybrid Petri Nets", *Proc. European Control Conference*, Grenoble (FR), July 1991.

[LeAlDa 92a] J. Le Bail, H. Alla, and R. David, "Asymptotic Continuous Petri Nets: An Efficient Approximation of Discrete Event Systems", *Proc. 1992 IEEE Int. Conf. on Robotics & Automation*, Nice (FR), pp. 1050–6, May 1992.

[LeAlDa 92b] J. Le Bail, H. Alla, and R. David, "Réseaux de Petri Hybrides", *Technique et Science Informatiques*, vol. 11, n°5, pp. 95–120, 1992.

[LeAlDa 93] J. Le Bail, H. Alla, and R. David, "Asymptotic Continuous Petri Nets", *J. of Discrete Event Dynamic Systems: Theory and Applications*, n°2, pp.235–263, 1993.

[Le *et al.* 03] D. Lefebvre, E. Leclercq, F. Druaux, and P. Thomas, "Source and Sink Transitions Controllers for Continuous Petri Nets: A Gradient-Based Approach", *IFAC Conf. on Analysis and Design of Hybrid Systems* (ADHS 03), Saint-Malo (FR), pp. 229–234, 2003.

[Le *et al.* 86] P. Le Guernic, A. Benveniste, P. Bournai, and T. Gauthier, "SIGNAL : A Data Flow Oriented Language for Signal Processing", *IEEE Trans. Acoust., Speech, Signal Processing*, vol. ASSP-34(2), pp. 362–374, April 1986.

[Li 76] Y. E. Lien, "A Note on Transition Systems", *J. Information Science*, Vol. 10, n° 4, pp. 251–265, 1976.

[Ma 97] O. Maler (Coordinator), *Hybrid and Real Time Systems*, Lecture notes in Computer Science 1201, Springer-Verlag, 1997.

[MaMaPn 92] O. Maler, Z. Manna, and A. Pnueli, "From Timed To Hybrid Systems", *Proc. REX Workshop "Real-Time: Theory and Applications"*, Lecture notes in Computer Science 600, Springer-Verlag, pp. 447–484, 1992.

[MaCh 91] A. Mandelbaum and H. Chen, "Discrete Flow Networks: Bottleneck Analysis and Fluid Approximations" *Math. Operations Research*, vol. 16, pp. 408–446, 1991.

[MaMuSi 87] J. Martinez, P. Muro, and M. Silva, "Modeling, Validation and Software Implementation of Production Systems using High Level Petri Nets", *Proc. of IEEE International Conference on Robotics and Automation*, Raleigh (US), March 1987.

[MaSi 82] J. Martinez and M. Silva, "A simple and Fast Algorithm to obtain all Invariants of a Generalized Petri Net", *2nd European Workshop on Petri Nets Theory and Applications*, Informatik Fachberichte 52, pp. 301–310, Springer Verlag, Berlin, 1982,.

[Ma 81] E. W. Mayr, "An Algorithm for the General Petri Net Reachability Problem", in *Proc. 13th Annual Symposium Theory Computing*, pp. 238–246, 1981.

[Me 78] G. Memmi, *Fuites et semi-flots dans les réseaux de Petri*, Docteur-Ingénieur thesis, University Pierre-et-Marie-Curie, Paris, Dec. 1978.

[Me 81] G. Memmi and R. Martin, "Specification and Validation of Sequential Processes Communicating by FIFO Channels", *4th Int. Conference on Software Engineering for Telecommunication Switching Systems*, I.E.E. Worwick, July 1981.

[MeRo 80] G. Memmi and G. Roucairol, "Linear Algebra in Net Theory" *Proc. of Advanced Course on General Net Theory of Processes and Systems*, Hambourg 1979; W. Bauer Ed., Springer Verlag LNCS 84, 1980.

[Me 76] P. M. Merlin, "A Methodology for the Design and Implementation of Communication Protocols", *IEEE Trans. on Communication,* vol. 24, n° 6, pp.614–621, 1976.

[MeFa76] P. M. Merlin and D. J. Farber, "Recovery of Communication Protocols — Implication of a Theoretical Study", *IEEE Trans. on Communication,* vol. 24, n° 9, pp.1036–1043, 1976.

[Me 94] M. Mezghani, *Modélisation et évaluation de performances d'un réseau du réseau de distribution de la ville d'Aix-Les-Bains*, Student Report, Laboratoire d'Automatique, INP Grenoble, June 1994.

[Mo 81] M. Moalla, *Spécification et conception sûre d'automatismes discrets complexes,basées sur l'utilisation du Grafcet et des réseaux de Petri,* Thèse d'Etat, Grenoble University (INPG), July 1981.

[Mo 85] M. Moalla, "Réseaux de Petri interprétés et Grafcet", *Technique et Science Informatiques*, Vol. 14, n° 1, pp. 17–30, 1985.

[Mo 01] M. Moalla, Personal communication, 2001.

[MoDa 81] M. Moalla and R. David, "Extension du Grafcet pour la représentation de systèmes temps réel complexes", *Journal RAIRO, Automatic Control Series*, vol. 15, n°2, pp. 159–192, 1981.

[MoPuSi 78] M. Moalla, J. Pulou, and J. Sifakis, "Réseaux de Petri synchronisés", *Journal RAIRO, Automatic Control Series*, vol. 12, n°2, pp. 103–130, 1978.

[MoCo 99] S. Mocanu and C. Commault, "Sparse Representations of Phase-Type Distributions", *Commun. Stat., Stochastic Models* vol. 15, n°4, pp. 759–778, 1999.

[Mo 82] M. K. Molloy, "Performance Analysis Using Stochastic Petri Nets", *IEEE Trans. on Computers*, vol. C-31, n° 9, pp. 913–917, March 1982.

[MüRa 00] C. Müller and H. Rake, "A Petri Net-State-Model for the Analysis and the Control Synthesis of Hybrid Technical Systems", *Proc. Int. Conf. on Automation of Mixed Processes: Hybrid Dynamic Systems* (ADPM). Dortmund (DE) , pp. 331–336, September 2000.

[MuDaAl 04] C. Munteanu, R. David and H. Alla, "Algorithmic Speed Calculation for a Timed Continuous Petri Nets", *WODES'04*, Reims (FR), September 2004.

[Mu89] T. Murata, "Petri Nets: Properties, Analysis and Applications", *Proceedings of the IEEE*, vol. 77, n° 4, pp. 541–580, 1989.

[Na80] S. Natkin, *Les réseaux de Petri stochastiques et leur application à l'évaluation de performances des systèmes informatiques,* Thèse de Doctorat, CNAM, Paris, June 1980.

[NeKo93] A. Nerode and Wolf Kohn, "Models for Hybrid Systems: Automata, Topologies, Stability", *Workshop on Theory of Hybrid Systems*, Lecture notes in Computer Science 736, Springer-Verlag, pp. 317–356, 1993.

[Ne81] M. F. Neuts, *Matrix Geometric Solutions in Stochastic Models*, Johns Hopkins University Press., Baltimore (US), 1981.

[NiSiYo 93] X. Nicollin, J. Sifakis, and S. Yovine, "From ATP To Timed Graphs and Hybrid Systems", *Acta Informatica,* vol. 30, pp. 181–202, 1993.

[Ol 93] G. J. Olsder, "Synchronized Continuous Flow Systems" in *Discrete Event Systems: Modeling and Control.* Editors S. Balemi, P. Kozak, & R. Smadinga, Birkhäuser Verlag, Basel, 1993.

[OrAb03] P. Orth and Dirk Abel, "Hybrid System Analysis Using a Parameterised Evolution Graph", *IFAC Conf. on Analysis and Design of Hybrid Systems* (ADHS 03), Saint-Malo (FR), pp. 343–348, 2003.

[Pe 81] J. L. Peterson, *Petri Net Theory and the Modelling of Systems*, Prentice Hall, Engelwood Cliffs, NJ, 1981.

[PeLe 95] S. Pettersson and B. Lennartson, "Hybrid Modelling Focused on Hybrid Petri Nets", *European Workshop on Hybrid Systems*, Grenoble (FR), pp.303–309, 1995.

[Pe 62] C. A. Petri, *Communication with Automata*, Supplement 1 to Technical Report RADC-TR-65-337, N.Y., 1965. Translation by C.F. Greene of *Kommunikation mit Automaten,* PhD Dissertation, Univ. of Bonn, 1962.

[Pe 76] C. A. Petri, *Interpretation of Net Theory,* Interner Bericht ISF-75-07, Saint Augustin, Bonn; Second Edition, December 1976.

[Pe 79] C. A. Petri, "Introduction to General Net Theory", in *Lectures Notes in Computer Science: Net Theory and Applications,* pp.1–19, 1979.

[PhHu 76] L. W. Philips and P. S. Hunger, "Mathematical Programming Solution of a Hoist Scheduling Program", *AIIE Transactions*, pp. 219–225, June 1976.

[QuGuBu94] Y. Quénec'hdu, H. Guéguen, and J. Buisson, "Les systèmes dynamiques hybrides : une nouvelle problématique", *Proc. Automation of Mixed Process: Dynamic Hybrid Systems* (ADPM), Brussels, pp. 1–8, 1994.

[RaHo80] C. V. Ramamoorthy and G. S. Ho, "Performance Evaluation of Asynchronous Concurrent Systems Using Petri Nets", *IEEE Trans. on Software Engineering,* vol. 6, n° 5, pp.440–449, 1980.

[Ra 73] C. Ramchandani, *Analysis of Asynchronous Concurrent Systems by Timed Petri Nets.* Ph.D., MIT (US), 1973.

[RaPh 84] R. R. Razouk and C. V. Phelps, "Performance Analysis Using Timed Petri Nets", in *Proc. Int. Conf. Parallel Processing*, pp. 126–129, August 1984.

[ReSi 01] L. Recalde and M. Silva, "PN Fluidification Revisited: Semantics and Steady State", *Journal Européen des Systèmes Automatisés*, vol. 35, n°4, pp. 435–449, 2001.

[ReTeSi99] L. Recalde, E. Teruel, and M. Silva, "Autonomous Continuous P/T Systems", *20th Int. Conference on Application and Theory of Petri Nets*, Williamsburg (US), Springer, pp. 108–126, June 1999.

[Re 85] W. Reisig, *Petri Nets: an Introduction,* Springer-Verlag, Berlin, 1985.

[Re 89] C. Reutenauer, *Aspects mathématiques des réseaux de Petri,* Masson, Paris, 1989.

[Ro 66] A. Robinson, *Non Standard Analysis*, North-Holland, Amsterdam, 1966. Printed again: Princeton University Press, 1996.

[Ro 79] G. Roucairol, "Vers une caractérisation de la synchronisation des processus parallèles sur les files et les piles", *1st European Conference on Parallel ans Distributed Processes*, Toulouse (FR), 1979.

[Sa 99] J.-M. Salanskis, *Le constructivisme non standard*, Presse Universitaires du Septentrion, Paris, 1999.

[ShNu89] G. W. Shapiro and L. W. Nuttle, "Hoist Scheduling for a PCB Electroplating Facility", *IIE Transactions*, vol. 20, n° 2, pp. 157–167, June 1989.

[Si 85] C. Sibertin-Blanc, "High Level Petri Nets with Data Structure", *6th European Workshop on Application and Theory of Petri Nets*, Finland, June 1985.

[Si 77] J. Sifakis, J., "Use of Petri Nets for Performance Evaluation". In *Measuring, Modelling and Evaluating Computer Systems*, H. Beilner and E. Gelenbe Eds., North-Holland Pub., Amsterdam, pp.75–93, 1977.

[Si 78] J. Sifakis, "Structural Properties of Petri Nets", *Mathematical Foundations of Computer Science*, J. Winkowski Ed., Springer Verlag, pp. 474–483, 1978.

[Si 79] J. Sifakis, *Le contrôle des systèmes asynchrones : concepts, propriétés, analyse statique*, Thèse d'Etat, Grenoble University (INPG), 1979.

[Si 85] M. Silva, *Las redes de Petri en la Automatica y la Informatica*, Ed. AC, Madrid, 1985.

[Si 93] M. Silva, "Introducing Petri Nets", in F. DiCesare, G. Harhalakis, J.-M. Proth, M. Silva, and F. Vernadat, *Practice of Petri Nets in Manufacturing Systems*, Chapman and Hall, London, 1993.

[SiCo 89] M. Silva and J. M. Colom, "On the Computation of Structural Synchronic Invariants in P/T Nets", *Advances in PN '88* (G. Rosenberg Ed.), *LNCS 340*, pp. 346–417, Springer-Verlag, Berlin, 1989.

[Si *et al.* 85] M. Silva, J. Martinez, P. Ladet and H. Alla, "Generalized Inverses and the Calculation of Symbolic Invariants for Coloured Petri Nets", *Technique et Science Informatiques*, vol. 14, n° 1, 1985.

[SiRe 02] M. Silva and L. Recalde, "Petri Nets and In tegrality Relaxations: A View of Continuous Petri Net Models", *IEEE Trans. on Systems, Man, and Cybernetics*, vol. 32, n° 4, pp. 314–327, 2002.

[SiRe 03] M. Silva and L. Recalde, "On Fluidification of Petri Nets: From Discrete to Hybrid and Continuous Models", *IFAC Conf. on Analysis and Design of Hybrid Systems* (ADHS 03), Saint-Malo (FR), pp. 9–20, 2003.

[SiTe 96] M. Silva and E. Teruel, "A System Theory Perspective of Discrete Event Dynamic Systems: The Petri Net Paradigm", *Proc. IMACS/IEEE SMC, CESA '96*, Lille (FR), pp. 1–30, 1996.

[SiTeCo 98] M. Silva, E. Teruel, and J. M. Colom, "Linear Algebraic and Linear Programming Techniques for the Analysis of the Net Systems", *Lectures in Petri Nets I: Basic Models*, Lecture notes in Computer Science 1491, Springer-Verlag, pp. 309–373, 1998.

[*SI*95] Special Issue on Hybrid Systems, *Theoretical Computer Science*, vol. 138, n° 1, A. Pnueli and J. Sifakis Eds., 1995.

[*SI*01] Special Issue on Hybrid Petri Nets, *J. of Discrete Event Dynamic Systems: Theory and Applications*, vol.11, A. Di Febbraro, A. Giua, and G. Menga Eds., 2001.

[SrKr91] R. S. Sreenivas, B. H. Krogh, "On Condition/Event Systems with Discrete State Realizations", *Discrete Event Dynamic Systems*, vol. 1, n° 2, pp. 209–236, 1991.

[SuMu83] I. Suzuki and T. Murata, "A Method for Stepwise Refinement and Abstraction of Petri Nets", *Journal of Computer and System Sciences*, n° 27, pp. 51–76, 1983.

[Ta87] L. Tavernini, "Differential Automata and their Discrete Simulation", *Nonlinear Analysis, Theory, Methods, and Applications*, vol. 11, n° 6, pp. 665–683, 1987.

[TeDa87] C. Terracol and R. David, "An Aggregation Method for Performance Evaluation of Transfer Lines with Unreliable Machines and Finite Buffers", *IEEE Int. Conf. on Robotics and Automation*, Raleigh (US), 1987.

[Te *et al.* 92] E. Teruel, P. Chrzatowski-Wachtel, J. M. Colom, and M. Silva, "On Weighted T-Systems," in *Proc. Int. Conf. of Applications of Petri Nets*, Sheffield, Lecture Notes in Computer Science, vol. 616, pp. 348–367, Springer, 1992.

[TeFrSi98] E. Teruel, G. Franceschinis, and M. Silva, "Untimed Petri Nets. Performance Models for Discrete Event Systems with Synchronization", in *HCM-MATCH Adv. Schools*, G. Balbo and M. Silva Eds., Dep. Leg. 2165/98, pp. 27–75, 1998.

[TeSi 93] E. Teruel and M. Silva, "Liveness and Home States in Equal Conflict Systems," in *Application and Theory of Petri Nets*, M. A. Marsan Ed., Lecture Notes in Computer Science, vol. 815, pp. 415–432, Springer, Berlin, 1993.

[TeSi 96] E. Teruel and M. Silva, "Structure Theory of Equal Conflict Systems", *Theoretical Computer Science*, vol. 153, pp. 271–300, Elsevier, Amsterdam, 1996.

[To 82] J. M. Toudic, "Algorithmes d'algèbre linéaire pour l'analyse structurelle des réseaux de Petri", *Revue technique Thomson-CSF*, vol. 14, n° 1, pp. 137–155, 1982.

[TrKu93] K. S. Trivedi and V. G. Kulkani, "FSPNs : Fluid Stochastic Petri Nets", 14th *Int. Conference on Application and Theory of Petri Nets*, Chicago (US), June 1993.

[TuChTr 01] B. Tuffin, D. S. Chen, and K. S. Trivedi, "Comparison of Hybrid Systems and Fluid Stochastic Petri Nets", *J. of Discrete Event Dynamic Systems: Theory and Applications*, vol.11, pp. 77–95, 2001.

[Va 98] C. Valentin-Roubinet, "Modelling of Hybrid Systems: DAE Supervised by Petri Nets, The Example of Gas Storage", *Proc. Int. Conf. on Automation of Mixed Processes: Hybrid Dynamic Systems* (ADPM), Reims (FR), pp. 142–149, 1998.

[Va 79] R. Valette, "Analysis of Petri Nets by Stepwise Refinements", *Journal of Computer and System Sciences*, n° 18, pp. 35–46, 1979.

[We 96a] R. Weiting, "Hybrid High-Level Nets", *Proceedings of the Winter Simulation Conference*, Coronado (US), pp.848–855, 1996.

[We 96b] R. Weiting, "Modeling and Simulation of Hybrid Systems Using Hybrid High-Level Nets", *Proceedings of the 8th European Simulation Symposium*, Genova (IT), pp.158–162, 1996.

[WeSo 95] R. Weiting and M. Sonnenschein, "Extended High-Level Nets for Modeling Hybrid Systems", *Proc. of the IMACS Symposium on Systems Analysis and Simulation*, Berlin, pp.259–262, 1995.

[Za 01] J. Zaytoon (Coordinator), *Systèmes Dynamiques Hybrides*, Hermès Science Pub., Paris, 2001.

[ZeAl 90] N. Zerhouni and H. Alla, "Dynamic Analysis of Manufacturing Systems Using Continuous Petri Nets", *IEEE International Conference on Robotics and Automation*, Cincinnati, vol. 2, pp. 1070–1075, 1990.

[Zi 56] B. Zimmern, "Etude de la propagation des arrêts aléatoires dans les chaînes de production", *Revue de Statistique Appli.*, vol.4, pp.85–104, 1956.

[Zu80] W. M. Zuberek, "Timed Petri Nets and Preliminary Performance Evaluation", in *Proc. 7th Ann. Symp. Computer Architecture*, La Baule (FR), pp. 88–96, May 1980.

[Zu85] W. M. Zuberek, "Performance Evaluation Using Timed Petri Nets", in *Proc. Int. Workshop on Timed Petri Nets*, Torino (IT), IEEE-CS Press, pp. 272–278, July 1985.

Index

Page numbers in **bold** characters correspond to pages where a word or concept is **defined** or **introduced**. The letter *n* following a page number refers to a footnote. *Italic* numbers correspond to *exercises* or *solutions to exercises* (only some of them are referred in this index).

Printing: Strauss GmbH, Mörlenbach
Binding: Schäffer, Grünstadt